Seismology of the Sun and the Distant Stars 2016

Using Today's Successes to Prepare the Future

TASC2 & KASC9 Workshop – SPACEINN & HELAS8 Conference

EPJ Web of Conferences Volume 160 (2017)

Acores Islands, Portugal
11 - 25 July 2016

Editors:

Mario J.P.F.G. Monteiro
Margarida S. Cunha
Joao Miguel T.S. Ferreira

ISBN: 978-1-5108-5055-2

Printed from e-media with permission by:

Curran Associates, Inc.
57 Morehouse Lane
Red Hook, NY 12571

proceedings.com

Some format issues inherent in the e-media version may also appear in this print version.

Printed by Curran Associates, Inc. (2017)

For additional information, please contact EDP Sciences – Web of Conferences at the address below.

EDP Sciences – Web of Conferences
17, Avenue du Hoggar
Parc d'Activité de Courtabœuf
BP 112
F-91944 Les Ulis Cedex A
France

Phone: +33 (0) 1 69 18 75 75
Fax: +33 (0) 1 69 28 84 91

contact-edps@webofconferences.org

Additional copies of this publication are available from:

Curran Associates, Inc.
57 Morehouse Lane
Red Hook, NY 12571 USA
Phone: 845-758-0400
Fax: 845-758-2633
Email: curran@proceedings.com
Web: www.proceedings.com

TABLE OF CONTENTS

edp sciences
EPJ Web of Conferences

Statement of Peer review

In submitting conference proceedings to EPJ Web of Conferences, I certify to the Publisher that I adhere to the **Policy on Publishing Integrity** of the journal in order to safeguard good scientific practice in publishing.

1. All articles have been subjected to peer review administered by the proceedings editors.
2. Reviews have been conducted by expert referees, who have been requested to provide unbiased and constructive comments aimed, whenever possible, at improving the work.
3. Proceedings editors have taken all reasonable steps to ensure the quality of the materials they publish and their decision to accept or reject a paper for publication has been based only on the merits of the work and the relevance to the journal.

Title, date and place of the conference

Title: Seismology of the Sun and the Distant Stars II

Conference: Seismology of the Sun and the Distant Stars 2016
Using Today's Successes to Prepare the Future
TASC2 & KASC9 Workshop – SPACEINN & HELAS8 Conference
11–25 July 2016, Angra do Heroísmo, Terceira, Açores Islands, Portugal

Proceedings editor(s):

Mário J. P. F. G. Monteiro (U.Porto)

Margarida S. Cunha (U.Porto)

João Miguel T. S. Ferreira (U.Açores)

Date and editor's signature

2017/10/02

Preface

Mário J. P. F. G. Monteiro[1,2], *Margarida S.* Cunha[2], and *João Miguel T. S.* Ferreira[2,3]

[1] *Departamento de Física e Astronomia, Faculdade de Ciências da Universidade do Porto, Portugal*
[2] *Instituto de Astrofísica e Ciências do Espaço, Universidade do Porto, Portugal*
[3] *Departamento de Ciências e Engenharia do Ambiente, Faculdade de Ciências Agrárias e de Ambiente, Universidade dos Açores, Portugal*

Abstract. This volume is a collection of original articles resulting from the contributions presented at the international conference:

SEISMOLOGY OF THE SUN AND THE DISTANT STARS 2016
USING TODAY'S SUCCESSES TO PREPARE THE FUTURE
TASC2 & KASC9 Workshop – SPACEINN & HELAS8 Conference
11–25 July 2016, Angra do Heroísmo, Terceira, Açores Islands, Portugal
Website: http://www.iastro.pt/research/conferences/spacetk16/

Rationale for the Conference

For the last 30 years, since the meeting on Seismology of the Sun and the Distant Stars held in Cambridge, in 1985, the range of seismic data and associated science results obtained has far exceeded the expectations of the community. The continuous observation of the Sun has secured major advances in the understanding of the physics of the stellar interiors and has allowed us to build and prepare the tools to look at other stars. Several ground facilities and space missions have completed the picture by adding the necessary data to study stars across the HR diagram with a level of detail that was in no way foreseen in 1985.

In spite of the great science successes, astero- and helioseismic data still contain many secrets waiting to be uncovered. The opportunity to use the existing data and tools to clarify major questions of stellar physics (mixing, rotation, convection, and magnetic activity are just a few examples) still needs to be further explored. The combination of data from different instruments and for different targets also holds the promise that further advances are indeed imminent. At the same time we also need to prepare the future, as major space missions and ground facilities are being built in order to collect more and better data to expand and consolidate the detailed seismic view of the stellar population in our galaxy.

This conference aimed at reviewing the major science achievements resulting from the extensive data collections available from space missions and ground facilities, as well as the tools the community has developed to study and explore those data. This also included the planning of asteroseismic activities of the TESS mission in preparation for its launch. The present view is fundamental to prepare the forthcoming facilities that will provide the data necessary to address challenging questions that have not yet been answered with the current data. The optimal way to achieve that goal was one of the expected outcomes of the conference, with an impact on how the astero- and helioseismology community prepares the future.

The meeting also cover major synergies with related fields, which benefit from a deeper understanding of the structure and evolution of the sun and stars across the HR diagram. These included topics covering the characterization of planet hosts, and their exoplanets, the description of stellar populations and impact on Galactic evolution, as well as star and planet formation and the early evolution of planetary systems.

Organization

The conference was jointly organized by;

- *Instituto de Astrofísica e Ciências do Espaço* (http://www.iastro.pt/),
- *Universidade do Porto* (http://www.astro.up.pt/) and
- *Universidade dos Açores* (http://www.uac.pt/).

The organization also involved *Fundação Gaspar Frutuoso* (http://fgf.uac.pt/), a private institution created by the *Universidade dos Açores*.

The scientific program occupied five working days from the 11th to the 15th (but with a social event taking place the 10th, together with the football celebrations of that day in Portugal!). It was a very dense program with a long list of communications (total of 82 oral communications and 83 posters), including:

- 2 Special Invited Reviews (40 min): Opening and closing presentations.
- 14 Invited Reviews (30 min): Review presentations on key topics of the conference.
- 14 Invited Contributions (20 min): Overview of specific work done of great impact.
- 6 Requested Oral Contributions (15 min): Reviews on the coordination of major actions done within TASC and SPACEINN.
- 43 Oral Contributions (15 min): Proposed oral presentations selected from the 76 proposals we received.
- 2 Poster sessions (1 min for 1 slide per poster): short report on the 83 posters displayed in the conference venue.
- 1 Public Talk (40 min): Talk presented at the site of the *Universidade dos Açores* for the general public.

The meeting was a great scientific success, with several new results being presented. Namely, the latest results obtained with KEPLER data and the major deliverables of the SPACEINN Project, as well as an update on the efforts towards the preparation of the asteroseismic component of the TESS mission. The organizers thank, in particular, the contribution of the speakers and all the participants, whose role was critical in making this conference such a scientific (and social) success.

DOI: 10.1051/epjconf/201716000001

Scientific Organizing Committee:

Conny Aerts	Belgium
Sarbani Basu	United States of America
Timothy Bedding	Australia
William Chaplin	United Kingdom
Jørgen Christensen-Dalsgaard	Denmark (*Co-Chair*)
Margarida S. Cunha	Portugal
João Miguel T. S. Ferreira	Portugal
Rafael García	France
Laurent Gizon	Germany
Hans Kjeldsen	Denmark
David W. Latham	United States of America
Eric Michel	France
Mário J. P. F. G. Monteiro	Portugal (*Co-Chair*)
Pere L. Pallé	Spain
Ennio Poretti	Italy
Markus Roth	Germany

Local Organizing Committee

Margarida S. Cunha	IA U.Porto (*Co-Chair*)
João Miguel T. S. Ferreira	U.Açores (*Co-Chair*)
Paulo Peixoto	IA U.Porto
Elsa Silva	IA U.Porto

Sponsors

The organisation of the conference and the publication of the present volume were supported by:

- The *Instituto de Astrofísica e Ciências do Espaço*, through the *Centro de Astrofísica da Universidade do Porto* (IA U.Porto), with funds from the European Commission's SPACEINN Project on *Exploitation of Space Data for Innovative Helio- and Asteroseismology* (FP7-SPACE-2012-312844),

- The *Fundação para a Ciência e a Tecnologia* (FCT), through IA U.Porto, with national funds from UID/FIS/04434/2013 and from FEDER through COMPETE2020 POCI-01-0145-FEDER-007672.

- The conference also received support from the *Governo dos Açores* (SRTT-DRT 81-9-168/1418).

- The *Câmara Municipal de Angra do Heroísmo* provided the venue – *Centro Cultural e de Congressos* – where the conference took place. The organization would like to thank the excellent technical support at the venue by Mr. Rúben Machado and his team.

Porto & Angra do Heroísmo – Portugal, 15 September 2017

DOI: 10.1051/epjconf/201716000001

Participants

There was a total of 169 participants in the conference, from 20 countries. The list includes:

1. Aarslev, Magnus Johan .... Aarhus University
2. Aerts, Conny .... Institute of Astronomy, University of Leuven
3. Agentoft, Camilla .... Stellar Astrophysics Centre, Aarhus University
4. Antoci, Victoria .... Stellar Astrophysics Centre, Aarhus University
5. Arentoft, Torben .... Stellar Astrophysics Centre, Aarhus University
6. Augustson, Kyle .... CEA/DRF/IRFU Service d'Astrophysique

7. Bachulski, Szymon .... Pedagogical University of Cracow, Poland
8. Ball, Warrick .... Institut für Astrophysik Göttingen
9. Ballot, Jérôme .... IRAP
10. Barban, Caroline .... Observatoire de Paris/LESIA
11. Basu, Sarbani .... Yale University
12. Batalha, Natalie .... NASA Ames Research Center
13. Baudin, Frederic .... IAS
14. Bazot, Michael .... New York University Abu Dhabi
15. Beck, Paul .... CEA, Paris-Saclay
16. Bedding, Tim .... University of Sydney
17. Belkacem, Kevin .... Paris Observatory
18. Bellinger, Earl .... Max-Planck-Institut für Sonnensystemforschung
19. Benomar, Othman .... New York University Abu Dhabi
20. Bigot, Leonel .... Observatoire de la Côte d'Azur
21. Bowman, Dominic .... University of Central Lancashire
22. Broomhall, Anne-Marie .... University of Warwick
23. Buldgen, Gaël .... AGO, Université de Liège
24. Böning, Vincent .... Kiepenheuer-Institut für Sonnenphysik

25. Campante, Tiago .... University of Birmingham
26. Cantiello, Matteo .... Kavli Institute for Theoretical Physics
27. Cassisi, Santi .... INAF - Astronomical Observatory of Teramo
28. Charpinet, Stephane .... bservatoire Midi-Pyrénées
29. Chiappini, Cristina .... Leibniz Institute for Astrophysics Potsdam (AIP)
30. Christensen-Dalsgaard, Jørgen .... Stellar Astrophysics Centre, Aarhus University
31. Colman, Isabel .... The University of Sydney
32. Corsaro, Enrico .... Service d'Astrophysique - CEA Saclay
33. Creevey, Orlagh .... Observatoire de la Cote d'Azur
34. Cunha, Margarida .... Instituto de Astrofísica e Ciências do Espaço, U.Porto

35. Daszyńska-Daszkiewicz, Jadwiga .... Uniwersytet Wrocławski
36. Davies, Guy .... University of Birmingham
37. De Cat, Peter .... Royal Observatory of Belgium
38. Deheuvels, Sébastien .... IRAP Toulouse
39. Derekas, Aliz .... ELTE Gothard Astrophysical Observatory
40. Di Mauro, Maria Pia .... INAF - Istituto di Astrofisica e Planetologia Spaziali

41. Dziembowski, Wojtek ... Copernicus Astronomical Center, Polish Academy of Sciences

42. Eff-Darwich, Antonio .. Universidad de La Laguna & Instituto de Astrofísica de Canarias
43. Elsworth, Yvonne .... University of Birminham
44. Eyer, Laurent .... Geneva Observatory, University of Geneva

45. Ferreira, João Miguel .... U. Açores & IA
46. Fournier, Damien .... Institute for numerical and applied mathematics Göttingen
47. Fu, Jianning .... Beijing Normal University

48. Gade Pedersen, May .... Institute of Astronomy, Leuven University
49. García, Rafael A. .... SAp/CEA-Saclay
50. García Hernández, Antonio ...... Instituto de Astrofísica e Ciências do Espaço, U.Porto
51. Gaulme, Patrick .... New Mexico State University
52. Gehan, Charlotte .... LESIA, Paris Observatory
53. Giammichele, Noemi .... IRAP-CNRS/OMP
54. Gizon, Laurent .... Max Planck Institute for Solar System Research
55. Goupil, Mariejo .... Observatoire de Paris
56. Guo, Kaiming .... Beijing Normal University

57. Handberg, Rasmus .... Stellar Astrophysics Centre, Aarhus University
58. Handler, Gerald .... Nicolaus Copernicus Astronomical Center Warsaw
59. Hanson, Chris .... Max-Planck-Institut für Sonnensystemforschung
60. Hartig, Erich .... Astrophysics, University of Vienna
61. Hekker, Saskia .... Max Planck Institute for Solar System Research
62. Hermes, J.J. .... University of North Carolina at Chapel Hill
63. Holdsworth, Daniel .... University of Central Lancashire
64. Hong, Kyeongsoo .... Korea Astronomy and Space Science Institute
65. Houdek, Günter .... Stellar Astrophysics Centre, Aarhus University
66. Huber, Daniel .... University of Sydney
67. Hypolite, Delphine .... CEA

68. Jannsen, Nicholas Emborg .... Physic and Astronomy, Aarhus University
69. Jiang, Chen .... Instituto de Astrofísica e Ciências do Espaço, U.Porto
70. Jimenez Mancebo, Antonio .... Instituto de Astrofísica de Canarias
71. Jones, Caitlin .... University of Birmingham

72. Kawaler, Steve .... Iowa State University
73. Ketzer, Laura .... Missouri State University
74. Kiefer, René .... Kiepenheuer Institute for Solar Physics
75. Kjeldsen, Hans .... Aarhus University
76. Kolenberg, Katrien .... Institute of Astronomy, KU Leuven
77. Koo, Jae-Rim .... Korea Astronomy and Space Science Institute
78. Kuehn, Charles .... University of Northern Colorado
79. Kurtz, Donald .... University of Central Lancashire
80. Kuszlewicz, James .... University of Birmingham

81. Lagarde, Nadège .... Institut UTINAM Observatoire de Besançon
82. Langfellner, Jan .... Max Planck Institute for Solar System Research Göttingen
83. Latham, David .... Harvard-Smithsonian Center for Astrophysics
84. Lebreton, Yveline .... Paris-Meudon Observatory
85. Lee, JaeWoo .... Korea Astronomy and Space Science Institute
86. Lehmann, Holger .... Thueringer Landessternwarte Tautenburg
87. Leibacher, John .... National Solar Observatory
88. Liang, Zhi-Chao .... Max Planck Institute for Solar System Research
89. Lignieres, François .... IRAP
90. Löptien, Björn .... Max-Planck-Institut für Sonnensystemforschung
91. Lund, Mikkel .... University of Birmingham
92. Lundkvist, Mia Sloth .... LSW, University of Heidelberg

93. Marcadon, Frédéric .... IAS
94. Mathur, Savita .... Space Science Institute
95. McKeever, Jean .... New Mexico State University
96. Metcalfe, Travis .... Space Science Institute
97. Michel, Eric .... Observatoire de Paris-LESIA
98. Miglio, Andrea .... University of Birmingham
99. Mirouh, Giovanni .... IRAP / OMP
100. Molnar, Laszlo .... Konkoly Observatory, MTA CSFK
101. Monteiro, Mário J P F G .... Instituto de Astrofísica e Ciências do Espaço, U.Porto
.... & Departamento de Física e Astronomia, Faculdade de Ciências, U.Porto
102. Moravveji, Ehsan .... Institute of Astronomy, KU Leuven
103. Mosumgaard, Jakob .... Aarhus University
104. Murphy, Simon .... SIfA, University of Sydney

105. Nasello, Guillaume .... UTINAM
106. North, Thomas .... University of Birmingham
107. Nsamba, Benard .... Instituto de Astrofísica e Ciências do Espaço, U.Porto

108. Østensen, Roy .... Missouri State University
109. Ouazzani, Rhita-Maria .... SAC - Aarhus University

110. Pallé, Pere L. .... Instituto de Astrofísica de Canarias
111. Paparo, Margit .... Konkoly Observatory, MTA CSFK
112. Papics, Peter .... Institute of Astronomy, KU Leuven
113. Papini, Emanuele .... Max-Planck-Institut für Sonnensystemforschung
114. Peralta, Raphaël .... Observatoire de Paris - LESIA
115. Pereira, Filipe .... Instituto de Astrofísica e Ciências do Espaço, U.Porto
116. Pinçon, Charly .... Paris Observatory
117. Pinsonneault, Marc .... Ohio State University
118. Plachy, Emese .... Konkoly Observatory, MTA CSFK
119. Poretti, Ennio .... INAF-OA Brera
120. Prat, Vincent .... CEA-Saclay
121. Prsa, Andrej .... Villanova University

DOI: 10.1051/epjconf/201716000001

122. Quitral, Paola .... Instituto de Astrofísica e Ciências do Espaço, U.Porto

123. Reed, Mike .... Missouri State University
124. Reese, Daniel .... LESIA, Paris Observatory
125. Ren, Anbing .... Beijing Normal University
126. Rieutord, Michel .... IRAP
127. Rogers, Tamara .... Newcastle University
128. Roth, Markus .... Kiepenheuer-Institut für Sonnenphysik
129. Roxburgh, Ian .... Queen Mary University of London
130. Rybicka, Monika .... Nicolaus Copernicus Astronomical Center

131. Samadi Ghadim, Aunia .... IVS KU Leuven
132. Santos, Ângela .... Instituto de Astrofísica e Ciências do Espaço, U.Porto
133. Schindler, Jan-Torge ... Steward Observatory, Dep. of Astronomy, University of Arizona
134. Schmid, Valentina .... Institute for Astronomy, KU Leuven
135. Schofield, Mathew .... University of Birmingham
136. Schou, Jesper .... Max Planck Institute for Solar System Research
137. Sekii, Takashi .... NAOJ
138. Serenelli, Aldo .... Instituto de Ciencias del Espacio (ICE/CSIC-IEEC)
139. Sharma, Sanjib .... University Of Sydney
140. Silva Aguirre, Victor .... Aarhus University
141. Slawinska, Joanna .... Center for Environmental Prediction, Rutgers University
142. Smolec, Radoslaw .... Nicolaus Copernicus Astron. Center, Polish Academy of Sciences
143. Sowicka, Paulina .... Nicolaus Copernicus Astronomical Center
.... & Isaac Newton Group of Telescopes
144. Stello, Dennis .... University of Sydney
145. Szabo, Robert .... Konkoly Observatory, MTA CSFK
146. Szewczuk, Wojciech .... Astronomical Institute of the Wroclaw University

147. Takata, Masao .... University of Tokyo
148. Themessl, Nathalie .... Max Planck Institute for Solar System Research (MPS)
149. Thompson, Michael .... NCAR / HAO
150. Thoul, Anne .... University of Liege
151. Tkachenko, Andrew .... Institute of Astronomy, KU Leuven
152. Trampedach, Regner .... Stellar Astrophysics Centre, Aarhus University
.... & Space Science Institute at Boulder
153. Turck-Chièze, Sylvaine .... CEA/Saclay

154. Van Eylen, Vincent .... Leiden University
155. Van Reeth, Timothy .... Instituut voor Sterrenkunde, KU Leuven
156. van Saders, Jennifer .... Carnegie Observatories
157. Viani, Lucas .... Yale University
158. Vidotto, Aline .... Trinity College Dublin
159. Vrard, Mathieu .... Instituto de Astrofísica e Ciências do Espaço, U.Porto

160. White, Tim .... Stellar Astrophysics Centre, Aarhus University
161. Wu, Yaqian .... Beijing Normal University

162. Wu, Tao .... Yunnan Observatories, Chinese Academy of Sciences

163. Yu, Jie .... Sydney Institute for Astronomy, School of Physics, University of Sydney

164. Zhang, Ruyuan .... Beijing Normal university
165. Zong, Weikai .... IRAP - CNRS & Université de Toulouse
166. Zwintz, Konstanze ..... Institute for Astro- and Particle Physics, University of Innsbruck

Figure 1. Group photo of the Conference, taken outside the conference venue by Victoria Antoci.

Foreword

Jørgen Christensen-Dalsgaard[1]

[1] *Stellar Astrophysics Centre, Department of Physics and Astronomy, Aarhus University, Ny Munkegade 120, DK-8000 Aarhus C, Denmark*

Abstract. N/A

The Kepler Asteroseismic Science Consortium (KASC) workshops have become an important meeting place for discussions of asteroseismology and related fields, from exoplanet research to Galactic archaeology, as well as the broader aspects of stellar physics. With the approaching launch, planned for March 2018, of the TESS mission activities within the TESS Asteroseismic Science Consortium (TASC) will take on increasing importance. The present workshop in addition marked the conclusion of the very productive SpaceInn project, organized under the HELAS collaboration and supported by the EU FP7 programme. And finally the meeting involved a look back to the conference '*Seismology of the Sun and the Distant Stars*' organized by Douglas Gough in Cambridge in 1985 and served to mark Douglas's 75th birthday in February 2016. Unfortunately Douglas was unable to attend, except via Skype, but it is excellent that he has contributed his remarks to these proceedings.

Each KASC meeting has its own highlights. It is difficult to beat the epiphany at KASC3 of realizing that the asteroseismic data that for decades we had been waiting for were now finally available. However, the present meeting was remarkable in the scope and physical understanding provided by the talks and posters, including fruitful discussions, and demonstrating the extreme richness still to be fully explored from the Kepler data. Much of this is reflected in the present volume. Also, the meeting provided a start of the planning of the target selection and data analysis for the TESS mission, which will greatly increase the volume of asteroseismic data and provide a huge set of data on other sorts of variability of stars and other objects.

I write this just a few days after the PLATO mission was adopted by ESA's Science Programme Committee, with planned launch in 2026. The future of our field looks very promising indeed!

Aarhus, 26 June 2017

EPJ Web of Conferences **160**, 00003 (2017) DOI: 10.1051/epjconf/201716000003
Seismology of the Sun and the Distant Stars II

Salvete

Douglas Gough[1,2,*]

[1] *Institute of Astronomy, Madingley Road, Cambridge, CB3 0HA, UK*
[2] *Department of Applied Mathematics and Theoretical Physics, Wilberforce Road, Cambridge, CB3 0WA, UK*

Abstract. N/A

It is an extremely great disappointment to me that I am unable to be with you physically at what I am sure will be a wonderful meeting. Electronic is inferior by far to physical presence. I have been temporarily banned from driving and flying, which is why I cannot be with you, although now I feel to be in fine fettle. I'm sure that the ban will be lifted after my next medical consultation, which has been deliberately delayed by the doctors until after this meeting; they seem to think that I cannot be trusted to stay home!

I have often been called the Father of Helioseismology, and I see from the list of participants that many of my offspring are present: three of my twenty-five academic children, particularly Jørgen Christensen-Dalsgaard and Margarida Cunha who have played a central role in the conception and organization of this conference, and Michael Thompson; also seven or so of my academic grandchildren, one of whom, Mário Monteiro, has also played a major role here; in addition, countless great-grandchildren, several former postdocs and many many friends. It may interest those of my mathematical offspring who are here that academically we are all directly descended from Isaac Newton, who received his MA from the University of Cambridge in 1668, and before him Galileo Galilei who graduated from the Università di Pisa in 1585. Doubtless at this meeting you will not be looking back quite so far as those great scientists, nor, I suspect, quite so far forward. In fact, ostensibly you will be looking back only as far as the original conference with essentially the same title as this, held in Cambridge in 1985 with the aim of broadening the foundations of the subject by bringing together theorists and observers who had either contributed directly to the subject already or had a mind to do so. So I shall just say a few words about what happened with helioseismology in the earliest days leading up to that conference – asteroseismology was little more than a dream at that time. Four of the participants of that conference are here today: Jørgen Christensen-Dalsgaard, Yvonne Elsworth, Don Kurtz and Ian Roxburgh. Of those I thank especially Yvonne, who has kindly stepped in to deliver the introduction that I am unable to give, on basically the development of the subject over the last three decades between the two like-titled conferences, yet no doubt from a perspective somewhat different from my own.

That first conference took place exactly a decade after the date on which I view helioseismology to have began, namely 24 June 1975. It was fifteen years to the day after I had first met my wife-to-be, Rosanne, and was at a workshop on astrophysical convection that I was organizing in Cambridge. At the suggestion of Ed Spiegel, who had recently heard a fascinating seminar by Henry Hill, I invited Henry at the last minute to present us with his news. Henry reported a discrete spectrum of oscillations in the apparent diameter of the Sun, with frequencies around 3 mHz (periods around 5 minutes) corresponding to high-order acoustic modes (p modes) of low degree. It was immediately clear to me that the spectrum was rich enough for one to be able to carry out inverse calculations of the kind that geoseismologists, then known simply as seismologists, were doing to infer the internal structure of the Earth. Jørgen Christensen-Dalsgaard and I had been computing model solar g modes, so I asked Jørgen to compute modes further up the spectrum; and the next day he was able to give a talk presenting a sequence of frequencies superficially similar to Henry's. With the inclusion of some intriguing observations of an apparent 160-minute oscillation (Severny, Kotov and Tsap, 1976; Brookes, Isaak and van der Raay, 1976), the way seemed clear to try to measure the inside of the Sun (Christenesen-Dalsgaard and Gough, 1976). Soon afterwards I became aware of Franz Deubner's (1975) pioneering observations of spatially resolved 5-minute oscillations of high degree which were so detailed as to render possible a calibration of the upper boundary layer of the convection zone. The 160-minute oscillation all but went away.

As we all know, unresolved five-minute oscillations had already been discovered by Bob Leighton, Bob Noyes and George Simon more than a decade earlier, and a degree of spatial resolution was subsequently achieved by Ed Frazier (1968) who realized that the oscillations are likely to be acoustic waves trapped in the outer layers of the convection zone. These observations had triggered theoreti-

*e-mail: douglas@ast.cam.ac.uk

cal analyses by John Leibacher and Bob Stein (1971) and by Roger Ulrich (1970), confirming Ed's interpretation. Then, soon after Franz's observations were announced, Hiroyasu Ando and Yoji Osaki (1976) carried out detailed p-mode calculations to compare with them, finding that their theoretical frequencies were systematically too high. However, until that time discussion was aimed simply at establishing the nature of the oscillations, rather than their implications concerning the structure of the Sun. It was Franz who pointed out that it was the solar models themselves that needed adjustment to fit his observations.

Fortunately Hiroyasu and Yoji's theoretical frequencies were close to the values that Franz had measured. That rendered it possible to apply perturbation theory to estimate how their equilibrium solar model needed to be adjusted to model the Sun more faithfully. It was not necessary for the adjustment to be extremely accurate for getting a first idea. I approximated the upper convective boundary layer with a polytrope, which is far from being accurate, but it gave me an analytical expression for the adjustment to the entropy jump that was required. Combining that with a relation to the depth of the convection zone, which I had recently established with Nigel Weiss (1976), I was able to estimate that the Sun's convection zone was some 50% deeper than the values preferred by solar modellers of the day (who were pushing parameters in such directions as to try to engineer agreement with the measured neutrino flux, thereby favouring a low initial helium abundance Y_0, which implied a shallow convection zone). That set the depth to be about 220 Mm, locating the transition to radiative equilibrium at $r = r_c \simeq 0.69R_\odot$. Soon afterwards, Roger Ulrich, Ed Rhodes and George Simon (1977) confirmed that result, setting r_c to lie between $0.62R_\odot$ and $0.75R_\odot$; and after some painstaking error analysis, reported only briefly in some cases, the robustness of the conclusion was established (e.g., Lubow et al., 1980; Berthomieu et al., 1980). By the time of the 1985 conference I had also made two direct estimates that were independent of any reference solar model, using an asymptotic analysis based on Duvall's Law (Duvall, 1982). The first (1984a) was made simply by inspection of the radial variation of an equatorially biased spherical average of the square of the sound speed c from data provided by Tom Duvall and Jack Harvey (1983), in which one could see the discontinuity in its second derivative. I recall getting the result one Friday morning, in the early hours – one could only compute effectively at night in those days – plotting c^2, holding the paper in such a way as to be able to squint almost in its plane, and locating the discontinuity. I grabbed a couple of hours sleep and then took the train to London where I was to present the result at a meeting of the Royal Astronomical Society later that day. The audience was incredulous, because the radiative-convective transition is difficult, almost impossible, to discern on a screen viewed at right-angles. However, I assured them that my identification had been confirmed by the only student still working at 4am in the computer room, and that he was certainly unbiased because, being a cosmologist, he had no idea where to look. The transition at the base of the convection zone appeared to be smoother than in theoretical models, indicating either a degree of material mixing or that the transition is aspherical: today Jørgen and I have seen tentative evidence that it might be both. The second estimate (1984b) relied on a separation between a (structural) function W of essentially sound speed alone and a purely thermodynamic function, Θ, the two being numerically equal in, and only in, adiabatically stratified regions. By comparing the two it was possible to identify more precisely where adiabatic stratification ceased. The method was blind-tested at the 1985 conference using theoretical eigenfrequencies kindly provided by Roger Ulrich, reproducing his depths to well within its estimated $0.005R_\odot$ accuracy. I had also carried out the analysis on new unpublished solar data obtained by Jack and Tom, the outcome of which I wanted to announce, but Jack and Tom didn't want to be included because the manner in which their data would have been presented looked, in Jack's words, "too ratty", errors in the observations having been magnified considerably in the inversion procedure, as we now know to be inevitable. So I had to hold back; all I could say was that the base of the convection zone was near to $0.7R_\odot$. The precise result was finally published six years later, after Jack and Tom had published their data (and after they had again refused an invitation to join in the publication of the consequences, this time on the grounds that they had contributed nothing beyond their own publication, an excuse which, though literally correct, was certainly not true of the overall sum of our interactions). By then I was joined by Jørgen and Michael Thompson in an extended investigation perturbing solar models. But that was after the 1985 conference, and is therefore not within the purview of this salutation.

Of the major unresolved issues concerning the Sun in those early days, there were two, perhaps one might say three, that stood head-and-shoulders above the others: the so-called solar neutrino problem, associated with which was the matter of the helium abundance, and the test of the theory of General Relativity from orbital measurements of Mercury and spacecraft, the latter by radar ranging. I shall discuss them in order.

Establishing that the convection zone was deep by the standards of the day implied, according to theoretical models of the Sun, that the zero-age-main-sequence helium abundance Y_0 was relatively high – of the order of 25 per cent by mass – which was comforting because an abundance low enough to reproduce Ray Davis's neutrino counts in a standard solar model would have required facing the problem of explaining how it could be substantially lower than what was believed to be the cosmic value immediately after the Big Bang. However, because there was evidently a serious problem with the production of neutrinos in the Sun, the theory of stellar evolution, at least as applied to the Sun, was evidently in some doubt. A deep convective envelope was only an indirect and questionable indicator of the conditions under which neutrinos are being produced in the energy-generating core. Consequently Jørgen and I (1981) undertook a calibration of solar models with the low-degree p-mode frequencies which penetrate into the core. At that time we made the most naive attempt imaginable, merely minimizing the difference between theoretical eigenfrequencies and full-disc observations by Claverie et al. (1980, 1982), Grec et al. (1980) and Deubner (1981), paying no regard to what aspects of the frequency spectrum pertained principally to the core. We found two promising fits, each having different mode orders n assigned to the peaks in the power spectrum, one fit with low Y_0, the other with high Y_0, the latter being marginally better, although not significantly so, given the coarseness of the procedure. We had earlier found evidence for Y_0 not being low from what are now called the large and small frequency separations observed by Gérard Grec, Eric Fossat and Martin Pomeranz (1980), the former being a measure of the acoustic radius of the Sun, the latter a signature of the core. To a first approximation, neither requires knowing n. We compared the observations with new theoretical values obtained in 1980 taking the influence of the Sun's atmosphere more carefully into account. Nevertheless, some people, such as George Isaak (1980), adhered to a former opinion to the contrary based on an earlier analysis of the Birmingham group's observations (Claverie et al., 1979) and earlier theory (Christensen-Dalsgaard et al., 1979). The reason that mode orders n could not be identified observationally is that the full-disc spectra did not extend to sufficiently low frequency for it to be possible to extrapolate reliably to $n = 0$; fortunately, at least the degrees l were known, for they could be inferred from the pattern of the large and small frequency separations.

It was Tom Duvall and Jack Harvey (1983) who came to the rescue: I had persuaded them to make crucial observations of modes of intermediate degree which connected the full-disc data to the high-degree modes whose lowest-order branch (f modes) is essentially independent of the internal structure of the Sun, and can therefore be identified unambiguously. The outcome completed the low-degree seismic calibration, selecting the solar model with the higher Y_0. That preference was reinforced by observations having low spatial resolution that were carried out by John Wilcox and Phil Scherrer (1983), which extended the range of the whole-disc observations of deeply penetrating modes up to $l = 5$. My interaction with Tom and Jack, and with John (until his untimely death) and Phil, continued. Those early days were the most exhilarating period in my scientific life, Jack and Tom, particularly, directing observations towards answering questions that I posed rather than pursuing what they could measure the most accurately, and I carrying out calculations stimulated by their new observations in the hope of shedding more light on the matters in hand, all of which, of course, raised new questions: it was textbook scientific method in action. It has been fascinating to investigate the ramifications of the new preliminary data, under the understanding, of course, that the outcome could not be relied upon until the analysis of the data had been confirmed. Working at the border of credibility, with appropriate caution, is essential for advancement. Indeed, I recall the exuberance that I experienced from one of Jack's tentative findings: it led to us entering into a serious wager about its veracity. The scientific details are of no concern here, except to say that I wagered that Jack was correct, and he that he was wrong. The matter would be decided by further observations, expected in a year or two. Jack and I regard the wager as being very serious, not just because of the scientific issue, but also because of the magnitude of the stake: it is defined operationally, the outcome of which depends on the circumstances in which the winner is determined. Unfortunately for one of us, the magnitude of the outcome is no doubt an increasing function of time, and the crucial observation is yet to be made. Perhaps a more serious matter is that Jack is now more reticent about revealing to me his preliminary observational results, on the ground that I might believe them.

The equivalence of W and Θ in the adiabatically stratified regions of the convection zone provided the means to measure the helium abundance Y directly, for in its ionization zones Θ is augmented significantly above its perfect-gas value. Werner Däppen and I (1984) made a preliminary analysis of that augmentation, finding that $Y \simeq 0.23$, which, after accounting for gravitational settling (which we did not do at first; it was nearly a decade before it was taken into account in this context), implies $Y_0 \simeq 0.25$, a result which has not changed significantly since, although there has been much more work on assessing its reliability. We also foresaw the possibility of measuring the individual abundances of the most abundant heavy elements.

I experienced an interesting consequence of the f-mode character. In view of the fact that the frequencies ω of the high-degree modes – which hardly sense the sphericity of the Sun – do not depend on stratification, but just on the surface gravity g and horizontal wavenumber k according to the relation $\omega^2 = gk$, I could assess the accuracy of published $k - \omega$ diagrams, both theoretical and observed. I recall one day finding in a new publication by Tom and Jack a plot that fitted the theoretical relation precisely. I was very impressed by the seemingly implied accuracy of the observations. I mentioned it to Jack when next we met, to which he responded with a smile, adding that on finding my paper he had realized that use of the formula was a more accurate way of gauging the image scale on his telescope than what was in use before. That formula had long been known to describe surface gravity waves in deep pure water. But it was not appreciated that the formula is more generally applicable. In fact, a different expression for the frequencies of surface gravity waves in terrestrial oceans, derived in the Boussinesq approximation, was presumed by oceanographers of the time to provide information about salinity stratification; little was it realized that the deviation from the simpler formula was actually a product of the inaccuracy of the Boussinesq approximation. The ubiquity of the formula came to the attention James Lighthill, the Lucasian Professor in the applied mathematics department in which I taught. One day at coffee time, he approached me for an explanation. It seemed to me that no sooner had I began to establish what I considered to be a fruitful way of thinking about the matter when James interrupted, saying: "Thank you; of course

 DOI: 10.1051/epjconf/201716000003

you're right". It's amazing how impressively rapidly some people think with such enormous perspicacity.

Jack and Tom refined their observations – or perhaps just the analysis of their data. With a view to measuring the acoustic stratification of the entire Sun, they had imaged the Sun through a cylindrical lens whose axis was aligned with solar latitude, thereby extracting just the zonal and near-zonal modes which provide a weighted latitudinal average concentrated somewhat in the polar regions. It was now possible to invert the data to obtain a direct estimate of the sound speed throughout all but the innermost regions of the Sun. With the help of Jørgen, who provided theoretical solar models with which to compare the seismological analysis together with considerable additional insight, a quite detailed picture of the solar interior was emerging (Christensen-Dalsgaard et al. 1985). In particular, it convincingly ruled out solar models with low Y_0, although, interestingly enough, that inference was not immediately appreciated by the solar-neutrino community who appeared to be unaware of the robustness of conclusions derived from such simple physics as acoustics. An interesting suggestion arising from the work – only a suggestion at the time because it concerned much more complicated physics – was that under conditions near the base of the convection zone the opacity incorporated in the theoretical models was about 20% too low. Five years later that suggestion was verified.

The body of helioseismological evidence now provided the most compelling evidence towards locating the resolution of the solar neutrino problem: it lay in nuclear or particle physics, and not in the theory of stellar structure and evolution.

In a stroke of genius Jack Harvey thought of rotating his cylindrical lens by $\pi/2$. Although that did not collapse the solar image precisely on lines of longitude, it almost did so principally in the equatorial regions where the sectoral modes were concentrated. In that way Tom and Jack were able to measure rotational splitting. They were joined by theorists to infer the Sun's angular velocity almost to the core, and, because the inversions, primitive as they were, were relatively straightforward, publication (Duvall et al., 1984) came before that of the sound speed. By far the dominant feature of the inference is that the interior of the Sun is rotating almost uniformly at roughly the photospheric equatorial rate, and not a great deal faster as most theorists of the time suspected. Indeed it immediately settled a dispute with Bob Dicke (1964, 1967; Bretherton and Spiegel, 1968), who had argued from his inference of the high oblateness of a measure of the visible solar disc that the Sun's core rotates with a period of no more than a day or so, and that the associated oblateness of the gravitational potential supported his scalar-tensor theory of gravity against General Relativity. A value of the quadrupole moment J_2 of the Sun's exterior gravitational potential of 2×10^{-7} was calculated from our seismological inference; that is consistent with General Relativity. Because the Sun's angular velocity is inferred from frequency splitting by advection of seismic waves, which depends linearly on the (small) angular velocity, and not on the much smaller centrifugal force, which is quadratic, the inference of the value of J_2 is much more robust than Bob's estimate, not to mention the difficulties that had confronted Bob in interpreting the apparent shape of the photosphere.

The 1985 conference also addressed, briefly, the seismology of other stars. At that time asteroseismology was only just emerging as a reality. There was just one definitive observation of a sequence of well defined p-mode oscillation frequencies, by Don Kurtz and Jan Seeman (1983) of the rapidly oscillating (ro) Ap star HR 1217, which heralded the beginning of the subject proper. In addition, there were strong hints of solar-like oscillations in α Centauri A (Fossat et al., 1984) and ϵ Eridani (Noyes et al., 1984), but the individual frequencies were not yet properly resolved. And certainly one should not forget that Don, the discoverer of roAp stars (1982), already had important observations of single modes of the significantly aspherical stars, which were to provide important asteroseismic diagnostics of not only the influence of asphericity on the dynamics of acoustic oscillations (Dziembowski and Goode, 1984, 1985; Gough and Taylor, 1984) but also, in due course, the means of addressing aspects of the nature of the large antipodal spots that are responsible for the asphericity. The state of the subject at that time is authoritatively reviewed by Don in the proceedings of the 1985 conference.

In addition to the issues that I have mentioned here, there was much work being undertaken to develop new instrumentation and advance the theory, the fruits from which I am sure Yvonne will address. The state of developments prior to the 1985 conference can be judged from reviews (Christensen-Dalsgaard & Frandsen, 1983; Christensen-Dalsgaard, 1984; Deubner & Gough, 1984; Christensen-Dalsgaard, Gough & Toomre, 1985) and the proceedings of conferences (Hill & Dziembowski, 1980; Belvedere & Paternò, 1984; Ulrich et al., 1984)

I have had a wonderful journey through the early days of the development of the subject, and I have enjoyed sharing it with so many of you. Many astronomers were stunned by what we had achieved, and there were some amongst them who even believed that we had accomplished all that there was to do. How wrong they were. The subject since has blossomed enormously, especially in the arena of asteroseismology. We are poised for further exciting advances, for which I'm sure this meeting will provide a solid foundation.

References

[1] Ando, H. & Osaki, Y., Publ. Astron. Soc. Japan, **27**, 581 (1976)

[2] Belvedere, G. & Paternò, L., Oscillations as a Probe of the Sun's Interior, Mem. Soc. Astron. Italiana, **55** (1984)

[3] Berthomieu et al., Nonradial and Nonlinear Stellar Pulsation (ed. H.A. Hill and W.A. Dziembowski), p.307 (1980)

[4] Bretherton, F.P. & Spiegel, E.A., Astrophys. J., **153**, L77 (1968)

[5] Brookes, J.R., Isaak, G.R. & van der Raay, H.B., Nature, **259**, 92 (1976)

[6] Christensen-Dalsgaard, J., Space Research Prospects in Stellar Activity and Variability (ed. A. Mangeney and F. Praderie, Springer-Verlag), p.11 (1984)

[7] Christensen-Dalsgaard, J. & Frandsen, S., Sol. Phys., **82,**, 165 (1983)

[8] Christensen-Dalsgaard, J. & Gough, D.O., Nature, **259**, 89 (1976)

[9] Christensen-Dalsgaard, J. & Gough, D.O., Nature, **288**, 544 (1980)

[10] Christensen-Dalsgaard, J. & Gough, D.O., Astron. Astrophys., **104**, 173 (1981)

[11] Christensen-Dalsgaard, J., Gough, D.O. & Morgan, J.G., Astron. Astrophys., **73**, 121 (1979)

[12] Christensen-Dalsgaard, J., Gough, D.O. & Toomre, J., Science, **229**, 923 (1985)

[13] Christensen-Dalsgaard, J. et al., Nature, **315**, 378 (1985)

[14] Claverie, A. et al., Nature, bf 282, 591 (1979)

[15] Claverie, A. et al., Astron. Astrophys. Lett., **91**, L9 (1980)

[16] Claverie, A. et al., Sol. Phys., **74**, 51 (1982)

[17] Däppen, W. & Gough, D.O., Theoretical problems in stellar stability and oscillations (ed. A. Noels & M. Gabriel, Univ. Li'ege), 264 (1984)

[18] Deubner, F-L., Astron. Astrophys., **44**, 371 (1975)

[19] Deubner, F-L., Nature, **290**, 682 (1981)

[20] Deubner, F-L. & Gough, D.O., Ann. Rev. Astron. Astrophys., **22**, 593 (1984)

[21] Dicke, R.H., Nature, **2902**, 432 (1964)

[22] Dicke, R.H., Astrophys. J., **49**, L121, (1967)

[23] Duvall Jr, T.L., Nature, **300**, 242 (1982)

[24] Duvall Jr, T.L. & Harvey, J.W., Nature, **302**, 24 (1983)

[25] Duvall Jr, T.L. et al., Nature, **310**, 22 (1984)

[26] Dziembowski, W.A. & Goode, P.R., Mem. Soc. Astron. Italiana, **55**, 185 (1984)

[27] Dziembowski, W.A. & Goode, P.R., Astrophys. J., **296**, L27 (1985)

[28] Fossat, E. et al., Comptes Rendus Acad, Sci Paris, Ser. II, **299**, 17, (1984)

[29] Frazier, E.N., Zeitschr. für Astrophys., **68**, 345 (1968)

[30] Gough, D.O., The Observatory, **104**, 118 (1984a)

[31] Gough, D.O., Mem. Soc. Astron. Italiana, **55**, 13 (1984b)

[32] Gough, D.O. & Taylor, P.P., Mem. Soc. Astron. Italiana, **55**, 215 (1984)

[33] Gough, D.O. & Weiss, N.O., MNRAS, **176**, 589 (1976)

[34] Grec, G., Fossat, E. & Pomerantz, M., Nature, **288**, 541 (1980)

[35] Hill, H.A. & Dziembowski, W.A. (ed.), Nonradial and Nonlinear Stellar Pulsation, Springer-Verlag, Berlin (1980)

[36] Iben Jr, I. & Mahaffy, J., Astrophys. J., **209**, L39 (1976)

[37] Isaak, G.R., Nature, **283**, 644 (1980)

[38] Kurtz, D.W., MNRAS, **200**, 807 (1982)

[39] Kurtz, D.W., Seismology of the Sun and the Distant Stars (ed. D.O. Gough), p417 (1985)

[40] Kurtz, D.W. & Seeman, J., MNRAS, **205**, 11 (1983)

[41] Leibacher, J.W. & Stein, R.F., Astrophys. Lett., **7**, 191 (1971)

[42] Leighton, R.B., Noyes, R.W. & Simon, G.W., Astrophys. J., **135**, 474 (1962)

[43] Lubow, S.H., Rhodes Jr, E.J. & Ulrich, R.K., Nonradial and Nonlinear Stellar Pulsation (ed. H.A. Hill and W.A. Dziembowski), p.300 (1980)

[44] Noyes, R.W. et al. Astrophys. J., **285**, L23 (1984)

[45] Scherrer, P.H., Wilcox, J.M. et al., Sol. Phys., **82**, 75 (1983)

[46] Severny, A.B., Kotov, V.A. & Tsap, T.T., Nature, **259**, 87 (1976)

[47] Ulrich, R.K., Astrophys. J., **162**, 993 (1970)

[48] Ulrich, R.K., Rhodes Jr, E.J., Harvey, J.W. & Toomre, J, Solar Seismology from Space, JPL Publication 84-84 (1984)

The BRITE-Constellation Nanosatellite Space Mission And Its First Scientific Results*

G. Handler[1,**], A. Pigulski[2], W. W. Weiss[3], A. F. J. Moffat[4], R. Kuschnig[3,5], G. A. Wade[6], P. Orleański[7], S. M. Ruciński[8], O. Koudelka[5], R. Smolec[1], K. Zwintz[9], J. M. Matthews[10], A. Popowicz[11], D. Baade[12], C. Neiner[13], A. A. Pamyatnykh[1], J. Rowe[14], and A. Schwarzenberg-Czerny[1]

[1]*Nicolaus Copernicus Astronomical Center, Bartycka 18, 00-716 Warsaw, Poland*
[2]*Instytut Astronomiczny, Uniwersytet Wrocławski, ul. Kopernika 11, 51-622 Wrocław, Poland*
[3]*Institute for Astrophysics, Universität Wien, Türkenschanzstrasse 17, A-1180 Wien, Austria*
[4]*Départment de physique, Université de Montréal, C. P. 6128, Succ. Centre-Ville, Montréal, QC H3C 3J7, Canada*
[5]*Institut für Kommunikationsnetze und Satellitenkommunikation, Inffeldgasse 12/I, 8010 Graz, Austria*
[6]*Department of Physics, Royal Military College of Canada, PO Box 17000, Stn Forces, Kingston, ON K7K 7B4, Canada*
[7]*Centrum Badań Kosmicznych, Polska Akademia Nauk, Bartycka 18A, 00-716 Warszawa, Poland*
[8]*Department of Astronomy and Astrophysics, University of Toronto, 50 St. George Street, Toronto, ON M5S 3H4, Canada*
[9]*Institut für Astro- und Teilchenphysik, Universität Innsbruck, Technikerstrasse 25/8, 6020, Innsbruck, Austria*
[10]*Department of Physics and Astronomy, University of British Columbia, Vancouver, BC V6T1Z1, Canada*
[11]*Silesian University of Technology, Institute of Automatic Control, Gliwice, Akademicka 16, Poland*
[12]*European Organisation for Astronomical Research in the Southern Hemisphere, Karl-Schwarzschild-Str. 2, 85748 Garching b. München, Germany*
[13]*LESIA, Observatoire de Paris, PSL Research University, CNRS, Sorbonne Universités, UPMC Univ. Paris 06, Univ. Paris Diderot, Sorbonne Paris Cité, 5 place Jules Janssen, F-92195 Meudon, France*
[14]*Institut de recherche sur les exoplanétes, iREx, Département de physique, Université de Montréal, Montréal, QC, H3C 3J7, Canada*

Abstract. The BRIght Target Explorer (BRITE) Constellation is the first nanosatellite mission applied to astrophysical research. Five satellites in low-Earth orbits perform precise optical two-colour photometry of the brightest stars in the night sky. BRITE is naturally well suited for variability studies of hot stars. This contribution describes the basic outline of the mission and some initial problems that needed to be overcome. Some information on BRITE data products, how to access them, and how to join their scientific exploration is provided. Finally, a brief summary of the first scientific results obtained by BRITE is given.

1 Introduction

A BRIght Target Explorer (BRITE) is a nanosatellite designed as a cube of 20 × 20 × 20 cm edge length that weighs approximately 7 kg. It carries a 3-cm telescope with a wide field of view (about 24° on the sky, nearly unvignetted) that feeds an uncooled 4008 × 2672-pixel KAI-11002M CCD. As such, it is predestined to obtain high-precision space photometry of apparently bright stars.

There is more than one BRITE. Each of the partner countries participating in the mission, Austria, Canada, and Poland, funded two satellites. One of each pair is equipped with a blue-sensitive, the other with a red-sensitive filter. The passbands are close to Johnson-Cousins B and R_c. The whole ensemble of satellites is called BRITE-Constellation and is therefore capable of multicolour time-resolved photometry.

The six satellites were launched into orbit between February 2013 and August 2014. One of the Canadian satellites, BRITE-Montreal, went astray during launch as it apparently did not separate from the last stage of the rocket. The remaining five satellites are called BRITE-Austria (BAb) and Uni-BRITE (UBr, both Austrian), BRITE-Lem (BLb) and BRITE-Heweliusz (BHr, both Polish), and BRITE-Toronto (BTr, Canadian), and are operating. The abbreviations of the satellites' names, that will be used in what follows, originate from their designations followed by the filter they carry, red (r) or blue (b).

More detailed accounts about the basic goals and outline of the BRITE-Constellation mission have been given by [1] and [2]. In this contribution, we briefly summarize these articles and the first science results, but also provide a glimpse into BRITE operations for prospective users.

*Based on data collected by the BRITE Constellation satellite mission, designed, built, launched, operated and supported by the Austrian Research Promotion Agency (FFG), the University of Vienna, the Technical University of Graz, the Canadian Space Agency (CSA), the University of Toronto Institute for Aerospace Studies (UTIAS), the Foundation for Polish Science & Technology (FNiTP MNiSW), and National Science Centre (NCN).
**e-mail: gerald@camk.edu.pl

2 Scientific goals and observations

As mentioned in Sect. 1, BRITE-Constellation obtains high-precision time-resolved two-colour photometry of bright stars. Apparently bright stars are often intrinsically luminous, which means that the stellar sample accessible to BRITE contains an overrepresentation of hot and evolved stars. As hot stars are short-lived, they are concentrated near the Galactic plane. This area is of prime interest for BRITE observations, because many target stars and bright guide stars used for tracking can be acquired. The primary science goal of the mission is thus easy to define: to study the variability of hot stars, be it caused by pulsation, rotation, wind variations, star-disk interactions, binarity etc. An HR Diagram of the stars observable by BRITE is shown in [1].

The observing strategy itself bears in mind that the longer the time base of observations, the more information on stellar variability can be obtained. Consequently, each BRITE observing run attempts to observe a given field as long as possible, for up to half a year. This goal is met in most cases. Given the low-Earth orbit of the satellites, occultations of the target field by the Earth or stray light imply that only a fraction of the ~ 100-min orbital period (up to 40%, depending on the satellite) can be used for observations.

The ensemble of satellites can be used to observe one field with all five of them (diminishing aliases at the satellites' orbital periods), or their number is split between two fields. A given satellite may also observe more than one field during its orbit. In the first few years of operation, it has been found most useful that two pairs of satellites (one with a blue filter, the other with a red one) each observe two different fields separated by about 6 h in right ascension, with the remaining fifth satellite used as a supplement.

The measurements of BRITE-Constellation are being retrieved by three ground stations, one in Graz (Austria), one in Warsaw (Poland), and the third in Toronto (Canada). Each country operates their satellites largely independently, with occasional help by the partners if need be. The downlink capacity is limited (also because of some interference issues at the European ground stations, see [2]), and therefore full-frame images are generally not downloaded. Instead, windows that contain target stars and a sufficiently large area of blank sky around them are downloaded for further processing. An overview of the BRITE observations so far is presented in Table 1, and the observed fields are depicted on a sky map in Figure 1.

As soon as the first science images had been obtained by the satellites launched earliest, it became clear that the CCDs suffered from radiation damages, causing hot pixels and columns, as well as Charge Transfer Inefficiency (CTI). Obviously, these defects would have a negative influence on the quality of the photometry to be obtained and needed to be mitigated. This was accomplished in three ways. First, some of the satellites were still on Earth when the problem was noticed. Therefore, BHr and BTr were equipped with better shielding, strongly decreasing the radiation damage. Second, a sophisticated data reduction pipeline has been developed that eliminates these adverse effects to a large extent [3].

Table 1: Target fields observed by BRITE-Constellation so far. Data taken from http://brite.craq-astro.ca/doku.php. The total number of stars observed is not the sum of the individual ones as some stars/fields have been re-observed.

Field	from civil date	to civil date	ΔT (d)	#stars
Ori I	01.12.2013	18.03.2014	108	15
Cen	25.03.2014	18.08.2014	147	32
Sgr I	29.04.2014	06.09.2014	42	19
Cyg I	12.06.2014	25.11.2014	167	36
Per	02.09.2014	18.02.2015	170	37
Orion II	24.09.2014	17.03.2015	175	35
Vel/Pup	11.12.2014	28.05.2015	169	52
Sco	18.03.2015	31.08.2015	167	26
Cyg II	01.06.2015	25.11.2015	178	34
Cas/Cep	23.08.2015	17.10.2015	55	25
CMa/Pup	18.10.2015	14.04.2016	180	32
Cru/Car	22.01.2016	22.07.2016	183	40
Sgr II	15.04.2016	23.09.2016	162	17
Cyg/Lyr	28.04.2016	21.10.2016	177	15
Total				350

Third, and most importantly, a "nodding" technique has been implemented in the observations, starting from early 2015. The BRITE satellites no longer stare at the same position. Instead, they are moved back and forth by about 0.2° every ≈ 20 s. Thus, the stellar images fall onto alternatingly different parts of their CCD windows, whereas the compromised pixels remain the same over such short time intervals. Consequently, differential images are computed, which are almost free of artifacts, and precise photometry can be obtained from these images. We refer to [2] for more information and illustrations of the process. Nodding mode is now the standard observing mode for BRITE. It also guarantees that the Constellation can be operated for much longer than the projected lifetime of two years because radiation damage of the CCDs is now a minor issue in this regard.

3 BRITE data products and policies

After the CCD data have been processed through the BRITE photometry pipeline [3], [4], the data for individual stars are assigned to a so-called "contact PI" who is responsible for their final reduction, analysis and publication. This assignment is based on calls for proposals in the years 2008, 2011 and 2016 (still open at the time of writing, see http://www.brite-constellation.at/).

If a given star has been proposed for observation by several researchers, these decide between themselves who will become contact PI. A contact PI is expected to provide a first report on data analysis three months after receiving the data (including a judgement whether or not

 DOI: 10.1051/epjconf/201716001001

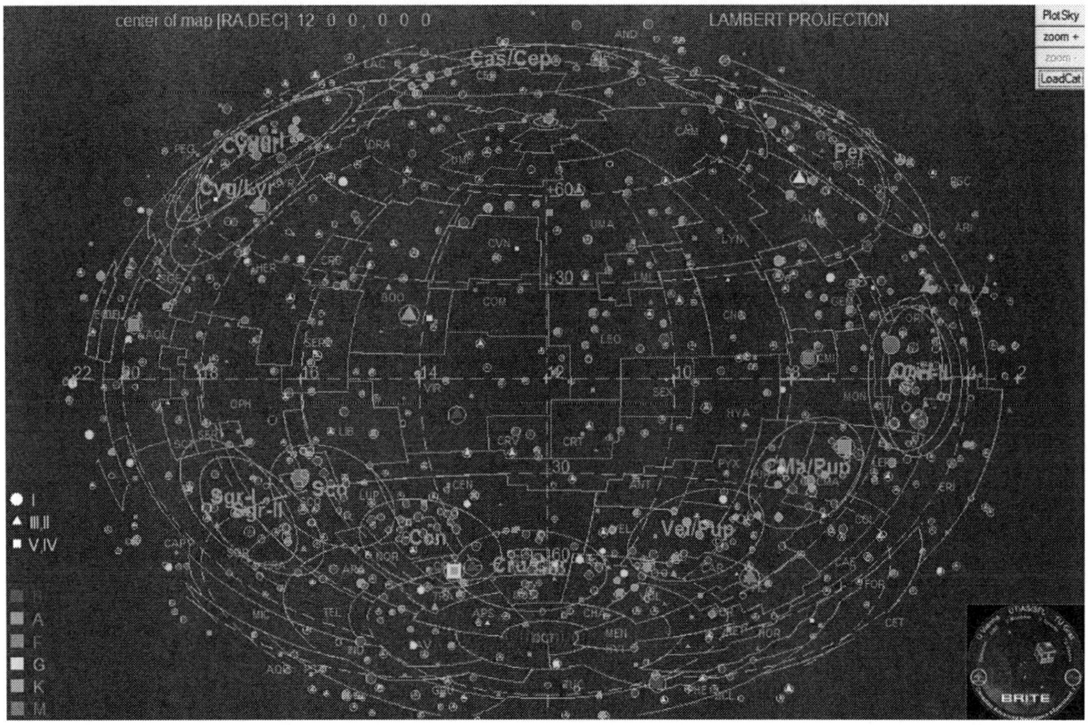

Figure 1: A sky map containing all stars brighter than $V = 4.5$. The colour code refers to spectral type, the symbols to luminosity class. Encircled stars have been proposed for BRITE observations. The sky areas labeled and encircled in pink have been observed by BRITE to date (cf. Table 1). Almost the whole galactic plane has been sampled.

the respective data set is suitable for stand-alone publication), and a final report after one year. After this one-year proprietary period has passed, the data will become public, unless an extension of the proprietary period of up to half a year has been approved in exceptional cases. A list of fields observed, stars measured therein, the respective contact PIs, and links to public data can be found at http://brite.craq-astro.ca/doku.php. Even if a data set is still in the proprietary phase, the contact PI can be approached for possible collaborations, as each contact PI is free to set up their own network of co-workers.

Light curves are supplied to scientists in the form of ASCII files. In the current format, these contain the timing in HJD, the measured target flux, and several additional parameters needed to decorrelate the light curves from remaining instrumental effects. These comprise the x/y position of the barycentre of the stellar PSF, the CCD temperature, and three further parameters describing the shape of the PSF (explained in detail by [4]).

Decorrelation of the BRITE photometry is required because the data orginate from an uncooled instrument (CCD and optics) and there is no possibility to acquire calibration images, such as bias, dark and flat field frames. Examples of the temperature changes occuring during a run on a given field, and within the orbit, are again given by [2]. Decorrelation with temperature is usually most important, as the individual CCD pixels' sensitivities depend on it, followed by decorrelation with x/y position to compensate for pixel-by-pixel sensitivity variations. Since temperature changes also change the PSF, decorrelation with the three PSF-related parameters may also be necessary, and sometimes the decorrelations must be performed in data subsets depending on average CCD temperature. In rare cases, stray light corrections need to be applied as well.

Several different approaches have been developed to carry out this decorrelation (e.g., [5], [6]). Sometimes it is necessary to first remove a target's variability from the observations to decorrelate optimally [7]). Even though these procedures may appear difficult, and several iterations may be required, the problems are understood and improved algorithms are permanently being developed.

Regarding the organization of the project, the scientific decisions such as target field selection, proposal evaluation, proprietary period extensions etc., are made by the BRITE Executive Science Team (BEST) whose current representatives are also the authors of the present manuscript. The scientific user community is organized as the BRITE-Constellation International Advisory Team (BIAST). Membership is granted after an informal request to one of us (GAW, `Gregg.Wade@rmc.ca`) and approval by BEST. BIAST members are informed about the actual status of the mission via a monthly email newsletter. Last but not least, the BRITE-Constellation Ground-Based Observing Team (GBOT) provides a plat-

DOI: 10.1051/epjconf/201716001001

form for BRITE scientists and observers worldwide to support collaborations and maximize the scientific output of BRITE-Constellation. The brightness of the targets facilitates ground-based support observations: even amateurs with modest-sized telescopes have obtained science-quality high-resolution spectroscopy.

4 First scientific results

The first three scientific papers have been published in a short series [5], [8], [9], and have examined three different types of object in the Centaurus field. At this early stage in the mission, only the Austrian BAb and UBr were available in full science mode, and were later joined by BLb and BTr after their commissioning had been completed.

First, the brightest rapidly oscillating Ap (roAp) star α Circini was observed over 33 of its 4.479-d rotational periods. The blue and red filter light curves were found to be qualitatively different, with the blue light curve showing a single-wave variation, whereas the red light curve exhibited a double-wave structure. Both were analysed using Bayesian photometric imaging. In addition, the strongest roAp-type oscillation were recovered in both filters, demonstrating the photometric sensitivity of BRITE data. We refer to [8] for details and the full analysis.

Together with the brightest roAp star, the brightest β Cephei star, β Centauri, was studied during the same observing campaign. This triple system contains two pulsating B-type stars, one of them with a measured magnetic field. Both stars exhibit pressure and gravity modes. Previous studies of the system revealed two oscillation modes spectroscopically; photometric studies remained inconclusive. BRITE observations revealed a completely different picture. A total of 17 independent frequencies (8 likely g mode pulsations and 9 p modes) and two combinations thereof were detected [5]. The orbital light-time effect may reveal which frequencies in the light variations originate from which star, but since the orbital period of the B-star binary is close to one year, more observations are needed to this end - and are certainly worth the effort.

The third of the initial three BRITE science publications reported a study of two Be stars, μ and η Centauri [9]. Whereas in μ Cen, strong light echoes from the inner circumstellar disk, which is viewed face-on, prevented new insights into the pulsational properties, the study of η Cen turned out to be more revealing. The modulation frequency of the star-disk mass transfer is equal to the frequency difference of two nonradial pulsation modes. This low frequency also modulates the (large) amplitude and the frequency of a circumstellar (so-called Štefl frequency), which is slightly lower than those of the stellar pulsations. The modulations in the mass loss are linked to variations in this Štefl frequency. This implies the presence of two engines causing the pulsation-mass loss interactions, a pulsational one from the star, and one related to viscosity in the circumstellar disk. In-depth arguments and discussions can be found in [9].

Another β Cephei studied by BRITE is ν Eridani [7], observed in the Orion II field by BAb, BLb, BTr and BHr. This star has been the subject of detailed observational and asteroseismic studies before, which have mostly concentrated on its p-mode spectrum. Earlier ground-based observations did detect two low-frequency g modes for ν Eridani, but it was the new BRITE observations that revealed six new gravity modes, firmly establishing it as a hybrid pulsator. The excitation of these g modes is still a problem for theoretical modelling. Whereas it is possible to excite a few g modes in a certain frequency range in stellar models, reproducing the whole extent of the observed gravity mode domain is extremely difficult. To this end, opacity-modified models of ν Eridani were computed [11]; only those can account for the pulsational mode instability. On the other hand, such models also have significantly modified interior structure and it remains to be seen how realistic these modifications are. Future analysis of BRITE measurements of additional β Cephei stars may provide an answer to this question.

When embarking on new types of observation, and achieving better precision than in the past, new scientific discoveries can result. Also BRITE has new discoveries to offer. In addition to the discovery of the role of difference frequencies in the mass loss of Be stars, BRITE has revealed the presence of massive heartbeat stars. These objects are close binaries in eccentric orbits, some of which also show tidally excited pulsations. The *Kepler* mission has discovered approximately two dozen such systems (e.g. [12]). However, all of these are composed of stars with masses below 2 $M_\odot$. With BRITE-Constellation, three more such systems have been discovered so far. ι Orionis [13] consists of an O9-type primary (23 $M_\odot$) and a B-type secondary (13 $M_\odot$), where the O star also shows tidally induced oscillations and is therefore the first of its kind. ϵ Lupi is the first doubly-magnetic massive star binary [14] with components of 8.7 and 7.3 $M_\odot$, respectively, and it shows peak flux modulation at periastron [15]. Finally, ζ Centauri ($M \approx 7\,M_\odot$) also exhibits a light curve strongly indicative of periastron brightening [16]; new radial velocities have been obtained to confirm this hypothesis.

5 Conclusions

Being the first nanosatellite mission carrying out astrophysical research, BRITE-Constellation had to overcome many "childhood diseases", including initially unexpected radiation damage of its CCD detectors, interference issues impeding the communication between the satellites and the European-based ground stations, and systematic effects occurring in the light curves. However, the child has grown up and developed antibodies successfully fighting these diseases. The mission has been in full science operation for about two years and is now in its presumably most fruitful phase of data acquisition.

The BRITE project is open for collaboration that can be established in several ways, be it through the submission of target proposals (a call is open at the time of writing), be it through joining the BRITE International Advisory Team (BIAST) or the Ground-Based Observing Team

DOI: 10.1051/epjconf/201716001001

(GBOT), or be it through attending the next BRITE science conference which will take place in Canada in August 2017.

For more information on the project itself, visit the BRITE home page http://www.brite-constellation.at/. Should you be interested in BRITE data, have a look at the BRITE photometry wiki (http://brite.craq-astro.ca/doku.php). In case you are a social media user, feel free to connect through the BRITE-Constellation facebook page.

Acknowledgments

The Polish contribution to the BRITE project is supported by the NCN grants 2011/01/B/ST9/05448, 2011/01/M/ST9/05914, 2011/03/B/ST9/02667 and 2015/18/A/ST9/00578. GAW and SMR acknowledge Discovery Grant support from the Natural Science and Engineering Research Council (NSERC) of Canada. AFJM is grateful for financial aid from NSERC and FQRNT (Quebec). KZ acknowledges support by the Austrian Fonds zur Förderung der wissenschaftlichen Forschung (FWF, project V431-NBL). BTr operations are supported through a Canadian Space Agency (CSA) Academic Development grant. RK and WW acknowledge financial support by the Austrian Space Application Programme (ASAP) of the Austrian Research Promotion Agency.

References

[1] Weiss W. W. et al., PASP **126**, 573 – 585 (2014)
[2] Pablo H. et al., PASP **128**, 125001 (20pp) (2016)
[3] Popowicz A., Proc. SPIE **99041R** (2016)
[4] Popowicz A., et al., in preparation
[5] Pigulski A., et al., A&A **588**, A55 (2016)
[6] Buysschaert B., et al., A&A, submitted
[7] Handler G., et al., MNRAS **464**, 2249 (2017)
[8] Weiss W. W., et al., A&A **588**, A54 (2016)
[9] Baade D., et al., A&A **588**, A56 (2016)
[10] Bruntt H., et al., MNRAS **396**, 1189 (2009)
[11] Daszyńska-Daszkiewicz J., et al., MNRAS **466**, 2284 (2017)
[12] Thompson S. E., et al., ApJ **753**, 86 (2012)
[13] Pablo H., et al., MNRAS, in preparation
[14] Shultz M., et al., MNRAS **454**, L1 (2015)
[15] Wade G. A., et al., Polish Astronomical Society Proceedings, submitted
[16] Handler G., et al., in preparation

Update on the APOKASC analysis of Red Giants

Yvonne Elsworth[1,2,★]

[1] *School of Physics and Astronomy, University of Birmingham, Birmingham B15 2TT, UK*
[2] *Stellar Astrophysics Centre, Department of Physics and Astronomy, Aarhus University, Ny Munkegade 120, DK-8000 Aarhus C, Denmark*

Abstract. APOKASC is a collaboration between APOGEE's spectroscopic survey in the near IR at Apache Point (SDSS) and Kepler Asteroseismic Science Consortium (KASC) for the study of red-giant stars. The aim of the collaboration is to take advantage of the good spectroscopic data of APOGEE together with the excellent KASC asteroseismic data from the *Kepler* mission to derive reliable and precise masses and ages for a large number of red-giant stars by combining data from the two sources. We report here on the progress to date on the just over 6600 stars being considered.

1 Introduction

The APOKASC consortium was formed to take advantage of the good spectroscopic data of APOGEE and the excellent asteroseismic data from the *Kepler* mission to derive reliable and precise masses and ages for a large number of red-giant stars. The first results on 1916 red-giant stars were published in 2014[1]. Since then, the number of stars observed has been considerably expanded, longer duration *Kepler* datasets are available and the spectroscopic pipeline of APOGEE has been improved. We report here on the progress with the grid-based modelling used to derive stellar masses and other parameters for just over 6600 red-giant stars. We take advantage of the latest advances in stellar modelling for stars on the upper portion of the red-giant branch where homologous scaling breaks down. Additionally, we are able to use the full precision of the asteroseismic data because data from several pipelines are provided as inputs to the grid-based modelling and it is now possible to provide information on the evolutionary state of individual stars based on the asteroseismic information and the effective temperature of the star.

2 The Asteroseismic Data

The average asteroseismic parameters of ν_{max} and $\Delta\nu$ have been provided by five different teams. The methods used by the teams are described in A2Z[2], CAN[3], COR[4], OCT[5], and SYD[6]. The *Kepler* data used in the analysis was prepared by various different preparation methods[7] [8][Mathur et al, in preparation] and it was checked that the preparation method had no significant impact on the derived parameters. From the asteroseismic spectra, it is possible to estimate the evolutionary state of the stars, i.e. has Helium started to be burnt in the core of the star? There are diverse ways of determining the evolutionary state and the detail of the methods used and the method to combine the different categorisation is given in Elsworth et al. (in preparation). In essence, the combination process looks for majority agreement between the different categorisation methods. For a summary of the different methods used see Elsworth et al.(submitted). In only about 8% of the stars is there not a reasonable consensus. These parameters, together with spectroscopic parameters from APOGEE DR13[9] are used as inputs to the grid-based modelling in order to determine the mass and other properties of the individual stars. The normal practice is to choose one preferred set of average asteroseismic parameters and to then inflate the uncertainties of that method to reflect the range of values provided by the other methods. We have chosen not to follow that route. Instead we have supplied all the data (after a small amount of extreme outlier rejection) to the grid-based modelling together with their original uncertainties. We do this to take full advantage of the precision in the average asteroseismic parameters and to explore the systematic effects of the different methods.

The asteroseismic ν_{max} values have an additional use in that they are believed to provide precise and accurate estimates of the surface gravity of the stars.

3 Goals of the Project

From the average asteroseismic parameters and the effective temperature, it is possible to use the principles of homology to compute mass and radius. For this we use the following scaling relations[10] which can be re-arranged to provide standard expressions for mass and radius. The constant of proportionality is provided by the solar reference values for $\Delta\nu$, ν_{max} and T_{eff} which are specific to the individual methods.

★e-mail: y.p.elsworth@bham.ac.uk

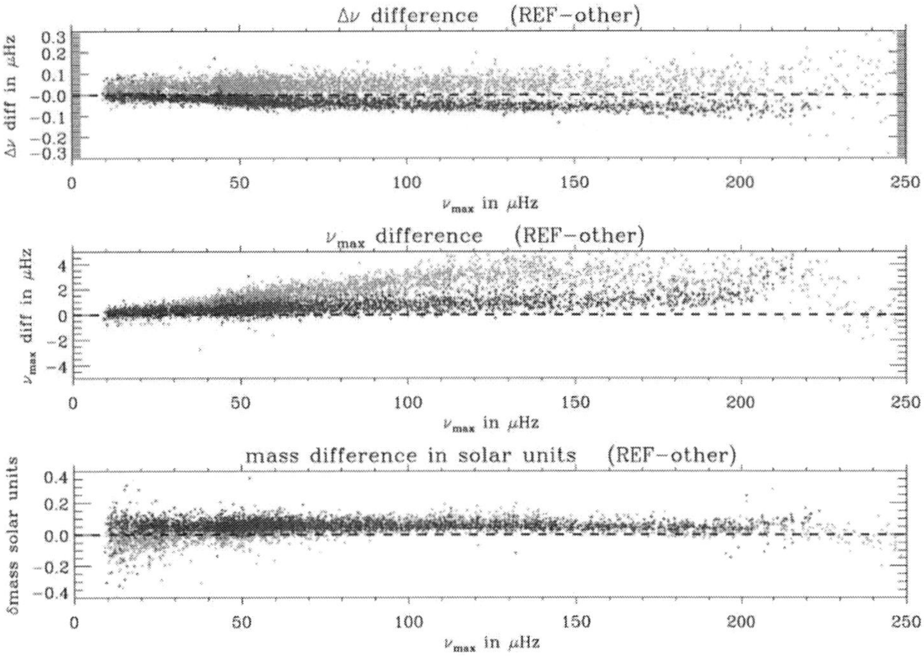

Figure 1: The average seismic parameters and the computed scaling-law mass are compared for 3 different methods (CAN, COR and OCT). In all cases the reference result is provided by CAN. In purple is CAN-COR and in light brown is CAN-OCT.

$$\Delta\nu \propto \sqrt{\frac{M}{R^3}} \tag{1}$$

$$\nu_{\max} \propto \frac{M}{R^2\sqrt{T_{\rm eff}}} \tag{2}$$

It is well-known that the different methods of measuring the average seismic parameters give slightly different values for the mass and the radius of stars. In other words, the solar reference values do not always fully compensate for the different approaches. Consider the data shown in figure 1. Here data from three methods are compared. The data from the CAN method are taken as the reference point. In the upper panel are shown, in light brown, the differences between $\Delta\nu$ from CAN and OCT as a function of the CAN $\nu_{\rm max}$ value. The sense of the comparison is CAN-OCT. In green are the differences for CAN-COR. As is evident from this plot, the $\Delta\nu$ values for the different methods do not precisely agree. The level of the disagreement is up to 10%. This is well known and has been extensively studied (see e.g.[11]). The second panel shows the comparison for $\nu_{\rm max}$ and the final one shows the scaling-law mass. The solar reference values have been used to mitigate the effects of the different seismic methods and it can be seen the OCT and COR now agree quite well but there are still small differences from the CAN result. Similar conclusions can be demonstrated for the other methods. The position is actually more complicated than shown here because the differences between the derived masses are actually somewhat dependent on the evolutionary states of the stars concerned.

We hope that the use of grid-based modelling will remove this disagreement. Hence, one of the major aims of the project is to measure the improvement that is gained for the giants from using grid-based modelling as opposed to the simple scaling laws. The ages of the stars, a very important parameter, can only be gained from the models. Furthermore, to give validity to the computed ages, we need to be sure that the mass calculation is precise and accurate and it is for that reason that we use both the simple scaling laws and the grid-based modelling.

4 Grid-based Modelling (GBM)

There are many teams involved in this project and here we show only a small number of interim results that illustrate progress to date. Within most grid-based models, the scaling laws are used to predict asteroseismic parameters from the mass, radius and $T_{\rm eff}$ of the model stars. However, it is becoming clear that it is not appropriate to use strict homology on the upper parts of the red-giant branch. Instead, the $\Delta\nu$ value for any given model star is calculated directly from the computed oscillation frequencies of the model star. For the results presented here, this method has been adopted. In figure 2, we show representative results from the Silva Aguirre GBM pipeline[12]. The difference between the grid mass and the scaling-law mass are plotted as a function of $\nu_{\max}$ for a variety of different seismic inputs. There is a clear trend in the difference between the two types of mass estimates. The scaling-law approach returns masses significantly higher than the GBM approach, particularly so for low $\nu_{\max}$ stars. This is due to the correction used within the models for the lack of homology high on the red-giant branch. The different seismic methods are giving similar but not identical results. Even tighter agreement between the different seismic inputs is seen in figure 3 where we show the results from Serenelli[13] extended to include the average seismic parameters. In this case, extra scaling is used to remove the effects of the different solar reference values for the different seismic methods. The level of the difference between the GBM mass and the scaling-law mass at low $\nu_{\max}$ is consistent with the results in figure 2. For the red-clump stars, the situation is less good. We do not show any results here, but there is disagreement and scatter in the results at the 15 to 20% level.

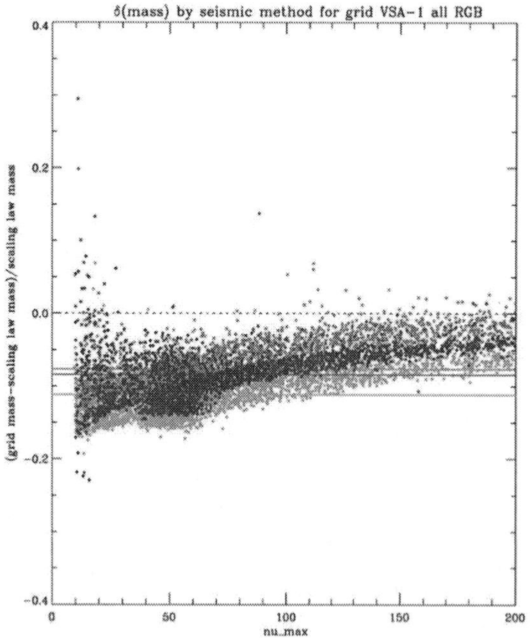

Figure 2: comparison between Mass results from the Silva Aguirre GBM pipeline and scaling-law masses for 3 different seismic input methods – CAN (green), COR(purple) and OCT(light brown)

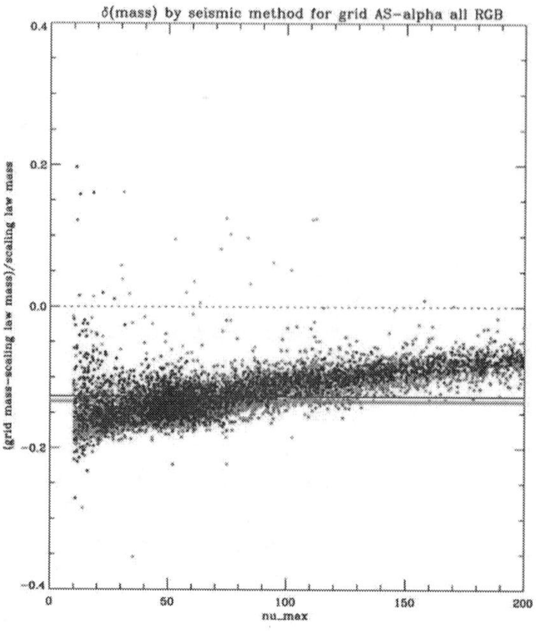

Figure 3: Comparison between Mass results from the Serenelli GBM pipeline and scaling-law masses for 3 different seismic input methods – CAN(green), COR(purple) and OCT(light brown)

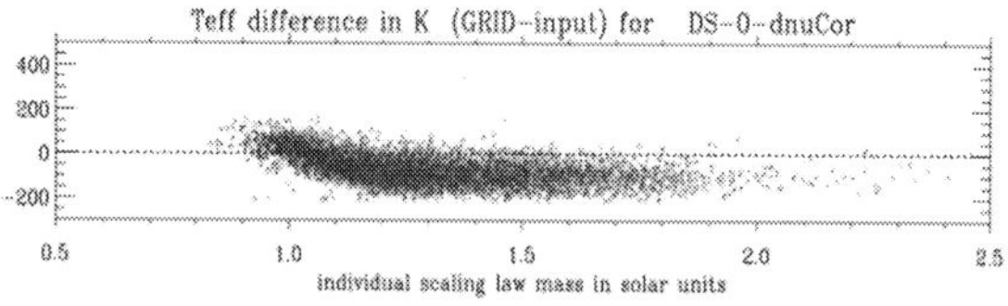

Figure 4: Comparison between the temperature of the chosen model and the observed input temperature from the Stello GBM pipeline for 3 different seismic input methods – CAN(green), COR(purple) and OCT(light brown). The independent variable is the scaling law mass.

There is one other very interesting effect that is seen in all the pipelines. The grid-based modelling process selects a model star that best fits the full set of input data. However, the parameters of the chosen model are never identically the same as those of the real, observed star. The input parameters are the average seismic parameters, $\nu_{\max}$ and $\Delta\nu$, plus the effective temperature $T_{\rm eff}$. No particular trends have been seen in the average seismic parameters but there is an effect for the temperature difference between the observed and selected model values. Here we illustrate the effect with results from the Stello GBM pipeline[14] [15] as shown in figure 4 where the temperature difference is plotted as a function of the scaling law mass. At scaling law masses above about $1.2M_{\odot}$ the chosen model has a lower temperature than the real star by about 100K. The size of this offset is dependent on the actual GBM pipeline and some have very small offsets. However, as the scaling law mass considered is reduced there is a clear trend in the temperature difference between the real star and the chosen model in the sense that the chosen model temperature gets closer to the observed one and then goes above it. This trend has the effect that the mass of the chosen is increased with respect to the scaling law mass. The GBM pipelines do not return very law mass

DOI: 10.1051/epjconf/201716001002

stars. We are currently exploring possible reasons for this effect.

5 Conclusions

This is an interim report on the first time that a significant range of average asteroseismic parameters have been used as inputs to grid-based modelling with corrections for the lack of homology for RGB stars. The work is undertaken in the framework of APOKASC. The initial results are very encouraging. One effect at low scaling law mass is curious and needs further investigation.

Acknowledgments

YE acknowledges the support of the UK Science and Technology Facilities Council (STFC). Funding for the Stellar Astrophysics Centre (SAC) is provided by The Danish National Research Foundation (Grant agreement no.: DNRF106).

References

[1] M.H. Pinsonneault, Y. Elsworth, C. Epstein, S. Hekker, S. Mészáros, W.J. Chaplin, J.A. Johnson, R.A. García, J. Holtzman, S. Mathur et al., ApJS, **215**, 19 (2014)

[2] S. Mathur, R.A. García, C. Régulo, O.L. Creevey, J. Ballot, D. Salabert, T. Arentoft, P.O. Quirion, W.J. Chaplin, H. Kjeldsen, A&A, **511**, A46 (2010)

[3] T. Kallinger, B. Mosser, S. Hekker, D. Huber, D. Stello, S. Mathur, S. Basu, T.R. Bedding, W.J. Chaplin, J. De Ridder et al., A&A, **522**, A1 (2010)

[4] B. Mosser, K. Belkacem, M.J. Goupil, E. Michel, Y. Elsworth, C. Barban, T. Kallinger, S. Hekker, J. De Ridder, R. Samadi et al., A&A, **525**, L9 (2011)

[5] S. Hekker, A.M. Broomhall, W.J. Chaplin, Y.P. Elsworth, S.T. Fletcher, R. New, T. Arentoft, P.O. Quirion, H. Kjeldsen, MNRAS, **402**, 2049 (2010)

[6] D. Huber, D. Stello, T.R. Bedding, W.J. Chaplin, T. Arentoft, P.O. Quirion, H. Kjeldsen, Communications in Asteroseismology **160**, 74 (2009)

[7] R. Handberg, M.N. Lund, MNRAS, **445**, 2698 (2014)

[8] R.A. García, S. Hekker, D. Stello, J. Gutiérrez-Soto, R. Handberg, D. Huber, C. Karoff, K. Uytterhoeven, T. Appourchaux, W.J. Chaplin et al., MNRAS, **414**, L6 (2011)

[9] SDSS Collaboration, F.D. Albareti, C. Allende Prieto, A. Almeida, F. Anders, S. Anderson, B.H. Andrews, A. Aragon-Salamanca, M. Argudo-Fernandez, E. Armengaud et al., ArXiv e-prints (2016), `1608.02013`

[10] H. Kjeldsen, T.R. Bedding, A&A, **293** (1995)

[11] S. Hekker, Y. Elsworth, B. Mosser, T. Kallinger, W.J. Chaplin, J. De Ridder, R.A. García, D. Stello, B.D. Clarke, J.R. Hall et al., A&A, **544**, A90 (2012)

[12] V. Silva Aguirre, G.R. Davies, S. Basu, J. Christensen-Dalsgaard, O. Creevey, T.S. Metcalfe, T.R. Bedding, L. Casagrande, R. Handberg, M.N. Lund et al., MNRAS, **452**, 2127 (2015)

[13] A.M. Serenelli, M. Bergemann, G. Ruchti, L. Casagrande, MNRAS, **429**, 3645 (2013)

[14] D. Stello, W.J. Chaplin, H. Bruntt, O.L. Creevey, A. García-Hernández, M.J.P.F.G. Monteiro, A. Moya, P.O. Quirion, S.G. Sousa, J.C. Suárez et al., ApJ, **700**, 1589 (2009)

[15] S. Sharma, D. Stello, J. Bland-Hawthorn, D. Huber, T.R. Bedding, ApJ, **822**, 15 (2016)

DOI: 10.1051/epjconf/201716001003

Expected asteroseismic performances with the space project PLATO

Mariejo Goupil[1] and the WP120[2] Members

[1] *Observatoire de Paris, LESIA, Observatoire de Paris, PSL Research University, CNRS, Université Pierre et Marie Curie, Université Paris Diderot, 5 place JUles Janssen 92190 Meudon, France*
[2] *PLATO STEllar and Seismology Inferences workpackage*⋆

Abstract. The PLATO (PLAnetary Transits and Oscillations of star) space project will observe about fifty percents of the sky with the main purpose of detecting, confirming and characterizing transiting exoplanets of (super)Earth sizes in the habitable zone of solar-like stars. Determining masses, radii and ages of exoplanets require the knowledge the masses, radii and ages of the host stars. We give a brief presentation of the main features of the mission. We then discuss some expected seismic performances of PLATO for characterizing bright solar-like stars, focusing on the challenging determination of accurate/precise stellar ages.

1 Introduction

The space project PLATO (PLAnetary Transits and Oscillations of stars) was selected as the M3 mission by ESA on February 2015. The mission is currently in phase B2 (consolidated definition phase). The adoption process is planned for February 2017 for a launch in 2025 by a Soyouz rocket (L2 orbit). The main scientific objective of PLATO is to answer the following questions

1. How do planets and planetary systems form and evolve?
2. Is our Solar System special or are there other systems like ours?
3. Are there potentially habitable planets?

In order to answer the above questions, the PLATO mission must :

- detect and characterize the (super) Earth-type planets, that-is determine the bulk properties (mass, radius, mean density) of planets in a wide range of systems, including terrestrial planets in the habitable zone of solar-like stars.

  By (Super)Earth-like planets, one means planets with masses, M_p, and radii, R_p, such that $M_p \leq 1M_E$; $R_p \leq 2R_E$ where M_E and R_E respectively are the Earth mass and radius. The habitable zone is defined here as the region around a star where liquid water on the surface of an Earth-like planet is in principle possible.

  It must be stressed that the space mission Kepler (and K2) does – and the future space mission TESS will – mostly provide transiting planets with orbital periods smaller than 80 days that-is planets with orbits not further than a Mercury-like orbit. Only PLATO will be able to provide a significant number of (super) Earth-type planets in the habitable zone of solar-like stars (an Earth-size planet at about 1 au from its host star).

  The expected output of PLATO is $\sim 100-400$ *precisely characterized exoplanets including* $\sim 5-30$ *(super-)Earths in the habitable zone of solar-like stars.*

- explore the diversity of the planets and planetary systems

  In order to determine accurately the mass and radius of a planet, one needs to determine with the same accuracy the mass and radius of the host star.

- study the evolution of planets and planet systems with age.

  The age of the planet is assumed to be the age of the host star.

- study the internal structure of stars and how it evolves with age.

  The accuracy/precision on the mass, radius and age determinations of the host stars depend on the quality of the stellar models (numerical precision, degree of realism of the physical description)

The above issues define the PLATO core programme. In addition, a complementary science programme is included. It is complementary because it has no influence on the design of the mission but will widely contribute to the output of the mission with a large number of astrophysical applications. To name a few, seismology of massive stars clusters, studies of galactic populations. This can be extended to any astrophysical open question which needs high-precision, continuous time series of photometric measurements for a large sample of stars.

⋆http://www.oact.inaf.it/plato/PPLC/PSM/Entries/2010/12/19_STELLAR_SCIENCE.html

The PLATO management consortium (PMC) is led by the Principal Investigator, H. Rauer (Germany). It includes more than 350 researchers from fourteen ESA Member States, as well as an Associate Member state, and several other nations across the world. The structure of the PLATO consortium is made of three main branches the payload, the PLATO Data Center (PDC) and the PLATO Science Management (PSM). The PLATO Data Center (PDC) is responsible for producing the data output of the mission. The PSM is in charge of specifying the algorithms ad tools which will be implemented and used by the PDC. The structure of the PSM is composed of several work-packages [Exoplanet science (WP110), Stellar science (WP120), Target/fields characterization (WP130), Follow up coordination (WP140), and Complementary Science (WP150)].

A full description of the Plato mission is detailed in [1] (see also PLATO Definition Study Report (2017)). We focused here on the expected performances of PLATO for the stellar part of the core program. The paper is organized as follows: Sect.2 explains how to reach the scientific objectives. Sect.3 presents several main issues of the scientific preparation. Sect.4 adresses some expected performances for the stellar part of the core programme. Sect.5 draws some conclusions.

2 Reaching the PLATO scientific objectives

Three techniques will be combined to reach the PLATO scientific objectives:

- Detection of exoplanets by photometric transits. This will provide the orbital parameters and the planet-to-star radius ratio R_p/R_*. The relative uncertainty on the planet radius must be smaller than 3% for the reference planet (an Earth-size planet orbiting at about 1 au from a G0V star with $V = 10$ i.e. orbital period of about 1 year). A driver for the mission efficiency is therefore the number of bright ($V \leq 11$) dwarfs and subgiants stars of spectral type F5-K7 which will be continuously monitored with a high signal-to-noise ratio ($\leq$ 34 ppm/h) over at least two years time periods. One must observe at least 15000 – 20000 such stars.
- Confirmation of the planet nature of the candidate by spectroscopic ground-based follow-up. Measurements of radial velocities will give the planet-to-star mass ratio M_p/M_* for transiting exoplanets. The relative uncertainty on the reference planet mass, M_p, must be smaller than 10% for a host star with an apparent magnitude $V = 10$. This will be obtain for the brightest planet candidates with the highest interest ranking.
- Precise seismic determinations of the mass, M_*, radius R_* and age for the host stars.

The specifications for the exoplanet part require a noise-to-signal ratio $NSR \leq 80$ ppm/h for detection and $NSR \leq 34$ ppm/h for the planet radius measurements for stars with $V \leq 11$ and 2-3 years of observation. The requirements for the host stars accordingly are that the relative uncertainties are no larger than 1-2% for the radius, 15 % for the mass and 10% for the age of the reference star at magnitude 10. The reference star is defined as a G0V star with the solar mass and radius and an effective temperature of 6000 K, see the document 'PLATO-DLR-MIS-LI-002_i1.1.pdf').

The reference star and any solar-like stars are expected to vibrate with solar-like oscillations. The above requirements then translate into the following specifications : the oscillation frequency uncertainties of the solar-like stars must be of the order or smaller than $\sim 0.1 - 0.2$ μHz for frequencies around ν_{max}, defined as the frequency at maximum amplitude in the frequency domain. These specifications lead to further specifications which are that the noise-to-signal level must be of the order or smaller than 34ppm/h for $V \leq 11$, the time series must be long and continuous with no periodic gaps of similar time scales than the oscillation ones.

2.1 Payload, observing scenario and data product

The payload is a multi-telescope concept for a wide-field-of view for each pointing: 24 «normal» 12cm cameras will collect white light (500-1050 nm) at a cadence 25 s for stars with apparent magnitude $8 < V < 16$. The field-of-view is set to $48.5° \times 48.5°$ ($2250 deg^2$). In addition, 2 «fast» 12cm cameras with a 2.5s cadence will observe in two colours (in a FoV 550 deg^2 (tot)) stars with apparent magnitude $4 \leq V \leq 8$. The daily downloaded telemetry will amount to about 435 Gb. Note that several samples of stars will be observed with different numbers of telescopes corresponding to different noise levels at a given apparent magnitude.

The observing scenario (Fig. 1) is based on four years of observation after commissioning and the possibility of a two-year extension. The choices are either two long runs of 2 years each and a sky survey in a step-and-stare mode (2 to 5 months) during the extension period or a three-year long run and one year for a sky survey in a step-and-stare mode (2 to 5 months). The spacecraft will operate a 90° rotation around the line-of-sight every 3 months so that the expected duty cycle is about 93%.

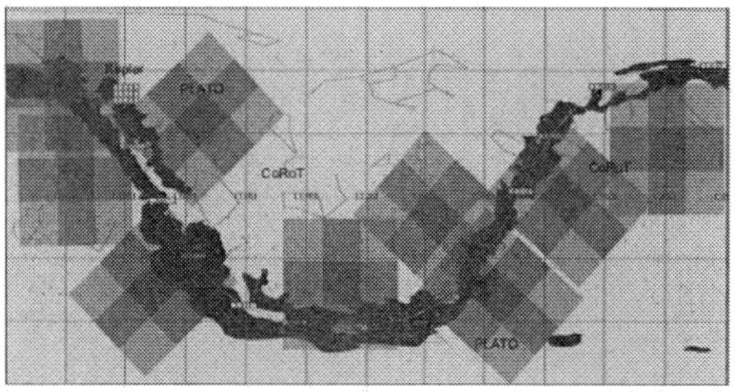

Figure 1: The reference observing scenario: 2 long runs and several short runs, covering 50% of the sky.

 DOI: 10.1051/epjconf/201716001003

Data product will be generated from the PLATO mission at different levels as follows:

- L0 : The validated light curves and centroid curves for all individual telescopes, housekeeping data, imagettes;
- L1 : The calibrated light curves and centroid curves for each star and corrected for instrumental effects e.g., jitter;
- L2 : Scientific output:
  - The planetary transit candidates and their parameters with formal uncertainties;
  - The asteroseismic mode parameters with formal uncertainties;
  - The stellar rotation periods and stellar activity models inferred from activity-related periodicities in the light curves, with formal uncertainties;
  - The seismically-determined stellar masses, radii and ages of stars, (and their formal errors), obtained from stellar model which oscillation frequencies fit best the observed ones;
  - Living catalogue of confirmed planetary systems and their characteristics using light curves and time transit variations.
- Lg : Follow up ground-based observation data;
- L3 : Living catalogue of confirmed planetary systems fully characterized by combining information from the planetary transits, the seismology of the planet-hosting stars, and the results of the ground-based observations; This represents the final and highest level PLATO science deliverable.

2.2 Samples of stars

The set of target stars in the core programme is divided into four samples. With 24 telescopes and the current baseline observing strategy,

- the P1 sample includes about 15000 (~ 20000) bright dwarfs and subgiants ($V \leq 11$) with spectral type F5-K7, a noise level ≤ 34 ppm/h, long observing run (LOP) with a 25s cadence. This sample will include the PLATO 'Rosetta stones';
- the P2 sample contains more than 1 000 dwarfs and subgiants ($V \leq 8.2$) (300 stars with 2 colours) with spectral type F5-K7, a noise level of ≤ 34 ppm/h, LOP with a 2.5s cadence. This sample will offer the possibility of caracterizing their atmospheres with ground (E-ELT) or space (JWST) facilities and the possibility of seismically caracterizing a large number of stars for improving the calibration and description of the stellar models;
- the P4 sample is made of more than 5000 M dwarfs $V \leq 16$, sampling time 25s, noise level 800 ppm/h.
- the P5 sample consists of ≥ 245,000 dwarfs and subgiants ($V \leq 13$) with spectral type F5-K7 with a lower SNR than P1, a 25s cadence. This sample is subdivided in two sub-samples:
  - P5-bright ($V \leq 11$) stars for which mass measurements from ground will be possible;
  - P5-faint ($11 < V \leq 13$) stars with a lower SNR but still with sufficiently precise characterization for statistical studies.

3 Scientific preparation

3.1 Input catalogue and target selection

Telemetry limitations impose the pre-selection of PLATO targets for the detection of planets. The optimal field selection is closely related to the target selection. The number of F5 or later spectral type dwarfs and sub-giants for which we can have photometry with the required SNR must be as large as possible. This led the WP130 workpackage to build a PLATO Input Catalogue (PIC) which includes P1-P5 targets, and provides their main parameters. The PIC is built using already existing star catalogues. As an illustration, this count yields for the fields of the reference observing scenario about 13 000 stars populating the P1 sample, about 29 000 stars in the P5 sample with $V \leq 11$, 34 ppm/h $< NSR \leq$ 80 ppm/h; about 80 000 stars for in the P5 sample with $V < 13$, $NSR <$ 80 ppm/h. Each step -and-stare can provide about 6100 stars in the P1 sample and about 160 000 stars in the P5 sample.

The Gaia release catalogues will also provide well-classified nearby dwarf/sub-giant stars to optimise the content of the PIC.

3.2 Estimation of planet candidates and needed resources for ground-based follow up

Over the whole space part of the PLATO mission, one expects the detection of about 13000 transiting planets of all sizes and periods out of which about 3400 transiting planets with host stars bright enough ($V \leq 11$) suitable for radial velocity follow up. From this set, the current counts foresee 2200 planets with radius $\leq 2R_E$ and which could be rocky in nature. These estimates by the WP110 workpackage are based on the stellar populations in the PIC and our current knowledge of the exoplanet planetary systems.

After identification of a planet candidate with PLATO data, ground-based follow up observations are needed to confirm and characterise the planetary nature of the companion. For most planet candidates, accurate masses will be derived from spectroscopic radial velocity observations. The PLATO mission therefore includes organized ground-based follow-up observations which will be carried out by the Ground-based Observations Programme (GOP) team. The needed ground-based resources considered two steps : first, observations must be designed to detect false positives and confirm the planet nature of the candidate. Second, the mass of the planet must derived from ground-based spectroscopic measurements. The necessary amount of resources has been estimated by the WP140 workpackage and found to be comparable to a large ESO programme. Then at least 150- 200 super-Earths are expected to be characterized out of which more than several dozen of planets are expected to seat in the

DOI: 10.1051/epjconf/201716001003

habitable zone of their host stars, representing the main targets for PLATO. The last figure remains however quite uncertain since it depends on the small planet occurrence rates in the habitable zone, a poorly known quantity that PLATO will contribute to establish on more solid grounds.

3.3 Plato noise level as a function of magnitude and number of telescopes

For performances studies, the stellar signal in the noise-to-signal ratio, NSR, is assumed constant. The PLATO noise is made of several components : the photon noise of the target star, the random noise from the instrument and the residual systematic noise from the instrument after correction. The residual systematic noise from the instrument after correction must not be larger than 1/3 of the total random noise. At this step of the performance studies, the signals considered as noises either for exoplanet or seismic studies (stellar activity, granulation, oscillatory components, transit effects in a power spectrum) are ignored.

The noise-to-signal ratio, SNR, for the reference star and for 24 telescopes was estimated by the performance working group on the basis of our knowledge of the instrument performance today and is found to be ~ 25 ppm/h at $V = 10$.

4 Expected performances for the stellar part of the core programme

For seismic performance estimations, only studies focusing on the P1 sample have been carried out so far [2]. The estimated number of stars passing the PLATO specifications is ~ 15000 out of which more than ~ 5000 stars will be seismically characterized with relative uncertainties $\Delta A/A \sim 10\%, \Delta R/R \sim 2\%, \Delta M/M \leq 15\%$ where A, $R_\star$ and $M_\star$ respectively mean the age, the radius and the mass of the star [3].

Assuming the PLATO noise level mentioned in the previous section and the case of the reference star, verification of the performances must then be carried out with two successive steps:

- step 1: reaching oscillation frequency uncertainties down to $\Delta\nu_{n,l} \sim 0.2\mu$Hz for several frequencies around ν_{max}

- step 2: assuming the step 1 frequency uncertainties, one must obtain the uncertainties for the stellar parameters at the level of 2% (radius), 15% (mass), 10% (age) for the reference star at magnitude $V = 10$.

Calculations show that the radius and mass uncertainties easily satisfy the PLATO specifications with the tools and methods currently used today for seismically characterizing CoROT and Kepler stars. On the other hand, the age specification is a challenge. From now on, we therefore focus on the age specification.

Age uncertainties can be seen as arising from several sources. One source is the propagation of the observational frequency uncertainties as estimated in step 1, we call it below *age random error*. A second source is referred to as *systematic error* and arises because the age must be determined with the use of stellar models which are not perfect. The physics implemented in evolutionary codes to compute stellar models remain by many aspects uncertain. Some processes such as convective energy transport are modelled with free parameters which values are currently set empirically. This can introduce large biases in stellar age derivation, everything else being equal [4]. Improvement in the modelling of these processes as well as modelling of missing processes are still in development and not yet implemented in stellar evolutionary codes. In the error budget, one must also include the impact of observational uncertainties on the non seismic constraints used in the stellar modelling such as the effective temperature, the surface gravity, the metallicity and when available the luminosity.

In the remaining of this section, we report first on the expected age random error for the reference star belonging to the P1 sample. A second part discusses the expected age systematic error for a star belonging to the P1 sample. For this second part, hare and hounds exercises ,(hereafter HH), were carried out over 2014-2015 within the workpackage WP120 (or STESSI for STEllar Seismic Inferences) using Plato simulated light curves.

4.1 Age random error for the reference star belonging to the P1 sample

As mentioned above, we estimate first the frequency uncertainties and second the age uncertainties that arise from the frequency uncertainty propagation for the reference star.

Step 1 : we use the [5]'s relation to compute the frequency uncertainties for the reference star.

$$\sigma(\nu) = \left[f(\beta) \; \frac{\Gamma}{4\pi T_{obs}} \right]^{1/2} \tag{1}$$

in Hz where $f(\beta) = (1+\beta)^{1/2} \, [(1+\beta)^{1/2} + \beta^{1/2}]^3$ and $\beta = NSR$ is the noise-to-signal ratio, Γ (in Hz) the width of the mode and T_{obs}, the observing time (in s).

This relation is valid for radial modes ($l = 0$). For non-radial modes, one must take into account the mode visibility and the inclination angle. The validity of the Libbrecht'relation is well established [6]. Nevertheless, we also checked its validity with the results of the PLATO HH exercises (not shown here).

The target is the reference star with an apparent magnitude $V = 10$ and the observing time, T_{obs}, is assumed to be 2 years. We evaluate the frequency uncertainties at ν_{max} i.e. $\sigma(\nu_{max})$ which then involves $\beta_{max} = NSR_{max}$ at ν_{max}, Γ the width of the modes at ν_{max} and $NSR_{max} = B_{max}/(V_\ell H_{max})$ where H_{max} is the mode height for radial modes at ν_{max}. B_{max} is the PLATO for the reference star at $V = 10$ (see [7]). V_ℓ is the mode visibility coefficient for mode degrees ℓ.

Fig.3 shows the frequency uncertainties computed according to Eq.1 as a function of the frequency for the reference star. For modes $l = 0$ and $l = 1$, the uncertainties are of the order of 0.15 μHz for several frequencies around

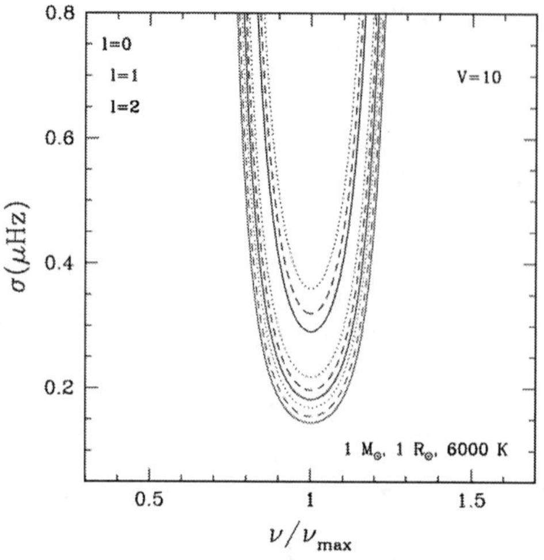

Figure 2: Frequency uncertainties (μHz) for different mode degrees ($l = 0, 1, 2$) as a function of the frequency normalized to ν_{max}. The uncertainty σ is computed according to the Libbrecht's relation for the reference star with an apparent magnitude $V = 10$, assuming two years of observation and 32 telescopes (solid line), 28 telescopes (dashed line) and 24 telescopes (dotted line).

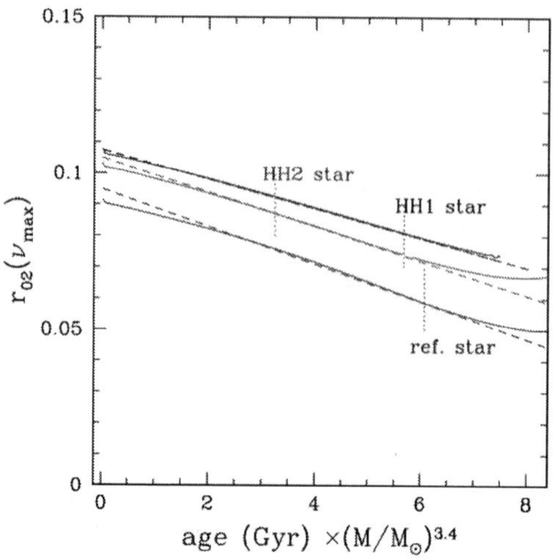

Figure 3: The seismic diagnostic r_{02} evaluated at ν_{max} as a function of age $\times M^{3.4}$ where M is the mass of the model in solar unit and age of the model in Gyr. Each solid curve is the result of a numerical evolution of a stellar model with a given mass. Three cases were considered corresponding to the HH target case HH1 ($1.18 M_{\odot}$, red line) and HH2 ($1.12 M_{\odot}$, black line) and the reference star ($1 M_{\odot}$, blue line). The stellar models which were input to the HH exercises and the reference model are located as vertical bars with the same colour than the corresponding lines. The dashed lines are linear fits to the solid lines.

ν_{max} i.e. satisfying the PLATO specification. For $l = 2$, modes the uncertainty amounts to $\sim 0.4\,\mu$Hz.

Step 2 : Let denote the oscillation frequencies, $\nu_{n,\ell}$, where n, ℓ represent the radial order and the degree of the mode, respectively. It is well accepted that the seismic diagnostic r_{02} defined as ([8])

$$r_{02}(\nu) = \frac{\nu_{n,0} - \nu_{n-1,2}}{\nu_{n,1} - \nu_{n-1,1}}, \qquad (2)$$

provides information on the age of the star [9]. Given the input physics, the computation of stellar models and provides $r_{02}(\nu)$ and the age of the model. As already stressed by [10], one finds that $r_{02}(\nu)$ evaluated at $\nu = \nu_{max}$ varies linearly with the age of the stellar model (Fig.4).

$$r_{02}(\nu_{max}) = c_0 \; age \; M^{3.4} + c_1 \qquad (3)$$

with the age in Gyr and the mass in solar unit. The mass dependence of the slope $c_0 = -6.\ 10^{-3} Gyr^{-1}$ is negligible. The quantity c_1 varies with the mass of the star but here we are interested only on the slope c_0 which is involved in the age uncertainties for the reference star. One can then compute the age uncertainties from the r_{02} uncertainties for the reference star with $\delta age = \delta r_{02}/(c_0)$ while the r_{02} uncertainties are obtained from the observational frequency uncertainties $\delta\nu_{n,l}$. The validity of the relation between the age uncertainty and the frequency uncertainties was assessed by using the results of the PLATO HH exercises (not shown here).

Fig.4 shows the relative age uncertainty $\delta age/age$ as a function of the assumed apparent magnitude of the star. One finds that with an observing time of 2 years, 24 telescopes and the associated NSR, the age specification for the reference star at $V = 10$ is easily fulfilled .

However for realistic age uncertainty estimates, we must now take into account the various biases due to stellar modelling.

4.2 Age systematic error and hare and hound exercices

The objective of these blind tests was to assess which accuracy/precision is currently obtained in the seismic determination of the age, mass and radius of a star similar- but not identical to the Sun- with the expected Plato data. In the error budget of these HH, were then included the propagation of frequency random errors, the biases due to the number of detected modes which depends on the type of star (mass, effective temperature) and the noise-to-signal ratio, biases due to the optimisation methods used to derive the age from oscillation frequencies, the assumptions made by the hounds for the various physical inputs, the chemical composition and the number and values of the free parameters entering the stellar modelling.

For each hare and hound exercise, the input stellar model and its oscillation frequencies were used to compute a PLATO simulated light curve. This simulated light curve was then analysed to derive the frequencies and their

 DOI: 10.1051/epjconf/201716001003

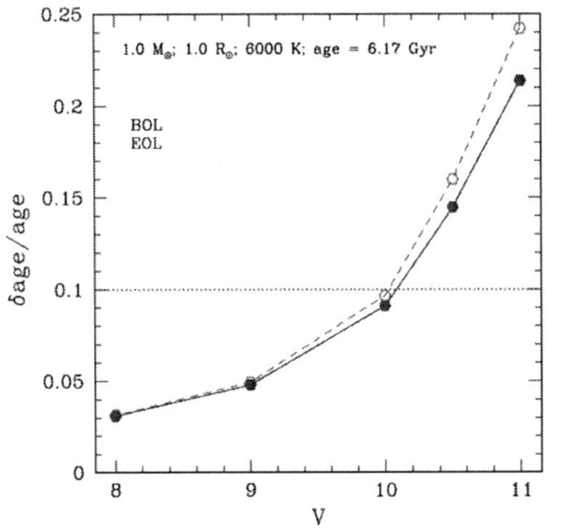

Figure 4: Relative age uncertainties due to the propagation of observational frequency errors as a function of the apparent magnitude of the reference star for 24 telescopes at the beginning (BOL) and end of the mission (EOL).

uncertainties and the parameters identifying the associated oscillation modes (degree, radial order). In a second step, the set of derived frequencies, their uncertainties and mode identification were given to hounds who used them for deriving a set of optimal stellar models, that-is stellar models with frequencies and/or frequency combinations best fitting the corresponding observed quantities. Non seismic constraints such as the effective temperature and optionally the luminosity were also included in the fitting processes. From the set of masses, radii and ages of these optimal models, one derives the estimated mass, radius and age of the 'target star' (i.e. input stellar model) and their uncertainties.

We briefly report here the main results for one such PLATO hare and hound exercise (HH2): the mass and radius of the target (input) stellar model were chosen to be $1.12M_{\odot}$, $1.20R_{\odot}$. The adopted physical assumptions for the stellar modelling led to an age = 3.44 Gyr (corresponding to a hydrogen central content $Xc = 0.30$) for an effective temperature of $T_{\text{eff}} = 6130K$ for this model. We adopted $Y_{ini} = 0.26, (Z/X)_{ini} = 0.019$ for the initial helium abundance in mass and the ratio metallicity over hydrogen.

A more detailed presentation of the expected PLATO seismic performances and results of the PLATO hare and hound exercises will be published a forth coming paper.

Step 1: frequency uncertainties from PLATO simulated light curves

The transit signature will be remove from the PLATO light curves before any seismic analysis. When computing a simulated light curve, the transit signature is therefore ignored. The stellar activity was assumed to be negligible over the frequency domain of interest for the oscillations of solar-like stars. On the other hand, the impact of the granulation was taken into account.

The simulated light curves were first built in the frequency domain where the oscillation and granulation signals, the target photon noise and the PLATO estimated noise were added together. The stochastically excited oscillating signals were composed of modes with degree $l = 0 - 3$ over a frequency range spanning several radial order around the radial order of the frequency at maximum power. The apparent magnitude of the simulated stars were set to $V = 8$ to 11. The observing time was assumed to be two years and the noise level was set for 32 telescopes (corresponding to a noise-to-signal level of 36 ppm/h at $V = 11$). For the target photon noise, we considered several apparent magnitudes $V = 9, 10$ and 10.5. Note that the noise level of a $V = 11$ star observed with 32 telescopes is the same than for a star with $V = 10.6$ with 24 telescopes. Details on the PLATO simulated light curves for seismology can be found in [7].

The simulated light curves were then passed on two groups who carried out independent data analyses: the Birmingham group with B. Chaplin, T. Campante and G. Davies (UK) on one hand and O. Benomar (United Arab Emirates and Germany) on the other hand. They provided sets of 'detected' oscillation frequencies with their uncertainties as well as mode widths and amplitudes with their uncertainties for each simulated light curves (i.e. each assumed apparent magnitude). The degree and radial order were also attributed. As a result, the frequencies are recovered with a good accuracy, indeed the differences between the true frequencies and the frequencies estimated from the data analyses are found to be of the order of $\leq 0.1-0.2\mu$Hz around ν_{max}. The frequency uncertainties as estimated by the two groups are of the same order of magnitude. Similar results are found for Kepler solar-like stars [11], [12].

Additional studies by M. Lund (UK, Denmark) investigated the evolution of frequency uncertainties with increasing observing time. Pieces of the $V = 10.5$ HH2 light curves successively increased by three months periods were analyzed, starting with an observing time of three months and ending with the full 24 months light. As expected, the frequency uncertainty decrease with time as $1/\sqrt{time}$. With three months data, the frequency uncertainties are at the level of $0.5 - 1.5\,\mu$Hz for $\ell = 0, 1$ modes and no $l = 2$ mode frequency was detected for this specific studied case. The $l = 2$ modes near ν_{max} were detected after 6-9 months observations. With 24 months data, the frequency uncertainties are found in the range $0.2 - 0.4\,\mu$Hz depending the mode degree, satisfying the PLATO specification for several mode frequencies around ν_{max}.

Step 2 : Age systematic errors

In order to estimate the age systematic error due to uncertainties in stellar modelling, a second step in the HH exercises was performed. The 'observed' frequency sets derived during the first HH2 step (as described above) were

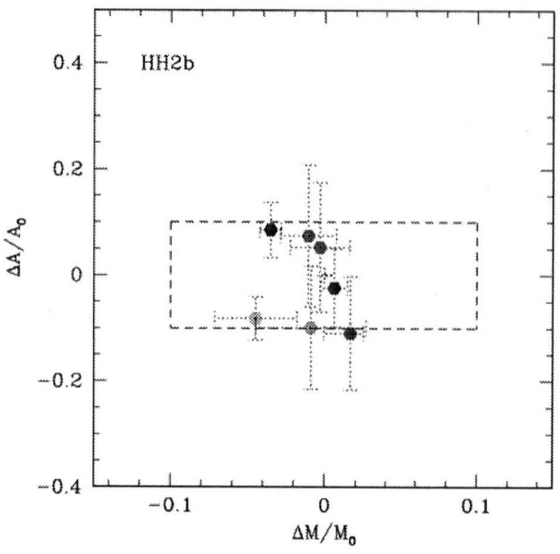

Figure 5: Relative differences between the age of the input model and the ages obtained by the hounds, $\Delta A/A$, obtained for the HH2 target as a function of the relative differences between the mass of the input model and the masses, $\Delta M/M$. The apparent magnitude is $V = 9$, the observing time was set to to years. Each colour corresponds to one hound. The uncertainties as estimated by the hounds are shown as error bar in dotted lines.

given to several 'hounds'. The non seismic observational constraints provided to the hounds were $T_{\rm eff} = (6100 \pm 80)$ K; $[Fe/H] = 0.04 \pm 0.05$ dex; $\log L/L_{\odot} = 0.22 \pm 0.03$. Independent modelling by S. Deheuvels (France), V. Silva-Aguirre (Denmark), J. Christensen-Dalsgaard (Denmark), Y. Lebreton (France), O. Creevey(France), A. Mazumdar & K. Verma (India), I. Roxburgh(UK), M. Bazot (Portugal) used different evolutionary codes with various assumptions for the physical description for the stellar models and their chemical compositions, adopting various seismic constraints and optimisation methods. The hounds provided the mass, radius and age of the HH2 target and their uncertainties.

Fig.5 shows the difference between the age of the input model and the age obtained by the hounds, $\Delta A/A$, as obtained for the HH2 target as a function the difference between the mass of the input model and the masses obtained by each hound, $\Delta M/M$. The target was the HH2 model with an apparent magnitude $V = 9$, the observing time was set to years. The derived values are within 10% from the exact values. The uncertainties estimated by the hounds are also of the order of 10%.

Similar exercises for the same HH2 target star but for a magnitude $V = 10$ and $V = 10.5$ yielded relative uncertainties for the mass in the range: [0.8 – 2.5]% and for the age in the range [25 – 27]%. For a star at the solar age, the relative age uncertainty then lies the range [17 – 19]% i.e. a few percents up to about 10 percents larger than the PLATO specification.

These results are in full agreement with previous similar studies ([13]; [14], [15] and [16]). For instance, the uncertainties obtained for Kepler host stars (with about four years of observations) yielded $(\Delta A/A)_{stat} \sim 14$ % + several % (systematics) ([15]).

5 Conclusion

The PLATO space mission will detect terrestrial exoplanets up to the habitable zone of solar-type stars and characterize their bulk properties. Several thousands of dwarfs and subgiants with F5 and later spectral type with apparent magnitudes less than 11 will be monitored during long uninterrupted observing runs with a high-precision photometry. Precise and accurate seismic properties will be obtained for a large sample of these stars. Their seismic characterisation will provide masses, radii with respective uncertainties as small as 2%, 15% for stars similar to the reference star with magnitude 10. Some corrections for systematics errors are needed to reach the PLATO specification for stellar ages (10% age uncertainties for the $V = 10$ reference star after corrections for biases and systematic effects). For that purpose, improvements in stellar modeling are currently being carried out and will continue in the coming years

The full seismic potential of the stellar part of the core programme of the PLATO mission has not been quantified yet. Indeed a large number of stars will have a noise level much smaller than 34 ppm/h. In addition, many other stars with a noise level 34 ppm/h $< NSR \leq 36$ ppm/h will be seismically characterized with uncertainties barely higher than for 34 ppm/h.

Studies of massive stars, clusters, red giants will also benefit from the wealth of high quality PLATO data.

More generally, the PLATO legacy database will provide a unique and essential resource for testing our models of planets and stars, their evolution and their interactions.

To become a consortium member, a way is to participate to the PLATO scientific preparation and be part of one workpackage. More details on the stellar part of the PLATO project can be found on the web sites:

http://www2.warwick.ac.uk/fac/sci/physics/research/astro/plato-science/

https://www.ias.u-psud.fr/PLATO_STESCI/PLATO_STESCI_OBJ.html

Acknowledgements

This work benefited from the financial support provided by the Centre national d'études spatiales (CNES) in the framework of the PLATO project.

References

[1] Rauer, H. and the PLATO team, Experimental Astronomy **38**, 249 (2014)

[2] Goupil, M.J. and the WP120 group, **PLATO-LESIA-PSPM-TS-01**, (issue 1.0) (2015)

 DOI: 10.1051/epjconf/201716001003

[3] Appourchaux, T., **PLATO-IAS-SCI-AN-001**, Issue 5 Revision 0, (2016)

[4] Lebreton, Y. and Goupil, M. J. and Montalbán, J., EAS Publications Series **65** , 99 (2014b)

[5] Libbrecht, K.G., ApJ **387**, 712 (1992)

[6] Toutain, T. and Appourchaux, T., A&A **289**, 649 (1994)

[7] Samadi, R., **PLATO-LESIA-PSPM-TN-014**, (Issue 1.0 Revision 1.0), (2015)

[8] Roxburgh, I. W., Vorontsov, S. V., A&A **411**, 215 (2003)

[9] Christensen-Dalsgaard, J., Space Research in Stellar Activity and Variability, ed. A. Mangeney & F. Praderie, 11 (1984)

[10] Appourchaux, T. and Antia, H. M. and Ball, W. et al., A&A **582**, 25 (2015a)

[11] Appourchaux, T. and Chaplin, W. J. and García, R. A. et al., A&A **543**, 54 (2015b)

[12] Davies, G. R. and Silva Aguirre, V. and Bedding, T. R. et al., MNRAS **456**, 2183, (2016)

[13] Lebreton, Y. and Goupil, M. J., A&A **569**, 21 (2014a)

[14] Reese, D. R. and Chaplin, W. J. and Davies, G. R. et al., A&A **592**, 14 (2016)

[15] Silva Aguirre, V. and Davies, G. R. and Basu, S. et al., MNRAS **452**, 2127, (2015)

[16] Silva Aguirre, V. and Lund, M. N. and Antia, H. M., ApJ **835**, 173 (2017)

DOI: 10.1051/epjconf/201716001004

LAMOST observations in the K2 fields: project and observation progress

Jian Ning Fu[1,⋆], *Martin C.* Smith[2], *Ru Yuan* Zhang[1], *An Bing* Ren[1], *Jian Rong* Shi[3], *A Li* Luo[3], and *Hao Tong* Zhang[3]

[1]*Department of Astronomy, Beijing Normal University, Beijing 100875, China*
[2]*Shanghai Astronomical Observatory, Chinese Academy of Sciences, Shanghai 200030, China*
[3]*National Astronomical Observatories of China, Chinese Academy of Sciences, Beijing 100012, China*

Abstract. To complement the time-series observations of the Kepler space mission in the K2 fields, spectroscopic observations for hundreds of thousands of stars in these fields are hugely important. LAMOST, a Chinese 4-meter class telescope equipped with 4000 fibers on the focal plane, is an ideal facility to fulfill this task. The LAMOST-*K*2 project was approved in the autumn of 2015 and observations commenced during the 2015-2016 winter season. The project will initially cover 6 of the first 10 observed K2 campaigns, and may be extended to include later *K*2 campaigns. We describe the project and introduce the current progress of observation with this poster presentation.

1 Motivation

After the failure of the *Kepler* spacecraft for continuously monitoring the original *Kepler* field in May 2013, the *K*2 mission enables continued scientific observations for a series of sequential observing "campaigns" of fields distributed around the ecliptic plane. Each campaign is limited to a duration of approximately 80 days. *K*2 became fully operational in June 2014 and is expected to continue operating until 2017 or 2018 (http://keplerscience.arc.nasa.gov/objectives.html). The fields of *K*2 from 2014 to 2016 are shown in Figure 1.

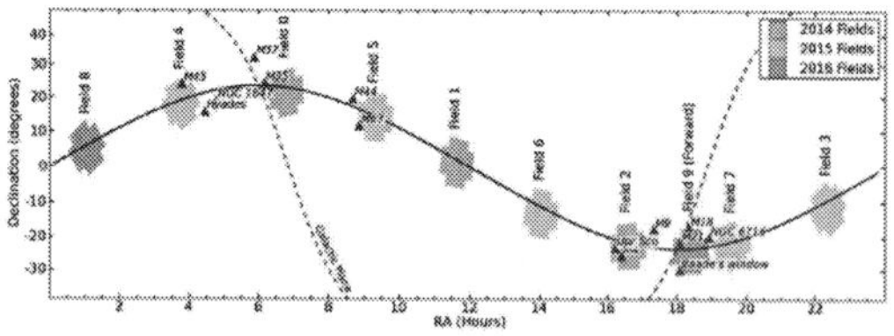

Figure 1: *K*2 fields from 2014 to 2016

In order to best exploit the *K*2 data, ground-based spectroscopic observations are needed to provide the atmospheric parameters for the large amount of stars in the *K*2 fields.

2 LAMOST and the LAMOST-*K*2 Project

The Large Sky Area Multi-Object Fiber Spectroscopic Telescope (LAMOST), as one of the National Major Scientific Projects undertaken by the Chinese Academy of Science, is a special quasi-meridian reflecting Schmidt telescope located in Xinglong Station of national Astronomical Observatory, China. It is a 4-m class telescope with the field of view of 5deg in diameter, equipped with 4000 fibers on the focal plane. The 16 fiber-fed spectrographs provide spectra of R=1800 for the spectral ranges of 370-900 nm (Su et al. 1998; Zhao et al. 2012). Figure 2 shows a picture of LAMOST.

Figure 2: The LAMOST telescope.

With the powerful capacity of spectra acquisition for a large amount of targets simultaneously, LAMOST becomes an ideal facility of observing the available *K*2 fields efficiently. In September of 2015, our proposal of observing 6 available *K*2 fields (DEC>-10deg) with LAMOST was approved by the LAMOST Science Committee. The observations were commenced December 30 of 2015.

⋆e-mail: jnfu@bnu.edu.cn

3 Observations and Stellar Parameter Calculation

Table 1 lists the observation log and the obtained data. The LASP pipeline (Luo et al. 2015) was applied to calculate the radial velocity and the stellar parameters ($T_{\rm eff}$, log$g$, $[Fe/H]$). The latter were calibrated following the formulae 1-6 of Ren et al. (2016) for the dwarf and giant stars, respectively. Figure 3 shows the observed plates of LAMOST for 5 $K2$ fields.

Table 1: Observation log and obtained data

K2-field	Date	#	Spectra	Parameters
Field0	2016/01-02	9	25833	18176
Field1	2016/01-02	17	25752	14679
Field4	2016/01-02	9	14566	10837
Field5	2015/12 2016/01-02	12	29269	19658
Field6	2016/01	1	813	259
Total		48	96233	63609
Unique			88694	59213
1×			81649	55061
2×			6554	3909
3×			488	242
4×			3	1

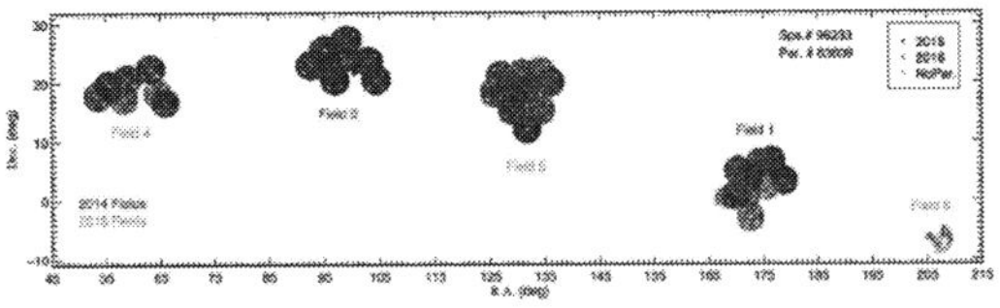

Figure 3: Observed plates of LAMOST for 5 K2 fields.

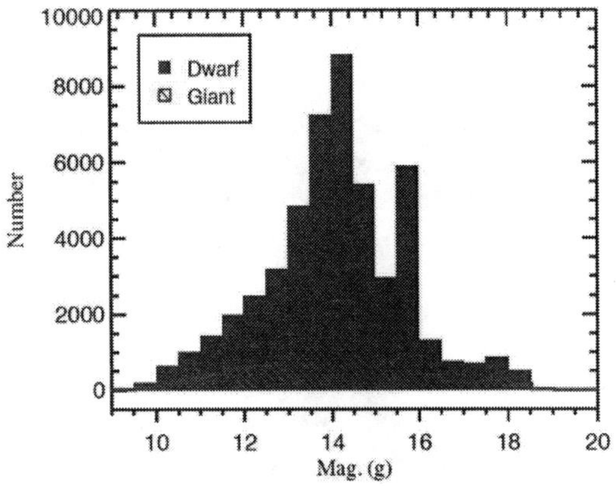

Figure 4: Magnitude distribution.

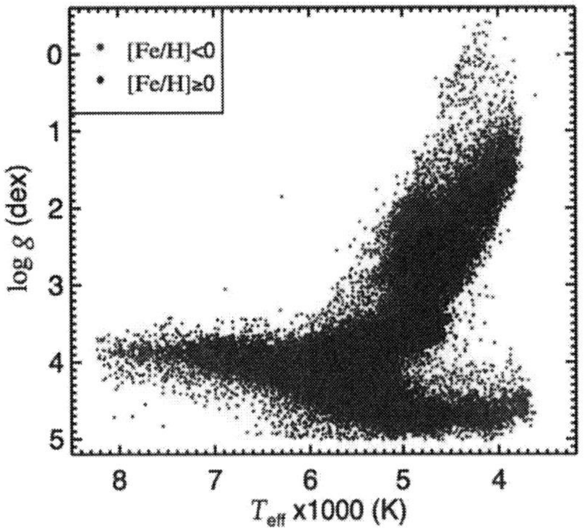

Figure 5: H-R diagram.

4 Parameter Analysis

The distribution of magnitude of the observed stars is plotted in Figure 4. In Figure 5, the H-R diagram is shown for the observed targets. Figure 6 shows the distribution of the parameters while Table 2 lists the particular stars: (V)MP - (Very) Metal Poor stars; HRV - High Radial Velocity stars.

References

[1] Luo,A.L.; Zhao,Y.H.; Zhao,G.; et al. RAA, 15, 1095 (2015)

[2] Ren,A.B.; Fu,J.N.; De Cat,P.; et al. ApJS, 225, 28 (2016)

[3] Su,D.Q.; Cui,X.Q.; Wang,Y. & Yao, Z.; Proc. SPIE, 3352, 76 (1998)

[4] Zhao,G.; Zhao,Y.H.; Chu,Y.Q.; et al. RAA, 12, 723 (2012)

 DOI: 10.1051/epjconf/201716001004

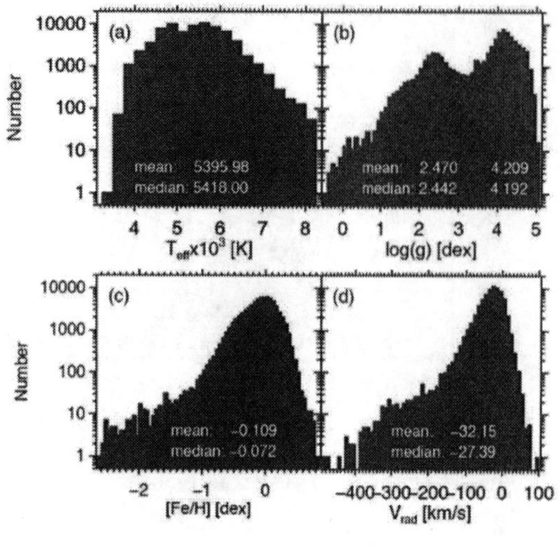

Figure 6: Distribution of parameters.

Table 2: Particular stars.

Field	MP	VMP	HRV
Field0	72	11	4
Field1	927	93	18
Field4	63	6	4
Field5	259	22	11
Field6	3	1	0

DOI: 10.1051/epjconf/201716001005

Data preparation for asteroseismology with TESS

Mikkel N. Lund[1,2]*, *Rasmus* Handberg[2]**, *Hans* Kjeldsen[2], *William J.* Chaplin[1,2], and *Jørgen* Christensen-Dalsgaard[2]

[1]*School of Physics and Astronomy, University of Birmingham, Edgbaston, Birmingham, B15 2TT, United Kingdom*
[2]*Stellar Astrophysics Centre, Department of Physics and Astronomy, Aarhus University, Ny Munkegade 120, DK-8000 Aarhus C, Denmark*

Abstract. The Transiting Exoplanet Survey Satellite (TESS) is a NASA Astrophysics Explorer mission. Following its scheduled launch in 2017, TESS will focus on detecting exoplanets around the nearest and brightest stars in the sky, for which detailed follow-up observations are possible. TESS will, as the NASA *Kepler* mission, include a asteroseismic program that will be organized within the TESS Asteroseismic Science Consortium (TASC), building on the success of the *Kepler* Asteroseismic Science Consortium (KASC). Within TASC data for asteroseismic analysis will be prepared by the TASC Working Group 0 (WG-0), who will facilitate data to the community via the TESS Asteroseismic Science Operations Center (TASOC), again building on the success of the corresponding KASOC platform for *Kepler*. Here, we give an overview of the steps being taken within WG-0 to prepare for the upcoming TESS mission.

1 Introduction

The Transiting Exoplanet Survey Satellite (TESS) is a NASA Astrophysics Explorer mission [1], scheduled for launch at the end of 2017 and with a nominal mission duration of 2 years. TESS may be seen as the successor to the NASA *Kepler* mission [2], and will as *Kepler* search for exoplanets using the transit method — here, a planet is identified from the dimming produced when it passes in front of its host star. Different from *Kepler*, TESS will focus on the nearest and brightest stars in the sky, allowing for detailed follow-up observations, and will over its nominal mission nearly cover the full sky. The primary science goal of *Kepler* was to determine the frequency of Earth-like planets in and near the habitable zone of solar-type stars [2]; TESS will instead focus on finding exoplanets smaller than Neptune where a detailed characterization is possible from follow-up observations.

With the advent of the space-based missions CoRoT [3] and *Kepler*, the field of asteroseismology has flourished over the last decade [4]. The reason for this advancement is that the photometric requirements needed for detecting transiting exoplanets coincide with those needed for asteroseismology, to wit, photometric observations of long duration and high precision. This synergy was realized early on for both the CoRoT and *Kepler* missions, and led for *Kepler* to the formation of the *Kepler* Asteroseismic Investigation (KAI). Via the *Kepler* Asteroseismic Science Consortium [KASC; 7] this provided direct access to the data from *Kepler* and helped to organize the work within the broad asteroseismic community.

Building on the success of KASC, the asteroseismic studies in TESS will be organized in the TESS Asteroseismic Science Consortium [TASC; 8]. In the following we will focus on the preparation of data from TESS for the sake of asteroseismology.

2 The TESS mission

Over its nominal mission TESS will observe the full sky, starting in the southern hemisphere. The total field of view (FOV) of the four cameras of TESS (each with 4 CCDs) will cover a rectangular slap of the sky spanning $24°\times96°$, starting from an ecliptic latitude of $\sim6°$. A given $24°\times96°$ field will be observed for ~27-days, corresponding to two orbits of the TESS spacecraft in its highly elliptical 13.7-day Lunar resonances orbit — we refer to such a field as an observing 'Sector'. Given the observing strategy adopted in TESS, some regions will be observed for longer than ~27-days. Most notable are the regions within 12° of the ecliptic poles that will be observed continuously, these are the so-called continuous viewing zones (CVZs).

Observing cadences will come at 20 and 120 seconds, and full-frame-images (FFIs) will be obtained every 30 minutes. Over the course of the nominal 2 year mission the number of stars observed in 20-sec and 120-sec cadences will exceed 200,000, and data for >20,000,000 stars are predicted from the 30-min FFIs. The pixels in TESS are, with a size of 21.1″, significantly larger than those of *Kepler*, which measured 3.98″. However, the pixel response function in TESS is very similar to that of *Kepler*, with

*e-mail: lundm@bison.ph.bham.ac.uk
**e-mail: rasmush@phys.au.dk

 DOI: 10.1051/epjconf/201716001005

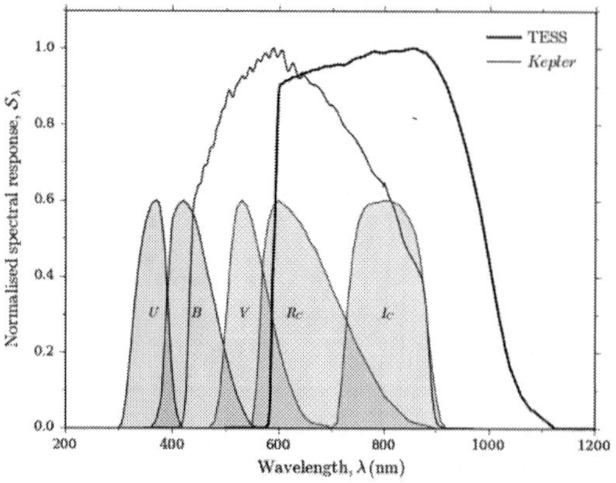

Figure 1: Spectral response functions S_λ for *Kepler* [5] and TESS [1], normalised to a maximum of 1. Shown are also the standard Johnson-Cousins $UBVR_CI_C$ photometric systems from [6], normalised to maximum values of 0.6.

~50% of light contained within 1 pixel, and ~90% contained within 4×4 pixels. The band-pass of TESS, roughly spanning the interval from 600 – 1000 nm and centred on the I_C band, is redder than that of *Kepler* which was centred on the R_C band (see Figure 1). At short wavelengths the TESS spectral response function is dominated by a long-pass filter transmission, and by the CCD quantum efficiency at long wavelengths.

Considering the number of stars observed and the larger number of pixels on average devoted to each of these (~100 pixels vs. ~32 in *Kepler*), the data rate for TESS from 120-sec cadence data will be a factor of ~13 that of *Kepler*. If FFIs are included the data rate rises to a factor of ~25 that of *Kepler* (Jenkins et al., in prep.). Data will be down-linked every 13.7-days when the TESS spacecraft reaches the perigee of its orbit. Here data will be transferred from TESS to the Deep Space Network (DSN), which will act as the relay for the TESS observations.

3 The TESS Asteroseismic Investigation

As mentioned in Section 1, the *Kepler* Asteroseismic Investigation (KAI) was organized within the broad international community in the KASC. Building on this, the TESS Asteroseismic Investigation (TAI) will be organized in the TESS Asteroseismic Science Consortium [TASC; 8]. Like KASC, the investigations within TASC will be divided between a number of Working Groups (WGs), each of which deals with the utilization of data for a specific group of objects. Each WG will have two co-chairs who will have the overall responsibility for the running of the WG, and these will be members of the TASC steering committee (SC). The TASC-SC, including also the TASC Board, is responsible for the overall running of TASC and will reports to the TESS team on issues pertaining to target selection. TASC will furthermore organize workshops aiming at target selection, science collaboration and data analysis.

Data and communication platforms for the WGs will be facilitated for TASC via the TESS Asteroseismic Science Operations Center (TASOC)[1], hosted at the Stellar Astrophysics Centre (SAC) at Aarhus University, Denmark. TASOC will furthermore provide long-term storage of all data products. By and large, TASOC will copy the facilities of the *Kepler* Asteroseismic Science Operations Centre (KASOC)[2]. Membership of TASC is open and any member of TASC can apply to become a member of a given WG. The WG-0 "*TASOC - Basic photometric algorithms and calibration of time / TASC data products*" will, as the name suggests, be responsible for maintaining the TASOC portal and the timely provision of data products for the whole of TASC. In Section 4 below we outline the different main tasks and responsibilities of WG-0.

4 WG-0 tasks

WG-0 will have the overall responsibility for delivering analysis-ready data for asteroseismology to TASC in a timely fashion. For each 27-day pointing, ~750 targets at 120-sec cadence, and ~60 targets at a 20-sec cadence, will be available for asteroseismology. WG-0 is, however, committed to the preparation of data for all targets with 120-sec and 20-sec cadences, not only those designated for asteroseismology. Additionally, WG-0 will analyse the 30-min FFIs in order to facilitate the detection of oscillations in red giants, SPBs, RR Lyraes, β Cep stars, Cepheids, etc., and will also produce light curves for eclipsing binaries. To produce optimally prepared data for the many different types of studies conducted within TASC, WG-0 will maintain close collaborations with the other WGs of TASC.

The TESS Science Processing Operations Center (SPOC) will process all 120-sec targets in the same manner as done by the Science Operations Center (SOC) for *Kepler*. This includes, for instance, the calibration of pixels, extraction of photometry and astrometry, definition of optimal pixel masks for aperture photometry, correction for systematic errors, etc. — i.e., an end-to-end analysis. For FFIs, SPOC is only committed to calibrating and archiving the pixels, while no corrections will be done at all for 20-sec data products (see Section 4.1). Data products from both WG-0 and SPOC will be modelled after those from *Kepler* (Jon Jenkins, private comm.).

4.1 20-sec-specific data correction

The 20-sec cadence data have been included amongst the cadences employed by TESS primarily for the sake of asteroseimology. The 20-sec cadence will be especially useful for studies of high-frequency oscillators, such as white dwarfs and some main-sequence solar-like oscillators. Because this sampling has been introduced for asteroseismology, only fully raw data will be delivered by the TESS team. WG-0 will then be responsible for the full calibration and analysis of these data, including basic corrections

[1] tasoc.dk
[2] kasoc.phys.au.dk

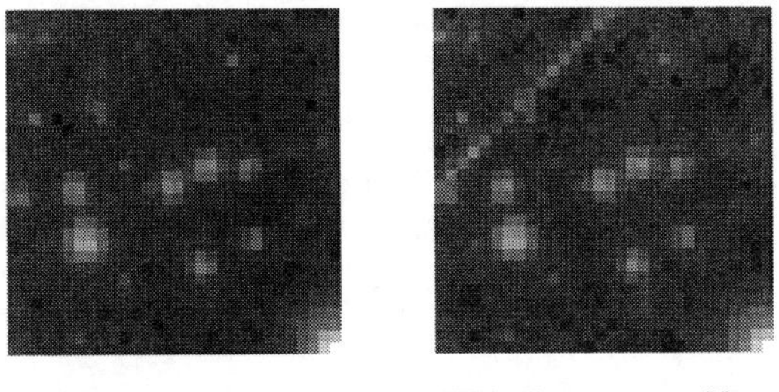

Figure 2: Potential effect of CRs in TESS. Shown are two 20-sec cadences of the same simulated pixel-field, one of which (right) are affected by a CR producing a trail impacting many pixels.

for 2D black levels; detector gain/linearity; smear; flat-fielding; and the removal of cosmic rays.

4.1.1 Cosmic rays

For 120-sec data and the 30-min FFIs, cosmic-ray (CR) signals will be mitigated on-board before the cadences are created from the 2-sec integrations in TESS. The idea for this mitigation is, at the time of writing, to identify outliers in the 2-sec light curves of individual pixels. If a given pixel is found to be affected by CRs, the identified 2-sec samplings are removed before the data are co-added to the 120-sec and 30-min cadences. Given that the 20-sec data will only consist of 10 such 2-sec integrations, it has been decided that removing the CRs from the co-added data on ground is more optimal. For every 20-sec cadence there is a ~1.7% chance per pixel for a CR hit. WG-0 will before launch need to identify suitable methods for such a correction. It is worth noting that CRs in TESS will impact the photometry in a manner quite different to that in *Kepler*, because of the difference in the pixels between TESS and *Kepler*. In TESS, the pixels have a width of 15μm and a depth of 100μm, whereas *Kepler* use pixels with a width of 27μm and a depth of 15μm. The reason for this choice is the desire for a high spectral response at long wavelengths (Figure 1), which requires significantly deeper pixels due to the quantum efficiency of the detector material. The deeper pixels, however, means that the cross-section of the detector for an incoming CR is much larger than in *Kepler*. Figure 2 shows a simulated pixel field at two different 20-sec cadences, where one (right panel) is affected by a CR. Where such an event in *Kepler* would likely only have affected a single pixel, it can in TESS produce a trail which impacts many pixels.

4.2 Sky backgrounds

For 20-sec data and FFIs WG-0 will need to estimate sky-background (SB) levels. The non-instrumental SB is mainly composed of the contribution from the diffuse background of unresolved stars and galaxies and the sky glow from Zodiacal light, which depends especially on ecliptic latitude [see, e.g., 9]. Before launch, WG-0 will work towards a proper and robust estimation of the SB for the highly diverse fields covered by TESS, going from near-ecliptic to polar and from very sparse to very dense (including regions containing stellar clusters, see Figure 3).

4.3 Extracting photometry

WG-0 is committed to extracting light curves for all possible sources in the 20-sec, 120-sec, and 30-min FFI data. As mentioned in Section 1, this will over the course of the nominal 2-year mission amount to >200,000 star from 20-sec and 120-sec cadences, and >20,000,000 stars from the 30-min FFIs. This number of targets, coupled with the requirement of a timely processing, means that the pipeline constructed for this task will need to be both fast and robust. The pipeline will also have to be flexible in terms of its ability to process very diverse fields, including dense fields close to the ecliptic, nebulous regions with high contamination from the SB, and open as well as globular clusters (Figure 3). It will be especially interesting in the pre-flight tests (Section 5) to see what can be expected for studies of star clusters given the relatively large TESS pixels.

Many methods exist for extracting photometry from CCD images, including aperture, point-spread-function (PSF), and so-called optimal photometry [10–13]. Some of these have already been adapted, or extended upon, for the *Kepler* and K2 missions [14–18]. Each of the methods have their pros and cons — aperture photometry is by far the simplest and fastest method, but deciding the optimal size and shape of the aperture is not always straightforward, and it is far from optimal for dense and crowded regions; optimal photometry can provide a more accurate extraction, but it is slower, requires knowledge (albeit not particularly accurate) of the PSF, and is still not optimal for dense and crowded regions; PSF photometry is optimal for dense and crowded regions, but requires accurate knowledge of the PSF and is again slower than aperture photometry. Concerning the PSF, it is worth noting that the TESS PSF will include both off-axis aberrations and chromatic aberrations arising both from the refractive elements of the TESS camera and from the deep-depletion CCDs, absorbing redder photons deeper in the silicon.

All these aspects of the different possible methods must be considered in a final pipeline — ideally, each method should be thoroughly tested on realistic simulated data, considering here also the hardware requirements that will be needed to keep up with the high data rates of TESS. In the end, light curves may well have to be extracted with a range of different methods, depending on the type or crowding of the field under study. Another option might be to run several methods for all fields, with the optimum choice of extracted photometry being made only after the fact.

4.4 Light curve preparation

Following the extraction of raw light curves from pixel data, WG-0 will for each star produce an analysis-ready light curve for asteroseismology, corrected for any instrumental features. From *Kepler* we know that instrumen-

DOI: 10.1051/epjconf/201716001005

tal features can come in many forms [19–21], including jumps from drops in pixel sensitivity, or from differences in sensitivities between the CCDs that a given star might land on. Such shifts in CCD position happened every Quarter in *Kepler*, and will also occur with TESS for stars with observing durations exceeding the ~27 days of an observing Sector; secular changes from variations in focus (e.g. from a change in solar heating of the spacecraft), or drifts either in pointing or from differential velocity aberrations; abrupt changes after safe-mode events or data down-links (which will happen every 13.7-days with TESS); transient events such as the Argabrightening events found in *Kepler* [22], CRs, or from momentum dumps in the reaction wheels orienting the spacecraft.

Currently, we can only speculate about the instrumental features that will be found in TESS, but it is near certain that some features will be found. The instrumental features that might be found cannot simply be rectified in the same manner for all types of stars under study by TASC (including solar-like oscillators, RR Lyraes, white dwarfs, eclipsing binaries, etc.). When observing a given star, the observed signal will be a mix of physical and instrumental contributions. Given that the time scales, amplitudes, and phase stability of the physical component will depend on the type of star observed, and thus also on its overlap with the instrumental signals, the method for isolating the instrumental contribution and preserving the astrophysical signal will in effect also depend on the stellar type.

The idea in WG-0 is to build on the collective knowledge of the community by bringing together people with expertise on the data preparation for different stellar types [see, e.g., 20, 21, 23–26]. Many methods for rectifying light curves for analysis were developed during the *Kepler* mission, and more recently for the re-purposed K2 mission [see, e.g., 15, 27–31]. WG-0 will develop a data-correction pipeline that adopts a star-based approach to the mitigation of instrumental effects; this will build on pipelines developed during the *Kepler* mission for specific types of stars. For the pipeline it is worth keeping the high data rate of TESS in mind — not only should the pipeline be robust and able to handle a diverse range of stellar types, it should also be fast enough to allow for a timely facilitation of processed data. Several versions of light curves will be available via TASOC for a given star, including a raw uncorrected light curve; a 'standard' light curve where the correction method adopted is the same for all stars; and a star-type customized light curve (based on the inputs and request of the TASC community).

4.5 Absolute timing

The TESS on-board clock should be accurate and stable to better than ~5 msec. To obtain a similar accuracy on the time stamps in Barycenter Julian Days (BJD) in the Earth frame, the correction to the light travel time between the spacecraft and the DSN should be accurate to the same level. This will be achieved from knowing the 3D-position of the TESS spacecraft in space to a high level of accuracy (1500 km, corresponding to a light travel time of 5 msec). However, delays may occur in the ground system (e.g. after data down-links or safe mode event) that cannot be accounted for without an independent assessment of any temporal shifts.

For the sake of ground-based follow-up observations, e.g. of transiting exoplanet hosts, it is naturally worth knowing the absolute time stamps of the data. Requirements on the accuracy of the absolute timing comes also from asteroseismology [32]:

- To reach the highest possible photometric quality from 120-sec observations, and the photon noise limit for the brightest stars, the absolute photometry needs to be accurate and stable to better than 5 msec.
- To reach the theoretical accuracy of high-amplitude coherent oscillations one needs the time at which each exposure is obtained to be very accurate over the period of an (27-day) observing Sector. For coherent pulsation modes this requires that the length of exposure is accurate over an observing Sector to better than 5 msec.
- To allow comparisons between ground-based observations with those from TESS, one needs to be able to estimate the absolute time of a given photometric data point and establish a stable reference (e.g. central time of a given observation). For coherent pulsation modes the absolute time (in HJD/BJD) should be known to better than 0.5 sec; for solar-like oscillations the required accuracy is better than 1 sec over a ~10 day period.

For the calculations leading to these estimates see [32].

The TESS team will make the corrections based on calculated light travel times; WG-0 is then committed to making independent checks of the absolute time stamps. The regular calibrations will be achieved by performing contemporaneous observations between TESS and ground-based facilities of several objects with photometry varying rapidly in time, such as bright, deep, detached eclipsing binaries. The absolute time shift, if any, can then be determined by cross-correlating the contemporaneous time series. The ideal objects for these checks will be found in the CVZs of TESS.

The work on the absolute timing issue will be handled by a dedicated sub-group of WG-0. As the checks of absolute times should be done regularly, and possibly after any data down-link or safe-mode event, the sub-group will have to be able to respond and obtain ground-based data on short notice. WG-0 will here depend on members of the TASC community with access to ground-based facilities.

4.6 Stellar classification

An additional sub-group will be formed under WG-0 to perform stellar classification of stars observed with TESS. The classification is important to select the proper course of action in rectifying a given light curve for asteroseismic studies (Section 4.4). WG-0 will conduct studies of the best classification of stars from the raw photometric data from TESS — this will be achieved using techniques from machine learning [see, e.g., 33–36], which will be tested on simulated TESS data before launch.

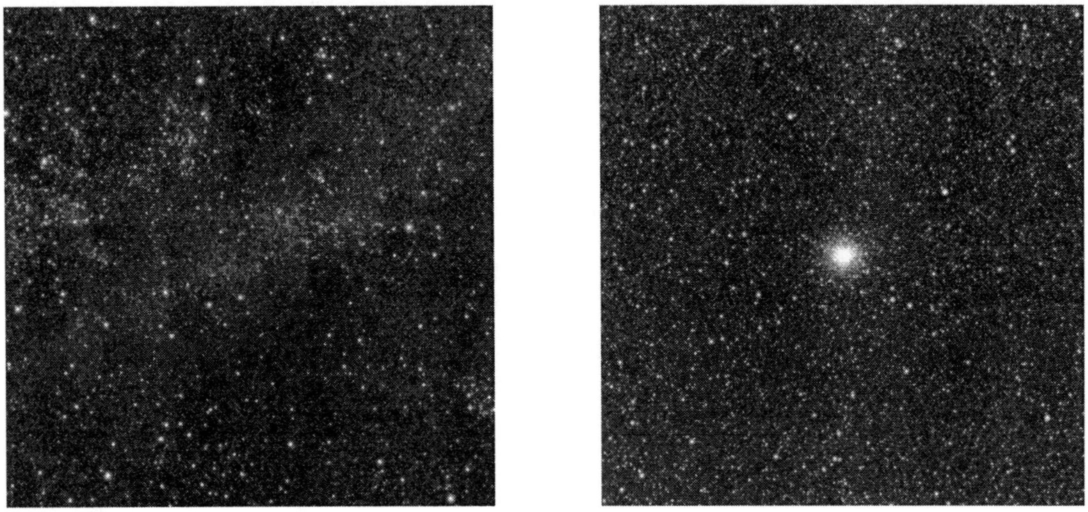

Figure 3: Simulated pixels fields from SPyFFI of the Large Magellanic Cloud (LMC; left) and the globular cluster ω Centauri (NGC 5139; right).

5 Pre-flight tests

In order for WG-0 to be able to construct a data processing pipeline that is ready when the first data from TESS are received, numerous tests will be conducted on simulated data (Section 5.1).

5.1 Pixel-data simulation

Pre-flight analysis will be performed on simulated TESS pixel data made using the "Spiffy Python for Full Frame Images" (SPyFFI) simulator. The simulator was created at the Massachusetts Institute of Technology (MIT) by Zachory K. Berta-Thompson (private comm.). As the name suggests, SPyFFI is a Python-based code for simulating TESS pixel data, including FFIs.

To simulate a given field, SPyFFI uses a user-specified input catalogue with stellar positions and magnitudes. The UCAC4 [37] catalog is currently used, but eventually the TESS Input Catalog [TIC; 38] will be adopted. SPyFFI includes realistic models for the TESS pixel response, differential velocity aberration, cosmic rays, spacecraft jitter, focus changes, and sky backgrounds (and the parameters of all of these contributions can be adjusted to test methods from best- to worst-case scenarios). Figure 3 gives examples of two simulated TESS pixel fields, one of the Large Magellanic Cloud (LMC) and one of the ω Centauri globular cluster.

SPyFFI furthermore has the option of assigning a simulated light curve to a given star in a given field. These light curves can include transits, eclipses, spot modulations, and/or oscillations. The light curves with solar-like oscillations and granulation signals are produced using the asteroFLAG simulator [39]; light curves for classical oscillators have been constructed with frequencies, phases, and amplitudes from such stars observed by *Kepler* (Vichi Antoci and Steven Kawaler, private comm.).

5.2 T'DA workshop series

To address the issues of TESS data preparation for asteroseismology, WG-0 is organizing the workshop series "TESS Data for Asteroseismology" (T'DA). The idea is to bring together people from the broad community, who either have expertise from missions such as *Kepler* or CoRoT, or who are students planning to work on data analysis issues. The T'DA series is planned to include, at least, workshops dedicated to (1) extracting light curves from pixel data; (2) correcting light curves for the optimal output from asteroseismic analysis; and (3) stellar classification. The first workshop (T'DA1), entitled "From Pixels to Light Curves", will be held at the University of Birmingham, UK, from 31st Oct. to 2nd Nov. 2016.

Acknowledgements

The authors would like to thank the organisers of the "Seismology of the Sun and the Distant Stars 2016 — Using Today's Successes to Prepare the Future" (SpaceTK16) conference, a joint TASC2-KASC9 Workshop – SPACEINN & HELAS8 Conference, where MNL presented a talk on the contents of these proceedings. Funding for the Stellar Astrophysics Centre (SAC) is provided by The Danish National Research Foundation (Grant DNRF106). MNL acknowledges the support of The Danish Council for Independent Research | Natural Science (Grant DFF-4181-00415). WJC acknowledges the support of the UK Science and Technology Facilities Council (STFC).

References

[1] G. R. Ricker, J. N. Winn, R. Vanderspek, D. W. Latham, G. Á. Bakos, J. L. Bean, Z. K. Berta-Thompson, T. M. Brown, L. Buchhave, N. R. Butler

et al., *Transiting Exoplanet Survey Satellite (TESS)*, in *Society of Photo-Optical Instrumentation Engineers (SPIE) Conference Series* (2014), Vol. 9143 of *Society of Photo-Optical Instrumentation Engineers (SPIE) Conference Series*, p. 20

[2] W. J. Borucki, D. Koch, G. Basri, N. Batalha, T. Brown, D. Caldwell, J. Caldwell, J. Christensen-Dalsgaard, W. D. Cochran, E. DeVore et al., Science **327**, 977 (2010)

[3] A. Baglin, M. Auvergne, P. Barge, J. T. Buey, C. Catala, E. Michel, W. Weiss, COROT Team, *COROT: asteroseismology and planet finding*, in *Stellar Structure and Habitable Planet Finding*, edited by B. Battrick, F. Favata, I.W. Roxburgh, D. Galadi (2002), Vol. 485 of *ESA Special Publication*, pp. 17–24

[4] R. L. Gilliland, *Prelude to, and Nature of the Space Photometry Revolution*, Vol. 101 of *European Physical Journal Web of Conferences* (2015), p. 1, arXiv eprint 1411.2531

[5] J. E. Van Cleve & D. A. Caldwell, *Kepler Instrument Handbook* (2009), (KSCI-19033-002; http://archive.stsci.edu/kepler/documents.html)

[6] M. S. Bessell, PASP, **102**, 1181 (1990)

[7] H. Kjeldsen, J. Christensen-Dalsgaard, R. Handberg, T. M. Brown, R.L. Gilliland, W. J. Borucki, D. Koch, AN, **331**, 966 (2010)

[8] H. Kjeldsen, J. Christensen-Dalsgaard, W. J. Chaplin, *TESS Asteroseismic Science Consortium* (2016), (SAC-TESS-0003-6; https://tasoc.dk/info/docs.php)

[9] P. W. Sullivan, J. N. Winn, Z. K. Berta-Thompson, D. Charbonneau, D. Deming, C. D. Dressing, D. W. Latham, A. M. Levine, P. R. McCullough, T. Morton et al., ApJ, **809**, 77 (2015)

[10] P. B. Stetson, PASP, **99**, 191 (1987)

[11] S. B. Howell, PASP, **101**, 616 (1989)

[12] T. Naylor, MNRAS, **296**, 339 (1998)

[13] D. Makovoz, F. R. Marleau, PASP, **117**, 1113 (2005)

[14] S. T. Bryson, P. Tenenbaum, J. M. Jenkins, H. Chandrasekaran, T. Klaus, D. A. Caldwell, R. L. Gilliland, M. R. Haas, J. L. Dotson, D. G. Koch et al., ApJL, **713**, L97 (2010)

[15] M. N. Lund, R. Handberg, G. R. Davies, W. J. Chaplin, C. D. Jones, ApJ, **806**, 30 (2015)

[16] S. Aigrain, S. T. Hodgkin, M. J. Irwin, J. R. Lewis, S. J. Roberts, MNRAS, **447**, 2880 (2015)

[17] M. Libralato, L. R. Bedin, D. Nardiello, G. Piotto, MNRAS, **456**, 1137 (2016)

[18] D. Nardiello, M. Libralato, L. R. Bedin, G. Piotto, L. Borsato, V. Granata, L. Malavolta, V. Nascimbeni, MNRAS, **463**, 1831 (2016)

[19] J. M. Jenkins, D. A. Caldwell, H. Chandrasekaran, J. D. Twicken, S. T. Bryson, E. V. Quintana, B. D. Clarke, J. Li, C. Allen, P. Tenenbaum et al., ApJL, **713**, L120 (2010)

[20] R. A. García, S. Hekker, D. Stello, J. Gutiérrez-Soto, R. Handberg, D. Huber, C. Karoff, K. Uytterhoeven, T. Appourchaux, W. J. Chaplin et al., MNRAS, **414**, L6 (2011)

[21] R. Handberg & M. N. Lund, MNRAS, **445**, 2698 (2014)

[22] F. C. Witteborn, J. Van Cleve, W. Borucki, V. Argabright, P. Hascall, *DEBRIS sightings in the Kepler field*, in *Techniques and Instrumentation for Detection of Exoplanets V* (2011), Vol. 8151 of *Society of Photo-Optical Instrumentation Engineers (SPIE) Conference Series*, p. 815117

[23] K. Kolenberg, S. Bryson, R. Szabó, D. W. Kurtz, R. Smolec, J. M. Nemec, E. Guggenberger, P. Moskalik, J. M. Benkő, M. Chadid et al., MNRAS, **411**, 878 (2011)

[24] K. Kinemuchi, T. Barclay, M. Fanelli, J. Pepper, M. Still, S. B. Howell, PASP, **124**, 963 (2012)

[25] C. Danielski, T. Kacprzak, G. Tinetti, *Detrending the long-term stellar activity and the systematics of the Kepler data with a non-parametric approach*, at *European Planetary Science Congress 2013* (http://meetings.copernicus.org/epsc2013)

[26] B. Kirk, K. Conroy, A. Prša, M. Abdul-Masih, A. Kochoska, G. Matijevič, K. Hambleton, T. Barclay, S. Bloemen, T. Boyajian et al., AJ, **151**, 68 (2016)

[27] S. B. Howell, C. Sobeck, M. Haas, M. Still, T. Barclay, F. Mullally, J. Troeltzsch, S. Aigrain, S. T. Bryson, D. Caldwell et al., PASP, **126**, 398 (2014)

[28] A. Vanderburg, J. A. Johnson, PASP, **126**, 948 (2014)

[29] B. J. S. Pope, T. R. White, D. Huber, S. J. Murphy, T. R. Bedding, D. A. Caldwell, A. Sarai, S. Aigrain, T. Barclay, MNRAS, **455**, L36 (2016)

[30] J. E. Van Cleve, S. B. Howell, J. C. Smith, B. D. Clarke, S. E. Thompson, S. T. Bryson, M. N. Lund, R. Handberg, W. J. Chaplin, PASP, **128**, 075002 (2016)

[31] S. Aigrain, H. Parviainen, B. J. S. Pope, MNRAS, **459**, 2408 (2016)

[32] H. Kjeldsen, J. Christensen-Dalsgaard, W. J. Chaplin, *Requirement Specifications for TESS: Timing requirements for Asteroseismology* (2013), (SAC-TESS-0002-5; https://tasoc.dk/info/docs.php)

[33] J. Debosscher, L.M. Sarro, M. López, M. Deleuil, C. Aerts, M. Auvergne, A. Baglin, F. Baudin, M. Chadid, S. Charpinet et al., A&A, **506**, 519 (2009)

[34] D. J. Armstrong, J. Kirk, K. W. F. Lam, J. McCormac, H.P. Osborn, J. Spake, S. Walker, D. J. A. Brown, M. H. Kristiansen, D. Pollacco et al., MNRAS, **456**, 2260 (2016)

[35] G. Bass, K. Borne, MNRAS, **459**, 3721 (2016)

[36] A. D'Isanto, S. Cavuoti, M. Brescia, C. Donalek, G. Longo, G. Riccio, S. G. Djorgovski, MNRAS, **457**, 3119 (2016)

[37] N. Zacharias, C. T. Finch, T. M. Girard, A. Henden, J. L. Bartlett, D. G. Monet, M. I. Zacharias, AJ **145**, 44 (2013)

[38] D. W. Latham, *TESS Input Catalog* (2015), (EXP-TESS-SAO-TSO-0001; Rev. 1.3.1;

DOI: 10.1051/epjconf/201716001005

https://cftest.stsci.edu/download/attachments/50632373/TICv1.3.1.pdf)

[39] W. J. Chaplin, T. Appourchaux, T. Arentoft, J. Ballot, J. Christensen-Dalsgaard, O. L. Creevey, Y. Elsworth, S. T. Fletcher, R. A. García, G. Houdek et al., AN, **329**, 549 (2008)

EPJ Web of Conferences **160**, 01006 (2017)
Seismology of the Sun and the Distant Stars II
DOI: 10.1051/epjconf/201716001006

On the detectability of solar-like oscillations with the NASA *TESS* mission

Tiago L. Campante[1,2,⋆]

[1] *School of Physics and Astronomy, University of Birmingham, Edgbaston, Birmingham B15 2TT, UK*
[2] *Stellar Astrophysics Centre (SAC), Department of Physics and Astronomy, Aarhus University, Ny Munkegade 120, DK-8000 Aarhus C, Denmark*

Abstract. The upcoming NASA *TESS* mission will perform an all-sky survey for planets transiting bright nearby stars. In addition, its excellent photometric precision will enable asteroseismology of solar-type and red-giant stars. We apply a newly developed detection test along a sequence of stellar evolutionary tracks in order to predict the detectability of solar-like oscillations with *TESS*.

1 Introduction

The *Transiting Exoplanet Survey Satellite*[1] [*TESS*; 1] is a NASA-sponsored Astrophysics Explorer mission that will perform an all-sky survey for planets transiting bright nearby stars. Its launch is currently scheduled for December 2017, with science operations due to start in March 2018. During the 2-year primary mission, *TESS* will monitor the brightness of several hundred thousand main-sequence, low-mass stars over intervals ranging from one month to one year, depending mainly on a star's ecliptic latitude. Monitoring of these pre-selected target stars will be made at a cadence of 2 min. Additionally, full-frame images (FFIs) will be recorded every 30 min, which will allow a wider range of stars to be searched for transits. Being 10–100 times brighter than *Kepler* targets and distributed over a solid angle that is nearly 300 times larger, *TESS* host stars will be well suited for follow-up radial-velocity/spectroscopic observations from the ground. *TESS* is expected to detect approximately 1700 transiting planets from $\sim 2 \times 10^5$ pre-selected target stars [2]. Analysis of the FFIs will lead to the additional detection of several thousand planets orbiting stars that are not among the pre-selected targets.

Furthermore, *TESS*'s excellent photometric precision, combined with its fine time sampling and long intervals of uninterrupted observations, will enable asteroseismology of solar-type and red-giant stars, whose dominant oscillation periods range from minutes to hours. In Section 2, we apply a newly developed detection test along a sequence of stellar evolutionary tracks in order to predict the detectability of solar-like oscillations with *TESS* across the Hertzsprung–Russell (H–R) diagram.

⋆e-mail: campante@bison.ph.bham.ac.uk
[1] https://tess.gsfc.nasa.gov

2 Detectability of solar-like oscillations across the H–R diagram

New images will be acquired by each of the four *TESS* cameras every 2 seconds. However, due to limitations in onboard data storage and telemetry, these 2-sec images will be stacked (before being downlinked to Earth) to produce two primary data products with longer effective exposure times: (i) subarrays of pixels centered on several hundred thousand pre-selected target stars will be stacked at a 2-min cadence, while (ii) FFIs will be stacked every 30 min.

Up to 20,000 2-min-cadence slots (or the equivalent to $\sim 10\,\%$ of the pre-selected target stars) will be allocated to the *TESS* Asteroseismic Science Consortium (TASC) over the course of the mission. In addition, a number of slots (notionally 1500) with faster-than-standard sampling, i.e., 20 sec, will be reserved for the investigation of asteroseismic targets of special interest (mainly compact pulsators and main-sequence, low-mass stars).

In [3] we developed a simple test to estimate the detectability of solar-like oscillations in *TESS* photometry of any given star. This test is in turn based on the test previously developed by [4] to estimate the detectability of solar-like oscillations in any given *Kepler* target, which looked for signatures of the bell-shaped power excess due to the oscillations. In what follows, we apply the newly developed detection test along a sequence of stellar evolutionary tracks in order to predict the detectability of solar-like oscillations.

Figures 1–3 illustrate the detectability of solar-like oscillations with *TESS*. We focus on that portion of the H–R diagram populated by solar-type and (low-luminosity) red-giant stars, bound at high effective temperatures by the red edge of the δ Scuti instability strip. The detection code was applied along a sequence of solar-calibrated stellar evolutionary tracks spanning the mass range 0.8–$2.0\,M_\odot$ (in steps of $0.2\,M_\odot$).

In Figure 1 we consider two different observing lengths and a cadence of $\Delta t = 2$ min. Further assuming a systematic noise level[2] of $\sigma_{\rm sys} = 60$ ppm hr$^{1/2}$, detection of solar-like oscillations in main-sequence stars will not be possible for $T = 27$ d. Increasing the observing length to $T = 351$ d (relevant for stars near the ecliptic poles) may lead to the marginal detection of oscillations in (very bright) main-sequence stars more massive than the Sun. In both cases, detection of oscillations in subgiant and red-giant stars is nonetheless made possible, owing to their higher intrinsic amplitudes. As one would expect, this situation is significantly improved as the systematic noise level is brought down to $\sigma_{\rm sys} = 0$ ppm hr$^{1/2}$, with detections now being made possible for the brightest main-sequence stars over a range of masses.

The longer 30-min cadence is considered in Figures 2 and 3, where we have assumed a systematic noise level of $\sigma_{\rm sys} = 60$ ppm hr$^{1/2}$. FFIs will allow detecting oscillations in red-giant stars down to relatively faint magnitudes. Furthermore, it becomes apparent from Figure 3 that it should be possible to detect oscillations in the super-Nyquist regime for the brightest red giants.

In Figure 4 we provide a sneak peek at a power spectrum of a low-luminosity red-giant star as it would be observed by *TESS* at the 30-min cadence. The two panels in this figure show the power spectra of Kepler-56 as observed both by *Kepler* (in its long-cadence mode) and *TESS* based on 1 year of observations. As can be seen, oscillations are easily identified in either power spectrum. Notice, however, that while the granulation power and shot noise levels are commensurate for the *Kepler* power spectrum, the shot noise level is one order of magnitude higher than the granulation power in the case of the *TESS* power spectrum. This is mostly due to the smaller (by 2 orders of magnitude) effective collecting area of the individual *TESS* cameras.

[2]This is an engineering requirement imposed on the design of the *TESS* photometer and not an estimate of the anticipated systematic noise level on 1-hour timescales. In the actual data it is expected that this systematic term will be rather small in the frequency range occupied by the oscillations. Herein, we explore the implications of having $\sigma_{\rm sys} = 0$ ppm hr$^{1/2}$ (ideal case) and $\sigma_{\rm sys} = 60$ ppm hr$^{1/2}$ (regarded as a worst-case scenario).

3 Summary and discussion

In [3] we developed a simple test to estimate the detectability of solar-like oscillations in *TESS* photometry of any given star. We applied this detection test here along a sequence of stellar evolutionary tracks in order to predict the detectability of solar-like oscillations across the H–R diagram (Section 2). Importantly, we note that bright subgiants are attractive targets for the 2-min-cadence slots reserved for asteroseismology, even though they are far from optimal for transiting planet detection. We should be able to use parallaxes from the ongoing *Gaia* mission [5] to deliberately target these bright subgiants.

Acknowledgements

The authors would like to thank the organizers of the "Seismology of the Sun and the Distant Stars 2016" conference, a joint TASC2/KASC9 workshop – SPACEINN & HELAS8 conference, where the author presented an invited talk on the contents of these proceedings. The author acknowledges the support of the UK Science and Technology Facilities Council (STFC). Funding for the Stellar Astrophysics Centre is provided by The Danish National Research Foundation (Grant DNRF106).

References

[1] G.R. Ricker, J.N. Winn, R. Vanderspek, D.W. Latham, G.Á. Bakos, J.L. Bean, Z.K. Berta-Thompson, T.M. Brown, L. Buchhave, N.R. Butler et al., Journal of Astronomical Telescopes, Instruments, and Systems **1**, 014003 (2015)

[2] P.W. Sullivan, J.N. Winn, Z.K. Berta-Thompson, D. Charbonneau, D. Deming, C.D. Dressing, D.W. Latham, A.M. Levine, P.R. McCullough, T. Morton et al., ApJ **809**, 77 (2015)

[3] T.L. Campante, M. Schofield, J.S. Kuszlewicz, L. Bouma, W.J. Chaplin, D. Huber, J. Christensen-Dalsgaard, H. Kjeldsen, D. Bossini, T.S.H. North et al., ApJ **830**, 138 (2016)

[4] W.J. Chaplin, H. Kjeldsen, T.R. Bedding, J. Christensen-Dalsgaard, R.L. Gilliland, S.D. Kawaler, T. Appourchaux, Y. Elsworth, R.A. García, G. Houdek et al., ApJ **732**, 54 (2011)

[5] Gaia Collaboration, ArXiv e-prints (2016), 1609.04153

DOI: 10.1051/epjconf/201716001006

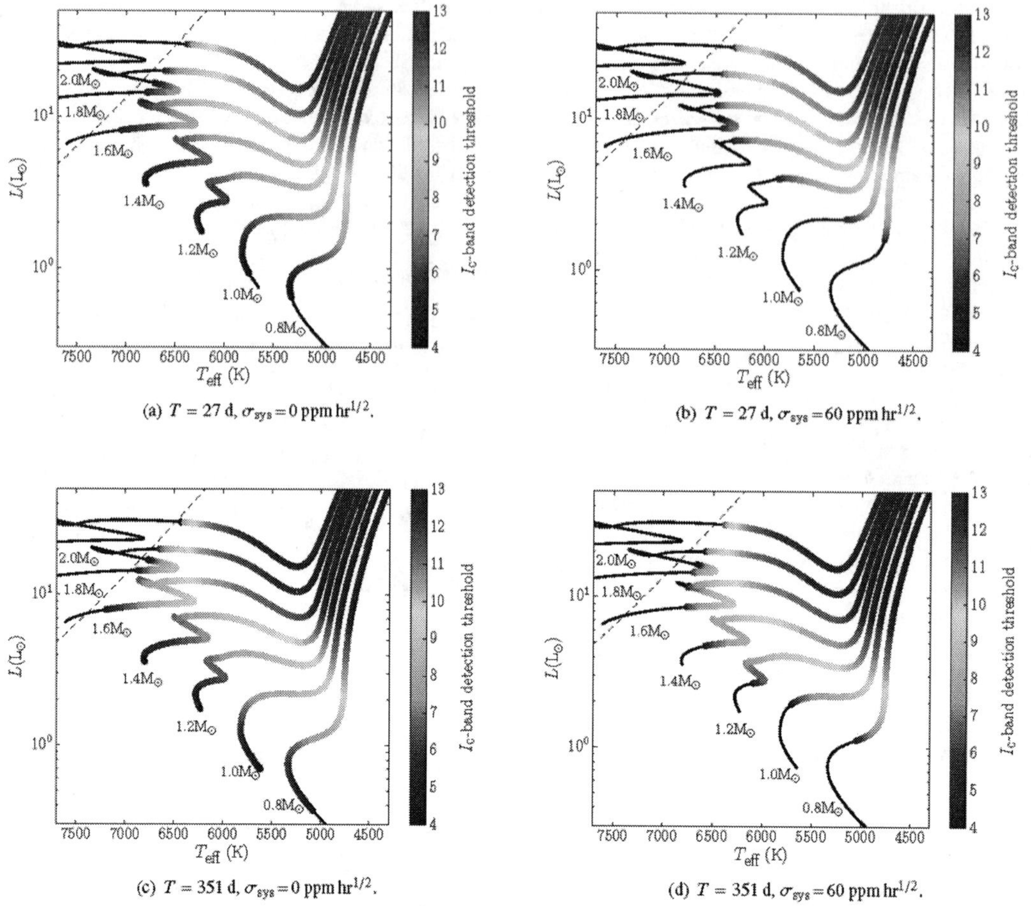

(a) $T = 27$ d, $\sigma_{\rm sys} = 0$ ppm hr$^{1/2}$.

(b) $T = 27$ d, $\sigma_{\rm sys} = 60$ ppm hr$^{1/2}$.

(c) $T = 351$ d, $\sigma_{\rm sys} = 0$ ppm hr$^{1/2}$.

(d) $T = 351$ d, $\sigma_{\rm sys} = 60$ ppm hr$^{1/2}$.

Figure 1: Detectability of solar-like oscillations with *TESS* across the H–R diagram for a cadence of $\Delta t = 2$ min. I_C-band detection thresholds are color-coded (no detection is possible along those portions of the tracks shown as a thin black line). The slanted dashed line represents the red edge of the δ Scuti instability strip. The several panels consider different combinations of the length of the observations (T) and systematic noise level ($\sigma_{\rm sys}$), as indicated. This figure was originally published in [3].

DOI: 10.1051/epjconf/201716001006

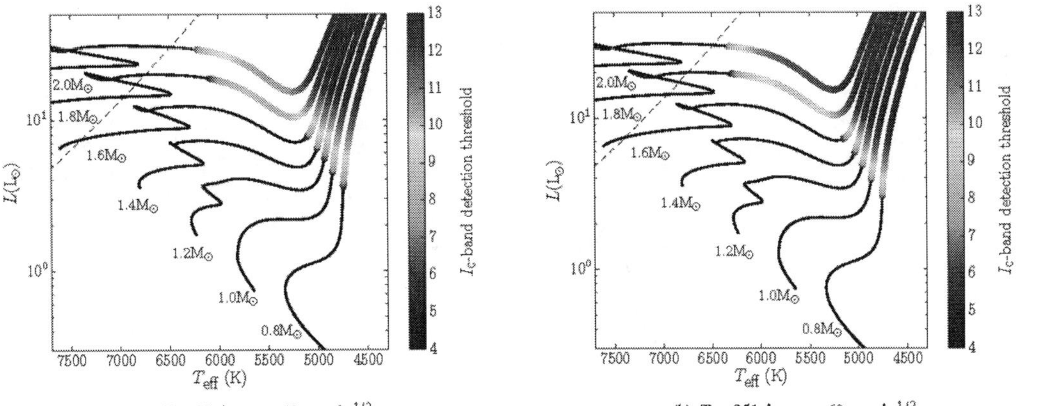

(a) $T = 27$ d, $\sigma_{\rm sys} = 60$ ppm hr$^{1/2}$.

(b) $T = 351$ d, $\sigma_{\rm sys} = 60$ ppm hr$^{1/2}$.

Figure 2: Detectability of solar-like oscillations with *TESS* across the H–R diagram for a cadence of $\Delta t = 30$ min. I_C-band detection thresholds are color-coded (no detection is possible along those portions of the tracks shown as a thin black line). The slanted dashed line represents the red edge of the δ Scuti instability strip. The two panels consider different lengths of the observations (T) and a systematic noise level of $\sigma_{\rm sys} = 60$ ppm hr$^{1/2}$, as indicated. This figure was originally published in [3].

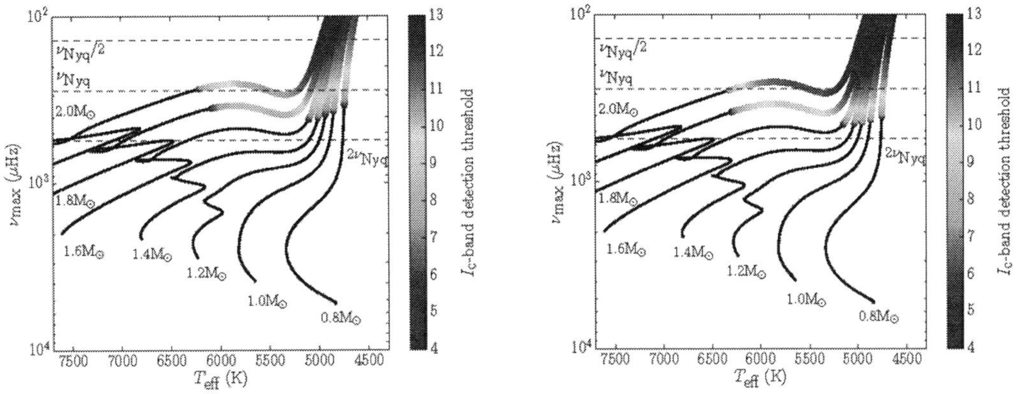

(a) $T = 27$ d, $\sigma_{\rm sys} = 60$ ppm hr$^{1/2}$.

(b) $T = 351$ d, $\sigma_{\rm sys} = 60$ ppm hr$^{1/2}$.

Figure 3: Detectability of solar-like oscillations with *TESS* across an asteroseismic H–R diagram for a cadence of $\Delta t = 30$ min. Note that the frequency of maximum oscillation amplitude, $\nu_{\rm max}$, is now plotted along the vertical axis and not luminosity. Horizontal dashed lines indicate $\nu_{\rm Nyq}/2$, $\nu_{\rm Nyq}$ and $2\nu_{\rm Nyq}$, where $\nu_{\rm Nyq}$ is the Nyquist frequency. I_C-band detection thresholds are color-coded (no detection is possible along those portions of the tracks shown as a thin black line). The two panels consider different lengths of the observations (T) and a systematic noise level of $\sigma_{\rm sys} = 60$ ppm hr$^{1/2}$, as indicated. This figure was originally published in [3].

 DOI: 10.1051/epjconf/201716001006

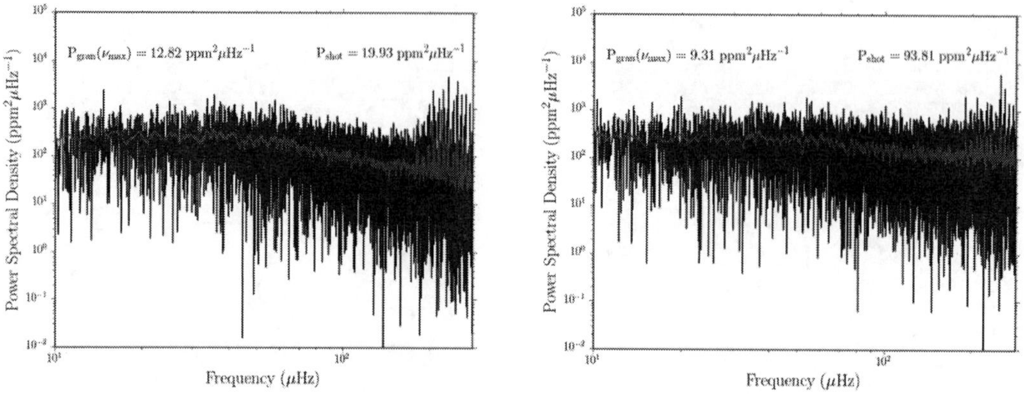

(a) Kepler-56 as observed by *Kepler*.

(b) Kepler-56 as observed by *TESS*.

Figure 4: Power spectra of Kepler-56 as seen both by *Kepler* (left) and *TESS* (right) based on 1 year of observations. A smoothed version of the power spectra is shown in red. Granulation power (at ν_{max}) and shot noise levels are indicated.

EPJ Web of Conferences **160**, 01007 (2017)
Seismology of the Sun and the Distant Stars II

DOI: 10.1051/epjconf/201716001007

Effect of *Kepler* calibration on global seismic and background parameters

Preliminary study on a sample of seismic solar analogs

David Salabert[1,★], *Rafael A.* García[1], *Savita* Mathur[2], and *Jérôme* Ballot[3,4]

[1] *Laboratoire AIM, CEA/DRF-CNRS, Université Paris 7 Diderot, IRFU/SAp, Centre de Saclay, 91191, Gif-sur-Yvette, France*
[2] *Center for Extrasolar Planetary Systems, Space Science Institute, 4750 Walnut street Suite#205, Boulder, CO 80301, USA*
[3] *CNRS, Institut de Recherche en Astrophysique et Planétologie, 14 avenue Edouard Belin, 31400, Toulouse, France*
[4] *Université de Toulouse, UPS-OMP, IRAP, 31400, Toulouse, France*

Abstract. Calibration issues associated to scrambled collateral smear affecting the *Kepler* short-cadence data were discovered in the Data Release 24 and were found to be present in all the previous data releases since launch. In consequence, a new Data Release 25 was reprocessed to correct for these problems. We perform here a preliminary study to evaluate the impact on the extracted global seismic and background parameters between data releases. We analyze the sample of seismic solar analogs observed by *Kepler* in short cadence between Q5 and Q17. We start with this set of stars as it constitutes the best sample to put the Sun into context along its evolution, and any significant differences on the seismic and background parameters need to be investigated before any further studies of this sample can take place. We use the A2Z pipeline to derive both global seismic parameters and background parameters from the Data Release 25 and previous data releases and report on the measured differences.

1 Introduction

The study of the characteristics of solar analogs [1] of different ages is a very promising way to understand the evolution of the Sun. It offers as well the possibility to infer new constraints on the variability of the solar magnetic activity and its associated dynamo processes. However, the identification of solar-analog stars depends on the accuracy of the estimated fundamental stellar parameters. The unprecedented quality of the continuous four-year photometric observations collected by the *Kepler* satellite [2] allowed the measurements of acoustic oscillations in hundreds of solar-like stars [3]. The addition of asteroseismic data combined with high-resolution spectroscopic observations was proven to provide the most accurate fundamental properties that can be derived from stellar modeling today, either from global oscillation properties [3] or from individual frequencies [4–7] compared to empirical scaling-law relations [8]. In consequence, the selection of stars analog to the Sun will benefit from the inclusion of the detection of solar-like oscillations as an additional selection criterium. This is what we called a seismic solar-analog star [9], thus extending the classical definition [1]. Among the *Kepler* sample of oscillating solar-like stars, we identified 18 seismic solar analogs [9].

However, calibration issues[1] present since the launch of *Kepler* were discovered in the Q2–Q16 Data Release 24 (DR24) short-cadence (SC) data, affecting approximately half of all *Kepler* SC targets. We note that the long-cadence (LC) data are not affected. The origin of the problems was traced to an accounting error which scrambled much of the SC collateral smear data used to correct for the effects of *Kepler*'s shutterless readout. As a result, the available SC calibrated pixels, and the light curves based on these pixels, are noisier. The final reprocessing of the *Kepler* Q0–Q17 SC light curves corrected from these calibration problems was thus performed and made available with DR25 in summer 2016.

It is thus important to comprehend the impact of the calibration issues affecting the previous releases of *Kepler* data, and the differences and the possible bias with DR25 on the derived stellar parameters. Here, we present a preliminary study on the differences between DR25 and previous releases on the global seismic and background parameters applied for the sample of 18 solar-analog stars [9]. Although the studied sample is rather small to perform a thorough statistical analysis, we provide initial inputs on the expected effects. Furthermore, this sample constitutes the set of stars the most comparable to the Sun observed by *Kepler* and any possible inferences on the estimated stellar parameters due to issues in the data calibration should be known for each target before any further studies of this sample are performed.

★e-mail: david.salabert@cea.fr

[1] See detailed explanations at http://archive.stsci.edu/kepler/.

DOI: 10.1051/epjconf/201716001007

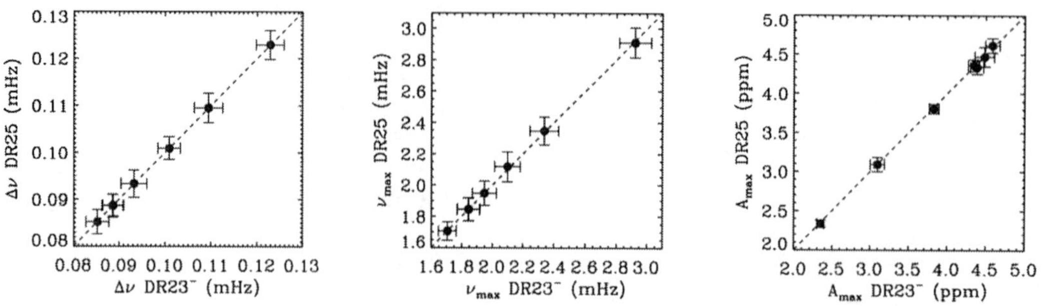

Figure 1: Comparison of the derived global seismic parameters between *Kepler* short-cadence observations from DR23⁻ and DR25. *From left to right*: the mean large separation $\Delta\nu$, the frequency of maximum power $\nu_{\rm max}$, and the maximum amplitude $A_{\rm max}$ at $\nu_{\rm max}$. In the three panels, the dashed lines correspond to the 1:1 relation. The associated uncertainties are also represented.

2 Global seismic and background parameters

Out of the 18 seismic solar analogs observed with *Kepler* [9], nine of them were observed in short cadence from Q5 to Q17: KIC 3656476, KIC 4914923, KIC 5774694, KIC 6116048, KIC 7296438, KIC 7680114, KIC 9098294, KIC 10644253, and KIC 11127479. We checked that all these targets were affected by the smear calibration issues from the list of affected stars available at the MAST archive[2]. We note that as the two stars KIC 5774694 and KIC 11127479 were observed only during one and three quarters respectively, they were disregarded for the following analysis. A total of seven solar analogs were then analyzed. DR25 and previous data releases (DR21 or DR23 depending on the star, denoted in the rest of the text DR23⁻) were corrected for the usual issues found in the *Kepler* light curves following the procedures explained in [10]. The gaps present in the data and affecting the stellar background were filled using inpainting methods [11, 12].

The global seismic parameters of our sample of stars were determined using the A2Z pipeline [13]. The mean large frequency, $\Delta\nu$, which is the distance between two modes of same angular degree and consecutive radial order, was measured by estimating the highest peak in the power spectrum of the power spectrum. The stellar background parameters were extracted by fitting the power spectrum with a function based on the results from [14]. In summary, five different components were fitted to the power spectrum: a white instrumental noise, two Harvey-like functions to mimic two granulation contributions [15], a power-law for the magnetic activity, and a Gaussian function for the p-mode power-excess envelop. The slopes of the Harvey-like functions were fixed to four as it was demonstrated to provide a better fit [14]. The parameters of the Gaussian function provide the center of the p-mode bell, which corresponds to the frequency of maximum power, $\nu_{\rm max}$. The associated maximum amplitude, $A_{\rm max}$, is also returned.

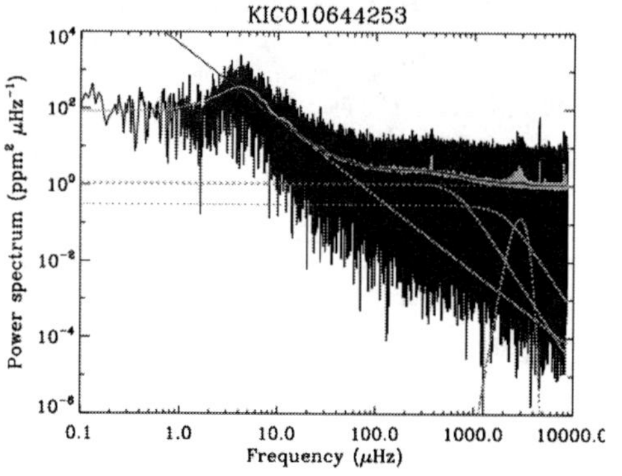

Figure 2: Power spectrum of the solar analog KIC 10644253 observed by *Kepler* DR23 (in black without smoothing, in grey with smoothing). The green dashed lines correspond to the different fitted components of the background model. The red solid line corresponds the fitted background model.

2.1 Global seismic parameters

Figure 1 shows the global seismic parameters ($\Delta\nu$, $\nu_{\rm max}$, and $A_{\rm max}$) extracted from both DR23⁻ and DR25 for the seven analyzed solar analogs. The corresponding uncertainties are also represented. No systematic bias can be seen between DR23⁻ and DR25, and the differences in the global seismic parameters are minimal (less than 1%) and well within the uncertainties.

2.2 Background parameters

An example of the background model used here is illustrated on Fig. 2 in the case of the 1-Gyr-old solar analog KIC 10644253 from DR23. This star is a particularly interesting target as it is the youngest G-type star within the

[2]See list of affected targets at https://archive.stsci.edu/missions/kepler/catalogs/kepler_scrambled_short_cadence_collateral_target_list.csv.

 DOI: 10.1051/epjconf/201716001007

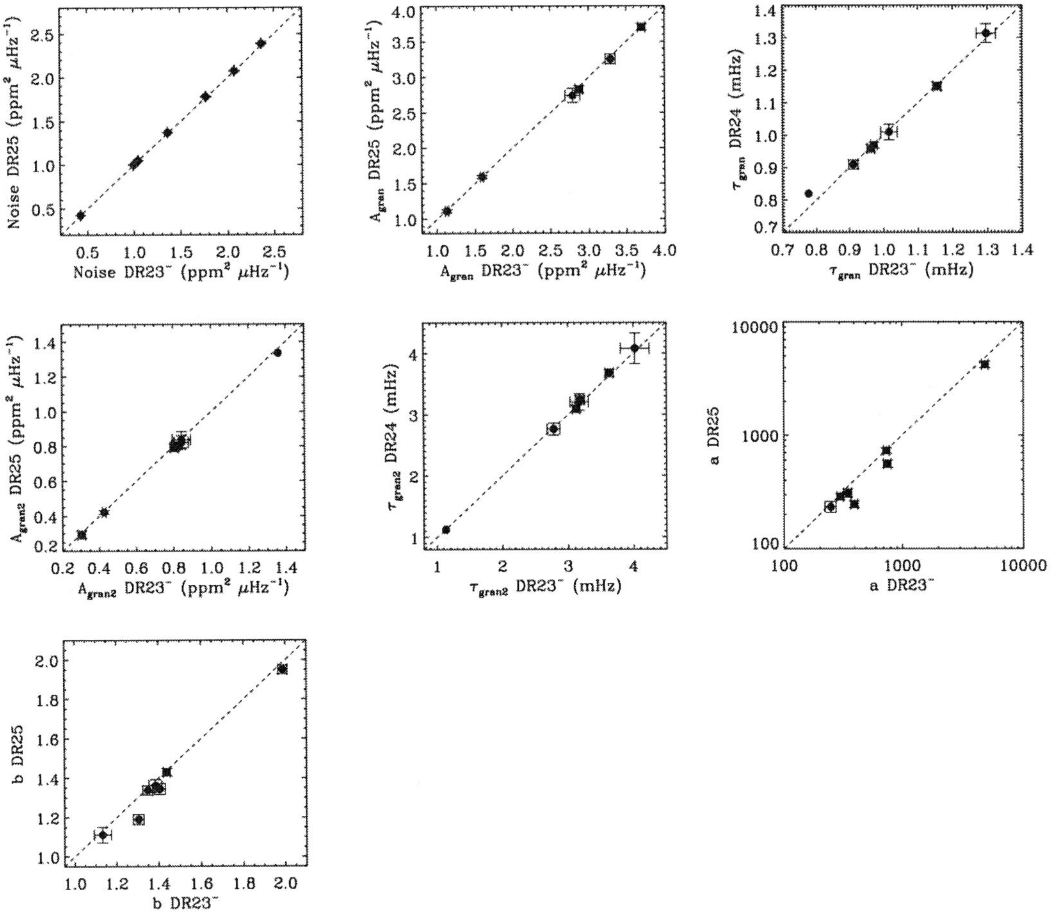

Figure 3: Comparison of the derived stellar background parameters between *Kepler* short-cadence observations from DR23⁻ and DR25. In the three panels, the dashed lines correspond to the 1:1 relation. The associated uncertainties are also represented.

solar-like oscillators observed by *Kepler* [5, 9]. Furthermore, it exhibits temporal variations of the oscillation frequencies associated to its magnetic variability [16]. Such temporal variability of the acoustic oscillation parameters are well studied in the case of the Sun over to date two 11-year solar cycles [see e.g., 17, and references therein], but has been observed so far in only three stars: the F-type stars HD 49933 [18, 19] and KIC 3733735 [20], and the solar-analog G-type star KIC 10644253 [13].

The different components of the background model (i.e., the two Harvey's laws, the p-mode gaussian envelop, the power law, and the white instrumental noise) are represented individually as well on Fig. 2. The background fitting was performed for the sample of solar-analog stars and each of the background components were compared between DR23⁻ and DR25 as shown on Fig. 3. The corresponding uncertainties are also represented. The granulation components, the amplitudes, $A_{\rm gran}$ and $A_{\rm gran2}$, and the characteristic frequencies, $\tau_{\rm gran}$ and $\tau_{\rm gran2}$, and the instrumental noise do not show significant differences between DR23⁻ and DR25 (respectively 3% and 1%). However, the two parameters, a and b, of the power law $a\nu^{-b}$ describing the stellar activity and low-frequency trends show significant differences for some of the stars between DR23⁻ and DR25 up to 60% and 10% respectively. Furthermore, all the parameters a and b are smaller when derived with DR25 compared to DR23⁻. It appears then from this small sample that the principal and significant impact of the smear calibration issues concerns the parameters of the power law in the background model.

3 Conclusions

Calibration problems associated to collateral smear were detected in the *Kepler* DR24 and reprocessed corrected light curves were provided in DR25. These issues were actually found to affect all data releases since launch. We performed here a preliminary study to evaluate the impact on the global seismic parameters and on the background parameters between the two data releases. We analyzed the sample of seven seismic solar analogs observed by *Kepler* in short cadence continuously between Q5 and Q17. We started with this set of stars as it constitutes the best sample to put the Sun into context along its evolution, and any significant differences on the seismic and background parameters need to be investigated before any further studies of this sample take place.

We used the A2Z pipeline to derive both global seismic parameters and background parameters with DR23$^-$ and DR25. We found that the global seismic parameters, i.e. the mean large frequency separation $\Delta\nu$, the frequency of maximum power $\nu_{\rm max}$, and the maximum amplitude $A_{\rm max}$, do not show significant differences (less than 1%) within the uncertainties. The components of the background model used to describe the granulation (the amplitudes $A_{\rm gran}$ and $A_{\rm gran2}$, and the characteristic frequencies $\tau_{\rm gran}$ and $\tau_{\rm gran2}$) with two Harvey's empirical laws are consistent between DR23$^-$ and DR25 within less than 3% as well as the white instrumental noise (1%). However, significant differences are observed for some of the stars in the power-law parameters, a and b, which are used to describe the response of stellar activity and low-frequency trends in the power spectrum. These parameters are smaller in DR25 by up to 60% and 10% for a and b respectively.

Although the number of analyzed stars is small, it appears that a particular attention is, at least, needed when studying the parameters extracted from the low-frequency power-law fitting. The analysis of a larger sample of stars is still necessary to get a better idea of the differences between DR25 and previous data releases, but this study presented here provides some initial elements on what can be expected. Furthermore, some additional individual oscillation modes are possibly observed at low frequency, which would provide additional constraints in modeling of stellar parameters [4–7]. Peak-fitting analysis of the individual modes with DR25 is thus necessary to confirm this possibility.

Acknowledgments

The authors wish to thank the entire *Kepler* team, without whom these results would not be possible. Funding for this Discovery mission is provided by NASA's Science Mission Directorate. DS and RAG acknowledge the financial support from the CNES GOLF and PLATO grants. SM acknowledges support from the NASA grant NNX12AE17G and NNX15AF13G.

References

[1] Cayrel de Strobel, G., A&ARv, **7**, 243 (1996)
[2] Borucki, W. J., Koch, D., Basri, G., et al., Science, **327**, 977 (2010)
[3] Chaplin, W. J., Basu, S., Huber, D., et al., ApJS, **210**, 1 (2014)
[4] Mathur, S., Metcalfe, T. S., Woitaszek, M., et al., ApJ, **749**, 152 (2012)
[5] Metcalfe, T. S., Creevey, O. L., Doğan, G., et al., ApJS, **214**, 27 (2014)
[6] Silva Aguirre, V., Davies, G. R., Basu, S., et al., MNRAS, **452**, 2127 (2015)
[7] Creevey, O. L., Metcalfe, T. S., Schultheis, M., et al., A&A, Under revision (2016)
[8] Kjeldsen, H., & Bedding, T. R., A&A, **293**, 87 (1995)
[9] Salabert, D., García, R. A., Beck, P. G., et al., arXiv:1608.01489 (2016)
[10] García, R. A., Hekker, S., Stello, D., et al., MNRAS, **414**, L6 (2011)
[11] García, R. A., Mathur, S., Pires, S., et al., A&A, **568**, A10 (2014)
[12] Pires, S., Mathur, S., García, R. A., et al., A&A, **574**, A18 (2015)
[13] Mathur, S., García, R. A., Régulo, C., et al., A&A, **511**, A46 (2010)
[14] Kallinger, T., De Ridder, J., Hekker, S., et al., A&A, **570**, A41 (2014)
[15] Harvey, J. W., Duvall, T. L., Jr., Jefferies, S. M., & Pomerantz, M. A., GONG 1992 *Seismic Investigation of the Sun and Stars*, **42**, 111 (1993)
[16] Salabert, D., Régulo, C., García, R. A., et al., A&A, **589**, A118 (2016)
[17] Salabert, D., García, R. A., & Turck-Chièze, S., A&A, **578**, A137 (2015)
[18] García, R. A., Mathur, S., Salabert, D., et al., Science, **329**, 1032 (2010)
[19] Salabert, D., Régulo, C., Ballot, J., García, R. A., & Mathur, S., A&A, **530**, A127 (2011)
[20] Régulo, C., García, R. A., & Ballot, J., A&A, **589**, A103 (2016)

EPJ Web of Conferences **160**, 01008 (2017)
Seismology of the Sun and the Distant Stars II

DOI: 10.1051/epjconf/201716001008

Kepler-K2: A Search for Very Red Variables

Erich Hartig[1,⋆], *Kenneth* Hinkle[2,⋆⋆], and *Thomas* Lebzelter[1,⋆⋆⋆]

[1] *University of Vienna, Department of Astrophysics*
[2] *National Optical Astronomy Observatories, P.O. Box 26732, Tucson, AZ 85726 USA*

Abstract. As a monitoring mission the NASA *Kepler* K2 mission offers the opportunity to search for long period variables, in this case along the ecliptic plane. The usability of K2 for this search is limited since the observing period of an individual target is restricted to about 80 days. However, this time scale allows detecting LPVs and making at least a rough estimate of the period length. Since no ready-to-use light curves were provided during the early phases of the K2 mission, we had to use a reduction algorithm developed by our group based on the experiences from the primary *Kepler* mission. Our results can now be nicely compared with the K2 'High Level Science Products' released in June 2016. The first results of our study up to campaign 3 are presented.

1 Introduction

Long period variables (LPVs) are highly evolved stars of low and intermediate mass. They are characterized by large amplitude variations in the visual, periods of a few 10 to several 100 days and very red colour due to both the low surface temperature and the circumstellar reddening by dust. As a consequence we chose stars with 2MASS colours J-Ks greater than 0.8 and Ks < 8.0 mag as targets. An additional selection criterion was the availability of basic information from Simbad, GSC 2.3, 2MASS, and AAVSO data bases to ensure we select stellar objects only. Figure 1 shows the interesting objects in the colour-magnitude diagram of campaign 1. In particular, we were interested in detecting highly reddened long period variables (LPVs), which is a main group of stars found with these selection criteria. To focus on these objects we set a lower period limit of 10 days. Most of the stars with very large amplitudes, usually dubbed 'Miras', have been detected as variables before, so the main interest goes into stars with small to moderate light amplitudes.

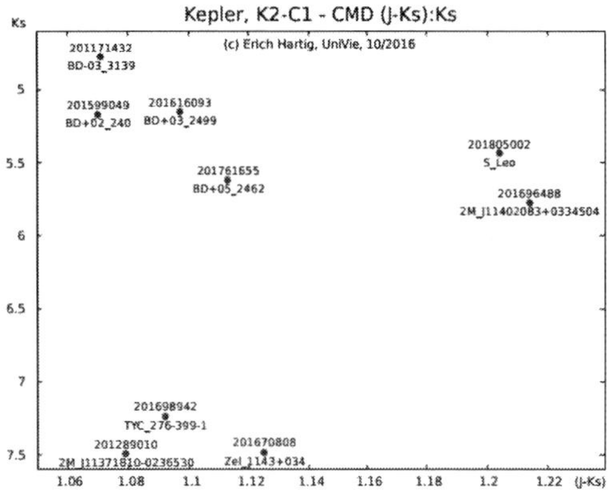

Figure 1: K2 - C1, Colour Magnitude Diagram (CMD) (J-Ks):J.

2 Data Preparation

Even more than for the *Kepler* primary mission (see Hartig et al. [4]) the data reduction had to be done with care to extract the true stellar signal. Dealing with LPVs we had to expect a variability on a time scale comparable to the length of the data set. Therefore, different to studies of other types of variables, it was a particular challenge to distinguish instrumental trends from long time variability.

For the data preparation we adjusted our pipeline software for TPD, which we had used for data from the *Kepler* primary mission, and modified it to the K2 requirements. While the 6 hours signal caused by the firing of the thrusters for correcting the telescope's pointing could be easily corrected, a non-linear flux increase over each observation period had to be removed up to campaign C2. Furthermore, it turned out that stars with magnitudes <8.5 Kepmag typically show elongated point spread functions. To circumvent this problem we applied a non-standard pixel aperture for these objects, which allowed to improve the light curve's signal-to-noise ratio significantly. Some examples are shown in Figures 4 to 10.

⋆e-mail: erich.hartig@univie.ac.at
⋆⋆e-mail: khinkle@noao.edu
⋆⋆⋆e-mail: thomas.lebzelter@univie.ac.at

DOI: 10.1051/epjconf/201716001008

3 Background Trend

We processed 18 targets of the target pixel files (TPF) from the 8 days two-wheel engineering test (K2-CEng) and observed an increase of the flux of a number of targets over time, which follows approximately the background increase, see Figure 2. We calculated the background from pixels around the mask and tested two strategies to remove this trend from the target pixel data (TPD):

- by using the median of the background
- by using the average of 6 stars with small amplitude variations showing the long-time trend (see Figure 2, EPIC 60020773).

Once released we compared our results with the 'K2 High Level Science Product' (HLSP) K2SFF.

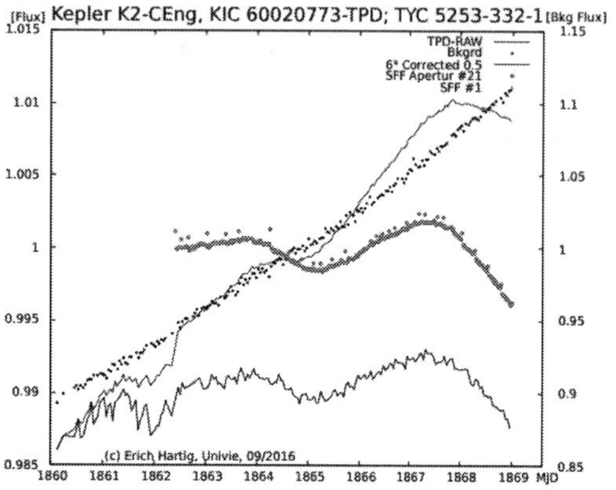

Figure 2: K2 - Engineering Test Data, EPIC 60020773, relative flux of raw TPD (red) compare with adjusted TPD (blue), K2SFF aperture 1 (green), aperture 21 (purple) and background (black).

Table 1: Number of Candidates

Campaign	Requested	Observed
C0	62	9
C1	68	34
C2	48	41
C3	35	12

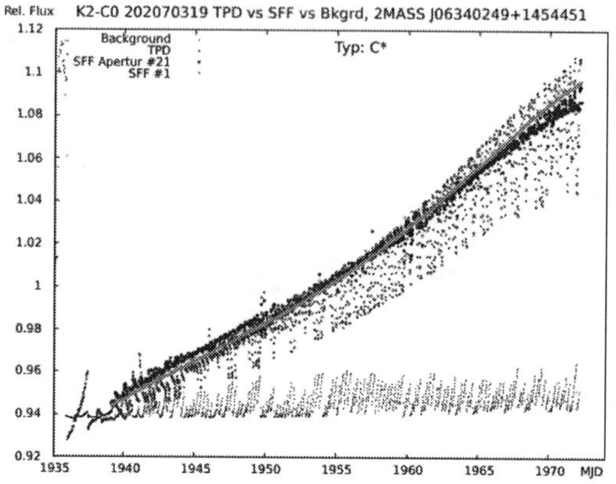

Figure 3: K2-C0, C-star EPIC 202070319, relative flux of background (black) compared with TPD (red), K2SFF aperture 1 and 21 (blue/green).

4 K2 Campaigns

4.1 K2 Campaign 0 (K2 - C0)

The K2-C0 campaign was a full length engineering test for the K2 concept but the spacecraft was not in fine-point mode for the first part of C0 and we treated these parts with care. Only 9 of our 62 requested targets were observed.

We find a number of stars, which, similar to the engineering test, show an increase of brightness by about 0.5 Kepmag, approximately following the increase of the background. In Figure 3 the TPD light curve retrieved for C-star EPIC 202070319, the corrected TPD light curve using the average of six reference stars, and the K2SFF results of aperture 1 and 21 are compared with the background.

Unfortunately, we found no unique relation of the adjustment factor for these long-time variations for all candidates, K2SFF again provides an acceptable solution. The exception is EPIC 202062457, SU Gem. K2SFF skips the first part of the data because *Kepler* achieved fine pointing control only for the last approximately 35 days; with the inclusion of the first part of the TPD data with our software we could improve the results. The TPD calculation gave a period of 25.158d, which fits better to the AAVSO result of 50.0d (half the period) than the K2FF period of 27.213d. We used the software package Period04 by Lenz P. and Breger M., [5] for their calculations.

In Figure 4 we show five light curves of our candidates, which indicates long periods.

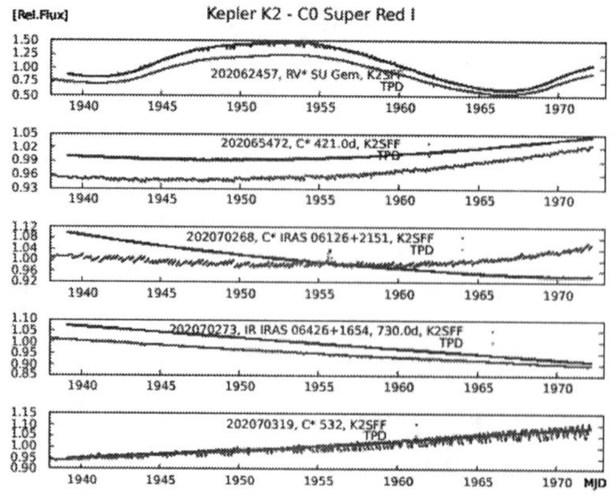

Figure 4: K2 - C0 LPVs, TPD (red), K2SFF aperture 21 (blue).

 DOI: 10.1051/epjconf/201716001008

4.2 K2 Campaign 1 (K2 - C1)

We applied the C0 correction also to the K2-C1 data to remove the long-time trend by using the average of six similar stars improving the determination of the amplitude and the long period. Only 3 of our requested objects are known LPVs, one Mira and two C-stars, showing periods longer than 10d. We additionally scanned the observed objects by (J-Ks) colour and Ks < 8.0 and found 12 more LPV candidates, the top five are shown in Figure 5. The last one, EPIC 201368935, TT Leo, does not show a significant variation in the SFF data.

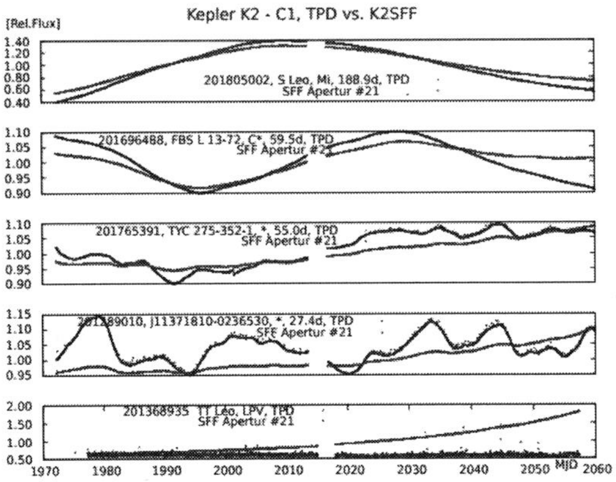

Figure 5: K2 - C1, Top 5 LPVs, TPD (red), K2SFF aperture 21 (blue).

4.3 K2 Campaign 2 (K2 - C2)

In K2-C2 the long-time disturbance still exists and has been corrected as described before. Some light curves derived with our own TPD analyses compared nicely with K2SFF, see Figure 6 and 7, some do not. From our 48 requested targets 41 have been observed. We found 21 possible candidates showing long periods, 5 of them are Miras, the rest are classified as infrared sources. Only V1158 Sco (EPIC 205240599) is a presently known LPV.

4.4 K2 Campaign 3 (K2 - C3) - K2SFF A new hope?

Finally, K2-C3 provided simple aperture photometry (SAP_FLUX) and a more processed version of SAP, with artifact mitigation included, called pre-search data conditioning flux (PDCSAP_FLUX) light curves (Smith, J. C. et al., [6], Stumpe, M. C. et al., [7]) for long cadence data. Three corrected light curves are generated: (a) a Bayesian MAP fit, (b) a Robust Least Squares fit and (c) no correction. PDC then selects the fit that results in the best photometric precision[1].

Background pixels were observed and used to model the background level across the field of view. The calibrated pixels available in the target pixel files now have this background level removed.

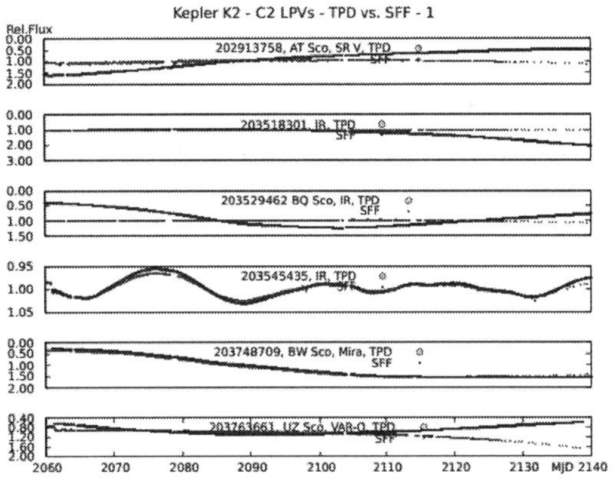

Figure 6: K2 - C2 sample 1 of LPVs, TPD (red), K2SFF aperture 21 (blue).

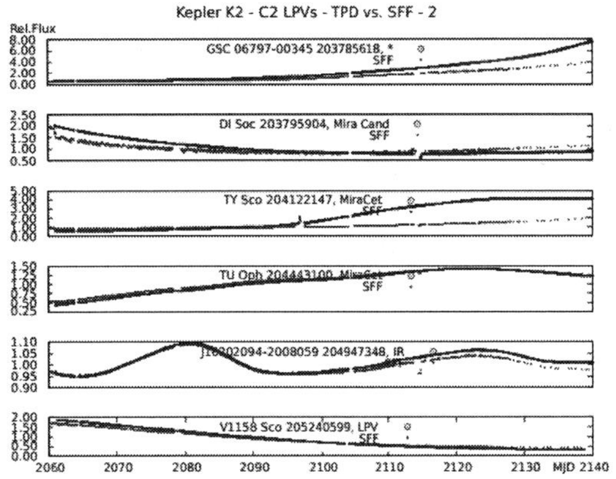

Figure 7: K2 - C2 sample 2 of LPVs, TPD (red), K2SFF aperture 21 (blue).

The new "High Level Science Products"[2] had been additionally tested and we found that K2SFF reproduces LPVs usually best, K2-*everest* did not provide data for this target yet. K2SFF is a technique for extracting and correcting K2 photometry similar to those developed for *Spitzer* but optimized to the peculiarities of the K2 mission, see Vanderburg A., and Johnson J. A. [8] and Vanderburg A [9]. In few cases we observe an amplitude jump at MJD 2189.052, see Figure 8, EPIC 206427151. The long-time trend of increasing brightness is no longer visible here and K2SFF preserved the long periods too. From our 35 requested objects 12 have been observed. In Figure 9 and 10 we show the 12 light curves and compare our TPD results with K2FF.

5 Conclusions

We noticed that stars with magnitudes < 8.5 Kepmag typically show elongated pixel configurations.

[1] https://keplerscience.arc.nasa.gov/K2/C3drn.shtml
[2] http://archive.stsci.edu/archive_news/2016/06-Jun/index.html

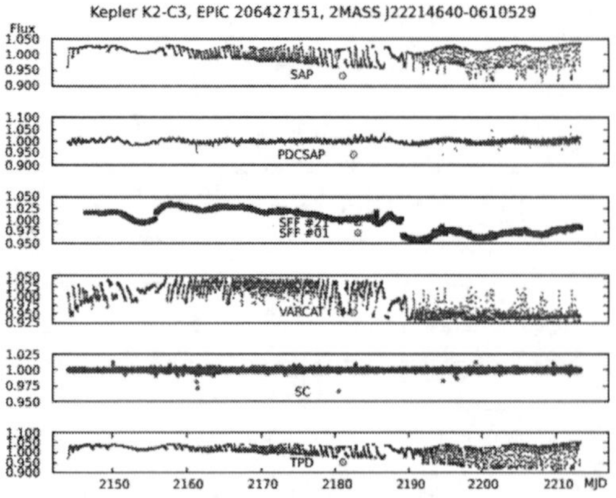

Figure 8: *Kepler* EPIC 206427151 Compare LCs with High Level Science Products like K2SFF, K2VARCAT, and K2SC.

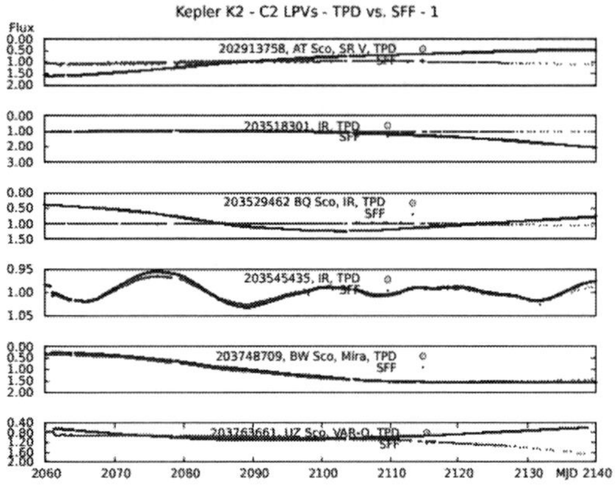

Figure 9: *Kepler* EPIC K2 - C3 sample 1 of LPV candidates.

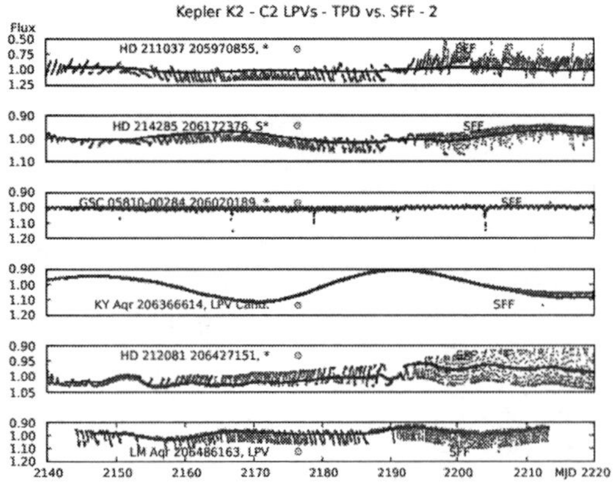

Figure 10: *Kepler* EPIC K2 - C3 sample 2 of LPV candidates.

Our software allows the specific selection of pixel masks, which gives good results but fails in the sample above. We will therefore continue all further analysis additionally to our own software package with K2SFF.

Acknowledgements

EH thanks the University of Vienna for support (FA538905). TL was funded by the FWF project P23737.

References

[1] Aigrain, S. et al., 2016, MNRAS, **459, 2408**

[2] Armstrong, D. J., Osborn, H. P., Brown, D. J. A., et al., 2014, e-prints arXiv:1411.6830

[3] Armstrong, D. J., Kirk, J., Lam, K. W., et al., 2016, MNRAS,**456, 2260**

[4] Hartig, E., Cash, J., Hinkle, K., Lebzelter, T., Mighell, K., and Walter, D., 2014, AJ, **148, 123**

[5] Lenz, P., Breger, M. 2005, Communications in Astrosysmologie,**146, 53**

[6] Smith, J. C., Stumpe, M. C., Van Cleve, J. E., et al., 2012, arXiv:1203.1382

[7] Stumpe, M. C., Smith, J. C., Van Cleve, J. E., et al., 2012, arXiv:1203.1383

[8] Vanderburg, A., Johnson, J. A. 2014, PASP,**126, 948**

[9] Vanderburg, A. 2014, arXiv:1412.1827v1

[10] Van Clevel, J. E., Howell, S. B., Smith J. C., 2015, arXiv: 1512.06162v3

Backup to the Future

Creating the *Kepler*/K2 long-term "living archive"

Rasmus Handberg[1,⋆], *Anders S.* Conrad[2], and *Michael* Svendsen[2]

[1] *Stellar Astrophysics Centre, Department of Physics and Astronomy, Aarhus University, Ny Munkegade 120, DK-8000 Aarhus C, Denmark.*
[2] *The Royal Library, Copenhagen, Denmark.*

Abstract. What if something horrible happens in the future? Great Scott! Do we have a backup? Is a backup even enough? At the Kepler Asteroseismic Science Operations Center (KASOC) we have a goal that all data and information from *Kepler* and KASC is preserved for the future. The benchmark is that the data should be useful for, at least, the next 50 years. But how do we ensure that hundreds of terabytes of data are understandable or even readable in half a century?

1 Introduction

Since the beginning of the Kepler Asteroseismic Investigation, the Kepler Asteroseismic Science Operations Center (KASOC) has been tasked with storing and archiving all Kepler data as well as distributing the data to the members of the Kepler Asteroseismic Science Consortium (KASC) [1]. This has been done through the KASOC website[1], which provides the members of KASC easy access to both the original Kepler data, but also to several unique data products created by the members of KASC. Records of all scientific publications by KASC are also stored as well as high-level data products like derived stellar properties and stellar models.

In terms of the future of this archive, it is important to realise that even though there are several ongoing and future missions and projects (e.g. SONG, BRITE, TESS, PLATO) devoted to time-domain astronomy, none of them is likely to achieve the same long time-coverage of many stars over many years. Therefore, *Kepler* datasets are in may ways unique and will be useful for active research for many decades to come.

As part of the SpaceINN work-package 3 ("Data Handling and Archiving"), KASOC has been tasked with constructing the long-term archive for the *Kepler*/K2 data. In this context we have started a close collaboration with The Royal Library in Copenhagen, Denmark, where we have created formal Data Management Plans, strategies and requirements for how such an archive should operate.

2 The "living" archive

Currently the KASOC is maintaining an archive of *Kepler*/K2 data as well as derived data, produced by KASC, and a database of derived stellar properties. The goal is that we should create a permanent archive of *all* these data.

Traditionally, a data archive can be thought of as the digital equivalent of putting all data, notes and publications in neatly labelled boxes and locking them in a storage room for safe keeping. This is not what we want! Instead we think that *the future is a living archive*. This means that we have the following list of requirements to the archive:

- The archive should, at a minimum, be available the next 50 years.
- Data are always freely available on-line.
- Data will continue to be used in active research.
- Extendable: New information should be added to the archive as our knowledge grows.
- Data should be stored in formats that are easily readable by both humans and computers.
- Understandable and useful for future researchers – No matter the science case of the researcher.

3 Discoverability

What will future researchers be interested in and how will they use *Kepler* data? Naturally, we have no idea about the answer to this question. But when we are thinking on timescales on more than 50 years, we have to take into account that all people having a direct hands-on experience with the data and worked on it when it was taken are no longer around. So we have to make sure that data is packaged and searchable in a way that makes it obvious to a future researcher what they are looking at.

But how will a future researcher, possibly with a very different research topic in mind, best *discover* the *Kepler*/K2 data? To try to get closer to an answer to this, we conducted some thought-experiments of how future

⋆e-mail: rasmush@phys.au.dk
[1] http://kasoc.phys.au.dk

researchers in astronomy would search for information. Which parameters are the most important in order to discover the data and their usefulness in future research? Besides general descriptions of *Kepler*/K2 and the different data products (measurements of stellar flux as a function of time), we found that the key parameters to search for is *the astronomical object*, in this case meaning the star, and/or its position on the sky. This is such a fundamental set of properties in astronomy, that this should be our main parameters for data discovery. This has an impact on the way we have chosen to bundle the information (see §4).

4 Data formats

The requirement of having data available on long timescales also requires that we ensure that the structure and formats of the data are readable, useful and extendable.

We have opted for using an XML (Extensible Markup Language) format for storing all results and meta-data for a star. One of the advantages is that format is easily readable both to humans and computers, and very wide-spread, meaning that parsers are available in most programming languages in use today. A very basic example of such a file is shown below.

```
<star kic="12345678">
  <numax value="3100" error="20" unit="uHz" />
  <mass value="1.0" error="0.01" unit="solar" />
  <radius value="1.0" error="0.01" unit="solar" />
</star>
```

Example 1: Example of XML format used to store results and meta-data.

The XML format will also be used on the existing KASOC websites for exchange of results between users and KASOC.

Files are finally put into a "bag" defined by the `BagIt`-format, a standard format defined by a collaboration between prominent libraries and universities, including California Digital Library and The Library of Congress (USA) [2]. A bag will be self-contained, holding the XML data, plus all data products for the star (incl. original *Kepler*/K2 data, processed data and power spectra), stellar evolution and structure models, auxiliary files, descriptions and documentation. The bag also contains a manifest of all files, including cryptographic file-hashes making it possible to verify the integrity of the bag.

5 The Future

First of all, it is important to clarify that KASOC will continue running, providing access to *Kepler*/K2 data. Regarding the actual implementation and hosting of a long-term archive we are currently in dialogue with national facilities in Denmark that can ensure its operation on timescales of many decades. Currently the most likely configuration looks like being a collaboration between Aarhus University, The Royal Library and Copenhagen University, utilizing different expertises in long-term file storage, discovery services and astronomical know-how. Another important issue that is currently under discussion, is who should be responsible for the long-term funding of such archives. This question is not yet fully resolved, and may require political decisions on faculty, university or even national level in Denmark.

On the long term (decades) the plan is that the *Kepler*/K2 data and the KASOC archives are copied into this new configuration, allowing researchers of the future who wants to continue to use the Kepler/K2 data in active research to discover new things we haven't even begun to think of...

Acknowledgements

Funding for the Stellar Astrophysics Centre is provided by The Danish National Research Foundation (Grant DNRF106). The research was supported by the ASTERISK project (ASTERoseismic Investigations with SONG and Kepler) funded by the European Research Council (Grant agreement no.: 267864).

References

[1] H. Kjeldsen, J. Christensen-Dalsgaard, R. Handberg, T.M. Brown, R.L. Gilliland, W.J. Borucki, D. Koch, Astronomische Nachrichten **331**, 966 (2010), `1007.1816`

[2] J. Kunze, J. Littman, L. Madden, E. Summers, A. Boyko, B. Vargas, Internet-Draft draft-kunze-bagit-13, IETF Secretariat (2016)

EPJ Web of Conferences **160**, 01010 (2017)
Seismology of the Sun and the Distant Stars II
DOI: 10.1051/epjconf/201716001010

Global helioseismology (WP4.1): From the Sun to the stars & solar analogs

Rafael A. García[1,⋆]

[1] *Laboratoire AIM Paris-Saclay, CEA/DRF-CNRS-Univ. Paris Diderot; IRFU/SAp, Centre de Saclay, 91191 Gif-sur-Yvette, France*

Abstract. Sun-as-a star observations put our star as a reference for stellar observations. Here, I review the activities in which the SPACEINN global seismology team (Working Package WP4.1) has worked during the past 3 years. In particular, we will explain the new deliverables available on the SPACEINN seismic+ portal. Moreover, special attention will be given to surface dynamics (rotation and magnetic fields). After characterizing the rotation and the magnetic properties of around 300 solar-like stars and defining proper metrics for that, we use their seismic properties to characterize 18 solar analogues for which we study their surface magnetic and seismic properties. This allows us to put the Sun into context compared to its siblings.

1 Introduction

The objective of the global seismology team inside the SPACEINN collaboration, namely Working Package 4.1, is to discuss the current problems in global seismology of the Sun and solar-analogue pulsating stars. Naturally, attending to the type of observations, i.e. Sun-as-a-star or imaged ones, the work and the discussions related were divided in low and high-degree modes. In this proceedings I will concentrate on the activities of the group concerning the low-degree modes obtained in Sun-as-a star observations and in the studies of solar analogues.

During the past three years, the discussion of high-degree global modes were focused on the use of different fitting codes and the differences obtained when several datasets of different instruments were used. Some of the works done in this direction are summarized in [e.g. 1, 2]. Briefly, it has been found that some of the p-mode parameters (for example the mode asymmetries) are inconsistent when different fitting methods are used, while they are consistent when a single methodology is used in different datasets, e.g., with GONG [3] or with MDI and HMI [4, 5]. Moreover, inversions of the internal rotation profile showed that the solar magnetic cycle 24 is different from cycle 23. The work with different datasets and methods evidenced that the twist appearing at high latitudes with time in several helioseismic variables is very likely to be a numerical artifact [for more details see 2].

Concerning the low-degree modes, the two main axes of the work in the WP4.1 were concentrated on the development of new statistical tools to study these modes and the study of the temporal evolution of their properties during the evolution of the solar magnetic activity cycle. The solar results were placed into a stellar context by comparing them to observations of other solar analogue stars observed by the NASA *Kepler* mission [6].

⋆e-mail: rafael.garcia@cea.fr

2 Real and simulated solar time series

A huge effort has been led by the WP4.1 in order to provide properly calibrated datasets obtained from the Sun-as-a-star instruments on board the Solar and Heliospheric Observatory (SoHO [7]), as well as from the Mark-I instrument, a solar spectrophotometer located and operated at Observatorio del Teide (Tenerife, Canary Islands, Spain). It provides precise radial velocity observations of the Sun-as-a-star[1] at the Potassium KI 7699 Å absorption solar line. Observations extend from 1976 to 2012 although only the summer campaigns of 1976 to 1983 are available.

Photometric light curves of the Sun Photometers (SPM) of the Variability of solar IRradiance and Gravity Oscillations instrument (VIRGO [8]) are now available at the SPACEINN portal[2]. These are 60 s cadenced time series starting on April 11, 1996 of the three channels of 5 nm bandwidth centered at 402 nm (blue), 500 nm (green) and 862 nm (red). These light curves are corrected from outliers and then filtered with a high-pass filter running mean to remove unwanted low-frequency trends. An additional correction is also applied whenever possible in order to correct for the so-called SPM "attractor" as explained in [9].

Doppler velocity time series observed by the Global Oscillations at Low Frequency (GOLF [10]) instrument are also available in the SPACEINN portal[3]. Observations obtained from the blue or the red wing of the Sodium doublet [11], each one with a different sensitivity to the solar disk [12], have been calibrated following the methods explained in [13] and the two independent channels have been averaged into one single time series. The starting

[1] http://www.spaceinn.eu/data-access/mark-i-data-archive/
[2] http://www.spaceinn.eu/data-access/calibrated-sohovirgospm-data/
[3] http://www.spaceinn.eu/data-access/calibrated-soho-golf-frequencies/

date is at midnight of April 11, 1996 and the sampling rate is 60 s in order to follow the same cadence than the other two helioseismic instruments on board SoHO.

Numerical simulations have also been created and are available on demand. All the information can be found in the SPACEINN portal[4]. Three different type of simulations were created:

1. Spherical-shell simulations of magnetoconvection.
2. Numerical Simulations of linear waves/modes in complex media.
3. Box simulations of compressible magnetoconvection.

3 Fitting low-degree solar modes

In helioseismology, it is common practice to characterize the acoustic modes by fitting single modes using a frequentist approach at low frequencies and pairs of even or odd modes at higher frequencies [e.g. 14–17]. But with the development of asteroseismology and due to the natural complications to properly identify the modes, global fittings were established as standard methods [e.g. 18–23]. Moreover, bayesian fitting techniques were soon adopted in asteroseismology [e.g. 24–29] and also tested on the Sun [30]. Although the difference between the bayesian and the frequentist approach in the case of the Sun could be neglected [e.g. 31], we decided to adapt the high-DImensional And multi-MOdal NesteD Sampling asteroseismic Bayesian global-fitting tool [DIAMONDS, 28] to the solar case[5]. An example of the application of the code to the GOLF velocity time series can be seen in figure 1. Because of the particularities of the GOLF measurements, the p-mode power excess has been fitted using a double Gaussian following [32]. A full analysis of the GOLF and VIRGO/SPM datasets using this tool will be soon available (Corsaro et al. in preparation).

A lot of discussions and efforts were employed to study low-frequency p modes and g modes in the Sun. Although the signature of the g-mode period spacing was found using GOLF observations [33, 34], it has been impossible to unambiguously detect individual g modes with a high confidence level [35–37]. Special effort has been spent to find better ways to combine contemporary datasets as it was already outlined in [30]. Proper covariant matrices have been developed (Broomhall et al. in preparation) but the results have not allowed us to find solar g modes so far.

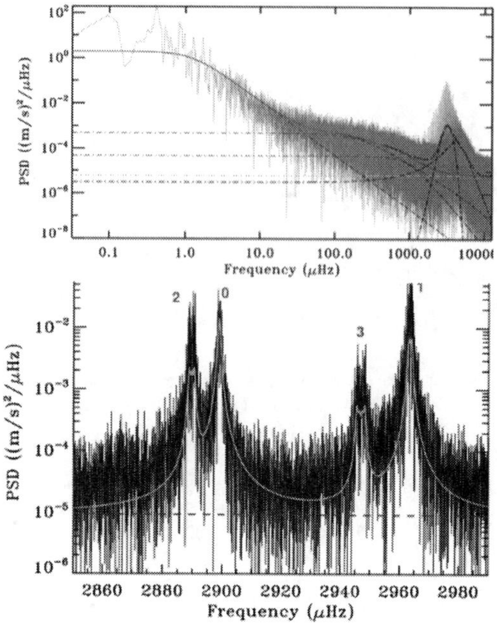

Figure 1: Application of the DIAMONDS asteroseismic Bayesian global-fitting tool [28] to the solar case. On the top panel, result of the fitting of the background and the power excess of GOLF velocity time series using a model of three Harvey-like profiles, 2 Gaussian functions and a flat photon-noise contribution. On the bottom panel, zoom of the region 2850 to 2990 μHz in black and the result of the fit of modes ℓ=0 to 3 in green. The horizontal dashed line represents the background level.

4 Photospheric magnetic activity proxy

The quasi continuous space observations required to perform helio- and astero-seismic studies can be used to track the evolution of the photospheric magnetic field [e.g. 38–42] as well as in the sub-photospheric layers through seismology [43–47]. In the SPACEINN framework, we have extensively studied the photospheric magnetic activity proxies obtained from photometric observations, $S_{\rm ph}$ [e.g. 48, 49] and Doppler velocity time series $S_{\rm vel}$. Briefly they track the standard deviation of subseries of a length proportional to the rotation period in order to ensure that the periodicities are related to stellar spots and hence, to magnetic activity. An example of the solar $S_{\rm vel}$ measured using the GOLF data is given in figure 2.

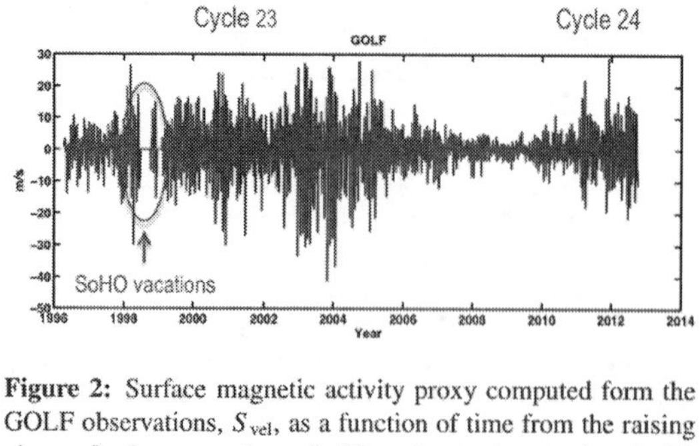

Figure 2: Surface magnetic activity proxy computed form the GOLF observations, $S_{\rm vel}$, as a function of time from the raising phase of solar magnetic cycle 23 to the maximum of cycle 24. The so-called SoHO vacation periods of 1998 and early 1999 are indicated.

[4]http://www.spaceinn.eu/data-access/wp4/simul/
[5]http://www.spaceinn.eu/data-access/wp4/tools-to-extract-low-frequency-mode-frequencies-d4-4/

In particular, we have used the Sun as a reference (Salabert et al. in preparation) and we have compared S_{ph} and S_{vel} to many other solar magnetic proxies [50]. An example of these comparisons is shown in figure 3.

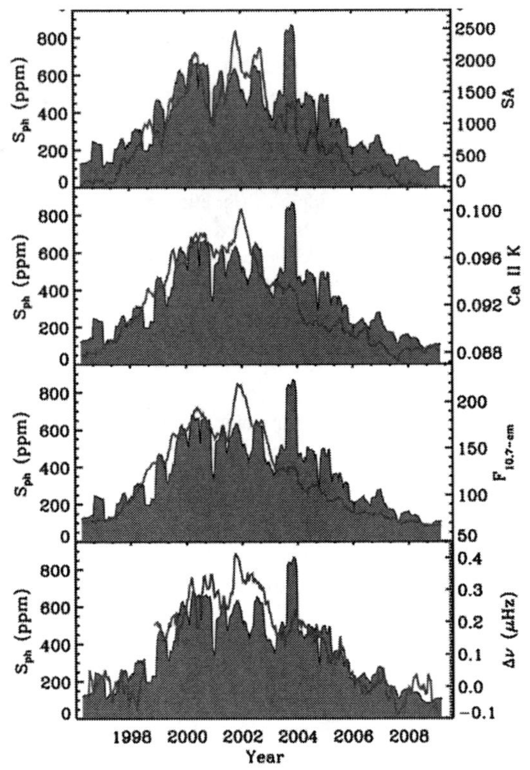

Figure 3: Evolution with time of the photospheric magnetic activity proxy S_{ph} measured from VIRGO/SPM (blue filled curve) compared to some standard solar magnetic activity proxies (red solid lines). From top to bottom: the right-hand y axis corresponds to the sunspot area (SA); the Ca II K-line emission index in Å; the 10.7-cm radio flux ($F_{10.7\text{-cm}}$) in 10^{-22} s^{-1} m^{-2} Hz^{-1}; and the frequency shifts $\Delta\nu$ of the acoustic low-degree p modes in μHz. Adapted from [51].

Four time series of the S_{ph} corresponding to the VIRGO/SPM blue, green, red channels, and the *Kepler*-like composite as well as one time series of the GOLF S_{vel} are available at the SPACEINN portal[6]. These files are updated in multiples of 5x27= 135 days because, as it has been demonstrated for the Sun and other stars [52], a length of five times the rotation period provide the best compromise to study the evolution of the magnetic activity cycle. These solar datasets have been computed from the raw time series of VIRGO/SPM and GOLF applying the calibration procedures developed for *Kepler* [53] using a high-pass filter with a cut-off period at 60 days.

[6]http://www.spaceinn.eu/data-access/photospheric-solar-activity-index-virgospm-sph/

5 Comparison of solar magnetic activity cycles 23-24

A particular effort has been done to study the magnetic activity cycle of the Sun and to better understand the differences between cycles 23 and 24 that we know had a particularly unexpected long magnetic activity minium [e.g. 54, 55]. Thanks to helioseismology, it si now possible to "see" inside the Sun below the photosphere and study the changes during the evolution of the solar cycle [e.g. 56]. An example of the frequency shifts of modes of degrees ℓ=0, 1, and 2 are shown in figure 4. They have been calculated at three different depths beneath the photosphere at (~2400, ~1300, and ~760 km) [57]. The low-frequency modes show nearly unchanged frequency shifts between Cycles 23 and 24, with a time evolving signature of the quasi-biennial oscillation [58, 59], which is particularly visible for the quadrupole component revealing the presence of a complex magnetic structure. The modes at higher frequencies show frequency shifts 30% smaller during Cycle 24, which is in agreement with the decrease observed in the surface activity between Cycles 23 and 24.

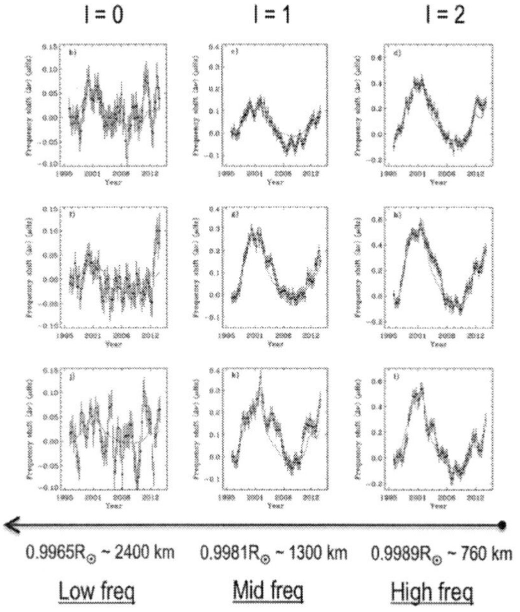

Figure 4: Temporal variations of the frequency shifts in μHz averaged over the modes ℓ= 0, 1, and 2, calculated for four different frequency ranges (low, middle and high frequencies), each one sensitive to a different depth. The associated 10.7-cm radio flux, F10.7, scaled to match the rising phase and the maximum of Cycle 23 is shown as a proxy of the solar surface activity (solid line). Adapted from [57].

The frequency tables associated with the work described in [57] are now available at the SPACEINN portal[7]. A total of 69 non-independent one-year frequency ta-

[7]http://www.spaceinn.eu/data-access/calibrated-soho-golf-frequencies/

bles of modes ℓ=0, 1, 2, and 3 are available. They were computed using 365 days time series (with a four-time overlap of 91.25 days) to avoid any perturbation induced by the 1-year orbital motion of the SoHO spacecraft and covering a total of 18 years starting in April 1996.

6 Sellar Activity: Solar analogues

One of the most important questions we have to answer in astrophysics is if the Sun is a typical star or not attending to its photospheric properties, in particular, to its activity. Indeed, the amount of magnetic activity in a star is crucial for the development of life. Therefore, it is important to be able to compare the Sun to its closest siblings as it has been shown that when a wider comparison is done in terms of the length of activity cycles and surface rotation rates, the Sun seems to be a particular star between the so-called active and inactive branches [60].

However, we know that stars exhibiting pulsations has low surface magnetism because the latter inhibits the oscillation modes [39, 40, 43, 61]. Therefore, the first step is to compare the Sun to solar-type pulsating analogs. This has been done by several members of the WP4.1 for 18 solar analogs observed by *Kepler* [62]. It has been found that the photospheric activity levels of 15 of the solar analogs are comparable to the range between the minimum and the maximum of the solar magnetic activity during the solar cycle (see figure 5).

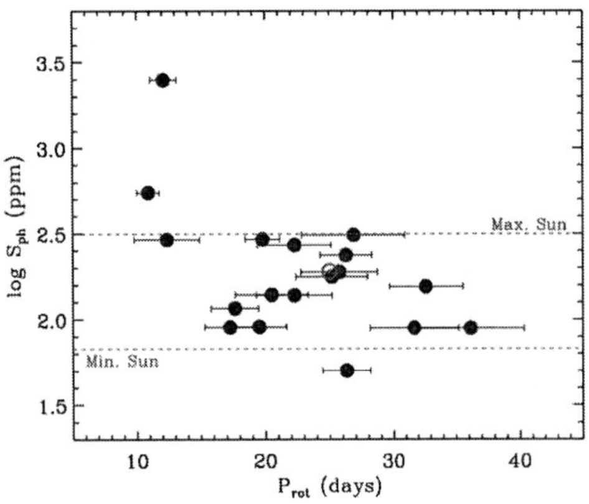

Figure 5: Photospheric magnetic activity index, S_{ph} (in ppm), as a function of the rotational period, P_{rot} (in days from [49]), of the 18 seismic solar analogs observed with the *Kepler* satellite. The mean activity level of the Sun, calculated from the VIRGO/SPM observations, is represented for a rotation of 25 days with its astronomical symbol, and its mean activity levels at minimum and maximum of the 11-year cycle are represented by the horizontal dashed lines. Adapted from [62].

The two stars in figure 5 with a higher S_{ph} correspond to the youngest stars in the sample. Hence, it is normal to have higher surface magnetic activity compared to the Sun. One star with a rotation period comparable to the Sun (i.e. with an age similar to the Sun as expected from gyrochronology), is observed to have a photospheric activity slightly lower than the Sun at its minimum. This could be a consequence of a tilted star compared to the line of sight as the photospheric activity proxy depends on the stellar inclination angle. A similar picture can be drawn when a chromospheric proxy, the Ca K lines, are used to study the magnetism of these stars. Again, inside the error bars, all apart for the two youngest ones have activity levels inside the solar range.

The youngest star in the sample, KIC 10644253 (1.1 ± 0.2 Gyr, [63]), has been extensively analyzed as a possible precursor of our Sun. A magnetic activity modulation of ~1.6 years has been measured in the S_{ph} as well as in the frequency shifts [45], and in the p-mode amplitudes [47]. This modulation could be analogous to what has been found by [64] in the Mount Wilson star HD 30495, having very close stellar properties and falling on the inactive branch reported by [60]. Interestingly, some discrepancies are seen when compared the seismic results with the photospheric proxy. Because this star seems to have a low inclination angle with a weighted average value of i = 23 ± 6°, it is conceivable that the regions of high activity are largely confined to the nearly out-of-sight hemisphere of the star where the discrepancy between the activity proxies are the largest [47].

Follow up observations have been engaged in the Hermes/Mercator telescope during the last two years for all these targets. An example is shown in figure 6 for KIC 10644253 and KIC 3241581. The chromospheric S index seems to follow the magnetic modulation depicted form the *Kepler* S_{ph} proxy. However, longer follow up observations will be required to confirm the periodicity of the magnetic activity cycle of those stars.

7 Conclusions

The global helioseismology working group (4.1) in the SPACEINN project has been a success because of the quality and quantity of the work done as well as for the tools and datasets delivered to the community through the SPACEINN web portal. But the work of this team will not finish at the end of 2016 with the official end of the project. WP4.1 is still working on several scientific papers (e.g. Broomhall et al., Corsaro et al., Salabert et al.) and it will continue working together in some other on-going analysis of helioseismic data. It is also important to mention that the deliverables provided go beyond what was originally proposed in the project. WP4.1 has also been a success reinforcing the synergies between helio- and asteroseismology with a continuous transfer from one to the other, in particular, on the studies of solar analogue stars, which are allowing us to better understand the Sun compared to its siblings.

Acknowledgements

The research leading to these results has received funding from the European Community's Seventh Framework Pro-

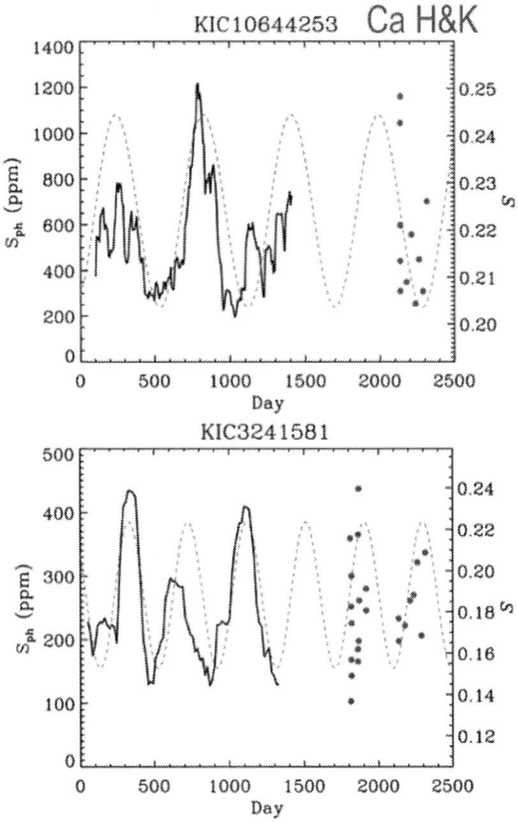

Figure 6: Time evolution of the S_{ph} from [45] obtained from the *Kepler* photometry (black continuous line) and Chromospheric S index obtained form the spectroscopic follow up observations obtained from the Hermes/Mercator telescope (blue dots) for two solar analogues, KIC 10644253 and KIC 3241581. The dashed curve represent the potential magnetic modulation plotted to guide the eye between the two sets of observations.

gramme ([FP7/2007-2013]) under grant agreement No. 312844 (SPACEINN).

References

[1] S.G. Korzennik, *Fitting medium and high degrees using GONG, MDI and HMI observations.*, in *American Astronomical Society Meeting Abstracts #224* (2014), Vol. 224 of *American Astronomical Society Meeting Abstracts*, p. 218.13

[2] S.G. Korzennik, *On the Temporal Changes of Helioseismic Properties Derived with Different Techniques*, in *AAS/Solar Physics Division Meeting* (2016), Vol. 47 of *AAS/Solar Physics Division Meeting*, p. 107.01

[3] J.W. Harvey, F. Hill, R. Hubbard, J.R. Kennedy, J.W. Leibacher, J.A. Pintar, P.A. Gilman, R.W. Noyes, A.M. Title, J. Toomre et al., Science **272**, 1284 (1996)

[4] P.H. Scherrer, R.S. Bogart, R.I. Bush, J.T. Hoeksema, A.G. Kosovichev, J. Schou, W. Rosenberg, L. Springer, T.D. Tarbell, A. Title et al., Sol. Phys., **162**, 129 (1995)

[5] P.H. Scherrer, J. Schou, R.I. Bush, A.G. Kosovichev, R.S. Bogart, J.T. Hoeksema, Y. Liu, T.L. Duvall, J. Zhao, A.M. Title et al., Sol. Phys., **275**, 207 (2012)

[6] W.J. Borucki, D. Koch, G. Basri, N. Batalha, T. Brown, D. Caldwell, J. Caldwell, J. Christensen-Dalsgaard, W.D. Cochran, E. DeVore et al., Science **327**, 977 (2010)

[7] V. Domingo, B. Fleck, A.I. Poland, Sol. Phys., **162**, 1 (1995)

[8] C. Fröhlich, J. Romero, H. Roth, C. Wehrli, B.N. Andersen, T. Appourchaux, V. Domingo, U. Telljohann, G. Berthomieu, P. Delache et al., Sol. Phys., **162**, 101 (1995)

[9] A. Jiménez, T. Roca Cortés, S.J. Jiménez-Reyes, Sol. Phys., **209**, 247 (2002)

[10] A.H. Gabriel, G. Grec, J. Charra, J.M. Robillot, T.R. Cortés, S. Turck-Chièze, R. Bocchia, P. Boumier, M. Cantin, E. Céspedes et al., Sol. Phys., **162**, 61 (1995)

[11] R.A. García, S.J. Jiménez-Reyes, S. Turck-Chièze, S. Mathur, *Helioseismology from the Blue and Red Wings of the NA Profile as Seen by GOLF*, in *SOHO 14 Helio- and Asteroseismology: Towards a Golden Future*, edited by D. Danesy (2004), Vol. 559 of *ESA Special Publication*, p. 432

[12] R.A. García, T. Roca Cortés, C. Régulo, A&AS, **128**, 389 (1998)

[13] R.A. García, S. Turck-Chièze, P. Boumier, J.M. Robillot, L. Bertello, J. Charra, H. Dzitko, A.H. Gabriel, S.J. Jiménez-Reyes, P.L. Pallé et al., A&A, **442**, 385 (2005)

[14] T. Appourchaux, L. Gizon, M.C. Rabello-Soares, A&AS, **132**, 107 (1998), astro-ph/9710082

[15] W.J. Chaplin, Y. Elsworth, G.R. Isaak, R. Lines, C.P. McLeod, B.A. Miller, R. New, MNRAS, **300**, 1077 (1998)

[16] R.A. García, C. Régulo, S. Turck-Chièze, L. Bertello, A.G. Kosovichev, A.S. Brun, S. Couvidat, C.J. Henney, M. Lazrek, R.K. Ulrich et al., Sol. Phys., **200**, 361 (2001)

[17] G.R. Davies, A.M. Broomhall, W.J. Chaplin, Y. Elsworth, S.J. Hale, MNRAS, **439**, 2025 (2014)

[18] T. Appourchaux, E. Michel, M. Auvergne, A. Baglin, T. Toutain, F. Baudin, O. Benomar, W.J. Chaplin, S. Deheuvels, R. Samadi et al., A&A, **488**, 705 (2008)

[19] R.A. García, C. Régulo, R. Samadi, J. Ballot, C. Barban, O. Benomar, W.J. Chaplin, P. Gaulme, T. Appourchaux, S. Mathur et al., A&A, **506**, 41 (2009), 0907.0608

[20] C. Barban, S. Deheuvels, F. Baudin, T. Appourchaux, M. Auvergne, J. Ballot, P. Boumier, W.J. Chaplin, R.A. García, P. Gaulme et al., A&A, **506**, 51 (2009)

[21] S. Mathur, R.A. García, C. Catala, H. Bruntt, B. Mosser, T. Appourchaux, J. Ballot, O.L. Creevey, P. Gaulme, S. Hekker et al., A&A, **518**, A53 (2010), 1004.4891

[22] S. Mathur, R. Handberg, T.L. Campante, R.A. García, T. Appourchaux, T.R. Bedding, B. Mosser, W.J. Chaplin, J. Ballot, O. Benomar et al., ApJ, **733**, 95 (2011), 1103.4085

[23] T.L. Campante, R. Handberg, S. Mathur, T. Appourchaux, T.R. Bedding, W.J. Chaplin, R.A. García, B. Mosser, O. Benomar, A. Bonanno et al., A&A, **534**, A6 (2011), 1108.3807

[24] T. Appourchaux, Astronomische Nachrichten **329**, 485 (2008), 0711.0435

[25] O. Benomar, T. Appourchaux, F. Baudin, A&A, **506**, 15 (2009)

[26] T.R. White, B.J. Brewer, T.R. Bedding, D. Stello, H. Kjeldsen, Communications in Asteroseismology **161**, 39 (2010), 1006.3992

[27] R. Handberg, T.L. Campante, A&A, **527**, A56 (2011), 1101.0084

[28] E. Corsaro, J. De Ridder, A&A, **571**, A71 (2014), 1408.2515

[29] G.R. Davies, V. Silva Aguirre, T.R. Bedding, R. Handberg, M.N. Lund, W.J. Chaplin, D. Huber, T.R. White, O. Benomar, S. Hekker et al., MNRAS, **456**, 2183 (2016), 1511.02105

[30] A.M. Broomhall, W.J. Chaplin, Y. Elsworth, T. Appourchaux, R. New, MNRAS, **406**, 767 (2010), 1004.4505

[31] R. Howe, G.R. Davies, W.J. Chaplin, Y.P. Elsworth, S.J. Hale, MNRAS, **454**, 4120 (2015), 1509.06894

[32] S. Lefebvre, R.A. García, S.J. Jiménez-Reyes, S. Turck-Chièze, S. Mathur, A&A, **490**, 1143 (2008), 0808.0422

[33] R.A. García, S. Turck-Chièze, S.J. Jiménez-Reyes, J. Ballot, P.L. Pallé, A. Eff-Darwich, S. Mathur, J. Provost, Science **316**, 1591 (2007)

[34] R.A. García, A. Jiménez, S. Mathur, J. Ballot, A. Eff-Darwich, S.J. Jiménez-Reyes, P.L. Pallé, J. Provost, S. Turck-Chièze, Astronomische Nachrichten **329**, 476 (2008), arXiv:0802.4296

[35] T. Appourchaux, K. Belkacem, A. Broomhall, W.J. Chaplin, D.O. Gough, G. Houdek, J. Provost, F. Baudin, P. Boumier, Y. Elsworth et al., A&ARv, **18**, 197 (2010), 0910.0848

[36] R.A. García, Highlights of Astronomy **15**, in press (2010)

[37] R.A. García, D. Salabert, J. Ballot, A. Eff-Darwich, R. Garrido, A. Jiménez, S. Mathis, S. Mathur, A. Moya, P.L. Pallé et al., Journal of Physics Conference Series **271**, 012046 (2011), 1012.0506

[38] S. Mathur, S.J. Jiménez-Reyes, R.A. García, *GOLF: A New Proxy for Solar Magnetism*, in *Solar-Stellar Dynamos as Revealed by Helio- and Asteroseismology: GONG 2008/SOHO 21*, edited by M. Dikpati, T. Arentoft, I. González Hernández, C. Lindsey, F. Hill (2009), Vol. 416 of *Astronomical Society of the Pacific Conference Series*, p. 529, 0810.1803

[39] B. Mosser, E. Michel, T. Appourchaux, C. Barban, F. Baudin, P. Boumier, H. Bruntt, C. Catala, S. Deheuvels, R.A. García et al., A&A, **506**, 33 (2009), 0908.2244

[40] W.J. Chaplin, T.R. Bedding, A. Bonanno, A. Broomhall, R.A. García, S. Hekker, D. Huber, G.A. Verner, S. Basu, Y. Elsworth et al., ApJL, **732**, L5 (2011), 1103.5570

[41] S. Mathur, H. Bruntt, C. Catala, O. Benomar, G.R. Davies, R.A. García, D. Salabert, J. Ballot, B. Mosser, C. Régulo et al., A&A, **549**, A12 (2013), 1209.5696

[42] S. Mathur, R.A. García, A. Morgenthaler, D. Salabert, P. Petit, J. Ballot, C. Régulo, C. Catala, A&A, **550**, A32 (2013), 1212.0630

[43] R.A. García, S. Mathur, D. Salabert, J. Ballot, C. Régulo, T.S. Metcalfe, A. Baglin, Science **329**, 1032 (2010), 1008.4399

[44] D. Salabert, C. Régulo, J. Ballot, R.A. García, S. Mathur, A&A, **530**, A127 (2011), 1104.5654

[45] D. Salabert, C. Régulo, R.A. García, P.G. Beck, J. Ballot, O.L. Creevey, F. Pérez Hernández, J.D. do Nascimento, Jr., E. Corsaro, R. Egeland et al., A&A, **589**, A118 (2016), 1603.00655

[46] C. Régulo, R.A. García, J. Ballot, A&A, **589**, A103 (2016), 1603.04673

[47] R. Kiefer, A. Schad, G. Davies, M. Roth, ArXiv e-prints (2016), 1611.02029

[48] S. Mathur, R.A. García, J. Ballot, T. Ceillier, D. Salabert, T.S. Metcalfe, C. Régulo, A. Jiménez, S. Bloemen, A&A, **562**, A124 (2014), 1312.6997

[49] R.A. García, T. Ceillier, D. Salabert, S. Mathur, J.L. van Saders, M. Pinsonneault, J. Ballot, P.G. Beck, S. Bloemen, T.L. Campante et al., A&A, **572**, A34 (2014), 1403.7155

[50] A.M. Broomhall, V.M. Nakariakov, Sol. Phys., **290**, 3095 (2015), 1507.02854

[51] D. Salabert, R.A. Garcia, P.G. Beck, C. Regulo, J. Ballot, O.L. Creevey, R. Egeland, J.D. do Nascimento, Jr., F. Perez Hernandez, L. Bigot et al., ArXiv 1610.00990 (2016), 1610.00990

[52] S. Mathur, D. Salabert, R.A. García, T. Ceillier, Journal of Space Weather and Space Climate **4**, A15 (2014), 1404.3076

[53] R.A. García, S. Hekker, D. Stello, J. Gutiérrez-Soto, R. Handberg, D. Huber, C. Karoff, K. Uytterhoeven, T. Appourchaux, W.J. Chaplin et al., MNRAS, **414**, L6 (2011), 1103.0382

[54] W. Livingston, M.J. Penn, L. Svalgaard, ApJ, **757** (2012)

[55] A.M. Broomhall, P. Chatterjee, R. Howe, A.A. Norton, M.J. Thompson, Space Science Reviews, **186**, 191 (2014), 1411.5941

[56] S. Basu, A.M. Broomhall, W.J. Chaplin, Y. Elsworth, ApJ, **758**, 43 (2012), 1208.5493

[57] D. Salabert, R.A. García, S. Turck-Chièze, A&A, **578**, A137 (2015), 1502.07607

[58] S.T. Fletcher, A. Broomhall, D. Salabert, S. Basu, W.J. Chaplin, Y. Elsworth, R.A. Garcia, R. New, ApJL, **718**, L19 (2010), 1006.4305

[59] R. Simoniello, W. Finsterle, D. Salabert, R.A. García, S. Turck-Chièze, A. Jiménez, M. Roth, A&A, **539**, A135 (2012), 1201.2773

[60] E. Böhm-Vitense, ApJ, **657**, 486 (2007)

[61] P. Gaulme, J. Jackiewicz, T. Appourchaux, B. Mosser, ApJ, **785**, 5 (2014), 1402.3027

[62] D. Salabert, R.A. Garcia, P.G. Beck, R. Egeland, P.L. Palle, S. Mathur, T.S. Metcalfe, J.D. do Nascimento, Jr., T. Ceillier, M.F. Andersen et al., ArXiv 1608.01489 (2016), 1608.01489

[63] T.S. Metcalfe, O.L. Creevey, G. Doğan, S. Mathur, H. Xu, T.R. Bedding, W.J. Chaplin, J. Christensen-Dalsgaard, C. Karoff, R. Trampedach et al., ApJS, **214**, 27 (2014), 1402.3614

[64] R. Egeland, T.S. Metcalfe, J.C. Hall, G.W. Henry, ApJ, **812**, 12 (2015), 1507.03611

DOI: 10.1051/epjconf/201716001011

Promoting access to and use of seismic data in a large scientific community

SpaceInn data handling and archiving

Eric Michel[1,*], *Kevin* Belkacem[1], *Reza* Samadi[1], *Raphael* de Assis Peralta[1], *Christian* Renié[1], *Mahfoudh* Abed[1], *Guangyuan* Lin[1], *Jørgen* Christensen-Dalsgaard[2], *Günter* Houdek[2], *Rasmus* Handberg[2], *Laurent* Gizon[3], *Raymond* Burston[3], *Kaori* Nagashima[3], *Pere* Pallé[4], *Ennio* Poretti[5], *Monica* Rainer[5], *Angelo* Mistò[5], *Maria Rosa* Panzera[5], and *Markus* Roth[6]

[1]*Observatoire de Paris – LESIA, UMR-8109, Université Pierre et Marie Curie, Université Paris-Diderot, PSL, Paris, France*
[2]*Stellar Astrophysics Centre, Department of Physics and Astronomy, Aarhus Univ., Aarhus, Denmark*
[3]*Max-Planck-Institut für Sonnensystemforschung, 37077 Göttingen, Germany*
[4]*Instituto de Astrofísica de Canarias/Department de Astrofísica, Universidad de La Laguna, La Laguna, Tenerife, Spain*
[5]*INAF-Osservatorio Astronomico di Brera, Merate, Italy*
[6]*Kiepenheuer-Institut für Sonnenphysik, Sconeckstr. 6, 79104 Freiburg, Germany*

Abstract. The growing amount of seismic data available from space missions (SOHO, CoRoT, Kepler, SDO,...) but also from ground-based facilities (GONG, BiSON, ground- based large programmes...), stellar modelling and numerical simulations, creates new scientific perspectives such as characterizing stellar populations in our Galaxy or planetary systems by providing model-independent global properties of stars such as mass, radius, and surface gravity within several percent accuracy, as well as constraints on the age. These applications address a broad scientific community beyond the solar and stellar one and require combining indices elaborated with data from different databases (e.g. seismic archives and ground- based spectroscopic surveys). It is thus a basic requirement to develop a simple and efficient access to these various data resources and dedicated tools. In the framework of the European project SpaceInn (FP7), several data sources have been developed or upgraded. The Seismic Plus Portal has been developed, where synthetic descriptions of the most relevant existing data sources can be found, as well as tools allowing to localize existing data for given objects or period and helping the data query. This project has been developed within the Virtual Observatory (VO) framework. In this paper, we give a review of the various facilities and tools developed within this programme. The SpaceInn project (Exploitation of Space Data for Innovative Helio- and Asteroseismology) has been initiated by the European Helio- and Asteroseismology Network (HELAS).

1 Introduction

With about 150,000 stars observed with CoRoT over periods going up to five months and about 200,000 stars observed with *Kepler* over periods up to 4 years, stellar seismology has known an unprecedented acceleration over the past decade. It has become common to use seismic data for hundreds or thousands of stars, and this information is fruitfully coupled with spectroscopic or astrometric informations available from various large surveys. This is true for studies addressing stellar structure and its evolution, but also for studies addressing galactic population formation and migration or for studies questioning planet formation. The stellar light-curves have also revealed signatures of other phenomena than just pulsation, like granulation, flares or activity, and the reference to the Sun is more than ever necessary to understand and interpret these phenomena which can be studied in great detail in the present wealth of solar data. Within the SpaceInn project, we have developed several facilities to make this information available for a large scientific community. Several data sources have been developed or upgraded and a portal has been created to help the access to this information.

*e-mail: Eric.Michelatobspm.fr

2 The Seismic Plus Portal

The Seismic Plus Portal (see figure 1) is intended to be a hub fostering access to existing seismic data sources and to the most relevant complementary data sources.

It offers three main functionalities:

- An extended census and an homogeneous and synthetic description of the most relevant data and data sources (see Sect. 2.1).
- localizing and quick-look of the data available for a given list of stars or for a given time period (see Sect. 2.2).
- Tools for data handling, including tools transmitting coordinated download queries to various data sources and

 DOI: 10.1051/epjconf/201716001011

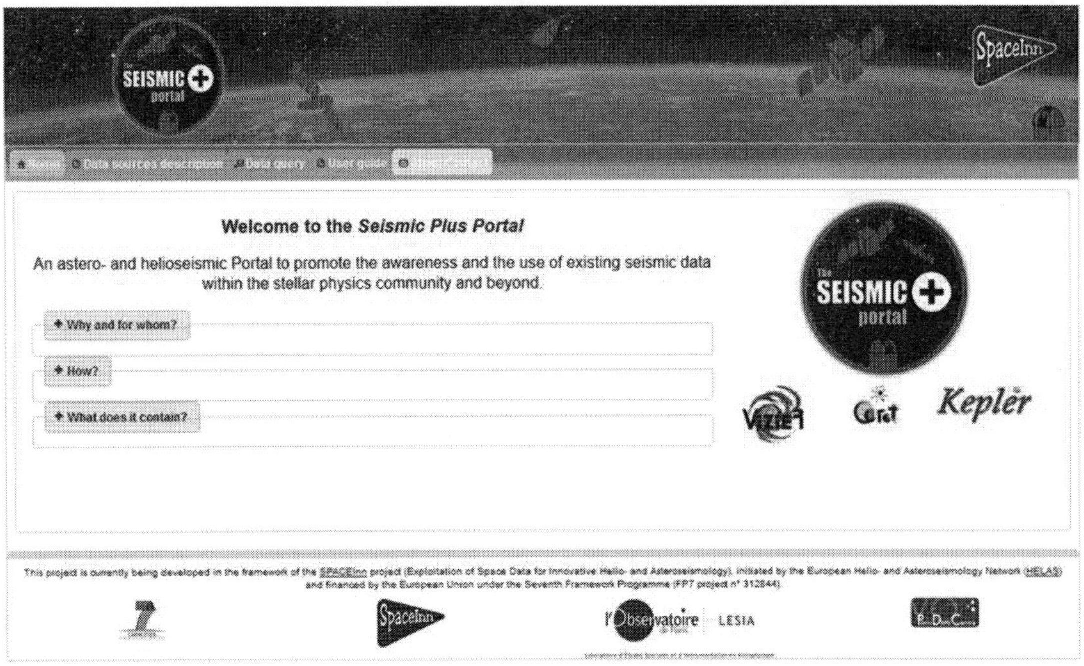

Figure 1: Home page of the Seismic Plus Portal, at http://voparis-spaceinn.obspm.fr/seismic-plus/

tools providing higher-level outputs combining information from various data sources (see Sect. 2.3).

2.1 The data and data sources

Beyond space and ground-based seismic data and data sources, the portal also addresses most relevant related sources of complementary data (e.g. astrometry, spectroscopy, ...) commonly used to complement seismic studies of stars.

For convenience, data sources and data products are distibuted under the following categories:

- Time series: they constitute the main observational material for seismology. It can be composed of photometric measurements, radial velocities, magnetograms, ... It can be 1D time series for integrated-light measurements but also 2D or 3D for time series of spectra or magnetograms. The total duration, the time sampling, and the duty cycle are the specific characteristic parameters which, along with the more classical S/N, characterize these data and show how they can be used.
- Spectroscopy/individual spectra: seismic studies often rely on stellar modeling which benefits from additional constraints brought by spectroscopic analysis (e.g. for [Fe/H] and $T_{\rm eff}$ estimates). On the other hand, it has become common, for red giant stars at least, to improve spectroscopic detailed analysis by using so-called seismic $\log g$ estimates as additional constraints. It is thus important to bridge the gap between these types of data in the context of seismology space missions and spectroscopic large programs.
- Stellar parameters: With space missions like CoRoT and Kepler, a new concept appeared, namely 'ensemble asteroseismology'. It deals with large samples of stars and allows stellar population studies. The characterization of these large samples relies on observational indices. In addition to the classical photometric color indices or parallaxes, new indices have emerged which characterize the stellar pulsation and the stellar structure to first order (large separation, frequency of maximum amplitude, ...).
- others: The use and application of seismic data is growing and evolving quickly. The list of data sources and data products considered at the Seismic Portal has to remain open to changes in the observational landscape and in current scientific issues.

The various data are distributed under these categories in a synthetic table (see figure 2). A brief complementary information is also available locally, and links toward the reference site or publication are also given.

2.2 Localizing and inspecting data available for a list of stars or a time period

Two modes of query exist at the moment. The first one, for a given list of stars, returns a localization table stressing which kind of data are available and where (see figure 3). This information is available for each star of the list and quick-look tools allow a first inspection of the data in order to determine to which extent they are appropriate for the anticipated application.

DOI: 10.1051/epjconf/201716001011

Figure 2: A few entries in the list of data sources referenced at the Seismic Plus.

A second mode of query exists for solar data. For a given time period, it returns a table stressing which kind of time series exists and where. It also provides information about occurence of solar events (essentially coronal mass ejections (CME) and flares so far) over the same period. The intention here is to help establishing connexion between solar events and their signatures in a time series, thus helping the interpretation of such signatures in the light-curve of other stars.

Figure 3: Result of a data query for a list of stars. Each line corresponds to a given star and green signs indicate existing data in the data source associated with the corresponding column.

2.3 Tools for coordinated download of the data and higher-level outputs

Once the result of the query is obtained, as in figure 3, it is possible to select a set of stars. The portal then offers the possibility to transmit a query to various data sources in order to download the corresponding data. Various formats are proposed. The portal is also intended to be the place where higher-level products can be built, combining several types of data from different sources. A first tool allows for instance to derive seismic estimates of the mass, radius and log g, for a selected list of stars and for various sources of seismic indices and effective temperatures. The procedures are based on scaling laws which are documented and the publication references are indicated.

3 Data sources developed or upgraded within the SpaceInn WP3 activities

Among the data sources referenced at the Seismic Plus Portal, several have been developed or upgraded within the SpaceInn activities. They are accessible via the Seismic Plus Portal but also via their dedicated web interface.

3.1 The SISMA spectral time series database

Thousands of high-resolution spectra were taken with the spectrograph HARPS mounted at the 3.6m ESO telescope at La Silla, Chile ([4]) during the ESO Large Programmes LP 182.D-0356 and LP 185.D-0056 as ground-based counterpart of the CoRoT ([1]) photometric measurements ([6]). The SISMA database (http://sisma.brera.inaf.it/ see figure 4) gathers 7103 fully reduced and normalized HARPS spectra, obtained for 261 stars, among which are 71 CoRoT bright targets observed simultaneously with the satellite. These spectroscopic time series allow to better exploit the photometric data and characterize the variability classes of CoRoT targets.

In addition to the spectra, SISMA offers mean line profiles (computed with the LSD software [3]), radial velocities, $V\sin i$, activity indices, differential rotation indicators, binarity and emmision flags. It also features the physical parameters of each target ($T_{\rm eff}$, log g, [Fe/H]) and, in the case of CoRoT targets, the relevant satellite light curves (see [5]).

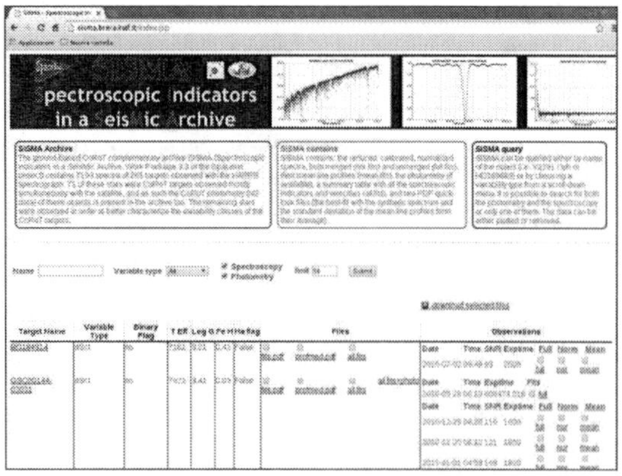

Figure 4: Home page of the SISMA database.

3.2 The KASOC and SpaceInn

The *Kepler* Asteroseismic Science Operations Center (http://kasoc.phys.au.dk/) provides asteroseismological data from the NASA *Kepler* mission to astronomers who are members of the *Kepler* Asteroseismic Science Consortium (KASC). In the framework of the SpaceInn project, efforts have been made to enhance reliability and efficiency of the access to *Kepler* data. A VO interface has been developed allowing for external access to KASOC database via the Virtual Observatory protocol. Various type of complementary information have been included in the database, like ground-based follow-up data and stellar models related to published studies of specific objects (figure 5).

In addition, a solution for long-term preservation of the data has been studied in collaboration with the Royal Library of Denmark.

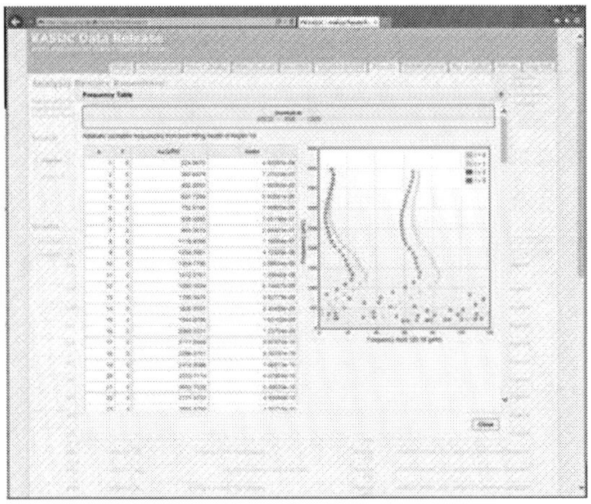

Figure 5: Interface allowing theoretical-frequency handling and visualisation at KASOC.

3.3 The MARK-I database

This historical archive (1976 till 2012) has been developed with the solar observations gathered at the Observatorio del Teide (IAC) with the pioneer Mark-I resonant-scattering spectrophotometer (http://svo2.cab.inta-csic.es/vocats/marki/). Observations extend from 1976 to 2012 with only summer campaigns from 1976 to 1983.

The data products are radial velocity measurements of the Sun-as-a-star at the Potassium KI 7699A absorption solar line.

3.4 The German Data Center for SDO and SpaceInn

The German Data Center for SDO (GDC-SDO) hosts the European depository for HMI data. The data processed and made available from the Joint Science Operations Center (JSOC) headquartered at Stanford University USA are immediatly transfered to GDC (several TBs per day). This constitutes the largest source of data available today for helioseismology (online storage capacity over 1 PB). The data products are line-of-sight Dopplergrams, line-of-sight magnetograms, continuum intensities, subset of Milne-Eddington inversions. They are available at various cadences.

Within the SpaceInn Project, a dedicated GDC-SDO Help Desk has been developed for SpaceInn (http://www2.mps.mpg.de/projects/seismo/GDC-SDO/spaceinn_access.htm). It contains information for setting up NetDRMS and receiving relevant data and users support.

3.5 The Stellar Seismic Indices database

The Stellar Seismic Indices (SSI) database (http://ssi.lesia.obspm.fr/ see figure 6) has been developed within the SpaceInn programme (see also [7]). It contains stellar seismic indices of solar-like oscillating stars as well as characteristic parameters of stellar granulation. These seismic indices obey characteristic scaling relations that depend directly on the radius, mass and effective temperature of the star. From the knowledge of these quantities it is possible to estimate the mass and radius of a star, and subsequently the surface gravity of the star (log g). The indices populating the current database were extracted from about 13,000 Kepler and 5,000 CoRoT red-giant stars using an automatic analysis pipeline [2].

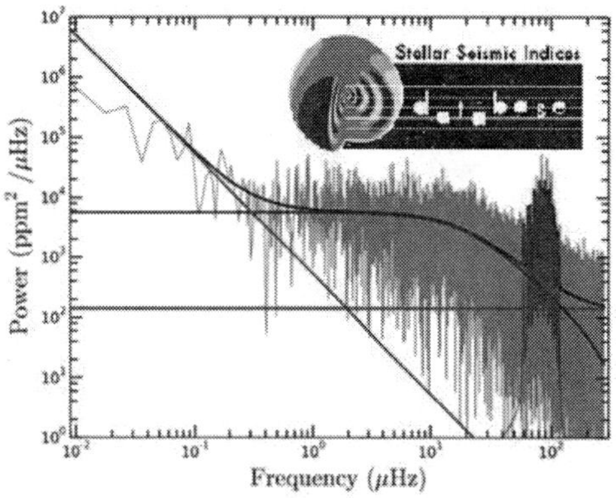

Figure 6: The SSI database provides seismic indices and granulation parameters characterizing solar-like pulsators.

4 Perspective

In the perspective of future projects like TESS (launch 2017) and PLATO (launch 2025) which will enrich considerably the seismic material available to the community and in the perspective of the Gaia wealth of information, it seems very relevant to foster as efficiently as possible the access to these data and techniques necessary to exploit them with the best outcome.

 DOI: 10.1051/epjconf/201716001011

Acknowledgements

This project has been developed within the European FP7 SpaceInn project (SpaceInn.eu). We acknowledge the great help from Spanish VO (E. Solano), Paris Astronomical Data Centre (P. Le Sidaner), Copenhagen Royal Library, INAF-Osservatorio Astronomico di Treste (M. Molinaro and F. Cepparo).

References

[1] Baglin, A. et al., AIPC **895**, 201 (2007)
[2] de Assis Peralta, R. et al. submitted
[3] Donati, J.-F. et al., MNRAS **291**, 658 (1997)
[4] Mayor, M. et al., The Messenger **114**, 20 (2003)
[5] Rainer, M. et al., Astronomical Journal **152**, 207 (2016)
[6] Rainer, M. et al., AN **333**, 1061 (2012)
[7] Samadi, R. et al., these proceedings

EPJ Web of Conferences **160**, 01012 (2017)
Seismology of the Sun and the Distant Stars II
DOI: 10.1051/epjconf/201716001012

Extraction of seismic indices and stellar granulation parameters for *CoRoT* and *Kepler* red giants using the MLEUP method. Main results and perspectives.

Raphaël de Assis Peralta[1,★], *Réza* Samadi[1], and *Éric* Michel[1]

[1] *LESIA, Observatoire de Paris, CNRS UMR8109, Université Pierre et Marie Curie, Université Paris Diderot, PSL*

Abstract. In the framework of the SPACEInn project, a Stellar Seismic Indices (SSI - http://ssi.lesia.obspm.fr/) database was developed in order to provide the scientific community with oscillations and granulation signatures for a large set of red-giant type stars. For this purpose, we have developed the method MLEUP able to extract simultaneously the seismic indices (the equidistance $\Delta\nu$, the frequency $\nu_{\max}$ and the height $H_{\rm env}$ of the maximum oscillation power) and granulation parameters (the *e-folding time* $\tau_{\rm eff}$ and the variance of the integrated brightness fluctuations σ^2). This method has been tested in terms of precision and accuracy, using Monte Carlo simulations. Then we applied it to all stars observed by CoRoT and all long-cadence *Kepler* light-curves. In total, we yield seismic indices and granulation parameters for about 5,000 stars for CoRoT and more than 13,000 for *Kepler*. In this paper, we focus on the main results for both seismic indices $\Delta\nu$ and $\nu_{\max}$ as well as for the stellar parameters (mass, radius and luminosity) seismically inferred. Then, in the perspective of Gaia, we discuss about the possibility to derive other seismic quantities like e.g. a seismic effective temperature.

1 Introduction

We have developed a new automatic method, called MLEUP, taking advantage of the MLE (Maximum Likelihood Estimate) algorithm combined with the parametrized representation of the red giants pulsation spectrum following the UP (Universal Pattern, cf. [1]) in order to measure simultaneously the oscillations and the granulation signatures. The full description of this method is in [2]. We describe briefly in Sect. 2 the mainlines of our method. In Sect. 3, we present the main results of the analysis of the CoRoT and *Kepler* datasets: PUBLI ! the mean large separation $\Delta\nu$, the peak frequency $\nu_{\max}$ as well as the "seismic" masses and radii. In Sect. 4, we discuss about the prospect, such as the possibility to derive the "seismic" effective temperature using the Gaia data. Finally, we conclude in Sect. 5.

2 Mainlines of the Method MLEUP

The model used to fit the power density spectrum is based on two Lorentzian-like functions for the granulation and activity component, a red giant parametric oscillations pattern based on the Universal Pattern [1] for the oscillations and a constant for the white noise (see Fig. 1). The oscillations are characterised by three seismic indices: The mean large separation, $\Delta\nu$, corresponding to the mean frequency spacing between two consecutive p-modes with same angular degree; $H_{\rm env}$ which is the maximum height of the oscillation envelope and $\nu_{\max}$, the corresponding frequency. Concerning the granulation, it is characterised by two parameters: the effective timescale $\tau_{\rm eff}$ (or the *e-folding time*), which measures the temporal coherence of the granulation in the time domain and the characteristic amplitude σ^2 which corresponds to the variance brightness fluctuation of the granulation. The method has been tested in order to characterize its bias and dispersion using Monte Carlo simulations. These simulations revealed that MLEUP presents low dispersions, especially for $\Delta\nu$ and $\nu_{\max}$, and that the internal errors are reasonably representative of the real dispersions

3 Mean large separation $\Delta\nu$ and peak frequency $\nu_{\max}$ – "seismic" mass and radius

We analysed all CoRoT data[1] with a duration of observation larger than 50 days and all long-cadence (dt = 29.42 min) *Kepler* data[2]. We yield 4783 CoRoT stars and 13,689 *Kepler* stars for which we get the seismic indices, and 1520 CoRoT stars and 12,879 *Kepler* stars for which we get also the granulation parameters. The results of the analysis of the large sets of *Kepler* and CoRoT stars are presented and discussed in [2]. Here, we focus on the results of both seismic indices $\Delta\nu$ and $\nu_{\max}$, and the stellar parameters inferred by them.

★e-mail: raphael.peralta@obspm.fr

[1] CoRoT data archive: http://idoc-corot.ias.u-psud.fr
[2] *Kepler* data archive: http://archive.stsci.edu/kepler/

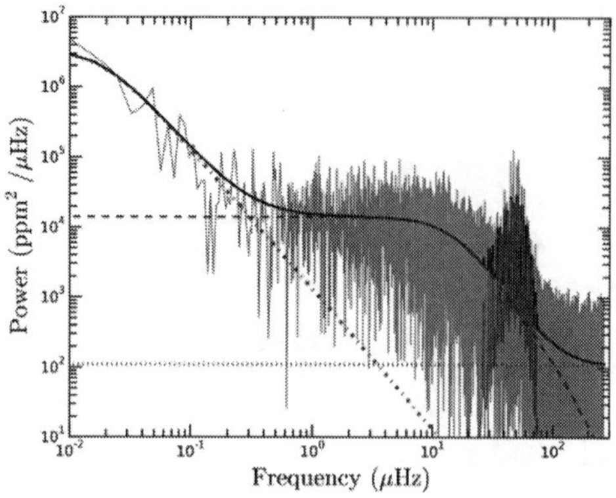

Figure 1: Results of the final adjustment of the background and oscillations of MLEUP (black line). In grey, the raw PSD of KIC 2850913. The dash-dot green line and the dashed blue one correspond respectively to the activity and granulation component. The dotted magenta line represents the white noise component and the solid red one is the Universal Pattern.

We plotted in figure 2 $\Delta\nu$ as a function of $\nu_{\max}$ from the 4783 CoRoT and 13,689 *Kepler* stars for which we extracted the seismic indices. We can see that the dispersion for CoRoT and *Kepler* is small, as well as the internal error (cf. Tab. 1). We deduced the two following scaling relations by adjusting the CoRoT and *Kepler* datasets:

- $\Delta\nu = (0.242 \pm 0.003)\ \nu_{\max}^{0.778\pm0.003}$ (CoRoT)
- $\Delta\nu = (0.282 \pm 0.001)\ \nu_{\max}^{0.7506\pm0.0010}$ (*Kepler*)

The *Kepler* scaling relation is consistent with the [3]'s scaling relation ($\Delta\nu = (0.274 \pm 0.002)\ \nu_{\max}^{0.751\pm0.002}$), determined by an average of scaling relations obtained by various different methods with the *Kepler* data, as well as the theoretical one [$\beta = 0.75$, e.g. 4]. Between our CoRoT and *Kepler* scaling relations, we note slopes significantly different. This difference is not yet understood. It could come from the difference in the stellar populations considered.

The previous seismic indices are very valuable to characterised large sets of stars. To illustrate this, we considered 13,408 of the *Kepler* selected dataset for which we have the effective temperature from [5]. Combining with $\nu_{\max}$ and $\Delta\nu$, one can estimate the mass, radius and luminosity [e.g. 6] seismically inferred with a high precision (cf. Tab. 1). The figure 3 shows the distribution of our results in the Hertzsprung–Russell (H-R) diagram where the red giant branch (RGB) and the red clump are well recognizable.

4 Perspective – The "seismic" effective temperature $T_{\rm eff}$

The arrival of Gaia data opens new perspectives. As an illustration, we introduce the "seismic" effective temperature. Indeed, if we combine the apparent magnitude m, the parallax p and the interstellar extinction $A_{\rm m}$ with both seismic indices $\nu_{\max}$ and $\Delta\nu$, it is possible to derive the $T_{\rm eff}$ via the following equation (cf. Sect. 5 for detailed derivation):

Table 1: Median of the relative internal errors given in percentage, for the results obtained with the *Kepler* and CoRoT selected datasets.

	δ_{err} *Kepler* (%)		δ_{err} CoRoT (%)	
	negative	positive	negative	positive
$\nu_{\max}$	−0.545%	+0.545%	−1.53%	+1.53%
$\Delta\nu$	−0.0386%	+0.0386%	−0.144%	+0.144%
$M/{\rm M}_\odot$	−3.98%	+3.98%	—	—
$R/{\rm R}_\odot$	−1.33%	+1.33%	—	—
$L/{\rm L}_\odot$	−11.6%	+11.6%	—	—

$$T_{\rm eff} = 10^{\frac{(\mathcal{M}_\odot - 5 + A_{\rm m} - m)}{12.5}}\, p^{-2/5} \left(\frac{\Delta\nu}{\Delta\nu_{\rm ref}}\right)^{4/5} \left(\frac{\nu_{\max}}{\nu_{\rm ref}}\right)^{-2/5} {\rm T}_{\rm eff,\odot}\,, \quad (1)$$

with $A_{\rm m}$ the interstellar extinction (mag), m the apparent magnitude (mag), p the parallax (arcsec), $\nu_{\max}$ the peak frequency (μHz) and $\Delta\nu$, the large separation (μHz). The constant ${\rm T}_{\rm eff,\odot}$ corresponds to the solar values and the constants $\nu_{\rm ref}$ and $\Delta\nu_{\rm ref}$, to the reference values defined by [7].

The corresponding error is given by:

$$\delta T_{\rm eff} = T_{\rm eff} \sqrt{\left(\frac{\ln 10}{12.5}\delta A_{\rm m}\right)^2 + \left(\frac{\ln 10}{12.5}\delta m\right)^2 + \left(\frac{2}{5}\frac{\delta p}{p}\right)^2 + \left(\frac{2}{5}\frac{\delta\nu_{\max}}{\nu_{\max}}\right)^2 + \left(\frac{4}{5}\frac{\delta\Delta\nu}{\Delta\nu}\right)^2}\,. \quad (2)$$

The MLEUP method provides the seismic indices $\nu_{\max}$ and $\Delta\nu$ with a high precision. The relative errors obtained with the selected *Kepler* dataset are 0.55% and 0.04% respectively (cf. Tab. 1). Regarding m, p and $A_{\rm m}$, they could be provided by Gaia. The precision that Gaia will soon provide should be of the order $\delta p \sim 10\ \mu$as for parallaxes, $\delta m \sim 3$ mmag for apparent magnitudes and $\delta A_{\rm m} \lesssim 0.1$ mag for extinctions[3]. Thereby, the relative incertitude on $T_{\rm eff}$ would be about 2%, mainly dominated by the interstellar extinction. For a star with $T_{\rm eff} = 5000$ K, we have $\delta T_{\rm eff} \sim 100$ K, which is very competitive with spectroscopic determination.

Alternatively, we could also imagine to derive the seismic interstellar extinction given the spectroscopic $T_{\rm eff}$.

5 Conclusion

The MLEUP method allowed us to extract simultaneously with a high precision the seismic indices and the granulation parameters from a large number of *Kepler* and CoRoT stars. We yield 4783 CoRoT stars and 13,689 *Kepler* stars for which we extracted the seismic indices, and 1520 CoRoT stars and 12,879 *Kepler* stars for which we get both the seismic indices and the

[3]For more details, see: http://www.cosmos.esa.int/web/gaia/science-performance

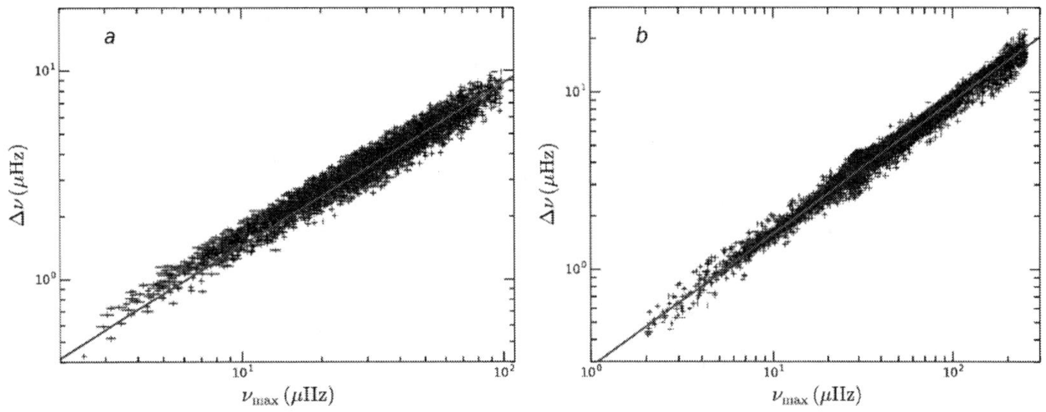

Figure 2: Mean large separation $\Delta\nu$ obtained with 4783 CoRoT stars (*a*) and 13,689 *Kepler* stars (*b*). Black crosses represent the values obtained, with their error bars in grey. The red line is the deduced scaling relations. Fig. from [2].

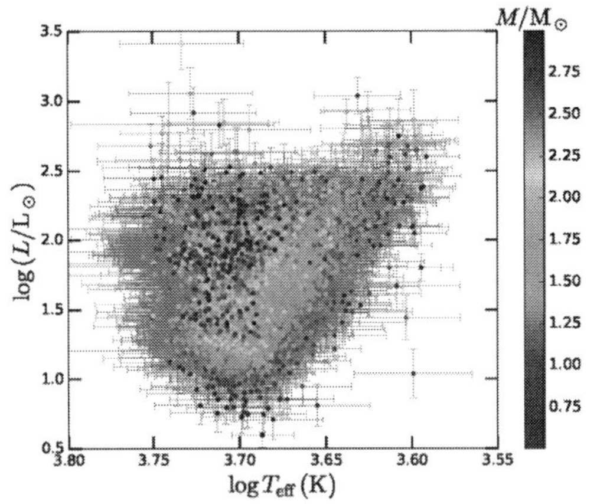

Figure 3: Hertzsprung–Russell diagram with 13,408 *Kepler* stars. The colour code represents the stellar mass. For better visibility of the mass variation, only stars with a mass included in the range $0.5 \leqslant M/\mathrm{M}_\odot \leqslant 3.0$ are plotted. Fig. from [2].

granulation parameters. Those indices and parameters are available in the *Stellar Seismic Indices* (SSI) database[4].

The measurements of $\nu_{\max}$ and $\Delta\nu$ obtained with MLEUP are very precise. Thereby, given an effective temperature, we could derive seismic masses and radii with a good precision. Thanks to the very precise measurements that the space mission Gaia will soon provide, we can plan to derive others parameters using the seismic indices. Indeed, it would be possible to derive a seismic effective temperature or the seismic interstellar extinction A_{m}.

[4]http://ssi.lesia.obspm.fr/

Acknowledgements

This paper is based on data from the CoRoT Archive.The CoRoT space mission has been developed and operated by CNES, with contributions from Austria, Belgium, Brazil, ESA (RSSD and Science Program), Germany, and Spain. This paper also includes data collected by the *Kepler* mission. Funding for the *Kepler* mission is provided by the NASA Science Mission directorate. The authors acknowledge the entire *Kepler* and CoRoT team, whose efforts made these results possible. We acknowledge financial support from the SPACEInn FP7 project (SPACEInn.eu) and from the "Programme National de Physique Stellaire" (PNPS, INSU, France) of CNRS/INSU.

Annexe: Detailed calculation of the "seismic" effective temperature T_{eff}

We detail in this annexe the calculation of the seismic effective temperature T_{eff}.

The Stefan-Boltzmann law, normalised with the solar values, gives:

$$\frac{L}{\mathrm{L}_\odot} = \left(\frac{R}{\mathrm{R}_\odot}\right)^2 \left(\frac{T_{\mathrm{eff}}}{\mathrm{T}_{\mathrm{eff},\odot}}\right)^4, \tag{3}$$

with L, the star luminosity (W) and R, its radius (m). The constants $\mathrm{L}_\odot$, $\mathrm{R}_\odot$ and $\mathrm{T}_{\mathrm{eff},\odot}$ correspond to the solar values.

The radius R can be expressed using the seismic indices $\nu_{\max}$ and $\Delta\nu$ via the following theoretical relation:

$$\frac{R}{\mathrm{R}_\odot} = \left(\frac{\nu_{\max}}{\nu_{\mathrm{ref}}}\right)\left(\frac{\Delta\nu}{\Delta\nu_{\mathrm{ref}}}\right)^{-2}\left(\frac{T_{\mathrm{eff}}}{\mathrm{T}_{\mathrm{eff},\odot}}\right)^{1/2}, \tag{4}$$

with $\nu_{\max}$, the frequency of the maximum height in the oscillation spectrum (μHz) and $\Delta\nu$, the large separation (μHz). The constants $\Delta\nu_{\mathrm{ref}}$ and ν_{ref} correspond to the reference values defined by [7].

Concerning the luminosity L, it can be expressed as a function of the stellar absolute magnitude $\mathcal{M}$ (mag):

$$\frac{L}{L_\odot} = 10^{\left(\frac{\mathcal{M}_\odot - \mathcal{M}}{2.5}\right)}, \tag{5}$$

with the constant $\mathcal{M}_\odot$ corresponding to the solar values.

And given that $\mathcal{M}$ is dependant of the apparent magnitude m (mag), the distance d (pc) and the extinction A_m (mag), as:

$$\mathcal{M} = m - 5\log(d) + 5 - A_\mathrm{m}\,. \tag{6}$$

One can finally deduce the seismic effective temperature as a function of A_m, m, d, ν_max and $\Delta\nu$, as:

$$\frac{T_\mathrm{eff}}{T_{\mathrm{eff},\odot}} = 10^{\frac{(\mathcal{M}_\odot - 5 + A_\mathrm{m} - m)}{12.5}} d^{2/5} \left(\frac{\Delta\nu}{\Delta\nu_\mathrm{ref}}\right)^{4/5} \left(\frac{\nu_\mathrm{max}}{\nu_\mathrm{ref}}\right)^{-2/5}, \tag{7}$$

with $T_{\mathrm{eff},\odot} = 5777$ K, $\mathcal{M}_\odot = 4.8$ mag, $\Delta\nu_\mathrm{ref} = 138.8\,\mu$Hz, $\nu_\mathrm{ref} = 3104\,\mu$Hz.

Furthermore, the stellar parallax p being small, one can approximate the distance d by $1/p$. Thus, the seismic T_eff can also be expressed with the parallax p (arcsec):

$$\frac{T_\mathrm{eff}}{T_{\mathrm{eff},\odot}} = 10^{\frac{(\mathcal{M}_\odot - 5 + A_\mathrm{m} - m)}{12.5}} p^{-2/5} \left(\frac{\Delta\nu}{\Delta\nu_\mathrm{ref}}\right)^{4/5} \left(\frac{\nu_\mathrm{max}}{\nu_\mathrm{ref}}\right)^{-2/5}. \tag{8}$$

The associated relative error of this seismic T_eff is then:

$$\frac{\delta T_\mathrm{eff}}{T_\mathrm{eff}} = \sqrt{\left(\frac{\ln 10}{12.5}\delta A_\mathrm{m}\right)^2 + \left(\frac{\ln 10}{12.5}\delta m\right)^2 + \left(\frac{2}{5}\frac{\delta p}{p}\right)^2 + \left(\frac{2}{5}\frac{\delta\nu_\mathrm{max}}{\nu_\mathrm{max}}\right)^2 + \left(\frac{4}{5}\frac{\delta\Delta\nu}{\Delta\nu}\right)^2}. \tag{9}$$

Note that the bolometric correction have to be taken into account in the calculation of the seismic effective temperature.

References

[1] B. Mosser, K. Belkacem, M. Goupil, E. Michel, Y. Elsworth, C. Barban, T. Kallinger, S. Hekker, J. De Ridder, R. Samadi et al., aa **525**, L9 (2011), 1011.1928

[2] R. de Assis Peralta, R. Samadi, E. Michel, submitted to mnras (2016)

[3] B. Mosser, Y. Elsworth, S. Hekker, D. Huber, T. Kallinger, S. Mathur, K. Belkacem, M.J. Goupil, R. Samadi, C. Barban et al., aa **537**, A30 (2012), 1110.0980

[4] B. Mosser, R. Samadi, K. Belkacem, *Red giants seismology*, in *SF2A-2013: Proceedings of the Annual meeting of the French Society of Astronomy and Astrophysics*, edited by L. Cambresy, F. Martins, E. Nuss, A. Palacios (2013), pp. 25–36, 1310.4748

[5] S. Mathur, D. Huber, N.M. Batalha, D.R. Ciardi, F.A. Bastien, A. Bieryla, L.A. Buchhave, W.D. Cochran, M. Endl, G.A. Esquerdo et al., ArXiv e-prints (2016), 1609.04128

[6] T. Kallinger, W.W. Weiss, C. Barban, F. Baudin, C. Cameron, F. Carrier, J. De Ridder, M.J. Goupil, M. Gruberbauer, A. Hatzes et al., aa **509**, A77 (2010)

[7] B. Mosser, E. Michel, K. Belkacem, M.J. Goupil, A. Baglin, C. Barban, J. Provost, R. Samadi, M. Auvergne, C. Catala, aa **550**, A126 (2013), 1212.1687

DOI: 10.1051/epjconf/201716001013

A new spectroscopic calibration to determine $T_{\rm eff}$ and [Fe/H] of FGK dwarfs and giants

G. D. C. Teixeira[1,2,⋆], *S. G.* Sousa[1], *M.* Tsantaki[1,3], *M. J. P. F. G.* Monteiro[1,2], *N. C.* Santos[1,2], and *G.* Israelian[4,5]

[1] *Instituto de Astrofísica e Ciências do Espaço, Universidade do Porto, CAUP, Rua das Estrelas, 4150-762 Porto, Portugal*
[2] *Departamento de Física e Astronomia, Faculdade de Ciências, Universidade do Porto, Rua do Campo Alegre, 4169-007 Porto, Portugal*
[3] *Instituto de Radioastronomía y Astrofísica, IRyA, UNAM, Campus Morelia, A.P. 3-72, C.P. J8089 Michoacán, Mexico*
[4] *Instituto de Astrofísica de Canarias, 38200 La Laguna, Tenerife, Spain*
[5] *Departamento de Astrofísica, Universidad de La Laguna, 38206 La Laguna, Tenerife, Spain*

Abstract. We present a new spectroscopic calibration for a fast estimate of $T_{\rm eff}$ and [Fe/H] for FGK dwarfs and GK giant stars. We used spectra from a joint sample of 708 stars, composed by 451 FGK dwarfs and 257 GK-giant stars with homogeneously determined spectroscopic stellar parameters. We have derived 322 EW line-ratios and 100 FeI lines that can be used to compute $T_{\rm eff}$ and [Fe/H], respectively. We show that these calibrations are effective for FGK dwarfs and GK-giant stars in the following ranges: $4500\,K < T_{\rm eff} < 6500\,K$, $2.5 < \log g < 4.9\ dex$, and $-0.8 < [Fe/H] < 0.5\ dex$. The new calibration has a standard deviation of 74 K for $T_{\rm eff}$ and 0.07 dex for [Fe/H]. We use four independent samples of stars to test and verify the new calibration, a sample of giant stars, a sample composed of Gaia FGK benchmark stars, a sample of GK-giant stars from the DR1 of the Gaia-ESO survey, and a sample of FGK-dwarf stars. We present a new computer code, GeTCal, for automatically producing new calibration files based on any new sample of stars.

1 Introduction

Deriving accurate and precise stellar parameters is a fundamental aspect of astrophysical studies. In order to determine the mass, radius, and age of stars it is necessary to measure stellar atmospheric parameters, such as effective temperature ($T_{\rm eff}$), surface gravity ($\log g$), metallicity ([Fe/H]), and microturbulence(v_{mic}), obtained mainly by spectroscopic or photometric methods (e.g. Casagrande et al. [3], Sousa et al. [12], Tsantaki et al. [20]).

The volume of data being produced by large-survey observational programs [2, 4] requires the existence of quick methods to obtain stellar parameters for a diverse range of stellar spectral types.

Methods such as line-strength ratios can be used to obtain $T_{\rm eff}$ [5, 6, 9], and equivalent widths (EWs) of FeI can be used to derive [Fe/H] [15]. The success of these methods is tied to the quality of the empirical calibrations.

One of the spectroscopic methods that can be used to determine spectral parameters is the ARES+MOOG method. This method makes use of the ARES code (to measure the EWs of a spectrum [11, 17]), and the MOOG code (to measure individual line abundances) combined with a minimization algorithm in order to derive the parameters of the stellar atmosphere [18]. For a more detailed description of this method see Sousa [16].

⋆e-mail: guilherme.teixeira@astro.up.pt

An automated tool, TMCalc, can be used to obtain $T_{\rm eff}$ and [Fe/H] using measurements of EWs of spectral lines [15]. It relies on EW line-ratios and on the EWs of FeI lines. The accuracy and precision of this TMCalc are mainly limited by the $T_{\rm eff}$ and [Fe/H] calibrations [15]. TMCalc is less accurate than the ARES+MOOG method but it is computationally more efficient.

The main body of this work was presented in detail in Teixeira et al. [19].

2 Stellar samples

We used six distinct stellar samples: two samples were used for calibration and four other samples were used for independent testing of the new calibration. A summary of each sample can be found in Table 1.

The parameters for each sample have been consistently determined homogeneously with ARES+MOOG. The samples used were:

- 451 FGK-dwarf stars described in Sousa et al. [12] that was revised in Tsantaki et al. [20], hereafter the So08 sample.
- 257 giant stars from Alves et al. [1], hereafter the Al15 sample.
- 44 giant stars from Santos et al. [10], hereafter the Sa09 sample.

Table 1: Summary of the stellar samples.

Sample	stars	stars_used	(S/N)	Spec type	$T_{\rm eff}$ (*K*)	$\log g$ (*dex*)	v_{mic} (kms^{-1})	[Fe/H] (*dex*)
So08	451	451	$[70, 2000]$	FGK	$[4400, 6431]$	$[3.60, 4.82]$	$[0, 2.1]$	$[-0.83, 0.36]$
Al15	257	257	~ 150	GK	$[4724, 5766]$	$[2.37, 3.92]$	$[1.08, 4.28]$	$[-0.75, 0.27]$
Sa09	56	44	~ 200	GK	$[4157, 6020]$	$[1.15, 4.82]$	$[0.86, 2.26]$	$[0, 0.32]$
Gaia	34	18	$[200, 400]$	FGK	$[3472, 6635]$	$[0.51, 4.67]$	$[0.89, 1.92]$	$[-2.64, 0.35]$
GES	36	36	~ 100	GK	$[4753, 5289]$	$[2.50, 3.86]$	$[1.17, 4.06]$	$[-0.51, 0.25]$
So11	582	582	$[100, 200]$	FGK	$[4487, 7212]$	$[3.61, 4.96]$	$[0, 2.87]$	$[-1.14, 0.55]$

Table 2: Limits of applicability of the Te16 calibration.

$T_{\rm eff}$ (K)	$\log g$ (*dex*)	[Fe/H] (*dex*)
$[4500, 6500]$	$[2.5, 4.9]$	$[-0.8, 0.5]$

- 18 benchmark stars of the Gaia survey, hereafter the Gaia sample [7, 8].
- 36 GK-giant stars from the Gaia-ESO survey DR1 [1] with $\log g < 3.9$, hereafter the GES sample.
- 582 FGK-dwarf stars from Sousa et al. [14] with with well-determined parameters, hereafter the So11 sample.

For calibration, we used a combination of the So08 and Al15 samples, hereafter the joint sample. The other sample were used to test our new calibration.

3 Calibration

This work aimed to build new $T_{\rm eff}$ and [Fe/H] spectroscopic calibrations for both FGK dwarfs and GK giants improving upon the work of Sousa et al. [15], hereafter the So12 calibration. The rationale for each step of the calibration procedure can be found in Teixeira et al. [19].

3.1 $T_{\rm eff}$ calibration

In order to obtain the $T_{\rm eff}$ calibration we used the established relations between EW line-ratios, R_EW[13]. This technique is based on the different sensitivity that metal lines have to $T_{\rm eff}$. EW line-ratios are more precise than the use of EWs of individual lines [5, 6, 9, 13].

To compute the $T_{\rm eff}$ calibration we:

- Selected lines with a difference in excitation potential greater than 3 eV.
- Selected only lines in close proximity, ($\Delta\lambda < 70$Å).
- Discarded ratios when $R\_EW \notin [0.01, 100]$.
- Fitted a linear and a third-degree polynomial function to the distributions of $T_{\rm eff}$ as a function of: a) R_EW; b) $1/R\_EW$; and c) $\log R\_EW$.
- Performed a cut-off based on the interquartile range method, IQR [21].
- Used a 2-σ cut on the distributions, reffited the functions and selected the fit with the smaller standard deviation.
- Only accepted a function that could fit 2/3 of the calibration sample.

3.2 Metallicity calibration

We calibrated the metallicity using only iron absorption lines, as iron abundance is a proxy for stellar metallicity [15]. We discarded ionized iron lines given their dependence on $\log g$.

We obtained the metallicity calibration by:

- Considering only lines with EW < 70 mÅ.
- Excluded lines with EW < 20 mÅ.
- Solved equation:

$$\begin{aligned} EW &= C_0 + C_1 \times [Fe/H] + C_2 \times T_{\rm eff} + \\ & C_3 \times [Fe/H]^2 + C_4 \times {T_{\rm eff}}^2 + \\ & C_5 \times [Fe/H] \times T_{\rm eff} \ . \end{aligned} \quad (1)$$

- Applied the IQR and a 2-σ to a linear fit made with the inversion of the previous equation.
- Made sure that the slope of the comparison between the calibrated and spectroscopic [Fe/H] was within 3% of the identity line.
- Only considered lines if the function had $\sigma < 0.06 dex$.

3.3 GeTCal

As a by-product of our work we created a Python code: GeTCal. This code is a pratical implementation of the methods described in Sects. 3.1 and 3.2 and is able to automatically produce $T_{\rm eff}$ and [Fe/H] calibrations. As input it requires three parameters: a line list, the stellar parameters (and errors) of a sample of stars, and the measured EWs for each star. The calibrations produced by GeTCal can then be used to compute the $T_{\rm eff}$ and [Fe/H] for stars of spectral classes similar to the calibration sample. The code can perform the $T_{\rm eff}$ calibration, the [Fe/H] calibration, or both simultaneously.

GeTCal is built in such a way as to produce output calibration files which are compatible with the TMCalc

[1] https://www.gaia-eso.eu/sites/default/files/file_attach/ESO-DR1-release-description.pdf

Table 3: Application of the Te16 calibration to the samples.

Sample	$\Delta T_{\rm eff}$ (K)	$\Delta T_{\rm eff}$ _median (K)	$\Delta T_{\rm eff}$ _σ (K)	Δ[Fe/H] (*dex*)	Δ[Fe/H]_median (*dex*)	Δ[Fe/H]_σ (*dex*)
So08	17	23	76	0.01	0.02	0.04
Al15	−32	−36	58	−0.07	−0.06	0.08
Joint	−1	−8	74	−0.02	0.00	0.07
GES	20	38	141	0.05	0.06	0.11
Sa09	−44	−41	75	−0.04	−0.06	0.10
Gaia	43	40	91	0.08	0.09	0.18
So11	47	46	89	0.02	0.02	0.05
Combined-validation	40	37	93	0.02	0.02	0.07

code [15]. This code is freely distributed and available for use by the community [2].

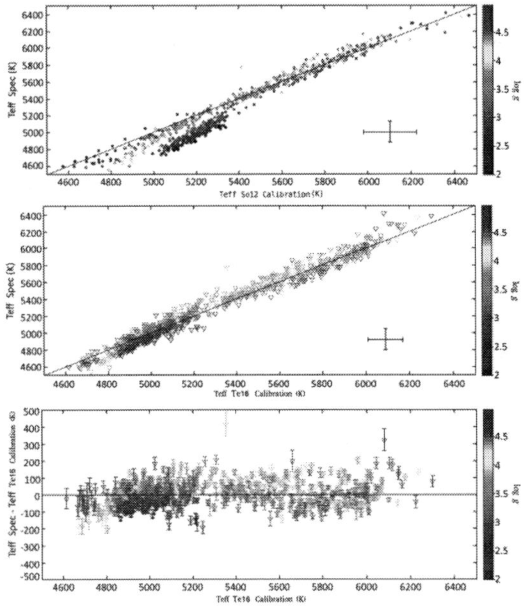

Figure 1: Comparison between the $T_{\rm eff}$ computed in this work and the spectroscopic values with the So12 calibration (top panel) and the Te16 calibration (middle panel) for the joint sample. The line in the two plots represents the identity line, the standard deviation is plotted as the cross in both panels. In the bottom panel we show the difference between the Te16 calibration and the spectroscopic $T_{\rm eff}$ as a function of $T_{\rm eff}$, the error bars represent the errors in our computation. The plots are colour-coded for $\log g$. Figure from [19].

4 Results

Figure 1 compares the computed $T_{\rm eff}$ with the spectroscopic $T_{\rm eff}$ using the different calibrations. The top panel of Fig. 1 shows the previous calibration, the middle panel shows the new calibration and the bottom panel shows the difference between the new calibration and the spectroscopic values. The Te16 calibration improves in the low $T_{\rm eff}$ and low $\log g$ regimes, i.e., the giants.

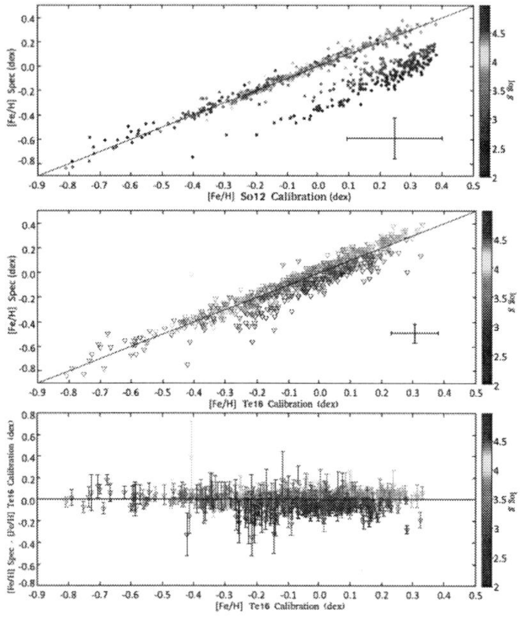

Figure 2: Comparison between the [Fe/H] computed in this work and the spectroscopic values for the So12 calibration (top panel) and the Te16 calibration (middle panel) for the joint sample. The black line represents the identity line, the standard deviation is shown as the cross. The bottom panel shows the difference between [Fe/H] from spectroscopy and the one in this work as a function of [Fe/H], the error bars are the errors resulting from this work. The plots are colour-coded for $\log g$. Figure from [19].

Figure 2 shows the same type of analysis but now for the computed [Fe/H]. The bottom panel shows most stars well within 0.2 dex from the zero value.

As shown in Table 3, for the joint sample, the $T_{\rm eff}$ standard deviation is 74 K, and the standard deviation of [Fe/H] is 0.07 dex for the Te16 calibrations.

The limits of the Te16 calibration reflect the parameters of the calibration sample and are presented in Table 2.

[2]The GeTCal code and the $T_{\rm eff}$ and [Fe/H] calibrations are available at http://www.astro.up.pt/exoearths/tools.html.

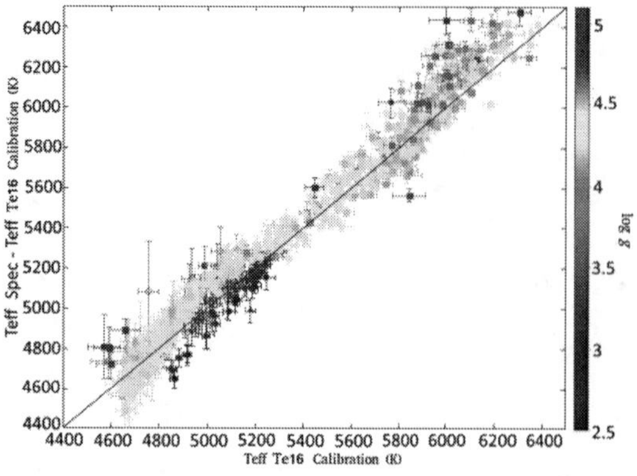

Figure 3: Comparison between the $T_{\rm eff}$ computed with TMCalc and the spectroscopic values with the Te16 calibration for the Gaia (closed triangles), Sa09 (circles), GES (open diamonds), and So11 (closed squares) sample. The colour-code represents the $\log g$ values and the black line represents the identity line. Only stars within the limits of applicability are plotted.

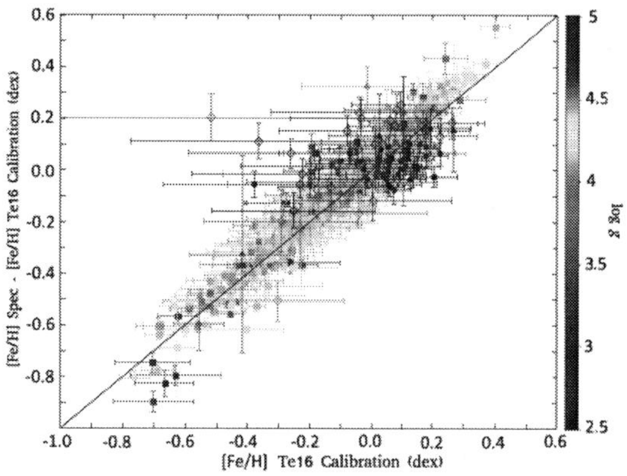

Figure 4: Comparison between the [Fe/H] computed with TMCalc and the spectroscopic values with the Te16 calibration for Gaia closed triangles), Sa09 (circles), GES (open diamonds), and So11 (closed squares) sample. The colour-code represents the $\log g$ values and the black line represents the identity line. Only stars within the limits of applicability are plotted.

Higher uncertainties will result from using the Te16 calibration outside the applicability limits [19].

After obtaining the Te16 calibration for $T_{\rm eff}$ and [Fe/H], we applied it to the four independent samples. Figure 3 shows the comparison of $T_{\rm eff}$ results between the Te16 calibration and the values from spectroscopy for the validation samples. Stars outside the limits of the Te16 calibration ($\log g < 2.5$) are not plotted.

The results for the [Fe/H] computation of the independent samples are shown in Fig. 4. The outliers are stars with low-$\log g$ values and $T_{\rm eff}$ and, therefore, very close to our applicability limits (see Table 2).

A summary of the application of the Te16 calibration to the various validation samples is provided in Table 3.

5 Conclusions

We presented new calibrations to obtain $T_{\rm eff}$ and [Fe/H] from the EWs of stellar spectra. These calibrations cover both FGK dwarfs and GK giants simultaneously [19]. The Te16 calibration is effective within the range of $4500K < T_{\rm eff} < 6500K$, $2.5 < \log g < 4.9$ dex, and $-0.8 < [Fe/H] < 0.5$ dex.

We built a Python code, GeTCal, capable of computing $T_{\rm eff}$ and [Fe/H] calibrations for any given sample of calibration stars. This program produces calibration files compatible with the existing TMCalc code.

This work provides a fast way to determine stellar atmospheric parameters from spectrographic observations of FGK-dwarf and GK-giant stars.

Acknowledgments

G.D.C.T. was supported by a research fellowship (CAUP2013-05UnI-BI), funded by the European Commission through project SPACEINN (FP7-SPACE-2012-312844) and by an FCT/Portugal PhD grant PD/BD/113478/2015. Part of this work was supported by FCT/Portugal (UID/FIS/04434/2013) and COMPETE2020 (POCI-01-0145-FEDER-007672).

References

[1] Alves, S., Benamati, L., Santos, N. C., et al. MNRAS, **448**, 2749 (2015)
[2] Bland-Hawthorn, J., Sharma, S., & Freeman, K. EAS Publications Series, **67**, 219 (2014)
[3] Casagrande, L., Ramírez, I., Meléndez, J., Bessell, M., Asplund, M., A&A, **512**, A54 (2010)
[4] Gilmore, G., Randich, S., Asplund, M., et al. , The Messenger, **147**, 25 (2012)
[5] Gray, D. F. Stellar Surface Structure, **176**, 227 (1996)
[6] Gray, D. F., *The Observation and Analysis of Stellar Photospheres*, 3rd Edition (ISBN 0521851866. , UK: Cambridge University Press, 2005)
[7] Heiter, U., Jofré, P., Gustafsson, B., et al. , A&A, **582**, A49 (2015)
[8] Jofré, P., Heiter, U., Soubiran, C., Blanco-Cuaresma, S. et al., A&A, **564**, A133 (2014)
[9] Kovtyukh , V. V., Soubiran, C., Belik, S. I., Gorlova, N. I., A&A, **411**, 559 (2003)
[10] Santos, N. C., Lovis, C., Pace, G., Melendez, J., Naef, D., A&A, **493**, 309 (2009)
[11] Sousa, S. G., Santos, N. C., Israelian, G., Mayor, M., & Monteiro, M. J. P. F. G., A&A, **469**, 783 (2007)
[12] Sousa, S. G., Santos, N. C., Mayor, M., et al., A&A, **487**, 373 (2008)
[13] Sousa, S. G., Alapini, A., Israelian, G., & Santos, N. C., A&A, **512**, A13 (2010)
[14] Sousa, S. G., Santos, N. C., Israelian, G., Mayor, M., & Udry, S. A&A, **533**, A141 (2011)
[15] Sousa, S. G., Santos, N. C., & Israelian, G., A&A, **544**, A122 (2012)

DOI: 10.1051/epjconf/201716001013

[16] Sousa, S., arXiv:1407.5817 (2014)

[17] Sousa, S. G., Santos, N. C., Adibekyan, V., Delgado-Mena, E., & Israelian, G., A&A, **577**, A67 (2015)

[18] Sneden, C. A. Ph.D. Thesis, The University of Texas at Austin (1973)

[19] Teixeira, G. D. C., Sousa, S. G., Tsantaki, M., et al. 2016, A&A, **595**, A15

[20] Tsantaki, M., Sousa, S. G., Adibekyan, V. Z., et al., A&A, **555**, A150 (2013)

[21] Upton, G., Cook, I., *Understanding Statistics* (Oxford University Press., 1996)

EPJ Web of Conferences **160**, 01014 (2017) DOI: 10.1051/epjconf/201716001014
Seismology of the Sun and the Distant Stars II

The SpaceInn SISMA archive

Monica Rainer[1,⋆], *Ennio* Poretti[1,⋆⋆], *Angelo* Mistò[1], and *Maria Rosa* Panzera[1]

[1] *INAF-Osservatorio Astronomico di Brera, via E. Bianchi 46, 23807 Merate, Italy*

Abstract. The *Spectroscopic Indicators in a SeisMic Archive* (SISMA) has been built in the framework of the FP7 SpaceInn project to contain the 7013 HARPS spectra observed during the CoRoT asteroseismic ground-based program, along with their variability and asteroseismic indicators. The spectra pertain to 261 stars spread around the whole Herztsprung-Russell diagram: 72 of them were CoRoT targets while the others were observed in order to better characterize their variability classes. The Legacy Data lightcurves of the CoRoT targets are also stored in the archive.

1 Introduction

A large number of high-resolution spectra have been taken in a six-year long ground-based observational campaign to complement the asteroseismic observations of the CoRoT satellite [1]. The main instrument used in this work was the HARPS spectrograph at the ESO-LaSilla Observatory [2], with which we collected 7103 good quality, high signal-to-noise ratio (S/N) spectra of 261 stars during the two Large Programmes LP 182.D-0356 and LP 185.D-0056, spanning 15 nights per semester from December 2008 to January 2013, for a total of 135 allocated nights. Seventy-two of the stars were CoRoT asteroseismic targets. Many other stars were observed to better define the variable classes of the CoRoT targets, in general by means of few spectra per star only. The coordination of the spectroscopic follow-up of the asteroseismic targets characterized the Italian participation to the scientific exploitation of the CoRoT mission [3], after the preparatory activities to select targets and fields [4, 5].

When the FP7-SPACE project *SpaceInn: Exploitation of Space Data for Innovative Helio-and Asteroseismology*[1] was set-up we took charge to create an online archive where not only the raw HARPS spectra could be found, but also the reduced ones. The goal was to make the HARPS spectra, in many cases already used in the first-analysis papers, directly exploitable for any further scientific study. To match this requirement we decided to compute specific spectral indicators, determine stellar parameters, and offer the possibility to compare straightly the spectroscopic timeseries with CoRoT ones. At the end of the *SpaceInn* project, we are proud to shortly present here the *Spectroscopic Indicators in a SeisMic Archive* (SISMA) database.

⋆e-mail: monica.rainer@brera.inaf.it
⋆⋆e-mail: ennio.poretti@brera.inaf.it

[1] http://www.spaceinn.eu/

2 Observations and reduction

We used HARPS mostly in the high-efficiency mode EGGS, with resolving power R=80,000, because most of our targets are hot stars (Fig. 1) with large $v \sin i$ values. As such, these targets are not suitable for the more precise radial velocities studies carried on using the high-accuracy mode HAM (R=115,000): using the EGGS mode, we were able to reduce the exposure times, and to increase the signal-to-noise ratios, without losing precision on the radial velocity estimation, which is limited by the rotational broadening of the lines and not by the instrument resolving power. Only the cooler, slowly rotating targets were observed in the HAM mode.

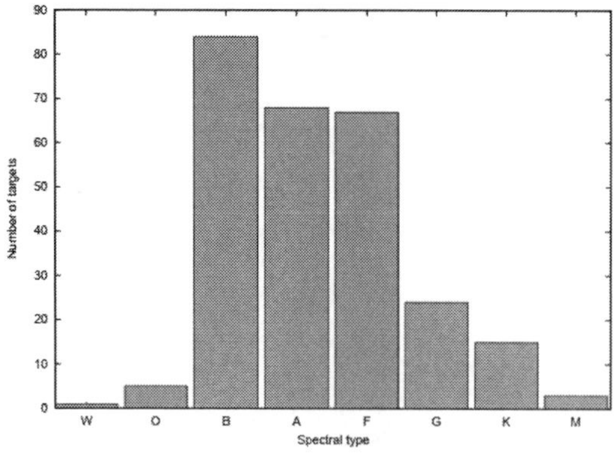

Figure 1: Histogram of the spectral types of the stars in the SISMA database.

Most of the spectra observed with EGGS have S/N around 200 at about 5800 Å, while the HAM spectra usually have a S/N around 150 in the same region (Fig. 2), with the exception of HD 46375, whose 1160 spectra cluster around 50, high enough to detect the solar-like oscillations in the radial velocity time series. The high number of spectra on a single target were acquired as a pilot study to support one of the scientific goals of the CoRoT second extension, i.e., to have a full asteroseismic picture of the stars hosting exoplanets. Though granted, this second extension did not take place due to fatal satellite failure in November 2012.

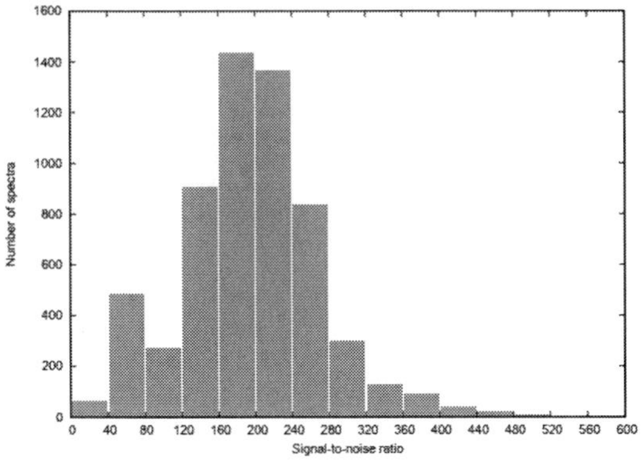

Figure 2: Histogram of the S/N values of the spectra in the SISMA database. The 1160 spectra of HD 46375 have been removed for easier reading.

The online ESO pipeline gave us as final outputs one-dimensional spectra with the echelle orders merged, which can be a problem when normalizing the whole spectrum because of the distorted continuum. In addition to that, we lose information on the pixel-by-pixel S/N and on the positions of the borders of the orders, which are important for detailed spectroscopic analysis: the S/N is very low on the borders of the orders and the merging may distort the spectral lines in these regions, particularly if the lines fall right on the border. In order to improve the quality of our spectra, we reduced them using a semi-automated pipeline developed at the Brera Astronomical Observatory. Our pipeline worked on the *e2ds* files given by the original DRS, i.e. the extracted, flat-field corrected spectra. For each observed spectrum, we obtain two output files: a five column ASCII file with wavelength, flux, normalized flux, S/N and number of the echelle orders, and a two column ASCII file with wavelength and normalized flux, with the echelle orders merged.

2.1 Instrumental problem in the spectra reduction

The intensive use of the HARPS instrument to study line-profile variations was very effective in detecting subtle instrumental effects. These effects could escape at the observer interested primarily at the value of the radial velocity. What happened in June 2009 was very instructive at this purpose. At that time the Be team noticed a strange feature in the HARPS spectra collected in the December 2008-January 2009 runs. The mean profile of the He I line at 4921 Å showed some "oscillations" in the spectra of the Be star HD 51452, both on the line profile and on the continuum. The peak-to-peak amplitude was 0.5% of the continuum. The origin was not stellar since it was detected in the spectra of the other stars observed in the same runs. These oscillations resulted to be strongly enhanced in the spectra obtained later in June 2009 since the amplitude raised to 2%. After interaction with the ESO staff, the problem was identified in the misalignement of a filter on the path of the flat-field lamp. Indeed, the "oscillations" were visible in the raw images of the flat field, but not in those of the stars (or, at least, they are not so evident). The strong misalignement occurred few weeks before, at the end of May 2009. The ESO staff corrected such misalignement and produced new spectra re-reduced using the new flat-field. The spurious oscillation was reduced in strength so as to be cancelled in the mean LSD profile, but still noticeable in the mean profile of a given line. This subtle effect constituted a serious problem for our line-profile studies. Therefore, the Brera team and the ESO staff continued to investigate the problem. In late July 2009, we solved it by removing the filter from the optical path of the calibration lamp. The ESO staff re-ran the complete reduction of the HARPS spectra collected in the previous runs of our Large Programme (i.e., December 2008, January, June and July 2009) and delivered new ones free from the instrumental effect. We re-processed these corrected spectra with our pipeline, and made them available in the SISMA database.

3 Variability indicators

In order to support the asteroseismic exploitation of the data, different kinds of indicators and secondary files have been computed to be provided alongside the spectra.

In particular, we used the LSD software [6] to compute mean line profiles for each spectrum. These profiles are stored in the archive and they were used to compute the radial velocities, the projected rotational velocities $v \sin i$, an indicator of the presence of differential rotation [7], and to amplify the effects of pulsation on the line profile variations. Just a quick look at the standard deviation between the mean line profiles and their average gives interesting information on the pulsational contents of the spectroscopic time series.

In addition to this, an activity index has been computed for each spectrum using the Ca II H and K lines, and the violet-to-red peak intensity ratio of the H_α is given in case of emission line stars.

The physical parameter $T_{\rm eff}$, $\log g$ and [Fe/H] of all the stars have been computed using the SME software [8] in the wavelength region 5160-5190 Å of an average spectrum in order ot lessen the effects of pulsations on the parameters determination.

DOI: 10.1051/epjconf/201716001014

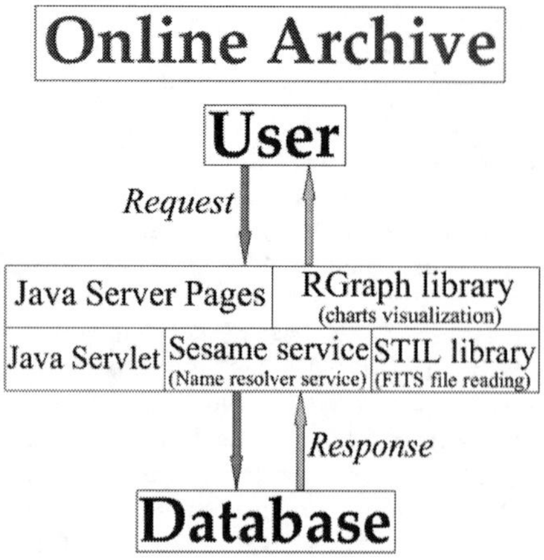

Figure 3: The software used in the interaction between user and database.

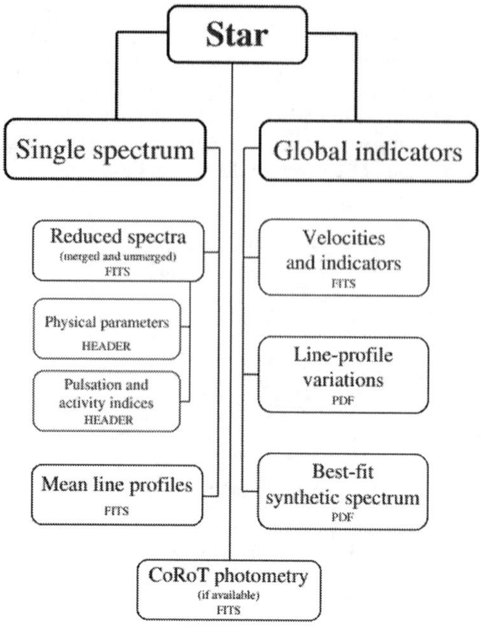

Figure 4: Structure of the HARPS archive, with the content and format of the available files.

4 Structure of the spectroscopic Archive

All the reduced spectra, indicators and lightcurves are stored in the online SISMA archive (http://sisma.brera.inaf.it/) and they can also be accessed through the Seismic Plus portal[2]. This portal is another product of the *SpaceInn* project and it was designed with the goal to coordinate the use of different helio- and asteroseismic databases.

4.1 Database infrastructure

The SISMA archive is hosted on a dedicated Linux server at the Brera Astronomical Observatory. The database management system used is the open-source PostgreSQL, while the web user interface is written in the Java/Servlet/JSP language. SISMA uses the name resolver service Sesame[3] in order to allow the user to query the database by target name. Otherwise it is possible to query the archive by variability class. The results can either be downloaded or displayed online using the Javascript library RGraph[4] in order to quickly create and plot inside the user browser all the required data (spectra, lightcurves or indicators). The archived FITS files are read using STIL (Starlink Tables Infrastructure Library) [5]. A schema of the software used in the interaction between the database and the user is shown in Fig. 3.

4.2 Scientific data

The SISMA archive contains several different kinds of data. For each of the observed 261 stars it is possible to find both single observations and time series and global data, as:

- the reduced spectra, which are the backbone of the archive. They are available in two format: unmerged with both normalized and non-normalized flux, and merged with normalized flux;
- the mean line profiles obtained by means of the LSD software;
- the time series of several spectroscopic indicators and quantities, i.e. the activity index, Hα violet-to-red peak-intensity ratio, the differential rotation indicator, the radial velocities and the projected rotational velocities $v \sin i$ with their relative errors;
- the estimated physical parameters $T_{\rm eff}$, $\log g$ and [Fe/H], which are stored in the header of all the FITS files;
- a PDF file with the best-fit synthetic and observed spectrum in the 5160-5190 Å region used to compute the physical parameters;
- a PDF with a quick look of the pulsational content of the mean-line profile time series;
- the CoRoT Legacy Data lightcurves (only for the 72 CoRoT targets stored in the archive).

Figure 4 summarizes how the data have been stored in the archive and how we structured the output of our reduction and analysis. Note that parameters and indices are properly stored in the FITS headers of all the reduced spectra. The graphical interface of the SISMA archive is shown in Fig. 5.

---
[2]http://voparis-spaceinn.obspm.fr/seismic-plus/
[3]http://cdsweb.u-strasbg.fr/doc/sesame.htx
[4]http://www.rgraph.net/
[5]http://www.star.bristol.ac.uk/~mbt/stil/

 DOI: 10.1051/epjconf/201716001014

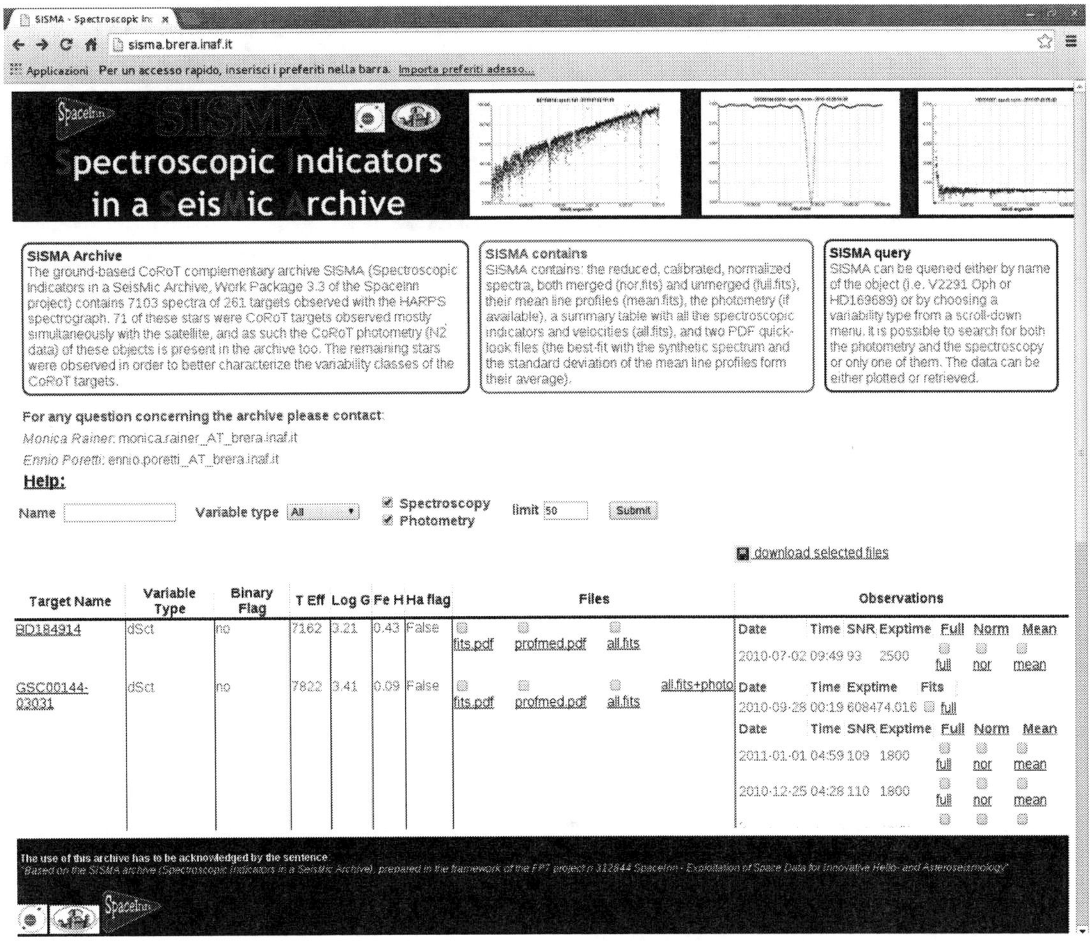

Figure 5: Screenshot of the SISMA web page.

5 Conclusions

The *SpaceInn* project aimed at coordinating several archives of helio- and astero-seismological data. We collected a large number of high-resolution HARPS spectra of the targets observed, often simultaneously, with the CoRoT satellite. Therefore, we considered very appropriate to make them available to the whole community. We strongly encourage more in-depth and statistical studies, since the reduced archival spectra are complemented with many indicators of variability and activity. Moreover, we suggest to look at the SISMA database as an opportunity to test new methods to extract physical information from high quality, high S/N spectra.

References

[1] Annie Baglin et al., AIPC **895**, 201 (2007)
[2] Michel Mayor et al., The Messenger, **114**, 20 (2003)
[3] Monica Rainer et al., AN **333**, 1061 (2012)
[4] Ennio Poretti et al., MemSAIt **78**, 624 (2007)
[5] Ennio Poretti, Francesco Borsa, Monica Rainer, Katrien Uytterhoeven, MemSAIt Suppl. **12**, 98 (2008)
[6] Jean-François Donati, Meir Semel, Brad D. Carter, Daniel E. Reese, Andrew Collier Cameron, MNRAS **291**, 658 (1997)
[7] Ansgar Reiners, Jürgen H.M.M. Schmitt, A&A **384**, 155 (2002)
[8] Jeff A. Valenti, Nikolai Piskunov, A&AS **118**, 595 (1996)

DOI: 10.1051/epjconf/201716001015

SIGS - Seismic Inferences for Glitches in Stars

L. Filipe R. Pereira[1,2], *João P. S.* Faria[1,2], and *Mário J. P. F. G.* Monteiro[1,2]

[1] *Departamento de F'ısica e Astronomia, Faculdade de Ciências da Universidade do Porto, Portugal*
[2] *Instituto de Astrofísica e Ciências do Espaço, Universidade do Porto, Portugal*

Abstract. The increased amount of high precision seismic data for solar-like stars calls for the existence of tools that can extract information from such data. In the case of the study of acoustic glitches there are no publicly available tools and most of the existing ones require a deep knowledge of their implementation. In this work a tool is presented that aims to both simplify the interaction with the user and also be capable of working automatically to determine properties of acoustic glitches from seismic data of solar-like stars. This tool is shown to work with both the Sun and other solar analogs but there are still some limitations to the methods being used when considering smaller frequency datasets.

1 Introduction

Acoustic glitches, such as the base of the convective zone (therein BCZ) and the helium second ionization zone (therein HeII), are regions in the interior of a star where the sound speed suffers an abrupt variation due to the sharp change in the internal structure. Since the eigenfrequencies of a star are directly related to the local sound speed, these glitches cause perturbations in thoem. These perturbations exhibit an oscillatory behavior when considering frequencies of sequential radial order [1–3]. It is thus possible to characterize some properties of these acoustic glitches by isolating and studying their contribution to the eigenfrequencies.

This oscillatory signature from the acoustic glitches has been extensively studied for the Sun to determine the location of the two previosly mentioned acoustic glitches, BCZ and HeII [4, 5] and it has also been applied to solar-like stars [6–9].

However, the existing tools to perform this study are not publicly available and require an intimate knowledge of the implemented methods in order to achieve good results. The work presented here aims to improve upon existing methods in order to make available a numerical procedure that is both more precise than the current methods and at the same time highly user-friendly so that it can be openly distributed to the community. Starting from the work developed by Faria [8], which in turn was adapted from the works of Monteiro et al. [4], Monteiro and Thompson [5] and Mazumdar et al. [9] the method was improved upon and an automatic pipeline was implemented to achieve robust results with minor user intervention.

The presented tool is validated initially by using observational data from the Sun and afterwards it is applied to both stars from the 16 Cygni binary and also to 10 solar-like stars chosen from the sample of Mazumdar et al. [9].

In the next section both methods implemented in the tool are described. Then in section 3 the code is validated using data from the Sun. In section 4 the automatic implementations of the code are described. In section 5 the results obtained for the selected stars are shown and finally in section 6 the results are discussed and some conclusions are drawn.

2 Methods

The code presented in this work is comprised of two similar but independent methods capable of isolating the oscillatory signature of the acoustic glitches and then determining some of their properties.

2.1 Method A – Frequencies

Method A isolates the signature in the frequencies themselves by iteratively removing a smooth component from the frequencies and then fitting the residuals to the free parameters from equation

$$\begin{aligned}\delta\nu \simeq \nu_s &+ \\ &+ A_{\rm BCZ}\left(\frac{\nu_r}{\nu}\right)^2 \cos\left(4\pi\tau_{\rm BCZ}\nu + 2\phi_{\rm BCZ}\right) + \\ &+ A_{\rm HeII}\left(\frac{\nu_r}{\nu}\right) \sin^2\!\left(2\pi\beta_{\rm HeII}\nu\right) \cos\!\left(4\pi\tau_{\rm HeII}\nu + 2\phi_{\rm HeII}\right),\end{aligned} \tag{1}$$

adapted from Faria [8] where $A_{\rm BCZ}$, $\tau_{\rm BCZ}$ and $\phi_{\rm BCZ}$ correspond, respectively, to the amplitude, acoustic depth and phase of the signal from the base of the convective zone. While $A_{\rm HeII}$, $\tau_{\rm HeII}$, $\phi_{\rm HeII}$ and $\beta_{\rm HeII}$ correspond, respectively, to amplitude, acoustic depth, phase and finally, the acoustic width of the signal, in this case due to the helium second ionization zone. The frequency ν_s is the smooth component removed from the frequencies. It is obtained independently for the N_l frequencies of each angular degree l

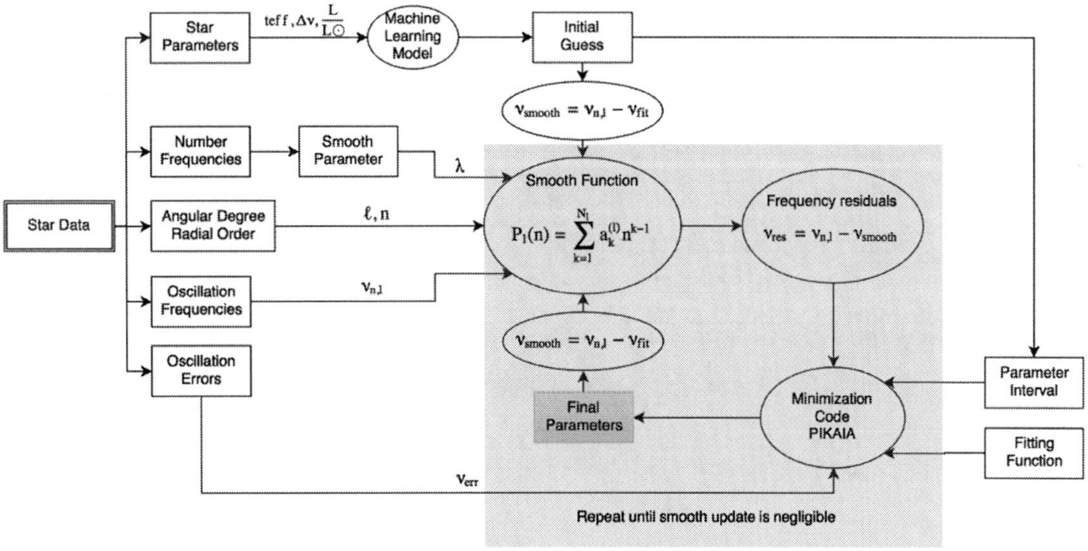

Figure 1: Diagram of the workflow of method A in SIGS, which determines the acoustic depths of the base of the convective zone and helium II ionization zone of a star by extracting the signal caused by these glitches directly from the oscillation frequencies of the star.

by fitting a polynomial of degree $N_l - 1$ with third derivative smoothing [4]. This smoothing can be controlled by defining a smoothing parameter λ, which can be selected by the user. By default, this parameter can be determined automatically in the code.

After removing the smooth function from the frequencies, the residuals are fitted to Equation 1 using the PIKAIA genetic algorithm [10] which is a global minimization algorithm that utilizes concepts inspired by the processes of evolution by natural selection. This method is advantageous compared to hill climbing methods because it doesn't require an initial guess and instead it explores the user-defined range of possible values for each of the free parameters and should, given enough time, converge to global minimum in the parameter space. Minimization is achieved by initiating a population of sets of possible solutions chosen randomly from the interval and then evolving then for a specified number of generations. Through the generations the best sets from the population are chosen to move forward by using a standard chi squared fitness function and after the evolution process the best set of parameters is chosen as the correct one.

The process of executing the smooth removal and fitting the residuals to the equation is repeated until the smooth function is perfectly characterized and convergence is said to be achieved. Figure 1 shows a diagram of the general workflow of this method

2.2 Method B – Second differences

In method B, instead of using the frequencies, the method searches for the signature of the glitches in the second differences [3]. In this method the smooth component is only removed once and the functional form of this component is a polynomial of a degree up to three depending on the amount of data points available for the fit.

After removing the smooth function the procedure is similar to the other method with the exception that there is no iteration on the smooth component and the functional form of the signal is taken to be

$$\begin{aligned} \delta\Delta_2\nu \simeq\ & \Delta 2\nu_s + \\ & + A^{*}_{\rm BCZ}\left(\frac{\nu_r}{\nu}\right)^2 \sin\left(4\pi\tau_{\rm BCZ}\nu + 2\phi_{\rm BCZ}\right) + \\ & + A^{*}_{\rm HeII}\left(\frac{\nu}{\nu_r}\right)\exp\left[-\beta^{*}_{\rm HeII}\left(\frac{\nu_r}{\nu}\right)^2\right]\sin\left(4\pi\tau_{\rm HeII}\nu + 2\phi_{\rm HeII}\right) \end{aligned} \quad (2)$$

which is very similar to the one from the frequencies, with the same seven free parameters, but a different smooth function as mentioned previously. This expression was adapted from the work of Mazumdar et al. [9], which in turn was adapted from the work of Houdek and Gough [7]. Figure 2 shows the workflow of method B.

2.3 Monte Carlo Simulations

To add robustness to both methods and estimate the errors in the measurements of the parameters we perform Monte Carlo simulations. We generate alternative data sets by taking the original frequencies and changing them by a value sampled from a normal distribution with standard deviation equal to the uncertainty of the original frequency. The methods are then used on all the new data sets to obtain a distribution of values for the final parameters.

3 Validation

To validate the implementations of both methods we apply them to both low-degree solar frequencies [11] and then

DOI: 10.1051/epjconf/201716001015

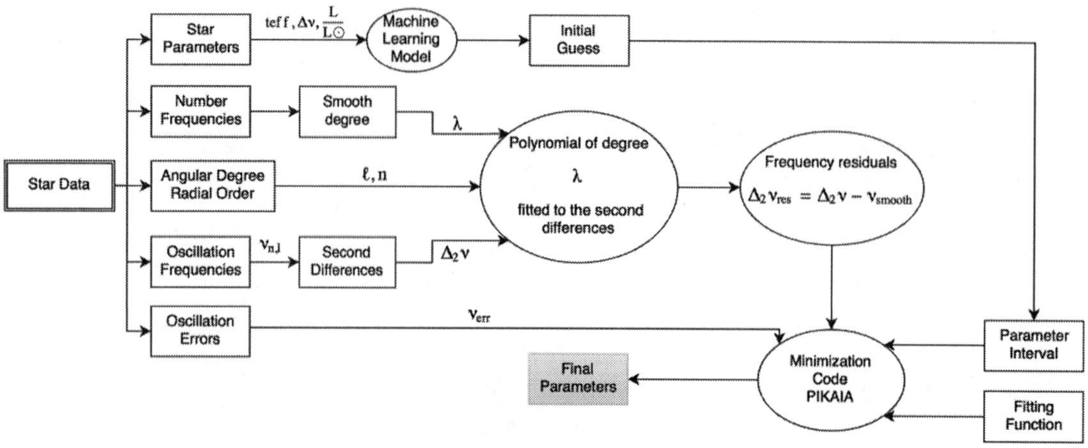

Figure 2: Diagram of the workflow of method B in SIGS, which determines the acoustic depths of the base of the convective zone and helium II ionization zone of a star by extracting the signal caused by these glitches from the second differences of the frequencies of the star.

compare the results with that of previous works. The results obtained for both methods are summarized in Table 1.

Table 1: Comparison of the acoustic location of the Sun's acoustic glitches, determined using observational data.

	This Work		Previous Works	
	$\delta\nu$	$\Delta 2\nu$	$\delta\nu$	$\Delta 2\nu$
$\tau_{\mathrm{BCZ}}(s)$	2282	2314	2337	2273
$\tau_{\mathrm{HeII}}(s)$	685	692	649	707

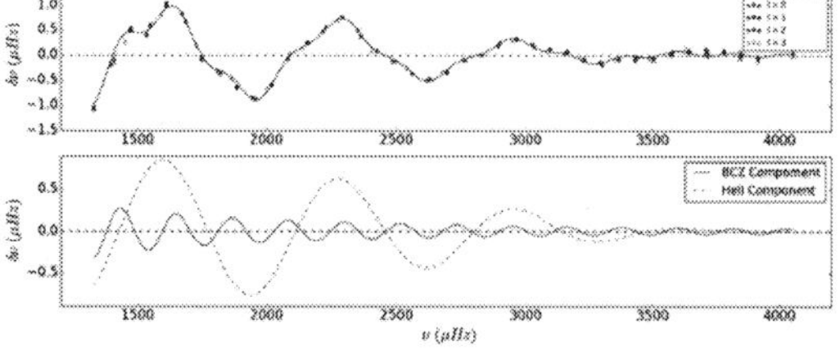

Figure 3: The top panel shows the fit of the residuals from the solar frequencies to Equation 1 using Method A. The color scheme detailed in the legend is used to differentiate between the degrees l of the frequencies. The bottom panel has the signal components from both the BCZ and HeII resulting from the fitting of the upper panel.

Figures 3 and 4 show fits to the best parameters and the oscillatory signature of both glitches for methods A and B respectively. These show how the code is working successfully in isolating the signals. The small difference in values compared to the literature is to be expected, considering the small changes made to the methods implemented and to the more recent data used.

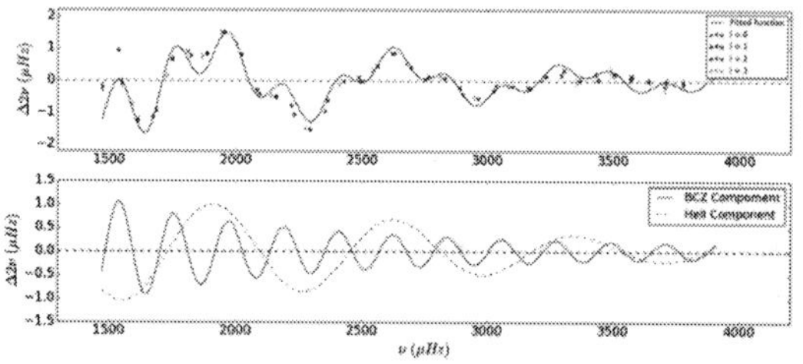

Figure 4: The top panel shows the fit of the second differences from the solar frequencies to Equation 2 using Method B. The color scheme detailed in the legend is used to differentiate between the degrees l of the second differences. The bottom panel has the signal components from both the BCZ and HeII resulting from the fitting of the upper panel.

4 Automatic Pipeline

After validating the code using the Sun, the program was used to determine the parameters of a grid of models and the results were used to build a pipeline that was capable of reducing the parameter interval automatically to improve robustness to the results without requiring user interaction.

A grid [12] of CESAM [13] models was considered with masses between $0.8M_{\odot}$ and $1.2M_{\odot}$ in intervals of $0.04M_{\odot}$ and each mass was considered with 5 different ages (1.5, 3, 4.5, 6 and 8 Gyr). The frequencies of the models were computed using the POSC Code [14]. The results from this run were then used to associate the determined values of the acoustic depths of both glitches to the values of effective temperature (T_{eff}) and large frequency separation (Large_sep) of the stars through a multivariate quadratic regression. The result of the regression is shown in Figure 5. Only the results using method A were shown, since the results for method B are very similar.

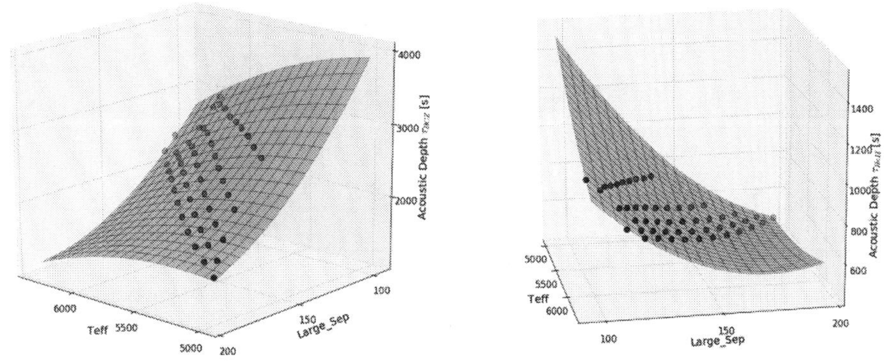

Figure 5: Multivariate regression applied to the values of effective temperature ($T_{\rm eff}$), and large frequency separation (Large_Sep), to the values of acoustic depth of the base of the convective zone (on the left) and the helium second ionization zone (on the right). The blue surface is the quadratic plane fitted.

The results from this regression allowed for the definition of a function that estimates the value of the acoustic depths of both glitches from the $T_{\rm eff}$ and Large_Sep of a star, improving the focus (and convergence) of the fitting procedure.

5 Results

To test the complete tool, which is to be available to the community, the program was applied to the stars of the 16 Cygni binary, with data from Verma et al. [15], and to 10 solar-like stars from the sample of Mazumdar et al. [9] with frequencies from Appourchaux et al. [16]. The errors determined for the results were obtained by executing the program for 500 realizations of the observational data of each star, using the aforementioned Monte Carlo simulations. The results for 16 Cyg A and B are presented in Table 2. Table 3 shows the results for the 10 selected stars.

Table 2: Results obtained in each method for 16 Cyg A (on the top) and 16 Cyg B (on the bottom).

16 Cyg A	Method A	Method B
$\tau_{\rm BCZ}(s)$	2904 ± 89	3038 ± 148
$\tau_{\rm HeII}(s)$	1178 ± 40	982 ± 14

16 Cyg B	Method A	Method B
$\tau_{\rm BCZ}(s)$	2538 ± 263	2452 ± 243
$\tau_{\rm HeII}(s)$	947 ± 31	902 ± 31

6 Discussion and Conclusions

This work presents a Code, to be made publicly available, that implements two alternative methods to measure the signature of acoustic glitches in the low-degree oscillation data of solar-like stars. The code was validated using solar seismic data and tested on an application to a sample of solar type stars with seismic data available from *Kepler* [16].

For the two components of the 16 Cygni stellar binary, the results reported here show precise parameters that are very similar to the ones found in other works [15]. For the 10 selected stars from [15], the results are compatible with the previous measurements in [9], although for some stars the code did not converge with any of the two methods.

The lack of convergence is most likely related to the amount of available data points for some of these 10 stars, and not to the procedures that were implemented, as these work very well for stars with a higher number of data points (either frequencies for method A or second differences for method B). This is a difficulty that has already be found with a the manual implementation of the methods (see [15]), which becomes worse when an automatic procedure is adopted, as done in this work. Indeed, the results reported here confirm that with the automatic pipeline, that reduces the parameter space for the fitting algorithm, the methods are still highly dependent on the amount and quality of the available data points.

The results for the Sun and the stars of the 6 Cygni binary show that *SIGS* is a tool with potential to be used in large amount of data sets, as it achieves the same results as other methods without the need for user intervention and guidance. Some improvements of the code are possible (and some are already ongoing regarding convergence). It will also be quite simple to add additional fitting functions (and number of free parameters) for the signature of the glitches in order to constrain the signal over a wider range of stellar masses and ages. Such a tool may be relevant for

Table 3: Results obtained from both methods for the 10 selected solar-like stars.

Star	Method A		Method B	
	$\tau_{BCZ}(s)$	$\tau_{HeII}(s)$	$\tau_{BCZ}(s)$	$\tau_{HeII}(s)$
KIC008006161	2140 ± 69	613 ± 27	2234 ± 51	—
KIC008379927	—	969 ± 31	—	841 ± 73
KIC008760414	—	—	—	—
KIC006603624	—	1084 ± 52	3066 ± 213	936 ± 50
KIC010454113	—	838 ± 10	2599 ± 90	751 ± 20
KIC006106415	—	—	—	963 ± 81
KIC010963065	2853 ± 101	1006 ± 33	2792 ± 139	1003 ± 73
KIC006116048	—	1166 ± 36	—	1071 ± 81
KIC004914923	3548 ± 37	1114 ± 26	3561 ± 55	—
KIC012009504	—	1139 ± 29	—	1116 ± 64

an efficient exploitation of the seismic data to be obtained by PLATO/ESA [17] for a wide range of main-sequence solar-like stars.

Acknowledgements

Part of this work had the support of the European Commission under the SPACEINN project (FP7-SPACE-2012-312844) and funds from FCT (UID/FIS/04434/2013 and PTDC/FIS-AST/7073/2014).

References

[1] D.O. Gough, M.J. Thompson, *Magnetic Perturbations to Stellar Oscillation Eigenfrequencies*, in *Advances in Helio- and Asteroseismology*, edited by J. Christensen-Dalsgaard, S. Frandsen (1988), Vol. 123 of *IAU Symposium*, p. 155

[2] S.V. Vorontsov, *A Search of the Effects of Magnetic Field in the Solar 5-MINUTE Oscillations*, in *Advances in Helio- and Asteroseismology*, edited by J. Christensen-Dalsgaard, S. Frandsen (1988), Vol. 123 of *IAU Symposium*, p. 151

[3] D.O. Gough, *Comments on Helioseismic Inference*, in *Progress of Seismology of the Sun and Stars*, edited by Y. Osaki, H. Shibahashi (1990), Vol. 367 of *Lecture Notes in Physics, Berlin Springer Verlag*, p. 283

[4] M.J.P.F.G. Monteiro, J. Christensen-Dalsgaard, M.J. Thompson, A&A **283**, 247 (1994)

[5] M.J.P.F.G. Monteiro, M.J. Thompson, MNRAS **361**, 1187 (2005)

[6] M.J.P.F.G. Monteiro, J. Christensen-Dalsgaard, M.J. Thompson, MNRAS **316**, 165 (2000)

[7] G. Houdek, D.O. Gough, MNRAS **375**, 861 (2007)

[8] J.P. Faria, Master's thesis, Masters Degree in Astronomy, Faculty of Sciences – University of Porto (2013), `http://hdl.handle.net/10216/69506`

[9] A. Mazumdar, M.J.P.F.G. Monteiro, J. Ballot, H.M. Antia, S. Basu, G. Houdek, S. Mathur, M.S. Cunha, V. Silva Aguirre, R.A. García et al., ApJ **782**, 18 (2014)

[10] P. Charbonneau, ApJS **101**, 309 (1995)

[11] A.M. Broomhall, W.J. Chaplin, G.R. Davies, Y. Elsworth, S.T. Fletcher, S.J. Hale, B. Miller, R. New, MNRAS **396**, L100 (2009)

[12] J.P. Marques, M.J.P.F.G. Monteiro, J.M. Fernandes, ApSS **316**, 173 (2008)

[13] P. Morel, Y. Lebreton, ApSS **316**, 61 (2008)

[14] M.J.P.F.G. Monteiro, ApSS **316**, 121 (2008)

[15] K. Verma, J.P. Faria, H.M. Antia, S. Basu, A. Mazumdar, M.J.P.F.G. Monteiro, T. Appourchaux, W.J. Chaplin, R.A. García, T.S. Metcalfe, ApJ **790**, 138 (2014)

[16] T. Appourchaux, W.J. Chaplin, R.A. García, M. Gruberbauer, G.A. Verner, H.M. Antia, O. Benomar, T.L. Campante, G.R. Davies, S. Deheuvels et al., A&A **543**, A54 (2012)

[17] H. Rauer, C. Catala, C. Aerts, T. Appourchaux, W. Benz, A. Brandeker, J. Christensen-Dalsgaard, M. Deleuil, L. Gizon, M.J. Goupil et al., Experimental Astronomy **38**, 249 (2014)

SSI: The Stellar Seismic Indices Data Base*

Reza Samadi[1,**], *Raphael* De Assis Peralta[1], *Eric* Michel[1], *Kévin* Belkacem[1], *Christian* Renie[1], and *Mahfoudh* Abed[1]

[1]*LESIA, Observatoire de Paris, CNRS UMR8109, Université Pierre et Marie Curie, Université Paris Diderot*

Abstract. The Stellar Seismic Indices database (hereafter SSI) has been developed within the SPACEInn project funded under the UE Seventh Framework Programme (FP7). This data base aims at providing stellar seismic indices of solar-like oscillating stars as well as other stellar indices such as the main characteristics of stellar granulation. These indices are extracted using an automatic method, which is able to derive simultaneously the seismic and granulation parameters. This method was applied to almost all *Kepler* long-cadence light-curves and to almost all CoRoT targets observed in the faint fields. A total of about 320 000 targets have been analyzed, among which seismic indices and granulation parameters have been extracted for about 18 000 redgiants. In this article, we present the data base content, its interface, as well as some illustrative results.

1 Stellar parameters extraction

The seismic indices and stellar granulation parameters are extracted for red-giant stars by fitting their power density spectrum (PDS) using a semi-empirical model, which is illustrated in Fig. 1. The latter is composed of an activity component (green line), a granulation component (blue line), and an oscillation pattern (red line). More details are provided in de Assis Peralta et al [1]. Subsequently, three seismic indices are extracted from the fitted PDS:

- The peak frequency, ν_{max}: it is defined as the frequency at which the oscillation spectrum maximum peaks in the PDS;
- The mean large separation, $\Delta\nu$: this quantity corresponds to the mean frequency spacing between two consecutive p-modes (with same angular degree). It is derived using the Universal Pattern as proposed by [2];
- The height of the oscillation envelope, H_{env}.

Together with the seismic indices, two parameters characterizing the stellar granulation background are also extracted from the PDS:

- The "e-folding time", τ_{eff}, which corresponds to the time the granulation signal loose its coherence [e.g. 3, 4]. This time is also believed to represent the typical granule lifetime ;
- The variance of the integrated brightness fluctuations, σ^2, which corresponds to the total integrated energy within the granulation background.

*http://ssi.lesia.obspm.fr

**e-mail: reza.samadi@obspm.fr

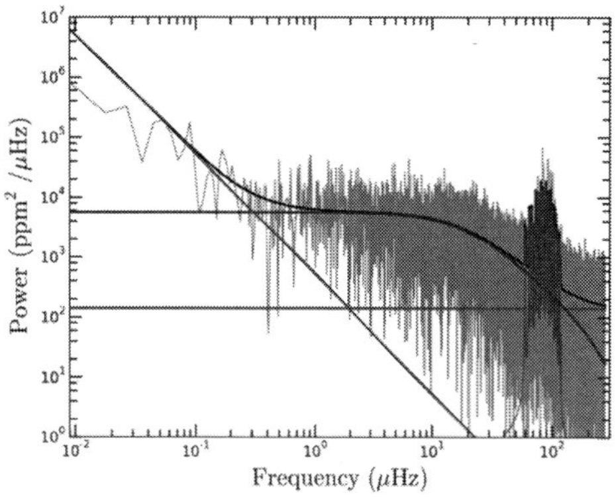

Figure 1: Result of the fit of the power density spectrum (PSD) associated with a *Kepler* red giant target (KIC 5527304).

2 Website and search interface

The SSI database and associated website is located at http://ssi.lesia.obspm.fr. It provides access to the data and a brief description of the database and its content. An User Guide is also available for download.

Data retrieval is possible through the 'Search interface'. It allows the users to perform queries using star ID, star properties (e.g. T_{eff}, log g, spectral type, ...), and finally stellar indices (seismic indices and granulation parameters). A list of star ID can be provided by downloading a file containing these ID. Access to the data is also possible through a VO-compatible web service as well as through the Seismic Plus portal (http://voparis-spaceinn.obspm.fr/seismic-plus/).

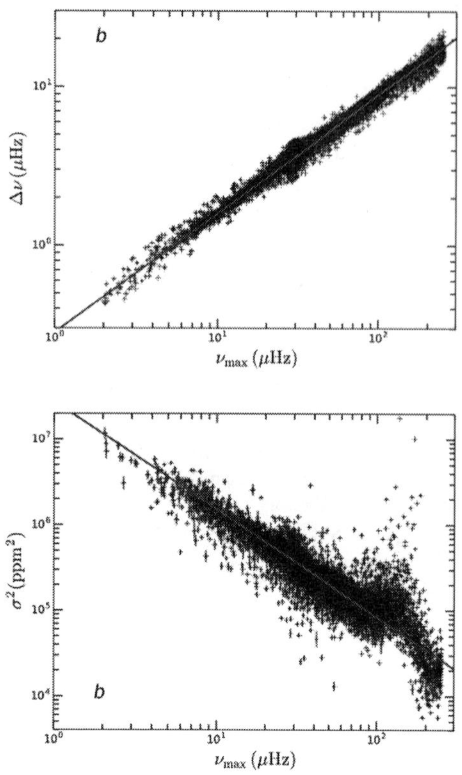

Figure 2: Top: $\Delta\nu$ as a function of $\nu_{\rm max}$ for about 13,000 *Kepler* red giants. **Bottom:** σ^2 as a function of $\nu_{\rm max}$.

3 Query results and some illustrations

The results of the request are displayed in the browser through a paginated data table (see the illustration of Fig. 3). They can be filtered and sorted by any parameters. All parameters are by default not displayed but by clicking on "Show/hide columns", it is possible to choose the parameters to be displayed. The results are displayed for a limited number of stars. The complete set can, however, be downloaded on demand in a CSV file.

The SSI pipeline was applied to about 114,000 stars observed by CoRoT in the exo- planet field and 210,000 stars observed during up to 4 years by *Kepler*. The corresponding light-curve were selected on the basis of their total duration (longer than 50 days) only, and accordingly without any prior concerning the evolutionary status of the targets. Stellar indices (seismic and granulation parameters) were then extracted for a total number of 18,000 red giants. Among the Kepler stars for which seismic indices were extracted, about 5,000 of them were not identified in the Kepler Input Catalog (KIC) as red giant. Some illustrative results are presented in Fig. 2.

The seismic indices $\Delta\nu$ and $\nu_{\rm max}$ obey characteristic scaling relations that depend directly on the radius, mass, and effective temperature of the star (for a review see e.g. [5]). From the knowledge of the effective temperature and these two seismic indices, it is then possible to estimate the mass and radius of the star. Note that a tool to derive stellar masses and radii from the seismic indices and effective temperatures is implemented in the Seismic Plus portal.

Concerning the granulation parameters, they also obey characteristic scaling relations that can provide informations about the stars in complement to the seismic indices [see e.g. 3, 4, 6, 7].

Acknowledgements

The Stellar Seismic Indices Data Base has been developed in the framework of the SPACEInn project (Exploitation of Space Data for Innovative Helio- and Asteroseismology), initiated by the European Helio- and Asteroseismology Network (HELAS) and financed by the European Union under the Seventh Framework Programme (FP7 project n. 312844). We acknowledge the Paris Data Centre for their financial and technical support. We thank S. Cnudde for the conception of the SSI logo and F. Henry for technical supports.
The CoRoT space mission has been developed and operated by CNES, with contributions from Austria, Belgium, Brazil, ESA (RSSD and Science Program), Germany, and Spain. Funding for the *Kepler* Discovery mission were provided by NASA's Science Mission Directorate.

References

[1] R.A. de Assis Peralta, R. Samadi, E. Michel, MNRAS, (submitted)
[2] B. Mosser, K. Belkacem, M.J. Goupil, E. Michel, Y. Elsworth, C. Barban, T. Kallinger, S. Hekker, J. De Ridder, R. Samadi et al., A&A, **525**, L9 (2011), 1011.1928
[3] S. Mathur, S. Hekker, R. Trampedach, J. Ballot, T. Kallinger, D. Buzasi, R.A. García, D. Huber, A. Jiménez, B. Mosser et al., ApJ, **741**, 119 (2011), 1109.1194
[4] T. Kallinger, J. De Ridder, S. Hekker, S. Mathur, B. Mosser, M. Gruberbauer, R.A. García, C. Karoff, J. Ballot, A&A, **570**, A41 (2014), 1408.0817
[5] K. Belkacem, *Determination of the stars fundamental parameters using seismic scaling relations*, in *SF2A-2012: Proceedings of the Annual meeting of the French Society of Astronomy and Astrophysics*, edited by S. Boissier, P. de Laverny, N. Nardetto, R. Samadi, D. Valls-Gabaud, H. Wozniak (2012), pp. 173–188, arXiv:1210.3505, 1210.3505
[6] R. Samadi, K. Belkacem, H.G. Ludwig, A&A, **559**, A39 (2013), 1309.1620
[7] R. Samadi, K. Belkacem, H.G. Ludwig, E. Caffau, T.L. Campante, G.R. Davies, T. Kallinger, M.N. Lund, B. Mosser, A. Baglin et al., A&A, **559**, A40 (2013), 1309.1488

 DOI: 10.1051/epjconf/201716001016

Download as csv file Show 10 entries Show / hide columns

StarID	Origin	alpha [deg]	delta [deg]	teff [K]	grav [cm/s2]	mag_v	mag_r	specType	lumClass	numax_up [μHz]		Deltanu_up [μHz]		v
										value	prec	value	prec	
100402467	corot	290.570746	1.644686	4329		-99	14.12	K	II	18.3	1	2.4912	0.012	8
100412751	corot	290.586053	1.634669	4343		-99	13.43	K	III	15.4118	0.586	1.9552	0.004	8
100422482	corot	290.600554	1.717	4549		-99	12.13	K	III	64.3284	0.963	6.0016	0.004	1
100434740	corot	290.618651	1.62033	5031		-99	12.87	G	V	47.502	0.39	4.1872	0.008	8
100440069	corot	290.626483	1.708309	4487		-99	12.87	K	II	51.6	1	5.605	0.005	8
100440565	corot	290.627243	1.647735	4507		-99	12.15	K	II	27.922	0.46	3.5088	0.004	1
100441421	corot	290.628492	1.636505	4753		-99	12.89	K	III	32.94	0.3	4.4695	0.005	1
100442460	corot	290.629967	1.714166	4549		-99	13.79	K	II	28.4526	0.958	3.5328	0.006	1
100448189	corot	290.638332	1.692805	4499		-99	13.53	K	II	16.3917	0.467	2.3356	0.004	1
100480626	corot	290.685877	1.647927	5673		-99	15.26	F	IV	19.1542	0.689	2.1554	0.002	2

Showing 1 to 10 of 4,903 entries First Previous 1 2 3 4 5 … 491 Next Last

Figure 3: Snapshot showing the query results

EPJ Web of Conferences **160**, 02001 (2017)
Seismology of the Sun and the Distant Stars II
DOI: 10.1051/epjconf/201716002001

Surface effects in solar-like oscillators

Warrick H. Ball[1,2,★]

[1] *Institut für Astrophysik, Georg-August-Universität Göttingen, Friedrich-Hund-Platz 1, 37077 Göttingen, Germany*
e-mail: wball@astro.physik.uni-goettingen.de
[2] *Max-Planck-Institut für Sonnensystemforschung, Justus-von-Liebig-Weg 3, 37077 Göttingen, Germany*

Abstract. Inaccurate modelling of the near-surface layers of solar models causes a systematic difference between modelled and observed solar mode frequencies. This difference—known as the "surface effect" or "surface term"—presumably also exists in other solar-like oscillators and must somehow be corrected to accurately relate mode frequencies to stellar model parameters. After briefly describing the various potential causes of surface effects, I will review recent progress along two different lines. First, various methods have been proposed for removing the surface effect from the mode frequencies and thereby fitting stellar models without the disproportionate influence of the inaccurate near-surface layers. Second, three-dimensional radiation hydrodynamics simulations are now being used to replace the near-surface layers of stellar models across a range of spectral types, leading to predictions of how some components of the surface effect vary between stars. Finally, I shall briefly discuss the future of the problem in terms of both modelling and observation.

1 Introduction

The era of space-based asteroseismology, driven chiefly by COROT [1] and *Kepler* [2], has provided observations of hundreds of cool main-sequence stars in which dozens of individual mode frequencies can be measured. To exploit this data, however, we need to correct for a systematic difference between observed and modelled mode frequencies caused by improper modelling of the near-surface layers of these stars: the so-called *surface term* or *surface effect*. Motivated by a newfound need to correct for the surface effect, significant progress has been achieved in the last few years and can be expected in the near future.

The purpose of this review is to first briefly recount our physical understanding of the surface effect (Sec. 2) and then review recent progress along two lines. First, several authors have proposed parametrizations of the surface effect (as a function of frequency) to suppress its influence when fitting stellar models to observed mode frequencies (Sec. 3). Second, a few research groups have begun replacing the near-surface layers of stellar models with average structures taken from detailed three-dimensional radiation hydrodynamic simulations (3D RHD, Sec. 4). Finally, I close with a few thoughts on how we might progress further on the problem of surface effects in the near future (Sec. 5).

I do not pretend that this review is exhaustive. Judging by the amount of material I excluded from my talk, it would be impossible to cover all the literature on the subject in 30 minutes. I apologize to anyone who feels their contribution has been omitted and seek to assure them that the cause is only brevity, not malice!

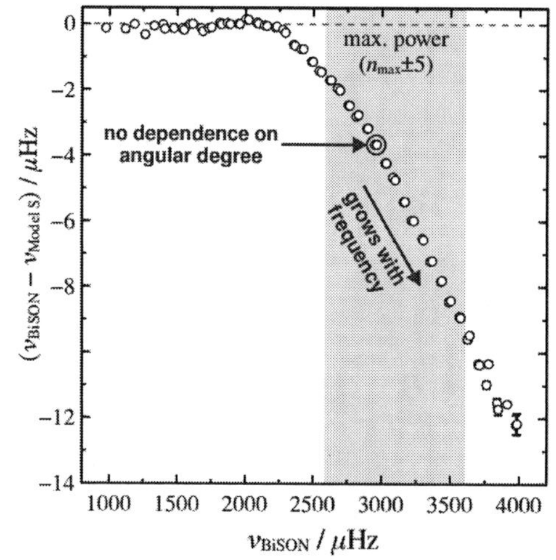

Figure 1: The basic problem of surface effects in the Sun. The white points show differences between low-degree ($\ell \leq 3$) mode frequencies observed by the Birmingham Solar Oscillation Network [BiSON; 3, 4] and linear adiabatic mode frequencies predicted for a standard solar model [Model S; 5]. The blue annotations show the two main features that lead us to believe that the systematic difference is a surface phenomenon: the differences grow with frequency and are largely independent of the angular degree ℓ (see text). The shaded region shows where the modes have their greatest power and are most easily observed, demonstrating that for most asteroseismic targets, *all* the observed frequencies are probably affected by the surface effect.

★e-mail: wball@astro.physik.uni-goettingen.de

2 The problem

2.1 Phenomenology of the surface effects

Suppose that one calibrates a solar model in the traditional sense by varying the mixing length parameter, initial helium abundance and initial metallicity to evolve a stellar model that matches the Sun's current radius, luminosity and surface metallicity, with the mass fixed at 1 $M_\odot$ and the age at the meteoritic age of the solar system. We take Model S [5] as an example. If we compute the linear adiabatic mode frequencies of this model and compare them with observed values [e.g. the low-degree data from the Birmingham Solar Oscillation Network, BiSON; 3, 4] we might hope that the differences between observed and modelled frequencies are randomly scattered about zero.

Instead, one gets the values plotted in Fig. 1. The white points indicate the differences between the mode frequencies predicted for Model S and those observed by BiSON. The discrepancy is much larger than the quoted uncertainties but it is also not random, and its structure tells us something about where the problem arises. First, the frequency differences do not depend on the angular degree ℓ, which suggests that the discrepancy lies well above the modes' lower turning points. The low-degree data alone only tells us that the cause is not very deep in the Sun but the frequency differences of the higher-degree modes are also ℓ-independent. Because they have shallower lower turning points, this suggests that the problem is quite close to the Sun's surface. Second, the frequency differences are close to zero at frequencies below about 2200μHz. Higher-frequency modes have shallower upper turning points, which again implies that the problem is somewhere near the Sun's surface. At 2200μHz, the modes' upper turning points are around 1 Mm below the surface, which implies that the effect really is confined to the near-surface layers.

The shaded region in Fig. 1 indicates the range of frequency covering about five radial orders either side of the frequency of maximum oscillation power ν_{max}. This is the range in which modes oscillate with the greatest power and are thus most easily observed. It shows that the surface effect probably affects nearly all the observed modes in distant Sun-like stars, unlike the Sun, in which we observe low-frequency modes that appear unaffected by the surface effect. The best targets from the nominal *Kepler* mission have lowest mode frequencies equivalent to about 2300μHz in the Sun, which is within the range of affected modes. For this reason, the surface effect is unavoidable: when fitting stellar models to individual mode frequencies, something *must* be done about the surface effect. In the case of the Sun, even the large separations of the modelled and observed frequencies differ by about 1μHz. When applied to the standard scaling relations [6], this bias in the large separation corresponds to biases in mass and radius of about 3 and 1.5 per cent, respectively.[1]

[1] The scaling relations are in essence empirical, which suppresses this effect. But if the surface effect varies significantly between different stars, it could be important.

2.2 The physical cause

It may come as a surprise that we have a fairly good idea about what causes the surface effect: improper modelling of near-surface convection. Most stellar models use some form of mixing-length theory (MLT), in which the convection zone is presumed to contain buoyantly-unstable rising and falling parcels of material (see Fig. 2, left). These parcels rise or fall by one mixing length, typically parametrized in terms of the local pressure scale height, $H_P = -\mathrm{d}r/\mathrm{d}\ln P$, after which they disperse, mixing the heat and composition of their origin into their new surroundings.

In reality, the flows are much more complicated, as is now understood from detailed 3D RHD simulations that accurately reproduce many observable features of convection at the Sun's surface [see 7, for an excellent review of the Sun's surface convection]. Let us start with one of the slow upflows. As it rises and the density decreases, so the flow expands horizontally and, to conserve mass, part of it must turn over and join whatever downflows exist (see Fig. 2, right). The rising plume ultimately appears as a granule at the surface, where the flows are chiefly horizontal. They then radiate heat to the vacuum of space before plummeting downward in narrow, turbulent intragranular lanes. Along the way back down, these downflows will draw material turning over from the widening upflows.

This is a very different picture from the calm rise and fall of MLT's parcels and it leads to a number of effects that affect the mode frequencies. Following the thorough discussion by Rosenthal [8], we can broadly divide these into two types of effect. *Model* physics includes everything that is wrong with the background model that we perturb. This includes, but is not limited to, MLT's incorrect temperature gradient, the incorrect atmospheric structure and the absence of turbulent pressure. *Modal* physics includes everything that is wrong with the calculation of the mode frequencies, which are affected by the perturbation to the turbulent pressure [e.g. 9], the modification of wave speeds when travelling with or against the flows [e.g. 10] and various effects of non-adiabaticity [e.g. 11]. All of these effects are most pronounced near the surface where convection becomes inefficient and the temperature gradient deviates furthest from the adiabatic value.

That so many physical effects contribute to the surface effect makes it a difficult problem to tackle piece by piece. In working on one component, one might think the problem is solved, only to find that another component returns you to square one. But it is not hopeless! We can learn how much each component might contribute and gradually add them up, bearing in mind that as our models improve, we might sometimes veer further from the observations before once again closing the gap.

3 Parametrizations

The surface effect in Fig. 1 appears to be a relatively simple function of mode frequency only. Thus, several groups have proposed parametric forms for this function whose

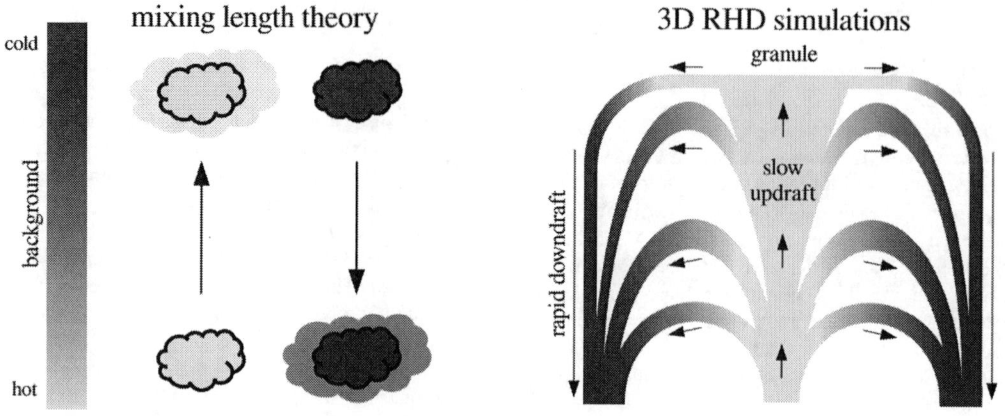

Figure 2: A crude illustration of the difference between the mixing-length theory of convection (MLT, left) and the structure of near-surface convection suggested by 3D RHD simulations (right). In MLT, buoyantly-unstable parcels of material retain their composition and heat content while floating upwards (or sinking downwards) by one *mixing length* l_{MLT} before dispersing their composition and heat into their new surroundings. In the 3D RHD simulations, slow, broad upflows expand as they rise through layers of decreasing density, ultimately reaching the surface and manifesting as granules. At the surface, material cools and sinks back down between the granules in rapid, cool downdrafts that we see as intragranular lanes.

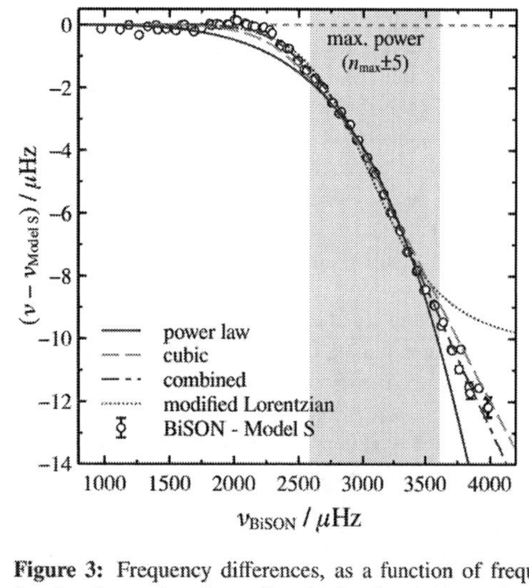

Figure 3: Frequency differences, as a function of frequency, for the observed surface effect in the Sun (white points) and several parametrizations (see Sec. 3). The shaded region shows the frequency range over which the modes have their greatest power, as in Fig. 1. The parametrizations shown are a power-law [solid blue, 12], the cubic and combined formulae [dashed orange and dash-dotted red, 13] and a modified Lorentzian [dotted green, 14].

parameters can be fit when comparing stellar models to observations. Here I shall review the best known and compare them for the Sun.

First, Kjeldsen et al. [12] proposed that the surface effect can be described as a power law with an index fixed to a solar-calibrated value (usually around 5 but slightly dependent on the precise physics of the stellar model). They also proposed that the magnitude of the power law be fit after rescaling the frequencies so that the stellar model being compared has the same mean density as the observed star. To rescale the frequencies so, they propose using the ratio of the large separations. This simple parametrization has been widely used since its publication [e.g. 15].

More recently, Ball & Gizon [13] proposed parametrizations based on surface perturbations and the asymptotic behaviour of the eigenmodes. Roughly speaking, the displacement eigenfunctions are exponentially decaying functions near the photosphere and, combining them with the variational principle for the linear, adiabatic oscillation equations [16], one finds that, for a sound speed perturbation or pressure scale height perturbation near the surface, the frequency shifts go either like $\nu^3/\mathcal{I}$ or $\nu^{-1}/\mathcal{I}$, where ν is the mode frequency and $\mathcal{I}$ the normalized mode inertia. These parametric forms, which Ball & Gizon [13] refer to as the *cubic* and *inverse* terms, respectively, were originally derived by Gough [17][2] in a discussion of the Sun's frequency shifts over the magnetic activity cycle. The inverse term alone does not fit the data well so Ball & Gizon [13] proposed to combine it with the cubic term, giving what they call the *combined* surface correction.

Most recently, Sonoi et al. [14] proposed to describe the surface effect as a modified Lorentzian function and calibrated its parameters to frequency shifts induced by replacing the near-surface layers of stellar models with averaged data from hydrodynamics simulations (see Sec. 4,

[2] The cubic term is also mentioned by Libbrecht & Woodard [18] and Goldreich et al. [19].

below). This parametrization is very new and has not yet been tested on observed data.

Fig. 3 shows the same data as Fig. 1 (BiSON against Model S), along with the above-mentioned parametrizations. The power law fit performs reasonably well in the shaded range around $\nu_{\max}$ but overestimates the surface effect both where it begins to rise and at higher frequencies. The cubic and the combined terms fare better. Though the improvement by using the combined term (rather than just the cubic term) is significant for the Sun, this was not the case for the COROT target HD 52265 studied by Ball & Gizon [13]. Finally, the modified Lorentzian captures most of the low-frequency behaviour but underestimates the difference at high frequencies.

Though different in principle, it is worth mentioning several methods proposed by Roxburgh (and Vorontsov in earlier work) [20–22]. These are all based on representing the oscillation modes as simple oscillations with phase shifts at the inner and outer boundaries. The outer phase shift contains the undesired and presumably ℓ-independent surface term whereas the inner phase shift is related to the structure of the stellar core. One can combine the frequencies into ratios of differences or so-called *separation ratios* that are nearly independent of the near-surface layers [20]. Otí Floranes et al. [23] computed kernels for these quantities and demonstrated that they are, indeed, largely insensitive to the near-surface layers and they have seen widespread use in asteroseismic modelling. From the same underlying principles, Roxburgh [21, 22] described methods to fit out a more general ℓ-independent component of the frequency differences. These are too new to have been used widely.

These various parametrized methods have not yet been systematically compared with observations, though the community's collective experience suggests that none generally leads to absurd results. Schmitt & Basu [24] conducted the most thorough study yet by inserting structural perturbations into stellar models across the HR diagram and then trying to fit the frequency differences using the solar-calibrated power law, the cubic and combined terms of Ball & Gizon [13] or the observed solar surface effect, rescaled by the large separation. The combined term by Ball & Gizon [13] appeared to fare best, although the scaled solar term also performed reasonably on the main sequence.

The parametrizations do not solve the problem of the surface effects but they at least allow us to exploit the reams of data already available while we work towards properly modelling the surface effects. The results should always be interpreted with the knowledge that the Sun remains the only star for which we can truly calibrate the frequency differences. Everything else depends on the confidence we place in how well our best-fitting models represent the stars under study.

4 Three-dimensional radiation hydrodynamics

I mentioned in Sec. 2 that the surface effect is chiefly caused by improper modelling of near-surface convection. So why not use better models of near-surface convection? This is the idea behind recent efforts to combine stellar models with 3D RHD simulations. Several groups have simulated near-surface convection from first principles in stars of various spectral types [e.g. 27–29]. These simulations are sufficiently realistic to reproduce most of the observed characteristics of the Sun's near-surface convection [again, see 7, for a review] and it is assumed that they are similarly realistic for other stars.

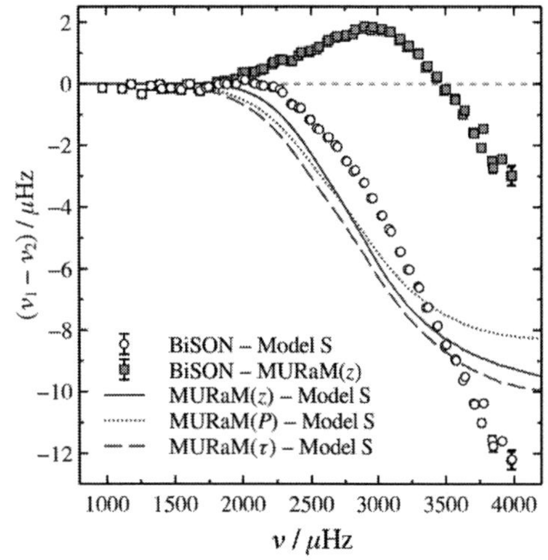

Figure 4: Frequency differences for the Sun between different combinations of models and data as a function of model frequencies. The lines show the frequency differences between the solar model before and after the near-surface layers are replaced by horizontally-averaged 3D RHD simulation data. The different lines correspond to different choices of averaging co-ordinate: geometric depth z (solid blue), pressure P (dotted green) or optical depth τ (dashed red). The white points show the same differences between observed and modelled frequencies as in Fig. 1. The orange squares show the frequency differences after the near-surface layers of the solar model have been replaced by 3D RHD simulation data averaged at constant geometric depth z. The overall extent of the surface effect is reduced to a few μHz but a clear systematic difference remains.

The process of replacing a stellar model's near-surface layers with averaged simulation data is becoming known as *patching*. The frequency differences are then computed between the *patched* model (the original stellar model) and the *patched* model (with the near-surface layers replaced).

The idea of patching is not new. Rosenthal et al. [9] restricted their study of solar oscillations to modes with angular degree $\ell > 60$. These modes are trapped within the solar convection zone, so they could compare their averaged simulation data with envelope models computed using MLT. Their early results showed that replacing the equilibrium structure of the stellar model with the simulation data, averaged at constant geometric depth, already introduced a surface effect of similar magnitude to the observed effect, although a significant systematic effect remained. More

DOI: 10.1051/epjconf/201716002001

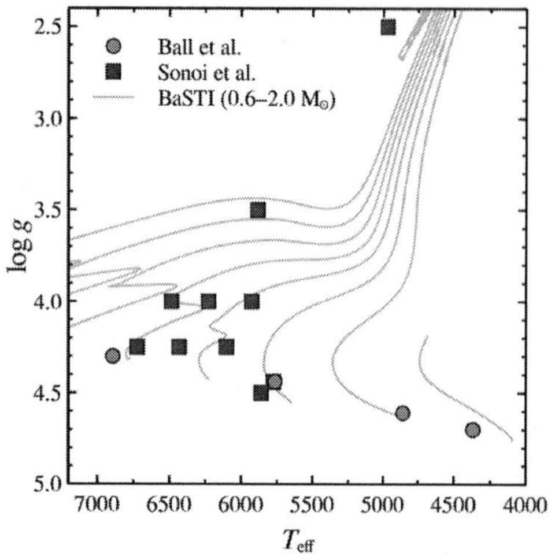

Figure 5: Kiel diagram (surface gravity against effective temperature) showing the parameters of the 3D RHD simulations used by [14, blue squares] and [25, orange circles]. The grey lines show solar-metallicity evolutionary tracks from BaSTI [26] for masses from 0.6 to 2.0 $M_\odot$ in steps of 0.2 $M_\odot$. Both sets of simulations include a solar model (models A and G2 in Sonoi et al. [14] and Ball et al. [25]) and the hottest models (models B and F3 in Sonoi et al. [14] and Ball et al. [25]) also have comparable parameters.

recently, Piau et al. [30] used 3D RHD simulation data to compute surface effects in a complete solar model (not just the convective envelope), finding that the structural component of the surface effect reduced the remain discrepancy to a few μHz. Finally, on the subject of the solar surface correction, Magic & Weiss [31] computed surface effects using simulations with different input magnetic field strengths and found that they could reproduce reasonably well the frequency shifts induced by the changing level of magnetic activity in the Sun.

The 12 months preceding this meeting saw the first papers to combine stellar models and 3D RHD simulations for the surface effects in other types of star. First, Sonoi et al. [14] combined stellar models from CESTAM [32, 33] with simulations from the CIFIST atmosphere grid [27]. Second, Ball et al. [25] combined stellar models from MESA [34–36] with simulations from the MURaM code [28]. The two groups independently performed nearly the same calculations using somewhat complementary sets of stellar models. The ten simulations used by Sonoi et al. [14] cover one red giant (around the red clump) and dwarfs and subgiants hotter than the Sun. The four simulations used by Ball et al. [25] span the main-sequence from spectral type F3 to K5. Fig. 5 shows the atmospheric parameters for the two groups' simulations.

These studies are not definitive. For a start, they only deal with the part of the surface effect caused by improving the structure of the equilibrium stellar model. The averaged simulation profiles include the turbulent pressure but it remains unclear what is the appropriate form of the perturbation to the turbulent pressure. Both Sonoi et al. [14] and Ball et al. [25] assume that the turbulent pressure varies with the total pressure: Rosenthal et al. [9] dubbed this the *gas gamma one* approximation. Changing this assumption potentially affects the results by a factor of about two [9]. Moreover, it is unclear exactly what is the appropriate horizontal average to take from the simulation data. Sonoi et al. [14] and Ball et al. [25] both used averages over constant geometric depth but different averages give surface effects that differ by a few μHz for the Sun (see Fig. 4).

Fig. 4 shows how the mode frequencies of a standard solar model [Model S again, 5] are changed by the modification of the near-surface equilibrium structure. The white points are the same differences shown in Figs 1 and 3. The solid blue, dotted green and dashed red curves are the differences in the model frequencies before and after patching with the G2-type MURaM simulation averaged over constant geometric depth, pressure or optical depth. The spread in the curves shows that uncertainty above the appropriate average introduces an uncertainty in the frequency shifts of about 0.5–1.0 μHz. Finally, the orange squares show the remaining difference between the patched model (with the simulation averaged at constant geometric depth) and the BiSON observations. The overall surface effect is reduced substantially, though clearly a large effect remains and the remaining difference is still a surface effect. It is not yet clear if the remaining trend is because the averaged near-surface structure is still not quite right, because non-adiabatic effects have been neglected, or (most likely) both.

With these uncertainties in mind, both teams found that the surface effect is larger in stars that are hotter. Based on their cooler dwarfs, Ball et al. [25] also noted that the overall shape of the frequency differences as a function of frequency is similar in the G2-, K0- and K5-type models, but some qualitative change sets in between the F3- and G2-type models. Sonoi et al. [14], with their greater coverage of surface gravity, also found that the surface effect increases with increasing surface gravity. Within their limitations, the two groups' results are mutually consistent. They have two simulations with similar parameters and the results agree well. Fig. 6 shows the frequency shifts for all the simulations by Ball et al. [25, left] and the simulations A and B of Sonoi et al. [14, right], which have similar parameters to models G2 and F3 of Ball et al. [25].

Both teams also compared the parametric fits described in Sec. 3, though Sonoi et al. [14] only compared their modified Lorentzian with a power law. Fig. 6 shows that a simple power law does not describe the differences between the patched and unpatched models very well. Ball et al. [25] found that the combined term was consistently superior, notably in their F3 model, but a scaled solar frequency correction fits reasonably well in their three cooler models. Sonoi et al. [14] provided simple fits to the best-fit parameters as a function of surface properties, though they did not consider a scaled solar correction or either of the corrections by Ball & Gizon [13].

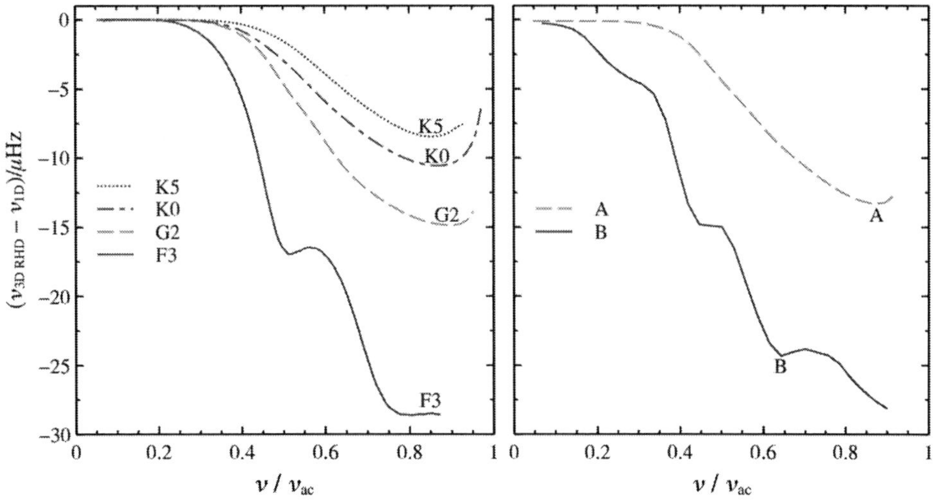

Figure 6: Frequency differences between various models before and after their near-surface layers are replaced with horizontally-averaged 3D RHD simulation profiles, as a function of frequency normalized to the acoustic cut-off frequency. The left panel shows models from Ball et al. [25]; the right panel for models from Sonoi et al. [14, adapted from their Fig. 3]. The models labelled A and B on the right correspond to the solar model and hottest model of Sonoi et al. [14] and can be compared to models G2 and F3 on the left, respectively.

Further exploitation of the 3D RHD simulations is underway, notably on non-adiabatic effects, but these early results already give some indication of how much of a surface effect is introduced by improving the background stellar model. It remains to be seen if the conclusions hold up as further surface effects are considered.

5 The future

To close, I briefly opine on how we might progress further on the problem of surface effects. The main theoretical path at this point is to further exploit the 3D RHD simulations. There is far more information available than simply the horizontally- and temporally-averaged profiles and this information can be used to investigate other components of the surface effect. But it should be remembered that even indirect conclusions drawn from the simulations can be useful. For example, the parametrizations of the surface effects tend to correlate with the mixing-length parameter in stellar models. There is good physical reason for this: both the surface effect and the mixing-length parameter are sensitive to the superadiabatic layer near the stellar surface. If the mixing-length parameter is constrained separately by the simulations [e.g. 37, 38] then the surface effect is also better constrained.

Progress is more difficult from the observational side. The best solar-like oscillators from the nominal *Kepler* mission show modes oscillating at frequencies nearly low enough that they are unaffected by the surface effect. If just a few more radial orders could be detected, these low frequencies could potentially be used to fit models without a surface term, though at the cost of discarding the many higher-frequency modes that are available. Alas, no imminent mission will provide such high-quality data for single targets, so we may have to wait until PLATO [39] for higher-quality data on single targets.

From the ground, however, there is tremendous potential from the Stellar Oscillation Network Group [see e.g. 40, these proceedings]. Because it observes in radial velocity, the background signal of granulation is weaker, which allows lower-frequency modes to be detected more easily. This could allow us to calibrate models directly to the unaffected frequencies and inspect the remaining frequencies to determine the surface effect after fitting the stellar model. One node of the network is fully operational and another partially so. The first results from the first node were reported at this meeting [40]. Adding nodes to the network probably represents our best chance of bringing tight observational constraints to bear on the problem of surface effects.

Acknowledgement

The author would like to thank the organizers for partial financial support to attend the meeting. He also acknowledges research funding by Deutsche Forschungsgemeinschaft (DFG) under grant SFB 963/1 "Astrophysical flow instabilities and turbulence", Projects A18.

References

[1] M. Auvergne, P. Bodin, L. Boisnard, J.T. Buey, S. Chaintreuil, G. Epstein, M. Jouret, T. Lam-Trong,

P. Levacher, A. Magnan et al., A&A, **506**, 411 (2009), `0901.2206`
[2] W.J. Borucki, D. Koch, G. Basri, N. Batalha, T. Brown, D. Caldwell, J. Caldwell, J. Christensen-Dalsgaard, W.D. Cochran, E. DeVore et al., Science **327**, 977 (2010)
[3] A.M. Broomhall, W.J. Chaplin, G.R. Davies, Y. Elsworth, S.T. Fletcher, S.J. Hale, B. Miller, R. New, MNRAS, **396**, L100 (2009), `0903.5219`
[4] G.R. Davies, W.J. Chaplin, Y. Elsworth, S.J. Hale, MNRAS, **441**, 3009 (2014), `1405.0160`
[5] J. Christensen-Dalsgaard, W. Dappen, S.V. Ajukov, E.R. Anderson, H.M. Antia, S. Basu, V.A. Baturin, G. Berthomieu, B. Chaboyer, S.M. Chitre et al., Science **272**, 1286 (1996)
[6] H. Kjeldsen, T.R. Bedding, A&A **293**, 87 (1995), `arXiv:astro-ph/9403015`
[7] Å. Nordlund, R.F. Stein, M. Asplund, Living Reviews in Solar Physics **6** (2009)
[8] C.S. Rosenthal, *Convective Effects on Mode Frequencies*, in *SCORe'96 : Solar Convection and Oscillations and their Relationship*, edited by F.P. Pijpers, J. Christensen-Dalsgaard, C.S. Rosenthal (1997), Vol. 225 of *Astrophysics and Space Science Library*, pp. 145–160
[9] C.S. Rosenthal, J. Christensen-Dalsgaard, Å. Nordlund, R.F. Stein, R. Trampedach, A&A, **351**, 689 (1999), `astro-ph/9803206`
[10] T.M. Brown, Science **226**, 687 (1984)
[11] G. Houdek, Ph.D. Thesis, Formal- und Naturwisseschaftliche Fakultät der Universität Wien, (1996) (1996)
[12] H. Kjeldsen, T.R. Bedding, J. Christensen-Dalsgaard, ApJL, **683**, L175 (2008), `0807.1769`
[13] W.H. Ball, L. Gizon, A&A, **568**, A123 (2014), `1408.0986`
[14] T. Sonoi, R. Samadi, K. Belkacem, H.G. Ludwig, E. Caffau, B. Mosser, A&A, **583**, A112 (2015), `1510.00300`
[15] V. Silva Aguirre, S. Basu, I.M. Brandão, J. Christensen-Dalsgaard, S. Deheuvels, G. Doğan, T.S. Metcalfe, A.M. Serenelli, J. Ballot, W.J. Chaplin et al., ApJ, **769**, 141 (2013), `1304.2772`
[16] P. Ledoux, T. Walraven, Handbuch der Physik **51**, 353 (1958)
[17] D.O. Gough, *Comments on Helioseismic Inference*, in *Progress of Seismology of the Sun and Stars*, edited by Y. Osaki, H. Shibahashi (1990), Vol. 367 of *Lecture Notes in Physics, Berlin Springer Verlag*, p. 283
[18] K.G. Libbrecht, M.F. Woodard, Nature, **345**, 779 (1990)
[19] P. Goldreich, N. Murray, G. Willette, P. Kumar, ApJ, **370**, 752 (1991)
[20] I.W. Roxburgh, S.V. Vorontsov, A&A, **411**, 215 (2003)
[21] I.W. Roxburgh, A&A, **574**, A45 (2015), `1406.6491`
[22] I.W. Roxburgh, A&A, **585**, A63 (2016)
[23] H. Otí Floranes, J. Christensen-Dalsgaard, M.J. Thompson, MNRAS, **356**, 671 (2005)
[24] J.R. Schmitt, S. Basu, ApJ, **808**, 123 (2015), `1506.06678`
[25] W.H. Ball, B. Beeck, R.H. Cameron, L. Gizon, A&A, **592**, A159 (2016), `1606.02713`
[26] A. Pietrinferni, S. Cassisi, M. Salaris, F. Castelli, ApJ, **612**, 168 (2004), `astro-ph/0405193`
[27] H.G. Ludwig, E. Caffau, M. Steffen, B. Freytag, P. Bonifacio, A. Kučinskas, Mem. Soc. Astron. Italiana, **80**, 711 (2009), `0908.4496`
[28] B. Beeck, R.H. Cameron, A. Reiners, M. Schüssler, A&A, **558**, A48 (2013), `1308.4874`
[29] Z. Magic, R. Collet, M. Asplund, R. Trampedach, W. Hayek, A. Chiavassa, R.F. Stein, Å. Nordlund, A&A, **557**, A26 (2013), `1302.2621`
[30] L. Piau, R. Collet, R.F. Stein, R. Trampedach, P. Morel, S. Turck-Chièze, MNRAS, **437**, 164 (2014), `1309.7179`
[31] Z. Magic, A. Weiss, A&A, **592**, A24 (2016), `1606.01030`
[32] P. Morel, A&AS, **124** (1997)
[33] P. Morel, Y. Lebreton, Ap&SS, **316**, 61 (2008), `0801.2019`
[34] B. Paxton, L. Bildsten, A. Dotter, F. Herwig, P. Lesaffre, F. Timmes, ApJS, **192**, 3 (2011), `1009.1622`
[35] B. Paxton, M. Cantiello, P. Arras, L. Bildsten, E.F. Brown, A. Dotter, C. Mankovich, M.H. Montgomery, D. Stello, F.X. Timmes et al., ApJS, **208**, 4 (2013), `1301.0319`
[36] B. Paxton, P. Marchant, J. Schwab, E.B. Bauer, L. Bildsten, M. Cantiello, L. Dessart, R. Farmer, H. Hu, N. Langer et al., ApJS, **220**, 15 (2015), `1506.03146`
[37] H.G. Ludwig, B. Freytag, M. Steffen, A&A, **346**, 111 (1999), `astro-ph/9811179`
[38] R. Trampedach, R.F. Stein, J. Christensen-Dalsgaard, Å. Nordlund, M. Asplund, MNRAS, **445**, 4366 (2014), `1410.1559`
[39] H. Rauer, C. Catala, C. Aerts, T. Appourchaux, W. Benz, A. Brandeker, J. Christensen-Dalsgaard, M. Deleuil, L. Gizon, M.J. Goupil et al., Experimental Astronomy **38**, 249 (2014), `1310.0696`
[40] J. Christensen-Dalsgaard, *200 nights with μ Herculis: early results from the SONG Hertzsprung telescope*, in *Seismology of the Sun and the Distant Stars 2016*, edited by J.P.F.G. Monteiro, M.S. Cunha, J.M.T. Ferreira (2017), this volume

DOI: 10.1051/epjconf/201716002002

Implications of the generation of internal gravity waves by penetrative convection for the internal rotation evolution of low-mass stars

Charly Pinçon[1,*], *Kévin* Belkacem[1], and *Marie-Jo* Goupil[1]

[1] *LESIA, Observatoire de Paris, PSL Research University, CNRS, Université Pierre et Marie Curie, Université Paris Diderot, 92195 Meudon, France*

Abstract. Due to the space-borne missions CoRoT and *Kepler*, noteworthy breakthroughs have been made in our understanding of stellar evolution, and in particular about the angular momentum redistribution in stellar interiors. Indeed, the high-precision seismic data provide with the measurement of the mean core rotation rate for thousands of low-mass stars from the subgiant branch to the red giant branch. All these observations exhibit much lower core rotation rates than expected by current stellar evolution codes and they emphasize the need for an additional transport process. In this framework, internal gravity waves (herefater, IGW) could play a signifivative role since they are known to be able to transport angular momentum. In this work, we estimate the efficiency of the transport by the IGW that are generated by penetrative convection at the interface between the convective and the radiative regions. As a first step, this study is based on the comparison between the timescale for the waves to modify a given rotation profile and the contraction/expansion timescale throughout the radiative zone of $1.3M_{\odot}$ stellar models. We show that IGW, on their own, are inefficient to slow down the core rotation of stars on the red giant branch, where the radiative damping becomes strong enough and prevent the IGW from reaching the innermost layers. However, we find that IGW generated by penetrative convection could efficiently modify the core rotation of subgiant stars as soon as the amplitude of the radial differential rotation between the core and the base of the convective zone is high enough, with typical values close to the observed rotation rates in these stars. This result argues for the necessity to account for IGW generated by penetrative convection in stellar modeling and in the angular momentum redistribution issue.

1 Introduction

Asteroseismology is a powerful tool to probe the internal structure of stars and the seismic data provided by the space-borne missions CoRoT and *Kepler* turn out to be a goldmine of information for the stellar evolution theory. The detection of mixed modes, which have amplitude both in the innermost layers and in the envelopes of stars, made the measurement of the mean core rotation possible for thousands of stars from the subgiant branch to the red giant branch. The observations clearly show that the mean core rotation moderately increases on the subgiant branch [1, 2] before it strongly drops along the red giant branch [3]. These observational facts are still unexplained by theoretical predictions. Indeed, transport of angular momentum by meridional circulation and shear-induced mixing as included in the current stellar evolution codes are unable to counteract the strong acceleration due to the core contraction [e.g. 4, 5]. Therefore, another mechanism must be found to efficiently slow the core rotation of these stars.

In this framework, internal gravity waves (hereafter, IGW), which are buoyancy waves propagating throughout the radiative zone of stars, may have a significative role to play. Indeed, in presence of differential rotation, IGW are radiatively damped, deposit the angular momentum that they carry into the medium and so can locally modify the internal rotation rate. Two main excitation mechanisms are usually invoked for these waves. Firstly, IGW can be generated by turbulent pressure in the convective bulk [6]. Turbulence-induced IGW have already been shown to be able to explain the quasi-flat rotation profile observed in the solar radiative zone [7, 8]. However, previous work [9] showed that the radiative damping drastically increases near the H-burning shell as the star evolves and should result in a rapid decoupling between these waves and the stellar core as soon as the beginning of the subgiant branch. Secondly, IGW can also be generated by the penetration of convective plumes into the top of the radiative zone. While penetrating by inertia, the convective plumes are slowed down by buoyancy braking and can transfer their kinetic energy into waves, as observed in geophysics or in numerical simulations [e.g. 10–12]. A semianalytical estimate of the generation of IGW by penetrative convection has already demonstrated the ability of these waves to efficiently affect the rotation of the solar radiative zone [13]. Nevertheless, an extension to more evolved stars is still missing.

*e-mail: charly.pincon@obspm.fr

Here, we investigate the ability of plume-induced waves to modify a given rotation profile in subgiant and red giant stars. As a first step in the investigation, our study is based on the comparison of the dynamical wave-driven timescale with the contraction/expansion timescale throughout 1.3M$_\odot$ stellar models. Moreover, the role of the radial differential rotation amplitude on the transport by IGW is stressed out.

2 Timescales and wave flux generated by convective plumes

2.1 Characteristic timescales

The local contraction/expansion timescale is equal to the characteristic timescale of evolution of the internal rotation rate in the case of local conservation of angular momentum. It is given by

$$t_{cont}(r) = \left|\frac{r}{2\dot{r}}\right| , \quad (1)$$

with r the radius and $\dot{r}$ the Lagrangian radial velocity of a mass shell at the radius r. The local timescale associated with the transport of angular momentum by IGW, given a rotation profile $\Omega(r)$ (within the shellular approximation), is equal to the ratio of the density of angular momentum in the star to the radial divergence of the angular momentum wave flux, i.e.

$$t_w(r) \sim \left|\frac{\rho r^2 \Omega}{\dot{J}}\right| , \quad (2)$$

with ρ the density at the equilibrium and $\dot{J}$ the divergence of the mean radial wave flux of angular momentum whose computation requires the knowledge of the wave energy flux, given below by Eq. (4). Therefore, IGW can be said to locally counteract the action of the contraction or the dilatation in the star if $t_w \leq t_{cont}$. Thus, the comparison of both timescales gives a first hint about the efficiency of the transport of angular momentum by IGW throughout the radiative zone of low-mass stars.

2.2 Wave flux of angular momentum

Convective plumes are strong coherent downwards flows that originate from the uppermost layers of low-mass stars and develop across their convective envelope by turbulent entrainment of matter. Once they reach the border with the underlying radiative zone, they penetrate by inertia into the stably stratified layers where they are slowed down by buoyancy braking. This process converts a part of the plume kinetic energy into wave energy, which can then propagate into the radiative zone. To model this excitation mechanism, [13] considered the pressure exerted by an ensemble of incoherent convective plumes in the penetration zone as the source term in the wave equation. By assuming a high Péclet number and a very sharp thermal transition at the base of the convective zone, the mean radial wave energy flux per unit of frequency, for an angular degree l and an azimuthal number m at the top of the radiative zone, can be written as

$$\mathcal{F}_{E,w}(r_t, \omega, l, m) \sim \frac{1}{4\pi r_t^2} \frac{\mathcal{A} \mathcal{S}_p}{2} \frac{\rho_b V_b^3}{2} \mathcal{F}_{R,l} \frac{e^{-\omega^2/4\nu_p^2}}{\nu_p} e^{-l(l+1)b^2/2r_t^2} , \quad (3)$$

where r_t is the radius at the top of the radiative zone, $\mathcal{A}$ is the plumes filling factor in the excitation region, $\mathcal{S}_p = \pi b^2$ is the horizonthal area occupied by one single plume, with b the plume radius, ρ_b and V_b are respectively the density and the plume velocity at the base of the convective region, $F_{R,l} = \sqrt{l(l+1)} V_b / r_t N_0$, with N_0 the Brunt-Väisälä frequency at the top of the radiative zone, and $\nu_p = 1/\tau_p$, with τ_p the plume lifetime.

Each spectral component of the total wave energy flux emitted from the top of the radiative zone are radiatively damped as it propagates through the radiative region and so contributes to the transport of angular momentum in presence of differential rotation. Then, the total wave energy flux at a radius r can be expressed following [7]

$$\mathcal{F}_{J,w}(r) = \sum_l \sum_{m=-l}^{m=+l} \int_{-\infty}^{+\infty} \frac{m}{\omega} \frac{r_t^2}{r^2} \mathcal{F}_{E,w}(r_t, \omega, l, m) e^{-\tau(r,\hat{\omega},l)} d\omega , \quad (4)$$

with

$$\tau(r, \hat{\omega}, l) = [l(l+1)]^{3/2} \int_r^{r_t} K \frac{N N_T^2}{\hat{\omega}^4} \left(\frac{N^2}{N^2 - \hat{\omega}^2}\right)^{1/2} \frac{dr}{r^3} , \quad (5)$$

where N is the Brunt-Väisälä frequency, with its thermal part N_T, K is the radiative diffusion coefficient, and

$$\hat{\omega}(r, \omega, m) = \omega - m\delta\Omega(r) \quad (6)$$

is the Doppler-shifted intrinsic frequency. Note that $\delta\Omega(r) = \Omega(r) - \Omega_t$ where Ω_t is the rotation rate at the top of the radiative zone. Near a critical layer (i.e. where $\hat{\omega} = 0$), we will suppose that the considered wave component is totally dissipated and deposits all the angular momentum that it carries into the medium, so that it cannot go deeper in the star.

3 Efficiency of the transport by IGW

3.1 Stellar models and assumptions on the rotation profile

We consider an evolutionary sequence of 1.3M$_\odot$ stellar models computed with the evolution code CESTAM [4] that covers the subgiant branch and the beginning of the ascent of the red giand branch. Their location in the Hertzsprung-Russel diagram is plotted in figure 1. The chemical composition follows the solar mixture as given in [14], with the initial helium and metal abundances $Y_0 = 0.25$ and $Z_0 = 0.013$. We used the NACRE nuclear reaction rates and the OPAL2005 equation of states and opacity tables. The convection was modeled by the mixing-length theory parametrized with $\alpha_{MLT} = 1.65$. We did not consider microscopic diffusion, overshooting nor rotation. To compute Eq. (3), we assume that the plume lifetime is

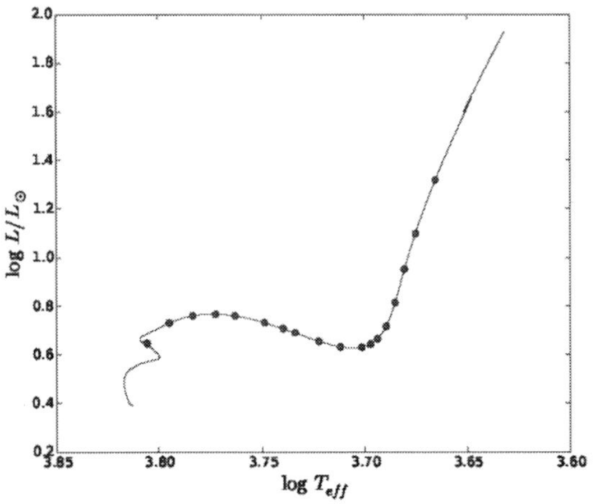

Figure 1: Location in the Hertzsprung-Russel diagram of the $1.3M_{\odot}$ stellar models used in this work (red filled circles). The evolutionary track, from the ZAMS to the top of the red giant branch, is represented by the solid blue line.

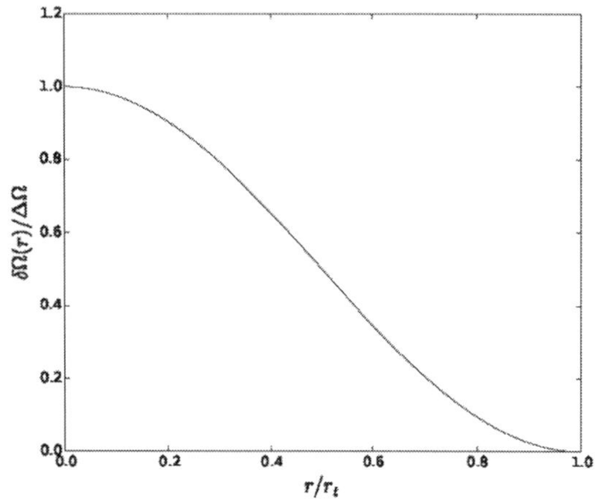

Figure 2: Normalized profile of the internal rotation rate as assumed in this work in the radiative zone of the stars (see Eq. (7)). It is represented as a function of the radius normalized by r_t, the radius at the top of the radiative zone.

close to the convective timescale at the base of the convective zone as given by the MLT, i.e. $v_p \approx \omega_{MLT} \sim 1$ μHz for all the considered models. In addition, we fix the plumes filling factor at a reasonable value $\mathcal{A} \approx 0.1$, as observed in the uppermost layers of numerical simulations of the Sun [e.g. 15]. All the other quantities are directly estimated using the equilibrium internal structure from the stellar models [see 13, for details].

As seen in Eq. (4), the total wave flux depends on the differential rotation in the radiative zone via Eq. (5). As shown by [13] in the solar case, its amplitude can have strong consequences on the transport by IGW. We will then assume a given rotation profile for each stellar model. Doing so, we assume that the rotation rate is low enough to have no effect on the stellar equilibrium structure. Nevertheless, little is known about the shape of the rotation profile in evolved stars. By assuming a priori, first, that a mechanism prevents the core contraction from developing a strong differential rotation in the radiative zone all along the star lifetime (as expected from the observations), and second, that this latter forces a quite smooth profile, it leads us to consider a rotation profile in the form

$$\delta\Omega(r) = \Delta\Omega \cos^2\left(\frac{\pi}{2}\frac{r}{r_t}\right) \quad \text{for} \quad r < r_t , \qquad (7)$$

with $\Delta\Omega$ the amplitude of the differential rotation between the center and the top of the radiative zone. The arbitrary shape of such a rotation profile is illustrated in figure 2. We thus assume a decreasing rotation rate from the core to the envelope and the $\cos^2$ function ensures a smooth profile at the center and near the base of the convective zone. Using such a synthetic profile is questionable, but it will give us a first hint about the efficiency of the transport by IGW in subgiant and red giant stars while considering different values for $\Delta\Omega$.

3.2 Timescale comparison and consequences

3.2.1 Stars at the beginning of the ascent of the RGB

As an example, we consider in the following the case of a model with $\log T_{\rm eff} = 3.68$, $\log L/L_{\odot} = 0.95$ and $\log g = 3.28$, which is located at the beginning of the ascent of the red giant branch. In figure 3 (*left panel*), the characteristic timescale of evolution of the rotation rate due to action of IGW, Eq. (2), is compared to the contraction/dilatation timescale, Eq. (1), as a function of the radius in the radiative zone. The values for $\Delta\Omega$ are chosen between 0 and 12 μrad s^{-1}, with the upper limit slightly above the maximum value given by the observations [2, 3]. As we can see, the wave-driven timescale drastically increases in average as the radius decreases and is still higher than the contraction/dilatation timescale below the H-burning shell regardless the value of $\Delta\Omega$ (in the considered range). Given that the result holds for values of the differential rotation that are consistent with the observations, we can conclude that IGW generated by penetrative convection are likely not to be responsible for the low rotation rates observed in red giant stars. Indeed, as showed in [9] in the case of the IGW generated by turbulent pressure, a strong radiative damping caused by the increase of the Brunt-Väisälä frequency as the stellar core contracts prevents IGW from reaching the innermost layers and modifying the core rotation. As we can note in figure 3, the wave-driven timescale reaches a local minimum just above the H-burning shell, which indicates a large deposit of angular momentum and a strong damping. Neverthless, we do not exclude that these waves damped near the H-burning shell may interact with the meridional circulation and boost the extraction of angular momentum in the core, but this will have to be tested with a more exhaustive calculation.

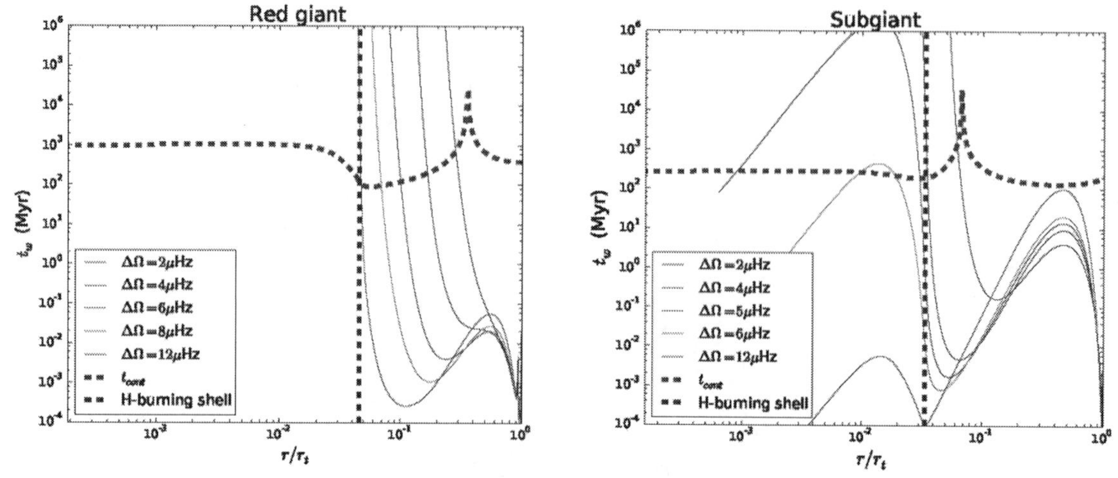

Figure 3: Left: Characteristic wave-driven timescale (solid lines), computed using Eq. (2), as a function of the normalized radius in the radiative zone of a 1.3M$_\odot$ red giant model, with $\log T_{\rm eff} = 3.68$ and $\log L/L_\odot = 0.95$. Colors correspond to different values for the amplitude of the differential rotation, $\Delta\Omega$. The red and the blue dashed lines represent the contraction/dilatation timescale and the location of the H-burning shell, repectively. **Right:** Same as the left panel, but for a 1.3M$_\odot$ subgiant model, with $\log T_{\rm eff} = 3.71$ and $\log L/L_\odot = 0.63$.

3.2.2 Subgiant stars

For the subgiant stars, we consider as an illustration the case of a model with $\log T_{\rm eff} = 3.71$, $\log L/L_\odot = 0.63$ and $\log g = 3.72$. The comparison of both timescales is plotted in figure 3 (*right panel*) throughout the radiative zone of the star. We can see that the situation is similar as the red giant case for low values of the differential rotation. Nevertheless, as $\Delta\Omega$ increases, t_w decreases and becomes smaller than the contraction timescale all over the region below the H-burning shell as soon as the amplitude of the differential rotation is higher than a threshold, noted $\widetilde{\Delta\Omega}$. In the example plotted in figure 3 (*right panel*), $\widetilde{\Delta\Omega} \approx 6$ µrad s^{-1}. This value is consistent with the mean core rotation rates observed in subgiant stars [2]. For this model, we can then conlude that IGW generated by penetrative convection are likely to play a role in the redistribution of angular momentum and that, at this stage, they have to be taken into account in stellar modeling. In fact, this result can be extended to the whole subgiant branch. In figure 4, the threshold value $\widetilde{\Delta\Omega}$, above which IGW can counteract the acceleration due to the core contraction in all the region below the H-burning shell, is plotted as a function of the $\log g$. We can see that IGW generated by penetrative convection are quite efficient at the beginning of the subgiant branch, even in the case of very small values for the differential rotation (lower than 0.1µrad s^{-1}), because of quite vigourous plume velocities. Further, $\widetilde{\Delta\Omega}$ increases with log g. Indeed, as the star evolves, the Brunt-Väisälä frequency increases, leading to an increase in the numerator of the damping depth Eq. (5). Nevertheless, this effect can be balanced by an increase of the amplitude of the differential rotation. Retrograde waves (m<0) are thus less damped, they can reach the innermost layers to deposit their negative angular momentum and slow down the core rotation. Figure 4 shows that the theoretical values are consistent with the observations in subgiants with values lower than about 10 µrad s^{-1}. The values found for stars at the beginning of the red giant branch ($\widetilde{\Delta\Omega} \approx 17$µrad s^{-1} for the model studied in Sect. 3.2.1) are much higher than the observations [3] and are the consequence of a stronger ra-

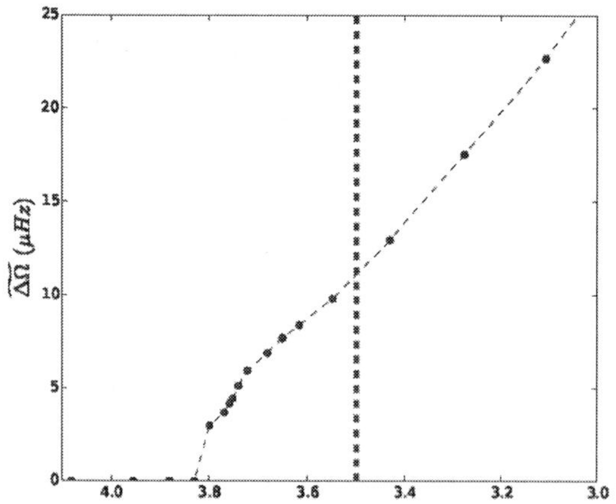

Figure 4: Theoretical threshold for the differential rotation as a function of the log g. The vertical red dashed line symbolizes the transition between the subgiant phase to the beginning of the RGB.

diative damping. As said before, another mechanism must be found for the more evolved red giant stars.

4 Conclusions

The comparison of the wave-driven timescale with the contraction/dilatation timescale provides us with a first indication about the efficiency of the transport of angular momentum by IGW generated by penetrative convection throughout the radiative zone of a sequence of $1.3M_\odot$ models. Plume-induced IGW are shown to be inefficient on their own to slow down the core rotation as soon as the star starts ascending the red giant branch because of a too strong radiative damping near the H-burning shell, in agreement with previous works [e.g. 9]. In contrast, we found that these waves are likely to play a major role on the subgiant branch as soon as the amplitude of the differential rotation is high enough. Indeed, the increase in the Brunt-Väisälä frequency as the star evolves, which should lead to a stronger radiative damping, can be balanced by the increase of the Doppler shift in the retrograde waves frequency. These waves can then deposit their negative angular momentum well below the H-burning shell. This scenario is supported by the strong similarities between the rotation rates observed in low-mass stars and the theoretical thresholds for the differential rotation. These preliminary results need for confirmation from a more thorough study accounting for the uncertainties on the wave energy flux parameters and the assumed shape of the rotation profile. This will be subject to a future paper (Pinçon et al., 2016, in preparation). In a further step, transport by IGW will have to be implemented in a stellar evolution code to model the evolution of the internal rotation along the stars lifetime by taking into account all the transport processes and their interaction with each other.

Acknowledgements

The authors acknowledge financial support from the EC Project SPACEINN (FP7-SPACE-2012-312844).

References

[1] S. Deheuvels, R.A. García, W.J. Chaplin, S. Basu, H.M. Antia, T. Appourchaux, O. Benomar, G.R. Davies, Y. Elsworth, L. Gizon et al., Astrophys. J. **756**, 19 (2012), 1206.3312

[2] S. Deheuvels, G. Doğan, M.J. Goupil, T. Appourchaux, O. Benomar, H. Bruntt, T.L. Campante, L. Casagrande, T. Ceillier, G.R. Davies et al., Astron. & Astrophys. **564**, A27 (2014), 1401.3096

[3] B. Mosser, M.J. Goupil, K. Belkacem, J.P. Marques, P.G. Beck, S. Bloemen, J. De Ridder, C. Barban, S. Deheuvels, Y. Elsworth et al., Astron. & Astrophys. **548**, A10 (2012), 1209.3336

[4] J.P. Marques, M.J. Goupil, Y. Lebreton, S. Talon, A. Palacios, K. Belkacem, R.M. Ouazzani, B. Mosser, A. Moya, P. Morel et al., Astron. & Astrophys. **549**, A74 (2013), 1211.1271

[5] T. Ceillier, P. Eggenberger, R.A. García, S. Mathis, Astron. & Astrophys. **555**, A54 (2013), 1305.5630

[6] P. Kumar, S. Talon, J.P. Zahn, Astrophys. J. **520**, 859 (1999), astro-ph/9902309

[7] J.P. Zahn, *Rotation and angular momentum transport*, in *SCORe'96 : Solar Convection and Oscillations and their Relationship*, edited by F.P. Pijpers, J. Christensen-Dalsgaard, C.S. Rosenthal (1997), Vol. 225 of *Astrophysics and Space Science Library*, pp. 187–200

[8] S. Talon, P. Kumar, J.P. Zahn, Astrophys. J. **574**, L175 (2002), astro-ph/0206479

[9] J. Fuller, D. Lecoanet, M. Cantiello, B. Brown, Astrophys. J. **796**, 17 (2014), 1409.6835

[10] R.B. Stull, Journal of Atmospheric Sciences **33**, 1279 (1976)

[11] B. Dintrans, A. Brandenburg, Å. Nordlund, R.F. Stein, Astron. & Astrophys. **438**, 365 (2005), astro-ph/0502138

[12] L. Alvan, A.S. Brun, S. Mathis, Astron. & Astrophys. **565**, A42 (2014), 1403.4052

[13] C. Pinçon, K. Belkacem, M.J. Goupil, Astron. & Astrophys. **588**, A122 (2016), 1512.07028

[14] M. Asplund, N. Grevesse, A.J. Sauval, P. Scott, Ann. Rev. Astron. Astrophys. **47**, 481 (2009), 0909.0948

[15] R.F. Stein, A. Nordlund, Astrophys. J. **499**, 914 (1998)

EPJ Web of Conferences **160**, 02003 (2017)
Seismology of the Sun and the Distant Stars II
DOI: 10.1051/epjconf/201716002003

Calibrating damping rates with LEGACY* linewidths

Günter Houdek[1,**]

[1] *Stellar Astrophysics Centre, Aarhus University, DK-8000 Aarhus C, Denmark*

Abstract. Linear damping rates of radial oscillation modes in selected *Kepler* stars are estimated with the help of a nonadiabatic stability analysis. The convective fluxes are obtained from a nonlocal, time-dependent convection model. The mixing-length parameter is calibrated to the surface-convection-zone depth of a stellar model obtained from fitting adiabatic frequencies to the LEGACY* observations, and two of the three nonlocal convection parameters are calibrated to the corresponding LEGACY* linewidth measurements. The atmospheric structure in the 1D stability analysis adopts a temperature-optical-depth relation derived from 3D hydrodynamical simulations. Results from 3D simulations are also used to calibrate the turbulent pressure and to guide the functional form of the depth-dependence of the anisotropy of the turbulent velocity field in the 1D stability computations.

1 Introduction

We use LEGACY [14] linewidths and frequencies of selected solar-type *Kepler* stars, together with 3D hydrodynamical simulation results [15, 16], for calibrating the global stellar and nonlocal convection parameters in 1D stability computations. Additionally to exploiting the seismic diagnostic of observed oscillation frequencies, linewidth measurements provide further diagnostic information about the physical processes prevailing in the outer superadiabatic boundary layers where convective transport becomes inefficient. It is in these outer stellar layers where solar-like oscillations are excited stochastically to the observed oscillation amplitudes and damped by various processes including the interaction between oscillations and convection. Therefore, a time-dependent convection model is required for describing these physical processes, including the pulsationally perturbations to both the convective heat (enthalpy) and momentum (turbulent pressure) fluxes. A consistent inclusion of turbulent pressure in the equilibrium structure is, however, only possible within the framework of a nonlocal formulation of convection [7]. Such a convection model is adopted here [7, 8]. Within the generally assumed approximations for constructing a 1D convection model, such as the Boussinesq approximation [17] (for a recent review see [10]), the turbulent fluxes are consistently estimated in both the equilibrium and nonadiabatic pulsation calculations. A nonlocal convection formulation typically has additional (nonlocal) parameters which need calibration. Here we use results from 3D hydrodynamical simulations and linewidth measurements from *Kepler* data for calibrating these additional (three) nonlocal convection parameters. We also include an analytical description for the variation of the anisotropy of the turbulent velocity field, the functional form of which being guided by 3D simulations [15].

2 Model computations

The 1D nonlocal model calculations are carried out essentially in the manner described by [9] and [2] but include, in addition, a description for the variation of the turbulent velocity anisotropy with stellar depth, and a temperature-optical-depth ($T-\tau$) relation derived from 3D simulations [16]. The convective heat flux and turbulent pressure are obtained from a nonlocal generalization of the mixing-length formulation [7, 8]. In this generalization three more (nonlocal) parameters, a, b and c, are introduced which control the spatial coherence of the ensemble of eddies contributing to the total convective heat flux (a) and turbulent pressure (c), and the degree to which the turbulent fluxes are coupled to the local stratification (b). The effects of varying these nonlocal parameters on the solar structure and oscillation properties were discussed in detail by [1].

The nonlocal parameter c is calibrated such as to have the maximum value of the turbulent pressure, $\max(p_t)$, in the 1D nonlocal model to agree with the 3D simulation result by [15]. The depth-dependence of the anisotropy $\Phi := \overline{\boldsymbol{u}\cdot\boldsymbol{u}}/\overline{w^2}$ of the convective velocity field $\boldsymbol{u} = (u, v, w)$ (an overbar denotes an ensemble average) is described by an analytical function, guided by 3D simulation results [15]. The remaining nonlocal parameters a and b cannot be easily obtained from the 3D simulations and are therefore calibrated such as to have a good agreement between calculated damping rates and LEGACY linewidths over the whole measured frequency range. The mixing length is calibrated such as to obtain the surface-convection depth of the frequency-calibrated evolutionary model calculated

*data provided by Mikkel N. Lund (e-mail: mikkelnl@phys.au.dk)
**e-mail: hg@phys.au.dk

DOI: 10.1051/epjconf/201716002003

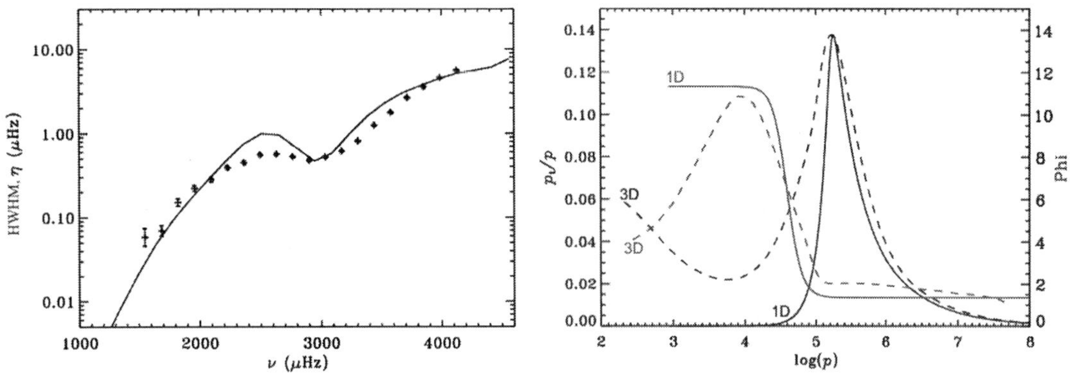

Figure 1: Comparison of solar stability computations with BiSON data and results from 3D hydrodynamical simulations. The **left panel** shows theoretical damping rates η, in units of cyclic frequency (solid curve), as functions of cyclic oscillation frequency ν, together with half-width-at-half-maximum (HWHM) measurements (symbols with error bars) of the spectral peaks in an acoustic power spectrum obtained by BiSON [2]. The **right panel** compares profiles of the turbulent pressure p_t over the total pressure p (black curves) and turbulent-velocity anisotropy Φ (red curves) between a 3D simulation (dashed curves) and a calibrated 1D solar model (solid curves).

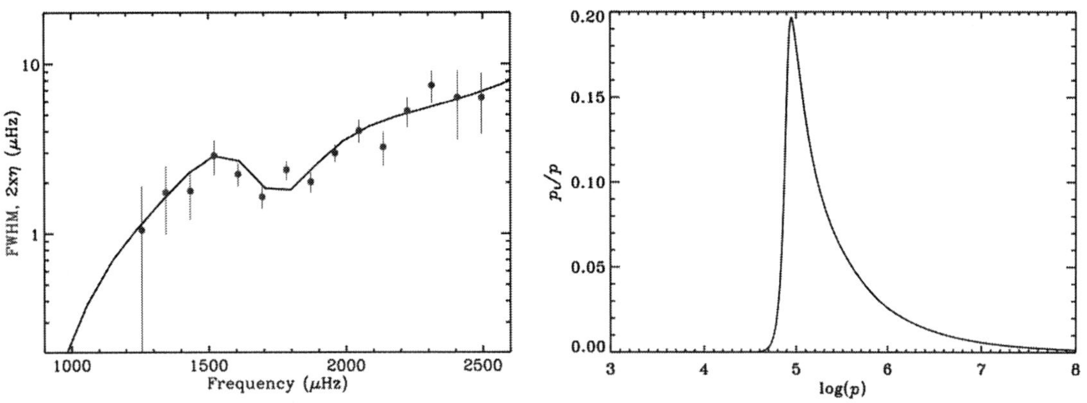

Figure 2: Results of a stability calculation for the *Kepler* star KIC12009504 (Dushera) with an effective temperature $T_{\rm eff}$=6217 K and surface gravity $\log g$ =4.214. The **left panel** compares LEGACY linewidths [14] (red symbols with 1 σ error bars) with twice the theoretical damping-rate estimates, $2\times\eta$ (solid curve), in units of cyclic frequency, as functions of cyclic oscillation frequency. The **right panel** shows the profile of (3D-calibrated) turbulent pressure p_t over the total pressure p in the 1D stellar model.

with ASTEC [4]. The ASTEC evolutionary model is obtained by minimizing the differences between observed (LEGACY) and adiabatically computed (ADIPLS [5]) oscillation frequencies, including corrections for surface effects according to [6].

Both the nonlocal envelope and pulsation calculations of the stability analyses assume the generalized Eddington approximation to radiative transfer [18]. The temperature gradient in the plane-parallel atmosphere is corrected by using a radially varying Eddington factor fitted to $T-\tau$ relations obtained from 3D simulation results [16]. The abundances by mass of hydrogen and heavy elements are adopted from the frequency-calibrated evolutionary calculations. The opacities are obtained from the OPAL tables [12], supplemented at low temperature by tables from [13]. The equation of state includes a detailed treatment of the ionization of C, N, and O, and a treatment of the first ionization of the next seven most abundant elements [3]. The integration of stellar-structure equations starts at an optical depth of $\tau = 10^{-4}$ and ends at a radius fraction $r/R_{\odot} = 0.2$.

The linear nonadiabatic pulsation calculations are carried out using the same nonlocal convection formulation with the assumption that all eddies in the cascade respond to the pulsation in phase with the dominant large eddies. A simple thermal outer boundary condition is adopted at the temperature minimum where for the mechanical boundary condition the solutions are matched smoothly onto those of a plane-parallel isothermal atmosphere. At the base of the model envelope the conditions of adiabaticity and vanish-

 DOI: 10.1051/epjconf/201716002003

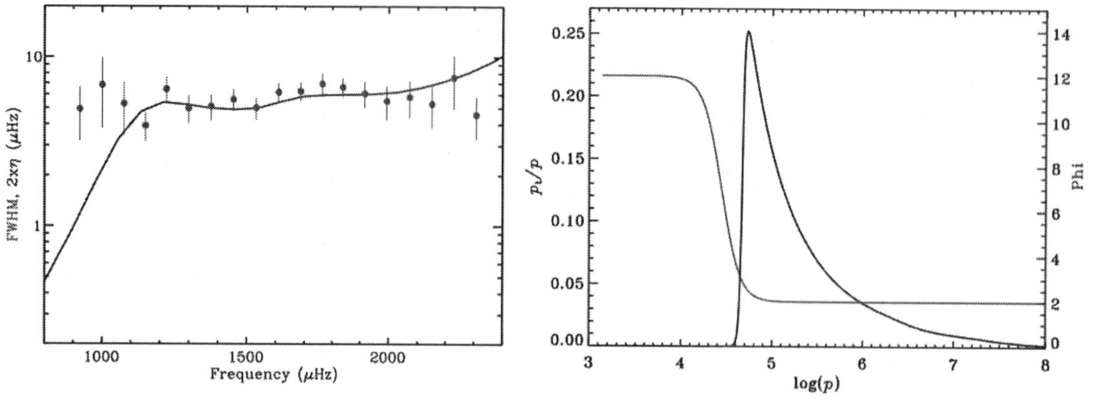

Figure 3: Results of a stability calculation for the *Kepler* star KIC11253226 (Tinky) with an effective temperature $T_{\rm eff}$=6715 K and surface gravity $\log g$ =4.171. The **left panel** compares LEGACY linewidths [14] (full-width-at-half-maximum FWHM, red symbols with 1 σ error bars) with twice the theoretical damping-rate estimates, $2\times\eta$ (solid curve), in units of cyclic frequency, as functions of cyclic oscillation frequency. The **right panel** shows the profiles of (3D-calibrated) turbulent pressure $p_{\rm t}$ over the total pressure p (black curve) and turbulent-velocity anisotropy Φ (red curve) in the 1D stellar model.

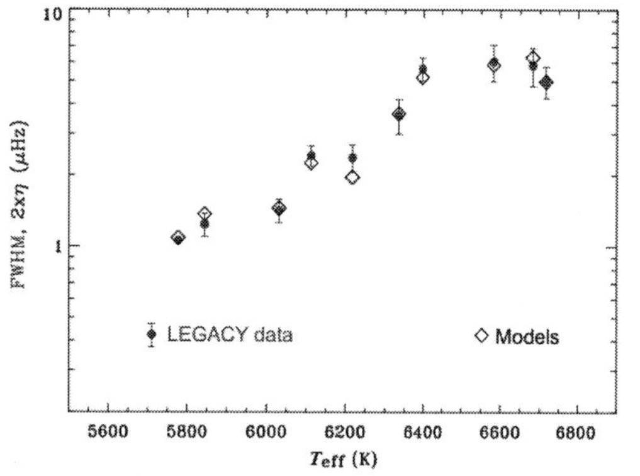

Figure 4: Comparison of measured LEGACY linewidths (full-width-at-half-maximum FWHM, red, filled circles with 1 σ error bars, [14]) with twice the estimated damping rates (black, open diamond symbols), in units of cyclic frequency, for 9 *Kepler* stars and the Sun (for the Sun BiSON linewidths [2] are shown). Results are plotted as functions of the model's effective temperatures and at the frequency $\nu \simeq \nu_{\rm max} \propto gT_{\rm eff}^{-1/2}$, assuming a solar $\nu_{\rm max} \simeq 3100\,\mu$Hz.

ing of the displacement eigenfunction are imposed. Only radial p modes are considered.

3 Results

We first tested the stability analysis against solar data provided by the BiSON group [2] (see also [11]). The results are illustrated in Figure 1. The left panel compares radial damping-rate estimates η, in units of cyclic frequency, with half-width-at-half-maximum (HWHM) measurements of the spectral peaks in the acoustic power spectrum. The power spectrum was obtained from a BiSON time-series collected over an epoch of 3456 days [2]. The right panel compares the profiles of turbulent pressure $p_{\rm t}$ and convective velocity anisotropy Φ as functions of the logarithm of the total pressure p in the outer stellar layers between a 3D solar simulation [15] and a calibrated 1D model. Only the maximum value of $p_{\rm t} := \overline{\rho w w}$ (ρ is density) in the 1D model was calibrated to the 3D simulation results by adjusting the nonlocal convection parameter c. The agreement between the model and simulation is particularly satisfactorily in the deeper stellar layers where the modes are propagating. Note that in the outer layers with $\log p \lesssim 5.3$ the modes are evanescent (e.g. [11]). The increasing difference in $p_{\rm t}$ between simulation (dashed curve) and model (solid curve) with decreasing total pressure p is predominantly a consequence of neglecting in a Boussinesq fluid the acoustic flux (H.-G. Ludwig, personal communication), generated by the convective fluctuations. These differences in stellar stratification have, however, little effect on the acoustic oscillation properties, for these differences are confined in the evanescent layers (see also [11]).

Next we estimated damping rates in 9 solar-type *Kepler* stars with effective temperatures 5844 K $\leq T_{\rm eff} \leq$ 6715 K. Results for the frequency-dependence of the estimated damping rates and profiles of turbulent pressure of the corresponding 1D stellar stratifications are illustrated in Figures 2 and 3 for two models with different effective temperatures. As in the solar case, max($p_{\rm t}$) in the 1D model was calibrated such as to match max($p_{\rm t}$) in the 3D simulations [15].

Figure 4 compares LEGACY linewidths [14] (red, filled circles with error bars) with estimated damping rates (black, open diamonds) for 9 selected *Kepler* stars and for the Sun. Results are plotted at the frequency $\nu_{\rm max}$ of maximum oscillation amplitude, determined from the scaling relation $\nu_{\rm max} \propto gT_{\rm eff}^{-1/2}$ (g being surface gravity) and assuming a solar $\nu_{\rm max} \simeq 3100\,\mu$Hz. The agreement between observations and models is very satisfactorily. Moreover, all 1D models have global stellar parameters determined from minimizing the differences between observed and adiabatically computed oscillation frequencies [14], calibrated max(p_t) values, depth-dependent functional forms of the turbulent-velocity anisotropy Φ and $T-\tau$ relations as suggested by 3D simulations [15, 16].

Acknowledgements

We thank Mikkel N. Lund for providing the LEGACY linewidths and Jørgen Christensen-Dalsgaard for the global stellar parameters of the frequency-calibrated ASTEC evolutionary models. Funding for the Stellar Astrophysics Centre is provided by The Danish National Research Foundation (Grant DNRF106).

References

[1] Balmforth N. J., MNRAS **255**, 603 (1992)

[2] Chaplin W. J., Houdek G., Elsworth Y., Gough D. O., Isaak G. R., New R., MNRAS **360**, 859 (2005)

[3] Christensen-Dalsgaard J., MNRAS **199**, 735 (1982)

[4] Christensen-Dalsgaard J., Ap&SS **316**, 13 (2008)

[5] Christensen-Dalsgaard J., Ap&SS **316**, 113 (2008)

[6] Christensen-Dalsgaard J., *ESF Conference: "The Modern Era of Helio- and Asteroseismology"* (M. Roth ed., Obergurgel), Astron. Nachr. **333**, 914 (2012)

[7] Gough D. O., *Lecture Notes in Physics* **71**, (Spiegel E. A., Zahn J.-P., eds, Springer, New York), 15 (1977a)

[8] Gough D. O., ApJ **214**, 196 (1977b)

[9] Houdek G., Balmforth N. J., Christensen-Dalsgaard J., Gough D. O., AA **351**, 582 (1999)

[10] Houdek G., Dupret M.-A., Living Rev. Sol. Phys., **12**:8 (2015)

[11] Houdek G., Trampedach R., Aarslev M. J., Christensen-Dalsgaard J., MNRAS **464**, 124 (2017)

[12] Iglesias C. A., Rogers F. J., ApJ **464**, 943 (1996)

[13] Kurucz R. L., *NATO Series C* **341**, (Crivellari L., Hubeny I., Hummer D. G., eds, D. Reidel Publishing Co., Trieste Italy), 441 (1992)

[14] Lund M. N., Silva Aguirre V., Davies G. R., Chaplin W. J., Christensen-Dalsgaard J. et al., submitted to ApJ (2016)

[15] Trampedach R., Asplund M., Collet R., Nordlund Å., Stein R. F., ApJ **769**, 18 (2013)

[16] Trampedach R., Stein R. F., Christensen-Dalsgaard J., Nordlund Å., Asplund M., MNRAS **442**, 805 (2014)

[17] Spiegel E. A., Veronis G., ApJ **131**, 442 (1960)

[18] Unno W., Spiegel E. A., PASJ **18**, 85 (1966)

Insights from Asteroseismology of Massive Stars: The Need for Additional Angular Momentum Transport Mechanisms

Ehsan Moravveji[1],*

[1] *Institute of Astronomy, KU Leuven, Celestijnenlaan 200D, B-3001 Leuven, Belgium*

Abstract. In massive stars, rotation and oscillatory waves can have a tight interplay. In order to assess the importance of additional angular momentum transport mechanisms other than rotation, we compare the asteroseismic properties of a uniformly rotating model and a differentially rotating one. Accordingly, we employ the observed period spacing of 36 dipole g-modes in the Kepler $\sim 3.2\,M_\odot$ target KIC 7760680 to discriminate between these two models. We favor the uniformly rotating model, which fully satisfies all observational constraints. Therefore, efficient angular momentum transport by additional mechanisms such as internal gravity waves, heat-driven modes and magnetic field is needed during early main sequence evolution of massive stars.

1 Introduction and objective

Massive stars are moderate to fast rotators [6]. Rotation modifies the thermal and structural equilibrium in stars, launches meridional circulation currents [29], and triggers a handful of hydrodynamical instabilities (Maeder 2009, [11]). Stars with masses exceeding $\sim 1.3\,M_\odot$ have diminishing outer convective envelopes, and above $\sim 2\,M_\odot$ the envelope becomes predominantly radiative (except for the narrow helium and/or iron opacity bumps). This radiative region provides a suitable propagation cavity for pressure (p-) and gravity (g-) modes to traverse it back and forth roughly on the dynamical timescale $\tau_{\rm dyn}$. Because the nuclear timescale $\tau_{\rm nuc}$ is much longer than the dynamical timescale, the oscillatory motions in stars have sufficient time to interact with the background star. One critical interaction is an efficient redistribution of angular momentum inside stars by non-radial oscillations [1, 8, 9, 25, 28], or through internal gravity waves [20, 21]. Townsend (2009, [25]) has demonstrated that in a $5M_\odot$ star, a very efficient angular momentum transport in the envelope takes place in less than 1 000 years. The coupling between the stellar structure and the pulsations is non-linear, and forms an interdependency: the oscillations can modify the structure on short timescales and the structure determines the oscillation properties. Therefore in massive stars, the coexistence of rotation and pulsation can have profound consequences and can impact the global observables (e.g. abundances, surface rotation, mass loss rate).

The conservation of angular momentum determines the coupling between the vertical component of the angular velocity $\Omega(r)$, the velocity of the advective meridional circulation currents U, and the angular momentum flux carried by waves $\mathcal{F}$ (see e.g. [4, 11, 29] for more details)

$$\frac{\partial}{\partial t}\left(\rho r^2\Omega\right) = \frac{1}{5r^2}\frac{\partial}{\partial r}\left(\rho r^4\Omega U\right) + \frac{1}{r^2}\frac{\partial}{\partial r}\left(\rho\, \nu\, r^4\frac{\partial\Omega}{\partial r}\right) - \frac{3}{2r^2}\frac{\partial}{\partial r}(r^2\mathcal{F}), \qquad (1)$$

where, ν is the vertical turbulent diffusivity. In some stellar structure and evolution models (e.g. Heger et al. 2000, [5]), the advective term is approximated as a diffusion process. Through Eq. (1), the angular momentum flux $\mathcal{F}$ can be extracted from one part of the star, and deposited in a different region. Consequently, some layers inside stars spin up, while some other regions spin down, trying to conserve the net amount of angular momentum. Eventually, some angular momentum will be lost through wind mass loss and/or via binary interaction.

How significantly can the waves contribute to the angular momentum transport in rotating stars? To answer this quantitatively, one must compare models that incorporate the last term in Eq. (1), and exploit the observable consequences [1, 8]. In this study, however, we present a preliminary and *qualitative* comparison between two models that represent two possible extremes, without any direct computation of extra angular momentum flux $\mathcal{F}$. One model is forced to rotate uniformly (abbreviated as UR), and it represents a case where waves contribute very efficiently to maintain rigid rotation. The other model has a differential rotation (abbreviated as DR). The latter represents ignoring the angular momentum flux carried by waves $\mathcal{F} = 0$, and obeys the standard angular momentum transport by rotation in a diffusion approximation [5, 19].

The high-precision Kepler observations of pulsating stars has opened a unique window to empirically constrain the angular momentum distribution inside heat-driven pulsators. This facilitates putting our knowledge of angular

*e-mail: Ehsan.Moravveji@kuleuven.be

momentum transport, and the underlying assumptions under scrutiny. As a demonstration, the radiative envelopes of three evolved γ Dor stars – near the end of core H burning phase – were shown to rotate almost uniformly [7, 14, 22]. The descendant of these stars are the red giants which are also shown to have a core-to-surface rotation rate slower than the model predictions [2].

In this paper, we concentrate on the moderately rotating and slowly pulsating B-type star KIC 7760680. Our stellar models are computed using MESA [17], which incorporate the effect of rotation [18]. The theoretical adiabatic frequencies are computed using GYRE [24], which treats the effects of rotation on pulsations through the Traditional Approximation of Rotation (TAR) [3]. Our inferences regarding model UR is heavily based on the in-depth seismic modeling presented by Moravveji et al. (2016, hereafter Paper1, [12]). We explicitly try to address two questions, in the light of strict observational (asteroseismic and spectroscopic) constraints that we place on this star. (Q1) Which of the rigidly rotating (UR) or differentially rotating (DR) models fulfil the observations? (Q2) What is the minimum required vertical diffusive mixing to maintain (near) rigid rotation?

2 Observational constraints

KIC 7760680 was thoroughly studied by Pápics et al. (2015, [15]). From spectroscopy the projected surface rotation velocity is $v \sin i = 62 \pm 5$ (km sec^{-1}). This is the first requirement that we try to fulfil in model UR and DR. In Paper1, stringent constraints were placed on the global and mixing parameters of this star through forward seismic modeling. In short, KIC 7760680 is a ~ 3.25 $M_\odot$ B8V star, with core hydrogen mass fraction $X_c = 0.503 \pm 0.001$. Assuming a solid body rotation, and trying to fulfil the observational constraints during asterseismic modeling, the surface rotation frequency is tightly bound to $\Omega_{surf} = 2\pi f_{rot} = 3.0096^{+0.0415}_{-0.0591}$ radian day^{-1}.

KIC 7760680 exhibits $N_{obs} = 36$ consecutive dipole prograde g-modes. The first and the last mode in this series have periods of $P_1 = 0.86930$ and $P_{36} = 1.46046$ days, respectively. The $N_{obs} = 36$ is the second asteroseismic constraint that we strictly require in this study. The motivation for this requirement is the high sensitivity of inertial g-modes to the rotation rate (see e.g. Fig. 2 in Townsend 2005, [26]).

3 Uniform versus differential rotation

To attack the first question Q1, we construct two MESA models (UR and DR) that have very similar stellar mass, radius, mass of the convective core M_{cc} and center hydrogen mass X_c. Table 1 summarises the parameters of both models. Based on Paper1, the model UR was computed from non-rotating MESA tracks, and the uniform rotation was only imposed during the GYRE computations of adiabatic frequencies. Model DR starts off the main sequence with an initial surface rotation velocity $v_{ini} = 75$ km sec^{-1}. The evolution of model DR is terminated at $X_c = 0.495$, and the surface rotation velocity has evolved to 68 km sec^{-1}. Thus, both UR and DR satisfy the first requirement in Section 2.

Fig. 1 compares model UR (dashed line) and model DR (solid line). The extent of the convective core (dark gray), and the core overshooting region (light gray) are

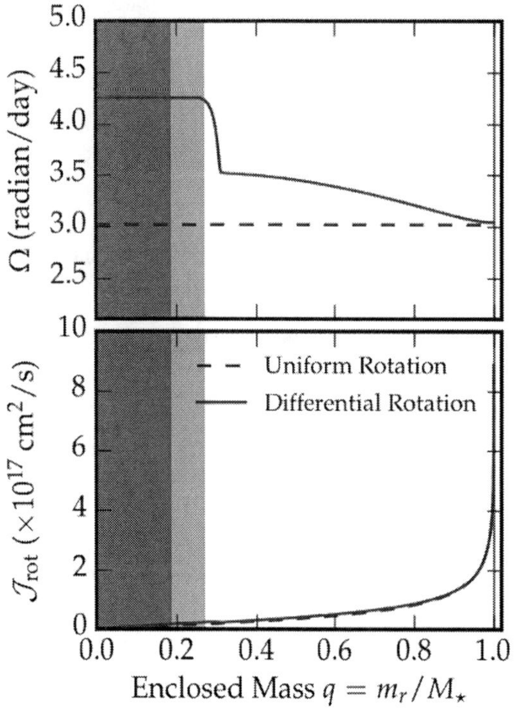

Figure 1: Comparison between the uniformly (UR) and differentially (DR) rotating models. The former is shown with a dashed blue line, and the latter with a solid red curve. Table 1 lists their global parameters. (Top) The internal angular velocity versus enclosed mass. Both models have identical radius and surface rotation velocity. (Bottom) The internal angular momentum distribution $\mathcal{J}_{rot}$ versus enclosed mass. The model DR has a slightly higher angular momentum across the interior.

Table 1: The physical parameters of the models UR and DR. OV, Diff and Rot stand for core overshoot mixing, vertical diffusive mixing and rotational mixing, respectively. v_{rot} is the initial rotation rate at zero-age-main-sequence.

	Model UR	Model DR
MESA	non-rotating	rotating
ZAMS v_{rot} (km sec^{-1})	0	75
GYRE	TAR	TAR
Mixing	OV + Diff.	OV + Rot.
Star mass ($M_\odot$)	3.25	3.20
Core mass ($M_\odot$)	0.621	0.606
X_c	0.503	0.495
Radius ($R_\odot$)	2.79	2.80

highlighted. The top panel shows the vertical angular velocity $\Omega(r)$. Model UR is assumed (and enforced) to rotate uniformly, whereas model DR rotates differentially. Both models have roughly equal surface angular velocity, and radius (Table 1), hence similar surface rotation velocity. Due to the efficient core boundary mixing by overshooting, this narrow region strongly contributes to the transport of angular momentum from the core region to the envelope. Thus, the narrow overshooting layer rotates synchronously with the core. The bottom panel shows the specific angular momentum $\mathcal{J}_{\rm rot} = i_{\rm rot}\Omega$, where $i_{\rm rot} \approx 2r^2/3$ is the specific moment of inertia. Given the faster inner rotation of the model DR, it has slightly higher angular momentum $\mathcal{J}_{\rm rot}$. This important feature is discussed in Section 5.

4 The asteroseismic test

To answer the first question Q1, we compute the period spacing for both models UR and DR, and verify if they fulfil the second requirement in Section 2, i.e. $N_{\rm obs} = 36$ between P_1 and P_{36}. Fig. 2 compares the period spacing $\Delta P_n = P_{n+1} - P_n$ from observation (top) with that of model UR (middle) and that of model DR (bottom). Model UR is the best seismic model of KIC 7760680 (Paper I), and it reproduces the number of observed modes $N_{\rm UR} = N_{\rm obs}$, in addition to the slope and the morphology of the observed period spacing. Thus, model UR satisfies both observational requirements in Section 2.

Model DR, however, has $N_{\rm DR} = 156$ modes between P_1 and P_{36}, which strongly contradicts with the observations (Fig. 2 top). The period spacing of the very high-order g-modes in this model approach zero, whereas the observed ΔP for the last mode in the series is 678 sec. Furthermore, the morphology of the period spacing pattern in this model does not resemble that of KIC 7760680, due to strong differences with the observed one. The fact that model DR does not fulfil the second observational requirement disqualifies it from representing KIC 7760680.

The current model of the diffusion of angular momentum employs various ad-hoc efficiency factors [5, 19]. These factors are poorly calibrated against observations, and are designed to suppress or boost the role of non-rotating and rotating mixing coefficients in transporting the angular momentum through the diffusion coefficient ν in Eq. (1). For choosing model DR, we varied most of the efficiency factors up to 5 orders-of-magnitude. However, the general conclusion did not significantly change; thus, we set the efficiency factors to the default value of unity. Furthermore, we also selected models closer to the zero-age-main sequence $X_c \gtrsim 0.60$ with slower initial rotation velocity $v_{\rm rot} \geq 60$ km sec^{-1}. For such models, $N_{\rm DR}$ approaches $N_{\rm obs}$. But on the flip side, the morphology of the period spacing (the number and position of dips) never resembled the observations. This is expected. The less-evolved models cannot develop extended μ-gradient layers on top of their receding convective cores; thus, the period spacings of models close to the zero-age-main-sequence hardly deviate from the asymptotic period spacing (straight line). In conclusion, we have so far not found any differentially rotating model that respects the observational constraints in Section 2, in addition to providing a reasonable match to the observed period spacing. As a result, we favor model UR, or any other model with near-uniform rotation as the more likely representative of the internal distribution of angular momentum in KIC 7760680.

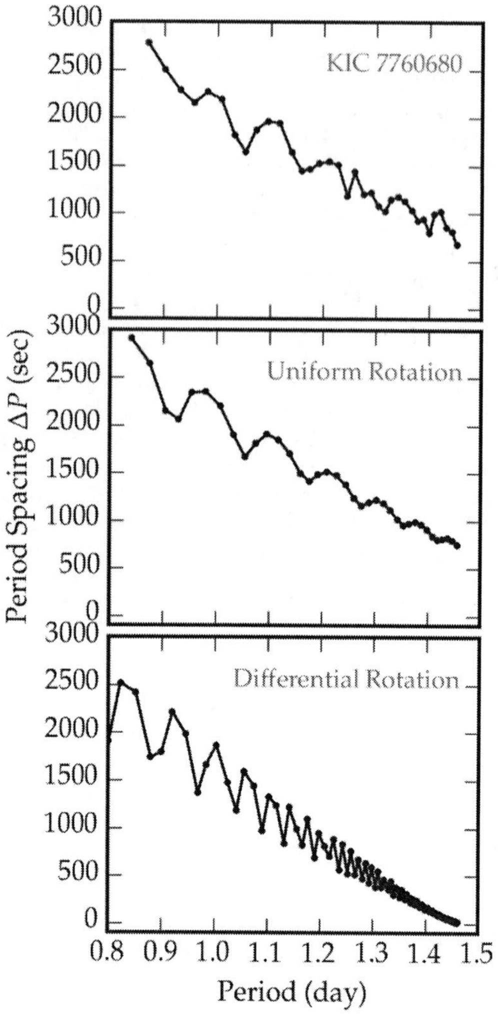

Figure 2: The period spacing versus mode period. (Top) The observed period spacing of KIC 7760680 forming a series of $N_{\rm obs} = 36$ consecutive g-modes with period in the range 0.86930 and 1.46046 days. (Middle) The period spacing from model UR with $N_{\rm UR} = 36$. The properties of this model are elaborated in [12]. This model satisfies both observational constraints (Section 2). (Bottom) The period spacing from model DR. This model has $N_{\rm DR} = 156$ modes in the observed period range, which is incompatible with observations.

DOI: 10.1051/epjconf/201716002004

5 Discussion and conclusions

Let's revisit Fig. 1. The difference between the angular momentum profile $\mathcal{J}_{\rm rot}$ in both models is minute, whereas their vertical angular velocities $\Omega(r)$ are different. In order to reduce the gradient of angular velocity in model DR, a small amount of additional angular momentum must be transferred from the near-core region to the envelope. Furthermore, small amount of angular momentum must be deposited from the star, to spin down the core to the same level of the envelope, and allow the total angular momentum $\mathcal{J}_{\rm rot}$ to approach that of model UR.

The need for additional angular momentum flux implies that $\mathcal{F} \neq 0$ in Eq. (1). Among possible mechanisms, other than rotation, that can actively redistribute angular momentum within short timescales (compared to nuclear timescale) are the transport by internal gravity waves [20, 21, 23] and heat-driven nonradial modes [1, 9, 25]. In fact, such additional angular momentum transport mechanisms are needed to explain the near-uniform rotation in the three slowly rotating Kepler γ Dor pulsators KIC 11145123 [7], KIC 9244992 [22] and KIC 7661054 [14], in addition to the slowly rotating and pulsating B star KIC 10526294 [13, 16, 27].

We now turn to the address how much turbulent diffusivity is needed in the radiative envelope to sustain uniform rotation (Q2 in Section 1)? In Paper I, we constrained the extra diffusive mixing above the overshooting zone to $\log D_{\rm ext} = 0.75 \pm 0.25$ cm^2 sec^{-1}. This value is between 100 to 1 000 times less than the rotationally-induced mixing coefficients due to e.g. shear, meridional circulation and other proposed mechanisms in [5] and [10]. Therefore, in massive stars, small amount of turbulent diffusivity is sufficient to maintain near-uniform rotation in their radiative envelopes. This is understandable in the light of Eq. (1); positive additional angular momentum flux $\mathcal{F} > 0$ would reduce turbulent diffusivity ν needed to sustain rigid rotation.

Asteroseismology of massive stars offers a great opportunity to exploit the inner workings of massive stars, which are inaccessible to us through other means. This encourages to direct the future research to provide quantitative answer to questions such as (a) how do oscillatory waves engage in angular momentum redistribution? and (b) how much turbulent diffusivity is needed to explain pulsation behaviour of stars, and maintain near-uniform rotation?

Acknowledgement

E. Moravveji is grateful to the organizers of the KASC9/TASC2 conference. The research leading to these results has received funding from the People Programme (Marie Curie Actions) of the European Union's Seventh Framework Programme FP7/2007-2013/ under REA grant agreement N° 623303 (ASAMBA).

References

[1] H. Ando, PASJ, **35**, 343 (1983)
[2] P.G. Beck, J. Montalban, T. Kallinger, et al., Nature, **481**, 55 (2012)
[3] M.P. Bouabid, M.A. Dupret, S. Salmon, et al., MNRAS, **429**, 2500 (2013)
[4] B. Chaboyer, J.P. Zahn, A&A, **253**, 173 (1992)
[5] A. Heger, N. Langer, S.E. Woosley, ApJ, **528**, 368 (2000)
[6] W. Huang, D.R. Gies, M.V. McSwain, ApJ, **722**, 605 (2010)
[7] D.W. Kurtz, H. Saio, M. Takata, et al., MNRAS, **444**, 102 (2014)
[8] U. Lee, S. Mathis, C. Neiner, MNRAS, **457**, 2445 (2016)
[9] U. Lee, H. Saio, MNRAS, **261**, 415 (1993)
[10] A. Maeder, G. Meynet, N. Lagarde, C. Charbonnel, A&A, **553**, A1 (2013)
[11] A. Maeder, *Physics, Formation and Evolution of Rotating Stars* (Springer Berlin Heidelberg, 2009)
[12] E. Moravveji, R.H.D. Townsend, C. Aerts, S. Mathis, ApJ, **823**, 130 (2016)
[13] E. Moravveji, C. Aerts, P.I. Pápics, et al., A&A, **580**, A27 (2015)
[14] S.J. Murphy, L. Fossati, T.R. Bedding, et al., MNRAS, **459**, 1201 (2016)
[15] P.I. Pápics, A. Tkachenko, C. Aerts, et al., ApJL, **803**, L25 (2015)
[16] P.I. Pápics, E. Moravveji, C. Aerts, et al., A&A, **570**, A8 (2014)
[17] B. Paxton, L. Bildsten, A. Dotter, et al., ApJS, **192**, 3 (2011)
[18] B. Paxton, M. Cantiello, P. Arras, et al., ApJS, **208**, 4 (2013)
[19] M.H. Pinsonneault et al. , S.D. Kawaler, S. Sofia, P. Demarque, ApJ, **338**, 424 (1989)
[20] T.M. Rogers, ApJL, **815**, L30 (2015)
[21] T.M. Rogers et al. ApJ, **772**, 21 (2013)
[22] H. Saio, D.W. Kurtz, M. Takata, et al., MNRAS, **447**, 3264 (2015)
[23] S. Talon, C. Charbonnel, A&A, **440**, 981 (2005)
[24] R.H.D. Townsend, S.A. Teitler, MNRAS, **435**, 3406 (2013)
[25] R.H.D. Townsend, American Institute of Physics Conference Series **1170**, 355 (2009)
[26] R.H.D. Townsend, MNRAS, **360**, 465 (2005)
[27] S.A. Triana, E. Moravveji, P.I. Pápics, et al., ApJ, **810**, 16 (2015)
[28] W. Unno, Y. Osaki, H. Ando, H. Saio, H. Shibahashi, *Nonradial oscillations of stars* (University of Tokyo Press, Tokyo, 1989)
[29] J.P. Zahn, A&A, **265**, 115 (1992)

DOI: 10.1051/epjconf/201716002005

An opaque Sun?

The potential for future, higher opacities to solve the solar abundance problem

Regner Trampedach[1,2,⋆]

[1] *Space Science Institute, 4750 Walnut Street, Suite 205, Boulder, CO 80301 USA*
[2] *Stellar Astrophysics Centre, Dept. of Physics and Astronomy, Ny Munkegade 120, Aarhus University, DK–8000 Aarhus C, Denmark*

Abstract. Last year Bailey et al. announced their measurement of iron opacity that increases the Rosseland mean at the base of the solar convection zone by 7%. I ask what happens if the absorption by other elements is also stronger than predicted so far.
Artificially increasing the absorption by other elements, proportional to the number of bound electrons in the absorber (reflecting our remaining ignorance of atomic physics) gives an opacity increase for a solar model, that has the potential to solve the long-standing solar abundance problem.
Conclusion: Opacities are the likely source of the solar abundance problem, and the solar abundances are likely closer to those of Asplund et al. (2009) than to the various classic sets of abundances.

1 Introduction

Solar abundances, as derived from spectra originating in the photosphere, are used as the reference for stellar abundances. Stars are usually assumed to have individual H and He abundances, but to have the same mixture of metals as does the Sun. This means that solar abundance determinations can affect our interpretation of a major part of astronomical observations.

A number of solar abundance analyses were published in the 1990's and form the *classic* solar abundances, e.g., [1–4] and [5] (the latter, from hereon GN93). They were based on 1D semi-empirical atmosphere models, e.g., [6].

The *modern* solar abundances, based on 3D simulations of deep, convective, radiation-coupled, solar atmospheres were first published by [7], challenged by [8] and [9] (from hereon CLSFB11) and updated by [10] (from hereon AGSS09). The latter were confirmed by improved analysis of the same solar 3D convection simulation, but with improved atomic physics and statistical analysis by [11–13] (from hereon SGA15). The crucial light elements are still awaiting a similar re-analysis. The differences from GN93 to AGSS09 and SGA15 are shown in Fig. 1.

The AGSS09 abundances constitute a reduction of C, N and O to about 70% of the classic GN93 values. In particular oxygen contributes significant opacity from the bottom of the solar convection zone and inward, so the reduction in [O/H] changes the interior structure of our solar models and moves the bottom of the convective envelope closer to the surface by many standard deviations.

This disrupts the previous agreement between models and helioseismic inversions, as is well documented by, e.g., [14, 15]. This is the so-called solar abundance problem. Many solutions to this problem have been proposed [16–20], but none of them have succeeded in achiev-

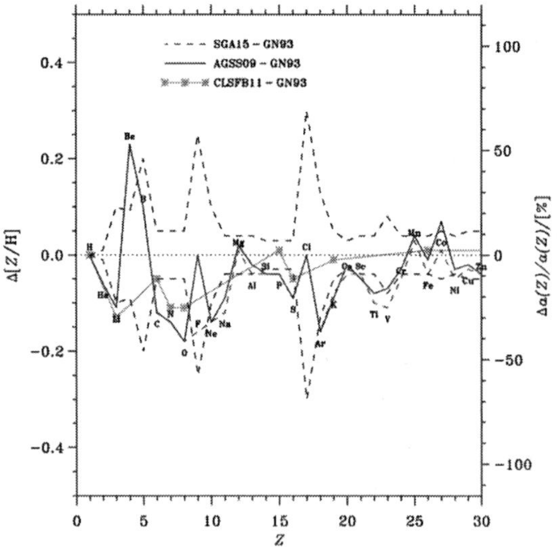

Figure 1: Some modern solar abundance determinations, relative to the 'classic' GN93 abundances. The dashed black lines show the error-bars of the SGA15 abundances, making everything outside, significantly different from the classic GN93 abundances, with C, N and O constituting the most significant difference. AGSS09 (blue) and SGA15 (red dashed) are very similar, whereas CLSFB11 (green) is closer to the classic mixture.

⋆e-mail: rtrampedach@SpaceScience.org

DOI: 10.1051/epjconf/201716002005

ing simultaneous agreement with the seismically inferred helium abundance *and* depth of the convection zone, as demonstrated by [21].

It is natural to question the 3D solar atmosphere simulations that precipitated this disagreement with helioseismology. We did [22], and found that the simulations exhibit an unprecedented simultaneous agreement with a wide range of solar observations, including continuum limb-darkening, spectral flux distribution, the Hα,β and Paβ,γ lines of hydrogen. Even the spatially (granule-scale) resolved oxygen lines (including non-LTE effects) match observations [23, 24]. This rather remarkable, considering the lack of parameters that are free to fit any of these results.

It should also be pointed out that [25] argue strongly against using the classic solar abundances, with Grevesse & Sauval being authors of both this rebuttal, many of the classic abundance determinations, as well as AGSS09. [25] found the main differences between the classic and AGSS09 abundances to be largely due to improved atomic line-parameters, the homogeneous analysis carried out by AGSS09, increased computer-power for non-LTE calculations and the much improved detection of blends due to the simulation's good agreement with the detailed C-shaped line-profiles. Direct 3D effects of the simulations are rather small except for molecular lines.

2 Opacities and Abundances

2.1 Opacity Solutions to the solar abundance problem

Only two viable solutions to the solar abundance problem, has bee found so far: Change the abundances back to the classic values, or increase the opacity below the solar convection zone. [26] determined the profile of opacity enhancements that would make a solar model with AGSS09 abundances have the same structure as Model S by [27]. This standard solar model was evolved with the classic GN93 abundances (dashed, black curve in Fig. 2, marked 'f_κ(AGSS09)'). The implication being that such a modified AGSS09 model would show similar helioseismic agreement as the highly successful Model S. A similar result was found by [28] (dot-dashed, black curve in Fig. 2, marked 'f_κ(AGS05)/2'). Whether such opacity increases were but wishful thinking, was not known at the time.

2.2 Laboratory Experiments

[29] (from hereon B15) carried out the impressive feat of measuring an iron plasma at conditions close to those at the bottom of the solar convection zone: a temperature of $T = 2.11 \times 10^6$ K and electron density of $N_e = 3.1 \times 10^{22}$ cm^{-3}. This N_e is still about three times lower than the solar value at that temperature, though. B15 measured the transmission spectra of the plasma (predominantly Fe^{16+}–Fe^{18+}) and computed the underlying absorption coefficient (ensuring that deviations from local thermodynamic equilibrium, homogeneity and stationarity were insignificant). This absorption coefficient is generally larger, and has broader features than any current opacity calculations. Specifically, compared to the [30] (OP05) Fe opacity at the same conditions, the observed ones are 1.5–1.8 times larger in the rather featureless, short-wavelength, continuum below 9.5 Å. In the minima between lines (the part that determines the harmonic Rosseland mean) the observed absorption is 3–5 times stronger. B15 did the experiment of calculating the OP05 Rosseland mean for a full solar mixture, and found that substituting the theoretical by the observed Fe absorption, increased this opacity by $7 \pm 3\%$. This is a substantial amount, and far from trivial to explain theoretically [31]. And it is about half of the opacity increase needed to reconcile an AGSS09 solar model with helioseismology.

2.3 OP05 Opacity Experiments

If the experimental results on iron are not somehow a special property of iron, then the absorption by other elements should also be larger than predicted by current calculations. Experiments are therefore carried out on the OP05 absorption coefficients of individual elements, summed up over the AGSS09 mixture and Rosseland averaged along the fixed ϱ, T-stratification of Model S [27].

The first step is to broaden the Fe absorption to make it qualitatively similar to Fig. 3 of B15, as shown with blue dotted line in Fig. 2. The next step is a wavelength independent enhancement factor on Fe, calibrated to give the observed 7% increase in Rosseland opacity of the full mixture (blue solid in Fig. 2). This is a crude but simple simulation of B15's results. More than qualitative agreement cannot be expected, as the present work does not offer an improved opacity calculation, but merely asks whether such improved opacity calculations are likely to solve the solar abundance problem, given the B15 experiment.

The next and crucial step is to extrapolate from iron to the 16 other elements included in OP05. We know the simple hydrogen- and helium-like ions very well, and it would be hard to argue for any phenomena that could increase their absorption. One- and two-electron systems are therefore kept at their current theoretical level of absorption. This already gives an electron dependency of the absorption enhancements, up to the 8–10 electron Fe ions, present in the plasma of the B15 experiment. A linear dependency is favoured for its simplicity. Whether to use the number of bound electrons in the absorber, or the number of closed electron shells (as motivated in Sect. 2.4 and shown with solid magenta in Fig. 2), has little effect on the final opacity increase. The opacity enhancements shown with solid lines in Fig. 2 also include a slight increase with $\log T$, designed to give no enhancement at $\log T = 5$. This is seen to give good agreement with the needed enhancement found by [26] (dashed black curve). Having no T-dependency makes it closer to the needed enhancement found by [28]. Either way, the artificial opacity enhancement, calibrated against the B15 experiment, has the correct shape and magnitude to solve the solar abundance problem.

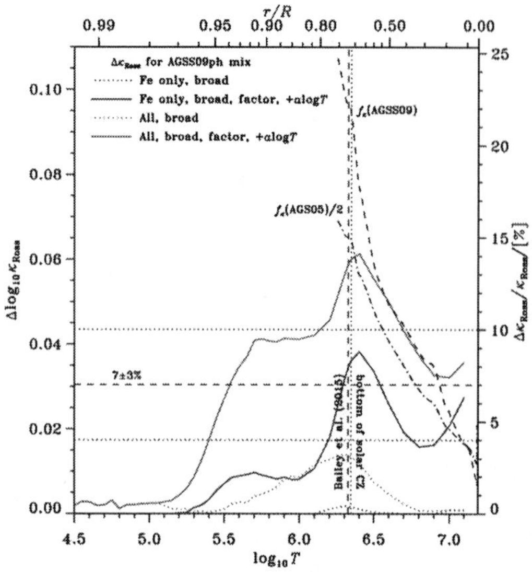

Figure 2: Opacity enhancements, $\Delta\kappa_{\mathrm{Ross}}$, along the fixed ϱ, T-stratification of standard solar Model S [27], as $\Delta \log \kappa$ on the left-hand scale and fractional change on the right-hand scale. Position in the model is shown both on the temperature scale, $\log T$ (bottom), and the fractional radius, r/R-scale (top). Blue curves shows only Fe enhanced, calibrated to the experimental result (horizontal and vertical dashed lines), and magenta curves show result of artificially enhancing the absorption by all elements. Dotted lines show the effect of broadening only, and solid lines also include the enhancement according to the number of closed electronic shells in the absorbing ions. The dashed, black curve marked 'f_κ(AGSS09)' is the opacity enhancement found by [26] to restore agreement with helioseismology, and the dot-dashed curve marked 'f_κ(AGS05)/2' is that found by [28].

2.4 New Opacity Calculations

Recently OPAS-team, [32], and the Next Gen Los Alamos opacity-team, [33], published their newest opacity calculations, aimed at solving the solar abundance problem. They only obtained 13–25% of the increase needed to restore agreement, however, again demonstrating how persistent the problem is.

New calculations by [34] has looked at how core electrons can increase the absorption. It has usually been assumed that only the outermost, valence electrons, interacts with the radiation field. Radiation is so intense at these temperatures, however, that the electrons of filled shells can be excited, often together with a valence electron. This gives rise to a whole new absorption system of continua and lines. Some of these processes were included in approximate ways for the OP05 calculations. The new calculation by [34] on the other hand, dispenses with some crucial approximations, and also check for completeness in terms of configurations included in the parent and target states. Their impressive calculation seems converged when including up to $n = 4$ complexes of the target ion, and gives continuum enhancements of orders of magnitude for absorption from individual states of the parent ion (although only a 9% increase for ground-state absorption), and in the right wavelength region to explain B15's results.

3 Conclusions

Solar abundance analysis based on realistic 3D convective atmosphere simulations, result in lower C, N and O abundances. This in turn makes for solar models that disagree with helioseismology by many standard deviations, known as the solar abundance problem (see Sect. 1). It has been shown that agreement can be restored if the opacities are artificially increased by 20% at the bottom of the convection zone, down to a few percent increase at the centre (see Sect. 2.1). Whether such increases in opacity were realistic was unknown at the time.

The experiments by B15 directly measured the iron absorption at conditions close to those at the bottom of the solar convection zone, and found larger and broader absorption than any theoretical calculations to date.

Performing experiments on the OP05 absorption coefficients, artificially enhanced and broadened to match the iron experiment, gives an increase of the Rosseland mean of the AGSS09 mixture, that has the potential to solve the solar abundance problem. The enhancement was extrapolated from iron, by scaling it by the number of filled electronic shells (see Sect. 2.3), reflecting our remaining ignorance of atomic physics, as well as the evidence for excited core electrons being the culprit.

And the theoretical work by [34], on absorption by core electrons in filled shells, does indeed point to this being the missing absorption in B15's experiments (see Sect. 2.4).

New physics in the opacity calculations therefore seems to be the likely solution to the solar abundance problem - a problem that turned out not to have much to do with abundances after all, but rather is an unfortunate conspiracy of nature and the historic events that lead to the classic abundances — confirming the old adage that you can have the correct results for the wrong reasons.

The artificial opacity enhancements, presented here, should not be used for computing new opacity tables to be applied to stellar evolution calculations. This is merely an exercise to answer the question, whether improved opacity calculations can realistically solve the solar abundance problem, and the affirmation will hopefully encourage the efforts of the various teams working on such opacity calculations. It will have a broad impact on a large section of solar and stellar modelling, asteroseismology, Galactic evolution, etc.

These changes to stellar opacities will fortunately not affect the atmospheres of solar-like stars, and there will therefore not be a feed-back into the solar abundance determination. The two are safely decoupled. The author recommends the AGSS09 abundances, supplanted by the [11–13] for all but the light elements, as the most likely, present day composition of the solar atmosphere and convective envelope.

Acknowledgements

RT acknowledges funding from NASA grant NNX15AB24G. Funding for the Stellar Astrophysics Centre is provided by The Danish National Research Foundation (grant DNRF106). This research has made extensive use of NASA's Astrophysics Data System.

References

[1] E. Anders, N. Grevesse, Geochim. Cosmochim. Acta **53**, 197 (1989)
[2] N. Grevesse, *Solar Abundances: the Reference System*, in *Evolution of Stars: the Photospheric Abundance Connection*, edited by G. Michaud, A.V. Tutukov (Kluwer, Dordrecht, 1991), Number 145 in IAU Symp., pp. 63–69
[3] N. Grevesse, A. Noels, A.J. Sauval, *Standard abundances*, in *Cosmic Abundances*, edited by S.S. Holt, G. Sonneborn, ASP (ASP, San Francisco, 1996), Vol. 99 of *Conf. Ser.*, pp. 117–126
[4] N. Grevesse, A.J. Sauval, Space Sci. Rev. **85**, 161 (1998)
[5] N. Grevesse, A. Noels, *Cosmic abundances of the elements*, in *Origin and Evolution of the Elements*, edited by N. Prantzos, E. Vangioni-Flam, M. Cassé (Cambridge Univ. press, 1993), pp. 15–25
[6] H. Holweger, E.A. Müller, Sol. Phys. **19**, 19 (1974)
[7] M. Asplund, N. Grevesse, J. Sauval, *The solar chemical composition*, in *Cosmic abundances as records of stellar evolution and nucleosynthesis*, edited by T.G. Barnes III, F.N. Bash (Astron. Soc. Pac., San Francisco, 2005), Number 336 in ASP Conf. Ser., pp. 25–38
[8] K. Lodders, ApJ **591**, 1220 (2003)
[9] E. Caffau, H.G. Ludwig, M. Steffen, B. Freytag, P. Bonifacio, Solar Physics **268**, 255 (2011)
[10] M. Asplund, N. Grevesse, A.J. Sauval, P. Scott, ARA&A **47**, 481 (2009)
[11] P. Scott, N. Grevesse, M. Asplund, A.J. Sauval, K. Lind, Y. Takeda, R. Collet, R. Trampedach, W. Hayek, A&A **573**, A25:1 (2015)
[12] P. Scott, M. Asplund, N. Grevesse, M. Bergemann, A.J. Sauval, A&A **573**, A26:1 (2015)
[13] N. Grevesse, P. Scott, M. Asplund, A.J. Sauval, A&A **573**, A27:1 (2015)
[14] J.N. Bahcall, S. Basu, M.H. Pinsonneault, A.M. Serenelli, ApJ **618**, 1049 (2005)
[15] H.M. Antia, S. Basu, ApJ **620**, L129 (2005)
[16] J.J. Drake, P. Testa, Nature **436**, 525 (2005)
[17] J.A. Guzik, L.S. Watson, A.N. Cox, ApJ **627**, 1049 (2005)
[18] A.C. Vincent, P. Scott, R. Trampedach, MNRAS **432**, 3332 (2013)
[19] J.A. Guzik, K. Mussack, ApJ **713**, 1108 (2010)
[20] A.M. Serenelli, W.C. Haxton, C. Peña-Garay, ApJ **743**, 24:1 (2011)
[21] F. Delahaye, M.H. Pinsonneault, ApJ **649**, 529 (2006)
[22] T.M.D. Pereira, M. Asplund, R. Collet, I. Thaler, R. Trampedach, J. Leenaarts, A&A **554**, A118:1 (2013), 1304.4932
[23] T.M.D. Pereira, D. Kiselman, M. Asplund, A&A **507**, 417 (2009)
[24] T.M.D. Pereira, M. Asplund, D. Kiselman, A&A **508**, 1403 (2009)
[25] N. Grevesse, M. Asplund, A.J. Sauval, P. Scott, *Why GN93 should not be used anymore*, in *Ageing Low Mass Stars: From Red Giants to White Dwarfs*, edited by J. Montalbán, A. Noels, V. van Grootel (2013), Vol. 43 of *European Physical Journal Web of Conferences*, pp. 01004:1–3
[26] J. Christensen-Dalsgaard, G. Houdek, Ap&SS **328**, 51 (2010)
[27] J. Christensen-Dalsgaard, W. Däppen, S.V. Ajukov, E.R. Anderson, H.M. Antia, S. Basu, V.A. Baturin, G. Berthomieu, B. Chaboyer, S.M. Chitre et al., Science **272**, 1286 (1996)
[28] A.M. Serenelli, S. Basu, J.W. Ferguson, M. Asplund, ApJ **705**, L123 (2009)
[29] J.E. Bailey, T. Nagayama, G.P. Loisel, G.A. Rochau, C. Blancard, J. Colgan, P. Cossé, G. Faussurier, C.J. Fontes, F. Gilleron et al., Nature **517**, 56 (2015)
[30] N.R. Badnell, M.A. Bautista, K. Butler, F. Delahaye, C. Mendoza, P. Palmeri, C.J. Zeippen, M.J. Seaton, MNRAS **360**, 458 (2005)
[31] C.A. Iglesias, High Energy Density Physics **15**, 4 (2015)
[32] M. Le Pennec, S. Turck-Chièze, S. Salmon, C. Blancard, P. Cossé, G. Faussurier, G. Mondet (2015), 1510.05600
[33] J. Colgan, D.P. Kilcrease, N.H. Magee Jr., M.E. Sherrill, J. Abdallah, Jr., P. Hakel, C.J. Fontes, J.A. Guzik, K. Mussack (2016), 1601.01005
[34] S.N. Nahar, A.K. Pradhan, Phys. Rev. Lett. **116**, 235003:1 (2016)

2D dynamics of the radiative core of low mass stars

Delphine Hypolite[1,*], *Stéphane* Mathis[1], and *Michel* Rieutord[2]

[1] *Laboratoire AIM Paris-Saclay, CEA/DRF - CNRS - Université Paris Diderot, IRFU/SAp Centre de Saclay, F-91191 Gif-sur-Yvette Cedex, France*

[2] *Institut de Recherche en Astrophysique et Planétologie, Observatoire Midi-Pyrénées, Université de Toulouse, 14 avenue Edouard Belin, 31400 Toulouse, France*

Abstract. Understanding the internal rotation of low mass stars all along their evolution is of primary interest when studying their rotational dynamics, internal mixing and magnetic field generation. In this context, helio- and asteroseismology probe angular velocity gradients deep within solar type stars at different evolutionary stages. Still the rotation close to the center of such stars on the main sequence is hardly detectable and the dynamical interaction of the radiative core with the surface convective envelope is not well understood. For instance, the influence of the differential rotation profile sustained by convection and applied as a boundary condition to the radiation zone is very important in the formation of tachoclines. In this work, we study a 2D hydrodynamical model of a radiative core when an imposed, solar or anti-solar, differential rotation is applied at the upper boundary. This model uses the Boussinesq approximation and we find that the shear induces a cylindrical differential rotation associated with a unique cell of meridional circulation in each hemisphere (counterclockwise when the shear is solar-like and clockwise when it is anti-solar). The results are discussed in the framework of seismic observables (internal rotation rate, core-to-surface rotation ratio) while perspectives to improve our modeling by including magnetic field or transport by internal gravity waves will be discussed.

1 Introduction

Asteroseismology has revealed the internal rotation profile of numerous stars over the past few years. On one hand, we now have partial access to the dynamical state of low mass stars on the main sequence and during sub giant and red giant evolution phases ([3], [12], [13], [14], [4]). On the other hand, the internal rotation profile of the Sun deep within the star, until $0.2R_\odot$, has been inverted ([11], [20]). Both helioseismology and asteroseismology show that a strong transport of angular momentum occurs in radiative zones. These constrains call for a new generation of stellar models since current 1D stellar models do not recover inverted quantities such as for instance the flat rotation profile of the radiative zone of the Sun (e.g. [46], [48]) and of more evolved stars ([15], [9], [32]).

Indeed, one who wants to model the evolution of stars gets immediately confronted with the complexity of stellar rotation ([29], [31]). It is known to have multiple strong consequences on stellar evolution since it drives rotational mixing and transport of angular momentum through the stable radiative zones ([50], [28], [34]) on secular timescale. In addition, it is connected with the generation of stellar magnetic fields and the presence of tachoclines ([43], [23], [6], [45]) requesting a deep understanding of the stellar rotation. One of the most challenging aspects is the multidimensional nature of rotating stars ([38], [16]). Indeed, the centrifugal acceleration leads to the rise of baroclinic flows which can not be fully described by low spherical harmonics expansion and/or latitudinal averages. For these reasons, we propose a 2D approach to describe the secular evolution of the internal dynamics of rotating radiation zones. In this framework, the convective envelope applies a latitudinal shear through differential rotation on the underlying radiative core in low mass stars. For example, in the Sun, the rotation profile of the envelope is conical with fast equatorial regions and slower poles ([42], [5]). Past studies have looked for the consequences of such a shear in the solar case ([17],[19]). However, it calls for a generalisation of the dynamical boundary conditions since 3D numerical simulations reveal that solar-like differential rotation is a function of stellar fundamental parameters and mean rotation rate while anti-solar differential rotation can also be expected in cool stars ([37], [26], [22], [49]).

In this framework, the question we address is: *How does the differential rotation of a convective envelope affect the dynamics* (differential rotation and meridional circulation) *of the radiative core in low mass stars?*

In this work, we use 2D Boussinesq modeling to perform a parametric study over the shear applied as a boundary condition on the radiative core and compute the latitudinally averaged core to the surface rotation ratio of the resulting models which can be compared with seismic observables.

* e-mail: delphine.hypolite@cea.fr

DOI: 10.1051/epjconf/201716002006

We describe the 2D modeling of the radiative core of low mass stars and its main assumptions in Sect. 2. In Sect. 3, we compute the core-to-surface rotation ratio numerically as a function of the shear applied at the top of the domain and discuss the fast rotating solar case. We give our conclusions and perspectives in Sect. 4.

2 Modeling low mass stars radiative core in 2D

We consider a viscous flow enclosed within a rotating spherical shell. We suppose the centrifugal acceleration does not deform it. We study the interaction of the combined effect of the inertia with rotation in a stably stratified environment, leaving aside magnetic fields ([18], [35], [45]), internal gravity waves ([10], [47]) and anisotropic turbulence ([50], [33], [27]) as a first step. Because we are interested in the evolution of the fluid on secular timescale, we solve as a first step the steady equation of the vorticity combined with the equations of energy and continuity using the Boussinesq approximation ([38], [40], [24]). This approximation retains density variations only in the buoyancy term taking into account both the gravity and centrifugal accelerations and implies that the stratification is only taken into account through the Brunt-Väisälä frequency profile (positive in a radiative zone). It is given as an input of the simulations using the 1D MESA stellar evolution code ([41]) to generate ZAMS low mass stars radiative core models (we use a metallicity of $Z = 0.02$, and a mixing length theory parameter of $\alpha_{\rm MLT} = 2$) and we seek the resulting velocity field, i.e. the differential rotation and the meridional circulation for a $1M_\odot$ model. As developed by [38], hereafter R06, and [25], the scaled equations we solve are

$$\left\{\begin{array}{l} \vec{\nabla}\times(\vec{e}_z\times\vec{u}-\theta_T\vec{r}-E\Delta\vec{u}) = -n^2(r)\sin\theta\cos\theta\,\vec{e}_\varphi \\ \left(\dfrac{n^2(r)}{r}\right)u_r = \dfrac{E}{Pr}\Delta\theta_T\ , \\ \vec{\nabla}\cdot\vec{u}=0\ , \end{array}\right. \quad (1)$$

where $\vec{u}$ is the scaled velocity field in the corotating frame, θ_T is the scaled temperature profile and n is the scaled Brunt-Väisälä frequency profile. We remove dimensions using R_c, the radius of the radiative core, as a length scale, V, the baroclinic velocity as a velocity scale, T' as a temperature scale and $\mathcal{N}$ the scale of the Brunt-Väisälä frequency profile both from R06 (We refer the reader to R06 for the detailed description of the adimensionalization).

The physical parameters that characterize the system are the Ekman number E which quantifies the importance of the viscosity (we note the kinematic viscosity ν) over the Coriolis effects

$$E = \frac{\nu}{2\Omega_0 R_c^2}\ , \quad (2)$$

where Ω_0 is the rotation rate of the corotating frame, and the Prandtl number Pr which is the ratio of the kinematic viscosity over the thermal diffusivity κ

$$Pr = \frac{\nu}{\kappa}\ . \quad (3)$$

The Ekman number is evaluated and found to be very small in radiative core of low mass stars (around 10^{-11}). We use here the smallest value that numerical resolution allows, i.e. $E = 10^{-6}$. Finally, we use a solar value for the Prandtl number, namely $Pr = 2.10^{-6}$ ([6]).

In this framework, 3D simulations ([37], [26], [22], [49]) found that the differential rotation is conical in convective envelopes of low mass stars. We therefore impose at the top of the radiative core

$$\Omega_{cz}(r = R_c, \theta) = \Omega_0 + \Delta\Omega\sin^2\theta\ . \quad (4)$$

In the corotating frame, the azimuthal dimensionless velocity reads

$$u_\varphi(r = 1, \theta) = b\sin^3\theta\ , \quad (5)$$

where the dimensionless number $b = \dfrac{R_c\Delta\Omega}{V}$ quantifies the shear at the top of the radiative core. The differential rotation is solar-like when the equatorial regions rotate faster than the pole ($b > 0$) and anti-solar when they rotate slower ($b < 0$).

At the radiative-convective envelope, the meridional components of the velocity are set to zero illustrating the non penetrative case $u_r = 0$ and assuming that the presence of a convective envelope induces $u_\theta = 0$ in the corotating frame.

The shear we impose by this dynamical boundary condition generates a geostrophic flow, i.e. a flow respecting the geostrophic balance where the pressure gradient compensates the Coriolis acceleration, characterized by a columnar differential rotation which depends only on the distance to the rotation axis as predicted by the Taylor-Proudman theorem. Without any shear or for a very weak shear $|b| < 10^{-2}$, the flow is driven by the baroclinic torque $-n^2(r)\sin\theta\cos\theta\vec{e}_\varphi$ in the vorticity equation. The solution generated and described by R06 in Eq. (7) is the thermal wind characterized here by a quasi shellular differential rotation as shown in Fig. 1 (left panel). The difference between the flow obtained by R06 and the one we show in Fig. 1 comes from the fact that the Brunt-Väisälä frequency profile used is different in R06 (polynomial interpolation) than the one we use (input of a $1M_\odot$ ZAMS MESA model) and the boundary conditions are stress-free in R06.

When $|b| > 10^{-2}$, the amplitude of the geostrophic flow arising from the shear applied at the upper boundary is larger than the baroclinic solution, which would dominate otherwise, and the Taylor-Proudman balance tends to be restored leading to a quasi-columnar structure. In Fig. 2, we show the resulting differential rotation $\delta\Omega$ for $b = 10$ and $b = -10$. The field is normalized by its extremum value and we subtract the rotation of the pole so it is zero there and the differential rotation in the bulk is relative to it. Therefore, the differential rotation tends towards a cylindrical profile which is very different from the shellular profile assumed by 1D stellar models. The associated meridional circulation is dominated by a single, global circulation cell in each hemisphere. For b positive (solar-like differential rotation), the meridional circulation is counter clockwise. At the surface of the model of the radiative

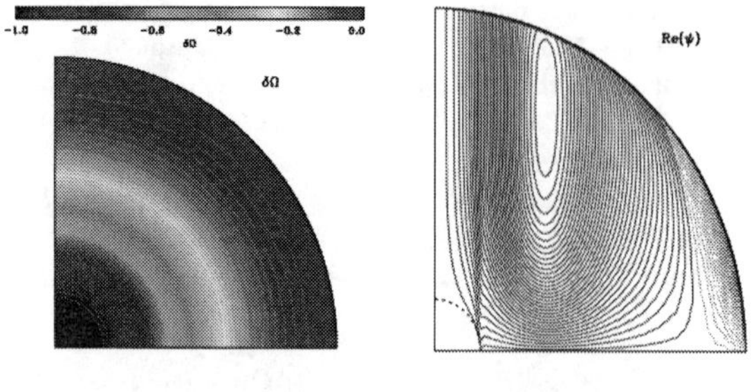

Figure 1: Differential rotation $\delta\Omega$ and meridional circulation stream function ψ (red : direct sens, blue: clockwise sens) shown in the meridional plane for an Ekman number $E = 10^{-6}$, a Prandtl number $\mathrm{Pr} = 2.10^{-6}$ and $b = 10^{-2}$. This is the pure baroclinic dynamics. The stellar rotation axis is vertical.

core, the fluid moves toward the pole which rotates slower than the equator. Conversely, for negative b, the meridional circulation is clockwise and towards the equatorial regions at the radiative-convective interface.

3 Seismic observables

The core-to-surface rotation ratio is the latitudinally averaged rotation rate at the top of the radiative core divided by the latitudinally averaged central rotation rate $\langle\delta\Omega_C\rangle_\theta/\langle\delta\Omega_S\rangle_\theta$ where $\langle\ldots\rangle_\theta = \int_0^{\pi/2}\ldots\sin\theta d\theta$. This ratio has been inverted in asteroseismology on several low mass stars. This only seismic diagnosis as of today is computed to characterize its behavior according to the shear strength we impose.

In our set-up, at the radiative-convective interface, an Ekman layer arises due to rotation and viscosity. This layer permits the velocity field (the sum of both the baroclinic flow and the geostrophic one) to match the boundary conditions by adding corrections whose amplitudes are important within the layer and weak outside of it. The Ekman layer has a characteristic thickness of $\sqrt{E}$, where we recall that E is the dimensionless Ekman number. We therefore compute the averaged surface rotation rate $\langle\delta\Omega_S\rangle_\theta$ just outside of it to avoid the boundary layer corrections. Moreover, we do not compute it at $r = 1$ since it would tend toward zero for small b and consequently the ratio $\langle\delta\Omega_C\rangle_\theta/\langle\delta\Omega_S\rangle_\theta$ would diverge. The averaged core rotation rate $\langle\delta\Omega_C\rangle_\theta$ is computed at the radius $r = 0.15R_c$ because g-modes, that would allow to probe central regions, are hardly identifiable closer to the center ([2], [21]).

We display the ratio $\langle\delta\Omega_C\rangle_\theta/\langle\delta\Omega_S\rangle_\theta$ as a function of b in Fig. 3 in logarithmic values. When $|b|$ tends towards zero, the ratio tends to a finite value. This is because, when the shear is very small, the dynamics is dominated by the baroclinic flow described by R06 visible in Fig. 1. With the Brunt-Väisälä frequency profile from MESA, the differential rotation is shellular and since we subtract the rotation of the pole, the differential rotation is negative in the entire zone. Therefore the ratio is positive that corresponds to negative averaged core rotation rate divided by a negative averaged surface rotation rate.

When b is negative, the ratio stays positive even when the shear becomes important and the geostrophic flow dominates. Moreover, we observe that when the absolute value of the shear increases, the ratio decreases. Indeed, the shear implies a transport of angular momentum, proportional to b both in the radial and latitudinal directions, which tends to accelerate the superficial regions more than the central regions.

If b is positive, the ratio displays a singularity around $b \sim 0.053$. This singularity is due to the change of sign of $\langle\delta\Omega_S\rangle_\theta$. Indeed for $b \geq 0.053$, the acceleration of the equator we impose through the shear boundary condition induces a positive $\langle\delta\Omega_S\rangle_\theta$. The ratio is then negative and has the same behavior than the branch of negative b. Namely, when b increases, the absolute value of the ratio decreases due to the enhanced transport of angular momentum. Roughly on the right side of the singularity, the solution is dominated by the geostrophic flow.

The position of the singularity depends on the zero we choose as a reference for the differential rotation; here, we recall that we choose the rotation of the pole to be zero. It also depends on the radius at which we compute $\langle\delta\Omega_S\rangle_\theta$. Indeed, if we compute it at $r = 1$, the ratio $\langle\delta\Omega_C\rangle_\theta/\langle\delta\Omega_S\rangle_\theta$ would diverge for small $|b|$ since the boundary conditions would reduce to no-slip ones. The singularity would disappear as it would coincide with the divergence at small b and the sign of the ratio $\langle\delta\Omega_C\rangle_\theta/\langle\delta\Omega_S\rangle_\theta$ would be the opposite of the sign of b for all b.

Since we work in the corotating frame at an unknown rotation rate and we set the rotation of the pole to be zero, this study provides only a qualitative behavior of the core-to-surface rotation ratio and the sign has a relative physical meaning. The important behavior to remember is the decrease of the ratio as the shear is strong.

In the derivation of the Eqs. (1), we assumed that the Brunt-Väisälä frequency N^2 is close to $4\Omega^2$ which is the fast rotating case where the transient baroclinic timescale is short enough for the steady state study to be relevant (R06, [39]). This means that a fast rotating solar case (a young Sun) is reachable for $b = 10$. For this value, the averaged core-to-surface rotation rate ratio is expected to be rather small, less than one in agreement with observations ([4]). But the internal rotation profile is far from flat. As we can see in Fig. 4, the rotation profiles for different colatitudes within the radiative core do not coincide as observed by heliosismology. Moreover, we do not produce a tachocline, pointing out that our current modeling is lacking a physical process responsible for the current rotation field of the solar radiative core (for reviews, see [36], [7]).

4 Conclusions

We build a 2D model of fast rotating radiative cores in low mass stars undergoing the shear of a differentially rotating convective envelope.

This shear is responsible for a redistribution of angular momentum deep within the star and gives rise to

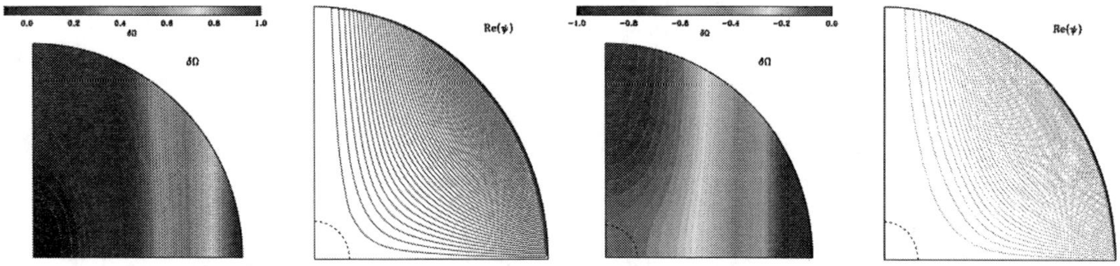

Figure 2: Differential rotation $\delta\Omega$ and meridional circulation stream function ψ (red : direct sens, blue: clockwise sens) shown in the meridional plane for an Ekman number $E = 10^{-6}$, a Prandtl number $\Pr = 2.10^{-6}$ with $b = 10$ (solar like rotation) on the left and $b = -10$ (anti-solar rotation) on the right. The stellar rotation axis is vertical.

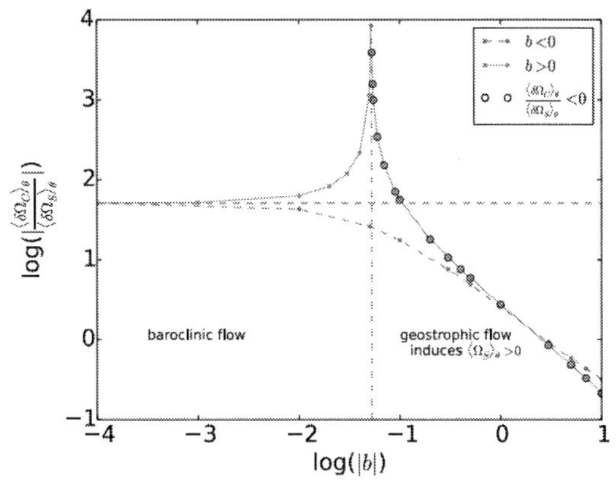

Figure 3: Logarithm of the averaged core-to-surface rotation ratio as a function of the logarithm of the shear parameter b for an Ekman number $E = 10^{-6}$ and a Prandtl number $\Pr = 2.10^{-6}$. Orange dots are for positive b, green dots for negative b. The circled dots are negative core-to-surface rotation ratio (non circled are positive). When $b \ll 1$ the dynamics is dominated by the baroclinic flow and the core-to-surface rotation ratio is constant (horizontal dashed line). For b positive, $\langle\delta\Omega_S\rangle_\theta$ changes sign inducing a singularity located at the vertical dashed line.

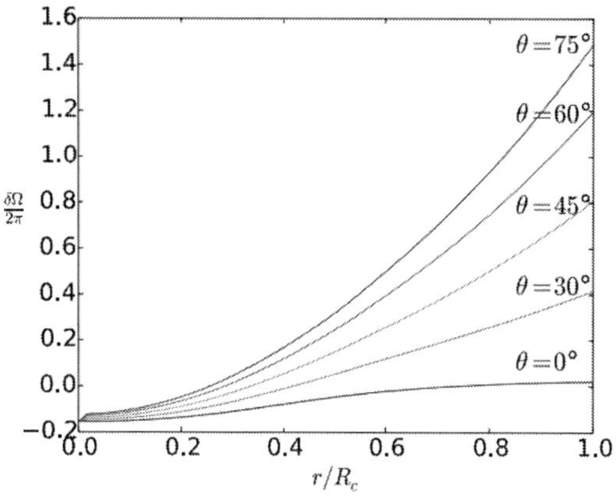

Figure 4: Relative radial differential rotation profiles within the radiative zone for $b = 10$ and $E = 10^{-6}$ at different colatitudes.

a geostrophic solution with a cylindrical differential rotation. Such a profile is very different from the shellular rotation profile assumed in 1D models. These models should therefore be used carefully in the radiative core of low mass stars when the shear applied by a convective envelope is important, i.e. $b > 10^{-2}$. Regarding the seismic observables such as the averaged core-to-surface rotation rate or the internal rotation profile, we conclude that we have to take into account other processes responsible for efficient transport of angular momentum both in the radial and in the latitudinal directions. Such processes may be internal gravity waves ([51]), magnetic fields ([23], [44], [45], [1]) or an anisotropic turbulence ([50], [30], [33]). These candidates are considered for implementation in our current 2D modeling of radiative zones.

Acknowledgments

D.H. and S.M. acknowledge funding by the European Research Council through ERC grant SPIRE 647383. The authors acknowledge funding by SpaceInn and PNPS (CNRS/INSU) and by CNES CoRoT/PLATO grant at SAp and IRAP.

References

[1] Acevedo-Arreguin, L. A. and Garaud, P. and Wood, T. S., MNRAS, **434** (2013), 720-741

[2] Appourchaux, T. and Belkacem, K. and Broomhall, A.-M. and Chaplin, W. J. and al., A&AR, **18** (2010), 197-277

[3] Beck, P. G. and Montalban, J. and Kallinger, T. and al., Nature, **481** (2012), 55-57

[4] Benomar, O. and Takata, M. and Shibahashi, H. and Ceillier, T. and García, R. A., MNRAS, **452** (2015), 2654-2674

[5] Brun, A. S. and Toomre, J., APJ, **570** (2002), 865-885

[6] Brun, A. S. and Zahn, J.-P., A&A, **457** (2006), 665-674

DOI: 10.1051/epjconf/201716002006

[7] Brun, A. S. and García, R. A. and Houdek, G. and Nandy, D. and Pinsonneault, M., Space Science Reviews, **196** (2015), 303-356

[8] Busse, F. H., Geophysical and Astrophysical Fluid Dynamics, **17** (1981), 215-235

[9] Ceillier, T. and Eggenberger, P. and García, R. A. and Mathis, S., A&A, **555** (2013), A54

[10] Charbonnel, C. and Talon, S., Science, **309** (2005), 2189-2191

[11] Couvidat, S. and García, R. A. and Turck-Chièze, S. and Corbard, T. and Henney, C. J. and Jiménez-Reyes, S. , APJ, **597** (2003), L77-L79

[12] Deheuvels, S. and García, R. A. and Chaplin, W. J. and al., APJ, **756** (2012), 19

[13] Deheuvels, S. and Doğan, G. and Goupil, M. J. and al., A&A, **564** (2014), A27

[14] Deheuvels, S. and Brandão, I. and Silva Aguirre, V. and Ballot, J. and Michel, E. and Cunha, M. S. and Lebreton, Y. and Appourchaux, T.,A&A, **589** (2016), A93

[15] Eggenberger, P. and Montalbán, J. and Miglio, A., A&A, **544** (2012), L4

[16] Espinosa Lara, F. and Rieutord, M., A&A, **470** (2007), 1013-1022

[17] Friedlander, S., Journal of Fluid Mechanics, **76** (1976), 209-228

[18] Garaud, P., MNRAS, **329** (2002) 1-17

[19] Garaud, P., MNRAS, **335** (2002) 707-711

[20] García, R. A. and Turck-Chièze, S. and Jiménez-Reyes, S. J. andal., Science, **316** (2007), 1591

[21] García, R. A., Highlights of Astronomy, **15** (2010), 345-346

[22] Gastine, T. and Yadav, R. K. and Morin, J. and Reiners, A. and Wicht, J., MNRAS, **438** (2014), L76-L80

[23] Gough, D. O. and McIntyre, M. E., Nature, **394** (1998), 755-757

[24] Hypolite, D. and Rieutord, M., A& A, **572** (2014), A15

[25] Hypolite, D. and Mathis, S. and Rieutord, M., To appear in the proceedings of "The 19th Cambridge Workshop on Cool Stars, Stellar Systems, and the Sun" (ed. G. A. Feiden), Uppsala, Sweden, 06-10 June 2016

[26] Käpylä, P. J. and Käpylä, M. J. and Brandenburg, A., A&A, **570** (2014), A43

[27] Kitchatinov, L. L. and Brandenburg, A., Astronomische Nachrichten, **333** (2012), 230

[28] Maeder, A. and Zahn, J.-P., A&A, **334** (1998), 1000-1006

[29] Maeder, A. and Meynet, G., ARAA, **38** (2000), 143-190

[30] Maeder, A., A&A, **399** (2003), 263-269

[31] Maeder, A., *Physics, Formation and Evolution of Rotating Stars* (Springer, 2009)

[32] Marques, J. P. and Goupil, M. J. and Lebreton, Y. and Talon, S. and al., A&A, **549** (2013), A74

[33] Mathis, S. and Palacios, A. and Zahn, J.-P., A&A, **425** (2004), 243-247

[34] Mathis, S. and Zahn, J.-P., A&A, **425** (2004), 229-242

[35] Mathis, S. and Zahn, J.-P., A&A, **440** (2005), 653-666

[36] Mathis, S., Lecture Notes in Physics, Berlin Springer Verlag, **865** (2013), 23

[37] Matt, S., Do Cao, O., Brown, B. P. and Brun, A. S., Astronomische Nachrichten, **332** (2011), 897

[38] Rieutord, M., A& A, **451** (2006), 1025-1036

[39] Rieutord, M., EAS Publications Series, **21** (2006), 275-295

[40] Rieutord, M. and Beth, A., A& A, **570** (2014), A42

[41] Paxton, B. and Bildsten, L. and Dotter, A. and Herwig, F. and Lesaffre, P. and Timmes, F., Astrophysics Source Code Library (2010)

[42] Schou, J. and Antia, H. M. and Basu, S. and al., APJ, **505** (1998), 390-417

[43] Spiegel, E. A. and Zahn, J.-P., A&A, **265** (1992), 106-114

[44] Spruit, H. C., A&A, **349** (1999), 189-202

[45] Strugarek, A. and Brun, A. S. and Zahn, J.-P., A&A, **532** (2011), A34

[46] Talon, S. and Charbonnel, C., A&A, **335** (1998), 959-968

[47] Talon, S. and Charbonnel, C., A&A, **440** (2005), 981-994

[48] Turck-Chièze, S. and Palacios, A. and Marques, J. P. and Nghiem, P. A. P., APJ, **715** (2010), 1539-1555

[49] Varela, J. and Strugarek, A. and Brun, A. S., "arXiv" (2016)

[50] Zahn, J.-P., A&A, **265** (1992), 115-132

[51] Zahn, J.-P. and Talon, S. and Matias, J., A&A, **322** (1997), 320-328

EPJ Web of Conferences **160**, 02007 (2017)
Seismology of the Sun and the Distant Stars II
DOI: 10.1051/epjconf/201716002007

Non-adiabatic pulsations in ESTER models

Daniel Roy Reese[1,⋆], *Marc-Antoine* Dupret[2], and *Michel* Rieutord[3,4]

[1] *LESIA, Observatoire de Paris, PSL Research University, CNRS, Sorbonne Universités, UPMC Univ. Paris 06, Univ. Paris Diderot, Sorbonne Paris Cité, 92195 Meudon, France*
[2] *Institut d'Astrophysique et Géophysique de l'Université de Liège, Allée du 6 août 17, 4000 Liège, Belgium*
[3] *Université de Toulouse, UPS-OMP, IRAP, 31028 Toulouse, France*
[4] *CNRS, IRAP, 14 avenue Edouard Belin, 31400 Toulouse, France*

Abstract. One of the greatest challenges in interpreting the pulsations of rapidly rotating stars is mode identification, i.e. correctly matching theoretical modes to observed pulsation frequencies. Indeed, the latest observations as well as current theoretical results show the complexity of pulsation spectra in such stars, and the lack of easily recognisable patterns. In the present contribution, the latest results on non-adiabatic effects in such pulsations are described, and we show how these come into play when identifying modes. These calculations fully take into account the effects of rapid rotation, including centrifugal distortion, and are based on models from the ESTER project, currently the only rapidly rotating models in which the energy conservation equation is satisfied, a prerequisite for calculating non-adiabatic effects. Non-adiabatic effects determine which modes are excited and play a key role in the near-surface pulsation-induced temperature variations which intervene in multi-colour amplitude ratios and phase differences, as well as line profile variations.

1 Introduction

The interpretation of pulsation spectra in rapidly rotating stars has proven to be a considerable challenge. Indeed, due to the lack of simple and easily identifiable pattern in the pulsation spectra, it has been very difficult to establish the correspondence between observed pulsation frequencies and theoretically calculated modes, i.e. to carry out mode identification [1]. This is extremely problematic since it means we do not know what regions of the star are being probed by individual modes, and cannot, accordingly, carry out meaningful inversions of the stellar structure. Nonetheless, [2] and [3] have recently shown the presence of patterns in the period spacings of rapidly rotating γ Dor stars, which agree with the most recent theoretical expectations [4–6] and allow us, in some cases, to determine the dominant harmonic degree and azimuthal order of these modes. For δ Scuti stars, progress remains more limited. In a recent paper, [7] derived an observational scaling relation between the large frequency separation and the mean density of several δ Scuti stars in eclipsing binary systems, in agreement with previous theoretical predictions [8]. Such large separations have also been used to create echelle diagrams [9, 10]. However, no firm identification has been obtained for individual modes.

As such, it is necessary to develop ways to observationally constrain mode identification, in addition to searching for patterns in the observed frequency spectra. In slowly rotating stars, amplitude ratios and phase differences based on multi-colour photometry, as well as line profile variations observed via spectroscopy, have proven to be popular methods for constraining mode identification [11, 12]. Various efforts have been made to extend these methods to rapid rotators, but it is only recently that all of the necessary ingredients are starting to come together. These include:

1. rapidly rotating models in which the energy conservation equation is satisfied. Indeed, this is important for carrying out self-consistent non-adiabatic pulsation calculations as will be described in the following point. Currently the only code able to produce such models is the ESTER[1] project[2] [13–15].
2. non-adiabatic pulsation calculations which fully take into account the effects of rapid rotation. Indeed, these can be used to predict which modes are excited by the κ-mechanism, the predominant mode-excitation mechanism in rapid rotators. Furthermore, only such calculations provide realistic pulsation-induced surface effective temperature variations, a key ingredient in photometric and spectroscopic mode signatures [11, 16]. Up until recently, the only non-adiabatic code which fully takes into account the effects of rotation was that of [17]. However, this code was generally applied with a limited number of spherical harmonics to models

⋆e-mail: daniel.reese@obspm.fr

[1] "Evolution STEllaire en Rotation"
[2] We note that the Self-Consistent Field method only solves the energy conservation equation after the averaging the model over level surfaces.

 DOI: 10.1051/epjconf/201716002007

based on a Chandrasekhar expansion, or even spherical models. This has, however, changed with the advent of the non-adiabatic version of the TOP code [18, 19], currently the only code which is set up to calculate pulsation modes in ESTER models.

3. detailed calculations of photometric mode visibilities and spectroscopic line profile variations (LPVs). Recently, [20] carried out detailed mode visibility calculations in rapidly rotating SCF models [21]. However these calculations, now need to be extended to the non-adiabatic case as described above. Probably the most realistic LPV calculations in rapidly rotating stars are those of [22], but were nonetheless carried out using the traditional approximation. Other calculations [23] fully include the effects of rotation on pulsations but use a simplified modelling of spectroscopic effects.

In the following, we describe the latest non-adiabatic pulsation calculations in ESTER models using the TOP code (Section 2), and some of the results for mode-excitation, mode visibilities, and line profile variations (Section 3).

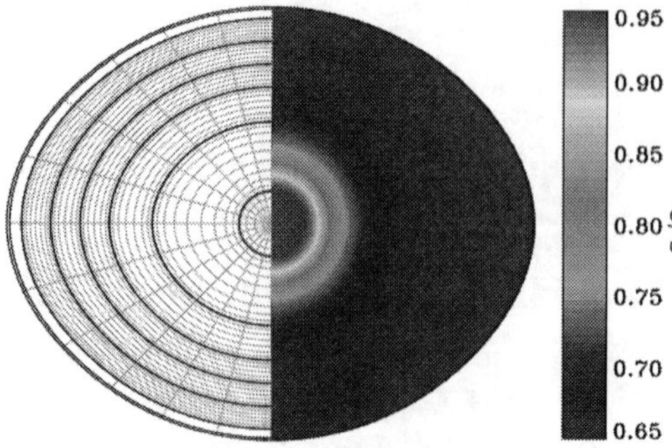

Figure 1: Multi-domain coordinate system and 2D rotation profile in an ESTER model.

2 Theory

2.1 ESTER models

The results presented in the following sections are based on 9 $M_\odot$ ESTER models with rotation rates ranging from 0.0 to 0.8 Ω_K, where Ω_K is the Keplerian break-up rotation rate. These models have a metallicity of $Z = 0.025$ and use OPAL opacities. They harbour β Cep type pulsations, in the p and g-mode range, which are excited by the iron opacity bump at $\log(T) = 5.3$.

Models from the ESTER project fully take into account the effects of rotation. These include: centrifugal deformation, gravity darkening, and baroclinic effects which result from the mismatch between isobars and isotherms (given that the temperature profile is determined via the energy conservation equation). These effects lead to meridional circulation as well as differential rotation which depends both on r and θ, as illustrated in Fig. 1. From a numerical point of view, the ESTER code uses a multi-domain spectral approach with Chebyshev polynomials in the radial direction, and spherical harmonics in the horizontal directions. The multi-domain approach is illustrated in Fig. 1.

2.2 Non-adiabatic pulsations

The set of non-adiabatic pulsation equations implemented in the TOP code are the perturbed continuity Equation, Euler's equation, Poisson's equation, the perturbed energy conservation equation, and the perturbed energy flux equation:

$$
\begin{aligned}
0 &= \frac{\delta\rho}{\rho_0} + \vec{\nabla}\cdot\vec{\xi} && (1)\\
0 &= [\omega + m\Omega]^2\vec{\xi} - 2i\vec{\Omega}\times[\omega + m\Omega]\vec{\xi} \\
&\quad -\vec{\Omega}\times\left(\vec{\Omega}\times\vec{\xi}\right) - \vec{\xi}\cdot\vec{\nabla}\left(\varpi\Omega^2\vec{e}_\varpi\right) - \frac{P_0}{\rho_0}\vec{\nabla}\left(\frac{\delta P}{P_0}\right) \\
&\quad +\frac{\vec{\nabla}P_0}{\rho_0}\left(\frac{\delta\rho}{\rho_0} - \frac{\delta P}{P_0}\right) - \vec{\nabla}\Psi + \vec{\nabla}\left(\frac{\vec{\xi}\cdot\vec{\nabla}P_0}{\rho_0}\right) \\
&\quad +\frac{\left(\vec{\xi}\cdot\vec{\nabla}P_0\right)\vec{\nabla}\rho_0 - \left(\vec{\xi}\cdot\vec{\nabla}\rho_0\right)\vec{\nabla}P_0}{\rho_0^2} && (2)\\
0 &= \Delta\Psi - 4\pi G\left(\rho_0\frac{\delta\rho}{\rho_0} - \vec{\xi}\cdot\vec{\nabla}\rho_0\right) && (3)\\
0 &= -i[\omega + m\Omega]\rho_0 T_0\delta S + \epsilon_0\rho_0\left(\frac{\delta\epsilon}{\epsilon_0} + \frac{\delta\rho}{\rho_0}\right) \\
&\quad -\vec{\nabla}\cdot\delta\vec{F} + \vec{\xi}\cdot\vec{\nabla}\left(\vec{\nabla}\cdot\vec{F}_0\right) - \vec{\nabla}\cdot\left[\left(\vec{\xi}\cdot\vec{\nabla}\right)\vec{F}_0\right] && (4)\\
\delta\vec{F}^{\mathrm{R}} &= \left[(1+\chi_T)\frac{\delta T}{T_0} + \chi_\rho\frac{\delta\rho}{\rho_0}\right]\vec{F}_0^{\mathrm{R}} - \chi_0\left[T_0\vec{\nabla}\left(\frac{\delta T}{T_0}\right)\right. \\
&\quad \left. +\vec{\xi}\cdot\vec{\nabla}\left(\vec{\nabla}T_0\right) - \vec{\nabla}\left(\vec{\xi}\cdot\vec{\nabla}T_0\right)\right] && (5)
\end{aligned}
$$

where ρ is the density, $\vec{\xi}$ the Lagrangian displacement, ω the pulsation frequency, m the azimuthal order, Ω the rotation profile (which depends on r and θ), $\vec{\Omega}$ the rotation vector (i.e. $\Omega\vec{e}_z$), ϖ the distance from the rotation axis, P the pressure, Ψ the gravitational potential, T the temperature, S the entropy, ϵ the nuclear energy generation rate, $\vec{F}$ the energy flux, $\vec{F}^{\mathrm{R}}$ the radiative energy flux, χ the opacity, χ_T the logarithmic derivative of χ with respect to T, i.e. $\frac{\partial\chi}{\partial T}$, and χ_ρ the logarithmic derivative of χ with respect to ρ. Quantities with the subscript "0" correspond to background equilibrium quantities whereas those preceded by a δ are Lagrangian perturbations. We note that the last term in Euler's equation is specific to non-barotropic models. In addition to the above equations, perturbed equations of state and opacities must be included, as well as a variety of boundary conditions which ensure regularity of the solution in the centre, a non-divergent gravitational potential perturbation at infinity, and the appropriate behaviour of

Table 1: Numerical cost for calculating non-adiabatic modes.

N_r	N_h	RAM (Gb)	Time (min)	Num. proc.
400	10	3.5		
400	15	7.9		
400	20	13.4	5	4
400	29	28.0	10	8
400	40	52.7	22	8
400	50	82.3	26	16

the pressure and temperature perturbations at the surface. Overall, the system comprises 10 equations with 10 unknowns. The numerical cost in solving this system is quite high as shown in Table 1 which gives the computational times and RAM for various radial and horizontal resolutions for given numbers of processors.

Using a similar approach as in the adiabatic case, it is possible to derive an integral expression for the frequency (but which will not benefit from a variational principle), as well as for the damping rate. The latter is known as the work integral and is used to increase the accuracy with which the damping rate is calculated. The relative difference between the numerical and integral values of the frequency is of the order of 10^{-4}, whereas it is around 10^{-2} to 10^{-1} for the damping rate given its smaller value.

3 Results

3.1 Frequencies and modes

Figure 2 shows a typical frequency spectrum at $0.5\,\Omega_K$. The hollow diamonds correspond to modes which are excited via the κ-mechanism. For the same model, Fig. 3 shows the difference between adiabatic and non-adiabatic frequencies as a function of the damping rate. As expected, the largest frequency differences occur for the largest damping rates, i.e. where non-adiabatic effects are the strongest.

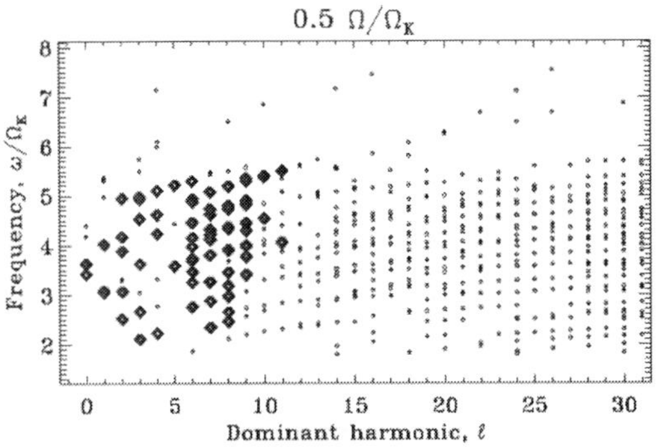

Figure 2: Frequency spectrum as a function of dominant spherical harmonic. The hollow diamonds correspond to unstable modes.

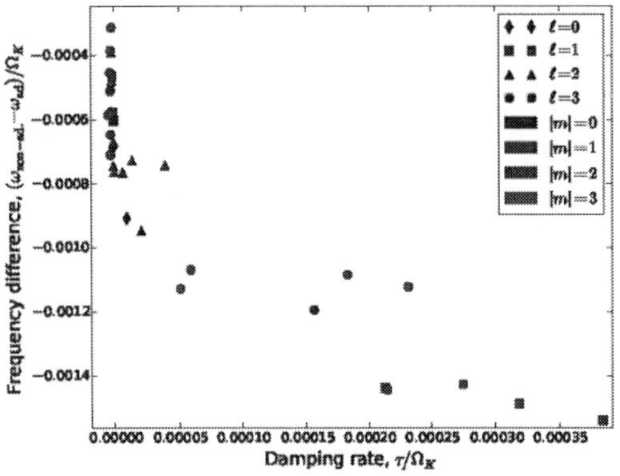

Figure 3: Difference between adiabatic and non-adiabatic frequencies as a function of the damping rate.

One can look more specifically at frequency multiplets. In Fig. 4, we show the evolution of various frequency multiplets as a function of the rotation rate. The thick lines correspond to excited modes. As can be seen, the prograde modes tend to remain excited longer, in agreement with the results in [24]. A look at the work integrals shows that modes are being stabilised by the very near surface layers, starting with the retrograde modes (see Fig. 2 of [25]).

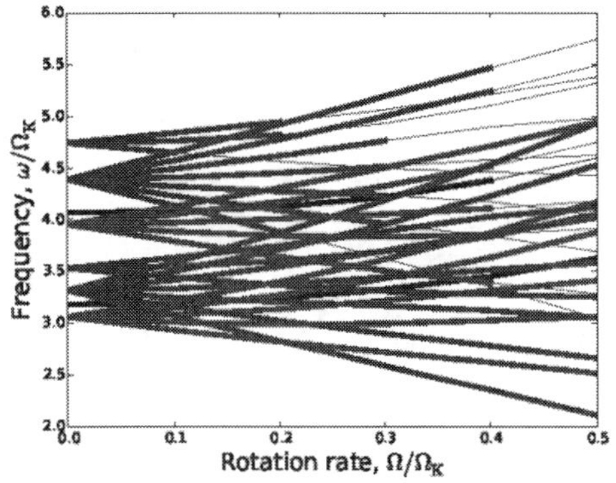

Figure 4: Frequencies multiplets as a function of the rotation rate. The thick lines corresponds to excited modes.

Different types of modes are excited. These include island modes and chaotic modes for the acoustic end of the spectrum [26], as well as mixed modes [27] and rosette modes [4], as illustrated in Fig. 5.

3.2 Amplitude ratios

Amplitude ratios were calculated for Strömgren and Johnson-Cousins photometric bands, and in the bolometric case using the formalism given in [20]. In order to calculate the intensity as a function of effective temperature,

DOI: 10.1051/epjconf/201716002007

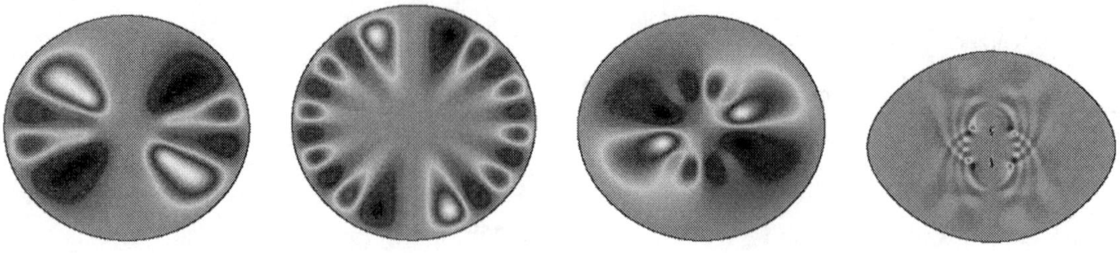

Figure 5: Different modes which are excited. These are, from left to right: an island mode, a whispering gallery mode, a mixed mode, and a rosette mode.

effective gravity, and orientation of the surface, we applied the following approximate formula:

$$I(T_{\text{eff}}, g_{\text{eff}}, \mu) = I_0(T_{\text{eff}})\, h(\mu, T_{\text{eff}}, g_{\text{eff}}) \quad (6)$$

where $I_0(T_{\text{eff}})$ was obtained from the integral of a black-body spectrum weighted by the various photometric filter profiles, and $h(\mu, T_{\text{eff}}, g_{\text{eff}})$ was obtained from [28], and $\mu = \cos\varphi$ where φ is the angle between the line-of-sight and the normal to the surface. Figure 6 shows the resultant function obtained for the Strömgren u filter for a star at $0.8\,\Omega_K$. The decrease as a function of colatitude corresponds to gravity darkening, whereas the increase as a function of μ values is caused by limb darkening.

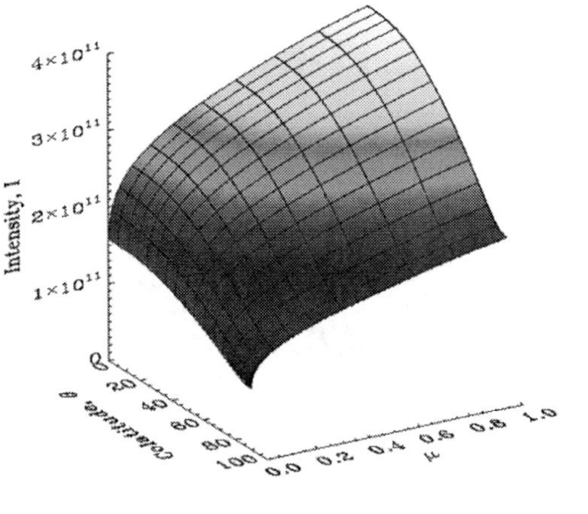

Figure 6: Intensity in the Strömgren photometric band as a function of the colatitude and μ for a model rotating at $0.8\,\Omega_K$.

Figure 7 shows the surface Lagrangian displacement, ξ_r, and surface $\delta T_{\text{eff}}/T_{\text{eff}}$ profiles for a given pulsation mode calculated with and without the adiabatic approximation. In contrast to ξ_r, the $\delta T_{\text{eff}}/T_{\text{eff}}$ profiles show a non-negligible imaginary part in the non-adiabatic case. This corresponds to a phase shift between the displacement and temperature variations, as expected in such a case. As also anticipated, the modulus of the effective temperature variations, $|\delta T_{\text{eff}}|$, is smaller in the non-adiabatic case. These illustrate the importance of taking non-adiabatic effects into account prior to calculate mode visibilities, amplitude ratios, and phase differences.

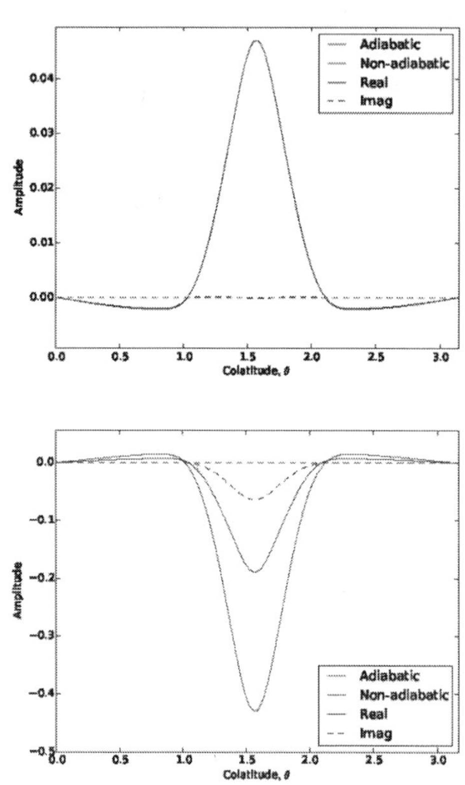

Figure 7: Surface vertical Lagrangian displacement (top panel) and surface $\delta T_{\text{eff}}/T_{\text{eff}}$ profiles (lower panel) for a given mode calculated with and without the adiabatic approximation.

Figure 8 shows a set of amplitude ratios for a given multiplet for a model with an inclination of $i = 30°$, and for a given mode at 4 different inclinations. In full agreement with previous results [29, 30], amplitude ratios become dependant both on the azimuthal order, m, and on the inclination as rotation is introduced. This, of course, complicates mode identification using amplitude ratios al-

DOI: 10.1051/epjconf/201716002007

though we do note that [31] proposed an alternate mode identification strategy in such stars.

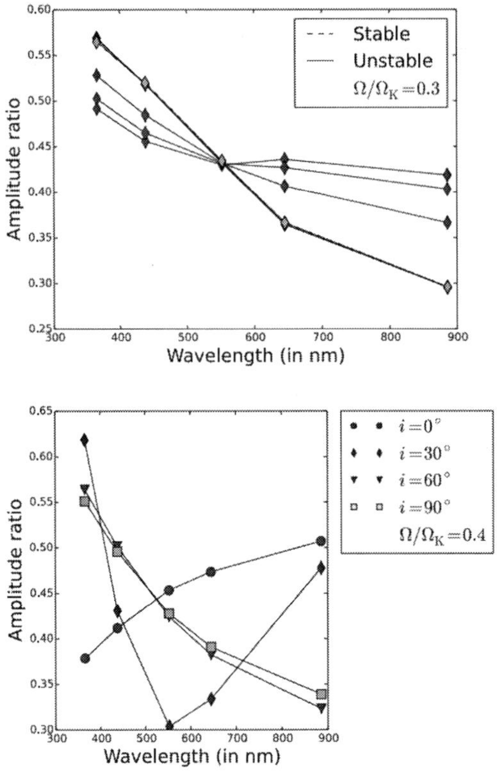

Figure 8: Amplitude ratios for an $\ell = 3$ multiplet at an inclination of $i = 30°$ (upper panel) and for an $(\ell, m) = (2, 0)$ mode at 4 different inclinations (lower panel).

3.3 Line profile variations (LPVs)

Another way of constraining mode identification is through LPVs. In the present case, we reuse some of the formulas from [20] with an additional Doppler term from both the stellar rotation and the oscillatory movements. The ray profile is modelled in crude way using a black-body spectrum (thereby accounting for gravity darkening) as a background and a Gaussian profile as an intrinsic absorption ray. Furthermore, the contributions of $\delta T_{\rm eff}$ and $\delta g_{\rm eff}$ are neglected as are any dependence of the absorption ray on $T_{\rm eff}$ and $g_{\rm eff}$. A rudimentary description of limb-darkening is also included.

Figures 9 and 10 show two examples of LPVs calculated in a model rotating at $0.5\,\Omega_K$. These plots also contain the first and second moments, the amplitudes and phases of the first three harmonics of the variations for a particular Doppler position, and a meridional cross-section of the modes. A note worthy feature is that most of the variations take place in the wings of the absorption ray, which is different than what takes place in the non-rotating case. This feature is fairly frequent at rapid rotations and has been suggested as a possible explanation for macro-turbulence in massive stars (Aerts, private communication).

4 Conclusions

The ability to carry out non-adiabatic pulsation calculations in rapidly rotating stars is an important step forward in being able to understand and interpret seismic data for such stars. Indeed, it is now possible to predict which modes are excited, and it will lead to more realistic predictions for amplitude ratios, phase differences, and line profile variations. The next steps forward include: gaining a better understanding of how rotation interacts with non-adiabatic effects, understanding what causes the differences between prograde and retrograde modes, including a realistic atmosphere in the stellar model when calculating amplitude ratios and line profile variations, and identifying modes in observed spectra using multi-colour photometry and spectroscopy.

Acknowledgements

DRR acknowledges financial support from the "Programme National de Physique Stellaire" (PNPS) of CNRS/INSU, France.

References

[1] M.J. Goupil, M.A. Dupret, R. Samadi, T. Boehm, E. Alecian, J.C. Suarez, Y. Lebreton, C. Catala, JA&A **26**, 249 (2005)

[2] T.R. Bedding, S.J. Murphy, I.L. Colman, D.W. Kurtz, *Échelle diagrams and period spacings of g modes in γ Doradus stars from four years of Kepler observations*, in *European Physical Journal Web of Conferences* (2015), Vol. 101 of *European Physical Journal Web of Conferences*, p. 01005

[3] T. Van Reeth, A. Tkachenko, C. Aerts, A&A **593**, A120 (2016)

[4] J. Ballot, F. Lignières, V. Prat, D.R. Reese, M. Rieutord, *2D Computations of g-modes in Fast Rotating Stars*, in *Progress in Solar/Stellar Physics with Helio- and Asteroseismology*, edited by H. Shibahashi, M. Takata, A.E. Lynas-Gray (2012), Vol. 462 of *Astronomical Society of the Pacific Conference Series*, p. 389

[5] M.P. Bouabid, M.A. Dupret, S. Salmon, J. Montalbán, A. Miglio, A. Noels, MNRAS **429**, 2500 (2013)

[6] R.M. Ouazzani, S.J.A.J. Salmon, V.L. Antoci, T.R. Bedding, S.J. Murphy, I.W. Roxburgh, MNRAS (2016), in press, astro-ph.SR:1610.06184

[7] A. García Hernández, S. Martín-Ruiz, M.J.P.F.G. Monteiro, J.C. Suárez, D.R. Reese, J. Pascual-Granado, R. Garrido, **811**, L29 (2015)

[8] D. Reese, F. Lignières, M. Rieutord, A&A **481**, 449 (2008)

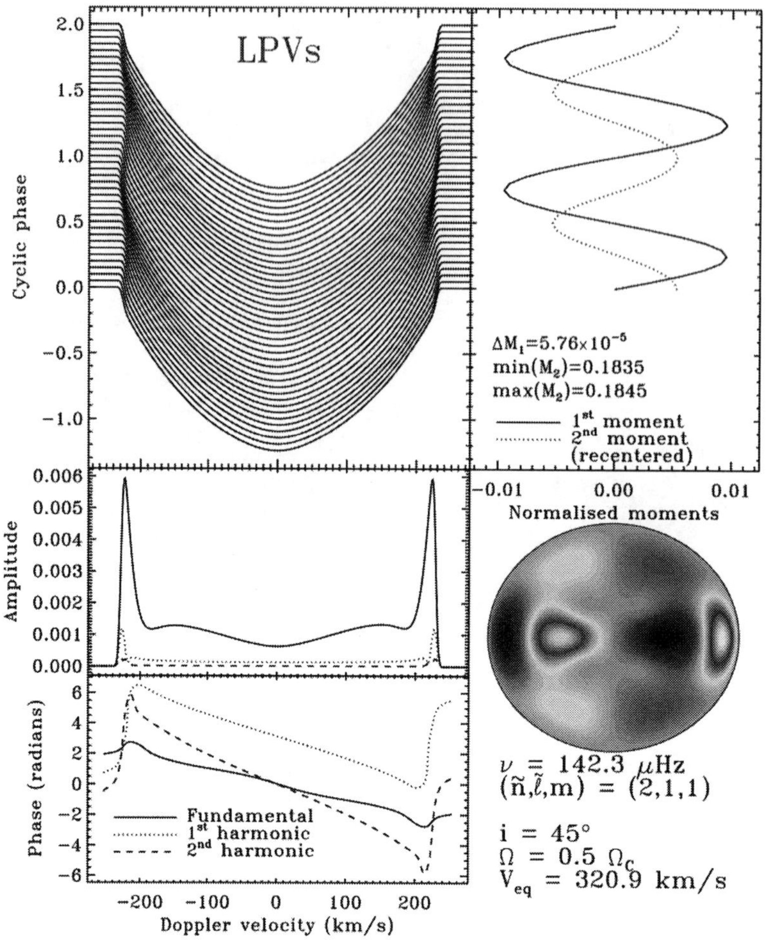

Figure 9: Line profile variations for an $(n, \ell, m) = (1, 3, 1)$ mode (top left panel), first and second moments (top right panel), amplitudes and phases of the first three harmonics of the variations across the line profile (lower left panels), and meridional cross-section of the mode (lower right panel).

[9] A. García Hernández, A. Moya, E. Michel, J.C. Suárez, E. Poretti, S. Martín-Ruíz, P.J. Amado, R. Garrido, E. Rodríguez, M. Rainer et al., A&A **559**, A63 (2013)

[10] M. Paparó, J.M. Benkő, M. Hareter, J.A. Guzik, ApJS **224**, 41 (2016)

[11] M.A. Dupret, J. De Ridder, P. De Cat, C. Aerts, R. Scuflaire, A. Noels, A. Thoul, A&A **398**, 677 (2003)

[12] W. Zima, D. Wright, J. Bentley, P.L. Cottrell, U. Heiter, P. Mathias, E. Poretti, H. Lehmann, T.J. Montemayor, M. Breger, A&A **455**, 235 (2006)

[13] M. Rieutord, F. Espinosa Lara, *Ab Initio Modelling of Steady Rotating Stars*, in *Lecture Notes in Physics, Berlin Springer Verlag*, edited by M. Goupil, K. Belkacem, C. Neiner, F. Lignières, J.J. Green (2013), Vol. 865 of *Lecture Notes in Physics, Berlin Springer Verlag*, p. 49

[14] F. Espinosa Lara, M. Rieutord, A&A **552**, A35 (2013)

[15] M. Rieutord, F. Espinosa Lara, B. Putigny, Journal of Computational Physics **318**, 277 (2016)

[16] M. Dupret, J. De Ridder, C. Neuforge, C. Aerts, R. Scuflaire, A&A **385**, 563 (2002)

[17] U. Lee, I. Baraffe, A&A **301**, 419 (1995)

[18] D. Reese, F. Lignières, M. Rieutord, A&A **455**, 621 (2006)

[19] D.R. Reese, K.B. MacGregor, S. Jackson, A. Skumanich, T.S. Metcalfe, A&A **506**, 189 (2009)

[20] D.R. Reese, V. Prat, C. Barban, C. van 't Veer-Menneret, K.B. MacGregor, A&A **550**, A77 (2013)

[21] K.B. MacGregor, S. Jackson, A. Skumanich, T.S. Metcalfe, ApJ **663**, 560 (2007)

[22] R.H.D. Townsend, MNRAS **284**, 839 (1997)

DOI: 10.1051/epjconf/201716002007

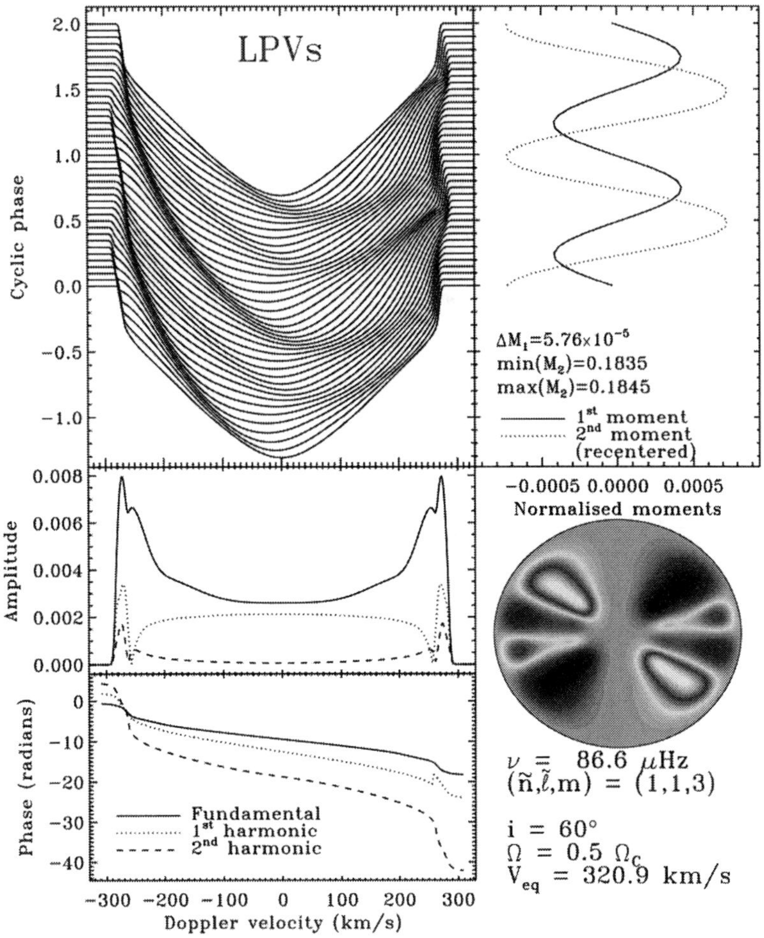

Figure 10: Same as Fig. 9 but for the $(n, \ell, m) = (0, 6, 3)$ mode.

[23] M.J. Clement, *Pulsation in Rapidly Rotating Stars*, in *IAU Symp. 162: Pulsation; Rotation; and Mass Loss in Early-Type Stars*, edited by L.A. Balona, H.F. Henrichs, J.M. Le Contel (1994), p. 117

[24] U. Lee, *Pulsational Stability of Rotating Stars*, in *American Institute of Physics Conference Series*, edited by V. Mioc, C. Dumitrche, N.A. Popescu (2008), Vol. 1043 of *American Institute of Physics Conference Series*, pp. 344–350

[25] D.R. Reese, *Internal rapid rotation and its implications for stellar structure and pulsations*, in *European Physical Journal Web of Conferences* (2015), Vol. 101 of *European Physical Journal Web of Conferences*, p. 05007

[26] F. Lignières, B. Georgeot, A&A **500**, 1173 (2009)

[27] R.M. Ouazzani, I.W. Roxburgh, M.A. Dupret, A&A **579**, A116 (2015)

[28] A. Claret, A&A **363**, 1081 (2000)

[29] J. Daszyńska-Daszkiewicz, W.A. Dziembowski, A.A. Pamyatnykh, M.J. Goupil, A&A **392**, 151 (2002)

[30] R.H.D. Townsend, MNRAS **343**, 125 (2003)

[31] D.R. Reese, F. Lignières, J. Ballot, V. Prat, C. Barban, C. van't Veer-Menneret, K.B. MacGregor, *Mode Visibilities and Frequency Patterns in Rapidly Rotating Stars*, in *Progress in Physics of the Sun and Stars: A New Era in Helio- and Asteroseismology*, edited by H. Shibahashi, A.E. Lynas-Gray (2013), Vol. 479 of *Astronomical Society of the Pacific Conference Series*, p. 545

Seismic diagnosis from gravity modes strongly affected by rotation

Vincent Prat[1,*], *Stéphane* Mathis[1], *François* Lignières[2,3], *Jérôme* Ballot[2,3], and *Pierre-Marie* Culpin[2,3]

[1] *Laboratoire AIM Paris-Saclay, CEA/DRF - CNRS - Université Paris Diderot, IRFU/SAp Centre de Saclay, F-91191 Gif-sur-Yvette, France*
[2] *Université de Toulouse; UPS-OMP; IRAP; Toulouse, France*
[3] *CNRS; IRAP; 14 avenue Édouard Belin; F-31400 Toulouse, France*

Abstract. Most of the information we have about the internal rotation of stars comes from modes that are weakly affected by rotation, for example by using rotational splittings. In contrast, we present here a method, based on the asymptotic theory of Prat et al. (2016), which allows us to analyse the signature of rotation where its effect is the most important, that is in low-frequency gravity modes that are strongly affected by rotation. For such modes, we predict two spectral patterns that could be confronted to observed spectra and those computed using fully two-dimensional oscillation codes.

1 Introduction

Stellar evolution plays a central role in astrophysics by linking observable quantities, such as effective temperature or surface gravity, to fundamental parameters of stars that can be used to constrain other objects such as galaxies or planets. In this context, angular momentum and chemical transport processes are crucial for stellar evolution, since they modify the structure of stars ([1]). For instance, rotation has three important effects: (i) it directly modifies the structure of stars because of the flattening due to the centrifugal acceleration, (ii) it induces many hydrodynamical instabilities, such as the shear instability, that create turbulent transport, and (iii) it drives large-scale flows such as meridional circulation ([2–5]).

Surface rotation can be investigated by different methods, but only asteroseismology gives access to internal properties of stars such as deep differential rotation and chemical stratification. For slowly rotating stars, perturbative methods can be used to estimate internal rotation by measuring the splittings between modes of different azimuthal orders. This has been done for the Sun (e.g. [6, 7]), subgiant and red giant stars (e.g. [8]), solar-type stars ([9]) and a few intermediate-mass (e.g. [10]) and massive stars ([11]). For rapidly rotating stars, because the problem of finding eigenmodes becomes intrinsically two-dimensional, especially for gravity modes, such methods fail (see e.g. [12]) and new diagnostic tools need to be developed. The traditional approximation, which neglects the latitudinal component of the rotation vector, allows for variable separation for all rotations (see e.g. [13]), thus providing such a tool, but its relevance for seismology is uncertain (see e.g. [14]).

*e-mail: vincent.prat@cea.fr

One way to go further is to use asymptotic theories that rely on some assumptions to gain physical insight. For example, it is possible to describe waves as rays in a way similar to geometrical optics by assuming that the wavelength is much smaller than the characteristic length of the variations of the background. These rays can be used to explore the properties of modes more efficiently than full numerical computations. Such an asymptotic theory has been built for acoustic waves ([15]) and has led to the prediction of regularities in corresponding spectra of fast-rotating stars ([16]). For gravito-inertial waves, a first study ([17]) has allowed us to constrain the structure of the phase space. In particular, we have shown that it is dominated by nearly integrable structures at low frequencies. This suggests the existence of a nearby integrable system that could be used to derive new diagnoses.

The aim of this article is to present a new low-frequency approximation (Sect. 2) that leads to an integrable and separable system. Then, we use semi-classical quantisation techniques to determine mode frequencies and resulting spectral patterns (Sect. 3). Finally, we conclude and discuss the prospects of this work in Sect. 4.

2 Low-frequency approximation

In [17], we applied the Jeffreys-Wentzel-Kramers-Brillouin (JWKB) small-wavelength approximation to the linearised equations describing the dynamics of gravito-inertial waves, and we obtained the general eikonal equation (i.e. local dispersion relation)

$$\omega^2 = \frac{f^2 k_z^2 + N_0^2\left(k_\perp^2 + k_\phi^2\right) + f^2 \cos^2\Theta k_c^2}{k^2 + k_c^2}, \qquad (1)$$

where ω is the wave pulsation, $f = 2\Omega$ with Ω the rotation rate, N_0 is the Brunt-Väisälä frequency, k_z, $k_\perp$ and k_ϕ are

 DOI: 10.1051/epjconf/201716002008

the components of the wave vector $\vec{k}$ parallel to the rotation axis, orthogonal to the effective gravity (in the meridional plane) and in the azimuthal direction, respectively; k is the norm of the wave vector, k_c is a term that accounts for the back-refraction of waves near the surface, and Θ is the angle between the rotation axis and the effective gravity (equal to the colatitude θ when centrifugal deformation is neglected).

As shown in [17], low-frequency gravito-inertial waves are trapped near the equatorial plane, but can propagate near the centre, which is not accounted for by the traditional approximation. In the limit of very low frequencies ($\omega \ll f$), the trapping implies that

$$\frac{N_0^2}{N_0^2 + f^2} \cos^2 \Theta \ll 1. \tag{2}$$

It is possible to use this condition to simplify the general eikonal equation in the vicinity of the equatorial plane. After some manipulations, one can finally write

$$\omega^2 = \frac{f^2 \cos^2 \delta \left(k_\beta^2 + k_c^2\right) + \left(N_0^2 + f^2\right) \frac{k_\delta^2 + m^2}{\zeta^2}}{k^2 + k_c^2}, \tag{3}$$

where

$$\cos^2 \delta = \frac{N_0^2}{N_0^2 + f^2} \cos^2 \Theta, \quad \zeta = \frac{r \sqrt{N_0^2 + f^2}}{N_0}, \tag{4}$$

m is the azimuthal order, β is the principal coordinate which is equivalent to $r \sin \theta$ near the centre and to r in the envelope, and k_β and k_δ are the natural components of $\vec{k}$ associated with β and δ, respectively.

The next step is to describe the ray dynamics using the Hamiltonian formalism given by

$$\frac{dx_i}{dt} = \frac{\partial \omega}{\partial k_i}, \tag{5}$$

$$\frac{dk_i}{dt} = -\frac{\partial \omega}{\partial x_i}, \tag{6}$$

where $k_i = \partial \Phi / \partial x_i$ is the natural component of the wave vector associated with the coordinate x_i. These equations ensure that the frequency ω is invariant. To go further, we make the simplification that k_c, N_0 and ζ depend only on β. This is rigourously valid only at zero rotation, but should be a good approximation in the vicinity of the equatorial plane, at least for moderate rotation rates. One then can show that the quantity

$$\chi = \frac{N_0^2 + f^2 \sin^2 \delta}{\zeta^2 (k^2 + k_c^2)} \tag{7}$$

is also invariant. Thus, the system has two invariants, ω and χ, which means it is integrable. Furthermore, the ray dynamics can be described in a separable way,

$$k_\beta^2 + k_c^2 = \frac{N_0^2 + f^2 - \omega^2}{\zeta^2 \chi}, \tag{8}$$

$$k_\delta^2 = \frac{\omega^2 - f^2 \cos^2 \delta}{\chi} - m^2, \tag{9}$$

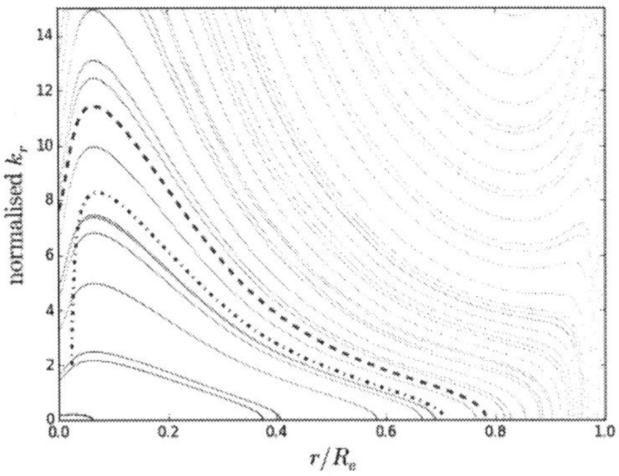

Figure 1: Poincaré surface of section at $\omega/f = 0.8$ of a polytropic model of star (with a polytropic index of 3) rotating at 38% of its critical velocity. The blue dash-dotted line and the green dashed line are computed using the traditional approximation and of the low-frequency approximation, respectively.

in the sense that each component of the wave vector can be expressed as a function of the corresponding coordinate only.

At the equator, $k_\beta = k_r$ and $\beta = r$, so Eq. (8) can be used to compare the present approximation with the full dynamics. This is illustrated in Fig. 1, where one can see that the low-frequency approximation reproduces well the structure of a Poincaré surface of section (see [17] for explanations) with the full dynamics, whereas the traditional approximation fails near the centre.

3 Seismic diagnoses

To construct modes from rays, one must satisfy quantisation conditions that ensure that when a ray crosses itself, its phase is the same. They read

$$\int_C \vec{k} \cdot d\vec{x} = 2\pi \left(p + \frac{\varepsilon}{4}\right), \tag{10}$$

where C is a closed contour, p and ε are integers, and ε accounts for phase-shifts due to boundary conditions. In our case, there are two independent sets of contours, which yield a non-linear system in ω and χ to solve:

$$\int_0^{r_s} \sqrt{\frac{N_0^2 + f^2 - \omega^2}{\zeta^2 \chi} - k_c^2}\, dr = \frac{\pi}{2} \left(\tilde{n} + \frac{1}{4}\right), \tag{11}$$

$$\int_{\delta_c}^{\frac{\pi}{2}} \sqrt{\frac{\omega^2 - f^2 \cos^2 \delta}{\chi} - m^2}\, d\delta = \frac{\pi}{2} \left(\tilde{\ell} + \frac{1}{2}\right), \tag{12}$$

where r_s is the stellar radius, $\delta_c = \arccos \frac{\sqrt{\omega^2 - m^2 \chi}}{f}$ the critical value of δ, $\tilde{n} = n - 1$ and $\tilde{l} = l_\mu - 1$ with n the radial order and l_μ the number of nodes in the latitudinal direction.

This system can be solved numerically, but it is possible to solve it analytically by making further assumptions, such as using the fact that $\omega \ll f$ and neglecting the influence of k_c in Eq. (11). This leads to

$$\omega^2 = \frac{(2\tilde{\ell}+1) f \int_0^{r_s} \frac{N_0}{r} \mathrm{d}r}{\pi\left(\tilde{n}+\frac{1}{4}\right)} + m^2 \frac{\left(\int_0^{r_s} \frac{N_0}{r} \mathrm{d}r\right)^2}{\pi^2\left(\tilde{n}+\frac{1}{4}\right)^2}, \tag{13}$$

expressed in the co-rotating frame.

For axisymmetric modes ($m = 0$), the period spacings between modes of equal $\tilde{l}$ and consecutive $\tilde{n}$ are

$$\Delta\Pi \simeq \frac{\pi^{3/2}}{\sqrt{2\left(\tilde{n}+\frac{3}{4}\right)(2\tilde{\ell}+1)\Omega \int_0^{r_s} \frac{N_0}{r} \mathrm{d}r}}. \tag{14}$$

They allow for measurements of the product $\Omega \int_0^{r_s} \frac{N_0}{r} \mathrm{d}r$. For non-axisymmetric modes, the period spacings in the inertial frame read

$$\Delta\Pi \simeq \frac{2}{m^2} \sqrt{\frac{\left(\tilde{\ell}+\frac{1}{2}\right)\pi \int_0^{r_s} \frac{N_0}{r} \mathrm{d}r}{\Omega^3\left(\tilde{n}+\frac{3}{4}\right)^3}}, \tag{15}$$

and they allow for measurements of $\left(\int_0^{r_s} N_0 \mathrm{d}r/r\right)/\Omega^3$.

4 Conclusion

Both kind of spacings contain information on the internal rotation and on the stratification of the star, but with different scalings laws. If both are measured, rotation and stratification can be constrained at the same time. When using stellar models in which stratification is given, only one kind of spacings is needed to constrain rotation.

These spacings could be used to interpret observed spectra of fast rotators exhibiting g modes with frequencies lower than their rotation frequency, such as γ Doradus, SPB or Be stars (see studies in the traditional approximation: [18] for γ Doradus stars, and [19] for a SPB star).

The new approximation on which is based our predicted spectral patterns is similar to the traditional approximation. However, our approximation is more accurate near the centre, as it allows waves to propagate there, and it is not limited to spherical models.

This work has been done under the assumption of uniform rotation. The next step is to investigate the case of differentially rotating stars (see [20, 21]).

Acknowledgement

V.P. and S.M. acknowledge funding by the European Research Council through ERC grant SPIRE 647383. The authors acknowledge funding by SpaceInn and PNPS (CNRS/INSU) and by CNES CoRoT/PLATO grant at SAp and IRAP.

References

[1] A. Maeder, *Physics, Formation and Evolution of Rotating Stars* (2009)
[2] J.P. Zahn, A&A **265**, 115 (1992)
[3] A. Maeder, J.P. Zahn, A&A **334**, 1000 (1998)
[4] S. Mathis, J.P. Zahn, A&A **425**, 229 (2004)
[5] F. Espinosa Lara, M. Rieutord, A&A **470**, 1013 (2007)
[6] J. Schou, H.M. Antia, S. Basu, R.S. Bogart, R.I. Bush, S.M. Chitre, J. Christensen-Dalsgaard, M.P. Di Mauro, W.A. Dziembowski, A. Eff-Darwich et al., ApJ **505**, 390 (1998)
[7] R.A. García, S. Turck-Chièze, S.J. Jiménez-Reyes, J. Ballot, P.L. Pallé, A. Eff-Darwich, S. Mathur, J. Provost, Science **316**, 1591 (2007)
[8] S. Deheuvels, J. Ballot, P.G. Beck, B. Mosser, R. Østensen, R.A. García, M.J. Goupil, A&A **580**, A96 (2015)
[9] O. Benomar, M. Takata, H. Shibahashi, T. Ceillier, R.A. García, MNRAS **452**, 2654 (2015)
[10] H. Saio, D.W. Kurtz, M. Takata, H. Shibahashi, S.J. Murphy, T. Sekii, T.R. Bedding, MNRAS **447**, 3264 (2015)
[11] S.A. Triana, E. Moravveji, P.I. Pápics, C. Aerts, S.D. Kawaler, J. Christensen-Dalsgaard, ApJ **810**, 16 (2015)
[12] J. Ballot, F. Lignières, D.R. Reese, M. Rieutord, A&A **518**, A30 (2010)
[13] R.H.D. Townsend, MNRAS **340**, 1020 (2003)
[14] T. Gerkema, J.T.F. Zimmerman, L.R.M. Maas, H. van Haren, Rev. Geophys. **46**, 2004 (2008)
[15] F. Lignières, B. Georgeot, A&A **500**, 1173 (2009)
[16] M. Pasek, F. Lignières, B. Georgeot, D.R. Reese, A&A **546**, A11 (2012)
[17] V. Prat, F. Lignières, J. Ballot, A&A **587**, A110 (2016)
[18] T. Van Reeth, A. Tkachenko, C. Aerts, ArXiv e-prints (2016)
[19] E. Moravveji, R.H.D. Townsend, C. Aerts, S. Mathis, ApJ **823**, 130 (2016)
[20] S. Mathis, A&A **506**, 811 (2009)
[21] G.M. Mirouh, C. Baruteau, M. Rieutord, J. Ballot, J. Fluid Mech. **800**, 213 (2016)

Seismological insights into solar and stellar magnetic activity cycles

Anne-Marie Broomhall[1,2,⋆]

[1] *Institute of Advanced Studies, University of Warwick, Coventry, CV4 7HS, UK*
[2] *Centre for Fusion, Space, and Astrophysics, Department of Physics, University of Warwick, Coventry CV4 7AL, UK*

Abstract. The Sun's magnetic activity cycle varies primarily on a time scale of approximately 11yrs from minimum to maximum and back again. It is well-known that the properties of the Sun's acoustic oscillations are affected by the near-surface internal magnetic field: Frequencies, damping rates, and powers are all known to vary systematically with solar cycle. Careful observation of these variations, therefore, allows aspects of the Sun's internal magnetic field to be inferred. However, the Sun is just one star and, with CoROT and Kepler in particular, oscillations can now be observed for thousands of other stars. However, despite many stars showing signs of magnetic activity in their lightcurves, to date, activity cycle-like variations in the properties of asteroseimic oscillations are sparse. Nevertheless, studying the solar-stellar connection advances our understanding of solar and stellar magnetic fields. For example, placing our Sun's activity cycle in a stellar context suggests that the Sun may be unusual. Conversely, observations of quasi-periodic pulsations in stellar flares suggest the underlying physics of solar and stellar flares are the same.

1 Introduction

The Sun's activity cycle varies primarily on a timescale of 11 years. However, each cycle is different. Indeed the last solar minimum was generally regarded as being unusually long and deep, while solar cycle 24 is the weakest seen since the beginning of the 20th century. One of the most obvious manifestations of solar activity on the solar surface is sunspots and the number of sunspots is known to track the solar cycle. Sunspot records extend back hundreds of years and records show that both the length and the amplitude of sunspot cycles vary from one cycle to the next, with lengths observed in the range of approximately 9-14yrs. There is also evidence that the amplitudes of the cycles themselves vary systematically over what is know as a Gleissburg cycle, which has a periodicity of about 70-100yrs, or around 9 cycles. The recent cycles are generally referred to as the Modern Maxima because, although the current cycle is small, that is just in comparison to recent cycles. There are also Grand Minima, the most famous of which is the Maunder Minimum, where almost no sunspots were observed for an extended period of time. Reconstructions based on radioisotopes of tree rings and icecores indicate other grand minima and maxima existed. These reconstructions also reveal a much longer periodicity of about 210yrs, which is known as the Seuss cycle.

There are also periodic variations associated with magnetic activity that have characteristic timescales shorter than 11yrs. Those shorter than 1yr are collectively referred to as Rieger-like periodicities [1]. While there is also strong evidence for what is loosely referred to as the quasi-biennial oscillation (QBO), although this term is generally attributed to any periodicity in the range 1-4yr. Both the Rieger and the quasi-biennial signal are fairly ubiquitous, being observed in activity proxies sensitive to the solar interior right out to the interplanetary medium [see 2, for a recent reviews]. The amplitude of the QBO varies as a function of time and is at a maximum at solar maximum and this feature is well-known and can be observed in all the vast majority activity proxies associated with the QBO.

To summarize the Sun's magnetic activity cycle is not just a simple 11yr variation but is far more complex. This paper describes helioseismic observations of the Sun's activity cycle (Section 2) and places these observations in a stellar context, discussing observations of stellar activity cycles and the potential for asteroseismology to reveal insights into the Sun's magnetic dynamo (Section 3). We finish, in Section 4, with a slight change of pace, switching to from stellar interiors to stellar atmospheres, and describing how stellar flare seismology can link magnetic features of the Sun and stars.

2 The helioseismic magnetic activity cycle

Properties of the Sun's acoustic (p mode) oscillations vary through the solar cycle. Figure 1 shows comparison of the profiles fitted to a p-mode spectrum at activity maximum and activity minimum. It is clear to see that the frequencies of the oscillations vary with time, as do the heights of the peaks. Less visible are variations in the widths of the profiles, which increase between cycle minimum and cycle maximum. We now describe the solar cycle varia-

⋆e-mail: a-m.broomhall@warwick.ac.uk

DOI: 10.1051/epjconf/201716002009

tions in these parameters in more detail, beginning with the mode frequencies.

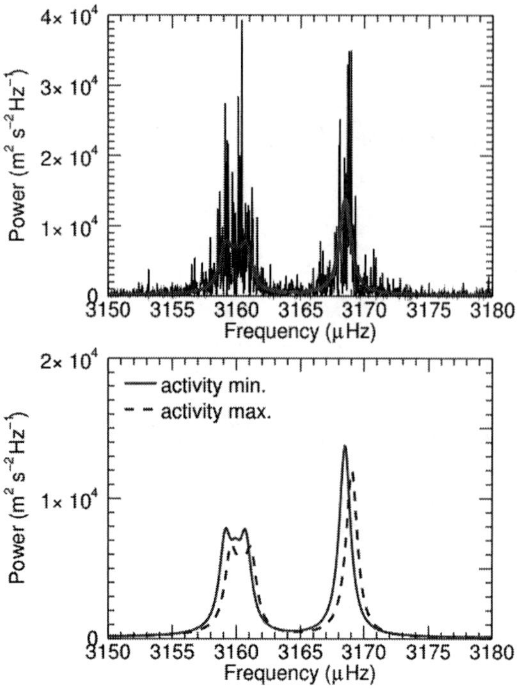

Figure 1: Top: Portion of a Sun-as-a-star power spectrum concentrating on an $\ell = 2$, $\ell = 0$ pair. Black fine line shows the data observed by Birmingham Solar Oscillations Network. Thick red line shows the asymmetric model fitted to the data. Bottom: A comparison of the asymmetric Lorentzian models fitted to an $\ell = 0$, $\ell = 2$ pair at solar maximum and solar minimum. The solid red line shows activity minimum. This is the same model fit as shown in the top panel. The blue dashed line shows a fit to data observed at activity maximum.

2.1 Solar cycle variations in p mode frequencies

There are two ways in which we generally determine the change in frequencies as a function of time and these methods have been employed to study variations in both solar and stellar p-mode frequencies. The first method determines the change in frequency directly from the fitted mode frequencies. Let $\delta\nu_{\ell,n}(t)$ be the change in frequency, or "frequency shift", as a function of time. The frequency shift is then given by

$$\delta\nu_{\ell,n}(t) = \nu_{\ell,n}(t) - \overline{\nu_{\ell,n}}, \tag{1}$$

where $\nu_{\ell,n}(t)$ is the frequency of a mode at time t and $\overline{\nu_{\ell,n}}$ is the "reference frequency" for that mode. This reference frequency is usually obtained by determining the average frequency of a the centroid of an individual mode over a longer period of time, say at solar minimum or the entire epoch under consideration. Although technically one can then consider frequency shifts for each individual mode, more frequently we determine averages over say a particular range of n and/or ℓ.

The second method determines the cross-correlation of time-lagged power spectra. To determine the frequency shift it is necessary to fit a function to the cross-correlation, where the average frequency shift is given by the frequency offset of the peak about zero. This method determines the average frequency shift for all modes in the frequency range for which the cross-correlation was determined. Therefore, it is possible to control the frequency range under consideration. However, it is not possible to isolate particular ℓ when using Sun-as-a-star data.

Chaplin et al. [3] compared the two methods for studying the solar cycle and found that the obtained frequency shifts were, for the most part, in very good agreement. However, they determined that the uncertainties associated with the cross-correlation method were systematically larger than those associated with the the frequency shifts obtained from the individual mode frequencies. In reality the decision over which method to utilise will depend on the quality and quantity of data available: The cross-correlation method may be a better option when considering low signal-to-noise data or short data series, such as those utilised in asteroseismic analyses, from which it is difficult to robustly fit profiles to the individual mode peaks.

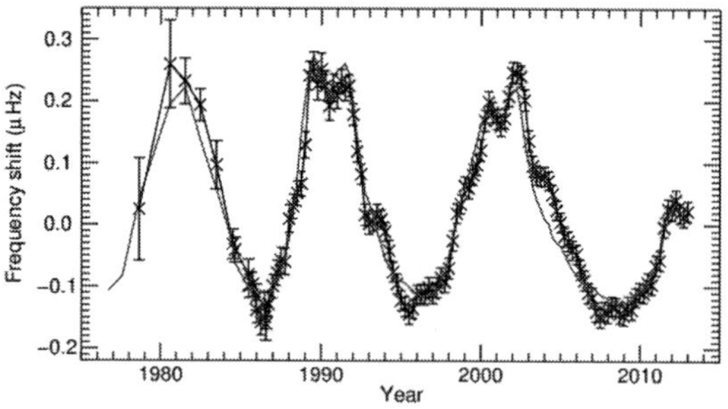

Figure 2: Average frequency shift of Sun-as-a-star oscillations as a function of time (black data points). The frequency shifts were obtained from Birmingham Solar Oscillations Network (BiSON) data. Frequency shifts were determined using Equation 1. Overplotted in red is a scaled and shifted version of the 10.7 cm flux, included for comparison purposes.

Figure 2 shows the average change in seismic frequency as a function of time. The frequency shifts were obtained using the individual mode frequencies (the first method described above). The frequencies were determined by fitting profiles, like those shown in Figure 1, to frequency-power spectra created from time-shifted 365 d time series observed by the Birmingham Solar Oscillations Network (BiSON). BiSON makes unresolved (Sun-as-a-star) Doppler velocity observations of the Sun. As such the BiSON data are only sensitive to those oscillations with the largest horizontal scales (or the lowest harmonic degrees, $\ell \leq 3$). The 11yr cycle is clear visible. The frequency shift, and therefore the mode frequencies, are at a maximum at cycle maximum, with modes at about

$3000\,\mu$Hz experiencing a shift of approximately $0.4\,\mu$Hz between cycle minimum and cycle maximum.

The exact causes of the frequency shifts are still somewhat uncertain. Broadly speaking there are both direct and indirect effects. The direct effects involve the Lorentz force, which provides an additional restoring force. While the indirect effects involve changes to the properties of the cavity in which the modes are trapped. For example, the presence of a magnetic field may alter the size of the acoustic cavity. The exact balance between direct and indirect effects is not clear. For example, Dziembowski & Goode [4] believe that the indirect effects are predominantly responsible for the frequency shifts. However, they do find that there is a frequency dependence, with the direct effect playing a more important role in shifting low-frequency modes. More recently, Santos et al., [5] find that approximately 30% of the frequency shifts can be accounted for by sunspots. We note that this proportion is dependent on the inclusion of a smooth component that represents long-term cycle variations.

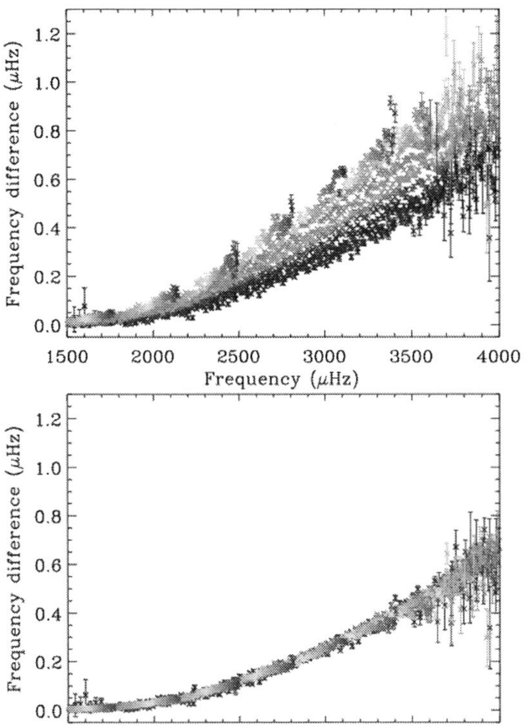

Figure 3: Shift in frequency observed between cycle minimum and cycle maximum as a function of mode frequency. The frequencies are obtained from the standard Global Oscillations Network Group (GONG) pipeline. The colours indicate the degree of the mode, with purple representing the lowest degree modes and red representing the highest degree modes. Figures taken from [6].

The top panel in Figure 3 shows the change in frequency between cycle minimum and cycle maximum as a function of frequency. It is clear that the size of the frequency shift depends both on the oscillation frequency and on the harmonic degree of the mode [also see e.g. 7–12]. The harmonic degree tells us about how deep in the solar interior the oscillations travel, with low degree modes travelling deeper in the solar interior than high degree modes. The degree dependence can be removed by normalizing by mode inertia [e.g. 11]. The normalized mode inertia is defined by Christensen-Dalsgaard [13] as

$$\begin{aligned} I_{n,l} &= M_\odot^{-1}\int_v |\xi|^2 \rho \mathrm{d}V, \\ &= 4\pi M_\odot^{-1}\int_0^{R_s} |\xi|^2 \rho r^2 \mathrm{d}r, \\ &= \frac{M_{n,l}}{M_\odot}, \end{aligned} \tag{2}$$

where ξ is the displacement associated with a mode, suitably normalized at the photosphere, V is the volume of the Sun, and $M_\odot$ is the mass of the Sun. $M_{n,l}$ is the 'mass' associated with a mode. The bottom panel of Figure 3 shows that the mode inertia normalization leaves a functional form that is dependent on frequency alone. The fact that the inertia-corrected frequency shifts show a strong frequency dependence but little or no dependence on degree implies that most of frequency changes are caused by effects confined to near-surface layers, specifically the upper few 100 km of the solar interior [e.g. 7, 14]. In fact, latitudinal inversions show that the frequency shifts are highly correlated with the observed surface magnetic field and even manage to replicate the well-known butterfly diagram [e.g. 15, 16].

In Sun-as-a-star data there is an additional dependence that occurs because of the latitudinal distribution of the magnetic field, as shown in Figure 4. The sectoral $|m| = \ell$ components are more sensitive to low latitudes than the zonal $m = 0$ components. Since the magnetic activity associated with the 11yr solar cycle is also concentrated near the solar equator this means that the frequencies of the sectoral components are shifted by a larger amount than zonal components [17]. This is important for Sun-as-a-star observations since the frequencies obtained for a $\ell = 2$ mode are dominated by the sectoral components as these are by far the most prominent components of an $\ell = 2$ mode observed in a frequency-power spectrum. More specifically, for an $\ell = 2$ mode 90% of the final frequency comes from the sectoral component and only 10% from the zonal component. Consequently, in Sun-as-a-star data, $\ell = 2$ modes experience a larger shift in frequency than $\ell = 0$ modes of approximately the same frequency. This could become important in asteroseismic studies of stellar activity cycles as it any degree dependence on the size of the shift can then be used to infer the latitudinal distribution of the star's activity.

2.2 Quasi-biennial Oscillation

Figure 5 shows the residuals of the frequency shifts once the 11 yr variation is removed. This was done by subtract-

DOI: 10.1051/epjconf/201716002009

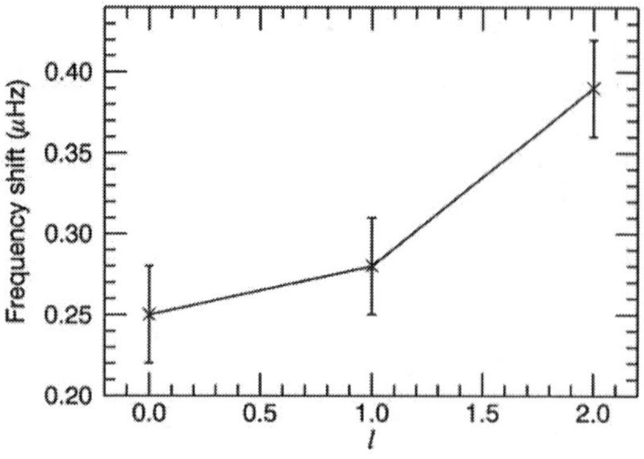

Figure 4: Frequency shift as a function of degree as seen in Sun-as-a-star data.

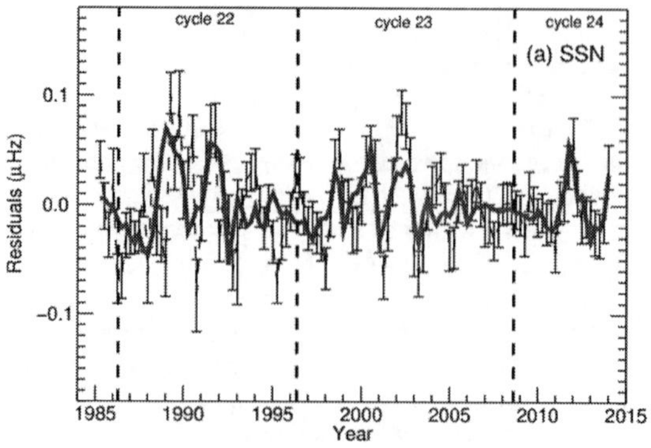

Figure 5: Residuals of frequency shifts once 11yr cycle has been removed (black data points, with dashed line). Overplotted in the red solid line is a scaled version of the sunspot number residuals. We note that the scaling has been made with respect to the 11yr cycle.

ing a smoothed component, where a boxcar smoothing of 3 yr was performed. Variations on timescales shorter than 11 yr are clearly visible. These variations are commonly referred to as a quasi-biennial oscillation (QBO) and Kolotkov et al., [18] demonstrated that the helioseismic QBO has an average periodicity of approximately 2.6 yr. The quasi-biennial variations seen in the frequency shifts are highly correlated with those seen sunspot number, and this is true for a large number of other activity proxies [19]. Furthermore, the QBO observed in different activity proxies even have the same amplitude relative to the 11yr variation. Another feature of the QBO common to helioseismic frequency shifts and other measures of activity is the observed amplitude variation, such that the amplitude of the QBO is at a maximum at solar maximum. However, mystery still surrounds the QBO: Whether or not it is a real phenomenon or an artifact of the data analysis techniques used is still debated, as is the potential physical mechanism responsible for the QBO. Current theories include a second, near-surface dynamo [20], the instability of magnetic Rossby waves in the solar tachocline [21], spatiotemporal fragmentation [22] and beating between dipole and quadropole magnetic configurations of a dynamo [23, 24].

2.3 Cycle-to-cycle variations

Although we don't have enough helioseismic data to look at the longer-term cycles mentioned in Section 1, we have been able to show that the cycles are different. Figure 6 shows the frequency shifts that have been obtained by averaging over 3 different frequency ranges. Notice that the 10.7 cm flux and sunspot number diverge in cycle 23, as is observed by, for example, Livingston, Penn, & Svalgaard [25]. The frequency shifts also appear to diverge during cycle 23 in all frequency ranges. Perhaps most eyecatching, though, is the behaviour of the low-frequency modes in cycle 23. These low-frequency modes appear to show a significant shift in cycle 22 but almost no solar-cycle shift in cycle 23.

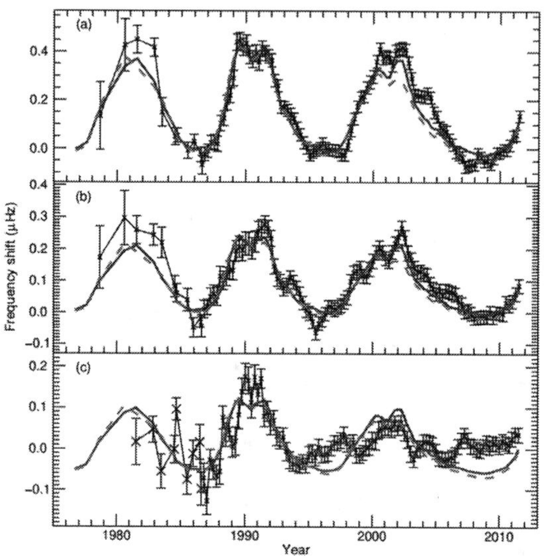

Figure 6: Average Sun-as-a-star frequency shifts as a function of time. Panel (a): Frequency shifts seen in modes with frequencies in the range $2920 < \nu \leq 3450\,\mu$Hz. Panel (b): Frequency shifts seen in modes with frequencies in the range $2400 < \nu \leq 2920\,\mu$Hz. Panel (c): Frequency shifts seen in modes with frequencies in the range $1860 < \nu \leq 2400\,\mu$Hz. Only modes with degrees in the range $0 \leq \ell \leq 2$ were considered. In all panels the red solid line represents a scaled and shifted version of the 10.7 cm flux and the blue line dashed line represents a scaled and shifted version of the sunspot number. Taken from [26].

The upper turning point of modes varies with frequency, with high-frequency modes turning at greater radii than low-frequency modes. So one possible explanation for the effect is that the magnetic layer responsible for the solar cycle shift has become thinner: In cycle 22 the magnetic layer is thick enough that the upper turning points of both the low and high frequency ranges lie within the

DOI: 10.1051/epjconf/201716002009

magnetic layer. However, in cycle 23 this layer is thinner so that the upper turning point of the high-frequency modes still lie within this layer but the upper turning point of the low frequency modes lies underneath the layer. An alternative explanation is that the cavity properties have changed, altering the upper turning point of the modes. To isolate the region responsible we can look at which parts of the Sun the modes are sensitive to using kernals. It can be shown that the changes between cycle 22 and 23 must occur above 2400km [26]. Salabert et al. [27] showed a similar result can be obtained using GOLF data.

2.4 Solar cycle variations in oscillation powers and damping rates

In addition to the mode frequencies, the damping rates and powers of the oscillations also show solar cycle variations. The damping rates are at a maximum at solar maximum while the powers are at a minimum [28]. The in-phase behaviour of the damping rates and anti-phase behaviour of the mode powers could be useful when searching for activity cycles in asteroseismic data, as any similar behaviour would add weight to any claimed detections. We note that the fractional variations in the mode widths appear to show double maxima, which are associated with the QBO. It is well known that active regions suppress mode powers and affect mode lifetimes but the mechanisms are still poorly understood [e.g. 29, 30]. We note that Houdek et al. [31] demonstrated that changes in the convection can affect mode damping rates and theorized that such changes could be brought about through the presence of magnetic structures.

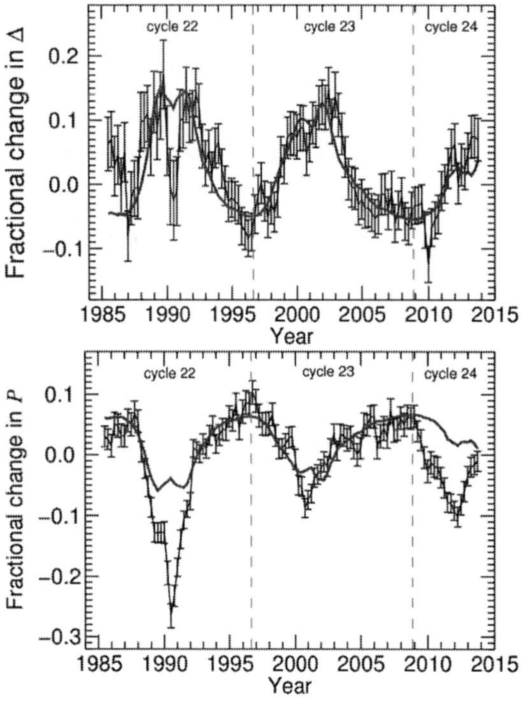

Figure 7: Top panel: Fractional variations in mode width as a function of time (black data points). Overplotted in red is a scaled and shifted version of the 10.7 cm flux for comparison. Bottom panel: Fractional variations in mode power as a function of time (black data points). Overplotted in red is an inversely scaled and shifted version of the 10.7 cm flux for comparison. The fractional variations in the mode widths and powers were averages over modes in the frequency range $2400 \leq \nu \leq 3500\mu$Hz. Mode parameters were obtained from Sun-as-a-star BiSON data and so only modes in the range $0 \leq \ell \leq 2$ were considered.

3 Stellar activity cycles

Evidence for stellar activity cycles has now been observed for some time, primarily because of the Mount Wilson Calcium HK survey [e.g. 32]. The calcium HK index is a measure of the activity in the chromosphere. The Mount Wilson survey showed that other stars exhibit activity cycles that are very reminiscent of our own Sun's. However, it also demonstrated that other stars show far more irregular activity-related variations, while others appear to show no variation whatsoever, with approximately 60% showing cyclic variability, 25% showing aperiodic variability, and 15% showing almost no variability (i.e. flat) [33].

Hopes were high that the long-term, high-precision asteroseismic observations of Kepler and CoROT might lead to observations of cycle-like variations in the oscillation parameters of other stars. The first evidence for such an asteroseismic activity cycle came from [34], who observed that the frequencies of the oscillations of HD49933, which was observed by CoROT, changed with time and that the amplitudes varied in anti-phase, as is observed for the solar data. Salabert et al. [35] examined the frequency dependence of the shifts, which increase with frequency. Again this is just as we see in the Sun. Salabert et al. [35] also noted that the shifts are larger than observed in the Sun. Observations of stellar activity induced frequency shifts in different stars will place additional constraints on predictions over the size of shift one might expect [36, 37], which in turn will improve our understanding of the underlying causes of the freqeuncy shift (see discussion in Section 2.1).

Other examples of asteroseismic cycles however are rare. Salabert et al. [38] and Régulo et al. [39] are two recent papers looking for such asteroseismic cycles that do appear to detect systematic variations in p mode frequencies. Furthermore, these systematic variations appear to be in phase with measures of activity based on the variations in the light curves. However, the variations are still small, especially when compared to typical uncertainties. Kiefer et al.[40] also finds evidence for cycle-like variability in asteroseismic data. However, the question still arises as to why evidence for asteroseismic activity cycles is not more abundant.

Essentially there is a catch 22 situation: Ideally we want to look at stars that are reasonably active so the activity-induced frequency shifts are large enough to be statistically significant. Since, even in the best case sce-

DOI: 10.1051/epjconf/201716002009

nario we only have approximately 4 yr of Kepler data available we also need stars with shorter activity cycles than the Sun: As can be seen from Figure 8 there is a well known relation between a star's rotation rate and the length of its cycle: Very broadly, cycle length increases with increasing rotation rate. However, a more careful examination reveals that the cycle periods increase proportionally with rotation period along two distinctly different sequences, known as the active "A" sequence and the inactive "I" sequence, with cooler and more slowly rotating stars generally lying on the I branch. Along each sequence the number of rotation periods per activity cycle is nearly the same, but the numbers are different for the different sequences. Bohm-Vitense [41] proposed that the branches are fed by different dynamo mechanisms with a near surface dynamo driving the A branch and a dynamo at the base of the convection zone responsible for the I branch. We can also see from Figure 8 that in terms of the primary cycles the younger stars generally lie on the A branch and the older stars like on the I branch.

There is also evidence that the faster a star rotates the more active it is: the X-ray luminosity ratio, $R_X = L_X/L_{\rm bol}$, is proportional to Rossby number (Ro $= P_{\rm rot}/\tau$, where τ is the convective turnover time at the base of the convection zone), until a saturation limit [see 42, and references therein]. The saturation limit occurs at Ro $= 0.13 \pm 0.02$ and for Rossby numbers less than this value R_X is constant and has a value of $\log R_X = -3.13 \pm 0.08$ [42]. Since X-rays are emitted from magnetically confined hot plasma in a star's corona an increase in R_X is believed to correspond to an increase in activity. The higher the X-ray luminosity is the hotter or denser the corona is and so the more active the star is. This means that faster rotators are more active and are expected to have shorter activity cycles.

However, we know from observations of the Sun that magnetic field suppresses the amplitude of acoustic oscillations (see Section 2.4). This result is validated for other stars in [43], who demonstrated that the fraction of stars in which one can detect oscillations decreases as a function of increasing activity. Therefore, the faster a star rotates the shorter its activity cycle, but the more active the star is and the less likely we are to be able to detect acoustic oscillations. It is, therefore, perhaps understandable that more asteroseismic activity cycles have not been detected to date.

The Sun's primary 11yr cycle and secondary quasi-biennial cycle have also been plotted in Figure 8. Neither match particularly well with the observations of the other stars and one proposed explanation for this is that the Sun is in transition. van Saders et al. [44] found a population of field stars in the Kepler field of view that were rotating faster than expected based on gyrochronology expectations. They explained this in terms of reduced magnetic braking efficiency beyond Ro ~ 2 i.e. the Sun's Rossby number. Metcalfe et al. [45] connect these results with chromospheric activity measures and conclude that the Sun's magnetic activity cycle really is unusual, being in an transitional state.

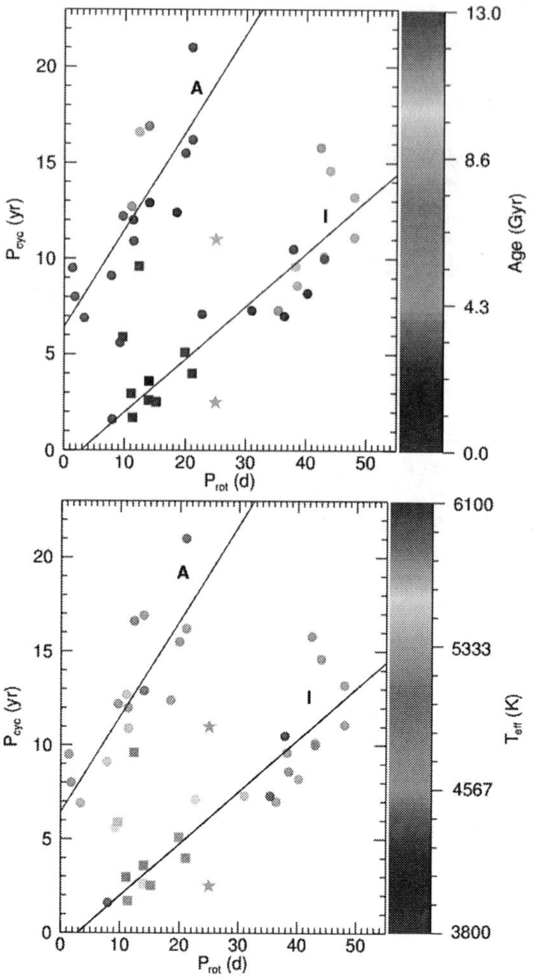

Figure 8: A Bohm-Vitense figure, adapted from [41], which plots cycle length as a function of a star's rotation rate. In the top panel colours indicate the age of the stars, where black indicates the age is as yet unknown. In the bottom panel colours indicate effective temperature. Circles represent primary cycles and squares represent secondary cycles. The two solid lines show best fits for the active "A" branch and the inactive "I" branch.

Finally, we recall from Section 1 that the variability in the Sun's magnetic activity is far more complex than even just the supposed primary 11yr and secondary quasi-biennial cycles. Oláh et al. [46] re-examined the Mount Wilson HK survey and found multiple cycles in a large number of stars. Figure 9 is a Bohm-Vitense diagram constructed with the Oláh et al. [46] results. In this figure there is no real separation of the active and inactive branches and the Sun no longer looks like an outlier, raising questions over whether the Sun's activity cycle is unusual after all.

DOI: 10.1051/epjconf/201716002009

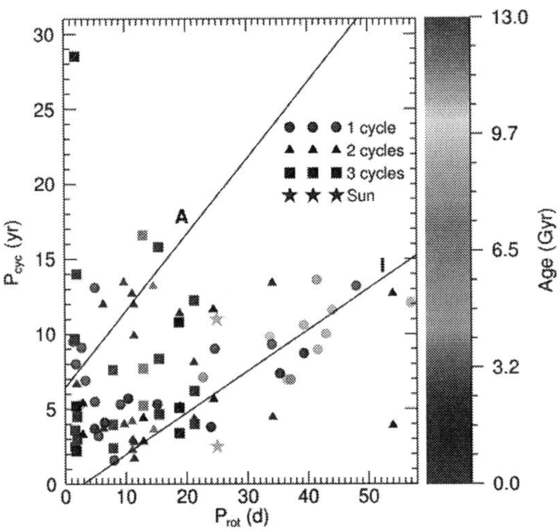

Figure 9: A Bohm-Vitense diagram constructed with the results of Oĺah et al. [46]. Here symbols indicate the number of cycles Oláh et al. found in the Mount Wilson data.

4 The seismology of stellar flares

Seismology is now providing evidence for synergies between solar and stellar flares, which are associated with magnetic activity. Quasi-periodic pulsations (QPPs) are time variations in the intensity of the light associated with the flare. They can have periods from fractions of a second to tens of minutes and the properties of QPPs can provide information about the associated active region [see 47, 48, for recent reviews]. QPPs have also been observed in stellar flares since the 1970s [49]. However, observations of stellar QPP flares are sporadic and often serendipitously discovered. The potential of QPPs to link solar and stellar flares has already been recognised: Pugh et al. [50] detected multiple periodicities in a stellar flare light curve, the most plausible explanation for which is magnetohydrodynamic oscillations, analogous to those observed in solar flares. This not only implies a link between stellar flares and stellar magnetic activity but that the same physical processes are involved in solar and stellar flares.

Balona et al. [52] demonstrated that QPPs can be observed in Kepler flare light curves. Pugh et al. [51] searched the entire Kepler short cadence database for flares with evidence of QPPs and Figure 10 shows an example of one such flare. Out of over 200 flares detected in the Kepler short cadence database we found statistically significant evidence for QPPs in only 56 flares and just 11 of these showed QPPs persistent enough to fit a model to (as shown in Figure 10). Based on these 56 QPP flares we performed a statistical study, comparing QPP period to various stellar parameters, including radius, effective temperature, rotation rate, and log g (see Figure 11). However, no significant correlations were found. This result is important as it implies that the QPP period is dependent upon the properties of the local active region from which it originated and not the global stellar parameters. This is exactly

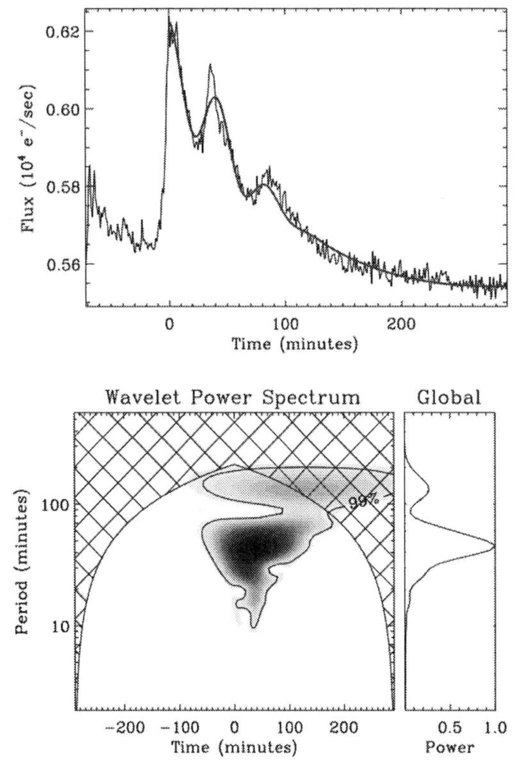

Figure 10: Top: Lightcurve of flare observed by Kepler on KIC12156549 (black). Overplotted in red is a fit to the data based on an exponential decay phase and a damped harmonic oscillator. Bottom: Wavelet spectrum of detrended light curve. Taken from [51].

what one might expect based on solar observations, since the periods of QPPs observed on the Sun themselves extend over a wide range.

5 Conclusions

The Sun's activity cycle is far more complex than a simple 11 yr (or even 22 yr) variation. Both the amplitudes and lengths change from one cycle to the next and numerous other timescales have been associated with it. Solar cycle variations in helioseismic parameters, including frequencies, damping rates and powers, are good proxies of solar activity. However, because the solar cycle is so complex observations of multiple cycles are required to gain insights into the chaotic nature of the Sun's dynamo.

We can search for similar variations in the oscillation parameters of other stars. However, to date only a few examples of asteroseismic activity cycles have been detected. Here we are limited by the length of data we have available and the fact that magnetic fields suppress the amplitudes of oscillations, meaning we can't perform asteroseismic studies on very magnetically active stars. We can, however, use seismology to study stellar flares. Although the

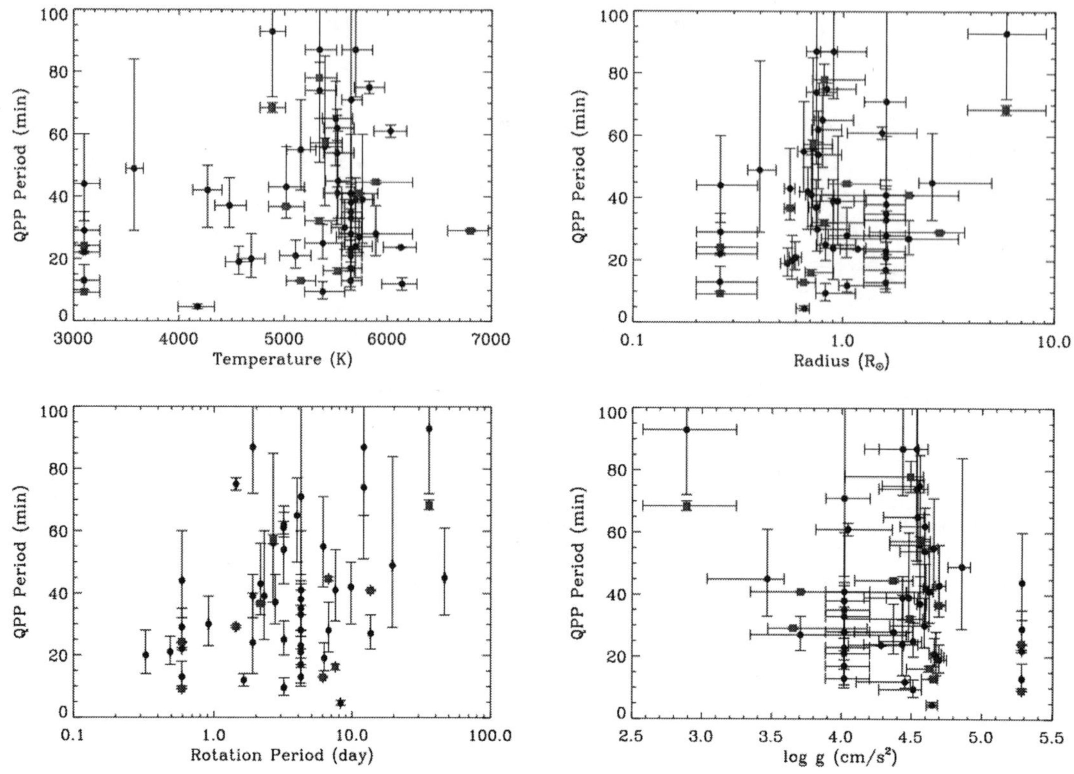

Figure 11: Scatter plots comparing the periodicity of the observed QPPs to various stellar parameters. Top left: QPP period as a function of effective temperature. Top right: QPP period as a function of stellar radius. Bottom left: QPP period as a function of stellar rotation period. Bottom right: QPP period as a function of log g. Black dots inditicate periodicities obtained from wavelet transforms. Red points indicate perioditicies obtained from fitting the decay light curve. Taken from [51].

field of stellar seismology is still very much in its infancy the first indications are that the physics of solar and stellar flares are the same. We can advance our understanding the Sun's dynamo by comparing parameters important to dynamo processes as observed in the Sun and other stars and by placing the Sun in a stellar context.

References

[1] E. Rieger, G. Kanbach, C. Reppin, G.H. Share, D.J. Forrest, E.L. Chupp, Nature, **312**, 623 (1984)

[2] G. Bazilevskaya, A.M. Broomhall, Y. Elsworth, V.M. Nakariakov, Space Science Reviews, **186**, 359 (2014)

[3] W.J. Chaplin, Y. Elsworth, B.A. Miller, G.A. Verner, R. New, ApJ, **659**, 1749 (2007)

[4] W.A. Dziembowski, P.R. Goode, ApJ, **625**, 548 (2005), arXiv:astro-ph/0503266

[5] A.R.G. Santos, M.S. Cunha, P.P. Avelino, W.J. Chaplin, T.L. Campante, MNRAS, **461**, 224 (2016), 1606.02133

[6] A.M. Broomhall, Sol. Phys., **292**, 67 (2017), 1702.03149

[7] K.G. Libbrecht, M.F. Woodard, Nature, **345**, 779 (1990)

[8] Y. Elsworth, R. Howe, G.R. Isaak, C.P. McLeod, B.A. Miller, R. New, C.C. Speake, S.J. Wheeler, ApJ, **434**, 801 (1994)

[9] W.J. Chaplin, Y. Elsworth, G.R. Isaak, R. Lines, C.P. McLeod, B.A. Miller, R. New, MNRAS, **300**, 1077 (1998)

[10] R. Howe, R. Komm, F. Hill, ApJ, **524**, 1084 (1999)

[11] W.J. Chaplin, T. Appourchaux, Y. Elsworth, G.R. Isaak, R. New, MNRAS, **324**, 910 (2001)

[12] S.J. Jiménez-Reyes, T. Corbard, P.L. Pallé, T. Roca Cortés, S. Tomczyk, A&A, **379**, 622 (2001), arXiv:astro-ph/0110364

[13] J. Christensen-Dalsgaard, G. Berthomieu, *Theory of solar oscillations*. (Solar Interior and Atmosphere, 1991), pp. 401–478

[14] D.O. Gough, *Comments on Helioseismic Inference*, in *Progress of Seismology of the Sun and Stars*, edited by Y. Osaki & H. Shibahashi (1990), Vol. 367 of *Lecture Notes in Physics, Berlin Springer Verlag*, p. 283

[15] R. Howe, R.W. Komm, F. Hill, ApJ, **580**, 1172 (2002)

[16] A.M. Broomhall, P. Chatterjee, R. Howe, A.A. Norton, M.J. Thompson, Space Science Reviews, **186**, 191 (2014), 1411.5941

[17] W.J. Chaplin, Y. Elsworth, G.R. Isaak, B.A. Miller, R. New, S. Thiery, P. Boumier, A.H. Gabriel, MNRAS, **343**, 343 (2003)

[18] D.Y. Kolotkov, A.M. Broomhall, V.M. Nakariakov, MNRAS, **451**, 4360 (2015)

[19] A.M. Broomhall, V.M. Nakariakov, Sol. Phys., **290**, 3095 (2015), 1507.02854

[20] E.E. Benevolenskaya, ApJL, **509**, L49 (1998), astro-ph/9810329

[21] T.V. Zaqarashvili, R. Oliver, J.L. Ballester, M. Carbonell, M.L. Khodachenko, H. Lammer, M. Leitzinger, P. Odert, A&A, **532**, A139 (2011), 1107.3395

[22] E. Covas, R. Tavakol, D. Moss, A&A, **363**, L13 (2000), astro-ph/0010491

[23] D. Moss, MNRAS, **306**, 300 (1999)

[24] R. Simoniello, K. Jain, S.C. Tripathy, S. Turck-Chièze, C. Baldner, W. Finsterle, F. Hill, M. Roth, ApJ, **765**, 100 (2013), 1210.6796

[25] W. Livingston, M.J. Penn, L. Svalgaard, ApJL, **757**, L8 (2012)

[26] S. Basu, A.M. Broomhall, W.J. Chaplin, Y. Elsworth, ApJ, **758**, 43 (2012), 1208.5493

[27] D. Salabert, R.A. García, S. Turck-Chièze, A&A, **578**, A137 (2015), 1502.07607

[28] A.M. Broomhall, C.E. Pugh, V.M. Nakariakov, Advances in Space Research **56**, 2706 (2015)

[29] D.T. Woods, L.E. Cram, Sol. Phys., **69**, 233 (1981)

[30] R. Howe, R.W. Komm, F. Hill, D.A. Haber, B.W. Hindman, ApJ, **608**, 562 (2004)

[31] G. Houdek, W.J. Chaplin, T. Appourchaux, J. Christensen-Dalsgaard, W. Däppen, Y. Elsworth, D.O. Gough, G.R. Isaak, R. New, M.C. Rabello-Soares, MNRAS, **327**, 483 (2001)

[32] S.L. Baliunas, R.A. Donahue, W.H. Soon, J.H. Horne, J. Frazer, L. Woodard-Eklund, M. Bradford, L.M. Rao, O.C. Wilson, Q. Zhang et al., ApJ, **438**, 269 (1995)

[33] S.L. Baliunas, R.A. Donahue, W. Soon, G.W. Henry, *Activity Cycles in Lower Main Sequence and POST Main Sequence Stars: The HK Project*, in *Cool Stars, Stellar Systems, and the Sun*, edited by R.A. Donahue, J.A. Bookbinder (1998), Vol. 154 of *Astronomical Society of the Pacific Conference Series*, p. 153

[34] R.A. García, S. Mathur, D. Salabert, J. Ballot, C. Régulo, T.S. Metcalfe, A. Baglin, Science **329**, 1032 (2010), 1008.4399

[35] D. Salabert, C. Régulo, J. Ballot, R.A. García, S. Mathur, A&A, **530**, A127 (2011), 1104.5654

[36] W.J. Chaplin, Y. Elsworth, G. Houdek, R. New, MNRAS, **377**, 17 (2007)

[37] T.S. Metcalfe, W.A. Dziembowski, P.G. Judge, M. Snow, MNRAS, **379**, L16 (2007), 0704.1606

[38] D. Salabert, C. Régulo, R.A. García, P.G. Beck, J. Ballot, O.L. Creevey, F. Pérez Hernández, J.D. do Nascimento, Jr., E. Corsaro, R. Egeland et al., A&A, **589**, A118 (2016), 1603.00655

[39] C. Régulo, R.A. García, J. Ballot, A&A, **589**, A103 (2016), 1603.04673

[40] R. Kiefer, A. Schad, G. Davies, M. Roth, A&A, **598**, A77 (2017), 1611.02029

[41] E. Böhm-Vitense, ApJ, **657**, 486 (2007)

[42] N.J. Wright, J.J. Drake, E.E. Mamajek, G.W. Henry, ApJ, **743**, 48 (2011), 1109.4634

[43] W.J. Chaplin, T.R. Bedding, A. Bonanno, A.M. Broomhall, R.A. García, S. Hekker, D. Huber, G.A. Verner, S. Basu, Y. Elsworth et al., ApJL, **732**, L5 (2011), 1103.5570

[44] J.L. van Saders, T. Ceillier, T.S. Metcalfe, V. Silva Aguirre, M.H. Pinsonneault, R.A. García, S. Mathur, G.R. Davies, Nature, **529**, 181 (2016), 1601.02631

[45] T.S. Metcalfe, R. Egeland, J. van Saders, ApJL, **826**, L2 (2016), 1606.01926

[46] K. Oláh, Z. Kővári, K. Petrovay, W. Soon, S. Baliunas, Z. Kolláth, K. Vida, A&A, **590**, A133 (2016), 1604.06701

[47] V.M. Nakariakov, V.F. Melnikov, Space Science Reviews, **149**, 119 (2009)

[48] T. Van Doorsselaere, E.G. Kupriyanova, D. Yuan, Sol. Phys., **291**, 3143 (2016), 1609.02689

[49] M. Rodono, A&A, **32**, 337 (1974)

[50] C.E. Pugh, V.M. Nakariakov, A.M. Broomhall, ApJL, **813**, L5 (2015), 1510.03613

[51] C.E. Pugh, D.J. Armstrong, V.M. Nakariakov, A.M. Broomhall, MNRAS, **459**, 3659 (2016), 1604.03018

[52] L.A. Balona, A.M. Broomhall, A. Kosovichev, V.M. Nakariakov, C.E. Pugh, T. Van Doorsselaere, MNRAS, **450**, 956 (2015), 1504.01491

EPJ Web of Conferences **160**, 02010 (2017)
Seismology of the Sun and the Distant Stars II
DOI: 10.1051/epjconf/201716002010

Dynamos and Differential Rotation:

Advances at the Crossroads of Analytics, Numerics, and Observations

Kyle Augustson[1,*]

[1] *CEA/DRF/IRFU Service d'Astrophysique, CEA-Saclay, 91191 Gif-sur-Yvette Cedex, France*

Abstract. The recent observational, theoretical, and numerical progress made in understanding stellar magnetism is discussed. Particularly, this review will cover the physical processes thought to be at the origin of these magnetic fields and their variability, namely dynamo action arising from the interaction between convection, rotation, radiation and magnetic fields. Some care will be taken to cover recent analytical advances regarding the dynamics and magnetism of radiative interiors, including some thoughts on the role of a tachocline. Moreover, recent and rapidly advancing numerical modeling of convective dynamos will be discussed, looking at rapidly rotating convective systems, grand minima and scaling laws for magnetic field strength. These topics are linked to observations or their observational implications.

1 Introduction

Whence came the crossroads? With HMI's new additions to helioseismology, Kepler's magnificent asteroseismic capabilities, and numerical breakthroughs with dynamo simulations, the last five years have been an exciting time during which the community has uncovered a few new plot twists in the mystery that is the dynamics occurring within stars.

One such mystery that has long dogged astrophysicists has been the source of the Sun's magnetism and its variability. While pieces of this mystery appear to be solved, there are still many vexing details that are left unexplained, such as why the cycle period is 22 years and why it varies in both duration and amplitude. Moreover, why surface magnetic features migrate toward the pole and equator as the cycle advances is still not clearly determined. Also, a long-running debate within the stellar community has been what is the precise role of a tachocline, which is the region where there likely is a distinct change in a star's rotation profile near the boundaries of its convective zones. In particular, the term tachocline originates from observations of the Sun's differential rotation [1], as shown in Figure 1(a), where the latitudinal differential rotation in its convective exterior is seen to transition to a latitudinally-uniform rotation in its convectively-stable, radiation-dominated interior. Part of the debate resides in a tachocline's role in a dynamo, being whether or not it influences the duration of the cycle period or the even-odd cycle parity [2]. Another piece of the debate regards the role of turbulent diffusion processes in homogenizing the rotation profile [3], as well as its apparent paradoxical ability to prevent the spread of the differential rotation into the stable region [4]. Some of these topics have begun to be addressed in global-scale dynamo simulations, as well as having been scrutinized both analytically and in local-scale simulations, as will be discussed in §4.

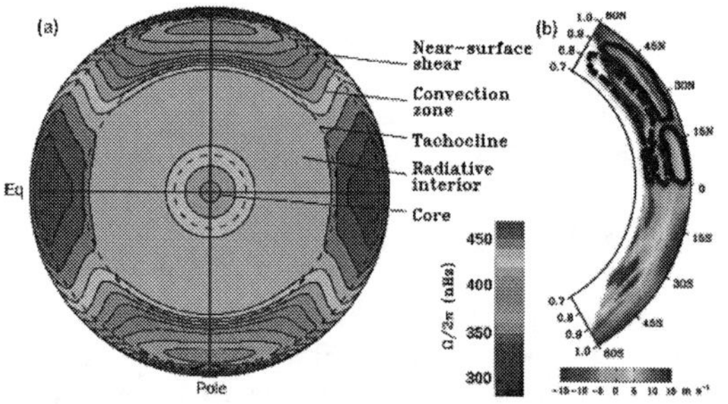

Figure 1: Helioseismically-determined large-scale flows in the Sun. (a) The solar differential rotation, with specific features highlighted [Adapted from 1]. (b) The meridional circulation, with the major circulation cells highlighted. Solid lines denote a counter-clockwise flow, whereas dashed lines denote a clockwise flow [Adapted from 5].

One such mystery was lurking in the large set of Kepler data for red giants. When parsed in the diction of asteroseismology, this data yields the measurement of many mixed modes. These modes behave like an acoustic wave (p-mode) near the stellar surface, but change their character to buoyancy-driven gravity waves (g-modes) within the convectively-stable core of those red giants [e.g., 11, 12], where these convectively-excited modes evanescently tunnel into the core. These modes can thus provide both information about the envelope and the core of the red giant, such as identifying the current evolutionary state of the star [e.g., 13, 14], as well as assessing the relative differential rotation of these two zones [e.g., 15, 16]. The enigma hidden in this data was the question: why are the

*e-mail: kyle.augustson@cea.fr

 DOI: 10.1051/epjconf/201716002010

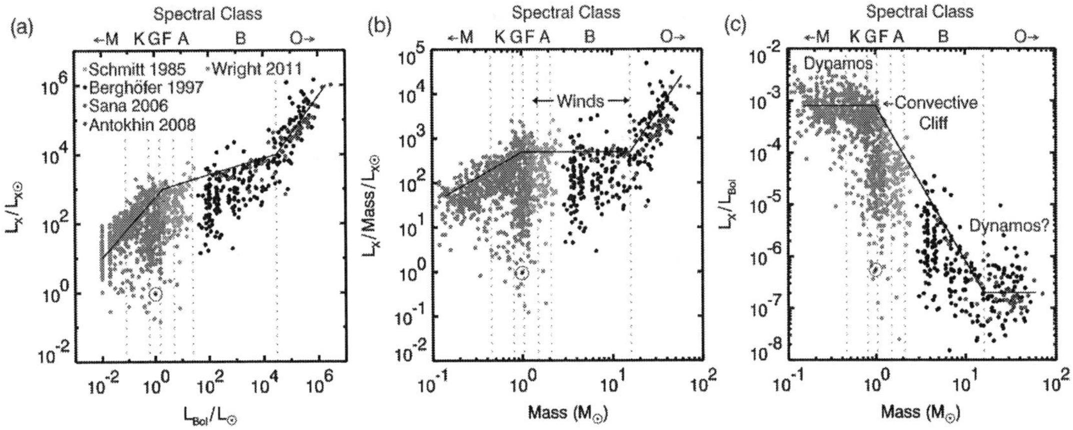

Figure 2: Stellar X-ray luminosity observations for a large range of stellar masses. (a) Average X-ray luminosity L_X versus bolometric luminosity $L_{\rm Bol}$ normalized by their respective solar values. Spectral type is indicated along the top abscissa and demarked with thin dotted lines. The black line illustrates the average behavior for three distinct mass ranges. The data sources are [6–10]. (b) L_X to mass ratio plotted versus mass, emphasizing a mass range with wind-driven emission where L_X depends linearly on mass. (c) The ratio $L_X/L_{\rm Bol}$ versus mass, which highlights the mass range where L_X is linearly related to $L_{\rm Bol}$. The location of the Sun is plotted with its symbol $\odot$, showing its relatively low X-ray luminosity.

dipole mixed modes suppressed for a some red giants and not for others? An astute prediction proposes that this effect is due to the magnetic field present within and near the core of red giants, being dubbed the magnetic "greenhouse effect" [17, 18]. Thus, Kepler observations have revealed the asteroseismic traces of the remnants of potentially very strong magnetic fields that were earlier built deep within the convective core of intermediate mass stars, such as the F-type and A-type stars. Such dynamos will be the topic of §7.

2 Magnetism Across the HR-Diagram

The physics underlying the generation of magnetic fields in astrophysical objects, including stars, appears to be quite generic, with an interaction of plasma motion and large-scale rotation playing key roles. In stars, the most vigorous motions occur in their convective regions. Massive stars have a convective core below a radiative exterior, whereas stars less massive than about 1.6 $M_\odot$ have an outer convective envelope above a radiative zone, and the lowest mass stars are fully convective. The magnetic field generated in those regions is largely responsible for the majority of the observed magnetic phenomena in, or above, the stellar photosphere such as starspots, a chromosphere, coronal plasma, flares, and coronal mass ejections. The detailed study of our nearest star, the Sun, has provided at least a framework for interpreting the X-ray emissions of other cool stars [19]. Indeed, most cool stars, except cool giant stars, have been found to exhibit characteristics grossly similar to what we know from the Sun. Most of these stars have variable X-ray emission at temperatures of at least 1-2 MK and occasional flaring.

Some of the behavior exhibited by low, intermediate, and high mass stars is displayed in Figure 2. There appear to be three distinct mass (or luminosity) ranges where there are changes in the power law relationship between X-ray luminosity (L_X) and bolometric luminosity as in Figure 2(a). The L_X-Mass ratio in Figure 2(b) shows that the intermediate mass range between the low-mass A-type stars and the higher-mass B-type stars has $L_X \propto M$, which may be due to strong shocks in their winds. When scaled by $L_{\rm Bol}$, as in Figure 2(c), two other regimes become evident, with one for the lower mass stars and another for the massive O-type stars where $L_X \propto L_{\rm Bol}$. The lower mass stars have a relatively high level of X-ray emission, whereas the relative X-ray emission declines significantly with higher mass and becomes nearly constant beyond about 16 $M_\odot$.

The precise nature of the physical processes driving these changes in X-ray emission is not known. Though, as always, there are a few hints about what might be occurring. For instance, the saturation of L_X at low masses is potentially a result of the strong magnetic confinement of their winds or a saturation of those stars dynamos, both of which could be due to the rapid rotation of the star. The saturation for higher mass stars may result from two mechanisms: one could be dynamo action in the Fe- and He-opacity bump convection zones; the other may be due to changing properties of the winds and mass loss for the O-type stars. It may be possible to disentangle these processes through asteroseismic measurements of these higher-mass stars, looking for a shift from a dominance of g-modes to p-modes.

3 Magnetic Activity-Rotation Correlation

William Herschel was the first to extend the contemporary idea of the rotation of the Sun and the planets to other stars [20]. In fact, he went so far as to attribute the periodic light curves that he observed for other stars to star spots

 DOI: 10.1051/epjconf/201716002010

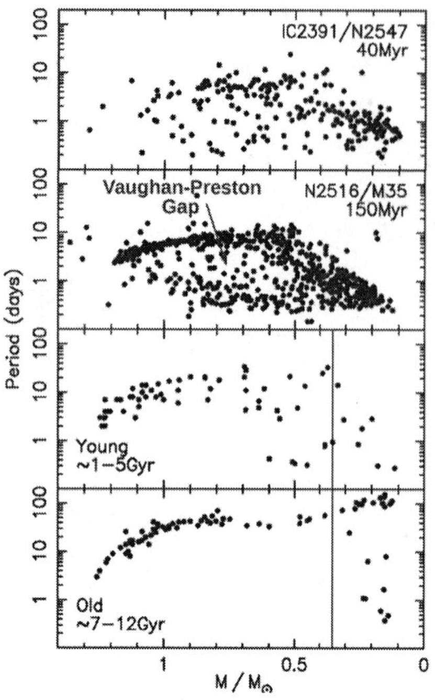

Figure 3: Rotation period distributions of stellar populations of differing ages and mass, showing the initially uniform distribution of a very young 1 Myr old cluster that evolves toward slower rotation with greater age. The panel in the upper right illustrates the Vaughan-Preston Gap, with a population of slow rotators, of fast rotators, and very few stars between them. The red line in the lower right two panels denotes the transition at 0.3 $M_\odot$ from stars with a radiative interior to those that are fully convective (Adapted from [25]).

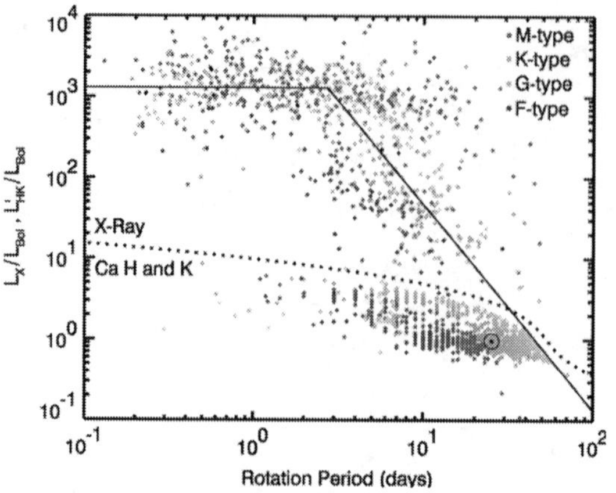

Figure 4: X-ray and Ca II observations of individual stars are shown, which cover a range of rotation rates and masses for lower mass stars. Ratios of X-ray luminosity L_X and chromospheric Ca II luminosity (L'_{HK}) to the bolometric value L_{Bol} (normalized so that these ratios are unity for the Sun; $L_X/L_{Bol} \approx 10^{-6}$ and $L'_{HK}/L_{Bol} \approx 10^{-5}$), displayed with rotational periods as abscissa. The color indicates a star's mass, with M-type in red, K-type in yellow, G-type in green, and F-type in blue. The dotted line demarks the boundary between X-ray and Ca II measurements. The solid line shows an estimated fit for the Rossby number scaling of L_X. The location of the Sun is shown with its symbol (⊙). Constructed with Ca II data from [30, 31] and X-ray data from [10].

rotating in and out of the line-of-sight [21]. Eighty years later, the first Doppler measurements of the Sun [22] were made confirming its rotation. Shortly thereafter and using similar techniques, studies of stellar rotation shed light on two regimes of stellar rotation along the main-sequence that depends upon mass [e.g., 23, 24]. In particular, they found that stars more massive than about 1.6 $M_\odot$ rotate fairly rapidly at an average speed of around 140 kms^{-1} with a slow increase with mass and reaching a plateau above about 10 $M_\odot$. Stars less massive than 1.6 $M_\odot$, on the other hand, show a steep decline in $v \sin i$ with decreasing mass. There is a distribution of their projected equatorial rotation speed ($v \sin i$) about this mean, as the statistics in the above references show and as can be seen in Figure 3.

Spectroscopic surveys of nearby field stars and clusters reveal a potential reason for the rapid decline of a star's rotational period with age. Based on observations of chromospheric lines in young clusters and old field stars [e.g., 26], it is apparent that magnetic activity declines with age. Indeed, the observational link between rotation, convection, and magnetic activity was described in [24], where an attempt is made to ascertain why stars less massive than about 1.5 $M_\odot$ rotate more slowly than more massive stars. A direct link between these physical processes is made when employing concepts pioneered in dynamo theory [2, 27]. As such, it is posited that the loss of an outer convection zone leads to an inability to generate a strong, large-scale magnetic field that would otherwise efficiently brake a star's rotation. Though, this logic would not hold for the most massive stars as they too have outer convection zones, however their relatively short lives lead to less time for wind-related breaking on the main-sequence even if their mass-loss rate is greater.

For lower-mass stars, the link between age, magnetic activity, and rotation rate was put into perspective and formulated into what is now widely accepted as the "Skumanich $t^{-1/2}$ law" for the decay of a star's rotation rate and chromospheric magnetic activity [28]. In particular, two regimes of dynamo behavior seem apparent along the main sequence. As the star spins down, there is a transition between two regimes of chromospheric activity, with a sparsely populated gap between them dubbed the "Vaughan-Preston Gap" [e.g., 29]. This can be seen most easily in Figure 3, which shows the rotation period versus mass for several clusters of increasing age.

The relationship between stellar rotation and proxies of magnetic activity remains an important diagnostic of the dynamo processes likely responsible for stellar magnetism in low mass stars. Chromospheric variability measured in Ca II, and other chromospheric lines, as well as the X-ray observations that probe stellar coronae, indicate that there is a monotonic increase in activity with rotation rate. This increase is halted as the rotation rate reaches a critical point, after which either these indicators of magnetic activity or these star's dynamos become saturated. Indeed, both coronal and chromospheric activity measures show such a saturation (Figure 4), though it is more prominent in the X-ray observations.

The first tentative connection between X-ray luminosity and rotation was established by [32], reporting that X-ray luminosity scaled as $L_X \propto (\upsilon \sin i)^{1.9}$ for low mass stars and that the X-ray luminosities of higher mass stars depends only on the bolometric luminosity. The scaling with rotation for low mass stars suggested that, like the Sun, X-ray emitting coronal magnetic activity is linked to the rotationally-constrained convection that drives the magnetic dynamo running in the outer convection zones of these stars, whereas more massive stars without such an outer convection zone likely produce X-rays primarily through shocks in their powerful winds. With more data, the nearly quadratic proportionality between L_X and $\upsilon \sin i$ was found to have a limited domain of applicability, since the ratio of X-ray to bolometric luminosity reaches a peak saturation of about 10^{-3} that is effectively independent of rotation rate and spectral type [e.g., 33–35]. The rotational period at which this saturation level is reached, however, does depend upon spectral type, with lower mass stars reaching this state at longer periods when compared to higher mass stars (Figure 4a). This can be attributed to the decrease in the vigor of the convection as one moves down the range of masses [36].

It is unclear whether the constancy of the X-ray luminosity below a threshold rotation period is derived of a fundamental limit on the dynamo itself [e.g. 33], a quenching arising from the pervasion of strong magnetic structures across much of the stellar surface [33], or a centrifugal stripping of the corona caused by the high rotation rates [37]. There could also be changes in the X-ray production mechanisms as well as their X-ray spectral content. However, recent work has aimed at modelling the coupling of a simple dynamo to a wind model, with the aim of describing this transition through fairly simple physics [38, 39].

Characteristics similar to the X-ray flux also hold for chromospheric measures of magnetic activity as can be seen in Figure 4. This figure shows the ratio of chromospheric Ca H and K luminosity $L'_{\rm HK}$ to bolometric luminosity for a sample of main-sequence stars ranging in mass between 0.3 and 1.6 $M_\odot$, with data from [30] and [31]. The ratio of X-ray luminosity and bolometric luminosity, also plotted with data from [10], has a much greater dynamic range, spanning six orders of magnitude compared to the roughly 1.5 orders of magnitude for the Ca H and K observations. Yet both show an increase of activity with decreasing rotation period and an eventual saturation. It is of note that the chromospheric activity seems to saturate at longer periods and that there are few stars that are rotating rapidly that show Ca H and K emission, at least in this sample.

If we now separate the X-ray observations by spectral type, the trends become more obvious as in Figure 5(a) for the G-type stars. Speaking broadly, the low mass stars of K, G, and F-type share similar characteristics in their X-ray emission involving increasing emission with more rapid rotation and a regime of saturated emission. What differs between them is the rapidity of the onset of this state of constant emission with rotation, the rotation period at which it saturates, and the maximum luminosity. The maximum luminosity seems to be determined by the bolometric luminosity, as surmised from Figure 4 where the data collapse onto the same fit. The rotation period at which it saturates may be related to the maximum achievable dynamo action, wherein the dynamo can no longer extract energy from the star's differential rotation due to strong Lorentz feedbacks.

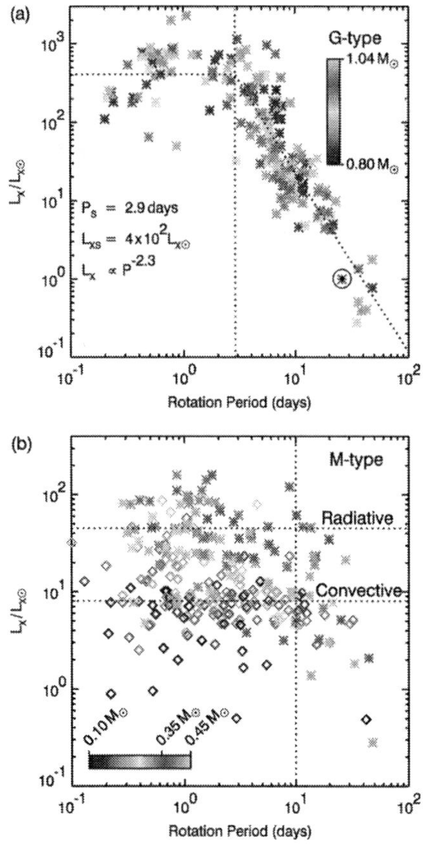

Figure 5: X-ray observations are shown for individual stars that are binned by their spectral type, which cover a range of rotation periods and masses. X-ray luminosity (L_X) normalized by the solar value and displayed with rotation period in days as abscissa. The color indicates a star's mass within the spectral type. (a) Spectral type G stars, with the Sun's position indicated. (b) Spectral type M stars, with stars above 0.3 $M_\odot$ as starred symbols and those below this mass as diamonds. The dotted lines indicate fits to the data for the scaling with period, the saturation value, and the rotation period corresponding to the beginning of the saturation (Constructed with data from [10]).

The rate of increase of X-ray emission with decreasing rotation period suggests a direct link to the Rossby number of the flows within these stars in that it measures the efficiency of the interaction of rotation and convection in building both the differential rotation and the magnetic field. In particular, there appears to be a change in the

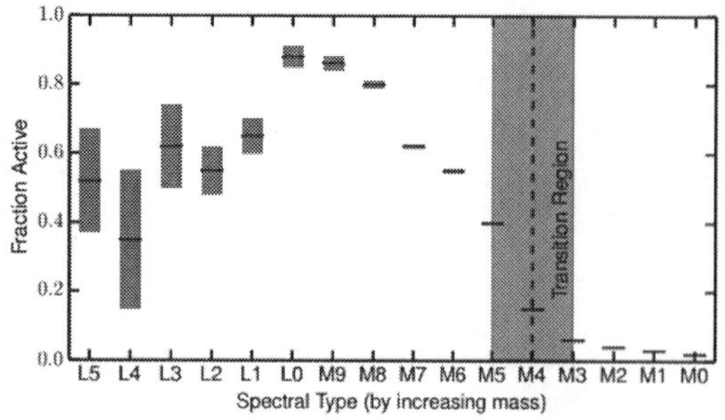

Figure 6: Fraction of active L and M dwarfs, showing the transition from a low to high activity for stars less massive than spectral type M4 [Adapted from data in 40]. The red line denotes the mean value and the blue boxes show the range of uncertainty.

slope of the decrease of X-ray emission with rotation period, from 1.7 to 2.5 between the F and K-type stars. This may indicate that the K-type stars have the most efficient conversion of kinetic to magnetic energy. An interesting feature of stars about the mass of the Sun, namely the G and F-type stars, is that there seems to be a regime of super-saturation with the X-ray luminosity decreasing at the rates of greatest rotation, which is a topic of ongoing research. What is also evident from all of these observations is that the Sun, while thought of as a magnetically active and robustly influential object, is in fact hundreds of times less active in both chromospheric and coronal proxies than some of its more rapidly rotating brethren, which serves to accentuate the overwhelming activity that must be present at the surfaces of these other stars.

4 Tiny Stars with Strong Magnetic Fields, and the Influence of a Tachocline

In the Sun, the tachocline is a site of strong radial shear revealed by helioseismology , which is a matching layer between the differentially rotating outer convection zone and an interior rotating as a solid body (see Figure 1). Other stars with a radiative interior may also have a tachocline. In an interface-type dynamo, the tachocline is thought to play a major role in the overall dynamo action. Such a layer below the convection zone can store and, through the radial differential rotation, comb small-scale magnetic field into a larger toroidal structure. When the magnetic field strength is sufficiently large, this field can become buoyant and thus could provide the seed magnetic field for the convective dynamo in the exterior convection zone [41, 42].

What is possibly more remarkable in the X-ray data is the transition between stars that possess a radiative core and those that are fully convective (Figure 5(b)). Indeed, most of the stars in our galaxy are less massive than our Sun, with more than half of all stars being M-dwarfs. These are stars that are less than half the mass of the Sun and thus between tens and thousands of times less luminous than it. These low-mass stars had been believed to harbor magnetic dynamos quite distinct from those in Sun-like stars. It seems appropriate that there should be a difference between stars that have an interface-type dynamo, with a region of penetration and convective overshoot, and those that are fully convective.

An intriguing suite of observations of low-mass stars in Hα indicates that there is a transition in magnetic activity lifetime from the more massive stars that become relatively inactive after about one billion years to the lower-mass fully-convective stars, which appear to remain active for much longer [43]. Indeed, the trend likely continues past the pp-chain fusion limit into the category of giant planets [40], with those objects potentially remaining active for their lifetimes. Moreover, as seen in Figure 6, the fraction of spectral-type L and M dwarfs that are active in Hα drops from a peak value of about 85% at spectral-type M9 to about 2% at spectral type M0.

A trend that is likely related can also be seen in open cluster data, wherein the variance in the distribution of the rotation rate for stars in the cluster decreases with time [25]. For stars with masses between about 0.3 and 1.3 solar masses that are evolving along the main sequence, the rotation-rate distribution slowly transits from a nearly uniform distribution to a tight distribution about a mean value. However, if the stars' masses fall below the fully-convective threshold value, the distribution of rotation rates appears to remain nearly uniform. This picture of long-term activity remaining even if the star spins down is largely consistent with observations both in X-ray, Hα, and Ca II spectral bands.

Hence, for these three observational measures of rotation rate, activity fraction, and activity lifetime, the behavioral transition occurs over a specific range of stellar mass. In particular, it occurs where main-sequence stars shift from having a convective envelope to being fully convective. Furthermore, such a transition may demark a region where a tachocline has a decreasing influence on the star's magnetic and rotational evolution, given that the radius of the convectively-stable radiative interior decreases for lower masses.

Stitching these pieces together hints at a question: it appears that stars with a tachocline, or at least with a radiative interior, spin down more rapidly than their fully-convective counter parts, so is the spin-down related to the internal dynamo? When comparing these two kinds of stars, the fully-convective stars tend to have stronger magnetic fields with simpler geometry [44, 45]. They are also more rapidly rotating. This all should lead to a more fierce wind and stronger angular momentum loss [46]. So, why is it that these stars remain active so long and retain much of their zero-age main-sequence angular velocity? It may be that the presence of a stable region is a critical component in the dynamo that inherently drives these stars' magnetic activity and hence impacts their rotational history. Yet precisely how a such a region could influence both of these facets of stellar evolution is unclear.

5 Prandtl Numbers and Large and Small-scale Dynamos

A powerful tool currently available in modern dynamo theory is 3D numerical simulations. In direct numerical simulations (DNS), all the turbulent processes are fully captured, from the driving scale to the diffusion scales. Although these simulations are typically far removed from the length scales important to astrophysical processes, they can shed light on the small-scale dynamics that strongly influence the transport properties of the larger-scale system. The aim of these simulations is usually to try to make contact with theoretical predictions of turbulent spectra, such as the Kolmogorov scaling. However, as with large-eddy simulations (LES) that parameterize the small-scale physics, they are limited by computational resources in obtaining ever larger resolution and are constrained in their ability to run large suites of simulations that accurately probe large swaths of parameter space.

There are many approaches to these problems: some that are true DNS in the sense that a particular physical system with its empirically determined diffusion coefficients are simulated, some that utilize other diffusive operators, and still others that employ only an implicit numerical diffusion. The first method assumes that the number of atoms within a grid cell is still large enough that the diffusion approximation to the collisional interactions is valid, whereas the last assumes effectively the same thing but with a numerically defined diffusion coefficient. This numerical diffusion could be considered a running coupling constant under the nomenclature of renormalization theory. The same can be said for LES schemes that explicitly and artificially increase the level of the Laplacian diffusion in the system. Thus there is some connection between these two schemes, though it may not be simple due to the anistropropies that are often present in LES of global-scale flows.

For simulations retaining the diffusion approximation to collisional interactions, there are two dynamo regimes one of low magnetic Prandtl number (*Pm*) and another of large *Pm*. Theoretically, one should expect that at large *Pm* the magnetic Reynolds number is much greater than that of the fluid Reynolds number, which leads to the length scales of the magnetic field being smaller than that of the velocity field. At low *Pm*, the expectation should be the opposite that the generated magnetic fields roughly tend to be of larger scale than the velocity field. This does not preclude smaller scale intermittent fields for the low-*Pm* regime or vice versa for the converse, rather it is more a statement about which components of the magnetic field contain the most energy: the large-scales for low-*Pm* dynamos and the small-scales for high-*Pm* dynamos. Using the Braginskii definitions of the viscous and magnetic diffusivities [47] and stellar models, one can determine the ranges of magnetic Prandtl numbers within a star. For dynamos in stars with convective exteriors, the low-*Pm* regime is of greatest interest. Using the Sun as an example, the molecular *Pm* ranges between about 10^{-1} in the core to roughly 10^{-6} at the photosphere, decreasing exponentially toward it. Other low-mass, main-sequence stars

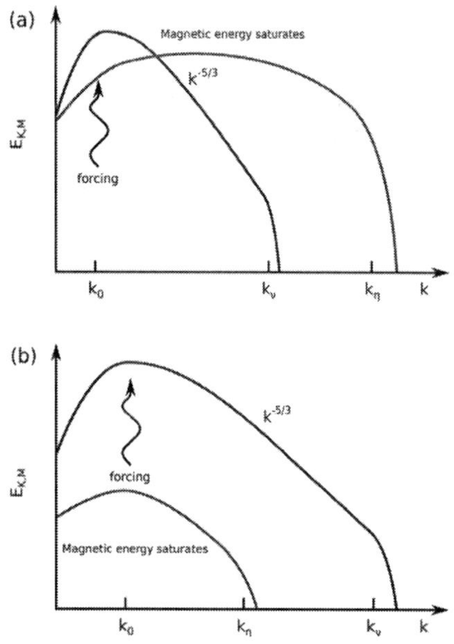

Figure 7: A sketch of energy spectra for low and high-*Pm* dynamos with wavenumber k. (a) Saturation phase of the high-*Pm* dynamo, with kinetic energy in blue and magnetic energy in red. (b) Saturation phase of the low-*Pm* dynamo. Here k_0 is the driving scale, k_ν the viscous dissipation scale, and k_η the magnetic energy dissipation scale.

are roughly similar, with a low *Pm* throughout the bulk of their domains. Higher-mass, main-sequence stars, in contrast, have relatively large molecular values of *Pm* ranging from about 10 throughout the convective core and remaining nearly unity until the near-surface region is reached at which point it drops to about 10^{-2}. So, the convective cores of massive stars and the convective envelopes of low mass stars represent very different dynamo regimes with respect to the scales possessing the greatest magnetic energy.

At small scales, the *Pm* parameter regimes have been studied extensively with numerical simulations. The primary results of these simulations are that, at large magnetic Prandtl number, the magnetic field generation is of the stretch-twist-fold variety where its structures are of a smaller scale than the viscous dissipation scale [e.g., 48]. On the other side of the Prandtl number coin, the low *Pm* regime has the stretch-twist-reconnect mechanism, where the flows act similar to the large *Pm* regime, but instead of the flows dissipating and the magnetic field folding on itself, the magnetic field dissipates leading to reconnection [e.g., 49, 50]. These two dynamo regimes are sketched in Figure 7, showing the saturated phase of the dynamo. While such simulations are instructive for creating subgrid-scale models and understanding the small-scale physics, they lack a direct connection to the global-scale phenomena that are more directly observable for astrophysical objects.

Likewise, there are many well-studied analytical or semi-analytical large-scale dynamos, for instance consider

DOI: 10.1051/epjconf/201716002010

those of [51], [52], [53], and [54]. These dynamos generate magnetic fields that have spatial scales greater than the velocity scales producing them. The Sun and the Earth, among many other astrophysical objects, seem to possess global-scale fields and as such they are likely large-scale dynamos with a cascade of energy moving from small-scale turbulence to large-scale magnetic fields. For stellar and planetary dynamos, the mechanisms building large-scale field is directly linked to their rotation. In these global-scale dynamos, one manner in which the rotation is crucial is that the Coriolis force induces a net kinetic helicity in the small-scale turbulent flows. Correlations between the fluctuating velocity components of the turbulence consequently yield Reynolds stresses that act to maintain the bulk differential rotation through an inverse cascade of kinetic helicity and energy. This inverse cascade is thus the indirect source of the differential rotation that fuels the generation of toroidal field, whereas direct action of the convection is largely to transform this toroidal field into poloidal field as in an α-effect. Thus the structure of the convection and the differential rotation that it supports is key to the overall nature of the global-scale stellar dynamo.

If the entire spectrum of the convective dynamo and its associated dynamics is resolved, the dissipation of the energy injected into the system is governed by the molecular values of the diffusivities. Using Braginskii plasma diffusivities for a solar-like metallicity [47] and a set of MESA stellar models to obtain the density and temperature profiles [55], one can define the average expected molecular magnetic Prandtl number in either the convective core of a massive star or the convective envelope of a lower mass star. To be specific, the equations of interest are

$$\nu = 3.210^{-5}\frac{T_e^{5/2}}{\rho\Lambda}, \quad (1)$$

$$\kappa_{\rm cond} = 1.810^{-6}\frac{T_e^{5/2}}{\rho\Lambda}, \quad (2)$$

$$\eta = 1.210^{6}\Lambda T_e^{-3/2}, \quad (3)$$

where ν is the kinematic viscosity, η is the magnetic diffusivity, $\kappa_{\rm cond}$ is the electron thermal diffusivity, $\kappa_{\rm rad}$ is the radiative thermal diffusivity, T_e is the electron temperature in electron volts, and Λ is the Coulomb logarithm. These were computed assuming the conditions of a MHD plasma. In particular, it is assumed that charge neutrality holds and that the temperatures are not excessively high, so the Coulomb logarithm is well-defined. Moreover, in the charge-neutral regime, the magnetic diffusivity is density independent because the electron collision time scales as the inverse power of ion density and the conductivity is proportional to the electron density times the electron collision time. So, one can find that the thermal Prandtl number Pr and magnetic Prandtl number Pm scale as

$$Pr = \frac{\nu}{(\kappa_{\rm cond} + \kappa_{\rm rad})}, \quad (4)$$

$$Pm = 2.710^{-11}\frac{T_e^4}{\rho\Lambda^2}. \quad (5)$$

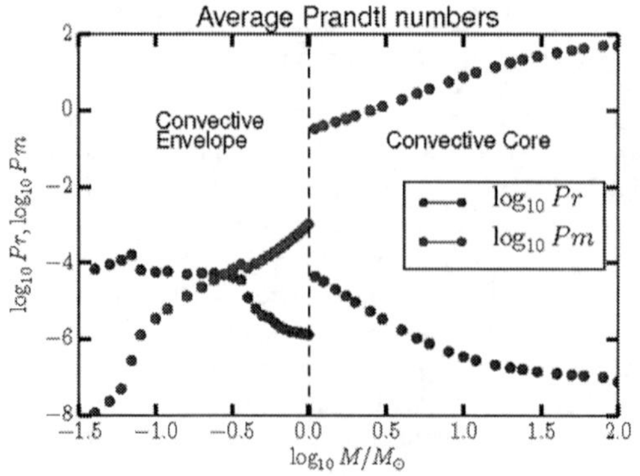

Figure 8: The thermal Prandtl number Pr (blue) and magnetic Prandtl number Pm (red) for stars with masses between 0.03 and 100 $M_\odot$. The convective region being averaged is the core for stars above about 1 $M_\odot$, and is either the convective envelope for stars below about 1 $M_\odot$ or full star for stars below 0.3 $M_\odot$.

In most cases, the radiative diffusivity is several orders of magnitude larger than the electron conductivity, so it dominates the thermal Prandtl number.

This prescription for the Prandtl numbers has been applied to MESA models of stars on the zero-age main-sequence in the mass range between 0.03 and 100 $M_\odot$ as shown in Figure (8). There, it is clear that all stars fall into the low thermal Prandtl number regime, whereas one can find two regimes of magnetic Prandtl number. This is directly related to where the convective region is located. For massive stars with convective cores, the temperature and density averaged over the convective volume are quite high when compared to lower-mass stars with a convective envelope. Such a high temperature leads to a large magnetic Prandtl number. This implies that there could be two fundamentally different kinds of convective dynamo action in low-mass versus high-mass stars.

6 Solar-like Stars and Their Cycles

From the perspective of dynamo theory, lower mass stars fall into the category of large-scale low-Pm dynamos. Nearly all such stars are observed to rotate to some degree and to have surface magnetic activity, as discussed above. Moreover, for sufficiently massive stars, they are likely to support a tachocline that separates their convective regions from the stable radiative zone below. Simulations of such stars have begun to show how the tachocline can influence the dynamo in ways that go beyond the original idea of the interface-type dynamo such as those of [64] and [42].

Having examined some of the basic principles of how differential rotation is built and maintained and examined how accurately global-scale models can capture such hydrodynamics, the topic of global-scale dynamo simulations in a solar setting must be addressed. To date, several different numerical models of solar convective dynamos have achieved cyclical polarity reversals, whether

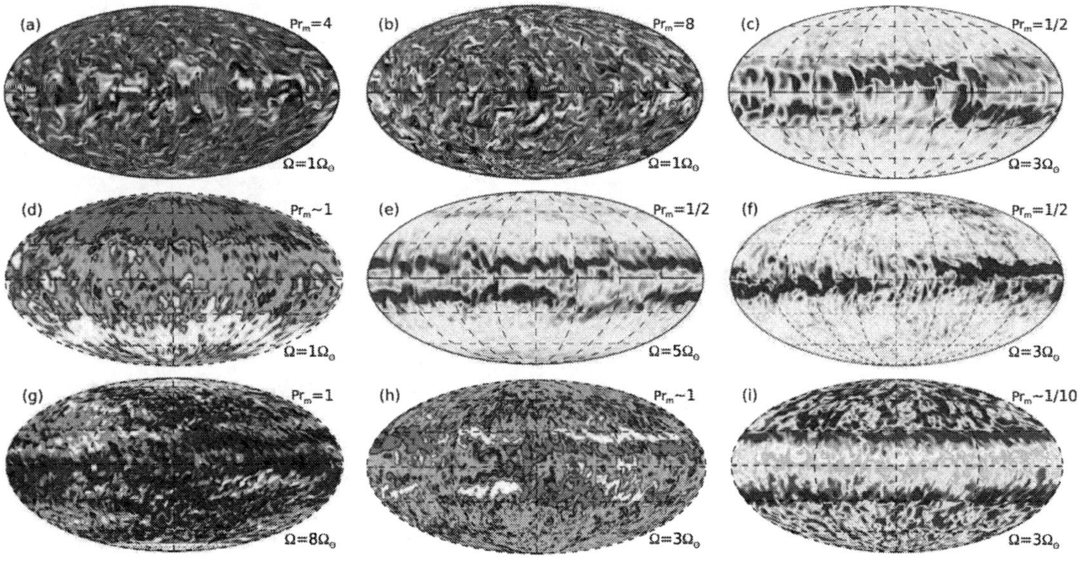

Figure 9: Modern 3-D global-scale convective dynamo simulations, showing snapshots of their toroidal magnetic fields in Mollweide projection. The effective magnetic Prandtl number (*Pm*) and rotation rate Ω of each simulation is indicated. (a) First dynamo simulation using ASH [56]. (b) ASH simulation with a tachocline [57]. (c) Large-scale, persistent toroidal structures dubbed wreaths maintained amidst moderately turbulent convection at lower *Pm* [58]. (d) EULAG solution with regular decadal-time-scale magnetic reversals [59]. (e) ASH simulation with quasi-periodic reversals [60]. (f) Recent continuation of (c) with lower diffusion and higher resolution. The greater levels of turbulence permits quasi-periodic cycles to emerge [61]. (g) Cyclic dynamo solution with the Pencil code, showing wreath formation in a spherical segment [62]. (h) EULAG solution rotating at 3 $\Omega_\odot$, for comparison with (c), (f), and (i) at the same rotation rate [63]. (i) Recent simulation using slope-limited diffusion in ASH exhibiting very regular cycles. Typical toroidal magnetic field strengths are between 1 and 10 kG in these simulations, with sense of fields suggested by color.

or not a tachocline is present. In all cases of such cycling global-scale dynamos, the interaction of convection, rotation, and magnetism are crucial ingredients. Though, precisely how the mechanisms are arranged so that a solar-like differential rotation can be maintained, and simultaneously yield a large-scale, periodic dynamo, has yet to be ascertained. Indeed, how the multiply-scaled convection of a thermally driven system, like the Sun, sempiternally maintains a global-scale kinetic helicity and sustains the massively disparate scales of the Sun's magnetism is seemingly a mystery.

The earliest attempts to address these issues with global-scale solar dynamo simulations were those of Gilman and Glatzmaier, who used an anelastic and spherical harmonic based spectral code to simulate the dynamics [65] rather than the finite difference Boussinesq scheme of [66]. A series of simulations were laid out in [67] that aimed at assessing how varying the magnetic Prandtl number *Pm* and the rotation rate modified the character of the resulting dynamo. In general, it was found that magnetic cycles are a common feature, provided that there is sufficient energy in the bulk toroidal fields and that the Lorentz force feedback on the established differential rotation was not too large. Moreover, Gilman also noted that there is little preference for symmetry or anti-symmetry about the equator and that the toroidal bands of magnetic field tended to propagate toward the poles. It was also found that, for a fixed Rayleigh number (ratio of thermal driving to diffusion), Taylor number (ratio of the Coriolis force to viscous diffusion), and thermal Prandtl number (ratio of viscous to thermal diffusion), there is a peak dynamo efficiency with $Pm \sim 1$. For *Pm* below unity, the efficiency reached a plateau and for those cases with *Pm* above unity it decreased.

Yet the mismatch between those early simulations and helioseismic inferences effectively ended that line of investigation for about 15 years. A similar approach to simulating global-scale dynamos was, however, restarted roughly around 2004 with the inclusion of magnetic fields into the anelastic spherical harmonic (ASH) code (see [56] for instance). These 21st century global-scale convective solar dynamo simulations were carried out at moderate *Pm*, ranging between 2 and 4. Hence it was much more difficult for a large-scale dynamo to take hold. In keeping with that notion, the fluctuating fields dominated the magnetic energy, accounting for most of the back-reaction on the flow via Lorentz forces, whereas the mean fields were relatively weak. There were also no hints of periodic reversals or dynamo wave behavior. Again as expected with a large *Pm* dynamo, the magnetic fields exhibit a complex and small-scale spatial structure and highly-variable time evolution. The radial magnetic field is swept into the downflow lanes, being stretched and folded there before reaching the resistive scale and reconnecting. The toroidal fields are organized into somewhat larger scales near the equator appearing as twisted ribbons, whereas they form

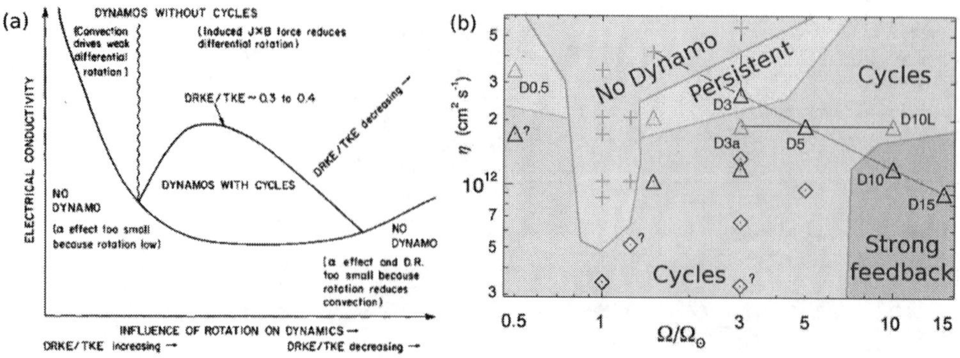

Figure 10: (a) A sketch of possible dynamo regimes with rotation rate and electrical conductivity, illustrating the cycling and persistent dynamo regimes as well as those without sustained dynamo action [adapted from [67]]. (b) A set of numerical dynamo experiments carried out with varying Prandtl numbers and rotation rates, exhibiting many of the behaviors predicted in [67] and some that were not, particularly at low rotation rate [adapted from [68]].

smaller scales at higher latitudes (Figure 9a). In [56], it was also found that the convection maintained a solar-like angular velocity profile despite the influence of Maxwell stresses, which tend to extract energy from the differential rotation and thus reduce its contrast throughout the convection zone.

The next logical step was to include a stable region and to emulate a tachocline in order to assess its impact upon the global-scale dynamo [57]. Both this simulation and that of [56] were fairly laminar with respect to the scales present in the velocity field. The dynamo action achieved in the convection zone of this simulation is also quite similar to the earlier calculation of [56], though with an even larger *Pm*. Thus, as one might expect the scales of magnetism become much smaller than the scales of the convection (Figure 9b), with the bulk of the magnetic energy appearing in the fluctuating magnetic fields and very little in the mean fields within the convection zone. The major difference between the dynamo action present in this simulation, as compared to that [56], is that the stable region provided a reservoir for storing magnetic energy and the tachocline may have provided a means of generation of large-scale magnetic field. Indeed, within the stable region, large-scale fields were able to persist, with the majority of the energy being in the mean magnetic fields. However, there was as yet no large-scale structure that might be able to appear at the surface (i.e. no magnetically buoyant structures) nor were there any cyclical polarity reversals.

Rapid progress over the last four years have confirmed the role that convection and rotation likely play in stellar dynamo action through global-scale dynamo simulations as well as suggested additional mechanisms that may be present in the solar dynamo. Several research teams have concurrently made inroads into solar dynamo theory utilizing three distinct numerical techniques, but nonetheless realizing similar dynamo action. The Anelastic Spherical Harmonic (ASH) code has provided a framework for constructing and understanding the dynamos that produce and sustain coherent toroidal fields exhibiting quasi-cyclic behavior. The Eulerian-Lagrangian (EULAG) code has shown that solutions with very regular cycles can be achieved within its implicit large eddy simulation (ILES) framework. Third, the Pencil code that simulates compressive convection has also found cyclical solutions with large-scale toroidal field structures within its spherical wedge geometry.

The first 3D nonlinear convective dynamo simulation, beyond those of Gilman and Glatzmaier in the 1980s, that indicated that sustained large-scale toroidal magnetic field structures could be built within the solar convection zone was presented in [58]. A snapshot of this solution's toroidal magnetic field is shown in Figure 9c and its evolution is shown in Figure 12a. The persistence of such structures had previously been thought impossible, as one might expect that the highly turbulent convection should rapidly eviscerate it, transporting the magnetic flux to the boundaries of the domain. What separated this simulation from those undertaken a few years earlier was its higher rotation rate (or lower Rossby number) and its lower *Pm*, being about 16 times smaller. The salient point that this simulation brings forth was that the rate of generation of magnetic field through the mean shear of the differential rotation could overpower the rate of its destruction by convection and dissipation. Indeed, the convection instead acts to sustain the differential rotation through Reynolds stresses, which indirectly aids in the generation of the toroidal fields. The convection then in turn converts the toroidal field into poloidal field, creating a positive feedback loop that sustains the wreaths against resistive decay. Although this simulation hinted that perhaps fields could be sustained within the solar convection zone, it still did not have cyclical behavior. Furthermore, it was still quite laminar, so the ability to sustain such large-scale magnetic field structures at greater levels of turbulence and increased complexity of the velocity field was still in question.

That question was partly answered in concurrent dynamo calculations utilizing the EULAG code. The ILES implementation of this code allows it to minimize the impacts of diffusion and achieve more complex velocity

 DOI: 10.1051/epjconf/201716002010

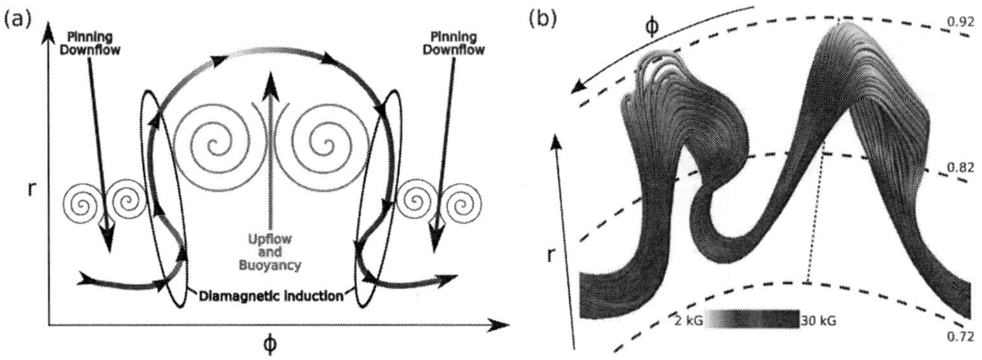

Figure 11: (a) A sketch of the processes leading to magnetic loop formation in ASH simulations with dynamic Smagorinski: downflows pin the foot points of the loops, shear between the upflow and downflow builds a large magnetic diffusivity gradient at the edge of the loop leading to diamagnetic induction that builds radial field of the correct sign, while advection and buoyancy lift the inner section of the toroidal field. (b) Two magnetic loops achieved in an ASH simulation utilizing a dynamic Smagorinski scheme shown in a field line rendering [adapted from [69]].

and magnetic fields at a given resolution than ASH can achieve. The solar dynamo modeled with this computational framework exhibited many multi-decadal cycles of roughly 40 years that were quite regular and persisted for roughly 500 years of evolution despite the fairly complex flow fields permitted by the low numerical diffusion [59, 70]. A snapshot of the toroidal magnetic field from this simulation is shown Figure 9d and its character in time and latitude is illustrated in Figure 12e. The turbulent EMF in these simulations is more amenable to a mean-field theory description than the earlier ASH simulations, which may in part be due to the greater complexity of the velocity field. Part of the reason for the greater agreement between these dynamos and mean-field theory may lie in the simple argument that as the number of convective cells becomes increasingly large the average poloidal magnetic field generation may approach a mean value. Such a characteristic might be expected if the averaging operator obeys the central limit theorem, or if the arguments given in [51] hold.

Further exploration of the rotation rate and *Pm* parameter space using the ASH code continued apace, with a quasi-periodic cycling dynamo solution published in [60] and a multitude of solutions explored in [68]. The toroidal fields of the cycling solution are shown in Figure 9, exhibiting the high degree of longitudinal connectivity that can be achieved within a low-*Pm* global-scale dynamo and also lending credence to their moniker of wreaths. These fields form at or near the regions of the greatest latitudinal shear in the differential rotation, which is why they tend to be generated closer to the equator. Such characteristics can be seen in several of the dynamo solutions from [68] as shown in Figures 12a–d. The dynamo behaviors exhibited within the menagerie of simulations in this latter paper seem to roughly correspond to the regimes described in [67], as is illustrated in Figure 10. The Gilman picture suggests that, as the magnetic diffusivity is reduced, dynamo action becomes increasingly likely and that cycles become more likely over a small region. Eventually, however, Lorentz-force feedback quenches both the convection and differential rotation. Similarly, as the rotation rate is increased, cycles become more likely to a point and then past this threshold they become less likely due to the rotational suppression of convection. Yet another set of simulations that sought to determine the influence of the density scale height and convective driving has recently been carried out that finds distinct transitions between dipolarity, multi-polarity, and cyclic dynamos [71], though in less of a solar context. Indeed, they found that their anelastic dynamo models produced a broad range of magnetic field geometries, with dipolar configurations for the least stratified and most rapidly rotating and a gradual transition to multipolar configurations if either the stratification or the rotation rate was increased. As the density stratification is increased the convective columns in their simulations become increasingly concentrated in a thin boundary layer near the upper boundary and at low latitudes. Such patterns of convection tended to be non-axisymmetric solutions. Furthermore, in multipolar solutions, zonal flows became more significant and led to the production of toroidal field through the Ω-effect. In those cases, dynamo waves seemed to then play an important role, leading to polarity reversing dynamos.

Continuing the rapid pace in solar dynamo modeling and under the assumptions of dynamic Smagorinski, mildly buoyant magnetic structures have been realized in a cycling dynamo simulation [Figure 11; 61, 69]. In this fortuitous simulation, the dynamic Smagorinski diffusion permits a greater degree of turbulence than can be achieved in the eddy viscosity model. It helps to establish a pattern of quasi-buoyant magnetic structures that form within the convection zone, which is in contrast to the common wisdom that they could only form in the tachocline as suggested in [42]. Regardless of how these structures come about, they tend to have statistics strikingly similar to the statistics of solar observations regarding Joy's and Hale's laws [72]. These magnetic structures arise from the structure of the flow field, natural buoyancy, and partly from

DOI: 10.1051/epjconf/201716002010

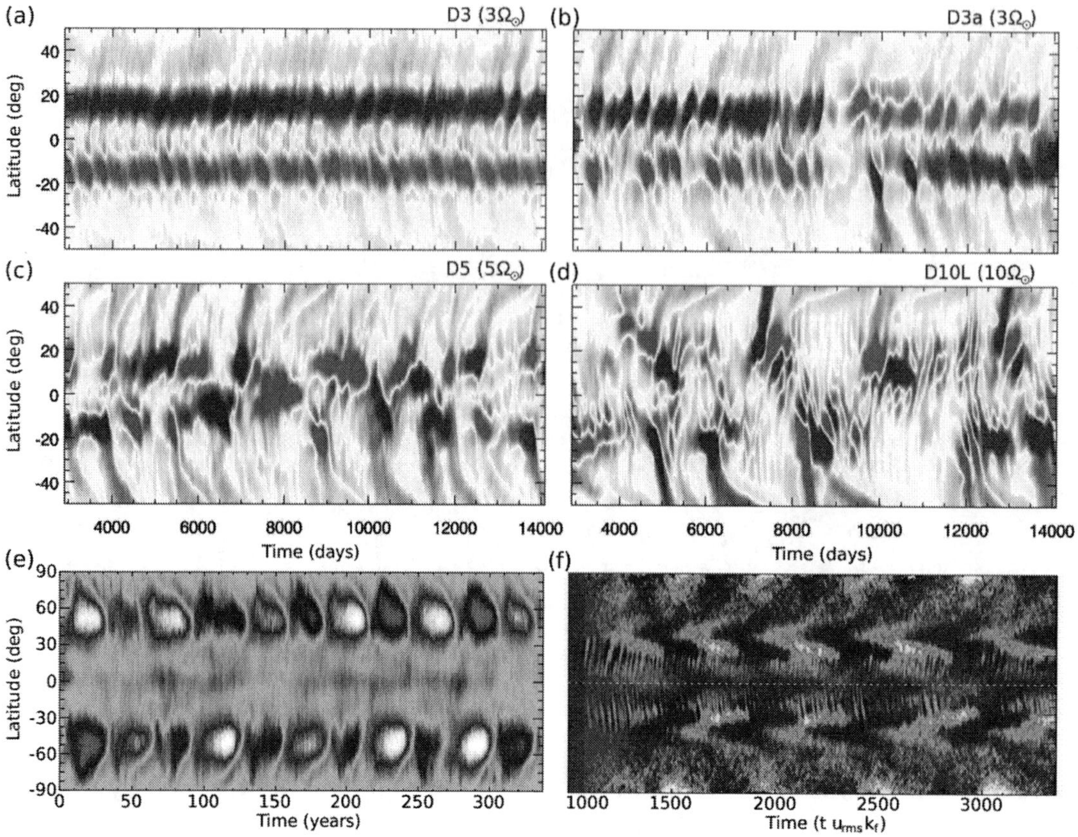

Figure 12: Time and latitude evolution of toroidal magnetic fields achieved in several global-scale convective dynamo solutions. (a-d) Dynamo calculations carried out with ASH showing four dynamo regimes arising from varying the rotation rate and magnetic diffusivity [adapted from [68]]. (e) Toroidal field evolution in a solar dynamo simulations utilizing EULAG, showing regular polarity reversals as in [59]. (f) A cycling case exhibiting polarward and equatorward branches of propagating dynamo waves in a Pencil code simulation described in [62].

diamagnetic processes. The latter process has not yet been fully considered, so the logic of it will be expanded upon a bit here. The downflows help to pin down the foot points of the loop-like magnetic fields, whereas the upflow between two downflows advects a portion of the magnetic field toward the upper boundary of the simulation. The natural magnetic and thermal buoyancy of the structure further boost the forces, compelling it to rise.

There have been recent simulations that also exhibit two branches of propagating nonlinear dynamo waves one that is equatorward at low latitudes and another that is poleward at higher latitudes [Figure 12f; 62]. As was also seen in [71], they find that the density stratification of the simulation has a substantial impact on the overall character of the dynamo. In particular, one of their solutions exhibits both equatorward and poleward propagating dynamo waves (Figure 12f), though the reason for this is unclear. The propagation characteristics of the dynamo are attributed to the stratification. Yet the number of density scale heights that are simulated in these calculations is only about 3 to 4.5 density scale heights, which is similar to those of previous ASH simulations. It may be that the differential rotation has a region of near-surface shear, which can promote equatorward propagation provided that the sign of the kinetic helicity does not reverse at the same time. One study has shed further light on this subject by assessing the role of the columnarity of the convective structures in determining the direction of propagation of a dynamo wave [73]. In that paper, below a threshold columnarity of about 0.3, an equatorward propagating dynamo wave is established. More precisely, in some of their simulations, the direction of propagation of dynamo waves is altered primarily by an inversion of the kinetic helicity in the deep interior relative to the near-surface kinetic helicity, rather than by changes in the differential rotation. This inversion tends to occur in cases with a low Prandtl number, internal heating, and in regions where the local density gradient is relatively small.

Several recent convective dynamo simulations carried out in spherical segments with the Pencil code have also manifested unprecedented magnetic self-organization, with strong mean field generation in nonlinear dynamo regimes, and cyclic activity on decadal time scales [Figure 9g; 74]. One simulation in particular, a global-scale

DOI: 10.1051/epjconf/201716002010

simulation using the ASH code and slope-limited diffusion, achieved a convective dynamo that exhibits many of these interesting dynamo features, generating magnetic fields with a prominent polarity cycle occurring roughly every 6.2 years for several hundred years of evolution [Figure 9i; 75, 76]. The polarity reversals seen in that simulation are related to the variation of the differential rotation and to a resistive collapse of the large-scale magnetic field. An equatorial migration of the magnetic field is also seen, which in turn is due to the strong, nonlinear modulation of the differential rotation rather than a dynamo wave. This simulation also enters an interval with reduced magnetic energy at low latitudes lasting roughly 16 years (about 2.5 polarity cycles), during which the polarity cycles are disrupted and after which the dynamo recovers its regular polarity cycles. An analysis of this grand minimum reveals that it likely arises through the interplay of symmetric and antisymmetric dynamo families. This intermittent dynamo state potentially results from the simulation's relatively low magnetic Prandtl number.

Another set of Pencil code simulations have begun to investigate how the upper boundary can influence the operation of the dynamo, such as through the release of magnetic helicity [e.g., 77]. The generation of large-scale fields in dynamos is often associated with an upscale transfer of magnetic helicity. In order to sustain this transfer, small-scale helicity must be either dissipated or removed from the system by passing through the boundaries. A possible manifestation of this on the Sun is that the magnetic helicity released in coronal mass ejections [78]. Indeed, these recent simulations have demonstrated that convective dynamos can eject helicity through CME-like eruptions and separately form bipolar magnetic features akin to sunspots [79–81], yet it is still unclear how such events effect the operation of the dynamo.

Finally, [82] presented a suite of global-scale simulations of rotating turbulent convective dynamos that focused on determining the role of the tachocline in setting the cycle period. The first set of their simulations considered only a stellar convective envelope, whereas the second set aimed at establishing a tachocline, and so they also included the upper part of a radiative zone. For the first set of models, either oscillatory or steady dynamo solutions are obtained, depending upon the global convective Rossby number of the simulation in question. The models in the second set naturally develop a tachocline, leading to the generation of a strong mean magnetic field. Since the field is also deposited in the stable deeper layer, its evolutionary timescale is much longer than in the models without a tachocline. Due to non-local dynamo effects, the magnetic field in the upper turbulent convection zone evolves on the same timescale as the deep field. These models result in either an oscillatory dynamo with a 30 yr period or a steady dynamo depending again on Rossby number. Hence, the presence of a tachocline appears to lengthen the cycle period.

7 Intermediate-mass Stars and Convective Core Dynamos

There may be the potential to identify a few regimes for which some global-scale aspects of stellar dynamos might be estimated with only a knowledge of the basic parameters of the system. For example, consider how the magnetic energy contained in the system may change with a modified level of turbulence (or stronger driving), or how does the ratio of the dissipative length scales impact that energy, or how does rotation influence it? Some of these questions will be addressed here. Establishing the global-parameter scalings of convective dynamos, particularly with stellar mass and rotation rate, is useful given that they provide an order of magnitude approximation of the magnetic field strengths generated within the convection zones of stars as they evolve from the pre-main-sequence to a terminal phase. This in turn permits the placement of better constraints upon transport processes, such as those for elements and angular momentum, most of which occur over structurally-significant evolutionary timescales. This could be especially useful in light of the recent evidence for magnetic fields within the cores of red giants, pointing to the existence of a strong core dynamo being active in a large fraction of intermediate mass stars [17, 18].

8 Scaling of Magnetic and Kinetic Energies

Convective flows often possess distributions of length scales and speeds that are peaked near a single characteristic value. One simple method to estimate these quantities is to divine that the energy containing flows have roughly the same length scale as the depth of the convection zone and that the speed of the flows is directly related to the rate of energy injection (given here by the stellar luminosity) and inversely proportional to the density of the medium into which that energy is being injected [83]. The latter is encapsulated as $v_{rms} \propto (2L/\rho_{CZ})^{1/3}$, where ρ_{CZ} is the average density in the convection zone. However, such a mixing-length velocity prescription only provides an order of magnitude estimate as the precise level of equipartition depends sensitively upon the dynamics [e.g., 84]. Since stars are often rotating fairly rapidly, taking for instance young low-mass stars and most intermediate and high-mass stars, their dynamos may reach a quasi-magnetostrophic state wherein the Coriolis acceleration also plays a significant part in balancing the Lorentz force. Such a balance has been addressed and discussed at length in [2, 85], and [86] for instance.

To characterize the force balance, consider the compressive MHD Navier-Stokes vorticity equation, wherein one has that

$$\frac{\partial \omega}{\partial t} = \boldsymbol{\nabla}\times\left[\mathbf{v}\times\omega_P + \frac{1}{\rho}\left(\frac{\mathbf{J}\times\mathbf{B}}{c} + \boldsymbol{\nabla}\cdot\sigma - \boldsymbol{\nabla}P\right)\right], \quad (6)$$

where $\mathbf{v}$ is the velocity, $\mathbf{B}$ is the magnetic field, ρ is the density, P the pressure. The following variables are also defined, with the current being $\mathbf{J} = c\boldsymbol{\nabla}\times\mathbf{B}/4\pi$, with c being

DOI: 10.1051/epjconf/201716002010

the speed of light, σ being the viscous stress tensor, and where $\omega = \nabla\times\mathbf{v}$ and $\omega_P = 2\mathbf{\Omega} + \omega$.

Taking the dot product of this equation with ω gives rise to the equation for the evolution of the enstrophy. Integrating that equation over the volume of the convective domain and over a reasonable number of dynamical times such that the system is statistically steady yields

$$\int d\mathbf{S}\cdot\left[\mathbf{v}\times\omega_P + \frac{1}{\rho}\left(\frac{\mathbf{J}\times\mathbf{B}}{c} + \nabla\cdot\sigma - \nabla P\right)\right]\times\omega$$
$$+\int dV\,(\nabla\times\omega)\cdot\left[\mathbf{v}\times\omega_P + \frac{1}{\rho}\left(\frac{\mathbf{J}\times\mathbf{B}}{c} + \nabla\cdot\sigma - \nabla P\right)\right] = 0. \tag{7}$$

If no enstrophy is lost through the boundaries of the convective domains, then the surface integral vanishes, leaving

$$\int dV\,(\nabla\times\omega)\cdot\left[\mathbf{v}\times\omega_P + \frac{1}{\rho}\left(\frac{\mathbf{J}\times\mathbf{B}}{c} + \nabla\cdot\sigma - \nabla P\right)\right] = 0. \tag{8}$$

This assumption effectively means that magnetic stellar winds will not be part of this scaling analysis. For timescales consistent with the dynamical timescales of the dynamos considered here, this is a reasonably valid assumption. Then, since $\nabla\times\omega$ is not everywhere zero, the terms in square brackets must be zero, which implies that

$$\rho\mathbf{v}\times\omega_P + \frac{\mathbf{J}\times\mathbf{B}}{c} + \nabla\cdot\sigma - \nabla P = 0. \tag{9}$$

Taking the curl of this equation eliminates the pressure contribution and gives

$$\nabla\times\left[\rho\mathbf{v}\times\omega + 2\rho\mathbf{v}\times\mathbf{\Omega} + \frac{\mathbf{J}\times\mathbf{B}}{c} + \nabla\cdot\sigma\right] = 0. \tag{10}$$

This, then, is the primary balance between inertial, Coriolis, Lorentz, and viscous forces. If one again scales the derivatives as the inverse of a characteristic length scale ℓ, and takes fiducial values for the other parameters, the scaling relationship for the above equation gives

$$\rho v_{\rm rms}^2/\ell^2 + 2\rho v_{\rm rms}\Omega_0/\ell + B^2/4\pi\ell^2 + \rho\nu v_{\rm rms}/\ell^3 \approx 0, \tag{11}$$

which when divided through by $\rho v_{\rm rms}^2/\ell^2$ yields

$$\mathrm{ME/KE} \propto 1 + Re^{-1} + Ro^{-1}. \tag{12}$$

Here the Reynolds number is taken to be $Re = v_{\rm rms}\ell/\nu$. However, the leading term of this scaling relationship is found to be less than unity, at least when assessed through simulations. Hence, it should be replaced with a parameter to account for dynamos that are subequipartition. This leaves the following

$$\mathrm{ME/KE} \propto \beta(Ro, Re) + Ro^{-1}, \tag{13}$$

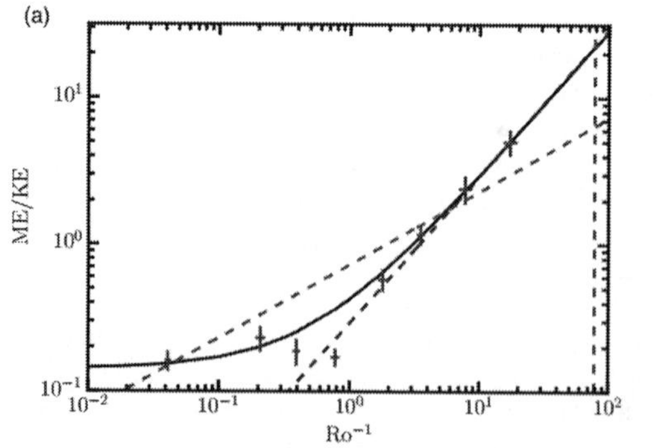

Figure 13: The scaling of the ratio of magnetic to kinetic energy (ME/KE) with inverse Rossby number (Ro^{-1}). The black curve indicates the scaling defined in Equation (13), with $\beta = 0.5$. The blue dashed line is for magnetostrophy, e.g. $\beta = 0$. The green dashed line is that for buoyancy-work-limited dynamo scaling, which requires ME/KE $\propto Ro^{-1/2}$. The red dashed line indicates the critical Rossby number of the star, correspoding to its rotational breakup velocity. The uncertainty of the measured Rossby number and energy ratio that arises from temporal variations are indicated by the size of the cross for each data point.

where $\beta(Ro, Re)$ is unknown apriori as it depends upon the intrinsic ability of the non-rotating system to generate magnetic fields, which in turn depends upon the specific details of the system such as the boundary conditions and geometry of the convection zone.

Thus, for a subset of dynamos such as those discussed in [87], Equation (13) may hold and the ratio of the total magnetic energy (ME) to the kinetic energy (KE) will depend upon the inverse Rossby number plus a constant offset, which will be sensitive to details of the dynamics. In some circumstances, the constant offset may also depend upon the Rossby number. Hence, such dynamos are sensitive to the degree of rotational constraint on the convection and upon the intrinsic ability of the convection to generate a sustained dynamo. Moreover, this line of argument indicates that the inertial term may give rise to a minimum magnetic energy state. This allows a bridge between two dynamo regimes, the equipartition slowly rotating dynamos and the rapidly rotating magnetostrophic regime, where ME/KE $\propto Ro^{-1}$. For low Rossby numbers, or large rotation rates, it is possible that the dynamo can reach superequipartition states where ME/KE > 1. Indeed, it may be much greater than unity, as is expected for the Earth's dynamo (see Figure 6 of [88]).

9 Comparison of Scaling Relationships

Since we wish to compare these three schemes, consider the data for the evolution of a set of MHD simulations using the Anelastic Spherical Harmonic code presented in [87]. These simulations attempt to capture the dynamics within the convective core of a massive star. In such stars, Pm is approximately four throughout the core, falls to about two at the core boundary, and to 1/10 nearer to the stellar surface in the radiative exterior. Hence, the convective core can be considered as a large magnetic Prandtl

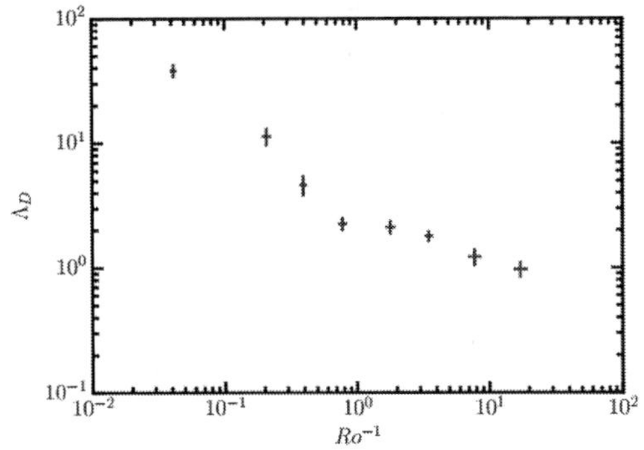

Figure 14: The scaling of the dynamic Elsasser number (Λ_D) with inverse Rossby number (Ro^{-1}). The uncertainty of the measured Rossby number and dynamic Elsasser number that arises from temporal variations are indicated by the size of the cross for each data point.

number dynamo. Such *Pm* regimes are accessible from a numerical point of view. These convective cores also present an astrophysical dynamo in which large-eddy simulations can more easily capture the hierarchy of relevant diffusive timescales. This is especially so given that the density contrast across the core is generally small, being a bit less than two. The heat generation from nuclear burning processes deep within the core and the radiative cooling nearer the radiative zone are also quite smoothly distributed in radius. So, there are no inherent difficulties with resolving internal boundary layers.

In the suite of simulations presented in [87], the rotation rates employed lead to nearly three decades of coverage in Rossby number, as shown in Figure (13). In that figure, the force-based scaling derived in §8 depicted by the black curve (where $\mathrm{ME/KE} \propto 0.5 + Ro^{-1}$) does a reasonably good job of describing the nature of the superequipartition state for a given Rossby number. Note, however, that the constant of proportionality in Equation (13) has been determined using the data itself, which is true also of the other two scaling laws shown in Figure (13).

These simulated convective core dynamos appear to enter a regime of magnetostrophy for the four cases with the lowest average Rossby number, where the scaling for the magnetostrophic regime is denoted by the dashed blue line in Figure (13). This transition to the magnetostrophic regime is further evidenced in Figure (14), which shows the dynamic Elsasser number

$$\Lambda_D = \mathrm{B_{rms}}^2/(8\pi\rho_0\Omega_0\mathrm{v_{rms}}\ell), \qquad (14)$$

where ℓ is the typical length scale of the current density **J**. Indeed, as Λ_D approaches unity, the balance between the Lorentz and the Coriolis forces also approaches unity, indicating that the dynamo is very close to magnetostrophy.

The scaling law derived in [89] and [86] suggests that there is a direct link between the buoyancy work and ohmic dissipation, whereas the Coriolis force only requires that there be two integral length scales: one for flows parallel and another for flows perpendicular to the rotation axis. Following this logic, one can show that this requires that the level of superequipartition of the magnetic energy relative to the convective kinetic energy is proportional to $Ro^{-1/2}$. This scaling does not capture the behavior of the set of dynamos in [87] very well, as indicated by the lack of agreement between the blue line and the data points in Figure (13). In contrast, for the many dynamo simulations and data shown in [90] and [85], it does perform well. This latter fact could be related to the fact that for large *Pm* the buoyancy work potentially has an additional *Ro* dependence.

10 Conclusions

This conference proceeding has attempted to shine some light on the link between many observations and dynamo theory. In particular, there was a focus on the changing nature of the observed magnetism as a function of mass and rotation rate, linking Ca II, H-α, and X-ray observations and magnetic activity. On the theory side, some effort was spent explaining the existence of two kinds of dynamos that depend upon magnetic Prandtl number and that they have relevance to stellar astrophysics. In particular, it was shown that there is likely a dichotomy in the kind of dynamo action taking place within stars that possess a convective core and those that possess an exterior convective envelope. Recent advances in describing the dynamos of solar-like stars was covered in a fair level of detail. Finally, there is a brief section discussing scaling laws for the level of the partitioning of magnetic energy and kinetic energy, which is relevant for intermediate and massive stars that possess a convective core.

In attempting to explain the great many observations of stellar magnetism, including its variability and intensity, dynamo theory has made some surprising advances in unexpected directions. Numerical simulations in particular have provided a fruitful path to understanding many of the observed features of the solar dynamo. While the exact processes that set the cycle period remain unknown, it has been shown that the tachocline can play a significant role in recovering longer cycle periods [82]. Most importantly, however, cyclic solutions seem now to be a more robust feature than in the past due to increased computational power and thus increased levels of turbulence. Further, the equatorward and poleward propagation of surface features, provided that they are tied to the deeper dynamics in a significant way, may follow a deeper dynamo wave, weather it is linear or nonlinear in origin [62, 73, 75]. Though promising, all of these recent convective dynamo simulations are still in the early stages of discovery. They are still quite sensitive to the treatment of diffusion where slight changes in the form and magnitudes of the diffusivities can lead to quite different behavior. Moreover, the physical mechanisms responsible for determining the cycle properties established within these various convection models remains unknown. What is clear is that these cyclic dynamos operate very differently than typical mean field dynamo models in that the meridional flow plays a minimal role in regulating the cycle period and flux emergence plays little or no role in poloidal field generation.

There appears to be two scaling laws for the level of equipartition of magnetic and kinetic energy that are applicable to stellar systems, one in the high magnetic Prandtl number regime and another in the low magnetic Prandtl number regime. Within the context of the large magnetic Prandtl number systems, the magnetic energy of the system scales as the kinetic energy multiplied by an expression that depends upon an offset, which depends upon the details of the non-rotating system, e.g. on Reynolds number, Rayleigh number, and Prandtl number, plus the inverse of the convective Rossby number, as was considered in [87]. For low magnetic Prandtl number and fairly rapidly rotating systems, such as the geodynamo and rapidly rotating low-mass stars, another scaling law that relies upon a balance of buoyancy work and magnetic dissipation as well as a balance between the buoyancy, Coriolis, and Lorentz forces may be more applicable. When focused on in detail, this yields a magnetic energy that scales as the kinetic energy multiplied by the inverse square root of the convective Rossby number, as has been shown to be widely applicable in such low Prandtl number and low Rossby number systems [89]. On the side of numerical experiments, a larger range of Reynolds number and level of supercriticality needs to be explored. Indeed, in [84], some work has already attempted to do this for the geodynamo.

Acknowledgement

K. C. Augustson acknowledges support from the the ERC SPIRE 647383 grant.

References

[1] R. Howe, Liv. Rev. Sol. Phys. **6** (2009)
[2] A.S. Brun, M.K. Browning, M. Dikpati, H. Hotta, A. Strugarek, Space Sci. Rev. **196**, 101 (2015)
[3] J.P. Zahn, A&A **252**, 179 (1991)
[4] D.O. Gough, M.E. McIntyre, Nature **394**, 755 (1998)
[5] J. Zhao, R.S. Bogart, A.G. Kosovichev, T.L. Duvall, Jr., T. Hartlep, ApJL **774**, L29 (2013)
[6] J.H.M.M. Schmitt, L. Golub, F.R. Harnden, Jr., C.W. Maxson, R. Rosner, G.S. Vaiana, ApJ **290**, 307 (1985)
[7] T.W. Berghöfer, J.H.M.M. Schmitt, R. Danner, J.P. Cassinelli, A&A **322**, 167 (1997)
[8] H. Sana, G. Rauw, Y. Nazé, E. Gosset, J.M. Vreux, MNRAS **372**, 661 (2006)
[9] I.I. Antokhin, G. Rauw, J.M. Vreux, K.A. van der Hucht, J.C. Brown, A&A **477**, 593 (2008)
[10] N.J. Wright, J.J. Drake, E.E. Mamajek, G.W. Henry, ApJ **743**, 48 (2011)
[11] J. Christensen-Dalsgaard, Sol. Phys. **220**, 137 (2004)
[12] M.A. Dupret, K. Belkacem, R. Samadi, J. Montalban, O. Moreira, A. Miglio, M. Godart, P. Ventura, H.G. Ludwig, A. Grigahcène et al., A&A **506**, 57 (2009)
[13] T.R. Bedding, B. Mosser, D. Huber, J. Montalbán, P. Beck, J. Christensen-Dalsgaard, Y.P. Elsworth, R.A. García, A. Miglio, D. Stello et al., Nature **471**, 608 (2011)
[14] B. Mosser, O. Benomar, K. Belkacem, M.J. Goupil, N. Lagarde, E. Michel, Y. Lebreton, D. Stello, M. Vrard, C. Barban et al., A&A **572**, L5 (2014)
[15] P.G. Beck, J. De Ridder, C. Aerts, T. Kallinger, S. Hekker, R.A. García, B. Mosser, G.R. Davies , Astro. Nachr. **333**, 967 (2012)
[16] S. Deheuvels, J. Ballot, P.G. Beck, B. Mosser, R. Østensen, R.A. García, M.J. Goupil, A&A **580**, A96 (2015)
[17] J. Fuller, M. Cantiello, D. Stello, R.A. Garcia, L. Bildsten, Science **350**, 423 (2015)
[18] M. Cantiello, J. Fuller, L. Bildsten, ApJ **824**, 14 (2016)
[19] G. Peres, S. Orlando, F. Reale, R. Rosner, H. Hudson, ApJ **528**, 537 (2000)
[20] W. Herschel, Roy. Soc. London Phil. Trans. Series I **85**, 46 (1795)
[21] W. Herschel, Royal Society of London Philosophical Transactions Series I **86**, 452 (1796)
[22] W.D.W. Abney, MNRAS **37**, 278 (1877)
[23] G. Shajn, O. Struve, MNRAS **89**, 222 (1929)
[24] O.C. Wilson, ApJ **144**, 695 (1966)
[25] K.G. Stassun, L. Hebb, K. Covey, A.A. West, J. Irwin, R. Jackson, M. Jardine, J. Morin, D. Mullan, I.N. Reid, 17th Cool Stars Proceedings, **448**, 505 (2011)
[26] O.C. Wilson, ApJ **138**, 832 (1963)
[27] H.K. Moffatt, *Magnetic field generation in electrically conducting fluids* (Cambridge, England, Cambridge University Press. 353 p., 1978)
[28] A. Skumanich, ApJ **171**, 565 (1972)
[29] A.H. Vaughan, G.W. Preston, Pub. Astron. Soc. Pac. **92**, 385 (1980)
[30] D.R. Soderblom, J.R. Stauffer, J.D. Hudon, B.F. Jones, ApJS **85**, 315 (1993)
[31] J.T. Wright, G.W. Marcy, R.P. Butler, S.S. Vogt, ApJS **152**, 261 (2004)
[32] R. Pallavicini, L. Golub, R. Rosner, G.S. Vaiana, T. Ayres, J.L. Linsky, ApJ **248**, 279 (1981)
[33] O. Vilhu, A&A **133**, 117 (1984)
[34] G. Micela, S. Sciortino, S. Serio, G.S. Vaiana, J. Bookbinder, L. Golub, F.R. Harnden, Jr., R. Rosner, ApJ **292**, 172 (1985)
[35] O. Vilhu, F.M. Walter, ApJ **321**, 958 (1987)
[36] N. Pizzolato, A. Maggio, G. Micela, S. Sciortino, P. Ventura, A&A **397**, 147 (2003)
[37] M. Jardine, Y.C. Unruh, A&A **346**, 883 (1999)
[38] E.G. Blackman, J.H. Thomas, MNRAS **446**, L51 (2015)
[39] E.G. Blackman, J.E. Owen, MNRAS **458**, 1548 (2016)
[40] S.J. Schmidt, S.L. Hawley, A.A. West, J.J. Bochanski, J.R.A. Davenport, J. Ge, D.P. Schneider, Astron.

J. **149**, 158 (2015)
[41] E.A. Spiegel, N.O. Weiss, Nature **287**, 616 (1980)
[42] E.N. Parker, ApJ **408**, 707 (1993)
[43] A.A. West, S.L. Hawley, J.J. Bochanski, K.R. Covey, I.N. Reid, S. Dhital, E.J. Hilton, M. Masuda, Astron. J. **135**, 785 (2008)
[44] M. Jardine, A. Collier Cameron, J.F. Donati, MNRAS **333**, 339 (2002)
[45] A.A. Vidotto, M. Jardine, J. Morin, J.F. Donati, M. Opher, T.I. Gombosi, MNRAS **438**, 1162 (2014)
[46] V. Réville, A.S. Brun, A. Strugarek, S.P. Matt, J. Bouvier, C.P. Folsom, P. Petit, ApJ **814**, 99 (2015)
[47] S.I. Braginskii, Reviews of Plasma Physics **1**, 205 (1965)
[48] A.A. Schekochihin, S.C. Cowley, S.F. Taylor, J.L. Maron, J.C. McWilliams, ApJ **612**, 276 (2004)
[49] A.A. Schekochihin, A.B. Iskakov, S.C. Cowley, J.C. McWilliams, M.R.E. Proctor, T.A. Yousef, New J. of Physics **9**, 300 (2007)
[50] A. Brandenburg, ApJ **697**, 1206 (2009)
[51] E.N. Parker, ApJ **122**, 293 (1955)
[52] G. Backus, Annals of Physics **4**, 372 (1958)
[53] A. Herzenberg, Royal Society of London Philosophical Transactions Series A **250**, 543 (1958)
[54] G.O. Roberts, Royal Society of London Philosophical Transactions Series A **266**, 535 (1970)
[55] B. Paxton, L. Bildsten, A. Dotter, F. Herwig, P. Lesaffre, F. Timmes, ApJS **192**, 3 (2011)
[56] A.S. Brun, M.S. Miesch, J. Toomre, ApJ **614**, 1073 (2004)
[57] M.K. Browning, M.S. Miesch, A.S. Brun, J. Toomre, ApJL **648**, L157 (2006)
[58] B.P. Brown, M.K. Browning, A.S. Brun, M.S. Miesch, J. Toomre, ApJ **711**, 424 (2010)
[59] É. Racine, P. Charbonneau, M. Ghizaru, A. Bouchat, P.K. Smolarkiewicz, ApJ **735**, 46 (2011)
[60] B.P. Brown, M.S. Miesch, M.K. Browning, A.S. Brun, J. Toomre, ApJ **731**, 69 (2011)
[61] N.J. Nelson, B.P. Brown, A.S. Brun, M.S. Miesch, J. Toomre, ApJ **762**, 73 (2013)
[62] P.J. Käpylä, M.J. Mantere, E. Cole, J. Warnecke, A. Brandenburg, ArXiv e-prints (2013)
[63] P. Charbonneau, Nature **493**, 613 (2013)
[64] R.B. Leighton, ApJ **156**, 1 (1969)
[65] G.A. Glatzmaier, J. Computational Physics **55**, 461 (1984)
[66] P.A. Gilman, G.A. Glatzmaier, ApJS **45**, 335 (1981)
[67] P.A. Gilman, ApJS **53**, 243 (1983)
[68] B.P. Brown, M.K. Browning, A.S. Brun, M.S. Miesch, J. Toomre, **448**, 277 (2011)
[69] N.J. Nelson, B.P. Brown, A.S. Brun, M.S. Miesch, J. Toomre, ApJL **739**, L38 (2011)
[70] M. Ghizaru, P. Charbonneau, P.K. Smolarkiewicz, ApJL **715**, L133 (2010)
[71] T. Gastine, L. Duarte, J. Wicht, A&A **546**, A19 (2012)
[72] N.J. Nelson, B.P. Brown, A. Sacha Brun, M.S. Miesch, J. Toomre, Solar Physics p. 20 (2013)
[73] L.D.V. Duarte, J. Wicht, M.K. Browning, T. Gastine, MNRAS **456**, 1708 (2016)
[74] P.J. Käpylä, M.J. Mantere, A. Brandenburg, ApJL **755**, L22 (2012)
[75] K. Augustson, A.S. Brun, M. Miesch, J. Toomre, ApJ **809**, 149 (2015)
[76] M.S. Miesch, M. Zhang, K.C. Augustson, ApJL **824**, L15 (2016)
[77] A. Brandenburg, K. Subramanian, Physics Reports **417**, 1 (2005)
[78] E.G. Blackman, A. Brandenburg, ApJL **584**, L99 (2003)
[79] J. Warnecke, P.J. Käpylä, M.J. Mantere, A. Brandenburg, Solar Physics **280**, 299 (2012)
[80] J. Warnecke, P.J. Käpylä, M.J. Mantere, A. Brandenburg, ArXiv e-prints (2013)
[81] J. Warnecke, I.R. Losada, A. Brandenburg, N. Kleeorin, I. Rogachevskii, ArXiv e-prints (2013)
[82] G. Guerrero, P.K. Smolarkiewicz, E.M. de Gouveia Dal Pino, A.G. Kosovichev, N.N. Mansour, ApJ **819**, 104 (2016)
[83] K.C. Augustson, B.P. Brown, A.S. Brun, M.S. Miesch, J. Toomre, ApJ **756**, 169 (2012)
[84] R. Yadav, T. Gastine, U. Christensen, S.J. Wolk, K. Poppenhaeger, Proc. Nat. Acad. Sci. **113** (2016)
[85] U.R. Christensen, Space Science Reviews **152**, 565 (2010)
[86] K. Augustson, S. Mathis, A.S. Brun, 19th Cool Stars Proceedings (2017)
[87] K.C. Augustson, A.S. Brun, J. Toomre, ApJ **829**, 92 (2016)
[88] P.H. Roberts, E.M. King, Reports on Progress in Physics **76**, 096801 (2013)
[89] P.A. Davidson, Geophys. J. International **195**, 67 (2013)
[90] U.R. Christensen, V. Holzwarth, A. Reiners, Nature **457**, 167 (2009)

DOI: 10.1051/epjconf/201716002011

On the oscillation spectrum of a magnetized core in a giant star

Michel Rieutord[1,2,⋆]

[1] *Université de Toulouse; UPS-OMP; IRAP; Toulouse, France*
[2] *CNRS; IRAP; 14, avenue Edouard Belin, F-31400 Toulouse, France*

Abstract. The spectrum of gravito-acoustic modes is depleted in dipolar modes for a significant fraction of the giant stars observed by the *Kepler* mission, a feature that has been explained by the presence of magnetic fields in the core of these stars (Fuller et al. 2015, Cantiello et al. 2016). We further investigate this possible scenario by considering first the oscillation spectrum of the core of a giant star modeled by a stably stratified, self-gravitating fluid of uniform density in a sphere pervaded by a uniform magnetic field. Our results show that the first effect of a magnetic field on the g-modes is to reduce their wavenumber and therefore reduce their damping. The magnetic effect, on this model, is therefore opposite Fuller's et al scenario. Moreover, the model shows that it is not possible to change the damping rate without changing the frequency of the modes and this latter change is not observed. Because of the simplicity of our model, the magnetized core scenario cannot be dismissed but further investigations are needed, and other ways of explaining the presence of depressed modes should also be considered.

1 Introduction

One of the important results obtained by the *Kepler* mission as far as asteroseismology is concerned, is the detection of mixed acoustic gravity modes in giant stars [1]. The detection of mixed acoustic-gravity modes is indeed a real opportunity to obtain informations on the core of giant stars and more generally on the core of stars. A particularly puzzling result of these observations has been the relatively slow rotation of the core when compared to the expected values given by models [2, 3]. Clearly data show that the mechanism, which controls the exchange angular momentum between a core and an envelope of a star, is not correctly implemented in stellar evolution models, if at all correctly understood.

However, the mysteries of the giant stars do not stop here. Seismic observations also show that a significant fraction of giant stars do not show all the modes they should. Dipolar ($\ell = 1$) modes are often depleted, with amplitudes strongly varying from one star to the next [1], while $\ell = 2$ or $\ell = 3$ modes are only slightly or not depleted [4]. Since the modes are most likely excited stochastically by the turbulent convection in the envelope (like solar acoustic modes), the selection of visible modes may be the result of some damping mechanism. Fuller et al. [5] suggested that magnetic fields in the radiative core could play an important role in this selection process. The idea is that g-modes of the core, which couple to p-modes of the envelope to make the mixed modes, are presumably strongly damped by their interaction with a residual magnetic field trapped in the core. While $\ell = 1$ modes are damped by such a mechanism, $\ell = 2$ (or higher) are much less affected because of a weaker coupling between their gravity and acoustic parts.

The scenario imagined by Fuller et al. [5] is appealing as it opens a window on the magnetic history of giant stars and more generally on dynamos in convective cores. However, these results are based on a local (WKB) analysis of the waves and assume that waves entering a magnetized core are fully dissipated. Hence, an open question is whether such a scenario is confirmed by a global analysis that takes the whole star into account. If confirmed, global inversion of the data may open a new window on the magnetic activity of convective core and on the dynamics history of intermediate mass stars. The road towards such a global model is however not a piece of cake since many parts of the problem are ill-known. In particular the way g-modes interact with magnetic fields has not been studied very much.

To progress on this question we consider a simplified model of the core of a giant star and study the effects of a uniform magnetic field on the damping rates of the g-modes. Results are surprising at first sight since magnetized g-modes are less damped than the non-magnetized ones. Such results are easily explained by the structure of the magnetized g-modes but they tease us to more in depth studies.

The paper is organized as follows: in the following section we give all the details of our simple model. In section 3 we give the results of the first investigations focusing on the modes that have the right frequency range. Some preliminary conclusions follow.

⋆e-mail: mrieutord@irap.omp.eu

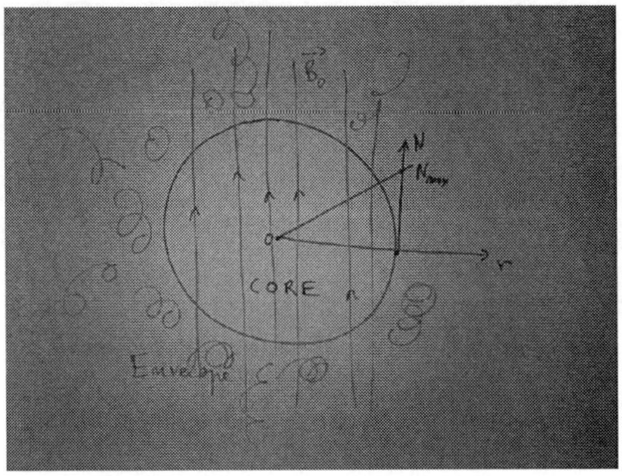

Figure 1: Schematic view of the simplified model of the magnetized radiative core of a giant star.

2 The simplified model

We consider an quasi-incompressible fluid filling a sphere of radius R, where cold sources are distributed, and bathed by a uniform magnetic field $\vec{B}_0 = B_0\vec{e}_z$ as sketched out in Fig.1. Cold sources, unlike heat sources, generate a stably stratified fluid where gravity modes can propagate. The equilibrium temperature field $T_{\rm eq}$ thus verify

$$\chi \Delta T_{\rm eq} + Q = 0 \tag{1}$$

where χ is the heat conductivity and $Q < 0$. Solving with imposed spherical symmetry leads to

$$T_{\rm eq} = -\frac{Qr^2}{6\chi} + T_c \tag{2}$$

where r is the radial coordinate and T_c the central temperature. The perturbed momentum and heat (linearized) equations read

$$\frac{\partial \vec{v}}{\partial t} = -\vec{\nabla}\delta P/\rho + \frac{\delta\rho}{\rho}\vec{g} + \nu\Delta\vec{v} + \overrightarrow{\mathrm{Rot}}\vec{b} \times \frac{\vec{B}_0}{\rho\mu_0} \tag{3}$$

$$\frac{\partial \delta T}{\partial t} + (\vec{v}\cdot\vec{\nabla})T_{\rm eq} = \kappa\Delta\delta T \tag{4}$$

where $\vec{v}$, $\vec{b}$, δP and δT are the velocity, magnetic field, pressure and temperature disturbances, respectively. ρ is the density, ν the kinematic viscosity, κ the heat diffusivity, μ_0 the vacuum permittivity and $\vec{g}$ the gravity. Obviously, we shall use the Boussinesq approximation, so mass conservation implies

$$\mathrm{Div}\vec{v} = 0 \tag{5}$$

We now rewrite the whole system of equations with scaled quantities, using the radius of the sphere R as the length scale, the inverse of the maximum Brunt-Väisälä frequency as the time scale, $|Q|R^2/\chi$ as the temperature scale and B_0 as the magnetic field scale. The Brunt-Väisälä frequency reads

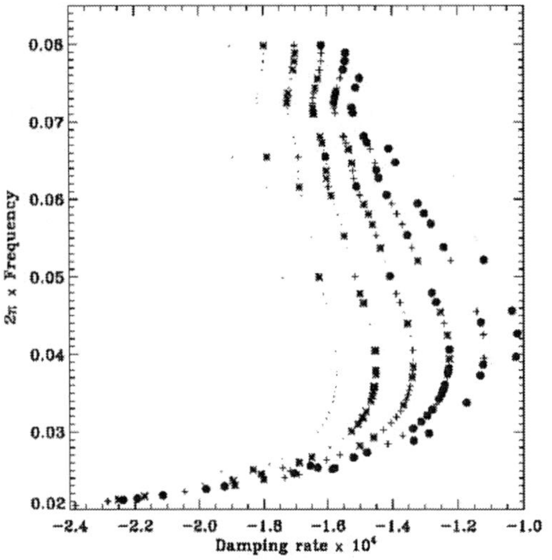

Figure 2: Evolution of four eigenvalues around $\omega = 0.01N$ when $10^{-8} \lesssim \mathcal{A} \lesssim 10^{-4}$. Here, $E_\nu = 10^{-10}$, $E_\eta = 10^{-10}$ and $E_t = 10^{-7}$.

$$N(r) = Nr \qquad \text{with} \qquad N^2 = \frac{\alpha|Q|Rg_s}{3\chi} \tag{6}$$

where r is the scaled radial coordinate, α the isobaric dilation coefficient and g_s the surface gravity. As the Boussinesq approximation is used $\vec{g} = -g_s\vec{r}$. Thus doing, we find

$$\vec{\nabla}\cdot\vec{u} = \vec{\nabla}\cdot\vec{b} = 0 \tag{7a}$$

$$\partial_\tau\vec{b} = \vec{\nabla}\times(\vec{u}\wedge\vec{e}_z) + E_\eta\Delta\vec{b} \tag{7b}$$

$$\partial_\tau\theta + ru_r = E_t\Delta\theta \tag{7c}$$

$$\partial_\tau\vec{u} = -\vec{\nabla}p + \theta\vec{r} + E_\nu\Delta\vec{u} + \mathcal{A}\,(\vec{\nabla}\times\vec{b})\times\vec{e}_z \tag{7d}$$

where we set

$$\mathcal{A} = \left(\frac{V_a}{NR}\right)^2, \quad E_\nu = \frac{\nu}{NR^2}, \quad E_t = \frac{\kappa}{NR^2}, \quad E_\eta = \frac{\eta}{NR^2} \tag{8}$$

Hence, the problem depends on three diffusion parameters (E_ν, E_t, E_η) and the coupling constant between gravity modes and Alfvén waves $\mathcal{A}$. With typical parameters of giant cores, we find that

$$E_\nu \sim 10^{-18}, \qquad E_\eta \sim 10^{-18}, \qquad E_t \sim 10^{-11} \tag{9}$$

Viscosity and magnetic diffusion are therefore extremely weak, and heat diffusion remains the basic source of damping. Since such numbers are not reachable numerically, we choose to consider higher values, namely

$$E_\nu \sim 10^{-10}, \qquad E_\eta \sim 10^{-10}, \qquad E_t \sim 10^{-7} \tag{10}$$

DOI: 10.1051/epjconf/201716002011

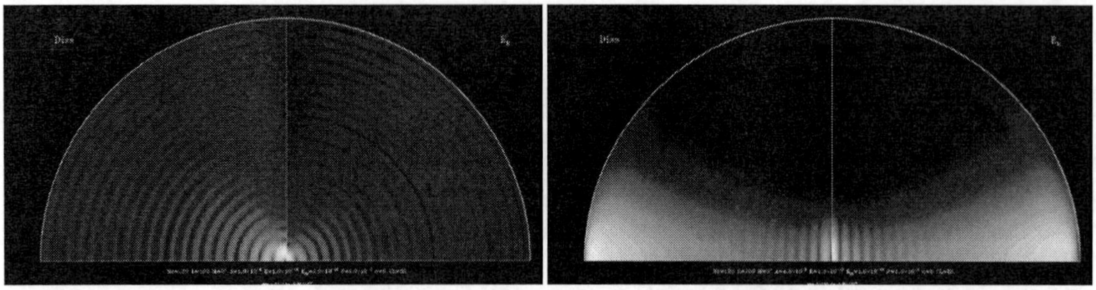

Figure 3: Left: Eigenfunction when the magnetic field is weak ($\mathcal{A} = 10^{-8}$). Right: Eigenfunction when the magnetic field is strong ($\mathcal{A} = 4 \times 10^{-5}$). On both plots we show on the right side the kinetic energy of the mode, while on the left side we show the viscous dissipation.

However, these higher values do not changed the qualitative results. [5] suggest that the maximum value of the Brunt-Väisälä frequency in the core is around $10^4 \mu$Hz. As for the magnetic field intensity we also follow [5] and take

$$2 \times 10^5 \mathrm{G} \quad \lesssim \quad B_0 \quad \lesssim \quad 2 \times 10^7 \mathrm{G} \tag{11}$$

Taking a typical core radius of a 1.5 $M_{\odot}$-giant of 4.2×10^8 m, and a mean density of 1.6g/cm^3 (from a MESA model [6]), we find that

$$10^{-8} \lesssim \mathcal{A} \lesssim 10^{-4} \tag{12}$$

so rather small values for the coupling coefficient.

3 First results

We solve the set of equations (7) by first assuming an exponential time-dependence, namely all perturbations are $\propto \exp(\lambda\tau)$ where λ is either the complex eigenvalue (free modes) or the complex frequency ($i\omega$ for forced oscillations). Then we discretize the partial differential equations on a spectral grid using spherical harmonics expansions, namely

$$\vec{u} = \sum_{l=0}^{+\infty} \sum_{m=-l}^{+l} u^{\ell}_m(r)\vec{R}^m_{\ell} + v^{\ell}_m(r)\vec{S}^m_{\ell} + w^{\ell}_m(r)\vec{T}^m_{\ell},$$

with

$$\vec{R}^m_{\ell} = Y^m_{\ell}(\theta,\varphi)\vec{e}_r, \qquad \vec{S}^m_{\ell} = \vec{\nabla} Y^m_{\ell}, \qquad \vec{T}^m_{\ell} = \vec{\nabla} \times \vec{R}^m_{\ell}$$

as in [7, 8]. Examples of the projected equations on the spherical harmonics basis may be found in [9–11]. The radial functions are then discretized on the Gauss-Lobatto grid leading to a generalized eigenvalue problem for the magneto-gravity modes (see Valdettaro et al. [12]).

3.1 Free modes

The typical frequency of the depleted mixed modes is around 100μHz so we need to consider gravity modes of frequency around $10^{-2}N$. We therefore monitored 4 eigenvalues associated with gravity modes with $\omega \sim 2 \times 10^{-2}$ for $B_0 = 0$, when the coupling parameter $\mathcal{A}$ is progressively increased.

We show in Fig. 2 the evolution of eigenvalues in the complex plane when $10^{-8} \lesssim \mathcal{A} \lesssim 10^{-4}$. From this plot we see that the damping rate first diminishes while the frequency slighly increases. Then, when $\mathcal{A} \gtrsim 1.1 \times 10^{-6}$, the damping rate approximately saturates, while the frequency continuously increases.

Hence, the first effect of magnetic fields is to reduce the damping of gravity modes. This is easily understood if we have a look at the eigenfunctions with and without magnetic fields. A meridional view of the eigenmodes is shown in Fig. 3. We note that at a frequency of a few percent of the Brunt-Väisälä frequency, $\ell = 1$-g-modes have a high radial wavenumber. On the contrary, in the low frequency range, Alfvén modes have a low wavenumber. The mixed magneto-gravity modes make a compromise between their double nature and show a typical wavenumber that is lower than their equivalent pure g-mode. Thus, the resulting wavenumber is decreased and the damping rate is decreased as well.

3.2 Forced modes

To view the impact of magnetic fields on g-modes from a different side, we also consider magneto-gravity modes forced by a deformation of the boundary. We also choose a frequency of $\sim 0.01N$ and now monitor the viscous dissipation as the strength of the magnetic field is increased. Such a forcing mimics the coupling between the acoustic modes of the envelope and the magneto-gravity modes of the core. The result is shown in Fig. 4 for two values of the parameter E_ν. In the two cases we note that the viscous dissipation decreases as B_0 increases, thus confirming the trend observed on free modes.

4 Discussion and conclusions

Alfvén modes are high frequency modes whose lowest frequency is of the order of V_A/R, V_A being the Alfvén speed and R the radius of the core. Besides, gravity modes are

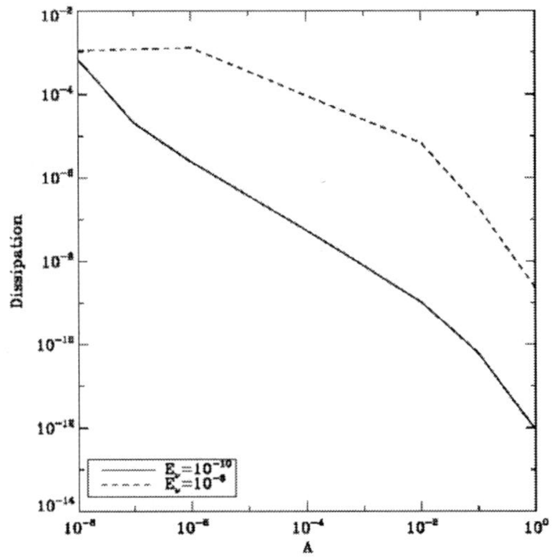

Figure 4: Evolution of the viscous dissipation with the coupling parameter $\mathcal{A}$ for $E_v = 10^{-6}$ and $E_v = 10^{-10}$.

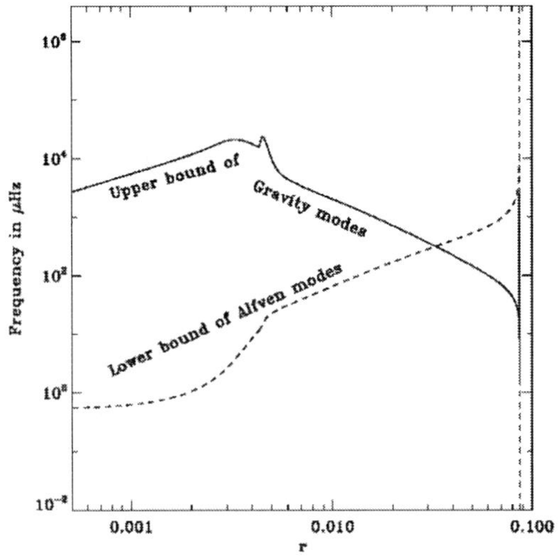

Figure 5: Propagation region for gravity and Alfvén waves in the core of giant star of mass 1.5 $M_\odot$. r is the radial coordinate scaled by the star radius $R_* = 4.87 \times 10^{11}$cm.

low frequency modes whose highest frequency is bounded by the Brunt-Väisälä frequency. For both of them, when the frequency is close to the limiting frequency, the wavelength is the largest possible. Magneto-gravity modes, for which magnetic field and stratification are both important, are thus expected to be a combination of long wavelength Alfvén and gravity modes as illustrate in Fig. 5. Such magneto-gravity modes should not therefore be very good at dissipating energy. The simplified Boussinesq model used to investigate the properties of such modes is indeed confirming that g-modes perturbed by a magnetic field have a reduced damping rate in the first place compared to the unperturbed modes.

One may argue that a uniform magnetic field might not be very realistic to represent magnetic fields in the core of giant stars. [5] indeed suggest that a highly disordered magnetic field is inducing the so-called magnetic greenhouse effect. The difficulty is that during the period between the end of the main sequence, where a dynamo could operate in the core, and the giant stage that is observed, small-scale magnetic fields are dissipated. For a typical 1.5$M_\odot$ giant star, the time lapse is ~700 Myrs. This is larger than the diffusion time of all scales larger than $0.035R_{\rm core}$. Moreover dynamos tend to build-up large-scale magnetic fields (e.g. Brandenburg et al.[13]). This let us think that even if the field is not uniform, it is expected to be dominated by the large-scales, in which case the magnetic perturbations should act in the same way as with our uniform field.

Using an approach based solely on observational data, Mosser et al. [14] recently showed that depressed modes are mixed modes, implying that the resonant cavity is similar for stars with and without depressed modes. Except the amplitudes, the spectrum of stars with depressed modes show no difference with the one with non-depressed modes. Hence, in view of the results of the foregoing model, it seems difficult to use magnetic fields to damp dipole modes without perturbing the frequencies as imposed by observations.

This does not exclude magnetic fields from being at the origin of the effect, but it is then a more subtle effect than just being an absorber. Our model is certainly very crude and one can think about the effect of density variations on the magnetic side of the magneto-gravity modes. Indeed, in the core of a giant the density varies by orders of magnitude. This leads to strong variations of the Alfvén speed that will modify the properties of the mixed magneto-gravity modes. This subject remains to be investigated.

Another possibility that should be studied is a mechanism that weakens the excitation of the modes. Such a mechanism would easily explain the point made by Mosser et al. (2016), but it remains to be found.

Acknowledgements

I would like to warmly thank Sébastien Deheuvels for drawing my attention to this problem and for enlightening discussions on the properties of giant stars.

References

[1] B. Mosser, Y. Elsworth, S. Hekker, D. Huber, T. Kallinger, S. Mathur, K. Belkacem, M.J. Goupil, R. Samadi, C. Barban et al., A&A **537**, A30 (2012)

[2] S. Deheuvels, R.A. García, W.J. Chaplin, S. Basu, H.M. Antia, T. Appourchaux, O. Benomar, G.R. Davies, Y. Elsworth, L. Gizon et al., ApJ **756**, 19 (2012)

DOI: 10.1051/epjconf/201716002011

[3] P. Eggenberger, J. Montalbán, A. Miglio, A&A **544**, L4 (2012)

[4] D. Stello, M. Cantiello, J. Fuller, R.A. Garcia, D. Huber, Pub. Astron. Soc. Austr. **33**, e011 (2016)

[5] J. Fuller, M. Cantiello, D. Stello, R.A. Garcia, L. Bildsten, Science **350**, 423 (2015)

[6] B. Paxton, L. Bildsten, A. Dotter, F. Herwig, P. Lesaffre, F. Timmes, Astrophys. J. Supp. Ser. **192**, 3 (2011)

[7] M. Rieutord, Geophys. Astrophys. Fluid Dyn. **39**, 163 (1987)

[8] M. Rieutord, L. Valdettaro, J. Fluid Mech. **341**, 77 (1997)

[9] F. Rincon, M. Rieutord, A&A **398**, 663 (2003)

[10] D. Reese, F. Rincon, M. Rieutord, A&A **427**, 279 (2004)

[11] S. Abbassi, M. Rieutord, V. Rezania, MNRAS **419**, 2893 (2012)

[12] L. Valdettaro, M. Rieutord, T. Braconnier, V. Fraysse, J. Comput. and Applied Math. **205**, 382 (2007), physics/0604219

[13] A. Brandenburg, D. Sokoloff, K. Subramanian, Space Science Rev. **169**, 123 (2012)

[14] B. Mosser, K. Belkacem, C. Pincon, M. Takata, M. Vrard, C. Barban, M. Goupil, T. Kallinger, R. Samadi, to appear in A&A, ArXiv e-prints/161003872 (2016)

Learning about the latitudinal distribution of starspots through the periodogram analysis of photometric data

Ângela R. G. Santos[1,2,3,⋆], *Margarida S. Cunha*[1,2], *Pedro P. Avelino*[1,2], *Rafael A. García*[4], and *Savita Mathur*[5]

[1] *Instituto de Astrofísica e Ciências do Espaço, Universidade do Porto, CAUP, Rua das Estrelas, PT4150-762 Porto, Portugal*
[2] *Departamento de Física e Astronomia, Faculdade de Ciências, Universidade do Porto, Rua do Campo Alegre 687, PT4169-007 Porto, Portugal*
[3] *School of Physics and Astronomy, University of Birmingham, Edgbaston, Birmingham B15 2TT, UK*
[4] *Laboratoire AIM, CEA/DRF-CNRS-Université Paris 7 Diderot; IRFU/SAp, Centre de Saclay, 91191 Gif-sur-Yvette Cedex, France*
[5] *Center for Extrasolar Planetary Systems, Space Science Institute, 4750 Walnut street Suite#205, Boulder CO 80301, USA*

Abstract. Starspots are cooler and darker than the stellar surface. Therefore, the emitted flux of a star changes when spots are visible on its surface. The presence of spots together with the stellar rotation leads to a periodic modulation on the light curve. By studying that modulation one can then learn about the stellar rotation and also magnetic activity. Recently, Reinhold & Arlt [1] proposed a method based on the analysis of the Lomb Scargle periodogram of the light curve to identify the sign of the differential rotation, i.e. whether the equator rotates faster than the poles or the opposite. We have been studying in detail the spots' impact on the light curve and on the resulting periodogram. We find that, under some conditions, the periodogram can actually provide an estimate of the true spot latitudes and/or the stellar inclination angle. Moreover, we find that the impact of the spot on the ratio between the heights of the second and first harmonics of the main peaks in the periodogram can be described by a single parameter, the visibility time of the spot. Finally, we also identify possible sources of false positives/negatives for the sign of the differential rotation.

1 Introduction

The light curves of active stars often show quasi-periodic modulations due to the presence of dark spots crossing the stellar disc. The study of such modulations provides insights on stellar rotation and magnetic properties [e.g 2].

The high-precision photometric data, such as those provided by the satellite *Kepler* [3], have allowed the detection of the rotation periods for a large number of stars [4], in particular through the periodogram analysis. If the star is differentially rotating, the periodogram of its light curve may show multiple peaks around the detected rotation period. Those peaks are associated to spots at different latitudes and, thus, with different rotation rates. For this reason, one can also use the periodogram analysis to measure the amplitude [e.g. 5–9] and, as proposed more recently, the sign [1] of the differential rotation.

The method proposed by Reinhold & Arlt [1] relies on the comparison of the peak-height ratios corresponding to different peaks (P_k with $k = 1, 2, ...$) in the periodogram related to the surface rotation. The peak-height ratios r_k are defined as the ratio between the heights of the second and first harmonics of P_k, h'_k and h_k respectively, i.e. $r_k = h'_k/h_k$. According to the authors, spots at higher latitudes lead to more sinusoidal light curves than spots at lower latitudes. At these lower latitudes, there is extra power in the second harmonic and, thus, larger ratios r_k. Therefore, following their argument, one can attribute a relative latitude ("low" and "high") to each peak P_k:

- if $r_k > r_{k+1} \Rightarrow P_{\rm low} = P_k$ and $P_{\rm high} = P_{k+1}$
- if $r_k < r_{k+1} \Rightarrow P_{\rm low} = P_{k+1}$ and $P_{\rm high} = P_k$.

Defining the observed relative differential rotation as

$$\alpha_{\rm obs} = \frac{P_{\rm high} - P_{\rm low}}{P_{\rm high}}, \quad (1)$$

$\alpha_{\rm obs}$ is positive/negative when the equatorial regions rotate faster/slower than the polar regions (solar/antisolar differential rotation).

In these proceedings, we show the main results from a detailed study of the peak-height ratios and their dependency on the spot and stellar properties (Sect. 2.1). We also present the first source for false-positives/negatives of $\alpha_{\rm obs}$ and for observational biases (Sect. 2.2).

2 Results

2.1 Peak-height ratios: 1-spot simulations

In this section, we investigate the dependency of the peak-height ratios on a number of stellar and spot parameters which affect the light curve modulation.

Taking the synthetic light curves from 1-spot simulations obtained with a tool that we developed, we start by

⋆e-mail: asantos@astro.up.pt

investigating the impact of the spot latitude and stellar inclination on the peak-height ratios. In this set of synthetic light curves, we fix the limb darkening, spot size and intensity contrast, and the rotation profile. We consider circular spots of radius $R_S \sim 5.7°$, intensity contrast $C_S = 0.67$, and infinite lifetime. We also consider solar differential rotation and a quadratic limb-darkening law adequate for solar-like stars.

For each synthetic light curve, we compute the Lomb Scargle periodogram and, then, the peak-height ratios associated to the rotation period. Figure 1 shows the peak-height ratios as a function of the spot latitude, L, for different stellar inclination angles, i. From this plot, one can already conclude that the relation between the peak-height ratios and the spot latitude claimed by Reinhold & Arlt [1] is not fully valid for $i \neq 90°$, being only correct for spots lying in the same hemisphere as the observer.

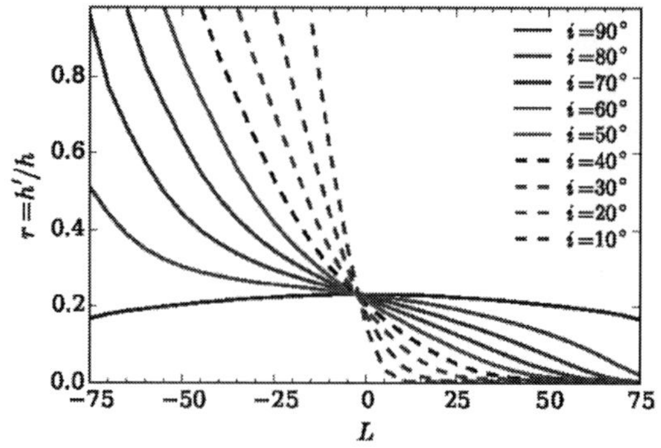

Figure 1: peak-height ratios as a function of the spot latitude for different stellar inclination angles.

From the results of figure 1, one can also anticipate the cases where the method proposed by [1] provides the correct sign of the differential rotation:

- when $i = 90°$ (note that the latitudinal dependency of the peak-height ratios for this inclination is weak, which hampers the determination of the sign of the differential rotation);
- when $i \neq 90°$ and the peak-height ratios are related to spots at the northern hemisphere (same as the observer);
- when $i \neq 90°$ and one of the spots is on the southern hemisphere but its absolute latitude is smaller than the latitude of the spots on the northern hemisphere.

For the remaining cases, we may expect to recover the wrong sign of the surface differential rotation.

We find that the peak-height ratios plotted against the spot latitudes follow a well defined relation for a given stellar inclination angle. Taking these results as reference for the dependency of the peak-height ratios on the latitude, if the inclination is known, one can estimate the latitude of the spot associated to a given rotation period. Even for the cases where the inclination is unknown, one may constrain the spot latitude and the stellar inclination.

But what determines the sinusoidality of the spot modulation on the light curve? Are the peak-height ratios a measure of that sinusoidality?

We find that the spot latitude and the stellar inclination can be translated into a single parameter related with the visibility of the spot. Figure 2 shows the peak-height ratios as a function of the ratio between the time the spot remains visible and the rotation period (t_{vis}/P_{rot}) for different spot latitudes and stellar inclination angles. The peak-height ratios tend to lie on the same curve, showing that they depend on a single parameter, the visibility time of the spot, and are indeed a measure of the sinusoidality of the light curve.

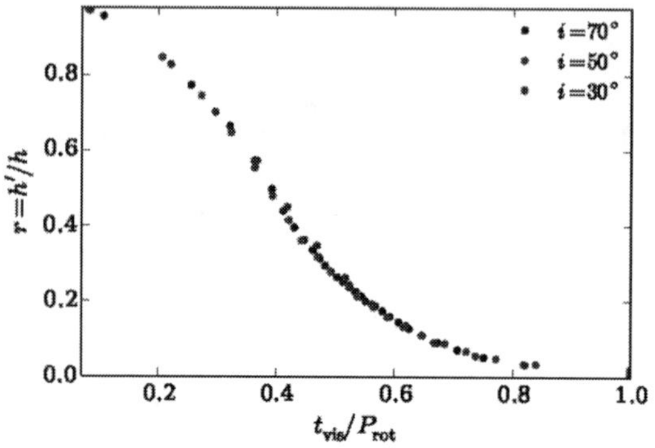

Figure 2: peak-height ratios as a function of the visibility time of the spot for different spot latitudes and the stellar inclination angles $i = 30°$, $i = 50°$, and $i = 70°$.

Besides the stellar inclination and latitudinal distribution of spots, the spot modulation also depends on the stellar rotation and limb-darkening, as well as, on the spot size and intensity contrast.

In the following simulations, the rotation profile is defined through

$$\Omega(L) = \Omega_{eq}(1 - \beta \sin^2 L), \tag{2}$$

where Ω and Ω_{eq} are the angular velocity at a given latitude L and at the equator ($L = 0$), and α is the parameter that determines the amount of surface shear. For the limb-darkening, we consider three different laws:

- linear limb-darkening law

$$\frac{I(\mu)}{I(1)} = 1 - u(1 - \mu) \tag{3}$$

- quadratic limb-darkening law

$$\frac{I(\mu)}{I(1)} = 1 - a(1 - \mu) + b(1 - \mu)^2 \tag{4}$$

- 3-parameter non-linear limb-darkening law

$$\frac{I(\mu)}{I(1)} = 1 - c_2(1 - \mu) - c_3\left(1 - \mu^{3/2}\right) - c_4\left(1 - \mu^2\right), \tag{5}$$

where $I(\mu)/I(1)$ is the relative photospheric intensity, μ is the cosine of the angle between the line of sight and the normal to a given surface element (at the center of the stellar disc $\mu = 1$ and at the limb $\mu = 0$), and u, a, b, c_2, c_3, and c_4 are the limb-darkening coefficients, which we take from [10].

By varying one of those parameters (rotation, limb-darkening, spot size and intensity contrast) at each set of simulations shown in figure 3, we find that the peak-height ratios are independent on stellar rotation and spot contrast (left panels of figure 3), as these parameters do not change the visibility time of the spot, measured as $t_{\rm vis}/P_{\rm rot}$, or the sinusoidality of the modulation. Although the effect is small, we find that the peak-height ratios depend on the limb-darkening (bottom right panel), due to its impact on the shape and sinusoidality of the spot modulation. Finally, the larger the spot the longer it remains visible, thus the spot size affects both peak-height ratios and visibility time of the spot (top right panel).

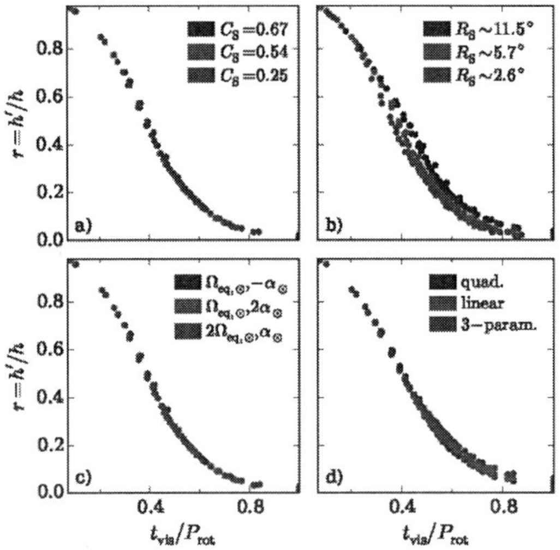

Figure 3: peak-height ratios as a function of the spot visibility time for different spot latitudes and stellar inclinations $i = 30°$, $i = 50°$, and $i = 70°$, and different: a) spot intensity contrast, C_S; b) spot radii, R_S; c) stellar rotation profiles; and d) limb-darkening laws.

2.2 Sign of the surface differential rotation: 2-spot simulations

This section focuses on the sign of the stellar surface differential rotation. In particular, we present one of the sources for false positives/negatives.

For each synthetic light curve obtained from 2-spot simulations, we compute the Lomb Scargle periodogram and the corresponding peak-height ratios. In order to avoid the contamination by nearby peaks, we impose an adequate criteria for the peak separation in the periodogram and only consider the peaks that obey that criteria.

Having the peak-height ratios and the rotation periods, we apply the method proposed by [1] and determine the observed relative differential rotation as defined in Eq. (1). Figure 4 shows the results for the sign of the differential rotation for different spot configurations (from a broad range of latitudes) and stellar inclination angles. The white regions in figure 4 correspond to the spot configurations where the rotation periods are not successfully recovered/distinguishable and, thus, no information on $\alpha_{\rm obs}$ is retrieved. Knowing the input differential rotation ($\alpha > 0$, solar differential rotation), the red dots show the cases where the wrong sign of $\alpha_{\rm obs}$ (false-negative) is determined, while the green dots correspond to the cases where solar differential rotation is correctly recovered. We note that we have considered steps of 5° for the spot latitudes and spots of equal size in these simulations.

Finally, for small stellar inclination angles, it is more likely to detect spot modulations on the light curve induced by spots at the same hemisphere as the observer than by spots at the opposite hemisphere. Therefore, as shown in figure 4, the rate of false negatives decreases as we study stars with lower inclinations.

3 Conclusions

The main goal of the work presented in these proceedings was to understand to what extent one can retrieve, from the light curve and consequent periodogram analysis, information about the latitudinal distribution of spots and stellar surface differential rotation.

We studied the dependency of the peak-height ratios on the spot and stellar properties. We found that those ratios are essentially a function of a single parameter, the spot visibility time, which is mainly determined by the stellar inclination and spot latitude.

We conclude that, despite the degeneracy between latitude and inclination, the periodogram analysis provides a simple and fast way to constrain these parameters. And if the inclination of the star is known, one can actually estimate the spot latitudes.

Our results also show that, for $i \neq 90°$ the method proposed by [1] only returns the correct sign of the differential rotation when:

- the rotation periods P_k and P_{k+1} are associated to spots at L_k, $L_{k+1} > 0$ (same hemisphere as the observer);
- the rotation peridos P_k and P_{k+1} are associated to spots at $L_k > 0$ and $L_{k+1} < 0$ (same and opposite hemispheres as the observer, respectively) and $|L_k| > |L_{k+1}|$.

Finally, we note that the modulation induced by spots at $L_k > 0$ will be preferentially observed, specially for small i. This will contribute for a low rate of false positives/negatives for the sign of $\alpha_{\rm obs}$.

For further details on this study, we recommend the reader to see [11].

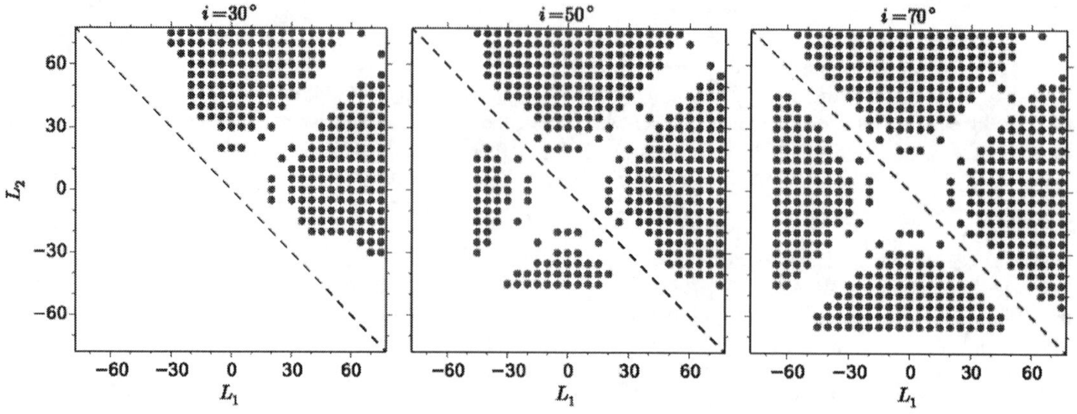

Figure 4: Sign of the differential rotation for different 2-spot configurations and inclination angles $i = 30°$ (left), $i = 50°$ (middle), and $i = 70°$ (right). L_1 and L_2 are the latitudes of each spot respectively. The red dots correspond to the cases of false negatives ($\alpha_{obs} < 0$), while the green dots represent the cases where solar differential rotation ($\alpha_{obs} > 0$) is correctly recovered. The spot configurations that are not marked in the plots correspond to the cases where no rotation periods are recovered or to the cases where the periods do not fulfill the peak separation criteria. The dashed line divides the regions where the correct (above) or wrong (bellow) sign of α_{obs} was expected from the results presented in figure 1.

Acknowledgements

This work was supported by Fundação para a Ciência e a Tecnologia (FCT) through the research grant UID/FIS/04434/2013. ARGS acknowledges the support from FCT through the Fellowship SFRH/BD/88032/2012 and from the University of Birmingham. MSC and PPA acknowledge support from FCT through the Investigador FCT Contracts No. IF/00894/2012 and IF/00863/2012 and POPH/FSE (EC) by FEDER funding through the programme Programa Operacional de Factores de Competitividade (COMPETE). RAG acknowledges the support of the GOLF and PLATO CNES grants. SM would like to acknowledge support from NASA grants NNX12AE17G and NNX15AF13G and NSF grant AST-1411685. The research leading to these results has received funding from EC, under FP7, through the grant agreement FP7-SPACE-2012-312844 and PIRSES-GA-2010-269194. ARGS, MSC, and PPA are grateful for the support from the High Altitude Observatory (NCAR/UCAR), where part of the current work was developed.

References

[1] T. Reinhold, R. Arlt, A&A **576**, A15 (2015)

[2] R.A. García, T. Ceillier, D. Salabert, S. Mathur, J.L. van Saders, M. Pinsonneault, J. Ballot, P.G. Beck, S. Bloemen, T.L. Campante et al., A&A **572**, A34 (2014)

[3] W.J. Borucki, D. Koch, G. Basri, N. Batalha, T. Brown, D. Caldwell, J. Caldwell, J. Christensen-Dalsgaard, W.D. Cochran, E. DeVore et al., Science **327**, 977 (2010)

[4] M.B. Nielsen, L. Gizon, H. Schunker, C. Karoff, A&A **557**, L10 (2013)

[5] T. Reinhold, A. Reiners, A&A **557**, A11 (2013)

[6] T. Reinhold, A. Reiners, G. Basri, A&A **560**, A4 (2013)

[7] T. Reinhold, L. Gizon, A&A **583**, A65 (2015)

[8] E. Nagel, S. Czesla, J.H.M.M. Schmitt, A&A **590**, A47 (2016)

[9] E. Distefano, A.C. Lanzafame, A.F. Lanza, S. Messina, F. Spada, A&A **591**, A43 (2016)

[10] D.K. Sing, A&A **510**, A21 (2010)

[11] A.R.G. Santos, M.S. Cunha, P.P. Avelino, R.A. García, S. Mathur, A&A **599**, A1 (2017)

On the relation between activity-related frequency shifts and the sunspot distribution over the solar cycle 23

Ângela R. G. Santos[1,2,3,⋆], *Margarida S.* Cunha[1,2], *Pedro P.* Avelino[1,2], *William J.* Chaplin[3,4], and *Tiago L.* Campante[3,4]

[1] *Instituto de Astrofísica e Ciências do Espaço, Universidade do Porto, CAUP, Rua das Estrelas, PT4150-762 Porto, Portugal*
[2] *Departamento de Física e Astronomia, Faculdade de Ciências, Universidade do Porto, Rua do Campo Alegre 687, PT4169-007 Porto, Portugal*
[3] *School of Physics and Astronomy, University of Birmingham, Edgbaston, Birmingham B15 2TT, UK*
[4] *Stellar Astrophysics Centre (SAC), Department of Physics and Astronomy, Aarhus University, Ny Munkegade 120, 8000 Aarhus C, Denmark*

Abstract. The activity-related variations in the solar acoustic frequencies have been known for 30 years. However, the importance of the different contributions is still not well established. With this in mind, we developed an empirical model to estimate the spot-induced frequency shifts, which takes into account the sunspot properties, such as area and latitude. The comparison between the model frequency shifts obtained from the daily sunspot records and those observed suggests that the contribution from a stochastic component to the total frequency shifts is about 30%. The remaining 70% is related to a global, long-term variation. We also propose a new observable to investigate the short- and mid-term variations of the frequency shifts, which is insensitive to the long-term variations contained in the data. On the shortest time scales the variations in the frequency shifts are strongly correlated with the variations in the total area covered by sunspots. However, a significant loss of correlation is still found, which cannot be fully explained by ignoring the invisible side of the Sun when accounting for the total sunspot area. We also verify that the times when the frequency shifts and the sunspot areas do not vary in a similar way tend to coincide with the times of the maximum amplitude of the quasi-biennial variations found in the seismic data.

1 Introduction

In the Sun, the activity-related frequency shifts are one of the numerous manifestations of the solar magnetic activity. As the activity level increases, the frequencies of the acoustic modes are observed to increase [e.g. 1–4]. In addition to this 11-yr, long-term variation, the solar acoustic frequencies also vary on a quasi-biennial timescale [e.g. 5–7].

The acoustic frequencies can be affected by different phenomena associated to both weak and strong components of the magnetic field [e.g. 7–10]. Those may contribute to the observed frequency shifts through indirect (thermal and structural) effects and direct (via Lorentz force) effects. Although the activity-related frequency shifts have been known for more than 30 years, the relative impact of the different contributions to the frequency shifts is not fully understood.

In the work presented in this proceedings, we have studied the short-term variations (on a timescale of days) in the solar acoustic frequencies. We may expect that the dominant contribution to such short-term variations is related to sunspots (strong component of the magnetic field), which, depending on their size, usually live and evolve in a few days. In a first study, we estimate the contribution from the spot component to the observed frequency shifts. To that end and following the approach of [11], we derived a model for the spot-induced frequency shifts (Sect. 2, see also [10]). Sect. 3 shows the results from a second study aimed at the characterization of the differences between the observed frequency shifts and the total area covered by sunspots. We have proposed a new observable, the weighted sum of the frequency-shift differences, to study the short- and, also, the mid-term variations in the acoustic frequencies [see 12]. This quantity is mostly insensitive to the long-term (11-yr) signal.

2 Model frequency shifts

By changing the medium where the acoustic waves propagate, sunspots can affect their propagation properties, in particular the oscillation frequencies. Sunspots contribute to the frequency shifts with both indirect and direct effects. By solving the incomplete wave equations, [13] studied the spots' indirect effect due structural changes within a sunspot. Here, we consider the total spot impact, by parameterizing the model frequency shifts.

⋆e-mail: asantos@astro.up.pt

One may quantify the spots' perturbation through the phase difference, $\Delta\delta$, between the magnetic (within a spot) and the non-magnetic wave solutions (in the quiet-Sun), at a depth R^* below the region of influence of the spot (here, we consider a depth of 10 Mm). In this work, we assume a characteristic phase difference, $\Delta\delta_{\rm ch}$, induced by sunspots. The comparison between the model and observed frequency shifts will allow to constrain this parameter.

Following a variational principle [11], taking the area, A_i, and central colatitude, θ_i, of a given spot i, we find the following relation for the spot-induced frequency shifts over time [for details, see 10]

$$\frac{\delta\omega_{lm}}{\omega_{lm}}(t) \simeq -\frac{\Delta\delta_{\rm ch}}{\omega_{lm}^2 R^2 \int_{r_1^l}^{R^*} c^{-2}\kappa^{-1} dr} \sum_{i=1}^{N(t)} \left[\left(P_l^m (\cos\theta_i) \right)^2 A_i \right], \tag{1}$$

where ω, l, and m are, respectively, the mode frequency, angular degree, and azimuthal order, R is the solar radius, r_1^l is the lower turning point of the mode, c is the sound speed, κ is the radial component of the acoustic wave number, r is the distance to the solar center, N is the number of spots at a given time t, and P_l^m are the normalized [see 10] Legendre polynomials.

Having the sunspot parameters, Eq. (1) can be used to estimate the spot-induced frequency shifts for different radial orders, angular degrees, and azimuthal orders.

Since, our goal is to determine the contribution from the spot component, which varies on a timescale of days, we shall consider frequencies obtained with a relative short cadence. We use the frequency shifts computed by [14] with a cadence of 36 days and an overlap of 18 days, constituting two samples of independent data points. The observed frequency shifts where obtained from GONG (Global Oscillation Network Group) data and includes modes of degree between $l = 0$ and 100 with frequencies ranging from 2000 and 3300 μHz. Within this frequency range, the frequency shifts increase almost linearly with the frequency [e.g. 2, 15]. Thus, we take the central frequency to estimate the mean frequency shifts. Using the sunspot daily records from the National Geophysical Data Center (NGDC/NOAA) database, the spot-induced frequency shifts, $\delta\nu_{lm} = \delta\omega_{lm}/(2\pi)$, are computed for $l = 0 - 100$ and corresponding azimuthal orders. The mean spot-induced frequency shifts, $\delta\nu_{\rm spots}$, are then obtained by combining $\delta\nu_{lm}$ as the observed frequency shifts [see 10, 14]. Finally, in order to obtain model frequency shifts consistent with the observational data, we average the model spot-induced frequency shifts, obtained from the daily sunspot records, over bins of 36 days with an overlap of 18 days.

To account for the weak component of the magnetic field we consider a global smooth component varying on the 11-yr timescale, given by

$$\delta\nu_{\rm global} = w f(t), \tag{2}$$

where w is the weight of the global component and

$$f(t) = \frac{a\,(t - t_0)^3}{\exp\left((t - t_0)^2 / b^2\right) - c}. \tag{3}$$

This functional form was proposed by [16] to describe the variation of the sunspot number over the cycle. The parameters a, b, c, and t_0 are related to the amplitude, starting time, size of the rising phase, and temporal asymmetry of the cycle, respectively, and are given by the best fit to the observed frequency shifts.

The total model frequency shifts are then

$$\delta\nu_{\rm model} = \delta\nu_{\rm global} + \delta\nu_{\rm spots}. \tag{4}$$

This way, the model frequency shifts contain two parameters, w and $\Delta\delta_{\rm ch}$, to be determined.

The best model is obtained through a χ^2 minimization between the observed and model frequency shifts. The value found for the characteristic phase difference induced by spots is $\Delta\delta_{\rm ch} \sim -0.44$. The weight of the global component, corresponding to the best fit, is $w \sim 0.71$, meaning that the global long-term component accounts for $\sim$ 70% of the total frequency shifts. The remaining $\sim$ 30% are related to the spot-induced component which varies on short timescales.

Figure 1 shows the comparison between the model predictions and the observational data for both samples of independent data points. Although the model and observed frequency shifts agree reasonably well and we have checked that our estimate of the relative contribution of each component is robust, the residuals shown in the bottom panels are found to vary in phase with the quasi-biennial variations in the frequency shifts [e.g. 5, 6]. This may suggest that our model is missing a mid-term contribution varying on a quasi-biennial timescale.

3 Weighted sum of the frequency-shift differences

The total area covered by sunspots varies on a timescale of days. Therefore, one may expect that sunspots induce frequency variations on the same timescale, being the main source for the 36-d frequency shifts. With this in mind, we propose a new quantity to investigate the short-term variations in the frequency shifts, being almost insensitive to the global long-term component of the frequency shifts [12]. This quantity combines the information contained in both observed frequency shifts and sunspot areas and is defined as the weighted sum of the frequency-shift differences ($W_{\rm D}$), i.e.

$$W_{\rm D} = \sum_n \Delta\delta\nu_n \times S_n, \tag{5}$$

where $\Delta\delta\nu_n$ is the frequency-shift difference at time n ($\Delta\delta\nu_n = \delta\nu_n - \delta\nu_{n-1}$ with $n = 2, 3, 4, ...$) and S_n is the weight to be determined by the variation in the total area covered by sunspots :

$$S_n = \begin{cases} 1, & \text{if } \Delta A_{{\rm T},n} > 0 \\ -1, & \text{if } \Delta A_{{\rm T},n} < 0 \end{cases} \tag{6}$$

DOI: 10.1051/epjconf/201716002013

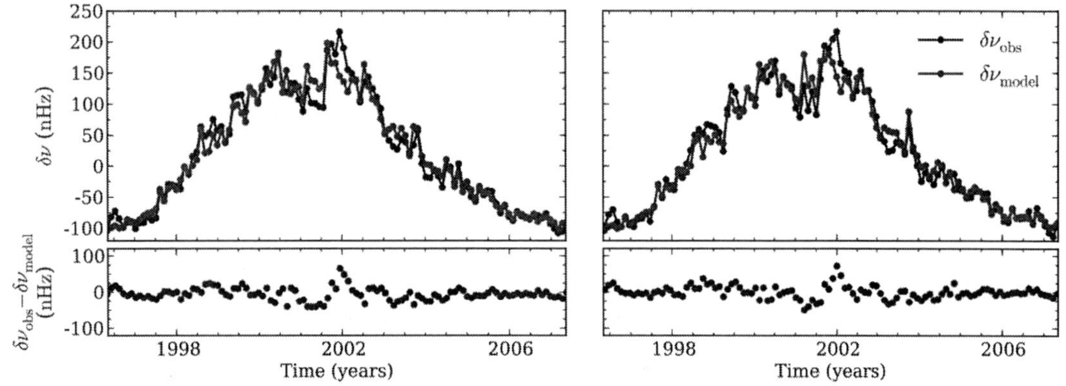

Figure 1: Top: Comparison between the observed (black) and model (blue) frequency shifts. The model frequency shifts include a global long-term component related with the overall solar magnetic field and a spot-induced component which varies on short timescales (i.e. $\delta\nu_{model} = \delta\nu_{global} + \delta\nu_{spots}$; Eq. (4)). Bottom: Residuals between the observed and model frequency shifts. Left and right panels concern each sample of independent data points.

where $\Delta A_{T,n} = A_{T,n} - A_{T,n-1}$ and $A_{T,n}$ is the total area covered by sunspots at a given time n. If the short-term variations in the frequency shifts and in the spot areas are completely correlated, $\Delta\delta\nu_n$ and $\Delta A_{T,n}$ will always have the same sign and W_D will correspond to the sum of the absolute (moduli) frequency-shift differences (M_D)

$$M_D = \sum_n |\Delta\delta\nu_n|. \tag{7}$$

The top panels of figure 2 compare the weighted sum of the frequency-shift differences, W_D, and the sum of the absolute values of the frequency-shift differences, represented by the quantity M_D. For the spot-induced frequency shifts obtained in Sect. 2, M_D and W_D overlap. This shows that the spot-induced frequency shifts, combined as done for the observed frequency shifts, are mostly proportional to the sunspot area, being weakly dependent on the sunspot latitudes. For the observed frequency shifts, M_D and W_D do not overlap and the maximum difference between them corresponds to the amount of correlation that is lost over the cycle, which cannot be fully explained by the sunspots on the far-side of the Sun [12]. Nevertheless, the comparison of W_D with the expected standard deviation for a random walk (at the end of the cycle), shows that the short-term variations in the frequency shifts and in the total area covered by sunspots are strongly correlated.

The middle panels of figure 2 show the difference between the quantities M_D and W_D. To better identify the times when frequency shifts and spot areas behave differently, the time derivative of that difference, $\partial(M_D - W_D)/\partial t$, is shown in the bottom panels of figure 2. The times of loss of correlation correspond to $\partial(M_D - W_D)/\partial t \neq 0$. We found that most of the loss of correlation between the short-term variations in the frequency shifts and in the sunspot areas occurs around epochs of maximum of quasi-biennial signal found by [6].

4 Conclusions

We investigated the short-term variations in the solar acoustic frequencies, which we expect to be mainly induced by the presence of sunspots on the solar surface.

We derived a parametric model for the spot-induced frequency shifts. The total model frequency shifts also account for a long-term component varying on the timescale of the solar cycle (11 yr). This component is assumed to be representative of the weak component of the magnetic field. The comparison between the model and observed frequency shifts indicates that the spot-induced contribution is ~ 30% of the total frequency shifts. The remaining ~ 70% corresponds to the long-term component. However, our results also suggest that there is a mid-term component varying on a quasi-biennial timescale that is not account for.

We also propose a new method for the analysis of the correlation between the frequency shifts and other activity proxies, such as the area coverage by spots. The method is based on the weighted sum of the frequency-shift differences, being the weight determined by the variation in the area covered by sunspots. This quantity amplifies the signal from the short-term component of the frequency shifts and is almost insensitive to the long-term component. We found a strong correlation between the short-term variations in the frequency shifts and in the sunspot areas. However, there is a significant loss of correlation at specific moments during the cycle, which tends to coincide with times of maximum of the quasi-biennial signal. These results further suggest that there is a mid-term contribution which has not been taking into account.

Acknowledgements

This work was supported by Fundação para a Ciência e a Tecnologia (FCT) through the research grant UID/FIS/04434/2013.

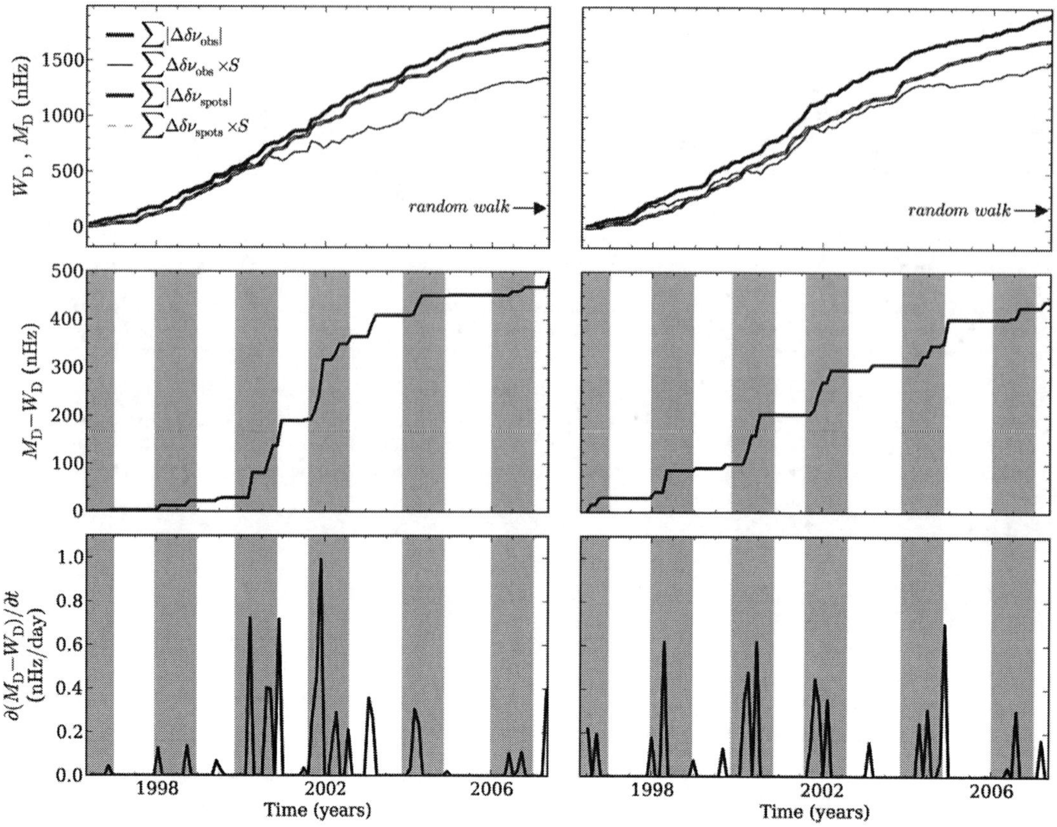

Figure 2: Top: Quantities M_D (thick solid lines) and W_D (thin solid and dashed lines) for the observed (black) and spot-induced (dark and light blue) frequency shifts. The arrow marks the standard deviation for a random walk. Middle: Difference between the quantities M_D and W_D. Bottom: Time derivative of the difference between M_D and W_D. The blue bars are centered around the maxima of the quasi-biennial signal, having a width of 1 yr. Left and right panels concern each sample of independent data points.

ARGS acknowledges the support from FCT through the Fellowship SFRH/BD/88032/2012 and from the University of Birmingham. MSC and PPA acknowledge support from FCT through the Investigador FCT Contracts No. IF/00894/2012 and IF/00863/2012 and POPH/FSE (EC) by FEDER funding through the programme Programa Operacional de Factores de Competitividade (COMPETE). TLC and WJC acknowledge the support of the UK Science and Technology Facilities Council (STFC). The research leading to these results has received funding from EC, under FP7, through the grant agreement FP7-SPACE-2012-312844 and PIRSES-GA-2010-269194. ARGS, MSC, and PPA are grateful for the support from the High Altitude Observatory (NCAR/UCAR), where part of the current work was developed.

References

[1] M.F. Woodard, R.W. Noyes, Nature **318**, 449 (1985)
[2] K.G. Libbrecht, M.F. Woodard, Nature **345**, 779 (1990)
[3] Y. Elsworth, R. Howe, G.R. Isaak, C.P. McLeod, R. New, Nature **345**, 322 (1990)
[4] W.J. Chaplin, Y. Elsworth, G.R. Isaak, R. Lines, C.P. McLeod, B.A. Miller, R. New, MNRAS **300**, 1077 (1998)
[5] S.T. Fletcher, A.M. Broomhall, D. Salabert, S. Basu, W.J. Chaplin, Y. Elsworth, R.A. Garcia, R. New, ApJ **718**, L19 (2010)
[6] A.M. Broomhall, W.J. Chaplin, Y. Elsworth, R. Simoniello, MNRAS **420**, 1405 (2012)
[7] A.M. Broomhall, V.M. Nakariakov, Sol. Phys. **290**, 3095 (2015), broomhall2015
[8] S.C. Tripathy, F. Hill, K. Jain, J.W. Leibacher, Sol. Phys. **243**, 105 (2007)
[9] K. Jain, S.C. Tripathy, F. Hill, ApJ **695**, 1567 (2009)
[10] A.R.G. Santos, M.S. Cunha, P.P. Avelino, W.J. Chaplin, T.L. Campante, MNRAS **461**, 224 (2016)
[11] M.S. Cunha, D. Gough, MNRAS **319**, 1020 (2000)
[12] A.R.G. Santos, M.S. Cunha, P.P. Avelino, W.J. Chaplin, T.L. Campante, submitted (2016)
[13] A.R.G. Santos, M.S. Cunha, J.J.G. Lima, Astron. Nachr. **333**, 1032 (2012)
[14] S.C. Tripathy, K. Jain, D. Salabert, R.A. García, F. Hill, J.W. Leibacher, J. Phys. Conf. Ser. **271**,

DOI: 10.1051/epjconf/201716002013

012055 (2011)
[15] W.J. Chaplin, T. Appourchaux, Y. Elsworth, G.R. Isaak, R. New, MNRAS **324**, 910 (2001)
[16] D.H. Hathaway, R.M. Wilson, E.J. Reichmann, Sol. Phys. **151**, 177 (1994)

EPJ Web of Conferences **160**, 03001 (2017) DOI: 10.1051/epjconf/201716003001
Seismology of the Sun and the Distant Stars II

What CoRoT tells us about δ Scuti stars

Existence of a regular pattern and seismic indices to characterize stars

Eric Michel[1,★], *Marc-Antoine* Dupret[2], *Daniel* Reese[1], *Rhita-Maria* Ouazzani[3], *Jonas* Debosscher[4], *Antonio* García Hernández[5,6], *Kevin* Belkacem[1], *Reza* Samadi[1], *Sébastien* Salmon[2], *Juan Carlos* Suarez[6], and *Sebastia* Barceló Forteza[7,8]

[1] *Observatoire de Paris, LESIA, UMR8109, Université Pierre et Marie Curie, Université Paris Diderot, PSL*
[2] *Institut d'Astrophysique et Geophysique, University of Liège, Belgium*
[3] *Stellar Astrophysics Centre, Department of Physics and Astronomy, Aarhus University, Ny Munkegade 120, DK-8000 Aarhus C, Denmark*
[4] *Instituut voor Sterrenkunde, KU Leuven, Celestijnenlaan 200B, 3001, Leuven, Belgium*
[5] *Instituto de Astrofísica e Ciências do Espaço, Universidade do Porto, CAUP, Rua das Estrelas, PT4150-762 Porto, Portugal*
[6] *Dept. Física Teórica y del Cosmos. University of Granada. 18071. Granada, Spain*
[7] *Instituto de Astrofísica de Canarias, 38200 La Laguna, Tenerife, Spain*
[8] *Departamento de Astrofísica, Universidad de La Laguna, 38206 La Laguna, Tenerife, Spain*

Abstract. Inspired by the so appealing example of red giants, where going from a handful of stars to thousands revealed the structure of the eigenspectrum, we inspected a large homogeneous set of around 1860 δ Scuti stars observed with CoRoT. This unique data set reveals a common regular pattern which appears to be in agreement with island modes featured by theoretical non-perturbative treatments of fast rotation. The comparison of these data with models and linear stability calculations suggests that spectra can be fruitfully characterized to first order by a few parameters which might play the role of seismic indices for δ Scuti stars, as $\Delta\nu$ and $\nu_{\max}$ do for red giants. The existence of this pattern offers an observational support for guiding further theoretical works on fast rotation. It also provides a framework for further investigation of the observational material collected by CoRoT ([1]) and *Kepler* ([2]). Finally, it sketches out the perspective of using δ Scuti stars pulsations for ensemble asteroseismology.

1 Introduction

δ Scuti stars constitute a large class of pulsating stars representative of chemically normal intermediate mass stars on and near the main sequence (see eg [3], [4]). Their seismic exploitation however meets major difficulties often refered to as 'the mode identification problem', 'the fast rotation treatment', and 'the selection effects'. These various expressions refer to two main difficulties. First, δ Scuti stars, as chemically normal A and early F stars are characterized by a large rotation rate ([5]) and the theoretical modelling of their pulsation spectra cannot rely on classical perturbative approaches ([6],[7]). Then, although we understand the process responsible for their pulsational instability, we have very little insight about the process responsible for the amplitude limitation (see however recent studies by [8] and [9]) and thus no clue about how amplitudes are distributed between modes and for different stars.

However, we do know a couple of things about these stars. First, there is increasing evidence that periodicities or regular spacings can be found in δ Scuti spectra ([10], [11], [12]) and recently, [13] demonstrated that this spacing is a good proxy of the mean density just as the large separation $\Delta\nu$ is for solar-like pulsators.

Then, linear stability calculations provide reliable results ([14]) and it is appealing to use them to characterize stars in terms of mass range or evolution stage (see e.g. [15] for stars in clusters and [16] in the case of pre-main sequence δ Scuti stars).

In the present paper, we use a large set of homogeneous spectra observed with CoRoT ([17]) to revisit these questions and see what CoRoT data tell us about δ Scuti stars.

2 The observational sample and the determination of $f_{\min}$, $f_{\max}$ and $a_{\max}$.

The automated supervised classification of variable stars in the CoRoT programme (ASCVC hereafter, [18]) brings about 1860 objects classified as δ Scuti stars with a probability higher than 80%. In comparison, catalogues before the space-photometry era gathered about 700 objects ([19]), among which a large fraction had been discovered by large surveys like the Hipparcos ([20]), OGLE ([21])

★e-mail: Eric.Michelatobspm.fr

and MACHO ([22]) projects. The present CoRoT sample is thus very valuable in terms of number of objects and also in terms of homogeneity.

We computed the Fourier spectrum for each of these light-curves and we set to zero-amplitude the parts of the spectra possibly hampered by intrumental/environmental artefacts induced by the orbital period (see [23]). It consists in narrow intervals around the frequency of the orbital period plus its harmonics, each of them associated with a few daily aliases as it can be seen in figure 6. For each spectrum, we also exclude from our study the part below f_{Lcut} = 25 μHz (also set to zero-amplitude) in order to avoid the influence on our analysis of possible power of instrumental or environmental origin at low frequency.

We define a limit amplitude criterion for peaks to be considered. Here we take the maximum between 10 times the mean amplitude level and the amplitude of the highest peak divided by 8. This last constraint aims at avoiding artefacts of large amplitude peaks convolved by the observational window.

Then, we determine for each spectrum the range of detected signal, noted $[f_{\min}, f_{\max}]$, as the frequency range encompassing all peaks satisfying the previous amplitude criterion.

We also use these spectra to produce an index characterizing the amplitude of the oscillations. We considered two versions of this index. One is simply the amplitude of the highest peak. The second is the square root of the quadratic sum of amplitudes of all peaks satisfying our amplitude criterion. Interestingly, the results were found to vary at most by a factor two from one version to the other. In the present work, all results for $a_{\max}$ refer to the square root of the quadratic sum, which we expect to be more stable a measurement and more representative of the energy involved in pulsation.

The results are presented in figure 1 and their interpretation in the light of theoretical models is discussed in Sect. 4.

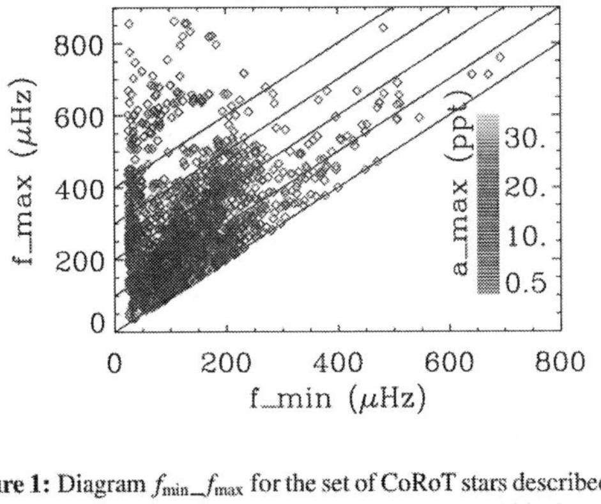

Figure 1: Diagram $f_{\min}$_$f_{\max}$ for the set of CoRoT stars described in the text. Oblique lines indicate equal frequency width $f_{\max} - f_{\min}$. The values mentioned on the y-axis indicate both the $f_{\max}$ value for a horizontal line and the $f_{\max} - f_{\min}$ width for an oblique line. Values of $a_{\max}$ are given by the colour code.

3 Theoretical estimates of $f_{\min}$ and $f_{\max}$

We used a grid of theoretical models representative of the whole δ Scuti stars instability strip for the main sequence evolution stage (see figure 2). This grid is the one used and described in [14]. The models have been computed with the code CLES ([24]). The physics of the models is standard and the only specific aspects of interest at the level of the present study are: the use of overshooting with $\alpha_{ov} = 0.2H_p$ and the mixing length parameter which has been set to the solar calibrated value 1.8. The metallicity (Z=0.02) had been chosen as representative of the solar one in [14]. Here again, the change for a more up-to-date value would not change significantly our results. The linear stability of the modes has been obtained following ([14], [25]).

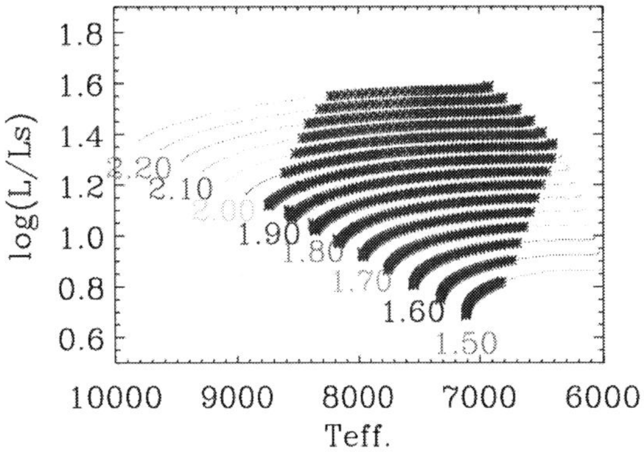

Figure 2: Hertzsprung-Russell diagram featuring models described in the text. Models showing at least one unstable mode are marked by red star symbols delimiting the theoretical instability strip. For each sequence, mass is indicated in solar units.

These linear stability calculations are used to derive theoretical counterparts of the $f_{\min}$ and $f_{\max}$ values obtained for observed stars in Sect. 2. They have been determined on modes of degree $l = 0, 1, 2$ and are illustrated in figure 3.

When looking at figure 3, it is clear that, for sequence of models of given mass, the position in the $f_{\min}$_$f_{\max}$ diagram changes under two effects. One is the decrease of mean density (or the decrease of $\Delta\nu_0 \propto (GM/R^3)^{1/2}$) with evolution. The second is associated with the gradual decrease of the radial orders of unstable modes, as the star crosses the instability strip from the blue to the red border, as described in Ref Dupret 2004. Both effects globally induce a decrease of $f_{\min}$ and $f_{\max}$ with age on the main sequence.

It is also worth noticing that, evolution sequences entering the instability strip through the so-called blue border (i.e. masses higher than $1.9M_\odot$) correspond, in the $f_{\min}$_$f_{\max}$ diagram, to sequences starting on the ($f_{\min} = f_{\max}$) central axis, i.e. with a very narrow frequency range of unstable modes.

On the contrary, evolution sequences starting on the ZAMS in the instability strip (i.e. masses between 1.5 and

DOI: 10.1051/epjconf/201716003001

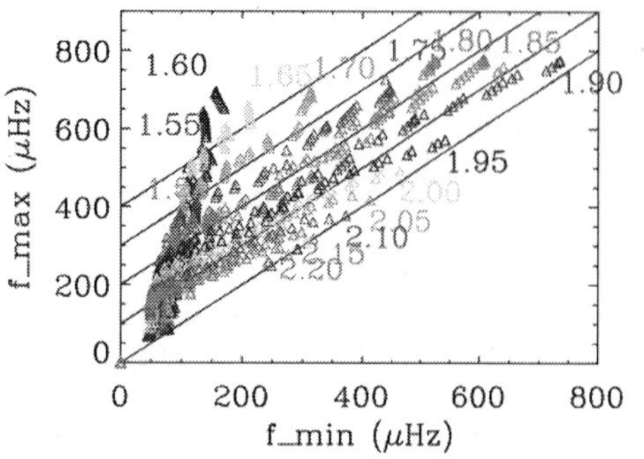

Figure 3: Diagram $f_{min}\_f_{max}$, as in figure 1, but for sequences of models illustrated in figure 2 and described in the text. The colours stand for sequences of models of a given mass.

$1.9 M_{\odot}$) correspond, in the $f_{min}\_f_{max}$ diagram, to sequences starting at high $f_{max} - f_{min}$ values, i.e. with a larger frequency range of unstable modes and thus farther from the ($f_{min} = f_{max}$) central axis.

Finally, we can also notice that, with the evolution, all sequences tend to converge at low f_{min} and f_{max} in the diagram, those going out of the instability strip through the red border corresponding to sequences terminating on the ($f_{min} = f_{max}$) central axis.

We take as a work hypothesis that the rotation will not impact drastically these theoretical estimates of f_{min} and f_{max} values. Rotational splitting is expected to extend the observed range of observed peaks by an amount which is difficult to estimate precisely in the present state of our knowledge, but should be, in the worst case, of the order of a few times the rotation frequency, i.e. a few times 10 to 20 μHz for the fastest rotating objects. This is not negligible, but if this extension remains below 50 μHz for most of the stars, the $f_{min}\_f_{max}$ diagram remains discriminent in terms of evolution stage and mass range, as can be seen in figure 3.

4 Comparison of observational and theoretical $f_{min}\_f_{max}$ diagrams

When comparing figure 1 and figure 3 in the light of previous remarks, we notice a few differences and similarities. The set of observed stars shows components in the $f_{min}\_f_{max}$ diagram which are not present for the models. The main difference consists in a vertival ridge observed at low f_{min} (below 50 μHz) in figure 1. This accumulation ridge could be due to an edge effect induced by the processing of spectra described in Sect. 2, when we fix the f_{Lcut} value to prevent our analysis from the influence of possible low frequency noise. However, we should keep in mind that this ridge is also the place where we should expect hybrid δ Scuti-γ Doradus stars. In addition to the modes characterizing δ Scuti stars, these objects present low-frequency pulsation modes not considered unstable in our models. A closer inspection of these spectra will be necessary to explore this possibility.

The second component presented by the observed set of stars and not by the models is much less important in numbers. It appears as a cloud of points around the upper left corner of figure 1, i.e. stars with f_{min} values bellow 200 μHz and f_{max} values between 500 and 900 μHz. As it reads from the oblique lines in figure 1, this corresponds to extremely large estimates of frequency range value ($f_{max} - f_{min} > 500$ μHz). The inspection of several spectra suggests that, to a large extent, these values could be due to the poor handling of particularly severe window-artefacts by the data processing presented in Sect. 2. This is supported by the presence of a significant fraction of spectra with high a_{max} values among those points.

Beside this, the $f_{min}\_f_{max}$ diagram for observed stars and the one for models look in reasonably good agreement, with a scarce distribution of points at high f_{min} and f_{max} values, where evolution sequences are highly spread in mass and age in figure 3, and a denser concentration of points at low values, where the evolution sequences tend to converge.

4.1 Amplitude of the oscillations versus evolution.

It is worth noticing that the distribution of a_{max} shows a clear gradient (from 0.1 to 40 part-per-thousand, ppt hereafter) with increasing amplitude values toward low values of f_{min} and f_{max}. As we already commented, according to the theoretical diagram, this domain of the $f_{min}\_f_{max}$ diagram is expected to host rather evolved stars.

The fact that evolved δ Scuti stars tend to show higher amplitudes than main sequence ones has been observed for long (see e.g. [4]), but to our knowledge, this trend has never been observed with such a resolution in amplitude and on such a large sample of objects. In the case of *Kepler* data, [26] considered about 1570 objects, but a large fraction of them (~1150) were observed in the so-called long cadence mode, i.e. with a 30 minutes sampling time which is not suited to address frequencies higher than 270 μHz. These data will thus be very helpful to study δ Scuti stars, but mostly evolved ones.

4.2 Evidence of a regular frequency pattern in early main sequence δ Scuti stars.

Models off the main sequence are not produced here, but we have inspected several such sequences, and they are always found in the domain of low f_{min} and f_{max}. Even with $\alpha_{ov} = 0$ (no overshooting), which corresponds to the shortest extension of the main sequence, post-main sequence models seem to remain below the $f_{max} = 400$ μHz limit.

We now consider stars with $f_{max} > 400$ μHz, i.e. for which, according to models, f_{min} and f_{max} values correspond unambiguously to main sequence models, whatever the amount of overshooting considered. As illustrated in figure 4 , we also avoid spectra with dubious values as discussed in Sect. 2. The domain of the HR diagram corresponding to this selection in the case of models presented here is illustrated in figure 5.

For this set of about 200 stars, we build an image presented in figure 6, where each amplitude spectrum is a line where amplitude is limited to an arbitrary value a_{thr} (here $a_{thr} = 0.2ppt$) and coded in grey scale. The full image thus presents more than 200 spectra, sorted by increasing $f_{\max}$ value from the bottom to the top.

In figure 6, we distinguish a few ridges approximately parallel to the ridge drawn by peaks associated with $f_{\max}$ values. These ridges separated by a few tens of μHz immediately recall the quasiregular pattern of axisymmetric island modes as described by nonperturbative calculation of fast rotators ([27] and [28]). They also recall the various detection of spacings of the order of the large separation in δ Scuti stars ([13],[11])

The pattern associated with these ridges has to be common to a sufficient fraction of our stellar sample to show up in figure 6. This suggests that this quasiregular pattern of peaks has to be, to some extent, independent of rotation, which necessarily varies from one star to another.

Axisymmetric modes are good candidates to explain this pattern as we will show hereafter.

An appealing idea is that the eigenspectra of axisymmetric modes of the different stars could be, at first order, homologous and thus distributed according to a common pattern of peaks, just multiplied by a different value of the large separation (or mean density).

In order to test a bit further this idea, we rescaled the (observational) spectra, taking in abscissae the logarithm of frequency instead of the frequency itself. If our hypothesis is correct, the rescaled eigenspectra should show a similar pattern, just shifted by an amount depending on (the logarithm of) the individual large separation value. These rescaled spectra are presented in figure 7 where they have been shifted according to their individual $f_{\max}$ value. Here the sequence of ridges is more visible even if the pattern remains blured at low abscissae.

We treated the same way the theoretical spectra associated with the set of models shown in figure 5 (models satisfying the same criterion $f_{\max} > 400$ μHz than our subsample of stars). Here again, we considered only unstable axisymmetric (m=0) l=0,1,2 modes. The result is presented in figure 8.

The figure 7 and figure 8 show great similarities with a few clear ridges near zero in abscissae followed by a less clear pattern. In fact, we have to pay attention to the fact that all these stars do not have the same range of unstable radial orders and this makes the superposition of the spectra less clear. This is demonstrated in figure 9, where the theoretical spectra this time are normalized by $\Delta\nu$ instead of $f_{\max}$. The ridges are much clearer now. This shows that (for models without rotation) a common pattern exists and is not too much distorted by the structural changes of models in the early main sequence.

In order to illustrate the anticipate impact of rotation on this comparison, we added at the bottom of figure 8 (between line 10 and 18), the pattern obtained for axisymmetric ($\tilde{l}$=0) island modes (see [29]) in 9 models based on the Self-Consistent Field (SCF) method ([30]). These models are ZAMS $2M_{\odot}$, ranging from 0 to 80% of the break-up rotation rate. These eigenfrequencies are shown to illustrate the effect of rotation on the pattern, but these modes are not necessary unstable.

We see that the regular pattern is preserved to a good extent over this large range of rotation. In fact, at low radial order, the distribution of the modes appears even more regular than the one of models without rotation.

To conclude with this point, it seems important to stress that the spectra considered in this study obviously show numerous peaks outside of the common quasiregular pattern revealed by figure 6 or figure 7. On the other hand, these figures do not suggest either that the modes associated to this pattern are systematically expressed (with detectable amplitude). However these results suggests that this pattern is characteristic of the eigenspectra of chemically normal early main sequence stars.

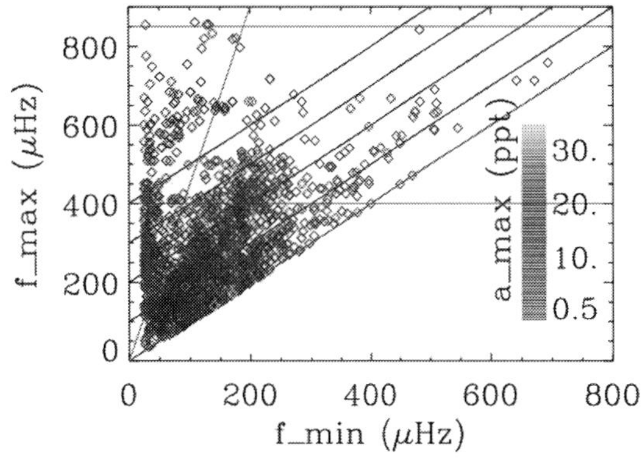

Figure 4: Same as in figure 1 but with purple lines marking the selection of the set of early main sequence objects with $f_{\max} > 400$ μHz and $f_{\max}/f_{\min} < 4.5$, as discussed in Sect. 4.2.

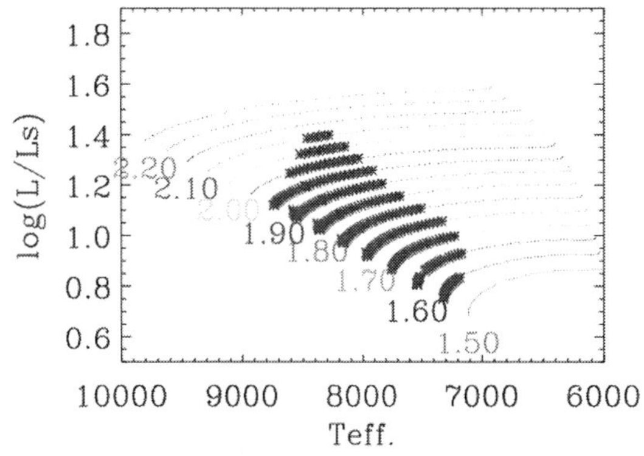

Figure 5: Same as in figure 2 but here only unstable models from the selection described in Sect. 4.2 and illustrated in figure 4 are marked by red symbols.

5 Conclusions

We used a homogeneous set of about 1860 stellar light curves collected with CoRoT for stars classified as δ Scuti

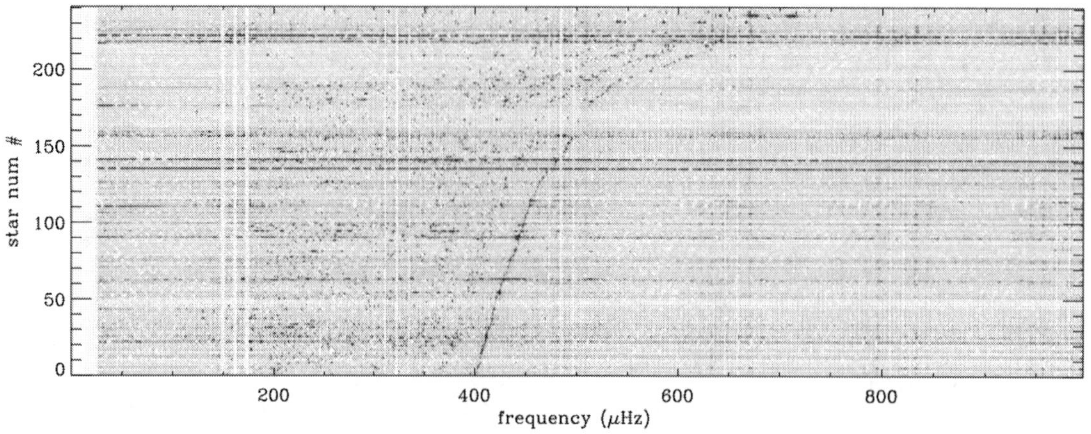

Figure 6: Each horizontal line of this image is an grey scale coded amplitude spectrum of one of the stars belonging to the set of early main sequence stars selected as described in Sect. 4.2 and illustrated in figure 4. Each spectrum is truncated in amplitude to values lower than a common limit value (here 2.10^{-4}). The spectra clearly show thin vertical dark ridges resulting from setting to zero-amplitudes parts of the spectra associated with the orbital period and its harmonics, plus day-aliases around them as explained in Sect. 2.

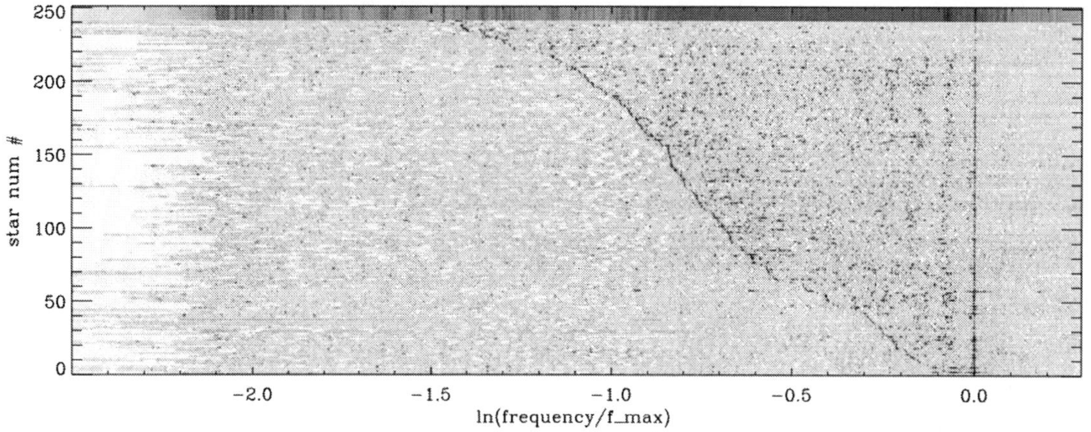

Figure 7: Same as in figure 6 but abscissae of the spectra have been normalized by f_{max} and converted in logarithm as described in the text. In addition, spectra have been ordered by increasing f_{max}/f_{min} value from the bottom to the top. The upper part of the image has been extended to host a few additional lines showing (also in grey scale) the mean of all individual spectra.

stars with a probability higher than 80% by the automated supervised classification of variable stars in the CoRoT archive.

We have characterized these spectra in terms of range of observed frequencies, defining two parameters, f_{min} and f_{max}. The distribution of our sample of stars in a f_{min}_f_{max} diagram appears to be consistent with the one obtained from a grid of theoretical models and linear stability calculations. This suggests that f_{min}_f_{max} values could be used to characterize a specific mass range or evolution stage.

Based on this criterion, we have selected stars that we consider to be on the early main sequence. We have shown that their spectra reveal a common pattern modulated by individual large separation values. The existence of regularities in the spectra of δ Scuti stars has already been demonstrated by several studies. It has even been demonstrated that this spacing is compatible with the classical large separation index used for solar-type pulsators and that rotation does not hamper the use of this index. Here the ridges we find for a large sample of stars confirm these results, but in addition, we show on a large set of objects that these regularties are due to a regular pattern of consecutive peaks which is compatible with patterns expected for axisymmetric island modes as suggested by recent non-perturbative modelling of fast rotating stars ([31]).

We still need to improve our knowledge and parametrisation of this pattern in synergy with further theoretical work on fast rotation. This study obviously would benefit from being extended to an even larger set of stars. The extension of this work to *Kepler* data is not expected to change considerably the case of the early main sequence stage since the high frequencies characterizing this domain

 DOI: 10.1051/epjconf/201716003001

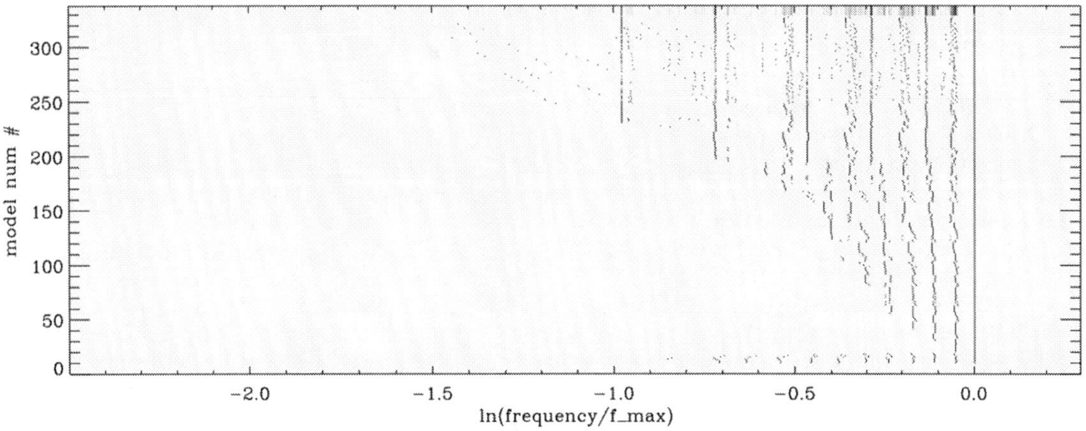

Figure 8: Same as in figure 7 (except for the 20 bottom raws), but for the set of models shown in figure 5, i.e. models satisfying the same criterion $f_{\max} > 400$ μHz than the subsample of stars considered in figure 7. The 20 bottom raws are dedicated to $\tilde{l}=0$ axisymmetric modes of models with rotation value going from zero to 80% of the breakup rotation rate, as described in Sect. 4.2.

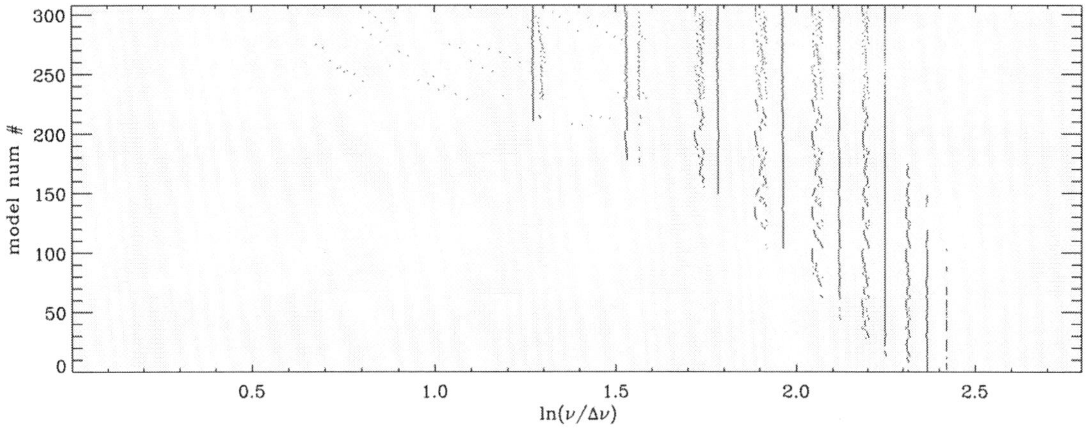

Figure 9: Same as in figure 8, but here the theoretical spectra are normalized by $\Delta\nu$ instead of $f_{\max}$.

are not accessible with the sampling time of long cadence data which constitute the essential of *kepler* data. However, in the difficult part of the $f_{\min}$_$f_{\max}$ diagram corresponding to evolved δ Scuti stars, the *Kepler* data might be of great help.

This suggests that $f_{\min}$ and $f_{\max}$ as well as large separation values might be used as seismic indices to characterize stars and this opens the perspective for ensemble seismology using δ Scuti stars.

Acknowledgements

The CoRoT space mission, launched on December 27th 2006, has been developed and is operated by the Centre National d'Etudes Spatiales (CNES), with contributions from Austria, Belgium, Brazil, the European Space Agency (RSSD and Science Programme), Germany and Spain. We acknowledge the support from the EC Project SpaceInn (FP7-SPACE-2012-312844). EM, KB, RS and DR acknowledge the support from the Programme de Physique Stellaire (PNPS). AGH acknowledges support from Fundação para a Ciência e a Tecnologia (FCT, Portugal) through the fellowship SFRH/BPD/80619/2011. JCS acknowledges funding support from Spanish public funds for research under project ESP2015-65712-C5-5-R (MINECO/FEDER), and under Research Fellowship program Ramon y Cajala (MINECO/FEDER).

References

[1] A. Baglin, M. Auvergne, P. Barge, M. Deleuil, C. Catala, E. Michel, W. Weiss, COROT Team, *Scientific Objectives for a Minisat: CoRoT*, in *The CoRoT Mission Pre-Launch Status - Stellar Seismology and Planet Finding*, edited by M. Fridlund, A. Baglin, J. Lochard, L. Conroy (2006), Vol. 1306 of *ESA Special Publication*, p. 33

[2] W.J. Borucki, D. Koch, G. Basri, N. Batalha, T. Brown, D. Caldwell, J. Caldwell, J. Christensen-

Dalsgaard, W.D. Cochran, E. DeVore et al., Science **327**, 977 (2010)
[3] M. Breger, *δ Scuti stars (Review)*, in *Delta Scuti and Related Stars*, edited by M. Breger, M. Montgomery (2000), Vol. 210 of *Astronomical Society of the Pacific Conference Series*, p. 3
[4] E. Rodríguez, M. Breger, A&A, **366**, 178 (2001)
[5] F. Royer, M. Gebran, R. Monier, S. Adelman, B. Smalley, O. Pintado, A. Reiners, G. Hill, A. Gulliver, A&A, **562**, A84 (2014), 1401.2372
[6] F. Lignières, M. Rieutord, D. Reese, A&A, **455**, 607 (2006), astro-ph/0604312
[7] D. Reese, F. Lignières, M. Rieutord, A&A, **455**, 621 (2006), astro-ph/0605503
[8] S. Barceló Forteza, E. Michel, T. Roca Cortés, R.A. García, A&A, **579**, A133 (2015), 1506.00543
[9] D.M. Bowman, D.W. Kurtz, M. Breger, S.J. Murphy, D.L. Holdsworth, MNRAS, **460**, 1970 (2016), 1605.03955
[10] A. García Hernández, A. Moya, E. Michel, J.C. Suárez, E. Poretti, S. Martín-Ruíz, P.J. Amado, R. Garrido, E. Rodríguez, M. Rainer et al., A&A, **559**, A63 (2013), 1307.2739
[11] M. Paparó, J.M. Benkő, M. Hareter, J.A. Guzik, ApJS, **224**, 41 (2016), 1603.09161
[12] M. Breger, L. Balona, P. Lenz, J.K. Hollek, D.W. Kurtz, G. Catanzaro, M. Marconi, A.A. Pamyatnykh, B. Smalley, J.C. Suárez et al., MNRAS, **414**, 1721 (2011), 1012.4373
[13] A. García Hernández, S. Martín-Ruiz, M.J.P.F.G. Monteiro, J.C. Suárez, D.R. Reese, J. Pascual-Granado, R. Garrido, ApJL, **811**, L29 (2015), 1509.01111
[14] M.A. Dupret, A. Grigahcène, R. Garrido, M. Gabriel, R. Scuflaire, A&A, **414**, L17 (2004)
[15] E. Michel, M.M. Hernández, G. Houdek, M.J. Goupil, Y. Lebreton, F. Pérez Hernández, A. Baglin, J.A. Belmonte, F. Soufi, A&A, **342**, 153 (1999)
[16] K. Zwintz, L. Fossati, T. Ryabchikova, D. Guenther, C. Aerts, T.G. Barnes, N. Themeßl, D. Lorenz, C. Cameron, R. Kuschnig et al., Science **345**, 550 (2014), 1407.4928
[17] A. Baglin, *1.1 The general framework* (2016), p. 5
[18] J. Debosscher, L.M. Sarro, M. López, M. Deleuil, C. Aerts, M. Auvergne, A. Baglin, F. Baudin, M. Chadid, S. Charpinet et al., A&A, **506**, 519 (2009)
[19] E. Rodríguez, M.J. López-González, P. López de Coca, A&AS, **144**, 469 (2000)
[20] M.A.C. Perryman, *The HIPPARCOS Catalogue Contents*, in *Hipparcos - Venice '97*, edited by R.M. Bonnet, E. Høg, P.L. Bernacca, L. Emiliani, A. Blaauw, C. Turon, J. Kovalevsky, L. Lindegren, H. Hassan, M. Bouffard et al. (1997), Vol. 402 of *ESA Special Publication*, pp. 1–4
[21] A. Udalski, A. Olech, M. Szymanski, J. Kaluzny, M. Kubiak, M. Mateo, W. Krzeminski, K.Z. Stanek, Acta Astron., **47**, 1 (1997), astro-ph/9704228
[22] C. Alcock, R.A. Allsman, D.R. Alves, T.S. Axelrod, A.C. Becker, D.P. Bennett, K.H. Cook, K.C. Freeman, M. Geha, K. Griest et al., ApJ, **536**, 798 (2000)
[23] M. Auvergne, P. Bodin, L. Boisnard, J.T. Buey, S. Chaintreuil, G. Epstein, M. Jouret, T. Lam-Trong, P. Levacher, A. Magnan et al., A&A, **506**, 411 (2009), 0901.2206
[24] R. Scuflaire, S. Théado, J. Montalbán, A. Miglio, P.O. Bourge, M. Godart, A. Thoul, A. Noels, Ap&SS, **316**, 83 (2008), 0712.3471
[25] A. Grigahcène, M.A. Dupret, M. Gabriel, R. Garrido, R. Scuflaire, A&A, **434**, 1055 (2005)
[26] L.A. Balona, W.A. Dziembowski, MNRAS, **417**, 591 (2011)
[27] D. Reese, F. Lignières, M. Rieutord, A&A, **481**, 449 (2008), 0801.4630
[28] R.M. Ouazzani, I.W. Roxburgh, M.A. Dupret, A&A, **579**, A116 (2015), 1505.01088
[29] F. Lignières, B. Georgeot, A&A, **500**, 1173 (2009), 0903.1768
[30] K.B. MacGregor, S. Jackson, A. Skumanich, T.S. Metcalfe, ApJ, **663**, 560 (2007), 0704.1275
[31] D.R. Reese, K.B. MacGregor, S. Jackson, A. Skumanich, T.S. Metcalfe, A&A, **506**, 189 (2009), 0903.4854

 DOI: 10.1051/epjconf/201716003002

The potential of space observations for pulsating pre-main sequence stars

Konstanze Zwintz[1,⋆]

[1] *Universität Innsbruck, Institute for Astro- and Particle Physics, Technikerstrasse 25, A-6020 Innsbruck*

Abstract. The first asteroseismic studies of pre-main sequence (pre-MS) pulsators have been conducted based on data from the space telescopes MOST and CoRoT with typical time bases of less than 40 days. With these data, a relation between the pulsational properties of pre-MS δ Scuti stars and their relative evolutionary phase on their way from the birthline to the zero-age main sequence was revealed. But it is evident from comparison with the more evolved pulsators in their main sequence or post-main sequence stages observed by the main Kepler mission, that many more questions could be addressed with significantly longer time bases and ultra-high precision.

Here, I will discuss the observational status of pre-MS asteroseismology and the potential of future space observations for this research field.

1 Introduction

A wealth of different types of pulsators has been studied extensively with data from the Kepler space telescope [1] and brought new insights into the physical processes acting inside these stars. As the field selected for the main Kepler mission pointed away from young (star forming) regions, we are lacking high-precision Kepler photometry for pulsating pre-main sequence (pre-MS) objects. Using space data obtained with the MOST [2] and CoRoT [3] satellites for a maximum of 39 days and the first Kepler K2 data [4] for a maximum of 80 days, we can illustrate the potential of space observations for the pre-MS stages of pulsating stars and the limitations of our current observational material.

2 Pre-main sequence evolution

Pre-MS stars do not have started nuclear fusions yet. They mainly gain their energy from gravitational contraction, and through this they get more compact and hotter on their way from the birthline to the zero-age main sequence (ZAMS). The speed of evolution in the pre-MS phase is quite rapid which is illustrated in Figure 1. According to our current models, a 1 $M_{\odot}$ star needs ~60 Myrs to reach the ZAMS. It takes only ~12 Myr for a 2 $M_{\odot}$ star, and ~4.5 Myr for a 4 $M_{\odot}$ star. At a mass of around 6 $M_{\odot}$ the birthline intersects the ZAMS, hence more massive stars do not have an optically visible pre-MS phase. Of course, here we are under several assumptions: first we assume our models are correct - but we know that we are lacking several physical ingredients and a better understanding of some quite crucial processes such as the influence of accretion from the birth environment, initial chemistry (on the surface and inside stars), convection that dominates young stars, magnetic fields or rotation. The evolutionary tracks used in Figure 1, for example, are calculated without rotation. Inclusion of rotation will shift the tracks significantly.

One way to approach answering some of these open issues, is using asteroseismology. To do so, we need to populate these regions with objects that are in their pre-MS phase and show pulsations.

3 Pulsating pre-main sequence stars

3.1 δ Scuti type

The firstly discovered and hence largest group of pre-MS pulsators are the δ Scuti type objects that span the mass range between about 1.5 and 3.5 $M_{\odot}$. The group of objects that we can use for the first detailed studies comprises of 34 objects that were observed photometrically from space by MOST [2] and CoRoT [3] and one object recently also by K2 [4]. To put these stars into a Kiel diagram, their fundamental parameters and chemical abundances were derived from dedicated high-resolution spectroscopic observations obtained with the McDonald Observatory 2.7m telescope (US), the ESO VLT (Chile), the Canada France Hawaii Telescope (CFHT; Hawaii, US) and the 1.2m Mercator telescope (La Palma, Spain). Using the pulsation frequencies from space photometry and the fundamental parameters from spectroscopy, we found an observational relation between the highest excited p-mode frequency, named f_{max}, and the evolutionary stage (see Figure 1 in [5]). Obviously such a relation could be the basis of an empirical scaling relation - similar as we know it for the solar-like oscillators.

For the classical post-MS δ Scuti pulsators this relation cannot be found. Figure 2 shows roughly 1000 δ Scuti

⋆e-mail: konstanze.zwintz@uibk.ac.at

DOI: 10.1051/epjconf/201716003002

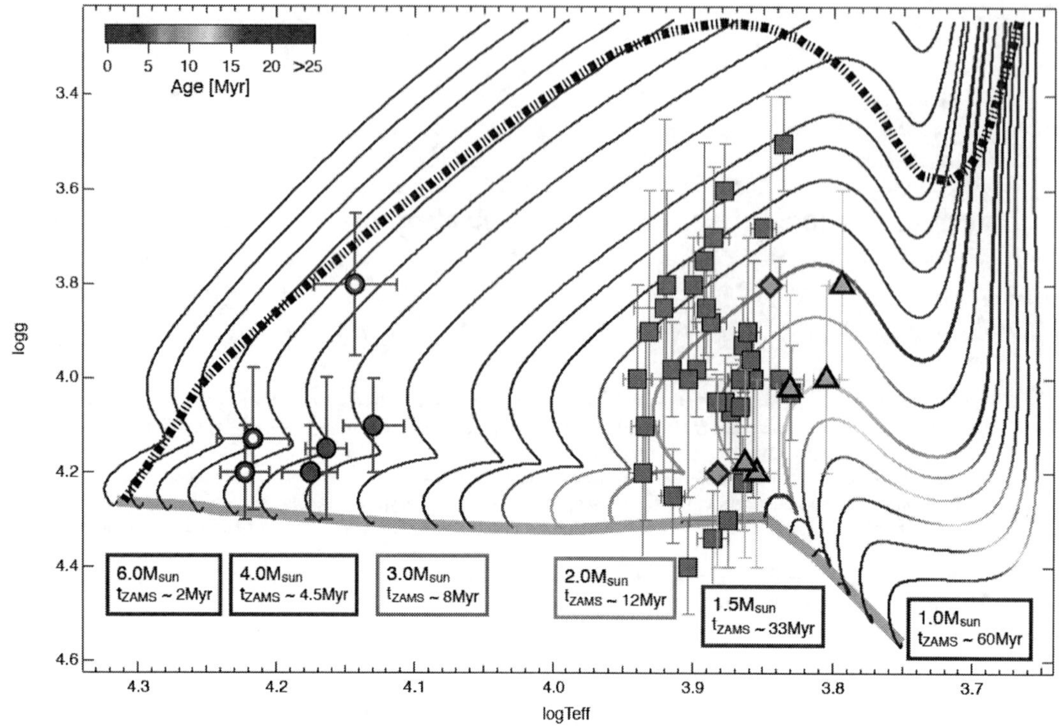

Figure 1: Kiel Diagram showing the currently known pre-MS pulsators. Evolutionary tracks from 1 to 6 $M_\odot$ are taken from Guenther et al. [7] and are color coded with age. The birthline is given as thick dash-dotted line, while the ZAMS is marked in light grey. Pre-MS δ Scuti stars are illustrated as grey squares, pre-MS γ Doradus stars as grey triangles, pre-MS δ Scuti – γ Doradus hybrid stars as grey diamonds with red borders and potential pre-MS SPB type stars as grey circles where stars in binary systems are additionally marked with a white dot.

stars observed by CoRoT in the IR and LRa01 for which spectroscopic observations were available from the Australian Astronomical Observatory [6]. A relation as described before is not seen.

The next observational step for pre-MS δ Scuti stars will be from the Kepler K2 campaign 9 data.

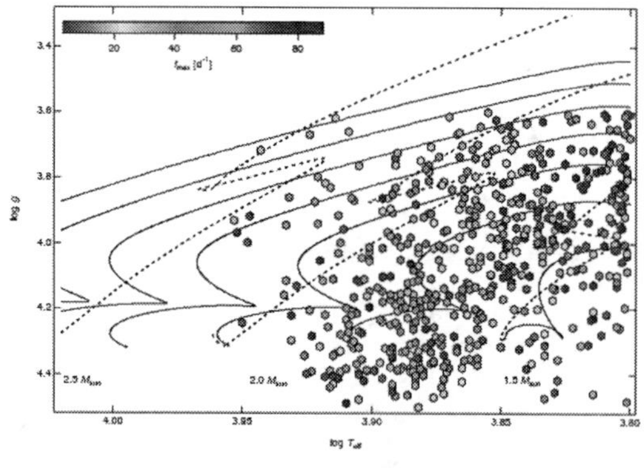

Figure 2: Kiel diagram showing ~1000 classical main sequence and post-main sequence δ Scuti stars observed by CoRoT in the IR and LRa01 runs with spectroscopic observations from the Australian Astronomical Observatory [6]. The symbols are color-coded with the highest excited p-mode frequency, $f_{\max}$. A relation like in the case for pre-MS δ Scuti stars cannot be seen.

3.2 γ Doradus and δ Scuti - γ Doradus hybrid types

Figure 1 shows that not many γ Doradus and δ Scuti - γ Doradus hybrid stars have been identified yet: Using CoRoT space photometry, Ripepi et a. [8] found one hybrid pre-MS pulsator, and Zwintz et al. [9] discovered two pre-MS γ Doradus stars. At least four more objects are currently under investigation (Zwintz et al., in preparation). This number is yet insufficient to conduct a study comparable to the one done for pre-MS δ Scuti stars [5]. Therefore, the first and immediate potential for future space missions is to discover more bona-fide members of the group of pre-MS γ Doradus stars to be able constrain their asteroseismic group properties.

Secondly, the positions of the borders of the instability region for γ Doradus pulsators in the HR diagram depend on how convection is treated in the theoretical models, and, hence, is quite uncertain [10]. With a statistically significant number of (pre-MS) γ Doradus stars whose location in the Kiel diagram is derived from high-resolution spectroscopy, we might be able to give an observational input to theory and put constraints on our theoretical concept for convection.

The ultimate goal for an observational study of pre-MS γ Doradus stars is the detection of prograde and retrograde period spacing patterns as they were discovered and analyzed in the hydrogen burning counterparts, e.g., in KIC 8375138 [11]. We would then be able to see if we find dips like for the (post-) main sequence objects, if we can identify signatures of mixing and rotation and really measure the internal differences of stars in both evolutionary stages that have similar atmospheric properties. We can only reach that by a combination of an excellent instrument in space, a suitable candidate target and a long enough time base.

BRITE-Constellation [12] is able to pick up period spacing patterns in a "classical", hydrogen core burning γ Doradus pulsator with photometric time series obtained within 156 days. This tells us that light curves of approximately that length are needed for pre-MS γ Doradus stars to clearly pick up period spacing patterns. Using 39-days light curves obtained with CoRoT on stars in NGC 2264 we detect first signs of period spacings in pre-MS γ Doradus stars (Zwintz et al., in prep.). But it is evident that we are missing several periods for a complete spacing sequence due to the insufficient time base of the observations.

3.3 Slowly Pulsating B (SPB) types

Pre-MS SPB stars are in the evolutionary transition phase from gravitational contraction to hydrogen core burning as their main energy source. Asteroseismic studies of pre-MS SPB stars would allow to measure the changes of their interior structures right before arrival on the ZAMS.

Observationally, pre-MS SPB stars are hard to find due to several reasons: (i) As their optically visible pre-MS phase is rather short lasting less than 3.6 million years for a 4 $M_{\odot}$ star and less than 1.6 Myrs for a 6 $M_{\odot}$ star, they are statistically relatively rare objects, (ii) a high percentage of B stars are in binary systems [13] which needs to be taken into account during the analysis, and (iii) rotation plays an important factor in B stars [14] which can complicate their analysis.

Currently only two candidates for pre-MS SPB pulsation have been reported [15] and four more candidates are identified in the young cluster NGC 2264 from MOST photometry (Zwintz et al. 2016, submitted). These six objects are marked as filled circles in Figure 1, where the pulsators identified as binaries from spectroscopy are identified with a white dot on top of the grey circle. For three of these objects first signs of period spacings are detected in photometric time series only ~ 22 and ~ 39 days long. Similar as in the case for pre-MS γ Doradus g-mode pulsators, the main aim is to find period spacings that lead to a better constraint on rotation and diffusive mixing in young stars as it was done already for main sequence stars (e.g., for KIC 7760680) [16, 17].

3.4 Solar like oscillators

The presence of solar like oscillations in pre-MS objects has been predicted by Samadi et al. 2005 [18]. The authors calculate the rate P at which the energy is injected into solar p-modes and determine that the maximum rate, P_{max}, scales with $(L/M)^{2.6}$ for main sequence stars, but follows a $(L/M)^{3.3}$ relation in the pre-MS phase, where L is the stellar luminosity and M is the stellar mass. Similar as in main sequence stars, also for solar-like pre-MS objects the p mode excitation occurs closer to the surface of the stars which is characterized by the effective temperature, T_{eff}, and gravity, $\log g$. These parameters then relate as T_{eff}^4/g to L/M. The authors describe that the difference in their computations lies in the usage of 3D simulations for the main sequence stars and mixing length theory (MLT; [19]) for the pre-MS phase, but they also note "However, the physical reasons [for the different relations] are not yet identified." [18].

Samadi et al. [18] also predict the maximum velocity amplitudes, v_{max} to lie between ~ 20 cms^{-1} and 120 cms^{-1} and the luminosity amplitudes, $(dL/L)_{max}$ between ~ 2 and 15 parts-per-million (ppm). This is yet again another observational challenge that might be tackled with future space missions like TESS [20] and Plato [21] because pre-MS solar-like oscillators are located in the same region of the Hertzsprung-Russel diagram (HRD) as the T Tauri stars and they also share the early evolutionary stage. T Tauri stars are characterized by a high degree of activity and show regular and irregular light variations of a magnitude and more (see, e.g., [22, 23]). Searching for pulsation amplitudes of a few ppm in stars with such a high level of activity requires the longest possible photometric time series obtainable from space. A first search for solar-like variability in the CoRoT NGC 2264 observations using a time base of 39 days did not reveal any signature of stochastic oscillations.

4 Conclusions

The main potential of future space photometry for pre-MS pulsators lies in a crucial increase of the photometric time bases from typically 40 days (with up to 80 days maximum from K2) to at least 150 days and more. With shorter light curves we will be able to identify new pre-MS δ Scuti, γ Doradus and SPB stars, but only with sufficiently long time series we will be able to detect long sequences of period spacing patterns for the g-mode pulsators. The discovery of solar-like oscillations in pre-MS stars – as they are predicted by theory – will also only be possible with long time base measurements which decrease the noise level in the Fourier domain sufficiently enough to detect signal at the ppm level.

Using the available data from MOST and CoRoT for pre-MS δ Scuti stars, we have first independent proof from asteroseismology that stars are formed sequentially in clusters [5], that giving one age for all cluster members is not sufficient and will always be affected by large errors in particular for the youngest objects. With new ultra-precise observations from space and time bases of

150 days and more, we would be able to seismically calibrate our pre-MS stellar structure models, and, hence, contribute to a better understanding of early stellar evolution.

The formation of stars is directly connected to the formation of planets. In their recent paper, David et al. [24] use K2 data from Upper Sco to study a Neptune-sized transiting planet closely orbiting a 5-10 million year old star. It is not unlikely that in our future space observations we will find a (young) pulsating star that hosts a planetary system. We will then be able to use asteroseismology to put constraints on the exoplanet host star. The first candidate for this analysis is the young ZAMS star β Pictoris which shows δ Scuti pulsations [25] and whose planet will cross the star's Hill sphere in 2017.

These examples illustrate that pre-MS asteroseismology also has a significant impact on other fields of stellar astrophysics, and that future space missions should put their focus also on this research area.

Acknowledgements

K.Z. acknowledges support by the Austrian Fonds zur Förderung der wissenschaftlichen Forschung (FWF, project V431-NBL).

References

[1] Koch, D. G., Borucki, W. J., Basri, G., et al., ApJ **713** L79 (2010)
[2] Walker, G., Matthews, J. M., Kuschnig, R., et al., PASP **115**, 1023 (2003)
[3] Baglin, A., *The CoRoT mission, pre-launch status, stellar seismology and planet finding*, (ESA SP-1306, ESA Publication Division, Noordwijk, The Netherlands, 2006)
[4] Howell, S. B., Sobeck, C., Haas, M., et al., PASP **126** 398 (2014)
[5] Zwintz, K., Fossati, L., Ryabchikova, T. A., et al., Science **345** 550 (2014)
[6] Kaiser, A., *Investigation of pulsation along the red edge of the classical instability strip using space photometry and ground based spectroscopy* (PhD thesis, University of Vienna, Austria, 2014)
[7] Guenther, D. B., Kallinger, T., Zwintz, K., et al., ApJ **704** 1710 (2009)
[8] Ripepi, V., Cusano, F., di Criscienzo, M., et al., MNRAS **416** 1535 (2011)
[9] Zwintz, K., Fossati, L., Ryabchikova, T., et al., A&A **550** 121 (2013)
[10] Bouabid, M., Montalban, J., Miglio, A., et al., A&A **531** 145 (2011)
[11] Van Reeth, T., Tkachenko, A., Aerts, C., et al., ApJs **218** 27 (2015)
[12] Weiss, W. W., Rucinski, S. M., Moffat, A. F. J., et al., PASP **126** 573 (2014)
[13] Sana, H., de Mink, S. E., de Koter, A., et al., Science **337** 444 (2012)
[14] Meynet, G., and Maeder, A., A&A **361** 101 (2000)
[15] Gruber, D., Saio, H., Kuschnig, R., et al., MNRAS **420** 291 (2012)
[16] Papics, P. I., Moravveji, E., Aerts, C., Tkachenko, A., Triana, S. A., Bloemen, S., Southworth, J., A&A **570**, A8 (2014)
[17] Moravveji, E., Townsend, R. H. D., Aerts, C., Mathis, S., A&A **823**, 130 (2016)
[18] Samadi, R., Goupil, M.-J., Alecian, E., et al., JApA **26** 171 (2005)
[19] Böhm-Vitense, E., Zeitschrift für Astrophysik **46** 108 (1958)
[20] Ricker, G. R., Winn, J. N., Vanderspek, R., et al., SPIE **9143** 20 (2014)
[21] Rauer, H., Catala, C., Aerts, C., et al., ExA **38** 249 (2014)
[22] Alencar, S. H. P., Teixeira, P. S., Guimaraes, M. M., et al., A&A **519** 88 (2010)
[23] Cody, A. M., Stauffer, J., Baglin, A., et al., AJ **147** 82 (2014)
[24] David, T. J., Hillenbrand, L. A., Petigura, E. A., et al., Nature **534** 658 (2016)
[25] Koen, C., Balona, L. A., Khadaroo, K., et al., MNRAS **344** 1250 (2003)

Accurate $\bar{\rho}$ and $\log g$ of δ Sct stars using Asteroseismology

Antonio García Hernández[1,*], *Mário J.P.F.G.* Monteiro[1,2], *Zhao* Guo[3], *Daniel R.* Reese[4], *Juan Carlos* Suárez[5,6], *Susana* Martín[6], *Javier* Pascual-Granado[6], *Andrés* Moya[7], and *Rafael* Garrido[6]

[1] *Instituto de Astrofísica e Ciências do Espaço, Universidade do Porto, CAUP, Rua das Estrelas, PT4150-762 Porto, Portugal*
[2] *Departamento de Física e Astronomia, Faculdade de Ciências da Universidade do Porto, Rua do Campo Alegre, 4169-007 Porto, Portugal*
[3] *Center for High Angular Resolution Astronomy and Department of Physics and Astronomy, Georgia State University, P. O. Box 5060, Atlanta, GA 30302-5060, USA*
[4] *LESIA, Observatoire de Paris, PSL Research University, CNRS, Sorbonne Universités, UPMC Univ. Paris 06, Univ. Paris Diderot, Sorbonne Paris Cité, 92195 Meudon, France*
[5] *Universidad de Granada. Dept. Física Teórica y de Cosmos. Fuentenueva Campus. 18007. Granada, Spain*
[6] *Instituto de Astrofísica de Andalucía (CSIC), Glorieta de la Astronomía S/N, 18008, Granada, Spain*
[7] *Universidad Internacional de Valencia - VIU. C/ Gorgos 5-7. Valencia. Spain*

Abstract. In this work, we present a new method to determine the surface gravity of δ Sct stars. We used a refined $\Delta\nu$-$\bar{\rho}$ relation and the stellar parallaxes or luminosities to determine their masses and radii. A comparison with the data obtained from the binary analysis, has shown that the values found by both methods are equivalent, within the uncertainties. Moreover, thanks to the refined relation, the uncertainties in $\log g$ are of the order of those usually estimated with high-resolution spectroscopy. Because of that, this new method to determine the surface gravity is an important step forward to break the degeneracy problem in the spectroscopic analysis.

1 Introduction

The pulsating spectrum of δ Sct stars is one of the most problematic to interpret, despite the large number of observed frequencies [7, 26]. This is because it is complicated to carry out a mode identification or find obvious patterns in their spectra, and because of the difficulties of the theoretical analysis, due to the rapid rotation of most of them.

Nevertheless, recent studies have found patterns in the periodograms of these pulsators [e.g., 2, 8, 11, 23, 33], pointing that they could be related to a large separation in the low radial order regime. In fact, a recent work has found a relation between this periodicity, pattern or large separation and the mean density of the star [9, from now GH15]. All these results are agree with theoretical predictions, both based on non-rotating models [29] and on 2D models using a non-perturbative approach [22, 27].

In the work we present here, we describe in detail the methodology to find such separation in the periodogram and we refine the relation found in GH15. The final aim is to use this relation to determine another stellar property, namely the surface gravity. Making use of the parallax or the luminosity, we estimated the stellar masses using the mass-luminosity relation. Based on the $\Delta\nu$-$\bar{\rho}$ relation, we found the radii and, thus, the surface gravity.

This article is divided into the following sections. In Sec. 2, we present the sample and its more relevant characteristics. Section 3 describes the properties of the large separation-mean density relation, the methodology carried out to determine $\Delta\nu$ and to fit the data. In Sec. 4, we described the procedure we followed to determine $\log g$. And Sec. 5 details the conclusions of this work.

*e-mail: agh@astro.up.pt

2 The sample

We used a sample of eclipsing binary systems with a δ Sct component. The study of these systems allows us to accurately determine the mass and radius, based only on the Kepler's laws and the transit analysis. In addition, the presence of pulsations in one of the components of the system allows us to check the pulsation theory.

We search in the bibliography for binary systems with such characteristics. We also imposed that they have precise space photometry. In this way, we found 7 systems fulfilling all the constraints. All these systems have precise fundamental parameters (masses and radii) and have been observe by MOST [31], CoRoT [1] or *Kepler* [15]. The sample is composed of 6 eclipsing binary systems and 1 binary system observed with optical interferometry, HD 159561. The δ Sct component of this system is a very fast rotator with $\Omega/\Omega_K = 0.88$.

With all this information, we computed the mean density of the stars. We have also used the frequencies derived from the light curve analysis to obtain the large separation. Table 1 shows the most relevant values used in the present

DOI: 10.1051/epjconf/201716003003

work. Additionally, it shows the bibliographic references from where we extracted those data.

3 The $\Delta\nu$-$\bar{\rho}$ relation

One of the most relevant results of the study in GH15 was the establishment of a relation between the pattern found in the frequency spectrum of a sample of δ Sct stars and their mean density. The method to obtain such a pattern is described in Sec. 3.1. The mean density was obtained from the values provided by the binary analysis. They even took into account the possible deviations from sphericity due to the rotation of the stars to compute their volumes. To that end, they used Roche's model [e.g., 19]. Following the same procedure and using the same sample of stars, we refined the relation between the mean stellar density and large frequency separation by implementing a Hierarchical Bayesian linear regression (see details in Sec. 3.2). The updated relation is the following:

$$\bar{\rho}/\bar{\rho}_{\odot} = 1.62(\Delta\nu/\Delta\nu_{\odot})^{2.085}. \quad (1)$$

A plot of the data, our fit and a comparison with the solar-like relation and the theoretical relation found by [29] can be seen in the Fig. 1. The simple dependency of the pattern with the mean density and the detection of a trend in the Échelle diagrams is the sign of a large separation (understood as a frequency spacing). However, this large separation is found in the non-asymptotic regime of low-radial order p-modes.

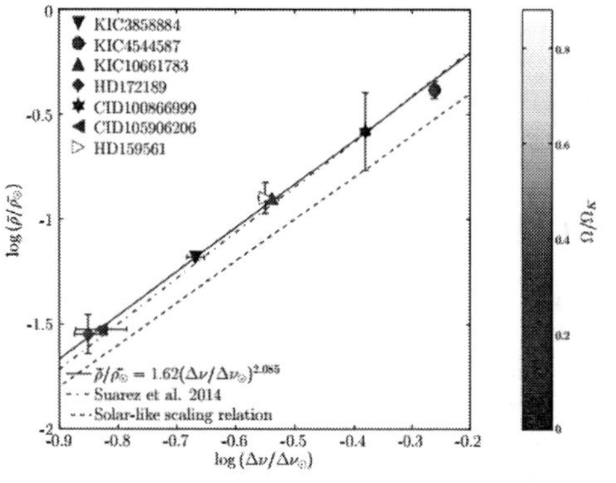

Figure 1: Large separation-mean density relation obtained for the seven binary systems of our sample. A linear fit to the points is also depicted, as well as the solar-like scaling relation [30] and the theoretical scaling relation for non-rotating models of δ Sct stars [29]. Symbols are plotted with a gradient colour scale to account for the different rotation rates.

Equation 1 fits those relations found in previous theoretical works, both using non-rotating models [29] and 2D models using a non-perturbative approach to compute the theoretical frequencies [27]. This relation is essentially the same obtained as that obtained in GH15, although the coefficients are slightly different.

The relation is parallel to that found for solar-type stars, once it is extrapolated to sub-solar densities. The observed shift between both relations is due to the different regimes where they are found. For the solar-like pulsators, this is the asymptotic regime, whereas for $\Delta\nu$ is around the fundamental radial mode.

But the most remarkable consequence, already suggested by previous non-perturbative studies [27], is the invariance of the relation with the rotation of the star. In our sample, we included a very fast rotating star, HD 159561. But, as can be seen in Fig. 1, this star also closely fits the relation. This invariance allows us to derive the mean density of a star independently of its rotation rate, making the large separation a powerful observable for δ Sct stars.

3.1 Obtaining $\Delta\nu$

Although the methodology is already described in GH15, it is convenient to provide some details to guide the reader. In order to obtain the pattern (large separation) from the frequency spectra of the stars in the sample, we followed the prescriptions given by [7, 8] and computed Fourier transforms. Additionally, we made a combined use of histograms of frequencies differences and Échelle diagrams, with the aim of searching for the most appropriated periodicity. An example of these three plots, can be seen in the Fig. 2, 3 and 4. The selected value of the large separation is marked with a dashed-line in the Fourier transform and in the histogram of frequencies differences. This value has then been used to build the Échelle diagram.

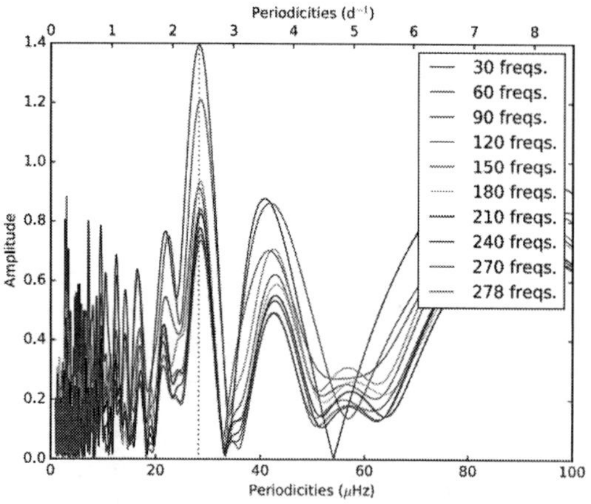

Figure 2: Fourier transform of different subsets of the highest-amplitude frequencies of KIC 3858884. The selected periodicity is marked with a dashed line (see text for details).

Essentially, the method lies in computing the Fourier transform of the frequencies, taking all the amplitudes equal to one (in whatever units). First, it is important to avoid harmonics or combinations between frequencies, frequencies driven by tidal effects, harmonics of the orbital period, and all frequencies above $\sim 50\ \mu$Hz, to avoid possible g modes.

Table 1: Characteristics of the systems taken from the literature (see the reference citation in the first column). The information corresponds to the pulsating component (which is not necessarily the primary).

System	$\Delta\nu$ (μHz)	P_{orb} (d) / ν_{orb} (μHz)	$\bar{\rho}$ ($\rho_\odot$)	v $\cdot$sin i (km s^{-1})	i (o)	ν_{rot} (μHz)	Ω/Ω_K
KIC 3858884 [20]	29 ± 1	25.953 / 0.44598	0.0657 ± 0.0021	25.7 ± 1.5	88.176 ± 0.002	1.93	0.0754
KIC 4544587 [10]	74 ± 1	2.1890951 / 5.289352	0.414 ± 0.039	75.8 ± 15	87.9 ± 3	10.99	0.172
KIC 10661783 [17]	39 ± 1	1.2313622 / 9.3981481	0.1255 ± 0.0039	78 ± 3	82.39 ± 0.23	6.99	0.200
HD 172189 [5]	19 ± 1	5.70198 / 2.03009	0.0283 ± 0.0061	78 ± 3	73.2 ± 0.6	4.63	0.281
CID 100866999 [4]	56 ± 1	2.80889 / 4.12037	0.262 ± 0.112	–	80 ± 2	–	–
CID 105906206 [28]	20 ± 2	3.6945708 / 3.1365741	0.02986 ± 0.00095	47.8 ± 0.5	81.42 ± 0.13	2.61	0.152
HD 159561 [13, 21]	38 ± 1	3148.4 / 0.0036762	0.123 ± 0.021	239 ± 12	87.5 ± 0.6	19.16	0.88

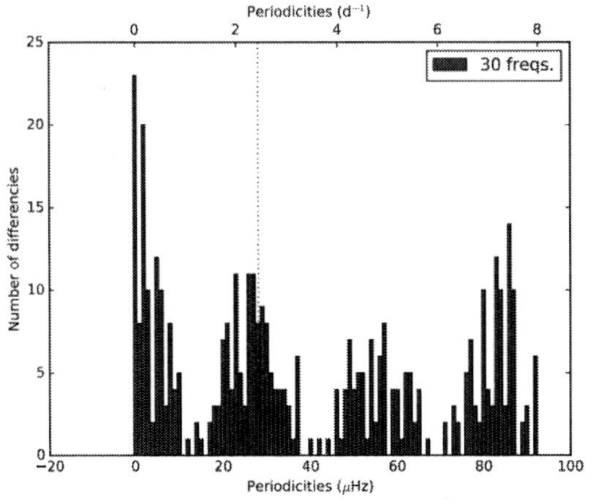

Figure 3: Histogram of frequencies differences for the 30 highest-amplitude frequencies of KIC 3858884. The selected periodicity is marked with a dashed line (see text for details).

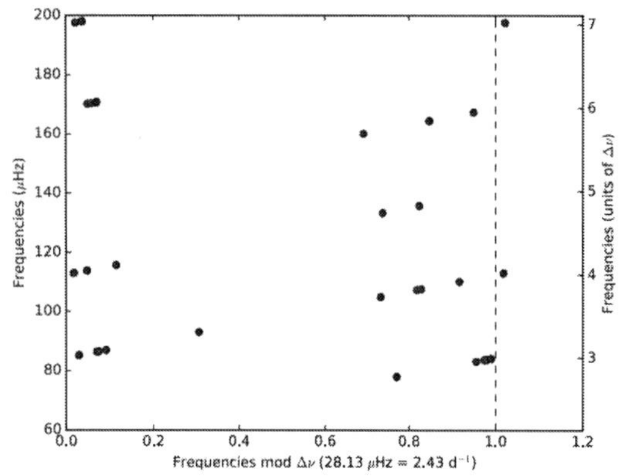

Figure 4: Échelle diagram for the 30 highest-amplitude frequencies of KIC 3858884. The large frequency separation used is that derived from the Fourier transform and the histogram of frequencies differences. Two clear ridges are visible around 0.8.

Second, the Fourier transform is computed with a certain number of the highest-amplitude frequencies. More frequencies are then added to the set and the Fourier transform is computed again for each subset of frequencies. As a first guess, we took the most prominent peak as the value of the periodicity. The sub-multiples of this peaks might be also visible.

The next step is to determine if the periodicity and its multiples appear in the histogram of frequencies differences. The final verdict is given by the Échelle diagram. The peak in the Fourier transform does not always coincide with the one from the histogram, but a pattern must be detected in the Échelle diagram.

Notice that the periodicities determined in this way are far from the orbital period of the system and from the rotational frequency of the star (see Table 1). The rotational velocity of the star has been calculated using the projected rotational velocity and the inclination angle of the orbital plane, assuming that equatorial and orbital planes are the same. This is a crude estimate of the rotational velocities and are just indicative, so we did not calculate their uncertainties.

One must also have in mind that sometimes the periodicity does not correspond to the value of the large separation. In some cases, the periodicity is a sub-multiple of the large separation, such a $\Delta\nu/2$ (as in the case of HD 172189 and HD 159561) or $\Delta\nu/4$ (as in the case of KIC 4544587). The $\Delta\nu/2$ possibility is expected from the theory [18, 27] and correspond to the cases in which ℓ =1 modes place between $\ell = 0$ modes at one half of the $\Delta\nu$ distance. The other possibility appears in the Fourier transform when a sub-multiple of $\Delta\nu$ is approximately equal to the rotational splitting or a multiple. To illustrate this, we can look at the case of KIC 4544587, for which $\Delta\nu/4 \sim 2\nu_{rot}$.

Recently, [24] determined the periodicity of a sample of 90 δ Sct stars observed by CoRoT. They estimated the masses, the radius and the rotational velocities of the stars and tried to place the sample over the $\Delta\nu$-$\bar{\rho}$ relation given in [29]. The conclusion was that it was necessary to assume that the periodicity was a combination of the large separation and the rotational splitting in most cases in order to put the data over the relation. One should point out that the method they used to determine the periodicity did not use the Fourier transform nor the histograms of frequencies differences.

3.2 Uncertainties of the coefficients

Despite the clear linear relation between the stellar density and the large separation observed in GH15, the uncertainties of coefficients in the relation are likely over-estimated in their weighted-least-square fitting procedure. We used a hierarchical Bayesian model to fit the relation, which fully takes account of the uncertainties in $\bar{\rho}$ and $\Delta\nu$. The calculation is implemented in the JAGS package [Just Another Gibbs Sampler, 25], and a similar application to fit the mass-radius relation of exoplanets can be found in [32]. We got the following result (C being the coefficient and λ being the exponent):

$$C = 1.62^{+0.13}_{-0.14}$$

$$\lambda = 2.085^{+0.052}_{-0.051}$$

The uncertainties obtained in this fit are one order of magnitude lower for the factor and half of the value for the exponent compared to GH15. The value of the coefficients have varied but they still are within the error-bars of both analysis.

4 Precise measurement of $\log g$ using Asteroseismology

To demonstrate the potential of the large separation as an observable, we tried to obtain the surface gravity of the star. Since most of the δ Sct stars are rapid rotators, we cannot determine their masses by just looking at the modelling. In addition, we wanted our calculations to be independent of any modelling. Since the Gaia mission [3] has just released its first series of data, we decided to use the stellar parallaxes or the luminosities (in case the first quantity is not available), to determine the masses. We thus used the mass-luminosity relation. With this value and the mean density provided by Eq. 1, we can derive the stellar radius, assuming spherical symmetry. We calculated the surface gravity of the stars in the sample and compared this value with the result of the binary analysis. A detailed discussion of the methodology and the results obtained are provided in the following sections.

4.1 Estimating masses

We have revised the bibliography again searching for independent measurements of the parallax, luminosity or absolute magnitude of each star. We just found information about the parallax of KIC 3858884, p = 2.49 ± 1.06 mas, which has been measured by the HIPPARCOS satellite [16]. We used this value, the apparent magnitude, m_v = 9.4297 mag, and the bolometric correction from [12], BC= 0.025. For the other cases, we used the true luminosity for four stars and the absolute magnitudes for other two, as provided by the main references. Table 2 shows a summary of the used parameters, as well as the bibliographic references where we extracted the values.

Once we have the stellar luminosities, we used a mass-luminosity relation to determine the masses. We thus selected the relation found by [14]. In that work, they used binary systems to derive the relation, expressed as: $L \propto M^{3.20\pm0.26}$ (in solar units).

The strongest assumption we made in this case is that this relation can be applied to stars off the main sequence (although not far) and that it is unaffected by the metallicity. Nevertheless, the mass estimates are not so critical for the calculation of $\log g$, as can be seen in Table 2.

4.2 $\log g$ from binarity vs $\log g$ from $\Delta\nu$ and L

Once we obtained the masses and densities from Eq. 1, it is straightforward to compute the radii. We assumed spherical symmetry, as usually assumed in binary analysis. Thus, we calculated the surface gravity, as shown in Table 2.

A few conclusions can be derived immediately. It is not possible to reach the precision level of the binary analysis, although those values are within the error-bars of the values derived in our work. Moreover, the uncertainties in the $\log g$ determined from $\Delta\nu$ and the stellar luminosities are of the order of those obtained from high-resolution spectroscopy.

At this point, one may see that both quantities correlate. We plotted the values to determine if there a equivalence between both. Fig. 5 shows the result, as well as a linear fit to the data and the relation "y = x". As expected, the equivalence is not perfect but pretty close. Indeed, as in Fig. 1, the dispersion of the points is significantly lower than their uncertainties for the case of the values computed with $\Delta\nu$ (x-axis). The key for obtaining such error-bars is the low uncertainty of the coefficients of the $\Delta\nu$-$\bar{\rho}$ relation, and, above all, of the exponent.

The only value of $\log g$ not obtained from the binary analysis is that of HD 159561. For this fast rotator, we decided to use the spectroscopic determination, since the most interesting question to solve is how spectroscopy deals with non-spherical stars. It is remarkable that the star still follows closely the identity relation.

Therefore, this methodology can help to obtain a precise surface gravity of δ Sct stars, breaking the degeneracy of this parameters in many spectroscopic analysis. It is remarkable the accuracy in the determination of $\log g$, even with the crude estimate of stellar masses and regardless of the rotation rate of the star.

5 Conclusions

In this work, we presented a methodology to derive the surface gravity of δ Sct stars using just the pulsation frequencies and an estimate of the luminosity. We demonstrated that it is possible to get $\log g$ with the same precision as in high-resolution spectroscopic analysis. The key point is the uncertainties of the $\Delta\nu$-$\bar{\rho}$ relation.

To achieve these results, we used the eclipsing binary systems from GH15, all of them with a δ Sct component. Thus, we determined the corresponding periodicity from the pulsation modes, using a Fourier transform of the frequencies, a histogram of frequencies differences and Échelle diagrams. We were able to correlate the periodicity and the mean densities of the stars in the sample, minimizing the uncertainties in the coefficients of the relation. This correlation proved that this periodicity is, indeed, a large separation in the low radial-order regime of

DOI: 10.1051/epjconf/201716003003

Table 2: Characteristics of the systems taken from the literature (named as *bin*) compared to those computed in this work (named as $\Delta\nu$). All the references are the same as in Table 1, except for HD 159561.

System	M_{bin} ($M_\odot$)	R_{bin} ($R_\odot$)	$\log g_{bin}$ (cgs)	For M-L relation	$M_{\Delta\nu}$ ($M_\odot$)	$\log g_{\Delta\nu}$ (cgs)
KIC 3858884	1.86 ± 0.04	0.0657 ± 0.0021	3.74 ± 0.01	$p = 2.49 \pm 1.06$ mas	1.82 ± 0.04	3.74 ± 0.18
KIC 4544587	1.61 ± 0.06	0.414 ± 0.039	4.33 ± 0.01	$M_b = 1.72 \pm 0.04$ mag	2.03 ± 0.03	4.32 ± 0.21
KIC 10661783	2.100 ± 0.028	0.1255 ± 0.0039	3.938 ± 0.004	$L= 21.62 \pm 1.03\ L_\odot$	2.19 ± 0.03	3.94 ± 0.19
HD 172189	1.78 ± 0.24	0.0283 ± 0.0061	3.48 ± 0.08	$L= 52.42 \pm 2.85\ L_\odot$	2.75 ± 0.04	3.54 ± 0.17
CID 100866999	1.8 ± 0.2	0.262 ± 0.112	4.1 ± 0.1	$M_b = 2.3 \pm 0.2$ mag	1.77 ± 0.05	4.13 ± 0.20
CID 105906206	2.25 ± 0.04	0.02986 ± 0.00095	3.53 ± 0.01	$L= 1.53 \pm 1.00\ L_\odot$	2.46 ± 0.02	3.56 ± 0.20
HD 159561 [6]	$2.40^{+0.23}_{-0.37}$	$R_{eq} = 2.858 \pm 0.015$ $R_{pol} = 2.388 \pm 0.013$	3.89 ± 0.20*	$L= 31.30 \pm 0.96\ L_\odot$	2.407 ± 0.019	3.94 ± 0.19

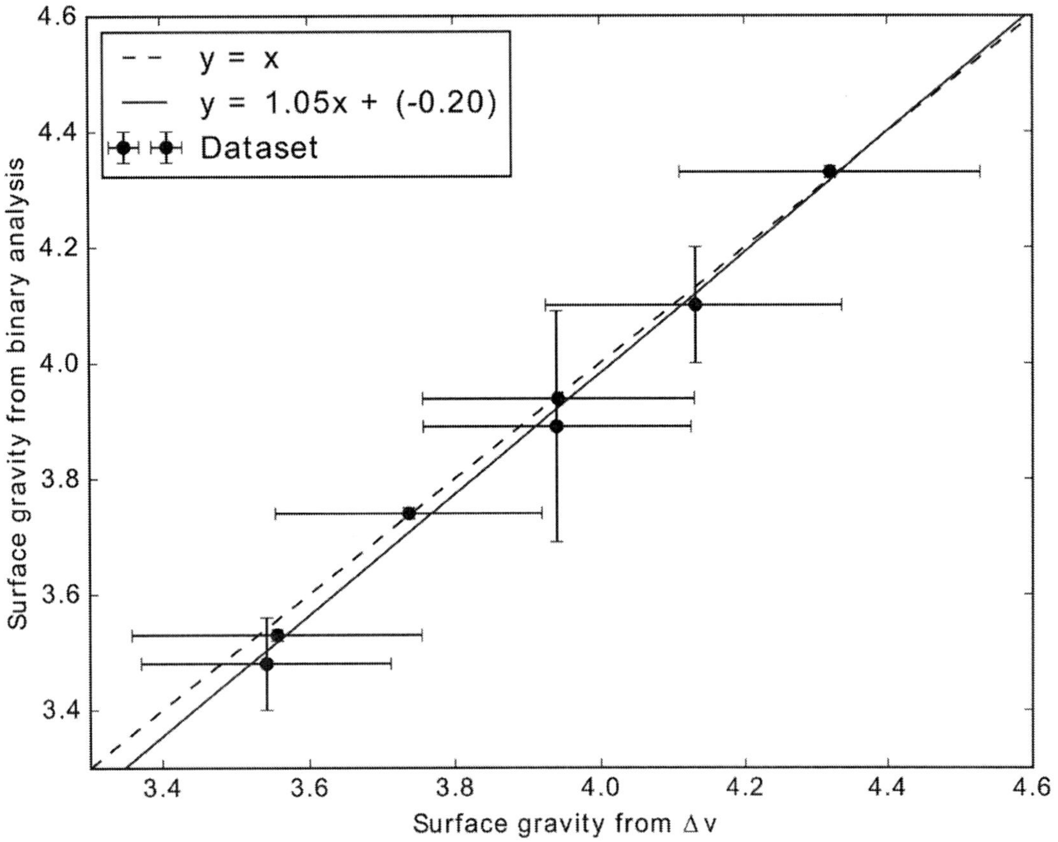

Figure 5: Surface gravity obtained from the analysis carried out in this work versus $\log g$ from the binary analysis (see text for details). A linear fit is also plotted as is the $y = x$ line for reference.

the δ Sct pulsators. The most remarkable characteristic of this relation is its independence with respect to the stellar rotation rate. That makes this large separation a promising observable for the study of δ Sct stars.

Pushing the analysis a step further, we used this relation to determine another physical quantity of the star, namely the surface gravity. We estimated stellar masses through their luminosities. Comparing our result with that obtained from the binary analysis, we found an almost identity relation between both quantities. Moreover, it is possible to determine $\log g$ with the same precision as in high-resolution spectroscopic analysis, with an uncertainty of around 0.2 dex.

The large separation, $\Delta\nu$, in δ Sct stars has the potential to break the degeneracy in $\log g$ determination from spectroscopic analysis. This will be possible thanks to missions such as Gaia, which will provide accurate measurements of the stellar parallaxes of the whole sky. This will become an important breakthrough in the study of δ Sct pulsations.

Acknowledgements

AGH acknowledges support from Fundação para a Ciência e a Tecnologia (FCT, Portugal) through the fellowship SFRH/BPD/80619/2011 and from grant UID/FIS/04434/2013. AGH, MJPFGM and JCS acknowledge support from the EC Project SPACEINN (FP7-SPACE-2012-312844). JCS also acknowledges funding support from the Spanish "Ministerio de Economía y Competitividad" under "Ramón y Cajal" subprogram. This research made use of the SIMBAD database and the VizieR catalogue access tool operated at CDS, Strasbourg, France, and the SAO/NASA Astrophysics Data System.

References

[1] Baglin, A., in The CoRoT Mission: Pre-Launch Status, ed. M. Fridlund, A. Baglin, J. Lochard, & L. Conroy, ESA Publications Division, Noordwijk, ESA SP, 1306 (2006)

[2] Breger, M., Pamyatnykh, A. A., Pikall, H., et al., A&A,, 341, 151 (1999)

[3] de Bruijne, J. H. J. 2012, ApJS,, 341, 31-41

[4] Chapellier, E., & Mathias, P., A&A,, 556, A87 (2013)

[5] Creevey, O. L., Uytterhoeven, K., Martín-Ruiz, S., et al., A&A,, 507, 901-910 (2009)

[6] Erspamer, D., & North, P., A&A,, 398, 1121 (2003)

[7] García Hernández, A., Moya, A., Michel, E., et al., A&A,, 506, 79G (2009)

[8] García Hernández, A., Moya, A., Michel, E., et al., A&A,, 559, A63 (2013)

[9] García Hernández, A., Martín-Ruiz, S., Monteiro, M. J. P. F. G. et al., ApJL,, 811, L29 (2015)

[10] Hambleton, K. M., Kurtz, D. W., Prša, A., et al., MNRAS,, 434, 925-940 (2013)

[11] Handler, G., Pikall, H., O'Donoghue, D., et al., MNRAS,, 286, 303 (1997)

[12] Hayes, D. S., IAU Symposium, 80, 65 (1978)

[13] Hinkley, S., Monnier, J. D., Oppenheimer, et al., ApJ,, 726, 104 (2011)

[14] Ibanoğlu, C., Soydugan, F., Soydugan, E., et al., MNRAS,, 373, 435-448 (2006)

[15] Koch, D. G., Borucki, W. J., Basri, G., et al., ApJ,, 713, L79 (2010)

[16] van Leeuwen, F., et al., å, 323, L61 (1997)

[17] Lehmann, H., Southworth, J., Tkachenko, A., et al., A&A,, 557, A79 (2013)

[18] Lignières, F., Georgeot, B., & Ballot, G., A&A,, 500, 1173 (2010)

[19] Maeder, A., Formation and Evolution of Rotating Stars (A&A, Library; Springer Berlin Heidelberg, 2009)

[20] Maceroni, C., Lehmann, H., da Silva, R., et al., A&A,, 563, A59 (2014)

[21] Monnier, J. D., Townsend, R. H. D., Che, X. et al., ApJ,, 725, 1192 (2010)

[22] Ouazzani, R. M., Roxburgh, I. W., & Dupret, M. A., A&A,, 579, A116 (2015)

[23] Paparó, M., Chadid, M., Chapellier, E. et al., A&A,, 531, A135 (2011)

[24] Paparó, M., Benkö, J. M., Hareter, M. et al., ApJS,, 224, 41 (2016)

[25] Plummer, M., Proceedings of the 3rd International Workshop on Distributed Statistical Computing (2003)

[26] Poretti, E., Michel, E., Garrido, R., et al., A&A,, 506, 85 (2009)

[27] Reese, D. R., Lignières & F., Rieutord, M., A&A,, 481, 449-452 (2008)

[28] da Silva, R., Maceroni, C., Gandolfi, D., et al., A&A,, 565, A55 (2014)

[29] Suárez, J. C., García Hernández, A., Moya, A., et al., A&A,, 563, A7 (2014)

[30] Tassoul, M., ApJS,, 43, 469-490 (1980)

[31] Walker, G., Matthews, J., Kuschnig, R., et al. 2003, PASP,, 115, 1023-1035

[32] Wolfgang, A., Rogers, L. A. & Ford, E. B, IAU General Assembly, 22, id.2258521 (2015)

[33] Zwintz, K., Lenz, P., Breger, M. et al., å, 533, A133 (2011)

 DOI: 10.1051/epjconf/201716003004

The first K2 roAp star: HD 24355 pulsating in a distorted quadruploe mode

Daniel L. Holdsworth[1,*], *Donald W.* Kurtz[1], *Barry* Smalley[2], *Hideyuki* Saio[3], *Gerald* Handler[4], *Simon J.* Murphy[5], and *Holger* Lehmann[6]

[1]*Jeremiah Horrocks Institute, University of Central Lancashire, Preston, PR1 2HE, UK*
[2]*Astrophysics Department, Keele University, Staffordshire, ST5 5BG, UK*
[3]*Astronomical Institute, School of Science, Tohoku University, Sendai 980-8578, Japan*
[4]*Copernicus Astronomical Center, Bartycka 18, PL-00-716 Warsaw, Poland*
[5]*Sydney Institute for Astronomy (SIfA), School of Physics, University of Sydney, NSW 2006, Australia*
[6]*Thüringer Landessternwarte Tautenburg (TLS), Sternwarte 5, D-07778 Tautenburg, Germany*

Abstract. We present an analysis of the first K2 observations of a rapidly oscillating Ap (roAp) star, HD 24355. The star was discovered to be a roAp star by [1], with a frequency of 224.31 d^{-1} (2596.18 μHz; P = 6.4 min) and an amplitude of 1.51 mmag in SuperWASP broadband photometry. Spectroscopic analysis of low-resolution spectra show the star to be an A5 Vp SrEu star. The high precision K2 data allow us to identify 13 rotationally split sidelobes to the main pulsation frequency. This is the largest number of sidelobes seen in a roAp star to date. We also see an unusual pulsational phase variation as the star rotates, showing this star to be the most distorted quadrupole roAp pulsator yet observed. Modelling of this star confirms its quadrupole nature, and allows us to constrain the magnetic field strength, angle of inclination and the angle of obliquity.

1 Introduction

The rapidly oscillating Ap (roAp) stars are a rare subclass of the chemically peculiar, magnetic, Ap stars. They show pulsations in the range of 6 – 23 min with amplitudes up to 18 mmag in Johnson B [2], and are found at the base of the classical instability strip on the Hertzsprung-Russell (HR) diagram, from the zero-age main-sequence to the terminal-age main-sequence in luminosity. Since their discovery by [3], only 61 of these objects have been identified (see [4] for a catalogue). The pulsations are high-overtone pressure modes (p modes) thought to be driven by the κ-mechanism acting in the H I ionisation zone [5]. However, [6] have shown that turbulent pressure in the convective zone may excite some of the modes seen in a selection of roAp stars.

The pulsation axis of these stars is inclined to the rotation axis, and closely aligned with the magnetic one, leading to the oblique pulsator model [e.g. 3, 7–11]. Oblique pulsation allows the pulsation modes to be viewed from varying aspects over the rotation cycle of the star, giving constraints on the pulsation geometry that are not available for any other type of pulsating star (other than the Sun, which is uniquely resolved).

The mean magnetic field modulus in Ap stars are strong, of the order of a few kG to 34 kG [12]. A strong magnetic field suppresses convection and provides stability to allow radiative levitation of some elements producing a stratified atmosphere with surface inhomogeneities. These inhomogeneities, or spots, are long lasting (decades in many known cases) on the surface of Ap stars, thus allowing for an accurate determination of the rotation period of the star. In the spots rare earth elements such as La, Ce, Pr, Nd, Sm, Eu, Gd, Tb, Dy and Ho, may be overabundant by up to a million times the solar value, leading to spectral line strength variations over the rotation period [e.g. 13]. Because of the complex atmospheres of the Ap stars, the roAp stars provide the best laboratory, beyond the Sun, to study the interactions between pulsations, rotation, and chemical abundances in the presence of magnetic fields.

Early photometric campaigns targeted known Ap stars in the search for oscillations [e.g. 14, 15], with later studies using high-resolution spectroscopy to detect line profile variations in Ap stars caused by pulsational velocity shifts [e.g. 16–21]. Most recently, the use of the SuperWASP [Wide Angle Search for Planets; 22] ground-based photometric survey led to the identification of 11 roAp stars [1, 2].

With the launch of the *Kepler* space telescope, the ability to probe to μmag precision has enabled the detection of four roAp stars with amplitudes below the ground-based detection limit [4, 23–25], and the analysis of a further known roAp star [26].

The re-purposing of the *Kepler* spacecraft after the failure of the second reaction wheel has opened up a host of new possibilities to observe variable stars at μmag precision. Here we present the first observations, in Short Cadence, of a classical, high-amplitude, roAp star.

*e-mail: dlholdsworth@uclan.ac.uk

HD 24355 is a bright ($V = 9.65$) rapidly oscillating Ap star, discovered by [1]. Their data show a pulsation at 224.31 d^{-1} (2596.18 μHz; $P = 6.4$ min) with an amplitude of 1.51 mmag in the WASP broadband filter. We present here spectral classification followed by a discussion of the SuperWASP discovery data, alongside further ground-based observations. We then present an analysis of the K2 Campaign 4 data.

2 Spectral Classification

We obtained two classification resolution spectra of HD 24355 (Fig. 1). The first with the Intermediate dispersion Spectrograph and Imaging System (ISIS) mounted on the 4.2-m William Herschel Telescope (WHT), at a resolution of $R \sim 2000$. The second spectrum was obtained using the Wide-Field Spectrograph (WiFeS) mounted on the 2.3-m Australian National Observatory (ANU) telescope at a resolution of $R \sim 3000$.

In addition, HD 24355 has been observed as part of the LAMOST campaign [27, 28] and was made public as part of Data Release 1 [DR1; 29]. The LAMOST spectrum was obtained at a resolution of $R \sim 1800$, and is shown in Fig. 1 alongside the other spectra. Furthermore, in addition to classification spectra, we have obtained high-resolution, $R \sim 58\,000$, spectra using the coudé échelle spectrograph mounted on the 2.0-m telescope at the Thüringer Landessternwarte (TLS) Tautenburg, Germany.

Following the classification process described by [30], we see that the Balmer line types are close matches to the A5 and A7 MK standard stars[1] (HD 23194 and HD 23156, respectively) with many of the metal lines being well matched by the standard A5 star. There are, however, peculiarities which do not match the standard stars. All spectra show strong enhancements of Sr II and Eu II. While the Eu II line at 4130 Å can be blended with a Si II doublet at 4128 and 4131 Å, we see no sign of Si II at 3856 and 3862 Å. Lines of Cr II are not enhanced when compared to the MK standard stars, ruling out Cr peculiarities. With the obvious presence of Sr II and Eu II in the spectra, we conclude that HD 24355 is an A5Vp SrEu star.

We derive an effective temperature for HD 24355 by fitting the Balmer lines of our spectra. Given their low resolution, we are unable to determine a $\log g$ of the star, so fix this value at 4.0 which is a reasonable approximation for this exercise. From the WiFeS spectrum we derive $T_{\rm eff} = 8200 \pm 250$ K and from the ISIS data we find $T_{\rm eff} = 8200 \pm 200$ K.

The LAMOST spectra are automatically analysed by instrument specific software to determine fundamental parameters. For HD 24355 the pipeline suggests $T_{\rm eff} = 8638 \pm 100$ K, $\log g = 4.15 \pm 0.41$ (cgs) and [Fe/H] $= 0.7 \pm 0.22$. However, when trying to fit these parameters to the spectrum, we find a large disparity. Given the resolution of the spectrum, we are unable to determine a revised $\log g$ (so we set this to 4.0 as above). We find that $T_{\rm eff} = 8150 \pm 250$ K fits the Balmer lines much better than the higher value of the LAMOST estimate, and 8150 ± 250 K is in agreement with the other two spectra.

For further details on the spectral analysis of HD 24355, we refer the reader to [31].

[1] Spectra of the standard stars were obtained from R.O. Gray's website: http://stellar.phys.appstate.edu/Standards/std1_8.html

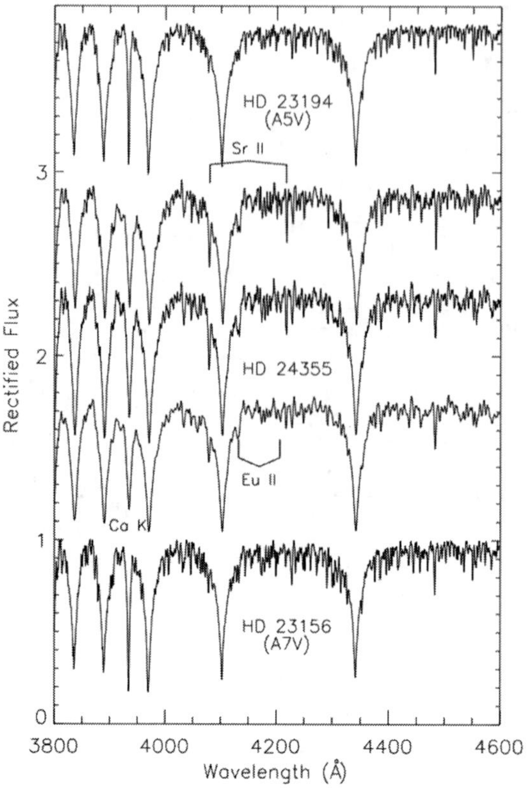

Figure 1: Spectra of HD 24355 (the three central spectra) compared with two MK standard stars of A5 and A7 type. Note the strong lines of Sr II at 4077 and 4216 Å and the enhanced lines of Eu II at 4130 and 4205 Å in the target star. The upper spectrum for HD 24355 was obtained with the ANU/WiFes instrument, the second with the WHT/ISIS instrument, with the lower obtained with the LAMOST instrument. The spectra have been ordered in rotation phase and offset for clarity.

3 Photometry

The pulsation in HD 24355 was discovered by [1]. Here we provide an overview of the discovery data, with the addition of further ground-based data obtained with the 0.75-m Automatic Photoelectric Telescope (APT) at Fairborn Observatory in Arizona, USA. Our main analysis is of the first Short Cadence (SC) data of a roAp star from the re-purposed *Kepler* space telescope, K2.

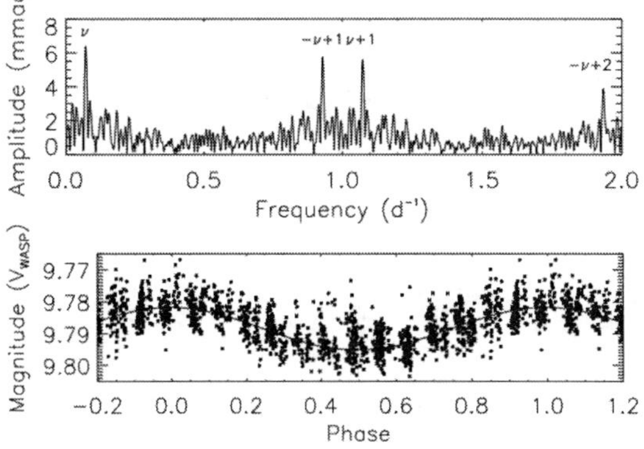

Figure 2: Top: low-frequency periodogram of the SuperWASP light curve from 2010 showing what was thought to be the rotation frequency of the star, as well as its positive and negative aliases. Bottom: SuperWASP phase folded light curve on the frequency determined in the upper panel, i.e. ν 0.07165 d^{-1}. The data have been binned to 5:1.

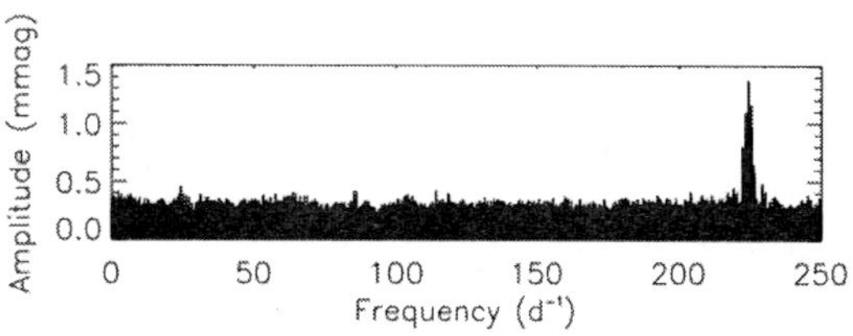

Figure 3: Periodogram of all seasons of WASP data. The data have been pre-whitened to 10 d^{-1} to the approximate noise level of the high-frequency range to remove the rotation signature and low-frequency noise. Note that the structure surrounding the pulsation signature is a result of daily aliases.

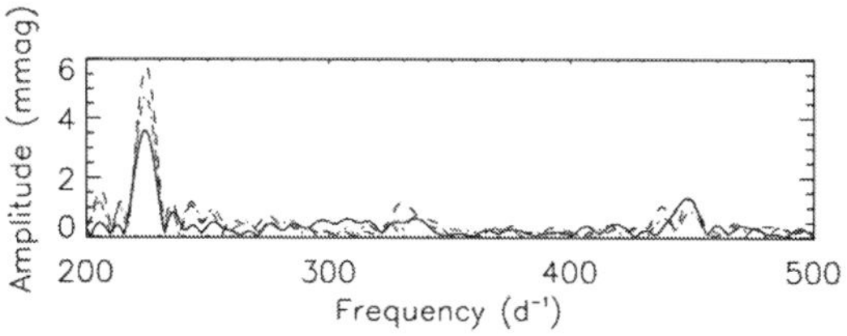

Figure 4: Periodograms of the three nights of APT data. There is a clear increase in the pulsation amplitude as the nights progress, from the solid black line, to the dot-dashed red line, to the dashed blue line. Also evident is the harmonic at twice the pulsation frequency.

3.1 Ground-based observations

3.1.1 SuperWASP

SuperWASP observed HD 24355 for three seasons, 2006, 2009 and 2010. The data were following the procedure described by [1].

The light curve shows a clear signature at low frequency that is indicative of rotation (Fig. 2). When all seasons of WASP data are combined to calculate the rotation period, there is evidence of a sub-harmonic to that frequency. We therefore simultaneously calculate the peak at 0.0716 d^{-1} and a frequency at half that value, namely 0.0358219 d^{-1}. In doing so, we are able to find a more precise measure of the rotation frequency and its harmonic. We find, $\nu = 0.0716439 \pm 0.0000063$d^{-1} and $\nu/2 = 0.0358220 \pm 0.0000032$, corresponding to periods of 13.9579±0.0011 d and 27.9158±0.0025 d, respectively. In this ground-based survey data, differentiating between the rotation frequency and any harmonics can be problematic. We therefore confirm the true rotation frequency with the K2 data, in section 3.2.

The pulsation signature of HD 24355 is apparent in all seasons of data, and is shown in Fig. 3. The pulsation is clearly seen at a frequency of 224.3071 d^{-1} with an amplitude of 1.51 mmag in the WASP passband.

Due to the survey nature of this ground-based data set, not much information can be extracted: the noise level in the high-frequency range is too great to discern, with confidence, sidelobes of the pulsation split by the rotation frequency. Analysis of the K2 data will provide more precise results on the pulsation and rotation frequencies.

3.1.2 0.75-m APT

We observed HD 24355 for three hours on each of three consecutive nights, 2013 November 6, 7, 8, with the 0.75-m APT installed at Fairborn Observatory, with integrations times of 20-s through a Johnson *B* filter.

The periodograms for each night of data presented in Fig. 4 show a clear increase in pulsation amplitude over the three nights that the target was observed. This is expected from the oblique pulsator model [3, 11], and is also observed in the K2 observations (see section 3.2). As the pulsation axis is often mis-aligned with the rotation axis in the roAp stars, the pulsation is viewed from varying aspects over the rotation period, giving rise to the amplitude variations we see in HD 24355.

3.2 K2 observations

HD 24355 was observed as part of Campaign 4 (proposal number GO4014) of the K2 mission. The observations covered a period of 70.90 d, and were reduced manually using the PyKE tools provided by the K2 Guest Observer office [32]. The final light curve is shown in the top panel of Fig. 5.

3.2.1 Rotation signature

Despite the thruster firings still being present in the data, the rotation signature is clearly seen in Fig. 5. The double-wave nature of the light curve suggests we see both magnetic poles of the star, where the chemically peculiar spots form. The K2 data allow us to unambiguously determine the rotation period of the star.

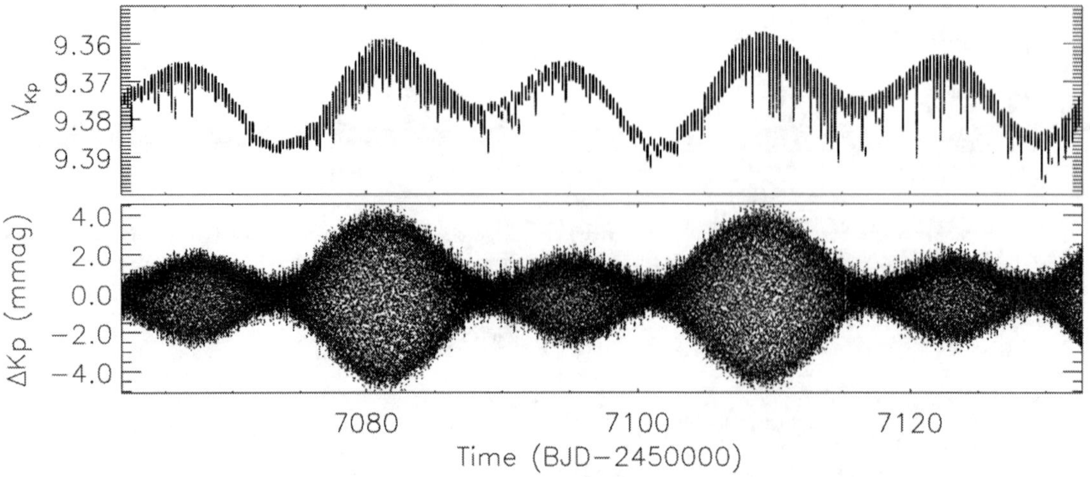

Figure 5: Top: full K2 light curve after data reduction and cleaning has been performed. Clearly evident is the rotational signal, as well as the spacecraft motions. Bottom: light curve of HD 24355 after removal of the rotation signature, and cleaning of low-frequency peaks from the periodogram; these include the harmonics of the thruster frequency. The pulsation envelope is evident, with pulsational maximum occurring at the same time as rotation light maximum within the errors.

Using Period04 we calculate an amplitude spectrum to $0.5\,d^{-1}$. The result shows two peaks, corresponding to the two consecutive dimmings in the light curve. The principal peak is found at a frequency of $0.071642\pm0.000011\,d^{-1}$, with the secondary at $0.036121\pm 0.000019\,d^{-1}$, as seen in the WASP data. Fig. 6 shows the light curve phased on the lower frequency. This is clearly the correct rotation frequency. Due to the relatively short length of the K2 data (covering ~ 2.5 rotation periods), we take the period derived from the WASP data (which covers ~ 57.3 rotation periods) as the rotation period for HD 24355, namely $P = 27.9158 \pm 0.0025$ d.

The rotation phases are calculated according to:

$$\phi(E) = 2457108.8867 + 27.9158 \times E, \qquad (1)$$

where E is the number of rotation cycles elapsed since the reference time, and the reference time is the second pulsational maximum in the K2 data set.

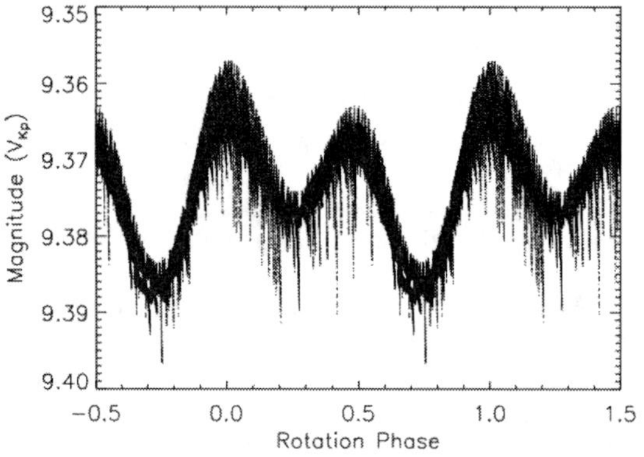

Figure 6: K2 phased light curve, phased on a period of 27.9166 d. The phases have been calculated according to equation (1).

3.2.2 Pulsation signature

Clearly evident in the K2 periodogram of HD 24355 is the pulsation detected by both the WASP and APT observations. To study the pulsation in detail, we remove the rotation signature from the light curve, and significant peaks in the low-frequency domain. This process results in the light curve shown in the bottom panel of Fig. 5.

Fig. 7 shows the pulsation at a frequency of $224.3043\,d^{-1}$ at an amplitude of 1.74 mmag, and four of the sidelobes of the principal peak after it has been removed. Removing the four peaks shown in Fig. 7 reveals further power at frequencies split by ever increasing multiples of the rotation frequency. In total, we identify 13 rotational sidelobes ($\nu + 7\nu_{rot}$ is below the noise level), the most ever seen in a roAp star. See [31] for a graphical representation of the pre-whitening of the sidelobes.

The oblique pulsation model expects that the frequencies of the sidelobes are exactly split by the rotation frequency. Therefore, using the rotation period derived from the spot variations, we force the frequencies to be equally split, and perform a linear least-squares fit to the data. When performing this test, we set the zero-point in time such that the phases of the first sidelobes are equal. When the exact splitting of the sidelobes is set, we see that the phases of the central five peaks are similar. For a pure quadrupole mode, we expect a quintuplet of frequencies with all phases equal at the time of pulsation maximum. Since this is not the case here for the phases of the quintuplet frequencies, and because there are further rotational sidelobes, we conclude that the mode is distorted.

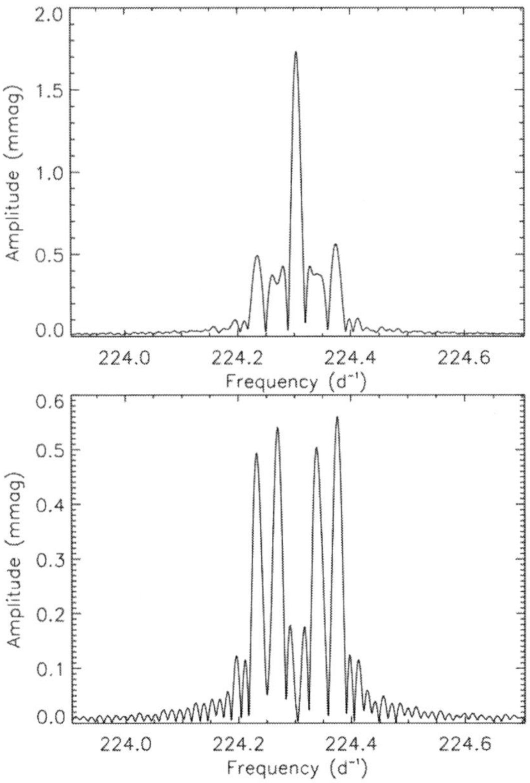

Figure 7: Top: the pulsation signature in the K2 SC data which shows a quintuplet separated by the rotation frequency. Bottom: the rotational sidelobes after the central peak, which is the actual pulsation frequency, has been removed. Analysis of the sidelobe amplitudes informs us on the geometry of the star (see text). Note the amplitude change between the two panels.

3.3 Frequency variability and a rotational view of the pulsation geometry

Many of the roAp stars that have been observed at such a high photometric precision as that afforded by the *Kepler* spacecraft have shown significant variations in their pulsation frequencies [e.g. 4, 26]. To this end, we investigate this phenomenon in HD 24355 by splitting the data into sections of 100 pulsation cycles, or about 0.45 d.

By calculating the frequencies, amplitudes and phases of the pulsation in these sections of data, we are able to monitor the variations of the pulsation over the rotation period of the star. The top-left panel in Fig. 8 shows the variation of the pulsation amplitude over the rotation period of the star. As is expected from the oblique pulsator model, the amplitude varies over the rotation cycle, as the pulsation mode is seen from varying aspects. The amplitude does not reach zero. If the mode were a pure mode, the amplitude would go to zero whenever the line-of-sight passes over a pulsation node; as this is not the case here, we can see that the mode is distorted.

The phase variations seen in the top-right panel of Fig. 8 are very different to those seen in other roAp stars (lower panels of Fig. 8). When the line-of-sight passes over a node, the oblique pulsator model predicts a π-rad phase shift for a pure dipole or quadrupole mode (as is seen in the other stars). This is not the case for HD 24355, supporting the conclusion that it is a distorted pulsator.

3.3.1 Mode geometry constraints

As the K2 data have a much higher precision than ground-based data, we can derive parameters of the geometry of the pulsation mode using the ratios of the sidelobes following the methodology of [33]. Initially, we test the phase relations between the pulsation mode and the first rotational sidelobes by forcing the sidelobes to have the same phase. We find that $\phi_{-1} = \phi_{+1} \neq \phi_0$, suggesting that the mode is distorted.

The clear frequency quintuplet strongly suggests that the pulsation is quadrupolar. We therefore examine the axisymmetric quadrupole case, where $l = 2$ and $m = 0$, applying a relationship from [33] for a non-distorted oblique quadrupole pulsator in the absence of limb-darkening and spots, i.e.

$$\tan i \tan\beta = 4\frac{A^{(2)}_{+2} + A^{(2)}_{-2}}{A^{(2)}_{+1} + A^{(2)}_{-1}}, \quad (2)$$

where i is the rotational inclination angle, β is the angle of obliquity between the rotation and magnetic axes, and $A^{(2)}_{\pm1}$ and $A^{(2)}_{\pm2}$ are the amplitudes of the first and second sidelobes of the quadrupole pulsation, respectively. Using the amplitudes from the periodogram, and substituting them into equation (2), we derive $\tan i \tan\beta = 3.950 \pm 0.012$ for this simplified case.

We investigate the values of i and β that satisfy the constraint of equation (2). All possible values sum to greater than 90°, which is consistent with the rotational light modulation under the assumption that there are spots at both magnetic poles. From our spectra we are able to place an upper limit of $v \sin i \leq 3.5\,\mathrm{km\,s^{-1}}$. Using this value, the rotation period of 27.9158 d and the derived radius of $R = 2.53\,\mathrm{R}_\odot$, we can estimate an upper limit of $i \lesssim 50°$.

4 Modelling the amplitude and phase modulation

Spectroscopic parameters of HD 24355 are $(T_{\rm eff}, \log g) = (8200\pm200\,\mathrm{K}, 4.0\pm0.2)$. Within the error-box we searched for models that have a frequency similar to $224.30\,\mathrm{d}^{-1}$ and reproduce the amplitude and phase modulation.

For each model we picked a non-magnetic quadrupole mode having a frequency similar to the observed frequency. Then, assuming a polar magnetic field strength, $B_{\rm p}$ of a dipole field, we obtained non-adiabatic eigenfrequency and eigenfunction using the method discussed in [34], in which the expansion was truncated at $J = 20$.

We obtain the amplitude modulation and the amplitudes of rotational sidelobes to compare with the observations. A good fit, as presented in Fig. 9, was sought by changing $B_{\rm p}, \beta$, and i assuming the limb-darkening parameter $\mu = 0.6$. Theoretical prediction depends weakly on

 DOI: 10.1051/epjconf/201716003004

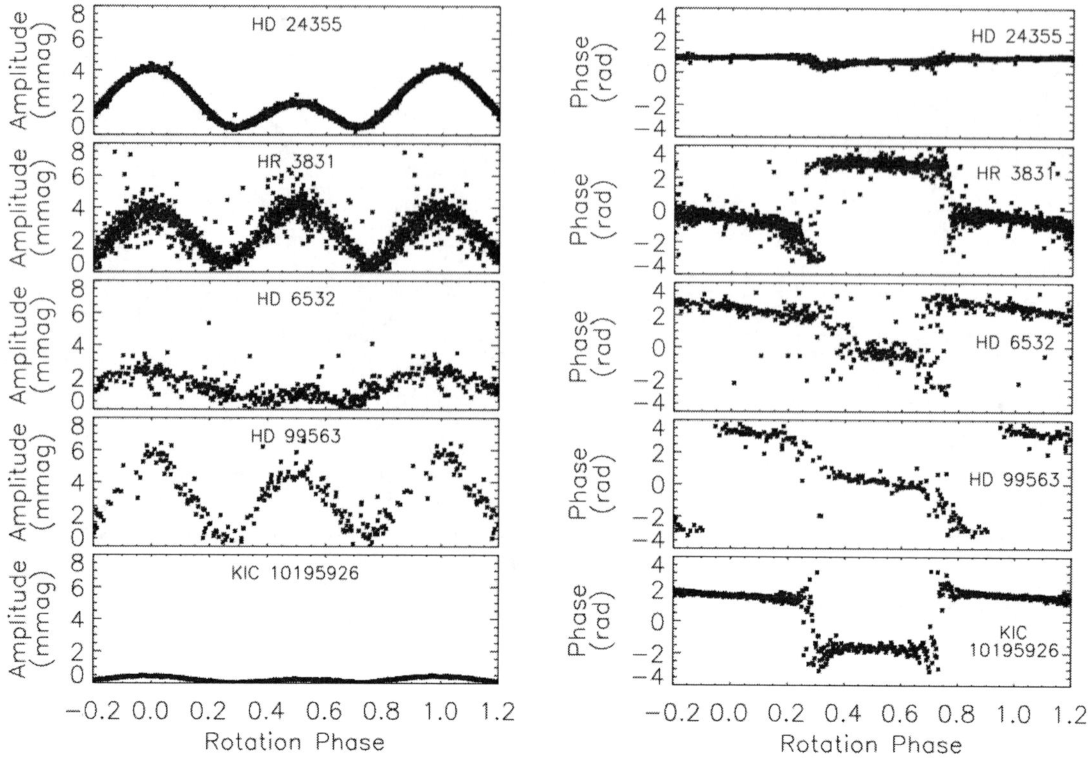

Figure 8: Pulsation amplitude and phase variations as a function of rotation phase for a set of roAp stars which are distorted dipole pulsators. All plots are on the same scale for easy comparison, and have been calculated in the same way. The stars are labelled in the plots. Despite all being distorted dipoles, there is a clear difference between the previously studied stars and HD 24355. Note, the phases are different from those previously published for the comparison stars as a different zero-point was chosen.

the inclination i. As our spectroscopic observations allow us to place an upper limit of $i < 50°$, we present models with $i = 45°$ as a standard choice, although we can obtain similarly good models in a wide range of i.

Although the amplitude modulation of HD 24355 can be reasonably reproduced by models with parameters around the spectroscopically derived values, qualitative properties of the phase modulation systematically change along evolutionary tracks (i.e., $T_{\rm eff}$), as shown in Fig. 10; there is a point where the direction of theoretical phase change switches; the "amplitude" of phase modulation decreases as $T_{\rm eff}$ decreases. In producing Fig. 10 we ensure that every model in each panel of the figure reproduces well the observed amplitude modulation, as in Fig. 9. We judge that the model in the second panel from the bottom at $\log T_{\rm eff} = 3.910$ best fits with the observed phase modulation among the 2.0 $M_\odot$ models. This model has a radius of 2.21 $R_\odot$, which gives an upper limit of $i \leq 48°$ consistent with $i = 45°$. We note that a slight asymmetry in observed phase modulation, which cannot be reproduced in our models, is probably caused by a small effect of Coriolis force [11] which we have neglected for simplicity.

Fig. 11 shows the position of HD 24355 in the $\log T_{\rm eff}$–$\nu L/M$ plane together with some of the known roAp stars. In this diagram, the mass dependence of the acoustic cut-off frequency as well as the range of excitation by the κ-mechanism in the H-ionisation zone [5] are partially compensated for. This figure shows that HD 24355 is a peculiar roAp star whose pulsation is highly super-critical – much more so than for the stars in the well known cooler roAp group consisting of HR 1217, α Cir and 10 Aql, etc., all of which are multi-periodic, while HD 24355 is a single-mode pulsator. Another single mode roAp star, HD 42659 [15], could have similar properties to HD 24355, although the position on the diagram depends on the uncertain quantity of L/M, and low-amplitude oscillations of HD 42659 are not well studied. The excitation mechanism of such super-critical high-order p-modes is not known.

5 Conclusions

We have presented the first *Kepler* spacecraft observations of the high-amplitude roAp star HD 24355. The K2 data have allowed us to unambiguously determine the rotation period of the star to be 27.9158 ± 0.0043 d, a parameter which was uncertain when considering the ground-based SuperWASP data alone. Classification dispersion spectra allowed us to classify this star as an A5 Vp SrEu star.

The high-resolution spectra also allowed us to estimate a mean magnetic field strength of 2.64 ± 0.49 kG; however,

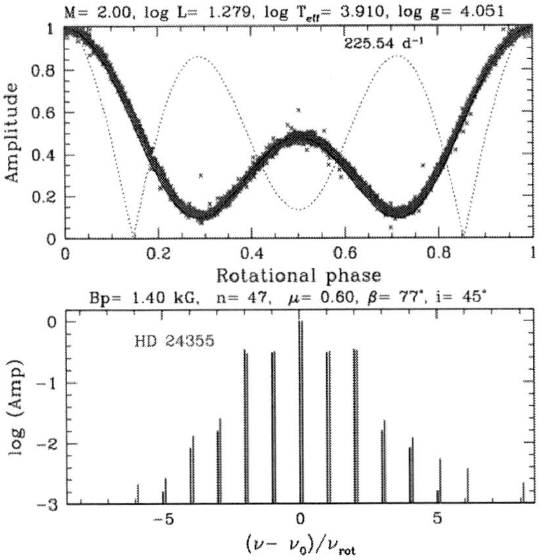

Figure 9: Amplitude modulation (top panel) and corresponding rotational sidelobes (bottom panel) of HD 24355 (red crosses and red bars) are compared with a model (black lines). The dotted line in the top panel shows amplitude modulation expected if the latitudinal distribution of pulsation were proportional to the Legendre function $P_2(\cos\theta_B)$. Note that bottom panel plots the logarithmic (base 10) values of sidelobes amplitude normalised by the amplitude of the central frequency.

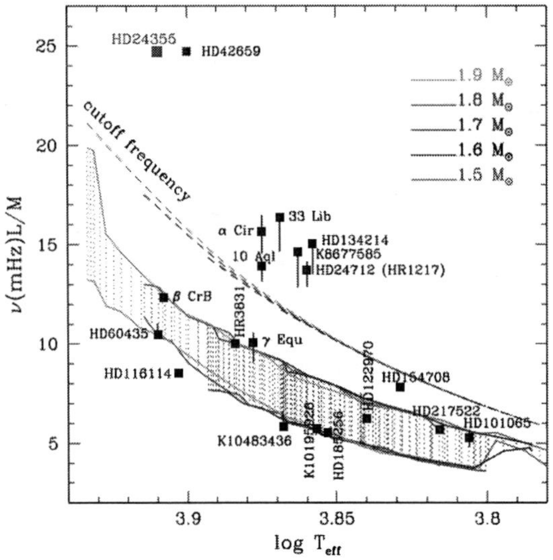

Figure 11: The position of HD 24355 in the log $T_{\rm eff}-\nu L/M$ plane is compared with some of the known roAp stars, in which L/M is in solar units. Main frequencies are shown by squares and frequency ranges by vertical bars. Shaded region indicates where high-order p-modes are excited by the κ-mechanism in the H-ionisation zone in non-magnetic models. Dashed lines show the acoustic cut-off frequencies. This figure is an extended version of figure 3 in [35].

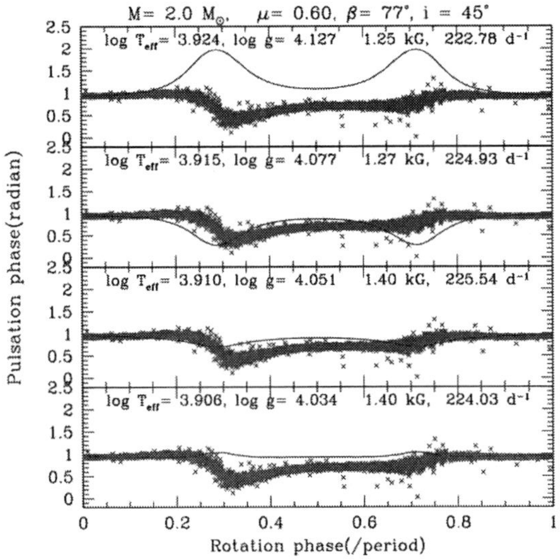

Figure 10: Pulsation phase modulation of HD 24355 (red crosses) with respect to the rotation phase is compared with theoretical ones obtained for some of the evolutionary models of 2 M$_\odot$. The effective temperature and surface gravity decrease from top to bottom.

we take this to be an upper limit on the value due to the lack of Zeeman splitting in the spectra, and the method used to derive that value.

There is a discrepancy between the $T_{\rm eff}$ of HD 24355 when using different methods to derive the parameter. Values from the literature, SED fitting abundance analysis and line fitting provide a wide range of $T_{\rm eff}$ values. However, we get the best agreement in results using solely the Balmer lines of both the low-resolution and high-resolution spectra, deriving 8200 ± 200 K, placing HD 24355 amongst the hotter roAp stars.

Analysis of the pulsation mode as detected in the K2 data have shown the characteristic signatures of a roAp pulsator as predicted by the oblique pulsator model [3, 11, 36]. The pulsational amplitude is modulated with the rotation period of the star. The behaviour of the pulsation phase is not as expected, however. The very small phase change at quadrature is a surprise for a quadrupole pulsator. Examples from the literature (cf. Fig. 8) show a clear π-rad phase change when a different pulsation pole rotates into view, but this is not the case for HD 24355. Here we see a shift of only 1-rad at most. This small 'blip' in the phase suggests that HD 24355 is pulsating in a very distorted mode, the most extreme case yet observed.

The rotationally split pulsation has provided us with the amplitudes to test the geometry of the star. As such, we modelled the system following the method of [34]. Changing values of the inclination and obliquity angles and the polar magnetic field strength, and searching the parameter space surrounding our observational constraints, we conclude that HD 24355 is a distorted quadrupolar pulsator, with a magnetic field strength of about 1.4 kG. The model accurately matches the observed amplitude modulation of the pulsation, and the amplitudes of the rotationally split sidelobes. The pulsational phase variations are a stronger function of the evolution of the star, and as such provide a slightly greater challenge to model. We believe, however, that the model presented is a satisfactory match to the data, given our current observational constraints on the evolutionary stage of the star. We determine that the pulsation seen in HD 24355 is super-critical, making it the most precisely observed super-critical roAp star to date. The driving mechanism for such a pulsation is currently unknown, thus making HD 24355 a highly important target in understanding how some roAp stars can pulsate with frequencies well above the critical cutoff frequency.

References

[1] D.L. Holdsworth, B. Smalley, M. Gillon, K.I. Clubb, J. Southworth, P.F.L. Maxted, D.R. Anderson, S.C.C. Barros, A.C. Cameron, L. Delrez et al., MNRAS, **439**, 2078 (2014)
[2] D.L. Holdsworth, Ph.D. thesis, Keele University (2015)
[3] D.W. Kurtz, MNRAS, **200**, 807 (1982)
[4] B. Smalley, E. Niemczura, S.J. Murphy, H. Lehmann, D.W. Kurtz, D.L. Holdsworth, M.S. Cunha, L.A. Balona, M. Briquet, H. Bruntt et al., MNRAS, **452**, 3334 (2015)
[5] N.J. Balmforth, M.S. Cunha, N. Dolez, D.O. Gough, S. Vauclair, MNRAS, **323**, 362 (2001)
[6] M.S. Cunha, D. Alentiev, I.M. Brandão, K. Perraut, MNRAS, **436**, 1639 (2013)
[7] H. Shibahashi, H. Saio, PASJ, **37**, 245 (1985)
[8] H. Shibahashi, H. Saio, PASJ, **37**, 601 (1985)
[9] H. Shibahashi, M. Takata, PASJ, **45**, 617 (1993)
[10] M. Takata, H. Shibahashi, PASJ, **47**, 219 (1995)
[11] L. Bigot, D.W. Kurtz, A&A, **536**, A73 (2011)
[12] H.W. Babcock, ApJ, **132**, 521 (1960)
[13] T. Lüftinger, O. Kochukhov, T. Ryabchikova, N. Piskunov, W.W. Weiss, I. Ilyin, A&A, **509**, A71 (2010)
[14] P. Martinez, D.W. Kurtz, G.M. Kauffmann, MNRAS, **250**, 666 (1991)
[15] P. Martinez, D.W. Kurtz, MNRAS, **271**, 118 (1994)
[16] I.S. Savanov, V.P. Malanushenko, T.A. Ryabchikova, Astronomy Letters **25**, 802 (1999)
[17] O. Kochukhov, T. Ryabchikova, A&A, **374**, 615 (2001)
[18] A.P. Hatzes, D.E. Mkrtichian, MNRAS, **351**, 663 (2004)
[19] D.E. Mkrtichian, A.P. Hatzes, H. Saio, R.R. Shobbrook, A&A, **490**, 1109 (2008)
[20] V.G. Elkin, D.W. Kurtz, G. Mathys, MNRAS, **415**, 2233 (2011)
[21] O. Kochukhov, D. Alentiev, T. Ryabchikova, S. Boyko, M. Cunha, V. Tsymbal, W. Weiss, MNRAS, **431**, 2808 (2013)
[22] D.L. Pollacco, I. Skillen, A. Collier Cameron, D.J. Christian, C. Hellier, J. Irwin, T.A. Lister, R.A. Street, R.G. West, D.R. Anderson et al., PASP, **118**, 1407 (2006)
[23] L.A. Balona, M.S. Cunha, D.W. Kurtz, I.M. Brandão, M. Gruberbauer, H. Saio, R. Östensen, V.G. Elkin, W.J. Borucki, J. Christensen-Dalsgaard et al., MNRAS, **410**, 517 (2011)
[24] L.A. Balona, M.S. Cunha, M. Gruberbauer, D.W. Kurtz, H. Saio, T.R. White, J. Christensen-Dalsgaard, H. Kjeldsen, J.L. Christiansen, J.R. Hall et al., MNRAS, **413**, 2651 (2011)
[25] D.W. Kurtz, M.S. Cunha, H. Saio, L. Bigot, L.A. Balona, V.G. Elkin, H. Shibahashi, I.M. Brandão, K. Uytterhoeven, S. Frandsen et al., MNRAS, **414**, 2550 (2011)
[26] D.L. Holdsworth, B. Smalley, D.W. Kurtz, J. Southworth, M.S. Cunha, K.I. Clubb, MNRAS, **443**, 2049 (2014)
[27] X.Q. Cui, Y.H. Zhao, Y.Q. Chu, G.P. Li, Q. Li, L.P. Zhang, H.J. Su, Z.Q. Yao, Y.N. Wang, X.Z. Xing et al., Research in Astronomy and Astrophysics **12**, 1197 (2012)
[28] G. Zhao, Y.H. Zhao, Y.Q. Chu, Y.P. Jing, L.C. Deng, Research in Astronomy and Astrophysics **12**, 723 (2012)
[29] A.L. Luo, Y.H. Zhao, G. Zhao, L.C. Deng, X.W. Liu, Y.P. Jing, G. Wang, H.T. Zhang, J.R. Shi, X.Q. Cui et al., Research in Astronomy and Astrophysics **15**, 1095 (2015)
[30] R.O. Gray, C. Corbally, J., *Stellar Spectral Classification* (2009)
[31] D.L. Holdsworth, D.W. Kurtz, B. Smalley, H. Saio, G. Handler, S.J. Murphy, H. Lehmann, MNRAS, **462**, 876 (2016)
[32] M. Still, T. Barclay, *PyKE: Reduction and analysis of Kepler Simple Aperture Photometry data*, Astrophysics Source Code Library (2012)
[33] D.W. Kurtz, H. Shibahashi, P.R. Goode, MNRAS, **247**, 558 (1990)
[34] H. Saio, MNRAS, **360**, 1022 (2005)
[35] H. Saio, *Pulsation of magnetic stars*, in *Precision Asteroseismology*, edited by J.A. Guzik, W.J. Chaplin, G. Handler, A. Pigulski (2014), Vol. 301 of *IAU Symposium*, pp. 197–204
[36] L. Bigot, W.A. Dziembowski, A&A, **391**, 235 (2002)

DOI: 10.1051/epjconf/201716003005

Asteroseismic inversions in the Kepler era: application to the Kepler Legacy sample

Gaël Buldgen[1,3,*], *Daniel* Reese[2,**], and *Marc-Antoine* Dupret[3,***]

[1] *Institut d'Astrophysique et Géophysique de l'Université de Liège, Allée du 6 août 17, 4000 Liège, Belgium*
[2] *LESIA, Observatoire de Paris, PSL Research University, CNRS, Sorbonne Universités, UPMC Univ. Paris 06, Univ. Paris Diderot, Sorbonne Paris Cité, 5 place Jules Janssen, 92195 Meudon Cedex, France*

Abstract. In the past few years, the CoRoT and Kepler missions have carried out what is now called the space photometry revolution. This revolution is still ongoing thanks to K2 and will be continued by the Tess and Plato2.0 missions. However, the photometry revolution must also be followed by progress in stellar modelling, in order to lead to more precise and accurate determinations of fundamental stellar parameters such as masses, radii and ages. In this context, the long-lasting problems related to mixing processes in stellar interior is the main obstacle to further improvements of stellar modelling. In this contribution, we will apply structural asteroseismic inversion techniques to targets from the Kepler Legacy sample and analyse how these can help us constrain the fundamental parameters and mixing processes in these stars. Our approach is based on previous studies using the SOLA inversion technique [1] to determine integrated quantities such as the mean density [2], the acoustic radius, and core conditions indicators [3], and has already been successfully applied to the 16Cyg binary system [4]. We will show how this technique can be applied to the Kepler Legacy sample and how new indicators can help us to further constrain the chemical composition profiles of stars as well as provide stringent constraints on stellar ages.

1 Introduction

In the last decade, the space photometry missions CoRoT and Kepler have provided us with a wealth of precise seismic data for a large number of stars. This revolution is now allowing us to carry out precise seismic modelling for these stars and to test the physical inputs of our stellar models. As a by-product of this modelling, asteroseismology also delivers precise and accurate fundamental parameters, which are essential to other fields such as galactic archeology or exoplanetology.

Amongst the observed targets, the Kepler Legacy sample contains the dwarfs that have the best frequency sets within the Kepler data (See [5] for the presentation of the dataset and [6] for forward modeling results of the Kepler Legacy sample). As such, they can be seen as benchmark stars to test our inputs of stellar models and the efficiency of seismic modelling. Due to their high number of frequencies and the precision of these determinations, they are also the ideal targets with which to carry out structural seismic inversions. In this paper, we will present forward modelling and inversion results for two targets of the Legacy sample: namely Doris (KIC8006161) and Saxo (KIC6603624). In Sect. 2, we present the targets and the forward modelling results for each one of them. In Sect. 3, we present the inversion procedure we used to further constrain the models of both stars and present the results of this process in Sect. 4. The implications of these results are then presented in Sect. 5.

2 Targets selection and forward modelling

The targets were selected by the number and precision of their observed frequencies as well as their estimated mass from a crude grid-based modelling approach. We specifically selected stars with masses close to $1M_{\odot}$. Firstly, stars closer to the sun are easier to model, and we know from helioseismology and from our theoretical test cases where it was possible to obtain a sufficiently good modelling that linear inversion techniques can be applied to constrain their structure (see [7] for an application to the solar case). This is potentially not the case for more massive stars for which the boundary of the convective core may not be reproduced accurately enough and thus the evolutionary stage of the target not very well constrained. In such cases, using a linear approximation may be questionable. The second reason is that we wanted to avoid convective cores in the reference models because we used specific indicators using radial derivatives to constrain the internal structure of the targets [3]. The thermodynamical quantities used in these indicators can have discontinuous derivatives which will appear in the target function of the SOLA method. Such targets are then impossible to fit with

*e-mail: gbuldgen@ulg.ac.be
**e-mail: daniel.reese@obspm.fr
***e-mail: ma.dupret@ulg.ac.be

DOI: 10.1051/epjconf/201716003005

Table 1: Forward modelling results for Doris and Saxo

	Doris	Saxo
Mass ($M_{\odot}$)	0.91 − 1.02	0.93 − 1.05
Age (Gy)	4.6 − 5.3	7.6 − 8.7

the continuous structural kernels built using the eigenfunctions of solar-like oscillation modes.

Before trying to carry out structural inversions, we obtained reference models for each target using a classical seismic forward modelling approach. This modelling was carried out using a Levenberg-Marquardt algorithm. The models were built using the Liège stellar evolution code (CLES, [8]) and the oscillations were computed using the Liège oscillation code (LOSC, [9]). We used the Ceff equation of state [10], the Opal opacities [11] supplemented at low temperatures by the opacities from [12] and the nuclear reaction rates from the NACRE project [13] including the updated reaction rate for the $^{14}N(p,\gamma)^{15}O$ reaction from [14]. Convection was implemented using the classical, local mixing-length theory [15]. We used the individual small frequency separations, d_{02} and d_{13}, the effective temperature, T_{eff}, and the metallicity, $[Fe/H]$ as constraints for our structural models. Some fits were also performed using the frequency ratios r_{01} and r_{02} [16]and this lead to very similar results. No surface correction was applied to the frequencies since we used quantities that are not strongly dependant on surface effects. The free parameters of the Levenberg-Marquardt algorithm were the mass of the star, M, its age, its hydrogen mass fraction, X, its metallicity, Z, and the mixing-length parameter, α_{MLT}. Both stars were modelled using the AGSS09 abundance tables [17], but tests were performed using the old GN93 abundances [18] to see whether some comments could be made on the metallicity scale used to relate stellar metallicity to solar metallicity.

2.1 Doris a.k.a KIC8006161

In Fig. 1, we illustrate the echelle diagram of Doris. This star has 54 observed individual frequencies with a mean 1σ uncertainty of 0.49 μHz, an effective temperature of $5488 \pm 77K$ and a $[Fe/H] = 0.34 \pm 0.10$. The initial mass estimate was $0.96M_{\odot}$. We illustrate in table 1 the results from the forward modelling process. These results are in agreement with the first estimate to within a reasonable accuracy. We note that most of the uncertainty stems from the well-known helium-mass degeneracy in seismic fitting and the intensity of diffusion. It seems also clear that extra-mixing processes acting inside the star would have an impact on these determinations. We only considered models with microscopic diffusion but without including turbulent diffusion. We note that it was possible to fit this target with both the $GN93$ abundances or the $AGSS09$ abundances within the same accuracy. Had we included extra-mixing, the scatter of the fundamental parameters would have been slightly larger.

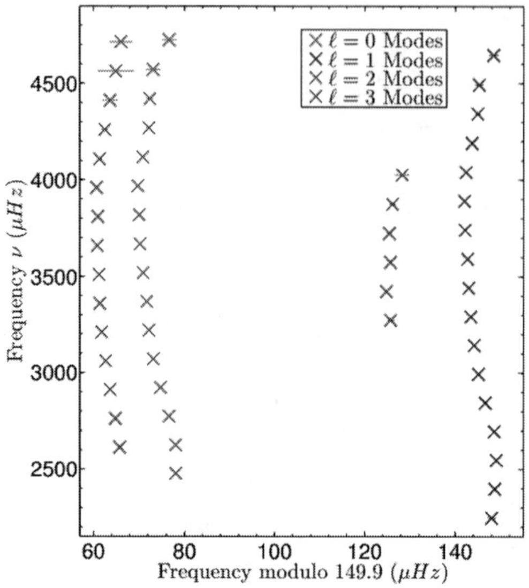

Figure 1: Echelle diagram of Doris ($KIC8006161$) illustrating the ridges of the $\ell = 0, 1, 2, 3$ modes.

2.2 Saxo a.k.a KIC6603624

In Fig. 2, we illustrate the echelle diagram of Saxo. This target has 44 individual frequencies with an average uncertainty of 0.336 μHz. Although the frequencies are accurately determined, we wish to point out a small deviation in the echelle diagram, the octupole mode of lowest frequency seems to be deviating from the ridge formed by the other octupole modes. This irregularity made this mode very difficult to fit and it seemed that it could be due to a miscalculation of the frequency value (Differences between different fitters have indeed been reported in [19]). It seemed unprobable that this difference could be physical so we eliminated this mode and used only 43 frequencies to calculate the small frequency separations used in our fits. In addition to the seismic data, we used the same constraints as for Doris, namely the effective temperature, at a value of 5674 ± 77 and the $[Fe/H] = 0.28 \pm 0.10$. Our results, given in table 1 are in agreement with the initial mass estimate of $1.01M_{\odot}$. Again, this target could be fitted using both the old GN93 and the new AGSS09 solar abundances.

3 Inversion Procedure

In this section we briefly present the inversion procedure we used for both targets. We followed the same methodology as in our study of the 16Cyg binary system [4] and carried out inversions for both the mean density and a core condition indicator based on the derivative of the squared isothermal sound speed, denoted t_u. The mean density inversions have been described in [2] and the t_u inversions have been presented in [3]. The SOLA inversion technique

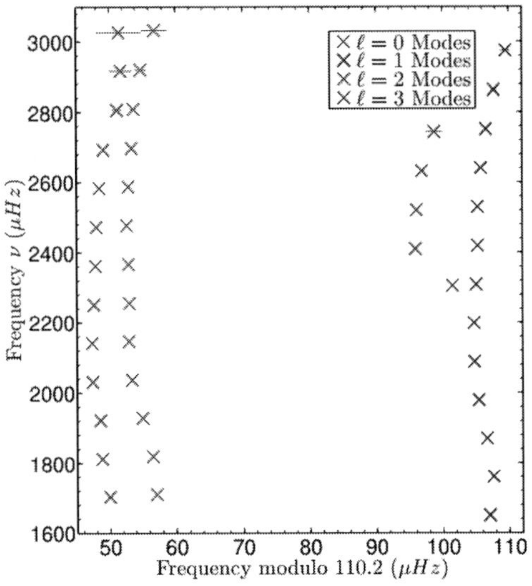

Figure 2: Echelle diagram of Saxo (*KIC*6603624) illustrating the ridges of the $\ell = 0, 1, 2, 3$ modes.

[1] was used to related the relative frequency differences to corrections of these integrated quantities, assuming the following relation

$$\sum_{i}^{N} c_i \frac{\delta \nu_i}{\nu_i} \equiv \left(\frac{\delta A}{A}\right)_{inv}, \quad (1)$$

where the c_i are the inversion coefficients, found by minimizing the cost function of the SOLA method and A is the integrated quantity whose correction is sought. Structural inversions in asteroseismology are carried out using a sample of reference models to analyze whether model dependency dominates the results or not. This is a consequence of the linearity hypothesis used to compute the structural corrections from seismic observations. To obtain the corrections of integrated quantities, we use the linear relations between frequency and structure from [20]

$$\frac{\delta \nu}{\nu} = \int_0^R K^{n,l}_{s_A,s_B} \frac{\delta s_A}{s_A} dr + \int_0^R K^{n,l}_{s_B,s_A} \frac{\delta s_B}{s_B} dr, \quad (2)$$

with s_A and s_B being structural quantities like the squared adiabatic sound speed, c^2, and the density ρ as in classical helioseismic inversions. The $K^{n,l}$ functions are the structural kernels, which depend only on the reference model and its eigenfunctions. The δ notation is related to the differences between observed and reference quantities following the convention

$$\frac{\delta x}{x} = \frac{x_{obs} - x_{ref}}{x_{ref}}, \quad (3)$$

where x can be a frequency or a structural quantity. The main problem in the context of asteroseismology is that the linearization hypothesis may not be valid. Thus, one can use multiple reference models to check for the robustness of the inference made with the inversion technique.

The definition of the t_u indicator is

$$t_u = \int_0^R f(r) \left(\frac{du}{dr}\right)^2 dr, \quad (4)$$

with $f(r) = r(r - R)^2 \exp(-7r^2)$, the weight function used for this inversion, R the stellar radius, r the radial coordinate associated with each layer inside the model, and u the squared isothermal sound-speed defined as $u = \frac{P}{\rho}$. This quantity is very sensitive to changes in the deep layers of stars and can be used to assess the reliability of the representation of the stratification of the deep layers by the models. In fact, one can show that $u \approx \frac{T}{\mu}$ in the core regions and thus, this quantity will be very sensitive to the effects of mixing processes that will change the temperature, T, and mean molecular weight, μ, gradients.

Besides this indicator, we also carried out inversions using a new core condition indicator based on an entropy proxy, denoted $S_{5/3} = \frac{P}{\rho^{5/3}}$. We consider the inversion results of this indicator to be preliminary and a follow up with the definition and limitations of this inversion will be presented in a future paper. This proxy is particularly well suited for stars with a convective core since it will show a plateau in regions where convection dominate. The width of the plateau is linked to the extent of the convective regions and the height is sensitive to the transition with radiative regions, thus to the temperature and chemical composition gradient. In Fig 3, we show the behaviour of $S_{5/3}$ for the convective envelope in the left panel and of $S^{-1}_{5/3}$ in the right panel for a convective core.

4 Inversion Results

In Fig. 4, we illustrate the inversion results for the t_u indicator and an indicator based on $S^{-1}_{5/3}$, denoted S_{Core}. The green crosses are the inverted results while the red circles illustrate the reference models computed using the Levenberg-Marquardt algorithm. As can be seen, some models are in agreement with the inversion results while others are not. There is also a clear trend showing that models not fitting the inverted mean density, given here in abscissa, are incompatible with the other indicators. As already seen in [4], the error bars on the t_u inversion are quite large, but there is a clear trend that can be deduced and used to select a subsample of models from our initial forward modelling. The numerical value obtained for the inverted $t_u/G^2R^6_{Tar}$ value is around 5.8 ± 0.7 g^2/cm^6, with R_{Tar} the target photospheric radius and G the gravitational constant. The S_{Core} indicator shows an inverted $S_{Core}GR^{1/3}_{Tar}$ value around 2.110 ± 0.005 $cm/g\,m^{1/3}$. Again only a subset of models are in agreement with these inversion results.

In fig. 5, we illustrate the inversion results for the t_u and the S_{Core} indicator. In this case, the results are stunningly different, since no model seems to agree with the inverted results, although the models were fitted to the observations with the same accuracy as for Doris. The difference is particularly significant for the S_{Core} indicators

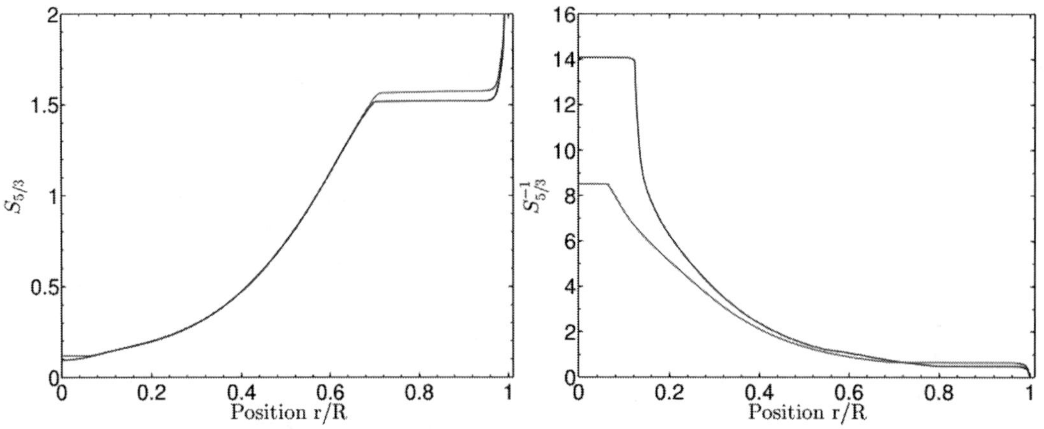

Figure 3: Left panel: radial profile of $S_{5/3}$ for two stellar models of various mass and age, illustrating the plateau in the convective region. Right panel: radial profile of $S^{-1}_{5/3}$ for two stellar models, illustrating the plateau of the proxy in convective cores.

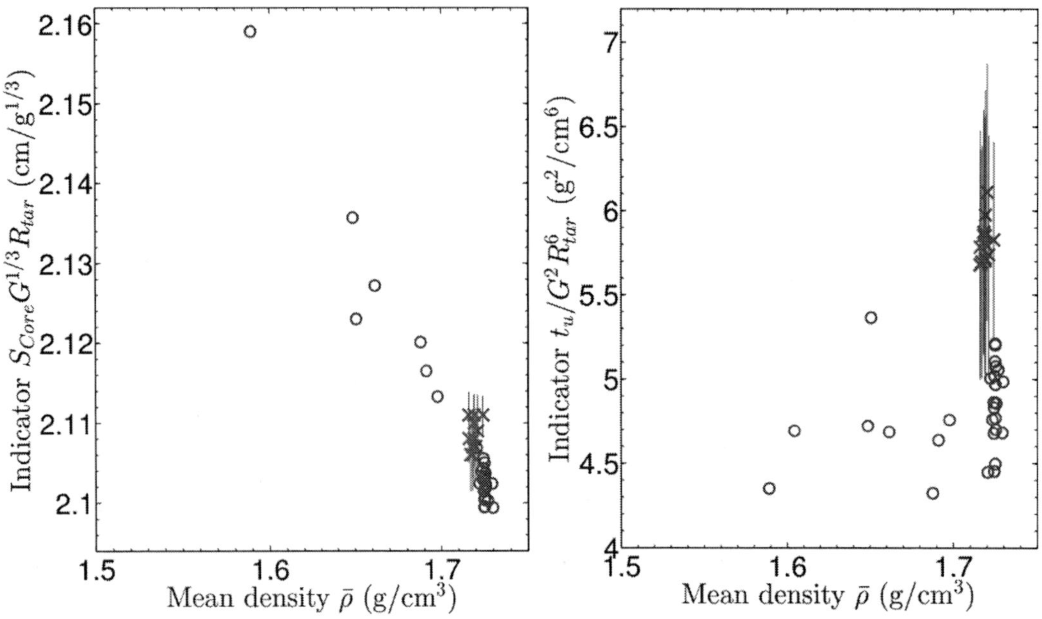

Figure 4: Inversion results for Doris. Left panel: $\bar{\rho} - S_{Core}G^{1/3}R_{tar}$ plane where the red circles illustrate the position of the reference models and the green crosses show the inversion results with their error bars. Right panel: $\bar{\rho} - t_u/(G^2R^6_{tar})$ plane following the same notations as the left panel.

and seems to indicate that something has been wrongly reproduced in the reference models. A small trend is seen with the mean density, as for the results for Doris, but the disagreement is still much larger. For the t_u indicator, the value found is around 1.5 ± 1 g^2/cm^6 while the S_{Core} inversion leads to a result around 1.7 ± 0.06 cm/g$^{1/3}$.

5 Implication for the modelling of Doris and Saxo

Using the inversion results for Doris, we could define a subsample of reference models, selected in agreement with the determinations of the mean density, the t_u and the S_{Core} indicators. These models are represented in Fig. 6, where the blue crosses represent the models of our sample which are in agreement with the inversion procedure, the purple crosses are models found in agreement with some

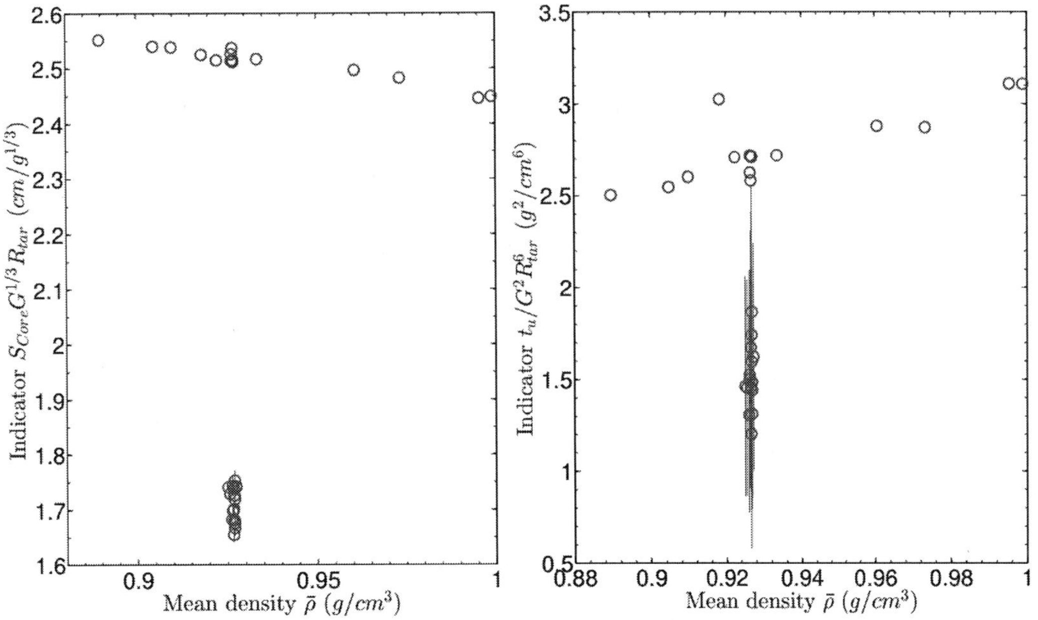

Figure 5: Inversion results for Saxo. Left panel: $\bar{\rho} - S_{Core}G^{1/3}R_{tar}$ plane where the red circles illustrate the position of the reference models and the green crosses show the inversion results with their error bars. Right panel: $\bar{\rho} - t_u/(G^2R_{tar}^6)$ plane following the same notations as the left panel.

but not all inversion results and the red dots are the models which disagree with the inversion results. We point out that all of these models included microscopic diffusion, so that the selection process is somewhat less spectacular than what is found in our study of 16Cyg. It should be noted however, that models without microscopic diffusion for Doris are systematically rejected by both inversions and have thus not been included in this study. Another point worth mentioning is that unlike 16Cyg, Doris has no determination of its surface helium abundance and thus further constraints could be brought on this model by determining the helium abundance of the convective envelope of this target. This could lead to further studies of this star and constrain the mixing inside this star, as is currently being done for 16Cyg.

However, the use of constraints from the inversions leads us to select a subsample of models with a 2.5% uncertainty in mass and 4.5% uncertainty in age, illustrating the diagnostic process of the inversion technique and its efficiency in constraining fundamental parameters. The selection effect in age is very small since we did not plot models without microscopic diffusion, which were systematically rejected by the inversion technique. These uncertainties are of course internal error bars and do not take into account possible inaccuracies in the physical ingredients of the model.

In the case of Saxo, no model seemed to fit the inverted constraints and thus a solution still has to be found. Moreover, the reliability of the frequencies has to be assessed before further inversions can be carried out. It is well known that the SOLA method is sensitive to outliers and it is also clear that any misfits to the frequencies would affect significantly the inversion results, and thus the diagnostic provided by the method. If these results are confirmed, then it would seem that something is clearly missing in the description of this target and further investigations would be needed.

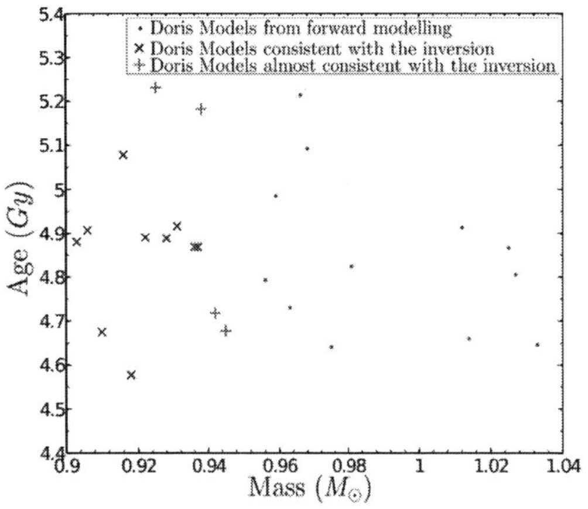

Figure 6: Illustration in terms of mass and age of the models validated by the inversions for Doris. The blue crosses are the models in agreement with the inversions, the purple crosses are at the borders of the intervals validated by the inversions and the red dots are rejected by the inversions.

6 Conclusions

In this paper, we applied the inversions of integrated quantities as defined in [2] and [3] to two targets of the Kepler Legacy sample. This sample of 66 dwarfs is considered to be the best sample of solar-like and F-type stars observed by Kepler. This means that these targets can be used as benchmark stars to constrain stellar physics and seismic modelling. Due to the high quality of the seismic data, they are very well suited for structural inversion techniques. We demonstrated this by applying our methods to Doris and Saxo and showed that there was indeed a diagnostic from the inversion technique.

For Doris, this diagnostic allowed us to further constrain the fundamental parameters of this star and lead to a final uncertainty of 2.5% in mass and 4.5% in age. These results are, of course, model-dependent and to be considered as internal error bars, not taking into account potential errors in the physical ingredients of the stellar models.

In the case of Saxo, a clear disagreement with the models was witnessed. The main problem with this result, but also with the seismic diagnostic for Doris, is linked to the uncertainties in the frequency determinations of the Kepler Legacy (see [19]). These results have then to be checked again once this problem will have been clarified.

References

[1] F.P. Pijpers, M.J. Thompson, A & A **281**, 231 (1994)

[2] D.R. Reese, J.P. Marques, M.J. Goupil, M.J. Thompson, S. Deheuvels, A & A **539**, A63 (2012)

[3] G. Buldgen, D.R. Reese, M.A. Dupret, A & A **583**, A62 (2015)

[4] G. Buldgen, D.R. Reese, M.A. Dupret, A & A **585**, A109 (2016)

[5] M.N. Lund, V. Silva Aguirre, G.R. Davies, W.J. Chaplin, J. Christensen-Dalsgaard, G. Houdek, T.R. White, T.R. Bedding, W.H. Ball, D. Huber et al., ApJ **835**, 172 (2017)

[6] V. Silva Aguirre, M.N. Lund, H.M. Antia, W.H. Ball, S. Basu, J. Christensen-Dalsgaard, Y. Lebreton, D.R. Reese, K. Verma, L. Casagrande et al., ApJ **835**, 173 (2017)

[7] S. Basu, W.J. Chaplin, Y. Elsworth, R. New, A.M. Serenelli, ApJ **699**, 1403 (2009)

[8] R. Scuflaire, S. Théado, J. Montalbán, A. Miglio, P.O. Bourge, M. Godart, A. Thoul, A. Noels, Ap& SS **316**, 83 (2008)

[9] R. Scuflaire, J. Montalbán, S. Théado, P.O. Bourge, A. Miglio, M. Godart, A. Thoul, A. Noels, Ap& SS **316**, 149 (2008)

[10] J. Christensen-Dalsgaard, W. Daeppen, A & AR **4**, 267 (1992)

[11] C.A. Iglesias, F.J. Rogers, ApJ **464**, 943 (1996)

[12] D.R. Alexander, J.W. Ferguson, ApJ **437**, 879 (1994)

[13] C. Angulo, M. Arnould, M. Rayet, P. Descouvemont, D. Baye, C. Leclercq-Willain, A. Coc, S. Barhoumi, P. Aguer, C. Rolfs et al., Nuclear Physics A **656**, 3 (1999)

[14] A. Formicola, G. Imbriani, H. Costantini, C. Angulo, D. Bemmerer, R. Bonetti, C. Broggini, P. Corvisiero, J. Cruz, P. Descouvemont et al., Physics Letters B **591**, 61 (2004)

[15] E. Böhm-Vitense, ZA **46**, 108 (1958)

[16] I.W. Roxburgh, S.V. Vorontsov, A & A **411**, 215 (2003)

[17] M. Asplund, N. Grevesse, A.J. Sauval, P. Scott, AR A & A **47**, 481 (2009)

[18] N. Grevesse, A. Noels, *Cosmic abundances of the elements.*, in *Origin and Evolution of the Elements*, edited by N. Prantzos, E. Vangioni-Flam, M. Casse (1993), pp. 15–25

[19] I.W. Roxburgh, ArXiv e-prints (2016)

[20] D.O. Gough, M.J. Thompson, *The inversion problem* (1991), pp. 519–561

DOI: 10.1051/epjconf/201716003006

A view into the core of α Cen A

Michaël Bazot[1,★], *Jørgen* Christensen-Dalsgaard[2], *Othman* Benomar[1], and *Laurent* Gizon[3,4]

[1] *Center for Space Science, NYUAD Institute, New York University Abu Dhabi, PO Box 129188, Abu Dhabi, United Arab Emirates*
[2] *Stellar Astrophysics Centre, Department of Physics and Astronomy, Aarhus University, Ny Munkegade 120, DK-8000 Aarhus C, Denmark*
[3] *Max-Planck-Institut für Sonnensystemforschung, 37077 Göttingen, Germany*
[4] *Institut für Astrophysik, Georg-August-Universität Göttingen, 37077 Göttingen, Germany*

Abstract. We present results of modelling of α Cen A. In order to estimate the physical parameters of this star, we modelled spectroscopic, interferometric, astrometric and asteroseismic data. To that effect we chose to use a Bayesian approach to parameter estimation, which allowed us, in particular, to define our prior knowledge on the parameters. An important question we wanted to address was to assess whether or not α Cen A has a convective core. We found that the data we used give indecisive results on this issue. If the star has a convective core, and provided that overshooting is taken into account, there is a possibility for the star to be in the peculiar state in which the ppII chain is the main driver of nuclear energy generation. We also found a non-negligible probability for α Cen A to be a very early subgiant.

1 Introduction

α Cen A has long been an attractive object for stellar physicists. It is the closest star to Earth and very precise data have been collected from interferometry or spectroscopy, in particular ground-based asteroseismology. Moreover, it is part of a relatively close binary, which gives us invaluable information on its mass. However, as of today, there were still uncertainties as to the state of its central regions. A outstanding issue is whether or not this star possesses a convective core. In this study, we aim at assessing rigorously our current state of knowledge about the central regions of α Cen A, given the current observational data, using numerical Bayesian statistical methods to conduct a (probabilistic) modelling of the star.

2 Method

2.1 Bayesian estimation

The goal of Bayesian Statistics is to obtain estimates of the parameters of some model, which we will call in the following a Bayesian statistical model. In mathematical terms, this model is described by what we will call a Posterior Density Function (PDF), $\pi(\theta|X)$, with θ the parameters of the model and X some observational data. It is key to understand that in a Bayesian statistical model the parameters are the random quantities, whereas the data are in a frequentist approach. The data are distributed according to a density $\pi(X|\theta)$, which is called a likelihood if we consider the data as a fixed quantity and let the parameters vary. Being given such a likelihood, one needs to define a prior density on the parameters, $\pi(\theta)$, in order to define completely a Bayesian statistical model. These densities are related through Bayes' formula

$$\pi(\theta|X) \propto \pi(\theta)\pi(X|\theta). \quad (1)$$

In general, whether one is using the frequentist or the Bayesian approach, observations and a parametric model are needed to define a statistical model. This is enough to construct a likelihood, but a Bayesian statistical model has to be supplemented by the definition of prior densities.

2.1.1 *Data*

We make use of various data collected throughout the literature. α Cen A being the closest star to our Sun, there exists a huge amount of high-quality observations, be it in astrometry, spectroscopy, interferometry, photometry or asteroseismology. In the present study we used the effective temperature and luminosity as referenced by Eggenberger, Charbonnel, Talon et al. [1], the radius published by Kervella et al. [2] and the metallicity adopted by Thoul et al. [3]. These are the quantities forming the vector X in the space of observations. We used the seismic data obtained from a 5-night HARPS run [4]. From the 34 frequencies, we were able to obtained 6 small separations $\delta_{n,l=0} = \nu_{n,l=0} - \nu_{n-1,l=2}$.

Measurements of the stellar mass also exist for α Cen A since it forms a close binary together with α Cen B. Having a period of ~80 years, the system has been observed for over a complete revolution and as such is known with good precision. The value of the mass used here is from

★e-mail: mb6215@nyu.edu

Pourbaix et al. [5]. This will enter our model through the prior density.

Table 1: Priors used for the Bayesian statistical model presented in this paper. The first column gives the parameter (mass, age, initial metallicity, initial hydrogen mass fraction, mixing-length parameter, overshoot parameter), the second column gives the functional form of the prior. The last column gives the relevant parameters of the prior (upper and lower bounds, the mean and standard deviation are given for the truncated Gaussian).

Parameter	Prior	Dist. parameters
M ($M_\odot$)	Truncated Gaussian	$[1.07, 1.13, 1.105, 7\times10^{-3}]$
$t_\star$ (Gyr)	Uniform	$[1, 8]$
Z_0	Uniform	$[0.01, 0.04]$
X_0	Uniform	$[0.6, 0.75]$
α	Uniform	$[1.2, 2.6]$
α_{ov}	Discrete/Uniform	$0/[0, 0.75]$

2.1.2 Physical model

To model these observations, we used the Aarhus Stellar Evolution Code [6]. It assumes spherical symmetry for the star and neglects the effect of magnetic fields, rotation, wave-induced mixing, mass-loss or radiative levitation.

We chose to use the OPAL 95 opacity tables supplemented by low-temperature values given by Ferguson et al. [7]. The OPAL 2005 equation of state [8] was adopted. The relative abundances from Grevesse & Noels [9] and the NACRE reaction rates [10] were selected. Convection is described using the classical framework of the mixing-length theory [11], which implies the tuning of a proportionality coefficient α between the mean-free path of convective elements and, as implemented in ASTEC, the characteristic pressure scale height.

In addition to this mixing-length parameter, other parameters have to be given as input to ASTEC. In our case, we will also consider the mass, the age, the initial hydrogen mass fraction and metallicity, and, in some cases, a core-overshoot parameter α_{ov} controlling the amount of additional mixing occurring above potential convective cores. These form the vector θ in the parameter space.

2.1.3 Bayesian statistical model

The last step in defining our Bayesian statistical model consists in defining our likelihood and priors on the parameters. The former is taken to be a multivariate Gaussian. Furthermore, we made the assumption of independent observations, hence reducing the non-zero entries of the covariance matrix to its diagonal components.

The type of prior chosen is given parameter by parameter in Table 1. For each of them we give the lower and upper boundaries of the definition domain. For the truncated Gaussian, these are followed by its mean and standard deviation.

Our model was then sampled using an adaptive Markov chain Monte Carlo (MCMC) algorithm [12].

2.2 Results

After sampling the PDF for θ, we were able to estimate the marginal densities of the individual parameters (by integrating over all the others) but also the PDFs for other functions of the parameters.

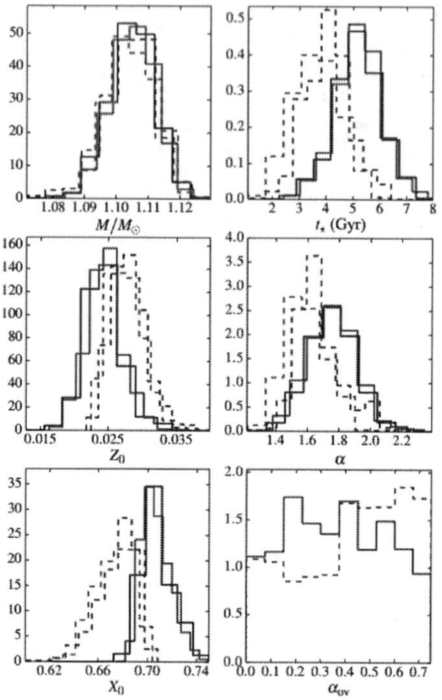

Figure 1: Marginal posterior densities for models without (full lines) and with (dashed lines) convective cores. In blue are shown the results for models without overshoot and in red with overshoot.

The estimates of the stellar parameters are given in Table 2. Interestingly enough the inclusion of overshooting does not impact much the estimates of the parameters. Since many models in our MCMC simulations show convective cores, which hence should be affected by overshoot processes, this hints to some liberty left by the seismic data on the constraining of the stellar innermost regions. To strengthen this picture, one can note that including overshoot does not dramatically increase the odds of α Cen A having a convective core. It simply increases its average radius. In terms of PDFs, this translate into the observation of two potential regimes of solutions that could both reproduce the observational constraints.

These are shown in Fig. 1. There we can see that the convective-core models are, on average, younger. They also have a lower initial hydrogen mass fractions and mixing-length parameters, and higher initial metallicities.

DOI: 10.1051/epjconf/201716003006

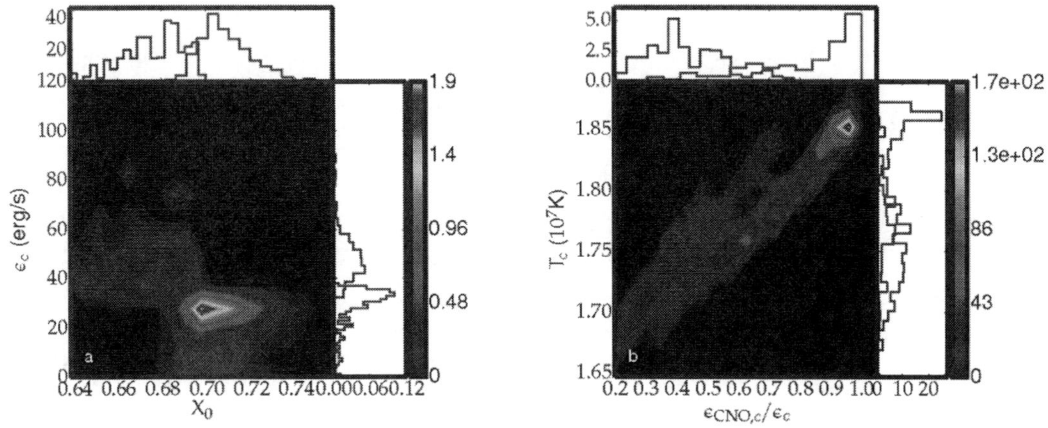

Figure 2: Posterior joint PDFs for (a) X_0 and ε_c the central energy generation rate, (b) $\varepsilon_{CNO,c}/\varepsilon_c$, the relative rate of CNO energy generation rate at the centre, and T_c, the central temperature. These distributions have been obtained from the second run in Table 2. The smaller panels display the marginal PDFs for each individual quantity. Two populations are distinguished, in red the radiative-core models and in green the convective-core models.

These two solutions obtained for α, X_0, Z_0 correspond to different positions on the zero-age main sequence. Basically, ZAMS progenitors for convective-core models of α Cen A are, again on average, more luminous and hotter than their radiative-core counterparts. The age is then simply an adjustment parameter mostly constrained by the need to reproduce the luminosity of the star.

A look at Fig. 2 gives a more detailed picture of the centre of α Cen A. In the left panel is plotted the joint probability density of the initial hydrogen mass fraction and the energy generation rate at the centre of the star. These are for models that include overshooting. As stated above, the models with and without convective cores represent two very distinct populations in terms of initial hydrogen content. This is equally the case for the energy generation rate, and we can clearly see the two above mentioned solutions appear. The models with convective cores have the largest energy production rate at the centre, as expected.

However, the overall picture of a potential convective core in models with overshoot does not exactly match the classical one, i.e. the CNO cycle becoming dominant and, due to the larger sensitivity of this process on temperature, giving rise to large energy flux near the centre. The right panel of Fig. 2 shows the joint probability density for the ratio of the CNO and total energy generation rates and the temperature, all evaluated at the centre of the star. Again, the two solutions to the estimation problem are seen clearly. However, the striking fact is that radiative-core models are, on average, hotter at the centre and dominated by the CNO cycle whereas the convective-core models are mostly powered by the ppII burning chain.

This phenomenon is not unknown and has already been noted by [13, p. 479], in the case of a 2 $M_\odot$ without overshooting for which a convective core develops before the CNO cycle takes over. Note, however, that this does not mean that the CNO cycle is unimportant, it must contribute non-negligibly to the energy generation rate in order for a convective core to develop.

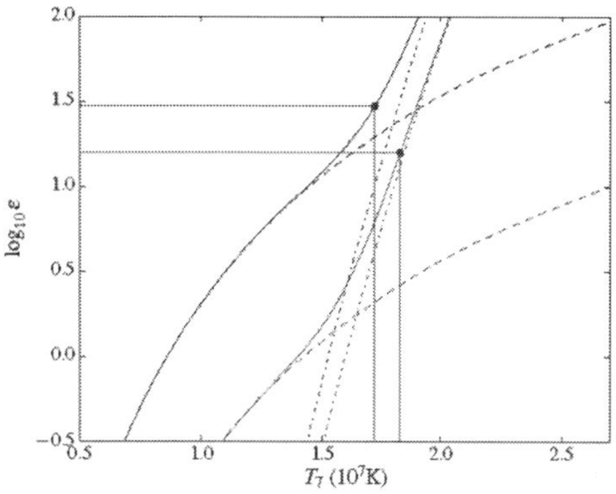

Figure 3: Energy generation rates as functions of the temperature computed for the best a posteriori models with a radiative core (blue lines) and convective core (black lines). The full lines represent the total energy generation rates, the dashed lines the pp-chain rates and the dot-dashed line the CNO rates. The green and red dots mark the position of the centre points of the best convective-core and radiative-core model in the plane.

A simple explanation in our case for this picture to occur is related to the fact that α Cen A, regardless of it having a convective core, lies in a region of the parameter space close to the transition from the pp to the CNO hydrogen-burning process. In this case, since these do not only depend on the temperature but also on the hydrogen abundances, a difference in chemical composition at the centre can therefore account for a hydrogen-rich ppII-dominated core with higher energy generation than

DOI: 10.1051/epjconf/201716003006

Table 2: Estimated parameters of α Cen A from different MCMC simulations without (top) and with (bottom) convective-core overshoot. Column 1-6 give the estimates as the posterior modes of the PDFs for the mass, age, initial metallicity and hydrogen mass-fraction, mixing-length and, if included, overshoot parameter. The upper and lower bounds of associated credible intervals are also given. Column 7 shows the odds for central convective onset. Column 8 gives the estimated fractional radius of the convective core (taken as the full extent of the mixing-homogenized region in case of overshoot) of α Cen A, if present.

M ($M_\odot$)	$t_\star$ (Gyr)	Z_0	X_0	α	α_{ov}	% Convective cores	$r_{cc}/R_\star$
$1.106^{+0.008}_{-0.008}$	$4.7^{+1.2}_{-1.0}$	$0.025^{+0.004}_{-0.003}$	$0.70^{+0.02}_{-0.02}$	$1.68^{+0.19}_{-0.13}$	-	37	$0.038^{+0.011}_{-0.013}$
$1.105^{+0.009}_{-0.007}$	$4.9^{+1.0}_{-1.5}$	$0.025^{+0.004}_{-0.003}$	$0.70^{+0.02}_{-0.02}$	$1.70^{+0.16}_{-0.20}$	$0.51^{+0.15}_{-0.35}$	41	$0.084^{+0.020}_{-0.025}$

a hydrogen-poor mostly powered by the CNO cycle. This is what is shown in Fig. 3 where simple formulations of the energy generation rates [see e.g. 14] have been plotted for the central hydrogen mass fraction of the most probable radiative-core and convective-core models. At their centres, these have, respectively, ρ = 209 g.cm^{-3}, $X = 0.11$, $X_{CNO} = 0.0062$, and ρ = 119 g.cm^{-3}, $X = 0.44$, $X_{CNO} = 0.0066$. These two models clearly fit the picture given above.

We note that this configuration can be related to the the behaviour seen in Fig. 2 where we see that the central temperature of the radiative-core models reach an upper limit. This is a "thermostat" effect that can be explained by the necessity for models fitting the data to reproduce the luminosity. This ensures that the central temperature of radiative-core models will never reach high levels. In turns, this allow for convective-core models with lower central temperature to have higher central energy generation rates, such as seen in Fig. 3.

Further studies including overshoot but also taking into account out-of-equilibrium nuclear reactions [15] also noticed that small convective cores could be sustained for long times by the pp chain. This is not our case here, since only equilibrium rates were considered.

The last point of interest is that approximately 30% of the models obtained from our MCMC simulations have almost completely exhausted their central hydrogen. The main region of hydrogen burning is a shell around the centre. As such they can technically be classified as subgiants. Their isothermal cores remain extremely small. However they mark a clear-cut possibility for α Cen A to have already exited the main sequence.

3 Conclusions

We focused here on the our latest findings concerning the nuclear reaction process at play at the core of α Cen A. Note that this depends on many other physical prescriptions such as the inclusion of diffusion or the precise formulation of the nuclear reaction rates. Changes in the formulation of these processes can lead to models either certainly convective or radiative in their cores. We refer to [16] for further details.

These results point however towards the need to obtain longer and better seismic data for α Cen A, which never ceases to be an intriguing target for stellar physicists. In particular, it will be necessary to obtain long, continuous and, if possible, precise data. This can be achieved either by the means of space missions (TESS potentially and PLATO) or ground-based networks of telescopes (SONG).

Acknowledgements

This material is based upon work supported by the NYU Abu Dhabi Institute under grant G1502. Funding for the Stellar Astrophysics Centre is provided by The Danish National Research Foundation (Grant DNRF106).

References

[1] Eggenberger P. et al., A&A, **417**, 235 (2004)
[2] Kervella P. et al., A&A, **404**, 1087 (2003)
[3] Thoul A. et al., A&A, **402**, 293 (2003)
[4] Bazot M. et al., A&A, **470**, 295 (2007)
[5] Pourbaix D., et al., A&A, **386**, 280 (2002)
[6] Christensen-Dalsgaard J., Ap&SS, **316**, 13 (2008)
[7] Ferguson J. W., ApJ, **623**, 585 (2005)
[8] Rogers F. J., Nayfonov A., ApJ, **576**, 1064 (2002)
[9] Grevesse, N., & Noels, Physica Scripta Volume T A. **47**, 133 (1993)
[10] Angulo C., et al., Nuclear Physics A **656**, 3 (1999)
[11] Böhm-Vitense E., Zeitschrift für Astrophysik **46**, 108 (1958)
[12] Haario H. et al., Bernoulli **7**, 223 (2001)
[13] Clayton D. D., *Principles of stellar evolution and nucleosynthesis* (The University of Chicago Press, Chicago, 1984)
[14] Kippenhahn, R., & Weigert, A. , Stellar Structure and Evolution (Springer-Verlag, Berlin Heidelberg New York, 1994) 468
[15] Deheuvels S., et al., A&A, **514**, A31 (2010)
[16] Bazot, M. et al., MNRAS, **460**, 1254 (2016)

EPJ Web of Conferences **160**, 03007 (2017) DOI: 10.1051/epjconf/201716003007
Seismology of the Sun and the Distant Stars II

Seismic inference of 57 stars using full-length *Kepler* data sets

Orlagh Creevey[1,*], *Travis S.* Metcalfe[2,3], *David* Salabert[4,**], *Savita* Mathur[2], *Mathias* Schultheis[1], *Rafael A.* García[4], *Frédéric* Thévenin[1], *Michaël* Bazot[5], and *Haiying* Xu[6]

[1] *Université Côte d'Azur, Laboratoire Lagrange, CNRS, Blvd de l'Observatoire, CS 34229, 06304 Nice cedex 4, France*
[2] *Center for Extrasolar Planetary Systems, Space Science Institute, 4750 Walnut St. Suite 205, Boulder CO 80301, USA*
[3] *Visiting Scientist, National Solar Observatory, 3665 Discovery Dr., Boulder CO 80303, USA*
[4] *Laboratoire AIM, CEA/DRF-CNRS, Université Paris 7 Diderot, IRFU/SAp, Centre de Saclay, 91191 Gif-sur-Yvette, France*
[5] *Center for Space Science, NYUAD Institute, New York University Abu Dhabi, PO Box 129188, Abu Dhabi, UAE*
[6] *Computational & Information Systems Laboratory, NCAR, P.O. Box 3000, Boulder CO 80307, USA*

Abstract. We present stellar properties of 57 stars from a seismic inference using full-length data sets from *Kepler* (mass, age, radius, distances). These stars comprise active stars, planet-hosts, solar-analogs, and binary systems. We validate the distances derived from the astrometric Gaia-Tycho solution. Ensemble analysis of the stellar properties reveals a trend of mixing-length parameter with the surface gravity and effective temperature. We derive a linear relationship with the seismic quantity $\langle r_{02} \rangle$ to estimate the stellar age. Finally, we define the stellar regimes where the Kjeldsen et al (2008) empirical surface correction for 1D model frequencies is valid.

1 Introduction

Time-series analysis of the full-length *Kepler* data sets of solar-like main sequence and subgiants stars is presented in [1]. They identify modes of oscillations for 66 stars. Using the *Asteroseismic Modeling Portal*, AMP[1], we analyse the individual frequency data from [1] for 57 of these stars and we supplement them with spectroscopic data from [2–7]. In this version of AMP, AMP 1.3, we fit the frequency separation ratios r_{01} and r_{02} [8] to determine the optimal models for each star in our sample.

The optimization method in AMP is genetic algorithm (GA) which efficiently samples the full parameter ranges without imposing constraints on unknown parameters, such as the mixing-length parameter or the initial chemical composition. The results of the GA is a dense clustering of models (1000s) around the optimal parameters. We analyse the distribution of these models to determine the stellar properties, such as mass, radius and age, along with their uncertainties. Details of the methods can be found in [9–11], with the most recent reference containing the tables of stellar parameters for the sample of 57 stars presented here. The results are validated using solar data and independent determinations of luminosity, radii, and ages.

In these proceedings we analyse the derived stellar properties of our sample. We compare their predicted distances with the recent solution provided by the Gaia-Tycho analysis (TGAS, [12], Sect. 2). In Section 3 we derive expressions for estimating the mixing-length parameter and the stellar age based on observed properties. Finally, in Section 4 we explore the range of parameters where the correction proposed by [13] to mimic the so-called *surface effect* related to seismic data is useful.

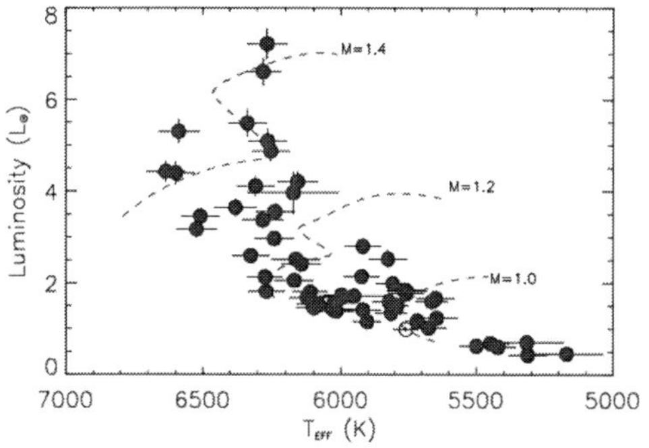

Figure 1: HR diagram showing the position of the 57 *Kepler* stars used in this work. Evolutionary tracks for solar-metallicity models with 1.0, 1.2, and 1.4 $M_\odot$ stellar masses are shown.

2 Asteroseismic distances

We use the stellar luminosity, L, constrained from the asteroseismic analysis to compute the stellar distance, as a parallax. The model surface gravity and the observed $T_{\rm eff}$ and [Fe/H] are used to derive the amount of interstellar absorption between the top of the Earth's atmosphere and the

*e-mail: ocreevey@oca.eu
**Visiting Scientist at 1
[1] https://amp.phys.au.dk/

DOI: 10.1051/epjconf/201716003007

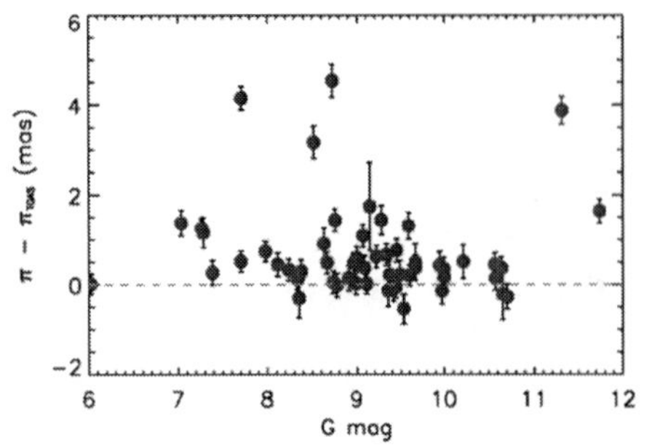

Figure 2: Comparison of the parallaxes derived in this work with those from the TGAS solution. Error bars are TGAS errors.

star, A_{Ks}, by applying the isochrone method described in [14]. Here, ths subscript Ks refers to the 2MASS K_s filter [15]. We compute the bolometric correction BC_{Ks}, using BC_{Ks} = 4.514650 − 0.000524 $T_{\rm eff}$ [16] where the solar bolometric magnitude is 4.72 mag. With L, K_s, A_{Ks}, and BC_{Ks} we derive the the distance to each of the stars in our sample, and then its parallax.

The TGAS catalogue of stellar parallaxes was recently made available through the first Gaia Data Release [17]. A comparison of the parallaxes we derive and these new values is shown in Figure 2. Apart from a few outliers there is an overall good agreement between the independent methods. However, the parallaxes that we derive are systematically larger, with a mean difference of +0.7 mas, just over 2σ the current uncertainties on the TGAS parallaxes. The cause of this systematic difference is currently under investigation. Nevertheless it is worthy to note that such comparisons of independent methods may allow us to reveal errors in the extinction measurements, the bolometric corrections, the $T_{\rm eff}$ or the underlying assumptions in the models which compute the stellar luminosities.

3 Trends in stellar properties

Performing a homogenous analysis on a large sample allows us to check for trends in some stellar parameters, and compare them to trends derived or established by other methods. We performed this check for two parameters: the mixing-length parameter and the stellar age.

3.1 The mixing-length parameter versus $T_{\rm eff}$ and $\log g$

The mixing-length parameter, α, is usually calibrated for a solar model and then applied to all models for a set range of masses and metallicities. However, several authors have shown that this approach is not correct [18, 19]. The values of α resulting from a GA offer an optimal approach to effectively test and subsequently constrain this parameter, since the only assumption is that α lies between 1 and 3.

The distribution of α with $\log g$ and $T_{\rm eff}$ (colour-coded, see figure caption) is shown in Fig. 3. We see that for a

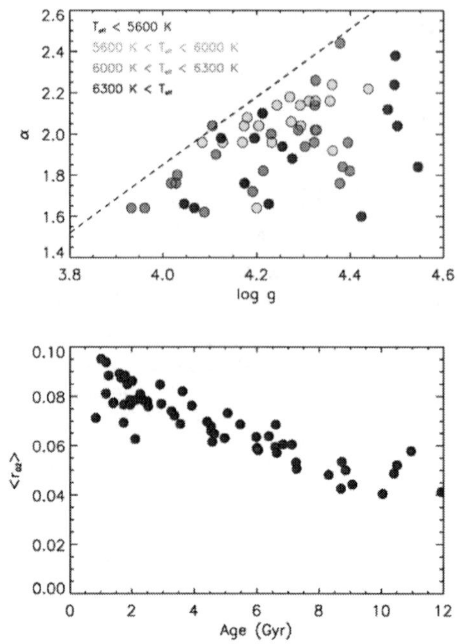

Figure 3: *Top:* Distribution of the $\log g$ and α for our sample. The colour coding is as follows: red – $T_{\rm eff}$ < 5600 K, yellow – 5600 K < $T_{\rm eff}$ < 6000 K, green – 6000 K < $T_{\rm eff}$ < 6300 K, blue – $T_{\rm eff}$ > 6300 K. *Lower:* The distribution of age and $\langle r02 \rangle$.

given value of $\log g$, the value of α has an upper limit, which we can approximate by $\alpha < 1.65 \log g - 4.75$, and this is denoted by the dashed line in the figure. A regression analysis considering the model values of $\log g$, $\log T_{\rm eff}$ and [M/H] yields:

$$\begin{aligned} \alpha &= 5.972778 + 0.636997 \log g \\ &\quad -1.799968 \log T_{\rm eff} + 0.040094\,[{\rm M/H}], \quad (1) \end{aligned}$$

with a mean and rms of the residual to the fit of −0.01 ± 0.15. This equation yields a value of $\alpha = 2.03$ for the known solar properties.

These results agree in part with those derived by [20], using full 3D radiative hydrodynamic calculations for convective envelopes. These authors also found that α increases with $\log g$ and decreases with $T_{\rm eff}$. Our fit indicates a very small dependence on metallicity while their results finds an opposite and more significant trend with this parameter. Our sample, however, does not span a very large range in [M/H], and the low coefficient is consistent with zero within the error bars.

3.2 Age and $\langle r_{02} \rangle$

The r_{02} frequency separation ratios are effective at probing the gradients near the core of the star [8]. As the core is most sensitive to nuclear processing, the r_{02} are a diagnostic of the evolutionary state of the star. Using theoretical models [21] showed a relationship between the mean

DOI: 10.1051/epjconf/201716003007

value of r_{02} and the stellar age. That relationship was recently used by [22] to estimate the age of KIC 7510397 (HIP 93511). Figure 3 shows the distribution of $\langle r_{02} \rangle$ versus the derived age for the sample of stars studied here. A linear fit to these data leads to the following estimate of the stellar age, τ, based on $\langle r_{02} \rangle$

$$\tau = 17.910 - 193.918 \langle r_{02} \rangle, \tag{2}$$

This is, of course, only valid for the range covered by our sample. Note that when inserting the solar value of $\langle r_{02} \rangle = 0.068\ \mu$Hz, Eq. 2 yields an age of 4.7 Gyr, in agreement with the Sun's age as determined by other means.

4 Surface Effects

The comparison of the observed oscillation frequencies from the Sun and other stars with model frequencies calculated from 1D stellar models reveals a systematic error in the models which increases with frequency. These are known as *surface effects*, which arise from incomplete modelling of the near-surface layers and the use of an adiabatic treatment on the stellar oscillation modes. Efforts to improve the stellar modelling is underway, but the application of improved models on a large scale is still out of reach. To alleviate this problem, several authors have advocated for the use of combination frequencies which are insensitive to this systematic offset [8], hence the exclusive use of r_{01} and r_{02} in the AMP 1.3 methodology. However, since individual frequencies contain more information than r_{01} and r_{02}, some authors have derived simple prescriptions to mitigate the surface effect. One such parametrization is that of [13] who suggest a simple correction to the individual frequencies $\delta\nu_{n,l}$ in the form of a power law, namely

$$\delta\nu_{n,l} = a_0 \left(\frac{\nu_{n,l}^{\mathrm{obs}}}{\nu_{\max}} \right)^b \tag{3}$$

where $b = 4.82$ is a fixed value, calibrated by a solar model, a_0 is computed from the differences between the observed and model frequencies [10, 23], $\nu_{n,l}^{\mathrm{obs}}$ is the observed frequency n, l mode and $\nu_{\max}$ is the frequency corresponding to the highest amplitude mode, see [1].

The AMP 1.3 methodology uses exclusively r_{01} and r_{02} as the seismic constraints, hence our results are insensitive to the surface effects. Using the model frequencies of the best-matched models, we can test how useful Eq. 3 is for different stellar regimes along the main sequence and early sub-giant phase. The interest in testing this becomes apparent when we consider that for many stars we do not have a high enough precision on the frequencies, or a sufficiently large range of radial orders, to use r_{01} and r_{02} to effectively constrain the stellar modelling. This is the case for some ground-based observations and for some stars that will be observed by TESS [24], and PLATO [25], where a limited time series of only one to two months may only be available.

In order to test where Eq. 3 is useful, we calculated the residual between the observed oscillation frequencies and the model frequencies corrected for the surface term (Eq. 3): $q_{n,l} = \nu_{n,l}^{\mathrm{obs}} - \nu_{n,l}^{\mathrm{mod}} + \delta\nu_{n,l}$,. Then we defined the metric Q as the median of the square root of the squared residuals:

$$Q = \mathrm{median}\left\{ \sqrt{q_{n,l}^2} \right\}, \tag{4}$$

for all n and all l defined in the region of $0.7 \leq \nu_{n,l}^{\mathrm{obs}}/\nu_{\max} \leq 1.3$. We find values of Q that vary between 0.3 and 10 and this variation is anti-correlated with a_0: as a_0 becomes more negative the match with the model becomes worse.

We define a subsample of 44 stars with the best fits to r_{01} and r_{02}. This subsample in shown in Fig. 4 as open circles for the observed values of $\langle\Delta\nu\rangle$ and T_{eff}, and the derived values of mass and radius. In these figures the stars represented by the dark blue filled circles have $Q \leq 1.0$ and those represented by the light blue filled circles have $1.0 < Q < 1.2$. It is evident from the figures that there exists regions of the parameter space where the surface correction proposed by [13] adequately corrects for the surface term. These regions are highlighted by the dashed lines, defined more conservatively here as $\log g > 4.2$, $T_{\mathrm{eff}} < 6250$ K, $\langle\Delta\nu\rangle > 80\mu$Hz and $\nu_{\max} > 1700\mu$Hz. In physical properties this corresponds to a star with $R < 1.6\ \mathrm{R}_\odot$, $M < 1.3\ \mathrm{M}_\odot$, and $L < 2.5\ \mathrm{L}_\odot$, with no apparent evidence of the age or the metallicity playing any role.

5 Conclusions

In these proceedings we used the derived properties of 57 *Kepler* stars presented in [11] to predict the star's parallaxes, to derive an expression relating the mixing-length parameter and the age to observed properties, and to explore the regions of parameters where the proposed surface correction to individual frequencies by [13] (Eq. 3) is useful. For this latter point, the aim of defining valid regions is to use the correction for data sets where the frequency precision or the range of radial orders is not sufficient for r_{01} and r_{02} to constrain the stellar models. This will be the case for some stars that will be observed in low ecliptic latitudes with TESS and for the *step-and-stare* pointing phases of the PLATO mission.

The parallaxes that we derived in this work were compared to the parallaxes derived from the Tycho-Gaia solution. We found in general a good agreement between our results. The mean differences between our results (ours – theirs) is +0.7 mas and the cause of this error is currently under study. Nevertheless we validate the new parallaxes from [12]. This comparison of parallaxes also highlights the ability of these external measurements to allow us to investigate the sources of systematic errors. We believe that in providing a prior on the luminosity, we may even be able to constrain the initial helium abundance in the star. We look forward to the wealth of forthcoming asteroseismic data on bright nearby stars with TESS and PLATO, and the forthcoming Gaia parallaxes with μas precision.

References

[1] Lund, M. N., Silva Aguirre, V., Davies, G. R., et al., ApJ, **835**, 172 (2017)

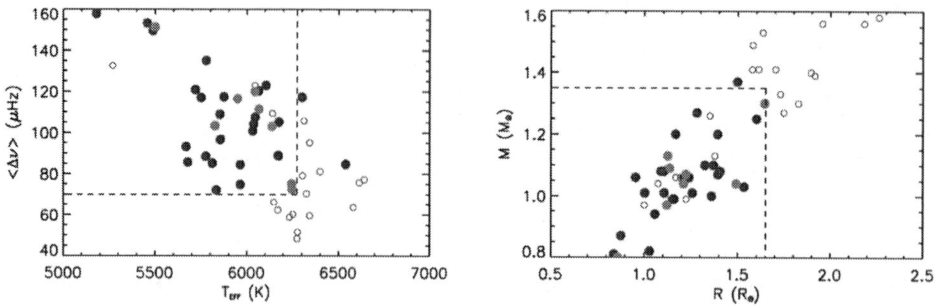

Figure 4: Distribution of observed (left) and derived parameters (right) for a subsample of our stars (open circles). The dark / light blue filled dots represent the stars with $Q \leq 1.0 / 1.2$. The prescription for correcting model frequencies is valid within the region delimited by dashed lines.

[2] I. Ramírez, J. Meléndez, M. Asplund, A&A, **508**, L17 (2009)

[3] L.A. Buchhave, D.W. Latham, A. Johansen, M. Bizzarro, G. Torres, J.F. Rowe, N.M. Batalha, W.J. Borucki, E. Brugamyer, C. Caldwell et al., Nature, **486**, 375 (2012)

[4] M.H. Pinsonneault, D. An, J. Molenda-Żakowicz, W.J. Chaplin, T.S. Metcalfe, H. Bruntt, ApJS, **199**, 30 (2012), 1110.4456

[5] D. Huber, W.J. Chaplin, J. Christensen-Dalsgaard, R.L. Gilliland, H. Kjeldsen, L.A. Buchhave, D.A. Fischer, J.J. Lissauer, J.F. Rowe, R. Sanchis-Ojeda et al., ApJ, **767**, 127 (2013), 1302.2624

[6] W.J. Chaplin, S. Basu, D. Huber, A. Serenelli, L. Casagrande, V. Silva Aguirre, W.H. Ball, O.L. Creevey, L. Gizon, R. Handberg et al., ApJS, **210**, 1 (2014), 1310.4001

[7] M.H. Pinsonneault, Y. Elsworth, C. Epstein, S. Hekker, S. Mészáros, W.J. Chaplin, J.A. Johnson, R.A. García, J. Holtzman, S. Mathur et al., ApJS, **215**, 19 (2014), 1410.2503

[8] I.W. Roxburgh, S.V. Vorontsov, A&A, **411**, 215 (2003)

[9] T.S. Metcalfe, P. Charbonneau, Journal of Computational Physics **185**, 176 (2003), astro-ph/0208315

[10] T.S. Metcalfe, O.L. Creevey, J. Christensen-Dalsgaard, ApJ, **699**, 373 (2009), 0903.0616

[11] O. Creevey, T.S. Metcalfe, D. Salabert, S. Mathur, M. Schultheis, R.A. García, F. Thévenin, M. Bazot, Z. Haiying, A&A, subm. (2017)

[12] D. Michalik, L. Lindegren, D. Hobbs, A&A, **574**, A115 (2015), 1412.8770

[13] H. Kjeldsen, T.R. Bedding, J. Christensen-Dalsgaard, ApJL, **683**, L175 (2008), 0807.1769

[14] M. Schultheis, G. Zasowski, C. Allende Prieto, F. Anders, R.L. Beaton, T.C. Beers, D. Bizyaev, C. Chiappini, P.M. Frinchaboy, A.E. García Pérez et al., AJ, **148**, 24 (2014), 1405.2180

[15] M.F. Skrutskie, R.M. Cutri, R. Stiening, M.D. Weinberg, S. Schneider, J.M. Carpenter, C. Beichman, R. Capps, T. Chester, J. Elias et al., AJ, **131**, 1163 (2006)

[16] P. Marigo, L. Girardi, A. Bressan, M.A.T. Groenewegen, L. Silva, G.L. Granato, A&A, **482**, 883 (2008), 0711.4922

[17] Gaia Collaboration, ArXiv e-prints (2016), 1609.04153

[18] A. Bonaca, J.D. Tanner, S. Basu, W.J. Chaplin, T.S. Metcalfe, M.J.P.F.G. Monteiro, J. Ballot, T.R. Bedding, A. Bonanno, A.M. Broomhall et al., ApJL, **755**, L12 (2012), 1207.2765

[19] O.L. Creevey, F. Thévenin, T.S. Boyajian, P. Kervella, A. Chiavassa, L. Bigot, A. Mérand, U. Heiter, P. Morel, B. Pichon et al., A&A, **545**, A17 (2012), 1207.5954

[20] Z. Magic, A. Weiss, M. Asplund, A&A, **573**, A89 (2015), 1403.1062

[21] Y. Lebreton, J. Montalbán, *Stellar ages from asteroseismology*, in *The Ages of Stars*, edited by E.E. Mamajek, D.R. Soderblom, R.F.G. Wyse (2009), Vol. 258 of *IAU Symposium*, pp. 419–430, 0811.2908

[22] T. Appourchaux, H.M. Antia, W. Ball, O. Creevey, Y. Lebreton, K. Verma, S. Vorontsov, T.L. Campante, G.R. Davies, P. Gaulme et al., A&A, **582**, A25 (2015)

[23] T.S. Metcalfe, O.L. Creevey, G. Doğan, S. Mathur, H. Xu, T.R. Bedding, W.J. Chaplin, J. Christensen-Dalsgaard, C. Karoff, R. Trampedach et al., ApJS, **214**, 27 (2014), 1402.3614

[24] G.R. Ricker, J.N. Winn, R. Vanderspek, D.W. Latham, G.Á. Bakos, J.L. Bean, Z.K. Berta-Thompson, T.M. Brown, L. Buchhave, N.R. Butler et al., Journal of Astronomical Telescopes, Instruments, and Systems **1**, 014003 (2015)

[25] H. Rauer, C. Catala, C. Aerts, T. Appourchaux, W. Benz, A. Brandeker, J. Christensen-Dalsgaard, M. Deleuil, L. Gizon, M.J. Goupil et al., Experimental Astronomy **38**, 249 (2014), 1310.0696

EPJ Web of Conferences **160**, 03008 (2017) DOI: 10.1051/epjconf/201716003008
Seismology of the Sun and the Distant Stars II

Amplitude modulation in δ Sct stars: statistics from an ensemble of *Kepler* targets

Dominic M. Bowman[1,*], *Donald W.* Kurtz[1], *Michel* Breger[2], *Simon J.* Murphy[3,4], and *Daniel L.* Holdsworth[1]

[1] *Jeremiah Horrocks Institute, University of Central Lancashire, Preston PR1 2HE, UK*
[2] *Department of Astronomy, University of Texas, Austin, TX 78712, USA*
[3] *Sydney Institute for Astronomy (SIfA), School of Physics, The University of Sydney, NSW 2006, Australia*
[4] *Stellar Astrophysics Centre, Department of Physics and Astronomy, Aarhus University, DK-8000 Aarhus C, Denmark*

Abstract. The results of a search for amplitude modulation of pulsation modes in 983 δ Sct stars, which have effective temperatures between $6400 \leq T_{\rm eff} \leq 10\,000$ K in the *Kepler* Input Catalogue and were continuously observed by the *Kepler* Space Telescope for 4 yr, are presented. A total of 603 δ Sct stars (61.3 per cent) are found to exhibit at least one pulsation mode that varies significantly in amplitude over 4 yr. Furthermore, it is found that amplitude modulation is not restricted to a specific region within the classical instability strip in the HR diagram, therefore its cause is not necessarily dependent on stellar parameters such as $T_{\rm eff}$ or log g. On the other hand, many δ Sct stars show constant pulsation amplitudes demonstrating that the cause of pulsational non-linearity in these stars is not well understood.

1 Introduction

The multiperiodic pulsators known as δ Sct stars are the most common group of variable A and F stars, and are found at the intersection of the classical instability strip and the main sequence in the HR diagram. On the main sequence, δ Sct stars typically range from A2 to F2 in spectral type [8] and within the effective temperature range of $6300 \leq T_{\rm eff} \leq 8600$ K [10]. Pulsations in δ Sct stars are excited by the κ-mechanism operating in the He II ionisation zone, which produces low radial order pressure (p) modes. Typical pulsation periods observed in δ Sct stars are of order a few hours, but can be as short as 15 min [10]. The *Kepler* Space Telescope provided a vast increase in the number of δ Sct stars observed at a high photometric precision and high duty cycle over a time span of 4 yr. In this article, the results from a search for amplitude modulation and non-linearity using 983 δ Sct stars that were continuously observed by *Kepler* for 4 yr are presented [4].

There are different mechanisms that can cause non-linearity in a pulsating star. For example, combination frequencies are mathematical sum and difference frequencies of pulsation mode frequencies, ν_i and ν_j, that have the form $n\nu_i \pm m\nu_j$, where n and m are integers. Possible mechanisms to explain combination frequencies are that the stellar medium does not respond linearly to a pulsation wave, or that the dependence of emergent flux variation is not a linear transformation from the temperature variation ($F = \sigma T^4$), which are often grouped into what is termed a non-linear distortion model [11]. These non-linear effects differ to families of pulsation modes that are resonantly excited due to the coupling of modes inside a star [5, 6]. This form of pulsational non-linearity gives rise to variable frequencies and amplitudes in pulsation modes over time, which will appear as a cluster of unresolved peaks in an amplitude spectrum if the length of observations is shorter than the modulation cycle [6].

1.1 Mode coupling in δ Sct stars

Coupled pulsation modes are grouped into a family containing one child mode with frequency ν_1, and two parent modes with frequencies ν_2 and ν_3, which satisfy a frequency resonance condition [5, 6] of

$$\nu_1 \simeq \nu_2 \pm \nu_3 , \tag{1}$$

but must also satisfy the phase relation of

$$\phi_1 = \phi_2 \pm \phi_3 . \tag{2}$$

However, equations 1 and 2 do not distinguish if ν_1 is a resonantly-excited pulsation mode frequency, or is a combination frequency that describes the non-linear interaction of the parent pulsation modes. To make this distinction, the amplitude of a child mode can be modelled as a product of the two parent mode amplitudes using

$$A_1 = \mu_c (A_2 A_3) , \tag{3}$$

where μ_c is defined as a coupling coefficient [4, 5]. The coupling coefficient μ_c in equation 3 represents the strength of non-linearity of pulsation modes in a star and thus how much coupling exists among the members of a

*e-mail: dmbowman@uclan.ac.uk

family [4, 5]. Therefore, the testable prediction for resonant mode coupling between a child and two parent modes in a δ Sct star is amplitude modulation in three similar-amplitude modes resulting in large values of μ_c [4, 5].

2 Searching for amplitude modulation in *Kepler* δ Sct stars

An ensemble of 983 δ Sct stars was comprised by selecting *Kepler* targets that met all of the following criteria:

1. $6400 \leq T_{\rm eff} \leq 10\,000$ K in the KIC;
2. Observed continuously from Q0 (or Q1) to Q17;
3. Contain peaks between $4 \leq \nu \leq 24$ d^{-1} with amplitudes greater than 0.10 mmag.

2.1 Identifying amplitude modulation

A tracking routine was used to determine amplitude and phase variations (at fixed frequency) for a maximum of 12 peaks in the amplitude spectrum using the 4-yr data set for each δ Sct star. The main motivation of using 12 peaks was to identify the dominant (changes in) pulsational behaviour in each star. For a star that pulsates in only a few modes, extracting 12 peaks was more than sufficient. On the other hand, for a star that pulsates in dozens of modes and has hundreds of combination frequencies in its amplitude spectrum, 12 peaks may not be enough to fully disentangle the star, but does provide vital information on the most dominant behaviour. Note that only peaks with amplitudes greater than 0.10 mmag were extracted in this analysis, so fewer than the maximum number of 12 peaks can be extracted for a star.

After extracting the frequencies by sequentially pre-whitening a star's amplitude spectrum, the frequencies were optimised using a simultaneous multi-frequency non-linear least-squares fit to the 4-yr data set, ensuring the highest possible frequency and amplitude precision were obtained. The data set of each star was divided into 30 time bins, each 100 d in length (except the last bin) with a 50-d overlap, and amplitude and phase were optimised at fixed frequency using linear least-squares in each bin for each peak. This approach has previously been used to study the amplitude and phase variability in the δ Sct star KIC 7106205 [2]. The amplitudes and phases of each time bin for all extracted peaks were plotted against time in what are termed tracking plots [4], with the amplitude spectrum and tracking plot for KIC 7106205 given in the top and middle panels of figure 1, respectively, as an example.

2.2 Defining significant amplitude modulation

A mean amplitude was calculated from the time bins for each peak and any bins that were more than $\pm5\sigma$ in amplitude away from this mean were flagged. The definition of significant amplitude modulation for each peak was defined as at least 15 (i.e., half) of its amplitude bins lying more than $\pm5\sigma$ from its mean value [4]. This is demonstrated graphically for the three highest-amplitude pulsation modes in the δ Sct star KIC 7106205 in the bottom panel of figure 1. Each δ Sct star in the ensemble of 983 stars was classified as either an AMod (**A**mplitude **Mod**ulated) star if at least a single peak exhibited significant amplitude modulation, or a NoMod (**No Mod**ulation) star if all the extracted peaks were constant in amplitude [4].

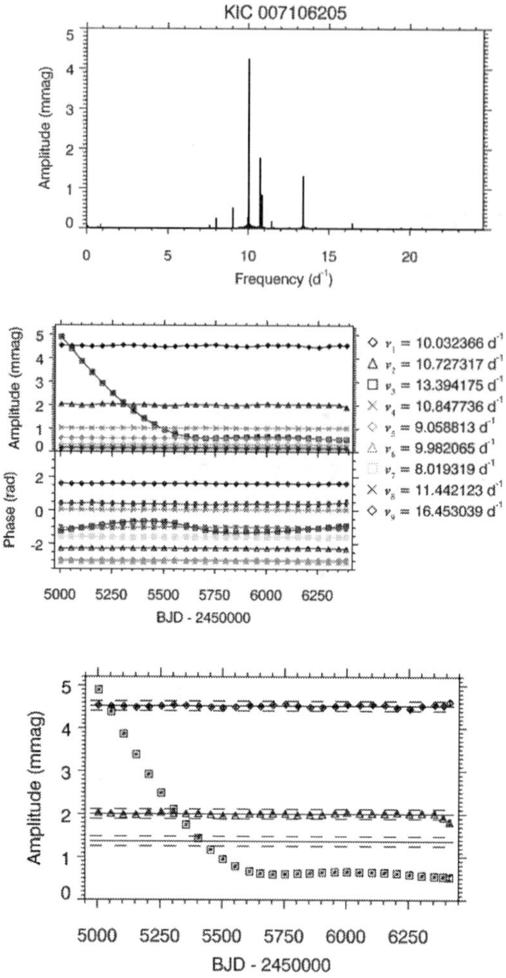

Figure 1: The top panel is the amplitude spectrum of the δ Sct star KIC 7106205, calculated using 4 yr of *Kepler* data. The middle panel is the amplitude and phase tracking plot for the nine significant peaks in KIC 7106205. The bottom panel is a demonstration of how a peak is defined as having significant amplitude modulation if 15 (half) of its bins lie outside $\pm5\sigma$ of the mean amplitude, where the solid and dashed lines represent the mean amplitude and its $\pm5\sigma$ significance interval, respectively, for each series of time bins.

 DOI: 10.1051/epjconf/201716003008

3 Results

In this section, the results for a thorough search for amplitude modulation using 983 δ Sct stars observed continuously by the *Kepler* Space Telescope for 4 yr are presented. The analysis of amplitude modulation in the ensemble of 983 δ Sct stars has been compiled into a single catalogue [4], which is available online as a PDF[1].

3.1 Constant pulsation mode amplitudes: NoMod stars

An example of a δ Sct star with constant pulsation mode amplitudes (i.e., a NoMod star) is KIC 2304168, which has been previously studied using *Kepler* data, asteroseismically modelled, and determined to pulsate in the fundamental and first overtone radial modes [1]. The amplitude spectrum and tracking plot of KIC 2304168 are shown in the top and bottom panels of figure 2, respectively. KIC 2304168 has an effective temperature of $T_{\rm eff}$ = 7220 ± 270 K and a surface gravity of log g = 3.67 ± 0.19 (cgs) [7], which are consistent with the results from stellar modelling [1]. The identification of radial pulsation modes in this star is also indicated by the period ratio of ν_1 = 8.107739 d^{-1} and ν_2 = 10.495495 d^{-1} being 0.7725, which is predicted by stellar models [9]. Therefore, the case study of KIC 2304168 demonstrates that it is possible for δ Sct stars to be coherent pulsators with constant pulsation mode amplitudes over a time scale of 4 yr.

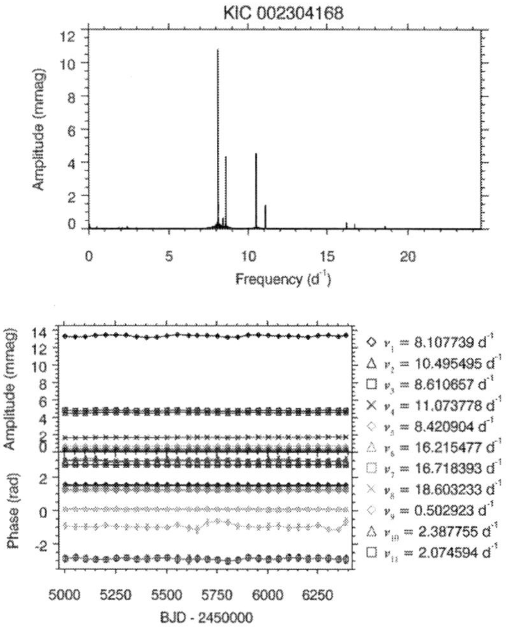

Figure 2: The amplitude spectrum and the tracking plot for the NoMod δ Sct star KIC 2304168 are shown in the top and bottom panels, respectively. There are eleven significant peaks in the amplitude spectrum of KIC 2304168, which all show constant amplitudes over the 4-yr *Kepler* data set using the $\pm 5\sigma$ definition of significant amplitude modulation.

3.2 Variable pulsation mode amplitudes: AMod stars

An archetypal example of amplitude modulation in a δ Sct star is KIC 7106205 [2, 3]. This star was discovered to contain only a single pulsation mode that varied significantly in amplitude, whilst all other pulsation modes stayed constant in amplitude and phase [2]. Later, ground-based data from WASP were used to extend the study of amplitude modulation 2 yr prior to the launch of the *Kepler* Space Telescope [3]. The amplitude spectrum and tracking plot for KIC 7106205 are shown in the top and middle panels of figure 1, respectively, which demonstrate that the visible pulsation energy budget is not conserved in this δ Sct star.

Another example of an AMod star within the ensemble of 983 δ Sct stars is KIC 4733344, with its amplitude spectrum and tracking plot shown in the top and bottom panels of figure 3, respectively. KIC 4733344 has an effective temperature of $T_{\rm eff}$ = 7210 ± 260 K and a surface gravity of log g = 3.50 ± 0.23 (cgs) [7]. KIC 4733344 pulsates with three relatively high-amplitude pulsation modes that exhibit significant amplitude modulation, which are labelled as ν_1, ν_2 and ν_3 in order of decreasing amplitude in figure 3. Frequency analysis of KIC 4733344 revealed that its pulsation mode frequencies ν_2 = 7.226764 d^{-1} and ν_3 = 9.412445 d^{-1} have a period ratio of 0.7678, which is typically associated with the fundamental and first overtone radial modes [9].

In this section, two δ Sct stars with similar $T_{\rm eff}$ and log g values have been presented, specifically the NoMod δ Sct star KIC 2304168 and the AMod δ Sct star KIC 4733344. However, it is not immediately obvious why these two stars can be similar in their stellar parameters, and yet be so different in their observed pulsational behaviour. Further work is needed to understand the diverse range of observed pulsations in δ Sct stars, and especially the mechanism that causes amplitude modulation.

3.3 Amplitude modulation across the classical instability strip

Using the described methodology, amplitude modulation was discovered to be common among δ Sct stars, with 603 stars (61.3 per cent of the 983 δ Sct stars in the ensemble) being classified as AMod [4]. However, amplitude modulation is not ubiquitous in δ Sct stars and further work is needed to address why this is the case. The distribution of AMod and NoMod stars on a $T_{\rm eff}$ − log g diagram is shown in figure 4, which shows that amplitude modulation is not restricted to a specific region within the classical instability strip in the HR diagram, and is unlikely to be related to these stellar parameters.

[1] http://uclandata.uclan.ac.uk/42/

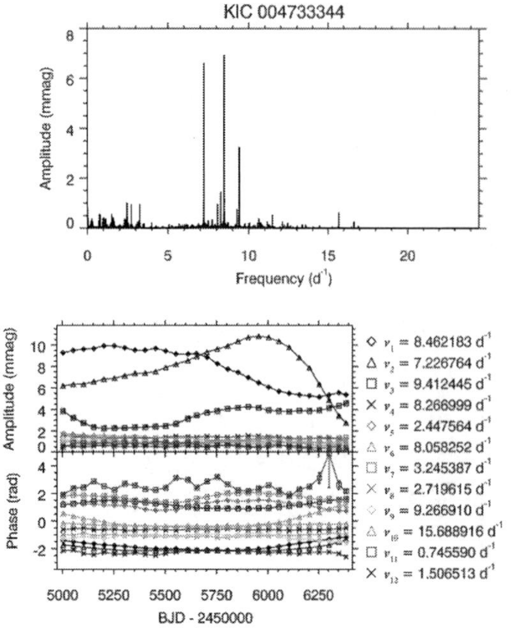

Figure 3: The amplitude spectrum and tracking plot for the AMod δ Sct star KIC 4733344 are shown in the top and bottom panels, respectively. There are many peaks in the amplitude spectrum of KIC 4733344, with only the twelve highest-amplitude peaks shown in the tracking plot.

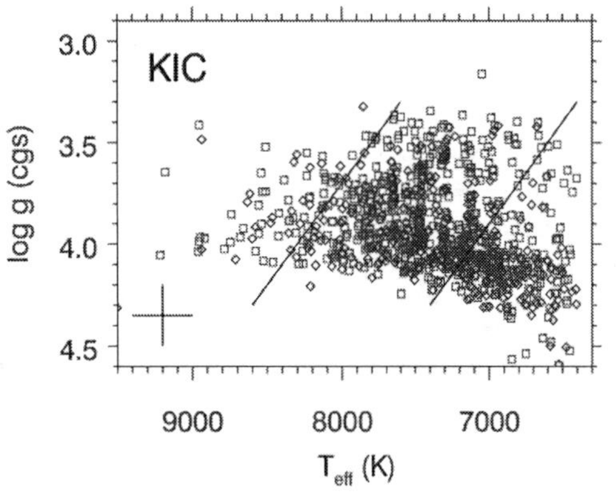

Figure 4: The ensemble of 983 δ Sct stars in a $T_{\rm eff}$ – log g diagram using KIC parameters, with each star's location given as a red square or a blue diamond for if it is AMod or NoMod, respectively. The cross in the bottom-left corner of the figure represents the typical uncertainty for each point, and the solid lines are the observed blue and red edges of the classical instability strip [8].

4 Conclusions

The *Kepler* data set provided extremely high frequency and amplitude precision, which was used to track amplitude and phase at fixed frequency in 100-d bins with a 50-d overlap, for a maximum number of 12 peaks with amplitudes greater that 0.10 mmag in 983 δ Sct stars [4]. These results were collated into an amplitude modulation catalogue and a selection of case study stars have been used to demonstrate the diversity in observed pulsational behaviour. A total of 603 δ Sct stars (61.3 per cent) exhibit at least one pulsation mode that varies significantly in amplitude over 4 yr, and so amplitude modulation is common, but not ubiquitous, in δ Sct stars [4].

For many δ Sct stars, the visible pulsation mode energy is not conserved in 4 yr of *Kepler* observations [4]. For example, the δ Sct star KIC 7106205 has only a single variable pulsation mode, which is shown in figure 1, and the δ Sct star KIC 4733344 contains many variable pulsation modes, which is shown in figure 3. Amplitude modulation in δ Sct stars may be caused by resonant mode coupling of parent modes to a high-degree child mode or internal g modes [6], but this is difficult to test using *Kepler* data [4]. This work has demonstrated that non-linearity is important when studying δ Sct stars on time scales as short as 4 yr [4].

The catalogue of 983 δ Sct stars using 4 yr of *Kepler* data demonstrates that observations spanning years (and longer) are often needed to study and resolve pulsational behaviour in these stars [4], which will be useful for comparison purposes when studying observations of δ Sct stars from K2, TESS and PLATO. Eventually, these missions will observe a large area of the sky but only for a short length of time. Therefore, *Kepler* may represent the best data set for studying δ Sct stars as 4 yr of continuous observations may not be surpassed for some time.

References

[1] Balona L. A., Dziembowski W. A., MNRAS 417, 591 (2011)
[2] Bowman D. M., Kurtz D. W., MNRAS 444, 1909 (2014)
[3] Bowman D. M., Holdsworth D. L., Kurtz D. W., MNRAS 449, 1004 (2015)
[4] Bowman D. M., Kurtz D. W., Breger M., Murphy S. J., Holdsworth D. L., MNRAS 460, 1970 (2016)
[5] Breger M., Montgomery M. H., ApJ 783, 89 (2014)
[6] Dziembowski W., Acta Astronomica 32, 147 (1982)
[7] Huber D., et al., ApJS 211, 2 (2014)
[8] Rodríguez E., Breger M., A&A 366, 178 (2001)
[9] Stellingwerf R. F., ApJ 227, 935 (1979)
[10] Uytterhoeven K. et al., A&A 534, A125 (2011)
[11] Wu Y., MNRAS 323, 248 (2001)

 DOI: 10.1051/epjconf/201716003009

Improving 1D Stellar Models with 3D Atmospheres

Jakob Rørsted Mosumgaard[1,⋆], *Víctor* Silva Aguirre[1], *Achim* Weiss[2], *Jørgen* Christensen-Dalsgaard[1], and *Regner* Trampedach[3,1]

[1] *Stellar Astrophysics Centre (SAC), Department of Physics and Astronomy, Aarhus University, Ny Munkegade 120, DK-8000 Aarhus C, Denmark*
[2] *Max-Planck-Institut für Astrophysik, Karl-Schwarzschild-Str. 1, D-85748 Garching, Germany*
[3] *Space Science Institute, 4750 Walnut Street, Suite 205, Boulder, CO 80301 USA*

Abstract. Stellar evolution codes play a major role in present-day astrophysics, yet they share common issues. In this work we seek to remedy some of those by the use of results from realistic and highly detailed 3D hydrodynamical simulations of stellar atmospheres. We have implemented a new temperature stratification extracted directly from the 3D simulations into the Garching Stellar Evolution Code to replace the simplified atmosphere normally used. Secondly, we have implemented the use of a variable mixing-length parameter, which changes as a function of the stellar surface gravity and temperature – also derived from the 3D simulations. Furthermore, to make our models consistent, we have calculated new opacity tables to match the atmospheric simulations. Here, we present the modified code and initial results on stellar evolution using it.

1 Introduction

Understanding stellar structure and evolution is one of the key ingredients in astrophysics. Perhaps the primary tool for doing so is evolutionary calculations of stellar structure, i.e. one-dimensional numerical models. These have been developed and tested through decades; as a result they are highly optimised and very efficient. However, in several aspects they are also highly simplified and can be improved. This work focuses on two of these: the use of a simple temperature stratification as part of an idealised atmosphere for the model and the treatment of stellar convection.

The artificial atmosphere is needed to supply the outer boundary conditions for stellar evolution codes in order to solve the differential equations of stellar structure (see e.g. [1]). The boundary conditions are obtained by integrating an analytical expression for the temperature as a function of optical depth – a so-called $T(\tau)$ relation. A general issue is that these relations are much too simple to capture the true stratification of stellar atmospheres.

In 1D models, convection is typically treated using a parametric description. The most common is the mixing-length theory (MLT) as introduced for stellar models by [2, 3] or some variant thereof. These parametric theories rely on rather crude assumptions, which are certainly not fulfilled in stars (see e.g. [4]). Furthermore, a free parameter (the mixing-length parameter, α_{MLT}) is "calibrated" against the Sun by some observational constraint, usually the solar radius. The solar value is then assumed to be valid for all stars across the HR diagram.

To improve the models, highly sophisticated three-dimensional simulations of stellar atmospheres (see the next section) can be used in different ways. One approach is to patch the averaged 3D structure on top of a 1D model, as described by [5]. This of course requires the 1D model to match the 3D simulation perfectly.

In this work, we employ a different approach, where condensed information – specifically targeted at the two issues mentioned above – is extracted from the simulations and interpolated to any point between them. The advantage of this approach is that it can be used in stellar structure models for full evolutionary calculations; the patched models are static by nature. An implementation similar to this work, however not fully consistent, was attempted by [6].

In section 2 we review the 3D simulations used in this work and how useful information is extracted from them. Our implementation is described in detail in section 3, and finally we present a few initial results in section 4.

2 Simulations of Stellar Atmospheres

Stellar atmospheres are very complex regions, where hydrodynamics and radiative transfer interact. For a better understanding of this important part of stars several authors have employed advanced 3D simulations, which account for both hydrodynamics and radiative transfer. These simulations have been used extensively to study the convection and granulation in the outer parts of our Sun (e.g. [7, 8]).

More recently, several groups have produced grids of 3D simulations to investigate the behaviour of stars different from our Sun. In this work, we will use the grid of simulations from Trampedach et al. [9], which consists of

⋆e-mail: jakob@phys.au.dk

37 simulations at solar metallicity – the physical parameters (effective temperature, $T_{\rm eff}$, and surface gravity, $\log g$) of the different simulations are shown in Figure 1, which is described in further details below.

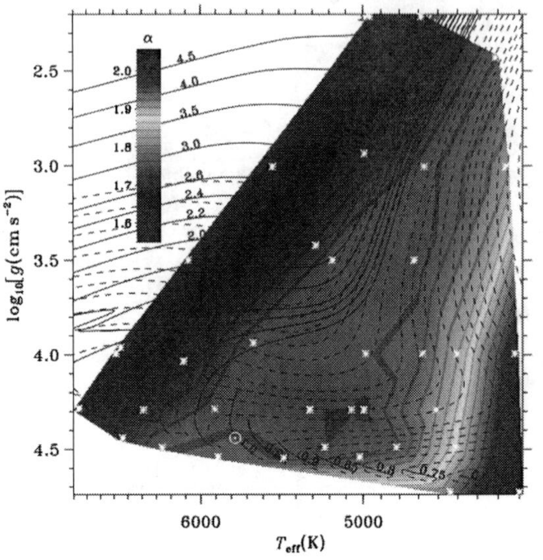

Figure 1: The grid of 3D simulations from [9] at solar metallicity. The simulation parameters are marked by white asterisks and the Sun by ⊙. The coloured background shows the behaviour of $\alpha_{\rm MLT}$ as a function of $T_{\rm eff}$ and $\log g$. The red lines represents evolutionary tracks, but they are unimportant for the present work. The figure is reproduced from [10] with permission.

2.1 Condensed 3D Simulations

For the purpose of stellar evolutionary calculations, the full (averaged) 3D structures cannot be used directly for several reasons. First of all, they only sample discrete points in the HR diagram, and it is not feasible to make the grids much denser since the simulations are extremely computationally expensive.[1] Secondly, due to differences in physics (e.g. treatment of convection and turbulent pressure) they are not directly compatible with the current 1D models.

For these reasons, Trampedach et al. [11] have devised a way of distilling the 3D simulations to be easily usable in stellar structure models. In short, the temperature stratification is extracted from the simulations in the form of the (generalised) Hopf function, q, as a function of Rosseland mean optical depth, τ. This is then used for reconstructing the photospheric transition, from optically thick to optically thin, of the 3D simulation in the 1D model.

Furthermore, it is possible to calibrate the mixing-length parameter, $\alpha_{\rm MLT}$, against the 3D simulations. This correctly connects the entropy of the adiabatic layers deep in the star with the surface layers in the 3D atmosphere. Besides, the need to assume a solar-calibrated value for all stars is eliminated. This is done by [10] for all simulations in the grid – the actual procedure is very similar to the patching of 3D simulations to 1D models mentioned earlier. The corresponding $\alpha_{\rm MLT}$'s are stored in the table with $q(\tau)$, and the interpolated behaviour as a function of $T_{\rm eff}$ and $\log g$ is shown in Figure 1. It should be noted that the calibrated values are of course only strictly correct, if the stellar evolution code uses the same MLT-formulation as the 1D (envelope) code used for the calibration.

[1]One could in principle interpolate between the full simulations, but such a scheme has yet to be developed.

3 Implementation

We have implemented the results from [10, 11] (in the form of a table containing $q(\tau)$ and $\alpha_{\rm MLT}$) into the Garching Stellar Evolution Code (GARSTEC, see [12]). A simple sketch of the implementation is shown in Figure 2.

The general principle is that the $q(\tau)$ and $\alpha_{\rm MLT}$ corresponding to $T_{\rm eff}$ and $\log g$ of the star are found by interpolation[2] in the grid and stored in the program. In each iteration of the code, these values are updated to always match the current stellar parameters.

The quantities are then supplied to the different parts of the code; the atmospheric integration (providing the outer boundary conditions) receives the $q(\tau)$ and the routine handling MLT-convection uses the new $\alpha_{\rm MLT}$. Moreover, for reasons explained below, the convection module also needs access to $q(\tau)$.

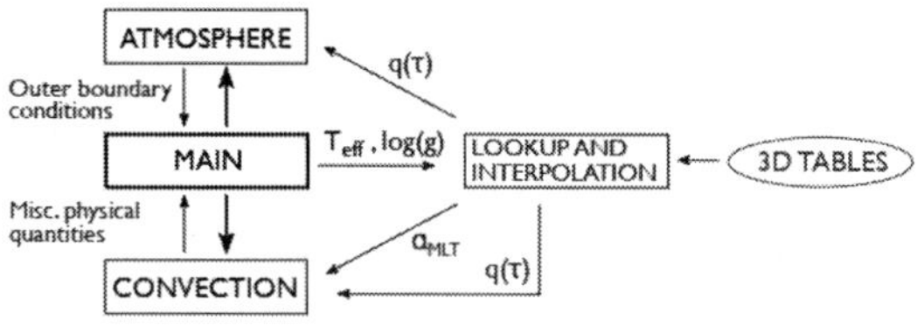

Figure 2: Schematic overview of our implementation marked in blue. The black represents modules in GARSTEC, while the green circle marks the tables from [10, 11]. Details are given in the text.

3.1 Changes to a Stellar Model

How a stellar structure model is modified by our implementation is outlined in Figure 3.

In the atmosphere, the generalised Hopf function from the simulations, $q(\tau)$, is used to provide the stratification instead of the standard analytical Eddington atmosphere, $T_{\rm edd}(\tau)$. Assuming radiative equilibrium, the new temperature structure is found from $q(\tau)$ using

$$\frac{4}{3}\left(\frac{T}{T_{\rm eff}}\right)^4 = q(\tau) + \tau\ , \tag{1}$$

as explained by [11, eq. 9].

[2]We use the interpolation scheme supplied by [10], which is a linear interpolation over a Thiessen triangulation of the irregular grid of simulations.

DOI: 10.1051/epjconf/201716003009

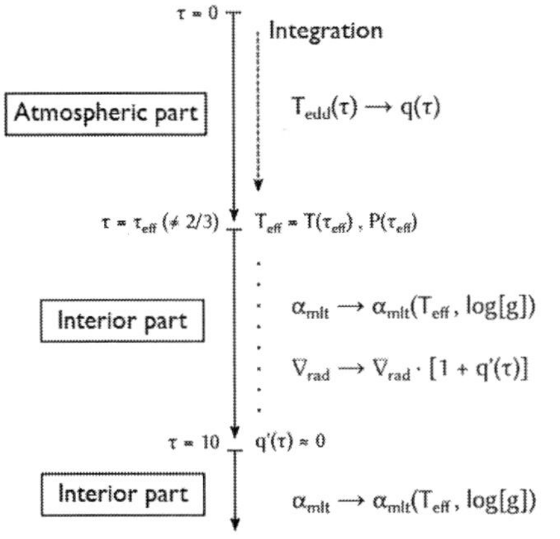

Figure 3: The changes to a stellar structure model as a result of our implementation. In the atmosphere, a different temperature stratification is used and the integration point is altered. In the outer, convective parts of the interior model, the radiative gradient is modified according to eq. (3). Everywhere in the interior of the star, the calibrated $\alpha_{\mathrm{MLT}}(T_{\mathrm{eff}}, \log g)$ is used.

Another alteration is the bottom point of the atmosphere, i.e. where the integration is stopped – note that in stellar structure models this point is usually defined to have $T = T_{\mathrm{eff}}$. The Eddington atmosphere is anchored at $\tau = 2/3$, but that is not the case for the relations extracted from 3D simulations. Thus before the actual integration, we find the point τ_{eff} in the interpolated temperature structure where $T = T_{\mathrm{eff}}$.[3]

Moving on to the interior part of the model, the first change is the use of the 3D-calibrated variable mixing-length parameter, $\alpha_{\mathrm{MLT}}(T_{\mathrm{eff}}, \log g)$, throughout the model. We are not using the $\alpha_{\mathrm{MLT,\,grid}}$ directly as provided from the tables, but rather using it differentially as recommended by [11, 13]. In other words, we introduce a scaling factor

$$\alpha_{\mathrm{MLT}}(T_{\mathrm{eff}}, \log g) = \frac{\alpha_{\odot}}{\alpha_{\odot,\,\mathrm{grid}}} \cdot \alpha_{\mathrm{MLT,\,grid}}(T_{\mathrm{eff}}, \log g)\,, \quad (2)$$

where $\alpha_{\odot}$ is what we obtain from a solar calibration in GARSTEC and $\alpha_{\odot,\,\mathrm{grid}}$ is the value given for the solar model in the grid. This ensures that the solar model is calibrated to the correct radius with this variable α_{MLT}.

When using Eq. (1) for the temperature structure, the temperature gradient for a stratification in radiative equilibrium (i.e. no convective flux), $\nabla_{\mathrm{rad}} \equiv (\partial \ln T/\partial \ln p)_{\mathrm{rad}}$, needs to be corrected from its optically deep value of $\tilde{\nabla}_{\mathrm{rad}}$, as

$$\nabla_{\mathrm{rad}} = \tilde{\nabla}_{\mathrm{rad}} \cdot [1 + q'(\tau)]\,, \quad (3)$$

according to [11, eq. 35]. Here $q'(\tau)$ is the derivative of the Hopf function with respect to τ and $\tilde{\nabla}_{\mathrm{rad}}$ is the usual expression for the radiative gradient, based on the diffusion approximation. This modification is applied before entering the MLT routine of GARSTEC; thus the resulting temperature gradient ∇ is properly corrected.[4] We apply this correction factor until $\tau = 10$, since $q'(\tau)$ at larger depth is always below 10^{-4} (and typically below 10^{-5}), where the adiabatic layers are reached.

3.2 Important Remarks

At this point it important to stress that it is crucial for the stellar evolution code to use the same microphysics as the 3D simulations for consistency.

To match the simulations, we use the MHD equation of state (EOS) [14–16]. To cover the necessary range of temperatures, we complement it with OPAL EOS [17] at higher temperatures.

For the 3D simulations, [11] calculated their own atmospheric (i.e. low-temperature) opacities, which in the envelope model used for the matching was merged with interior opacities from the Opacity Project (OP, see [18]). We have calculated similar opacity tables for the specific solar mixture used in the simulations (see [9]) using the atmospheric opacities provided by the author and the available OP data.

4 Initial Results

The natural test of our implementation is to calculate evolutionary tracks and compare to a *standard* evolution. In this context *standard* means: using Eddington atmosphere and a constant solar-calibrated α_{MLT}.

We did a solar calibration (using the opacities explained above) and obtained $\alpha_{\odot} = 1.742$ for the traditional Eddington atmosphere and $\alpha_{\odot} = 1.870$ for our new atmosphere. In the latter case, eq. (2) is utilised to obtain the scaled α_{MLT} used in the calculation.[5]

The results for a $1.0M_{\odot}$ and a $1.4M_{\odot}$ star are shown in Figures 4 and 5, respectively. It can be seen that for the first track the temperature differences are minuscule except at the RGB. For the heavier star, the differences are larger, but the tracks are actually closer to each other higher up the RGB.

Regarding the evolutionary pace of the models, it is also quite similar. At the end of the main sequence (i.e. where central hydrogen is exhausted) the age difference is less than 0.01 Gyr between the tracks – similar for the end of the calculation at $\log g = 2.4$. However, using a more observational "definition" of the turn-off – the point of the track where T_{eff} reaches maximum – there is a difference in age of ~ 0.5 Gyr for the $1.0M_{\odot}$ model. This is an interesting aspect of the new atmosphere, which we will investigate further.

We plan to next look for differences in interior structure (if any) between models with our implementation and the standard calculation. It will be very interesting to use asteroseismology (see e.g. [19]) and compare the oscillation frequencies between modified and unmodified models. We will follow this line of investigation in a coming paper.

[3]The value is typically $\tau_{\mathrm{eff}} \simeq 0.5$, and T_{eff} is here the parameter of the (interpolated) 3D simulation.

[4]The modified ∇_{rad} is of course also used when checking for convection with the Schwarzschild criterion.

[5]The solar value of the grid from [11] is $\alpha_{\odot,\,\mathrm{grid}} = 1.767$, which gives a scaling factor of 1.058 to be applied to the interpolated $\alpha_{\mathrm{MLT}}(T_{\mathrm{eff}}, \log g)$.

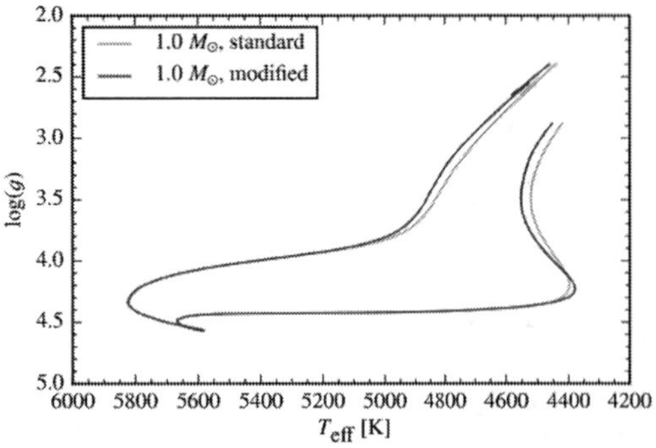

Figure 4: Evolutionary track of a $1.0M_\odot$ star from the pre main sequence to $\log g = 2.4$. The *standard* track uses a Eddington atmosphere and solar calibrated α_{MLT}. The *modified* track uses our implementation of $q(\tau)$ and $\alpha_{\text{MLT}}(T_{\text{eff}}, \log g)$.

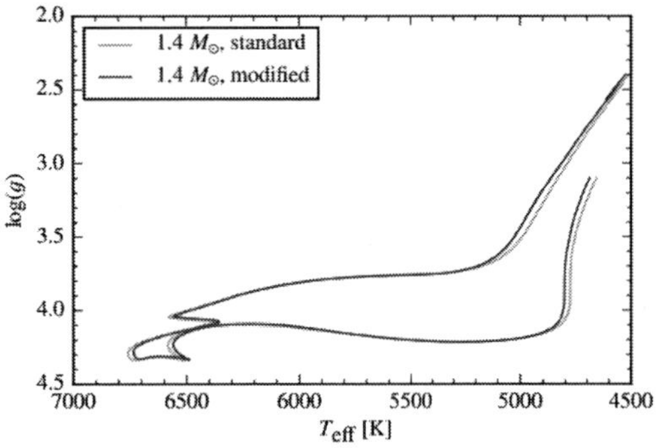

Figure 5: As Figure 4, but for a $1.4M_\odot$ star.

5 Conclusion

We have successfully implemented $T(\tau)$ relations extracted from 3D simulations and the associated modifications to the thermal gradients in the Garching Stellar Evolution Code. Furthermore, we have also modified the code to use a variable mixing-length parameter, $\alpha_{\text{MLT}}(T_{\text{eff}}, \log g)$, obtained from calibration against the 3D simulations. For consistency we have calculated new opacity tables to match those used in the simulations, and we use the same equation of state.

In a future work we will investigate the effect of our modifications on the interior structure of the models and more specifically the oscillation frequencies. We will also investigate the impact on the derived age when comparing models to observations.

Acknowledgements

Funding for the Stellar Astrophysics Centre (SAC) is provided by The Danish National Research Foundation (Grant agreement no.: DNRF106).

References

[1] R. Kippenhahn, A. Weigert, A. Weiss, *Stellar Structure and Evolution*, second edition edn. (Springer, 2012)

[2] E. Vitense, Zeitschrift für Astrophysik **32**, 135 (1953)

[3] E. Böhm-Vitense, Zeitschrift für Astrophysik **46** (1958)

[4] R. Trampedach, Astrophysics and Space Science **328**, 213 (2010)

[5] C.S. Rosenthal, J. Christensen-Dalsgaard, Å. Nordlund, R.F. Stein, R. Trampedach, Astronomy and Astrophysics **351**, 689 (1999)

[6] M. Salaris, S. Cassisi, Astronomy & Astrophysics **577**, A60 (2015)

[7] R.F. Stein, Å. Nordlund, The Astrophysical Journal **342**, L95 (1989)

[8] R.F. Stein, Å. Nordlund, The Astrophysical Journal **499**, 914 (1998)

[9] R. Trampedach, M. Asplund, R. Collet, Å. Nordlund, R.F. Stein, The Astrophysical Journal **769**, 18 (2013)

[10] R. Trampedach, R.F. Stein, J. Christensen-Dalsgaard, Å. Nordlund, M. Asplund, Monthly Notices of the Royal Astronomical Society **445**, 4366 (2014)

[11] R. Trampedach, R.F. Stein, J. Christensen-Dalsgaard, Å. Nordlund, M. Asplund, Monthly Notices of the Royal Astronomical Society **442**, 805 (2014)

[12] A. Weiss, H. Schlattl, Astrophysics and Space Science **316**, 99 (2008)

[13] H.G. Ludwig, B. Freytag, M. Steffen, Astronomy & Astrophysics **346**, 111 (1999)

[14] D.G. Hummer, D. Mihalas, The Astrophysical Journal **331**, 794 (1988)

[15] D. Mihalas, W. Dappen, D.G. Hummer, The Astrophysical Journal **331**, 815 (1988)

[16] W. Däppen, D. Mihalas, D.G. Hummer, B.W. Mihalas, The Astrophysical Journal **332**, 261 (1988)

[17] F.J. Rogers, F.J. Swenson, C.A. Iglesias, The Astrophysical Journal **456**, 902 (1996)

[18] N.R. Badnell, M.A. Bautista, K. Butler, F. Delahaye, C. Mendoza, P. Palmeri, C.J. Zeippen, M.J. Seaton, Monthly Notices of the Royal Astronomical Society **360**, 458 (2005)

[19] C. Aerts, J. Christensen-Dalsgaard, D.W. Kurtz, *Asteroseismology*, Astronomy and Astrophysics Library (Springer Science+Business Media, 2010)

 DOI: 10.1051/epjconf/201716003010

A fitting LEGACY – modelling *Kepler*'s best stars

Magnus J. Aarslev[1,⋆], *Jørgen* Christensen-Dalsgaard[1,⋆⋆], *Mikkel N.* Lund[2,1], *Victor Silva* Aguirre[1], and *Douglas* Gough[3]

[1] *Stellar Astrophysics Centre (SAC), Department of Physics and Astronomy, Aarhus University, Denmark*
[2] *School of Physics and Astronomy, University of Birmingham, UK*
[3] *Institute of Astronomy & Department of Applied Mathematics and Theoretical Physics, University of Cambridge, UK*

Abstract. The LEGACY sample represents the best solar-like stars observed in the *Kepler* mission[5, 8]. The 66 stars in the sample are all on the main sequence or only slightly more evolved. They each have more than one year's observation data in short cadence, allowing for precise extraction of individual frequencies. Here we present model fits using a modified ASTFIT procedure employing two different near-surface-effect corrections, one by Christensen-Dalsgaard[4] and a newer correction proposed by Ball & Gizon[1]. We then compare the results obtained using the different corrections. We find that using the latter correction yields lower masses and significantly lower χ^2 values for a large part of the sample.

1 Introduction

The use of helioseismology has been tremendously successful in mapping the solar interior and the general discipline of asteroseismology has revolutionized our understanding of the physics of stellar interiors. as well as enabled precise determination of crucial global stellar parameters such as the mean density directly inferable from the large frequency separation. The space-based *Kepler* mission has, through long, continuous time-series photometry, provided invaluable data for asteroseismic analysis. *Kepler* has been a huge success in general expanding our knowledge of many types of stars as well as the discovery and characterization of exoplanets. The LEGACY sample focuses on the very best observations of main-sequence (and a few subgiant) solar-like oscillators observed by *Kepler*[5, 8]. It features 66 stars all observed for at least 12 months in short cadence making them some of the highest signal-to-noise solar-like targets in the *Kepler* mission. None of the stars exhibit properties of mixed modes. The details of their oscillation properties are presented in [5], while radii, masses and ages are the topic of [8]. Here we consider one aspect of fitting stellar models to these phenomenal observations, namely the choice of surface-effect corrections.

One of the major challenges in stellar astrophysics has always been the determination of stellar ages. They are extremely relevant in many areas, for example in trying to understand the formation of stellar systems or the evolution of our galaxy. But the age of a star cannot be measured or inferred from observations. Rather it is derived by fitting models of stellar evolution to match asteroseismic and spectroscopic data. However, stellar evolution models are rather crude, neglecting or oversimplifying key physical processes, which makes the resulting ages somewhat unreliable. Silva Aguirre et al.[7] determine stellar ages of *Kepler* exoplanet host stars to within a median statistical error of 14 per cent - an order of magnitude larger than the corresponding errors on radius, density, mass and distance. Here we apply essentially the same model fitting procedure using both the same surface-effect correction as well as a more recent correction and compare the outcome of the two.

2 Model Fitting

We use a model-fitting procedure, which combines the Aarhus stellar evolution code[2] with the Aarhus adiabatic oscillation package[3] to fit individual model frequencies within a grid of stellar models. The grid was computed as described for ASTFIT in[7] with the following exceptions: The heavy-element mixture was based on Grevesse & Noels (1993), taking the solar surface ratio between the abundances of heavy elements and hydrogen as $Z_s/X_s = 0.0245$. Furthermore the grid extends in mass from 0.7 to 1.7 $M_\odot$, in steps of 0.01 $M_\odot$. The heavy-element abundance ranged from $Z = 0.0032$ to 0.059, relating Y and Z by $\Delta Y/\Delta Z$ values varying between 1 and 2, in steps of 0.2. Here ΔY and ΔZ are the differences in helium and heavy element abundance between today and the time of the big bang. Often this is assumed to be around 1.4. However, since the exact value is unknown, and subject to debate, it is allowed to vary here. Diffusion and settling were not taken into account. Two different surface-effect corrections were used to account for the fact that model frequencies are systematically overestimated:

⋆e-mail: mja@phys.au.dk
⋆⋆e-mail: jcd@phys.au.dk

a: A correction based on a scaled solar fit (see [4]). This works very well for other stars with masses close to $1M_\odot$, but falls increasingly short for heavier, and consequently hotter, stars as shown in Section 3.

b: A recent correction presented by Ball & Gizon (2014);

$$\delta\nu = (c_{-1}(\nu/\nu_{ac})^{-1} + c_3(\nu/\nu_{ac})^3)/\mathscr{I}, \quad (1)$$

where $\delta\nu$ is the frequency difference $\nu_{obs} - \nu_{model}$ for a given mode, ν_{ac} is the acoustic cut-off frequency, $\mathscr{I}$ is the mode inertia, and c_{-1} and c_3 are the fitting coefficients. The formula is based on calculations originally carried out to explain solar cycle frequency variations, and includes two terms: one proportional to ν^{-1}, which represents a frequency shift due to the increase in pressure scale height that would arise from better modelling of convection, and another term proportional to ν^3 correcting for a frequency shift caused by local changes to the sound speed without changing the density stratification[10]. The correction is weighted by individual mode inertia, for which we use the standard ASTEC calculation, which is normalized by the displacement at the photosphere. This correction formula seems to produce very reasonable fits for all the solar-like stars in the mass and temperature range considered here.

When combined with the aforementioned correction we produce frequency fits of unprecedented quality – even for stars where the surface term differs significantly from that of the Sun. The best fitting model is found by minimizing

$$\chi^2 = \chi^2_{spec} + \chi^2_\nu, \quad (2)$$

where χ^2_{spec} indicates the contribution from fitting spectroscopic data (T_{eff} and [Fe/H]) and χ^2_ν is the contribution from fitting the frequencies as defined in [7]. Our ability to model observations in a precise manner helps constrain the physical parameters of the observed stars as well as pinpoint the aspects of stellar models that still need improvement.

3 Results

Figures 1 and 2 show frequency fits for the best fitting models of two LEGACY stars: KIC 8006161 (top) and KIC 7940546 (bottom). Both **a** and **b** work well for the $0.97M_\odot$ star; in fact the scaled solar fit is slightly better here. For the second star the two approaches yield slightly different masses close to $\sim1.5M_\odot$ and effective temperatures of 6275.8K (**a**) and 6313.7 (**b**). The surface effect of this star is very different from the Sun and thus modelled poorly by the scaled solar fit **a**, whereas the inertia-scaled fit **b** captures the overall trends in the curve, which is evident from the much lower χ^2_ν value (see figure captions).

Figure 3 (top) shows the effect on obtained masses of using either approach for the whole LEGACY sample. This is shown as δM, the mean mass from approach **b** minus the mean mass from **a** divided by the combined standard deviation σ, i.e.

$$\delta M = \langle M\rangle_b - \langle M\rangle_a, \quad (3)$$

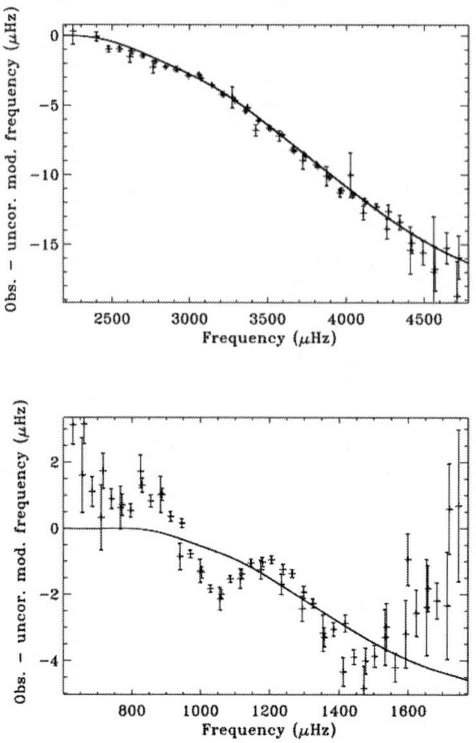

Figure 1: Scaled 'solar fit' (**a**) applied to models of KIC 8006161 and KIC 7940546. The fits pertain to the models with the lowest combined χ^2. **Top panel:**. Surface effect correction for a $0.97M_\odot$ model with T_{eff} = 5465.9K. The contribution from the frequency fit to χ^2 is χ^2_ν = 2.64. **Bottom panel:** Surface effect correction for a $1.47M_\odot$ model with T_{eff} = 6275.8K and χ^2_ν = 12.60.

and

$$\sigma(\delta M) = \sqrt{\sigma(M_b)^2 + \sigma(M_a)^2}. \quad (4)$$

The mean mass is obtained by averaging over many models within a suitable mass range in the grid, where each model is weighted by $\exp\left(-\chi^2/2\right)$. For $M \gtrsim 1.3M_\odot$ there is a significant difference between the two, where **b** consistently yields lower-mass stellar models, and as a consequence they are older[1].

Figure 3 (bottom) shows the χ^2 ratio between the two approaches. Here χ^2_{min}(a) is the best fitting model using the scaled solar fit, and (b) is using the Ball & Gizon (2014) surface effect correction. Except for some cases around $\sim1M_\odot$, model fits using the Ball & Gizon (2014) correction have lower χ^2 values. This is thus preferable to any previous correction formulae – even for stars where the surface term differ significantly from that of the Sun. It is worth noting that the lower masses obtained using surface fit **b** are accompanied by lower χ^2 values; not only are

[1] This has also been found explicitly in our analysis but is not shown here

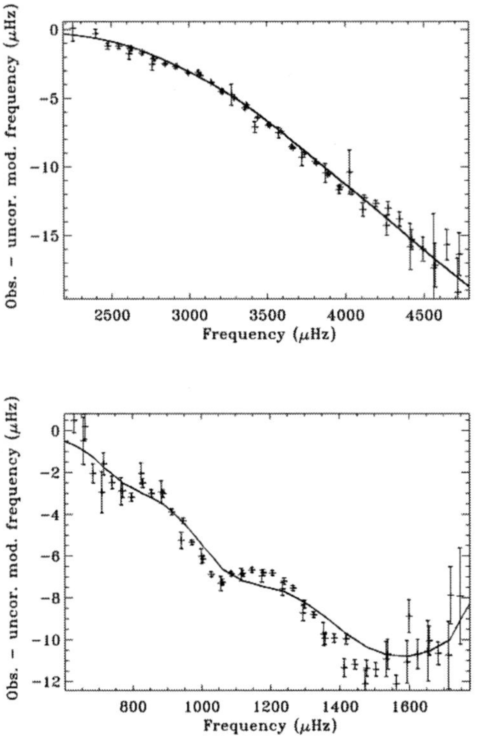

Figure 2: Inertia-scaled two-term fit (**b**) applied to models of the same stars as in Figure 1. Again, the fits shown are from the models with the lowest combined χ^2. **Top panel:**. Surface effect correction for a 0.97M$_\odot$ model with T_{eff} = 5465.8K and χ^2_ν = 2.90. **Bottom panel:** Surface effect correction for a 1.46M$_\odot$ model with T_{eff} = 6313.7K and χ^2_ν = 6.73.

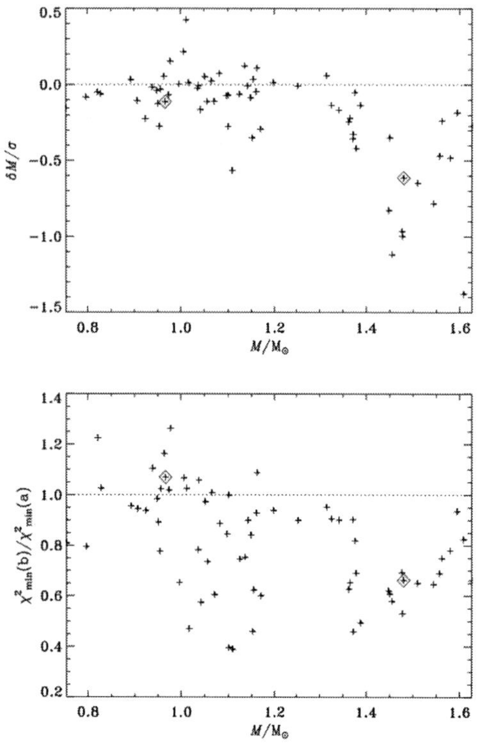

Figure 3: Top: Difference in mean mass obtained with either surface effect correction for all stars in the LEGACY sample. **bottom**: Ratio between χ^2 of the best fitting models. Diamonds highlight the stars in Figures 1 and 2.

the masses and ages different compared to the previous approach **a**, they are also determined with better precision.

3.1 Inertia scaling

We have established that the inertia-scaled two-term fit is superior for the more massive, and consequently hotter, stars in the sample. We can gain some insight into why this is by looking at the inverse mode inertia used for scaling the fitting terms shown in Figure 4. As has also been pointed out by Trampedach et al.[9], the peculiar features in the frequency differences between $\sim$ 1000μHz and $\sim$ 1500μHz coincide with an oscillatory contribution to the inverse mode inertia. We have identified the issue as arising from effects of the second helium ionization zone on the eigenfunctions. There is a good reason for this feature to only show up in the surface layers of hotter stars: There is a significantly larger drop in Γ_1 in the second helium ionization zone, possibly due to the gas density being lower here than in a corresponding zone in a cooler star. This will be elaborated on in an upcoming paper.

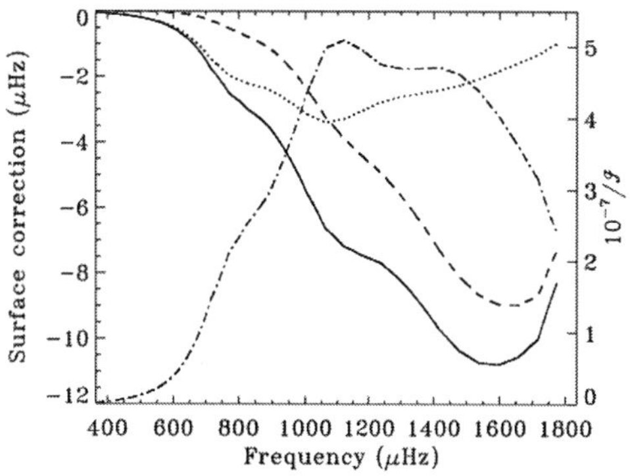

Figure 4: Surface-effect correction and inverse inertia of KIC 7940546. For the left y-axis we present the full line showing the full two-term fit (**b**), while the dashed and dotted lines show the contributions to the fit from the cubic and inverse frequency terms respectively. On the right y-axis we present the inverse mode inertia (dot-dashed line) used to scale the fitting terms

DOI: 10.1051/epjconf/201716003010

Table 1: Table of global parameters for the two stars presented in Figures 1 and 2 as well as the fit quality described by χ^2 (see eq. (2))

Correction	KIC ID	M [M$_\odot$]	R [R$_\odot$]	$T_{\rm eff}$ [K]	Age [Gyr]	χ^2
a	8006161	0.967 ± 0.018	0.923 ± 0.006	5465.9	5.008 ± 0.463	3.640
b	8006161	0.964 ± 0.018	0.922 ± 0.006	5465.8	5.093 ± 0.460	3.892
a	7940546	1.480 ± 0.026	1.990 ± 0.014	6275.8	2.325 ± 0.104	24.670
b	7940546	1.458 ± 0.025	1.971 ± 0.015	6313.7	2.356 ± 0.106	16.340

4 Conclusions

We have presented a comparison between two surface-effect corrections employed in a grid model fitting scheme. We find that there are significant differences in the obtained global stellar parameters for stellar masses $\gtrsim 1.1\mathrm{M}_\odot$, the largest differences occurring at $\gtrsim 1.4\mathrm{M}_\odot$. Here we see that not only are the masses we find using correction **b** systematically lower, they are also determined more accurately as reflected by their lower χ^2 values.

Trampedach et al[9] investigate the appropriateness of eq. (1) across the HR diagram using 1D envelope models patched with averaged 3D atmospheres from hydrodynamical simulations. Their findings support our notion that the fit is very versatile and can be used across the HR diagram for solar-type stars also including the pre-main sequence and red giant phase. The versatility is reflected by the fact that the ratio c_{-1}/c_3 changes, c_{-1} increasing relative to c_3, when going from cooler to hotter stars. Another study modelling the surface term across the HR diagram was performed by Schmitt & Basu[6]; they also found that, while various correction formulas might suit different types of stars, the two-term fit by Ball & Gizon[1] performs well across the whole range in question (from the main sequence to red giants of models ranging in mass from 0.8 to $1.5\mathrm{M}_\odot$). The results presented in this paper agree with other recent studies, which suggests that overall the physically motivated correction **b** is the best available surface term to be used for general-purpose model fitting. Of course it requires the structure of the model to be close to that of the observed star in order to get the appropriate mode inertia necessary for eq. (1). However, we do also show that for more specific cases, perhaps other surface terms might also be reasonable and in some cases even slightly better. As an example the surface term of KIC 8006161 is so close to that of the Sun that it is better described using the scaled solar fit yielding both lower χ^2_ν and χ^2.

Acknowledgements

Funding for the Stellar Astrophysics Centre is provided by The Danish National Research Foundation. The research was supported by the ASTERISK project (ASTERoseismic Investigations with SONG and Kepler) funded by the European Research Council (Grant agreement no.: 267864).

References

[1] Ball, W., & Gizon, L., A&A **568**, A123 (2014)
[2] Christensen-Dalsgaard, J., A&SS **316**, 13-24 (2008a)
[3] Christensen-Dalsgaard, J., A&SS **316**, 113-120 (2008b)
[4] Christensen-Dalsgaard, J., Astron. Nachr. **333**, 914-925 (2012)
[5] Lund, M. N., et al., ApJ, (2016 submitted)
[6] Schmitt, J.R., & Basu, S., ApJ **808**,123-135 (2015)
[7] Silva Aguirre, V., et al., MNRAS **452**, 2127-2148 (2015)
[8] Silva Aguirre, V., et al., ApJ, (2016 submitted)
[9] Trampedach, R., et al., MNRAS (2016 submitted)
[10] Gough, D. O., *Progress of Seismology of the Sun and Stars, Proceedings of the Oji International Seminar*, Osaki, Y., & Shibahashi, H., *Vol. 367* (Springer-Verlag, 1990) 283

DOI: 10.1051/epjconf/201716003011

The more the merrier: grid based modelling of Kepler dwarfs with 5-dimensional stellar grids

Aldo Serenelli[1,⋆], *William J.* Chaplin[2,3,], and *Daniel* Huber[3,4,5]

[1] *Institute of Space Sciences (IEEC-CSIC), Campus UAB, Barcelona, E-08193, Spain*
[2] *School of Physics and Astronomy, University of Birmingham, Edgbaston, Birmingham B15 2TT, United Kingdom*
[3] *Stellar Astrophysics Centre, Dept. of Physics and Astronomy, Aarhus Univ., Ny Munkegade 120, DK-8000 Aarhus C, Denmark*
[4] *Sydney Institute for Astronomy, School of Physics, University of Sydney, Sydney, Australia*
[5] *SETI Institute, 189 Bernardo Avenue, Mountain View, CA 94043*

Abstract. We present preliminary results of our grid based modelling (GBM) of the dwarf/subgiant sample of stars observed with Kepler including global asteroseismic parameters. GBM analysis in this work is based on a large grid of stellar models that is characterized by five independent parameters: model mass and age, initial metallicity ($Z_{\rm ini}$), initial helium ($Y_{\rm ini}$), and mixing length parameter ($\alpha_{\rm MLT}$). Using this grid relaxes assumptions used in all previous GBM work where the initial composition is determined by a single parameter and that $\alpha_{\rm MLT}$ is fixed to a solar-calibrated value. The new grid allows us to study, for example, the impact of different galactic chemical enrichment models on the determination of stellar parameters such as mass radius and age. Also, it allows to include new results from stellar atmosphere models on $\alpha_{\rm MLT}$ in the GBM analysis in a simple manner. Alternatively, it can be tested if global asteroseismology is a useful tool to constraint our ignorance on quantities such as $Y_{\rm ini}$ and $\alpha_{\rm MLT}$. Initial findings show that mass determination is robust with respect to freedom in the latter quantities, with a 4.4% maximum deviation for extreme assumptions regarding prior information on $Y_{\rm ini} - Z_{\rm ini}$ relations and $\alpha_{\rm MLT}$. On the other hand, tests carried out so far seem to indicate that global seismology does not have much power to constrain $Y_{\rm ini} - Z_{\rm ini}$ relations of $\alpha_{\rm MLT}$ values without resourcing to additional information.

1 Introduction

Grid based modelling (GBM) is now routinely employed in the determination of stellar parameters using global asteroseismic quantities, i.e. the large frequency separation $\Delta\nu$ and the frequency of maximum power $\nu_{\rm max}$ (see e.g. [1]). Many quantities need to be determined to construct the large grids of stellar models needed for GBM. Leaving aside variations in the input physics employed in stellar evolution calculations, the most important simplifications made in order to keep the problem tractable are the assumption of a fixed mixing length parameter $\alpha_{\rm MLT}$ and a uniquely relation between the initial helium $Y_{\rm ini}$ and metallicity $Z_{\rm ini}$ of the models. The latter, together with the normalization $X_{\rm ini}+Y_{\rm ini}+Z_{\rm ini} = 1$ and an adopted element mixture leave only one free parameter associated with the initial composition of models. This sets the minimal framework for GBM in which each stellar model in the grid is determined by three independent parameters: its age, the initial mass of the evolutionary sequence to which the stellar model belongs (might differ from its actual mass due to mass loss) and the initial $Z_{\rm ini}$ (or any other parameter defining the initial composition). Needless to say, such construction of models inherently include strong prejudices: that a solar $\alpha_{\rm MLT}$ is universal and that $Y_{\rm ini}$ and $Z_{\rm ini}$ are uniquely and universally related to each other. These are strong and limiting assumptions.

In the framework of GBM the way to break free is to use stellar grids where $\alpha_{\rm MLT}$ is also a free parameter, together with a second parameter linked to initial composition. Based on this necessity, in this work we present GBM results based on a newly computed grid of stellar models based on a five-dimensional parameter space: age, mass (mass loss is negligible for dwarfs and subgiant stars), $Z_{\rm ini}$, $Y_{\rm ini}$, and $\alpha_{\rm MLT}$ are the parameters that characterize a given stellar model in the grid. The grid of models is combined with BeSPP, a Bayesian algorithm [2] developed to determine stellar physical quantities from any appropriate combination of spectroscopic, photometric or asteroseismic quantities. Different assumptions regarding galactic chemical enrichment, e.g. by introducing a relation between $Y_{\rm ini}$ and $Z_{\rm ini}$, or regarding dependences of $\alpha_{\rm MLT}$ on surface stellar parameters ($T_{\rm eff}$, $\log g$ and [Fe/H]) can be tested by considering appropriate prior probabilities in BeSPP. We apply our methods to the Kepler sample from [1] in order to answer the following questions. How robust is the determination of stellar parameters given our lack of knowledge regarding $Y_{\rm ini}$ and $\alpha_{\rm MLT}$ in particular? Inverting the problem, can global seismology be used to

⋆e-mail: aldos@ice.csic.es

establish a history, e.g. in the form of a $Y_{\rm ini} - Z_{\rm ini}$ relation, of galactic chemical enrichement? Can it be used to empirically establish a relation between $\alpha_{\rm MLT}$ and stellar properties to, e.g. test recent results from three dimensional stellar atmosphere models [3]?

2 Methods and results

Stellar models have been computed using GARSTEC. For the present work, the grid spans a mass range between 0.7 and 1.8 $M_\odot$ with $\Delta M/M_\odot = 0.02$ and extends down to $\log g = 3.2$. $Y_{\rm ini}$ and $Z_{\rm ini}$ and $\alpha_{\rm MLT}$ are taken as three independent parameters and the qubic parameter space covered is illustrated in Fig. 1. Note we take $Z_{\rm ini}$, not [Fe/H] as independent parameter. However, the grid is uniformly spaced in $\log Z_{\rm ini}$.

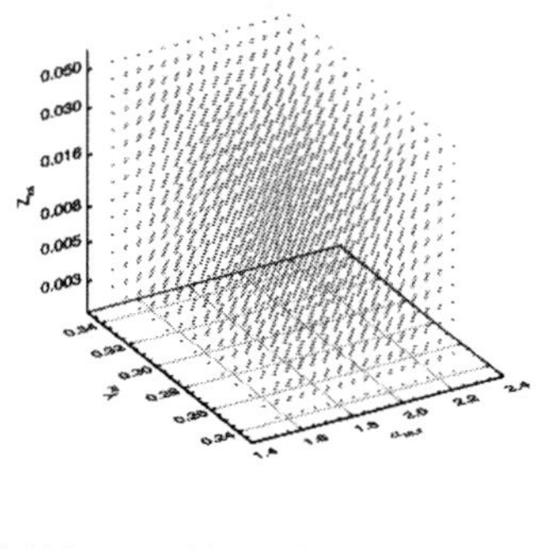

Figure 1: Initial composition and $\alpha_{\rm MLT}$ parameter space of the grid of stellar models used in the present work.

Depending on the problem at hand, priors are included in BeSPP that account for different assumptions regarding model parameters. In this work we consider different possibilities regarding priors for initial composition or $\alpha_{\rm MLT}$ as follows:

- **Flat**: flat priors are used for composition and $\alpha_{\rm MLT}$.
- $\mathbf{\Delta_{YZ} = 1.2}$: a Gaussian prior relates $Y_{\rm ini}$ and $Z_{\rm ini}$ such that

$$p(Y_{\rm ini}) \propto \exp\left[-\frac{(Y_{\rm ini} - Y_\Delta)^2}{2\sigma_Y^2}\right], \qquad (1)$$

with $Y_\Delta = Y_{\rm SBBN} + \Delta Z_{\rm ini}$, $\Delta = 1.2$ and $\sigma_Y = 0.01$.

- **3DMLT**: a Gaussian prior relating $\alpha_{\rm MLT}$ with 3D model atmosphere results such that

$$p(\alpha_{\rm MLT}) \propto \exp\left[-\frac{(\alpha_{\rm MLT} - \alpha_{\rm 3D})^2}{2\sigma_\alpha^2}\right], \qquad (2)$$

with $\alpha_{\rm 3D} \equiv \alpha_{\rm 3D}(T_{\rm eff}, \log g, [{\rm Fe/H}])$ from [3] (but shifted by a constant value to match our solar calibrated value $\alpha_{\rm MLT} = 1.802$) and σ_α=0.05.

- **3DMLT +** $\mathbf{\Delta_{YZ} = 1.2}$: combination of Eqs. 1 and 2

$$p(Y_{\rm ini}, \alpha_{\rm MLT}) \propto p(Y_{\rm ini}) \cdot p(\alpha_{\rm MLT}). \qquad (3)$$

- **SUN**: effectively constraints parameter space to the standard three dimensional space by using

$$p(Y_{\rm ini}) \propto \exp\left[-\frac{(Y_{\rm ini} - Y_\Delta)^2}{2\sigma_Y^2}\right] \qquad (4)$$

as before and the solar calibrated $\alpha_{\rm MLT} = 1.802$.

Input data in our analysis is taken from [1] and includes $T_{\rm eff}$, [Fe/H], $\Delta\nu$ and $\nu_{\rm max}$. In that work, [Fe/H] values for all stars were taken equal to a mean solar neighborhood value and do not reflect any spectroscopic or photometric determination.

Fig. 2 shows results for the mass determination of KIC 8547279 in the $Y_{\rm ini}-\alpha_{\rm MLT}$ plane. Mass is color coded, and at each point in the plot, the mass value marginalized over $Z_{\rm ini}$, and age is shown. Contours correspond to 0.9, 0.5, and 0.1 probability levels. Priors applied in each case are given in each plot title. Very little constraining power comes from the asteroseismic analysis for $Y_{\rm ini}$ and $\alpha_{\rm MLT}$ when flat priors are applied. This is probably due to a degenerate behavior than $\alpha_{\rm MLT}$ and $Y_{\rm ini}$ have on $T_{\rm eff}$, as can be seen by the diagonal orientation of the contours. Increasing either $Y_{\rm ini}$ or $\alpha_{\rm MLT}$ leads to higher $T_{\rm eff}$ in models, but given that $T_{\rm eff}$ is bound by observations, then our results show a compensation effect: higher $Y_{\rm ini}$ correspond to lower $\alpha_{\rm MLT}$ values and viceversa.

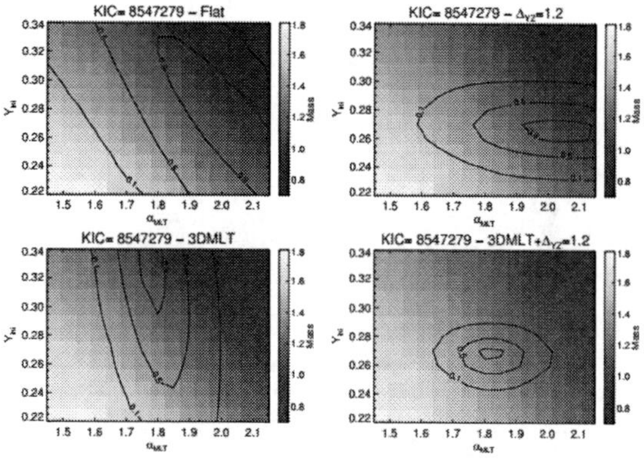

Figure 2: Impact of using different priors in the mass determination of KIC 8547279. For each $Y_{\rm ini} - \alpha_{\rm MLT}$ point, mass value is the mean value of the mass PDF after marginalizing the 5-dimensional PDF over age and metallicity. Contours depict isoprobability curves for the mass as a function of $Y_{\rm ini}$ and $\alpha_{\rm MLT}$.

As priors are added, results become more constrained for $Y_{\rm ini}$ and $\alpha_{\rm MLT}$, as expected. It is interesting to note, however, that the estimated mass determination is not too sensitive to the choice of priors. In fact, KIC 8547279 is one of the worst cases we have found in this study because the estimated mass varies by about 15% depending on the priors used. Note that the comparison between the flat and the 3DMLT + $\Delta_{YZ} = 1.2$ priors is an extreme one, so this level of variation in mass determinations due to ignorance regarding $Y_{\rm ini}$ and $\alpha_{\rm MLT}$ is probably an upper limit. In Fig. 3 we show one more example, this time for KIC 5629080, which shows almost no variation in its mass estimate. However, as for the previous case, constraints for $Y_{\rm ini}$ and $\alpha_{\rm MLT}$ come mostly from priors; asteroseismic data is in fact providing little or no information about these two quantities.

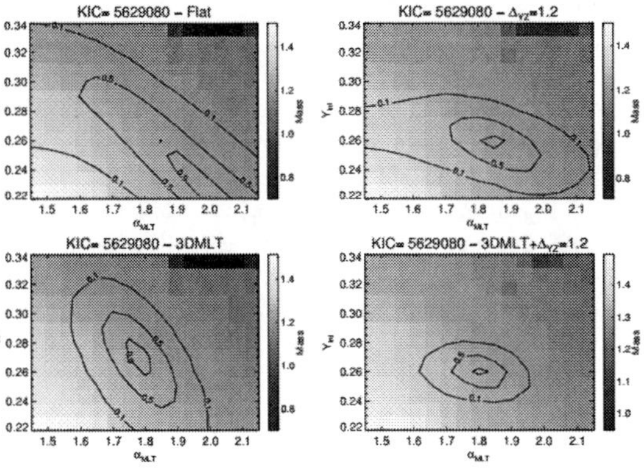

Figure 3: Same as Fig. 2 but for KIC 5629080.

Fig. 4 summarizes results for the mass determinations in the whole sample by comparing results obtained for the 5 different assumptions on priors given above. In all cases, the x-axis corresponds to priors in Eq. 4. The maximum dispersion is seen in the comparison Flat-Sun (top left panel) and it is at most 15% (except for a few cases). In fact, the fractional mass difference between the two cases has $\sigma_{\rm F,\odot} = 4.4\%$, which can be taken as an estimate of the uncertainty between assuming no prior info on initial composition and $\alpha_{\rm MLT}$. We believe this is a very encouraging result. The prior on $\alpha_{\rm MLT}$ (bottom left panel) leads to almost the same mass determination as the flat prior, i.e. constraints on $\alpha_{\rm MLT}$ do not affect mass determinations by themselves. On the contrary, relating $Y_{\rm ini}$ to $Z_{\rm ini}$ (Eq. 1) influences mass estimates. Interestingly, the interplay between $Y_{\rm ini}$ and $\alpha_{\rm MLT}$ due to their degeneracy in relation to $T_{\rm eff}$ values, as described before, make $\alpha_{\rm MLT}$ priors relevant if used in combination with constrained initial composition, as shown by comparision of the upper and lower right panels in Fig. 4.

Recently, 3-dimensional model atmospheres have been used to determine the dependence of $\alpha_{\rm MLT}$ on stellar surface properties [3, 4]. But it is not easy to test these results because observational tests have to rely on $T_{\rm eff}$ determinations and, in stellar models, the $T_{\rm eff}$ scale does not depend solely on $\alpha_{\rm MLT}$ but also, for example, on both the metal and helium abundance of stars. Here, we consider whether or not global seismic quantities, in combination with $T_{\rm eff}$ and [Fe/H] measurements, can constrain $\alpha_{\rm MLT}$ in the $T_{\rm eff} - \log g$ plane.

We use BeSPP to determine posterior values for $\alpha_{\rm MLT}$ for our whole sample, using as before different assumptions for priors. Ideally, we would like to obtain good constraints for $\alpha_{\rm MLT}$ in the case of flat priors that could be used to test theoretical determinations of $\alpha_{\rm MLT}$ based on stellar atmosphere models. This has been attempted before, with different data and methods, by [5]. Our initial results are

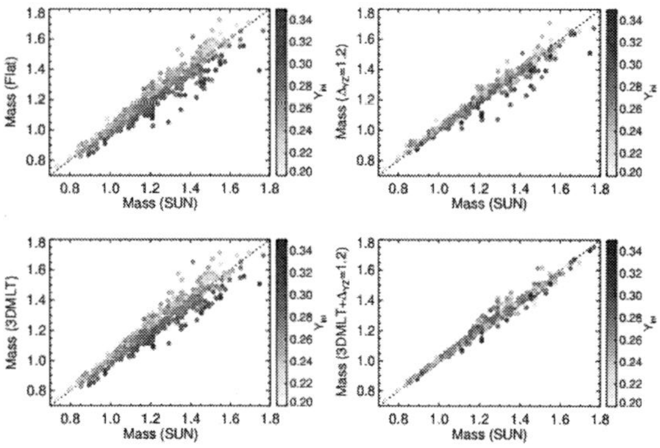

Figure 4: Mass determinations for the whole [1] sample depending on the different priors used in BeSPP as indicated in the axes.

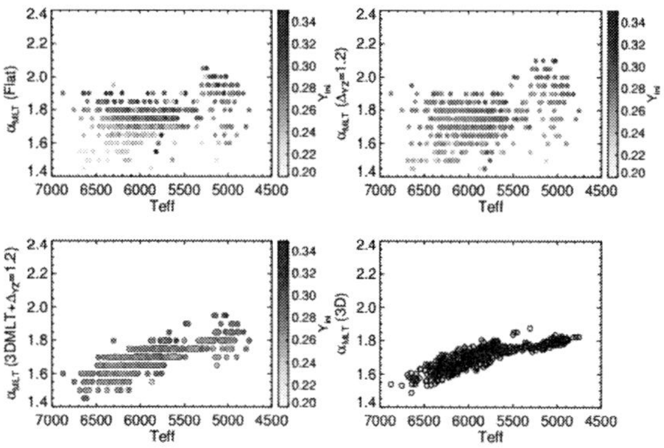

Figure 5: Determination of $\alpha_{\rm MLT}$ using different priors in BeSPP. Bottom right panel shows results directly obtained from $\alpha_{\rm MLT}$ values based on 3D stellar atmospheres [3].

shown in Fig. 5, where $\alpha_{\rm MLT}$ is shown as a function of $T_{\rm eff}$ and color-coded to represent $Y_{\rm ini}$. For the Flat case, there is a clear correlation between $\alpha_{\rm MLT}$ and $Y_{\rm ini}$, regardless of $T_{\rm eff}$. Although it is not impossible that there is a physical dependence of $\alpha_{\rm MLT}$ on the helium abundance of stars, i.e. on their composition, we are more inclined to believe that seismic data and the [Fe/H] from [1] are not enough to give too meaningful constraints on $\alpha_{\rm MLT}$. The results change partially once the Δ_{YZ} prior is added, but still a correlation between $\alpha_{\rm MLT}$ and $Y_{\rm ini}$ seems to be present. Only after adding the **3DMLT** prior this correlation is almost completely removed (bottom left panel). The lower right panel shows the $\alpha_{\rm MLT}$ values corresponding to our whole sample that correspond to [3] 3D atmosphere models, directly determined from the stars $T_{\rm eff}$, [Fe/H] and $\log g$ values using formulae given in that reference but with a shift applied so that $\alpha_{\rm MLT}$ values match for the Sun. Clearly, the bottom two panels look quite similar, because our **3DMLT** prior is constructed from [3] results. A more rigurous study is required, at this point, to better assess the possibilities that global seismology has to offer for determining α_{MLT} values and testing 3D hydrodynamic atmosphere models. Given the correlation with Y_{ini}, i.e. with the stellar composition, it might be crucial to use actual [Fe/H] values in the analysis before firm conclusions can be drawn.

The last exercise we present here relates to determination Y_{ini} from asteroseismic data, much as done above for α_{MLT}, but present results in terms of the enrichement parameter $\Delta = \Delta Y/\Delta Z$, where $\Delta Y = Y_{ini} - Y_{SBBN}$, $Y_{SBBN} = 0.2485$ is the standard Big Bang Nucleosynthesis (SBBN) value and $\Delta Z = Z_{ini}$ because $Z_{SBBN} = 0$. Attempts to constrain Δ from asteroseismology, based on modeling stars based on individual frequencies or combinations, have been plagued with difficulties linked to problematic determination of helium abundances. In fact, in many cases helium abundances are too low, below the SBBN value [6, 7], which makes us wary of the robustness of results for the whole sample. Asteroseismic modelling of stars using individual frequencies is known to introduce biases. This might also be the case when using frequency separation ratios. Is the situation different when global seismic quantities are used instead?

Unfortunately, our first tests show than in fact Y_{ini} values are also at odds with SBBN. Results for Δ are shown in Fig. 6 where, for different assumptions regarding priors, we show resulting Δ distributions. $\Delta < 0$ indicates $Y_{ini} < Y_{SBBN}$. For the flat or the 3DMLT priors, clearly there are many stars that lead to low unrealistic Y_{ini} values. Also, and despite the fact that most Δ values nucleate between 0.5, the distribution looks very broad. When adding the Δ_{YZ} prior, the Δ distribution is peaked around 1.2, but obviously this simply reflects the prior information. The combination of the Δ_{YZ} and 3DMLT priors (lower right panel) gives the best behaved results, actually improving over the Δ_{YZ}-alone case (top right panel). Of course, Δ depends critically on Z_{ini} and, as stated before, [Fe/H] values in our sample are just fiducial values. Only after we improve this aspect, we can make more firm statements about this.

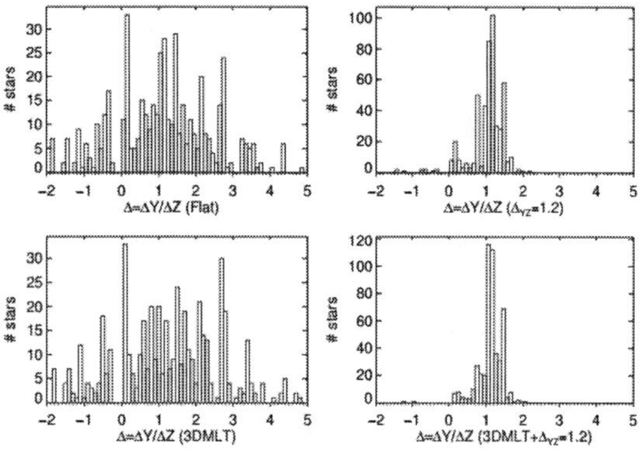

Figure 6: Histograms showing the distribution of $\Delta = \Delta Y/\Delta Z$ for different assumptions regarding priors.

3 Conclusions

We have extended the capabilities of BeSPP to determine stellar physical parameters by adding Y_{ini} and α_{MLT} as additional free parameters in its grid of stellar models. The fully new 5-dimensional grid of models relaxes the strong constraints imposed by: 1) the standard assumption that there is a one-to-one correlation between initial metal and helium abundance in stars; 2) that α_{MLT} takes a universal value that is solar calibrated. By adding priors in BeSPP, it is possible in principle to study different galactic chemical enrichment models, test α_{MLT} results from 3D model atmospheres and, very importantly, assess the impact of the usual assumptions (one-to-one $Y_{ini} - Z_{ini}$ relation and solar α_{MLT}) in determination of stellar parameters such as mass, radius and age.

In this preliminary exercise we have used [1] sample, so we do not use reliable [Fe/H] values. Therefore,results here are *experimental*. However, one interesting and probably robust conclusion is that mass (and radius - not shown here) determinations using global seismic quantities is robust with respect to our ignorance on α_{MLT} and Y_{ini}.An estimate of the uncertainty in mass determinations yield a typical 4.4% deviation between the two most extreme cases: flat priors on Y_{ini} and α_{MLT} and priors that mimic typical GBM results (e.g. [1]). This is an encouraging result. On the other hand, constraining $Y_{ini} - Z_{ini}$ relations or α_{MLT} values that can be used to test galactic chemical enrichment or stellar atmosphere models seems at the moment a very daunting task because either Y_{ini} or α_{MLT} can introduce changes in the model T_{eff} scale, leading to degenerate results. However, this needs further investigation, particularly by considering actual [Fe/H] in the analysis.

References

[1] W.J. Chaplin, S. Basu, D. Huber, A. Serenelli, L. Casagrande, V. Silva Aguirre, W.H. Ball, O.L. Creevey, L. Gizon, R. Handberg et al., ApJS, **210**, 1 (2014)

[2] A. Serenelli, M. Bergemann, G. Ruchti, L. Casagrande, MNRAS, **429**, 3645 (2013)

[3] Z. Magic, A. Weiss, M. Asplund, A&A, **573**, A89 (2015),

[4] R. Trampedach, R.F. Stein, J. Christensen-Dalsgaard, Å. Nordlund, M. Asplund, MNRAS, **445**, 4366 (2014)

[5] A. Bonaca, J.D. Tanner, S. Basu, W.J. Chaplin, T.S. Metcalfe, M.J.P.F.G. Monteiro, J. Ballot, T.R. Bedding, A. Bonanno, A.M. Broomhall et al., ApJL, **755**, L12 (2012)

[6] T.S. Metcalfe, O.L. Creevey, G. Doğan, S. Mathur, H. Xu, T.R. Bedding, W.J. Chaplin, J. Christensen-Dalsgaard, C. Karoff, R. Trampedach et al., ApJS, **214**, 27 (2014)

[7] V. Silva Aguirre, G.R. Davies, S. Basu, J. Christensen-Dalsgaard, O. Creevey, T.S. Metcalfe, T.R. Bedding, L. Casagrande, R. Handberg, M.N. Lund et al., MNRAS, **452**, 2127 (2015)

DOI: 10.1051/epjconf/201716003012

Deciphering the period spacing pattern in the oscillation spectrum of the SPB star KIC 7760680

Wojciech Szewczuk[1,*], *Jadwiga* Daszyńska-Daszkiewicz[1], and *Przemysław* Walczak[1]

[1]*Astronomical Institute of the Wrocław University, Kopernika 11, 51–622 Wrocław, Poland*

Abstract. We present the analysis of KIC 7760680, the rotating Slowly Pulsating B–type star identified in the Kepler photometry. The oscillation spectrum of the star exhibits a series of 36 frequencies which are quasi-equally spaced in period. We confirm that this series can be associated with prograde dipole modes of consecutive radial orders. In our studies, the effects of rotation were included in the MESA equilibrium models as well as in the puslational calculations in the framework of the traditional approximation. We find that pulsational models computed with the OPLIB opacities best reproduce the observed frequency range. The modified opacity data with an enhancement of the opacity at $\log T = 5.3$, 5.46 and 5.06 were tested as well. Increasing the OPLIB opacities by about 50% at $\log T = 5.3$ is sufficient to excite modes in the whole range of 36 frequency peaks of KIC 7760680.

1 Introduction

KIC 7760680 was classified as the Slowly Pulsating B-type star by Pápics et al. [1] who found a series of 36 frequencies quasi-equally spaced in period from the Kepler photometry. From spectroscopic observations the authors derived the effective temperature, $T_{\rm eff} = 11650 \pm 210$ K, surface gravity, $\log g = 3.97 \pm 0.08$, metallicity, $[M/H] = 0.14 \pm 0.09$ (which corresponds to about $Z = 0.017$) and the minimum value of the rotational velocity, $V_{\rm rot} \sin i = 61.5 \pm 5$ km s^{-1}. Based on the zero-rotation asymptotic theory, Pápics et al. [1] attributed these frequencies to the prograde dipole gravity modes of high consecutive radial orders. Recently, Moravveji et al. [2] made a detailed seismic modelling of the star taking into account effects of rotation in the framework of the traditional approximation [e. g., 3–9] applied to the non-rotating MESA [10–12] evolutionary models. The authors found that KIC 7760680 is a moderately rotating star with the rotational frequency amounting to 26% of its Roche break up frequency (0.4805 d^{-1}) and with a mass $\sim 3.25 M_\odot$. Furthermore, they constrained the exponentially decaying convective core overshooting parameter to $f_{\rm ov} \approx 0.024 \pm 0.001$. Additionally, their best seismic models employed extra diffusive mixing in the radiative envelope. Here, we repeat the Moravveji et al. [2] approach but our computations include the effects of rotation also in equilibrium models. In Sect. 2 we present results of our seismic modelling and conclusions ends the paper.

2 Seismic modelling

Knowledge of the geometry of pulsational modes is essential in seismic modelling. In the case of KIC 7760680, the quasi-equally spaced in period frequency peaks suggest that these are the modes with the same degree, ℓ, azimuthal order, m, and consecutive radial orders, n. According to Pápics et al [1], the frequencies are associated with prograde dipole modes ($\ell = 1$, $m = +1$).

In order to check whether this statement is valid if the effects of rotation are included, both in the equilibrium and pulsational calculations, we constructed a grid of rotating stellar models within the 3σ error box in $T_{\rm eff}$ and $\log g$. To this end we used the MESA code [10–12] considering a wide range of the rotational velocity, from $V_{\rm rot} = 62$ km s^{-1} to the upper limit of the validity of the traditional approximation. As the upper limit we adopted 200 km s^{-1} which is about a half of the break-up velocity. The step was $\Delta V_{\rm rot} = 1$ km s^{-1}. We included various rotational mixing mechanisms, i.e., dynamical shear instability, Solberg-Høiland instability, secular shear instability, Eddington-Sweet circulation and the Goldreich-Schubert-Fricke instability [see 11, 13, 14]. We assumed the initial hydrogen abundance, $X_0 = 0.7$, metallicity $Z = 0.017$, the chemical mixture of Asplund et al. [15], and the overshooting parameter in exponential description, $f = 0.02$. Three commonly used opacity data were adopted: OP [16], OPAL [17] and OPLIB [18, 19].

Then, we calculated nonadiabatic pulsation models in the framework of traditional approximation (here the step in $V_{\rm rot}$ was decreased to 0.1 km s^{-1}) for modes which seemed to be good candidates to reproduce observed series of frequencies, i.e., (1, +0), (1, +1), (2, +0) and (2, +1).

* e-mail: szewczuk@astro.uni.wroc.pl

DOI: 10.1051/epjconf/201716003012

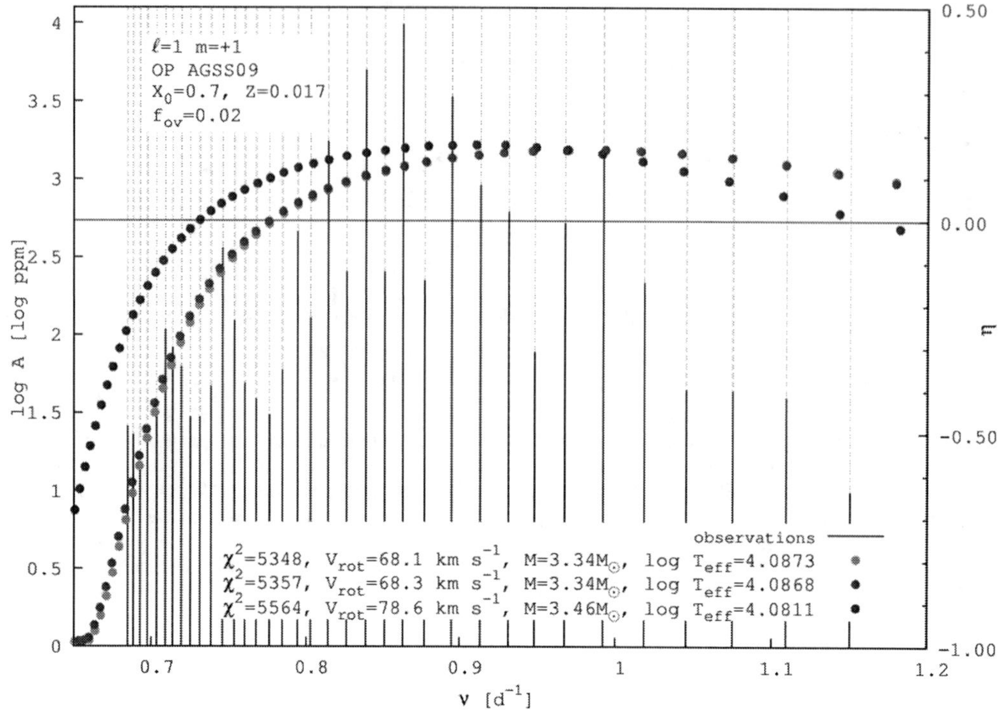

Figure 1: The best seismic models calculated with OP opacities. The observed frequencies and their amplitudes are plotted with vertical lines (left Y-axis). Colour circles mark theoretical frequencies and their instability parameter, η, can bee read on the right Y-axis.

Since for modes other than $(1, +1)$ the best models give worse quality of the fit, χ^2 (defined as in Moravveji et al. [2]), by the order of magnitude, we confirm that the frequencies in the series can be associated with the prograde dipole modes.

In Fig. 1, the theoretical frequencies of $(1, +1)$ modes for the best three OP models are compared with the observed ones. The parameters of these models are summarised in Table 1. Without a doubt, the quality of fitting needs improvement. For the best models, we obtained the values of χ^2 of the order of 5500 which is a value almost three times higher than the one obtained by Moravveji et al. [2]. This is because in our calculations we fixed the metallicity and the overshooting parameter compared to the cited authors. Moreover, Moravveji et al. fitted the coefficient of the extra diffusive mixing in the radiatively stable envelope whereas we employed rotational induced mixing with fixed coefficients. In Fig. 1 on the right Y-axis it is also shown the instability parameter η ($\eta \leq 0$ – stable mode, $\eta > 0$ – unstable mode). As one can see, modes with the lowest frequencies in our best models are predicted to be stable. This is obviously contrary to the observations and needs to be improved.

The quality of the fit is quantitatively the same for each considered opacity source but for models with OPLIB opacities the theoretical instability domain is in an excellent agreement with the observed frequencies. This is shown in Fig. 2, in which three best OPLIB models are plotted. As one can see only 3–4 modes are marginally stable, $\eta \approx 0$. Parameters of these models as well as of those obtained with the OPAL opacities are summarised in Table 1.

Then, we recalculated the best OPLIB model with different values of the metallicity and overshooting parameter as well as with the modified OPLIB opacity profile. The modifications of the OPLIB data consisted of an enhancement of the opacity at the depths corresponding to temperatures $\log T = 5.3$, 5.46 and $\log T = 5.06$. The increase of the opacity at $\log T = 5.06$ mimics the new opacity bump identified in Kurucz models by Cugier [20]. The modified models are shown in Fig. 3. The increase of the opacity at $\log T = 5.3$ by about 50% gives theoretical instability in the whole range of the observed frequencies. Moreover, for each observed frequency we found a corresponding unstable mode. Increasing opacity at $\log T = 5.46$ helps in exciting low frequency modes whereas the instability of high frequencies is unchanged. On the other hand, adding the Kurucz bump results in reducing the pulsational instability, both, at low and high frequencies. Larger value of the metallicity shifts the instability domain towards higher frequencies whereas more effective overshooting from the convective core acts in the opposite direction.

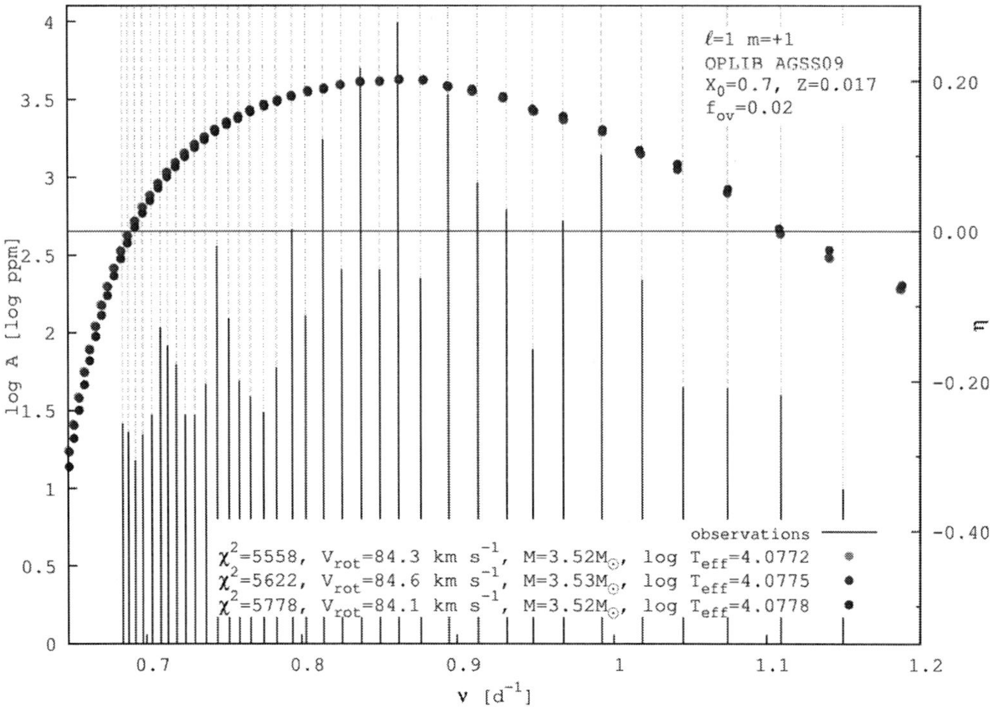

Figure 2: The same as in Fig. 1 but OPLIB models are shown.

Table 1: The parameters of the best seismic models. Column 1 – opacity data (#1 – OP, #2 – OPAL, #3 – OPLIB); Column 2 – mass (in the solar mass units) Column 3 – logarithm of the effective temperature; Column 4 – logarithm of the luminosity (in the solar luminosity units); Column 5 – central hydrogen abundance; Column 6 – rotational velocity (in km s^{-1}); Column 7 – measure of the fitting quality.

op	M	$\log T_{\rm eff}$	$\log \frac{L}{L_\odot}$	X_c	$V_{\rm rot}$	χ^2
#1	3.340	4.0873	2.222	0.42	68.1	5348
#1	3.340	4.0868	2.224	0.42	68.3	5357
#1	3.460	4.0811	2.327	0.35	78.6	5564
#2	3.510	4.0807	2.375	0.31	83.3	5786
#2	3.460	4.0834	2.336	0.34	78.8	5851
#2	3.510	4.0801	2.376	0.31	83.5	5998
#3	3.520	4.0772	2.371	0.30	84.3	5558
#3	3.525	4.0775	2.373	0.30	84.6	5622
#3	3.520	4.0778	2.370	0.31	84.1	5778

3 Conclusions

Our calculations confirmed Pápics et al. [1] finding that the 36 quasi-equidistant in period frequencies observed in KIC 7760680 are most probably associated with the consecutive prograde dipole modes. However, our best models have slightly higher masses (from 3.340 M$_\odot$ to 3.525 M$_\odot$ depending on the opacity data) compared to 3.25 M$_\odot$ obtained by Moravveji et al. [2]. The reason is that we used rotating evolutionary models which give lower effective temperature for the same mass. Rotational induced mixing can contribute to the differences as well. We got a little worse quality of the fit, $\chi^2 \sim 5500$, compared to the cited authors, $\chi^2 \sim 2000$ but in our calculations we have fixed the metallicity and the parameter of the core overshooting, which Moravveji et al. [2] set as free. It is prominent that the instability domain of the $(1, +1)$ modes obtained with the OPLIB data (see Fig. 2) nearly covers the range of the observed 36 frequency peaks. A slight increase of the opacity at $\log T = 5.3$ gives an excellent agreement. On the other hand, adding the Kurucz bump at $\log T = 5.06$ spoils the agreement. Finally, it should be added that series of frequencies equidistant in period may be accidental as they can be reproduced by modes of various angular numbers (ℓ, m; see e.g. [21]). Therefore, it is desirable to confirm the results of mode identification using another method, e.g., using multicolour time-series photometry.

Acknowledgements

This work was financially supported by the Polish National Science Centre grant 2015/17/B/ST9/02082. Calculations have been carried out using resources provided by Wrocław Centre for Networking and Supercomputing (http://wcss.pl), grant no. 265.

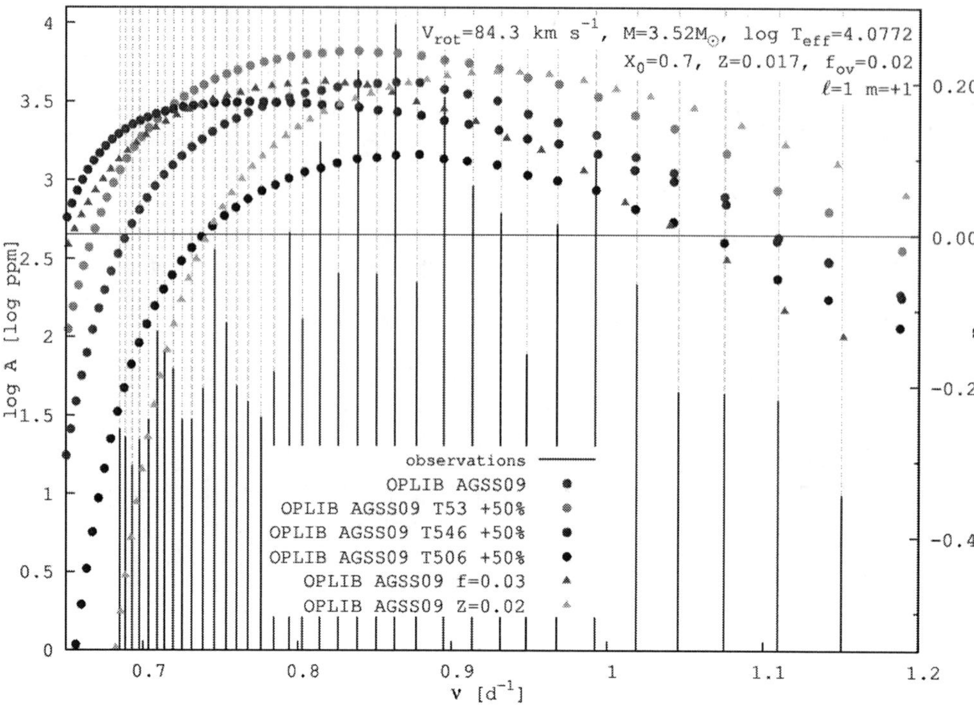

Figure 3: Best OPLIB model recalculated with increased OPLIB opacities by 50% at $\log T = 5.30$ (green circles), at $\log T = 5.46$ (blue circles) and at $\log T = 5.06$ (black circles) as well increased overshooting parameter (magenta triangles) and increased metallicity (cyan triangles). Original OPLIB model is marked by red circles. The coordinates meaning is the same as in Fig 1.

References

[1] P.I. Pápics, A. Tkachenko, C. Aerts, T. Van Reeth, K. De Smedt, M. Hillen, R. Østensen, E. Moravveji, Astrophys. J. **803**, L25 (2015)

[2] E. Moravveji, R.H.D. Townsend, C. Aerts, S. Mathis, Astrophys. J. **823**, 130 (2016)

[3] U. Lee, H. Saio, Astrophys. J. **491**, 839 (1997)

[4] R.H.D. Townsend, MNRAS **340**, 1020 (2003)

[5] R.H.D. Townsend, MNRAS **343**, 125 (2003)

[6] R.H.D. Townsend, MNRAS **360**, 465 (2005)

[7] R.H.D. Townsend, MNRAS **364**, 573 (2005)

[8] W.A. Dziembowski, J. Daszyńska-Daszkiewicz, A.A. Pamyatnykh, MNRAS **374**, 248 (2007)

[9] J. Daszynska-Daszkiewicz, W.A. Dziembowski, A.A. Pamyatnykh, Acta Astron. **57**, 11 (2007)

[10] B. Paxton, L. Bildsten, A. Dotter, F. Herwig, P. Lesaffre, F. Timmes, Astrophys. J. Suppl. **192**, 3 (2011)

[11] B. Paxton, M. Cantiello, P. Arras, L. Bildsten, E.F. Brown, A. Dotter, C. Mankovich, M.H. Montgomery, D. Stello, F.X. Timmes et al., Astrophys. J. Suppl. **208**, 4 (2013)

[12] B. Paxton, P. Marchant, J. Schwab, E.B. Bauer, L. Bildsten, M. Cantiello, L. Dessart, R. Farmer, H. Hu, N. Langer et al., Astrophys. J. Suppl. **220**, 15 (2015)

[13] A. Heger, N. Langer, S.E. Woosley, Astrophys. J. **528**, 368 (2000)

[14] A. Heger, S.E. Woosley, H.C. Spruit, Astrophys. J. **626**, 350 (2005)

[15] M. Asplund, N. Grevesse, A.J. Sauval, P. Scott, Ann. Rev. Astron. Astrophys. **47**, 481 (2009)

[16] M.J. Seaton, MNRAS **362**, L1 (2005)

[17] C.A. Iglesias, F.J. Rogers, Astrophys. J. **464**, 943 (1996)

[18] J. Colgan, D.P. Kilcrease, N.H. Magee, J. Abdallah, M.E. Sherrill, C.J. Fontes, P. Hakel, H.L. Zhang, High Energy Density Physics **14**, 33 (2015)

[19] J. Colgan, D.P. Kilcrease, N.H. Magee, M.E. Sherrill, J. Abdallah, Jr., P. Hakel, C.J. Fontes, J.A. Guzik, K.A. Mussack, Astrophys. J. **817**, 116 (2016)

[20] H. Cugier, Astron. & Astrophys. **565**, A76 (2014)

[21] W. Szewczuk, J. Daszyńska-Daszkiewicz, W. Dziembowski, *Interpretation of the oscillation spectrum of HD 50230 - a failure of richness*, in *Precision Asteroseismology*, edited by J.A. Guzik, W.J. Chaplin, G. Handler, A. Pigulski (2014), Vol. 301 of *IAU Symposium*, pp. 109–112

EPJ Web of Conferences **160**, 03013 (2017) DOI: 10.1051/epjconf/201716003013
Seismology of the Sun and the Distant Stars II

On possible explanations of pulsations in Maia stars

Jadwiga Daszyńska-Daszkiewicz[1,⋆], *Przemysław* Walczak[1], and *Alexey* Pamyatnykh[2]

[1] *Inytut Astronomiczny, Uniwersytet Wrocławski, ul. Kopernika 11, Wrocław, Poland*
[2] *Nicolaus Copernicus Astronomical Center, Bartycka 18, 00-716, Warsaw, Poland*

Abstract. The long-time photometric surveys in a few young open clusters allowed to identify the light variability in stars located on the HR diagram between the well defined δ Scuti variables and Slowly Pulsating B-type stars. Several objects of this type were suggested also from the analysis of the Kepler data. Assuming the pulsational origin of this variability, we try to explain the observed frequencies with pulsational models involving rotation and/or modification of the mean opacity profile.

1 Introduction

Pulsating variables between the instability strips of δ Scuti and Slowly Pulsating B-type (SPB) stars have been suggested many years ago by Struve [1]. Although later a member of the Pleiades Maia appeared to be a constant star (Struve [2]), the idea remained and the search have continued for many years (McNamara [3], Scholz [4]). Finally, the analysis of a 7-year observation campaign of the open cluster NGC 3766 led to the discovery of a large number of variable stars (Mowlavi et al. [5]). Amongst them the authors found 36 new variables, which according to the position on the color-magnitude diagram, have the spectral types between late type B and early type A. The periods of their light variability are from the range of about 0.1 - 0.7 [d], corresponding to the frequencies of about 1.4 - 10 [d^{-1}], and the photometric amplitudes are of the order of a few millimagnitudes. The very recent work by Mowlavi et al. [6] showed that most of them rotate fast (with $V_{rot} \sin i > 150$ km s^{-1}) which supports the hypothesis that Maia pulsators are fast rotating SPB stars (Salmon et al. [7].)

This new type of variables was also suggested from the analysis of space data. From the Kepler photometry, Balona et al. [8–10] found B-type stars with high frequencies and low amplitudes. These stars are located below the red edge of the β Cep instability strip and some of them can be considered as Maia candidates. Several candidates are also from the CoRoT observations (Degroote et al. [11]).

In this paper, assuming the pulsational origin of this variability, we test a few hypotheses. Because NGC3766 is a very young open cluster, we consider only Zero Age Main Sequence (ZAMS) models. The first and already studied hypothesis involves the very high rotation. In the second, new scenario, we modify the opacity profile in order to excite pulsational modes in the required frequency range in the rotating models with masses below $M \approx 3.0\ M_{\odot}$. The last hypothesis demands the modifications of the opacity profile and higher degree modes (up to 6-8) and works for both, low and fast rotating stars of B8-A2 spectral type. Sect. 2 presents the results of our modelling and conclusions are give in Summary.

⋆e-mail: daszynska@astro.uni.wroc.pl

2 The three explanations

The first hypothesis:
fast rotating stars with underestimated masses

In this scenario, Maia stars are linked to fast rotating Slowly Pulsating B-type stars. The fast rotation causes that masses of the stars seen close equator-on are underestimated (Salmon et al. [7]). We perform pulsational computations for models with masses $M = 3-4\ M_{\odot}$ in the framework of the traditional approximation (e.g. Lee & Saio [12], Townsend [13], Savonije [14]). As mentioned above, NGC3766 is a very young open cluster, thus we analysed only ZAMS models. The initial hydrogen abundance and metallicity were set at 0.7 and 0.015, respectively, and the chemical mixture was adopted from Asplund at al. [15]. Here, we show results obtained with the OPAL opacities (Iglesias & Rogers [16]) but with the OPLIB (Colgan et al. [17, 18]) and OP data (Seaton [19]), the results are qualitatively similar.

High order gravity modes as well as mixed gravity-Rossby modes are considered. In models with masses corresponding to B8-A2 type, these modes are stable. Their instability begins from the mass of about $M = 3M_{\odot}$. Moreover, including the effects of rotation on g modes, both, enhances the pulsational instability and causes the shift of modes to the higher frequency range. In Fig. 1, we plot the instability parameter, η, as a function of mode frequency for a ZAMS model with a mass $M = 3.2\ M_{\odot}$, effective temperature, $\log T_{\rm eff} = 4.119$. and rotational velocity $V_{rot} = 250$ km s^{-1}. The parameter η is the normalized work integral and for unstable modes it takes positive values. For clarity, we depicted only modes which

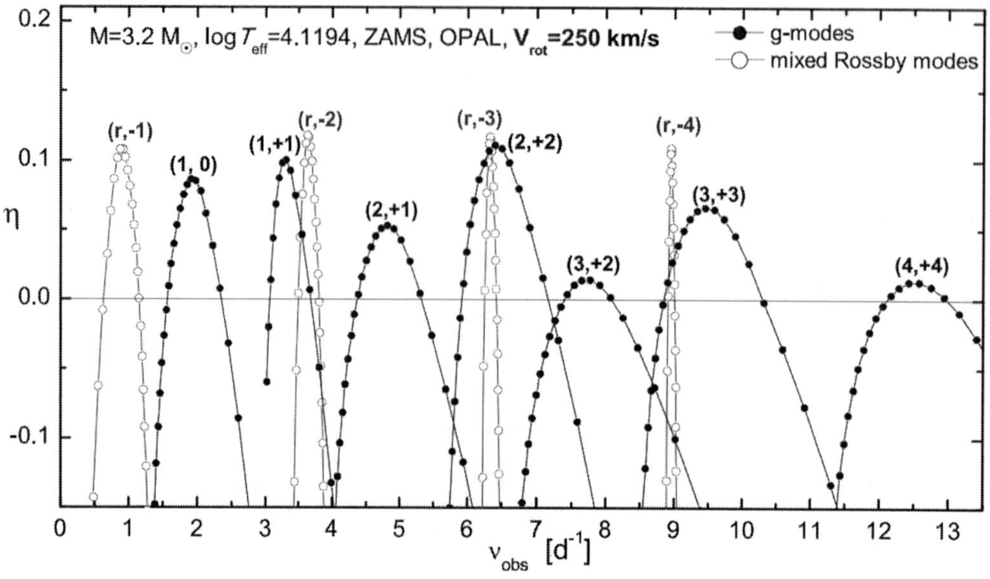

Figure 1: The instability parameter, η, as a function of the mode frequency for a ZAMS model with a mass $M = 3.2\ M_\odot$ and effective temperature, $\log T_{\rm eff} = 4.1194$, rotating with the velocity of $V_{\rm rot} = 250\ \ {\rm km\ s^{-1}}$. Only $(\ell,\ m)$ and $(r,\ -m)$ modes which reach instability $(\eta > 0)$ are shown.

become unstable $(\eta > 0)$ in some range of frequencies.

The second hypothesis:
fast rotating stars with non-standard opacities

In the second, new scenario, we continued pulsational computations with the traditional approximation but additionally we modified the mean opacity profile in models with masses below $M = 3.0\ M_\odot$. The standard opacities were increased at the depth corresponding to temperature $\log T = 5.1$. This modification mimics the new opacity bump identified in the Kurucz atmosphere models (Cugier [20]). This new bump was suggested, for example, as a possible cause of excitation of low frequency g-modes in δ Scuti stars as detected in the Kepler data (Balona et al. [21]). The effect of increasing the OPAL opacities by 100% at $\log T = 5.1$ is presented in Fig. 2. A model with the mass $M = 2.5\ M_\odot$ on ZAMS was considered. The rotational rate was $V_{\rm rot} = 200\ \ {\rm km\ s^{-1}}$. For models with such masses, pulsational computations with the standard opacities model do not predict any instability. As one can see, it is more than enough to consider modes with the degree $\ell \leq 3$ to cover the frequency range observed in the NGC3766 stars.

The third hypothesis:
slow rotating stars with non-standard opacities

The third explanation involves modifications of the opacity profile and including modes with the harmonic degrees up to 8. Here, we performed all pulsational computations assuming the zero-rotation approximation, i.e., all effects of rotation on pulsations were neglected.

We found that by increasing the mean opacity at the depth $\log T = 5.1$ in a $M = 2.5\ M_\odot$ model, it is possible to get the instability in the observed frequency range for modes with $\ell \geq 2$. Adding opacity at the Z-bump $(\log T = 5.3)$ makes also the dipole modes unstable. This scenario works for, both, slow and fast rotating stars of B8-A2 type. Below, we show the results obtained with the modified OPAL data. We added 150% of the opacity at $\log T = 5.1$ and 50% of the opacity at $\log T = 5.3$. As one can see, in the mass range corresponding to late B-type and early A-type stars, $M \in (2.5,\ 4)$, high degree modes $(\ell > 6)$ can reach frequencies as high as about 9 $\rm d^{-1}$. Considering modes with such high degrees is justified by the fact that the amplitudes of the new variable stars in NGC3766 are very low and only very long campaign (about 7 years) allowed to detect them.

3 Summary

Here, we discussed three possible explanations for pulsations in Maia stars. Each of them has its strengths and weaknesses.

Fast rotating stars seen equator-on can account for pulsations of the Maia type. This hypothesis has been already explored by Salmon et al. [7]. In this case, we demand the close equator-on orientation for all stars. As a consequence, all modes which have an equatorial node are not good candidates. It is worth to mention that prograde sectoral modes reach the highest visibility on the equator (e.g. Salmon et al. [7], Daszyńska-Daszkiewicz et al. [22]). On the other hand, given the fact that there are not many identified Maia variables this scenario is plausible. Moreover, it is strengthened by the recent result of Mowlavi et al. [6]

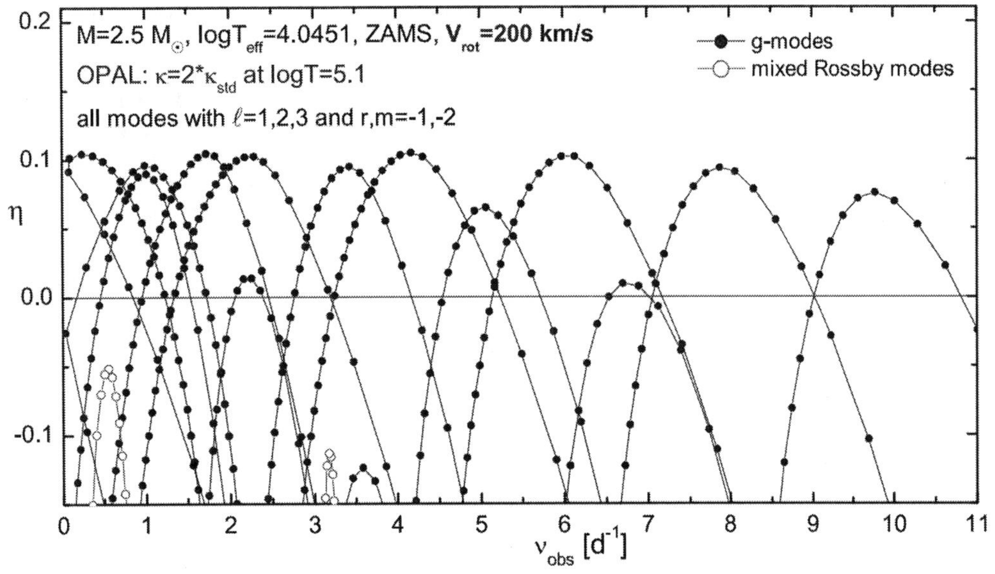

Figure 2: The instability parameter, η, as a function of the mode frequency for the ZAMS model with a mass $M = 2.5\ M_{\odot}$ and effective temperature, $\log T_{\rm eff} = 4.0451$, rotating with the velocity of $V_{\rm rot} = 200\ \ {\rm km\ s^{-1}}$. The model was computed with the modified OPAL opacities which were artificially increased by 100% at $\log T = 5.1$. Modes with the degree up to 3 and mixed-Rossby modes with $m = -1,\ -2$ are shown.

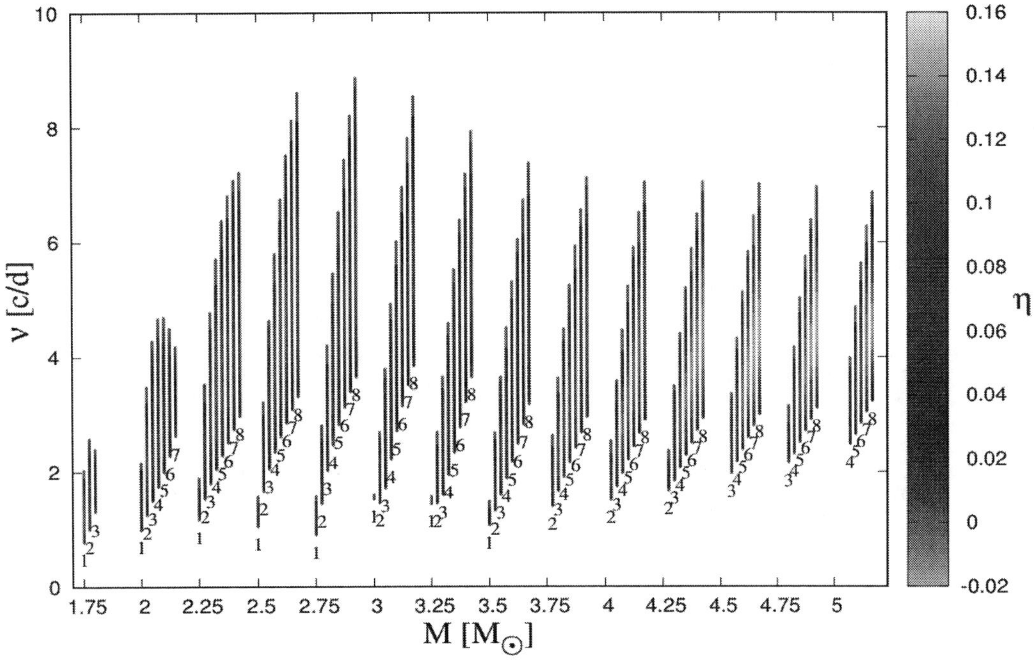

Figure 3: The frequency range of unstable modes as a function of mass for ZAMS models computed with the modified OPAL opacity profile. Modes with the harmonic degrees up to $\ell = 8$ were considered. The opacities were increases by 150% at $\log T = 5.1$ and by 50% at $\log T = 5.3$. The values of the instability parameter are coded by colours.

who showed that all new-class variables in NGC 3766 are fast rotators.

Two new scenarios involve modifications of the mean opacity profile. We showed that an increase of the opacity at $\log T = 5.1$ is indispensable to get the instability in the frequency range 1.4 – 10 d^{-1} in the models with masses $M < 3.0\,M_\odot$. The above conclusions are valid if the effects of rotation on pulsations are taken in to account. In the case of g-mode pulsations, we do this via the traditional approximation. If the effects of rotation on pulsation are neglected then, additionally to the opacity modifications, the higher ℓ modes have to be considered.

Here, we need quite significant modifications of the opacities which have to be justified at some point. There are some results that support such approach, e.g., by increasing the opacity at the depth $\log T = 5.1$ it is possible to explain low frequency g-modes in δ Scuti stars detected in the Kepler data (Balona et al. [21]). Given that there are still many uncertainties in computations of stellar opacities and, for example, in case of hybrid B-type pulsators significant modifications have been suggested (Daszyńska-Daszkiewicz et al. [23]), such hypothesis for Maia variables is reasonable. Additionally, these two hypothesis do not have any requirements for the inclination angle and all unstable modes can be considered. The third explanation needs the higher degree modes, up to 8, which have poorer visibility. However, given the low values of the observed amplitudes and the lack of knowledge of the selection mechanism, we should allow such a possibility. Moreover, the often forgotten fact is that while the mode visibility is getting worse with higher ℓ according to the disc averaging effects, it increases as ℓ^2 according to the geometrical factor $(\ell-1)(\ell+2)$ (e.g. Daszyńska-Daszkiewicz et al. [24]). The explanation with the opacity modifications and higher ℓ modes works for both fast and slow rotators.

To decide which scenario is preferred, firstly, more determinations of the rotational velocity of these stars are needed. Then, pulsational analysis of more Maia stars has to be performed taking into account various effects on pulsational instability. It cannot be ruled out that rotation and opacity modification can work together and hold the key to solving the Maia star problem.

Acknowledgements

This work was financially supported by the Polish National Science Centre grant 2015/17/B/ST9/02082. Calculations have been carried out using resources provided by Wrocław Centre for Networking and Supercomputing (http://wcss.pl), grant no. 265.

References

[1] O. Struve, S&T **14**, 492 (1955)
[2] O. Struve, J. Sahade, C.R. Lynds, S.S. Huang, Astrophys. J. **125**, 115 (1957)
[3] B.J. McNamara, Astrophys. J. **289**, 213 (1985)
[4] G. Scholz, H. Lehmann, G. Hildebrandt, K. Panov, L. Iliev, Astron. & Astrophys. **337**, 447 (1998)
[5] N. Mowlavi, F. Barblan, S. Saesen, L. Eyer, Astron. & Astrophys. **554**, A108 (2013)
[6] N. Mowlavi, S. Saesen, T. Semaan, P. Eggenberger, F. Barblan, L. Eyer, S. Ekström, C. Georgy, ArXiv e-prints (2016), 1610.01077
[7] S.J.A.J. Salmon, J. Montalbán, D.R. Reese, M.A. Dupret, P. Eggenberger, Astron. & Astrophys. **569**, A18 (2014)
[8] L.A. Balona, A. Pigulski, P.D. Cat, G. Handler, J. Gutiérrez-Soto, C.A. Engelbrecht, F. Frescura, M. Briquet, J. Cuypers, J. Daszyńska-Daszkiewicz et al., MNRAS **413**, 2403 (2011)
[9] L.A. Balona, A.S. Baran, J. Daszyńska-Daszkiewicz, P. De Cat, MNRAS **451**, 1445 (2015)
[10] L.A. Balona, C.A. Engelbrecht, Y.C. Joshi, S. Joshi, K. Sharma, E. Semenko, G. Pandey, N.K. Chakradhari, D. Mkrtichian, B.P. Hema et al., MNRAS **460**, 1318 (2016)
[11] P. Degroote, C. Aerts, M. Ollivier, A. Miglio, J. Debosscher, J. Cuypers, M. Briquet, J. Montalbán, A. Thoul, A. Noels et al., Astron. & Astrophys. **506**, 471 (2009)
[12] U. Lee, H. Saio, Astrophys. J. **491**, 839 (1997)
[13] R.H.D. Townsend, MNRAS **340**, 1020 (2003)
[14] G.J. Savonije, Astron. & Astrophys. **443**, 557 (2005)
[15] M. Asplund, N. Grevesse, A.J. Sauval, P. Scott, Ann. Rev. Astron. Astrophys. **47**, 481 (2009), 0909.0948
[16] C.A. Iglesias, F.J. Rogers, Astrophys. J. **464**, 943 (1996)
[17] J. Colgan, D.P. Kilcrease, N.H. Magee, J. Abdallah, M.E. Sherrill, C.J. Fontes, P. Hakel, H.L. Zhang, High Energy Density Physics **14**, 33 (2015)
[18] J. Colgan, D.P. Kilcrease, N.H. Magee, M.E. Sherrill, J. Abdallah, J., P. Hakel, C.J. Fontes, J.A. Guzik, K.A. Mussack, Astrophys. J. **817**, 116 (2016)
[19] M.J. Seaton, MNRAS **362**, L1 (2005)
[20] H. Cugier, Astron. & Astrophys. **565**, A76 (2014)
[21] L.A. Balona, J. Daszyńska-Daszkiewicz, A.A. Pamyatnykh, MNRAS **452**, 3073 (2015)
[22] J. Daszyńska-Daszkiewicz, W.A. Dziembowski, A.A. Pamyatnykh, AcA **57**, 11 (2007)
[23] J. Daszyńska-Daszkiewicz, P. Walczak, A.A. Pamyatnykh, MNRAS **subbmited** (2016)
[24] J. Daszyńska-Daszkiewicz, W.A. Dziembowski, A.A. Pamyatnykh, M.J. Goupil, Astron. & Astrophys. **392**, 151 (2002)

DOI: 10.1051/epjconf/201716004001

Asteroseismic Constraints on the Models of Hot B Subdwarfs: Convective Helium-Burning Cores

Jan-Torge Schindler[1,⋆], *Elizabeth M.* Green[1], and *W. David* Arnett[1]

[1] *Steward Observatory, University of Arizona, 933 North Cherry Avenue, Tucson, AZ 85721,USA*

Abstract.
Asteroseismology of non-radial pulsations in Hot B Subdwarfs (sdB stars) offers a unique view into the interior of core-helium-burning stars. Ground-based and space-borne high precision light curves allow for the analysis of pressure and gravity mode pulsations to probe the structure of sdB stars deep into the convective core. As such asteroseismological analysis provides an excellent opportunity to test our understanding of stellar evolution.
In light of the newest constraints from asteroseismology of sdB and red clump stars, standard approaches of convective mixing in 1D stellar evolution models are called into question. The problem lies in the current treatment of overshooting and the entrainment at the convective boundary. Unfortunately no consistent algorithm of convective mixing exists to solve the problem, introducing uncertainties to the estimates of stellar ages. Three dimensional simulations of stellar convection show the natural development of an overshooting region and a boundary layer. In search for a consistent prescription of convection in one dimensional stellar evolution models, guidance from three dimensional simulations and asteroseismological results is indispensable.

1 Hot Subdwarf B stars

Subdwarf B (sdB) stars are a class of hot ($T_{\rm eff}$ = 20,000–40,000 K) and compact ($\log g$ = 5.0–6.2) stars with very thin hydrogen envelopes ($M_{\rm H} < 0.01\,M_\odot$) [1, 2]. They form the so-called extreme horizontal branch (EHB) in the Hertzsprung-Russell diagram, where most of them quietly burn helium in their cores for $\sim 10^8$ years.

A violent process stripped them of most of their hydrogen envelope before they underwent the helium flash near the tip of the red giant branch (RGB). As a consequence they will not climb up the asymptotic giant branch (AGB) once their core helium is exhausted, but rather evolve directly to become white dwarfs.

Although some sdB stars appear to be single and others occur in wide binaries, surveys have concluded that the majority are in close binaries with white dwarf or low-mass main sequence companions [3–5]. A small fraction of the latter exhibit eclipses, offering insight into the orbital parameters and masses of the systems. These HW Virginis stars, named after the discovery system, play a crucial role in determining the mass distribution of sdB stars. The median mass of sdB stars estimated from these binaries is $0.469\,M_\odot$ [6].

The formation scenarios for sdB stars have to explain their existence in both single and binary systems as well as introduce a process by which the hydrogen envelope of the progenitor stars can be lost. Common envelope ejection (CE) or stable Roche lobe overflow (RLOF) during binary evolution result in systems with close and wide separations, respectively, [7]. Population synthesis models can well explain the occurrence of binary systems and the mass distribution of the primary and secondary components [8]. To explain the existence of single sdB stars, extreme mass loss on the RGB [9–11] as well as white dwarf mergers [12] were suggested. However, the masses from the white dwarf merger channel over-predict the median mass of observed sdB stars [6]. After the discovery of Earth-sized bodies around sdB stars [13–16], it seems possible that sub-stellar companions may play a role in shaping single sdB stars.

Our goal in this article is to discuss the latest results of asteroseismology and their implications for stellar evolution of core helium-burning stars. Therefore we will continue by introducing sdB asteroseismology in Section 1.1 and the state-of-the-art of sdB stellar evolution in Section 1.2. In Section 2 we focus on our main discussion of convection in helium-burning cores in light of asteroseismological results. We present a conclusion to the discussion in Section 3. For a full review of the field of Hot Subdwarfs we refer the interested reader to the publications of Heber [17, 18].

1.1 sdB Asteroseismology

Short period (80–600 s) and longer period (2000–14000 s) stellar pulsations are found in a significant fraction of the sdB star population. The shorter pressure (p)-mode pulsations were predicted by S. Charpinet et al. [20] and independently discovered by Kilkenny et al. [21] at nearly the same time. Currently more than 60 p-mode pulsators,

⋆e-mail: jtschindler@email.arizona.edu

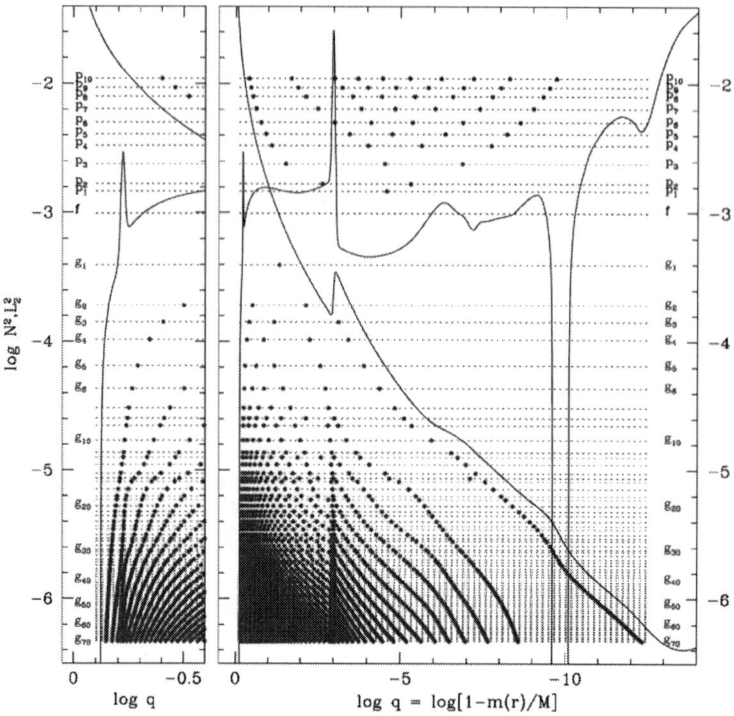

Figure 1: This illustration shows the l=2 modes in a representative sdB model. P-modes propagate where the mode angular frequencies exceed both the Lamb (L2) and Brunt-Väisälä (N) frequencies (blue filled circles show the nodes of modes with order up to k=10). G-modes are confined to the region where the mode angular frequencies are both below the L2 and N frequencies (red filled circles show the nodes of modes with order up to k=70). The left panel shows a close up view of the mode structure in the central region of the star. This Figure was adapted with permission from Charpinet et al. 2013 [19].

termed V361 Hya stars, have been found among the hotter sdB stars. These short pulsations are due to low-order, low-degree acoustic waves.

The longer pulsations have their origin in mid to high order, low degree gravity waves (g-modes [22]). There are about 50 known g-mode pulsators, or V1093 Her stars, on the cooler end of the sdB distribution. Furthermore, at the intersection of these two populations on the EHB exists a narrow range in which hybrid pulsators can be found [23], showing both p- and g-mode oscillations.

The non-radial pulsations in sdB stars are opacity driven (κ-mechanism) [24, 25] by partial ionization of iron group elements in their stellar envelopes. The balance between gravitational settling and radiative levitation creates a region with an overabundance of these iron group elements (especially Fe and Ni, [26]) in the envelopes of these stars, leading to an opacity bump. The inclusion of both diffusion processes is therefore not only important to explain the low atmospheric helium abundances, but also to create the driving region for the pulsations. Since the pulsation regions are by no means pure, inefficient stellar winds have been proposed to destroy the diffusive balance in the driving region [27].

As can be seen in Figure 1, p-modes can only propagate in the outer part of the star ($\log q < 0.4$). They are reflected back to the surface before reaching the convective core. G-modes on the other hand can penetrate into the deep interiors ($\log q \lesssim 0.1$). The steep chemical transitions between the hydrogen-rich envelope and the helium mantle and between the base of the mantle and the convective C-O-He core result in spikes in the Brunt-Vaisälä-frequency (see Figure 1). Pulsational modes having nodes (filled circles in Figure 1) close to these transition regions are either partially trapped above or confined to the lower part of the star. This leaves clear signatures on the period spacing of those modes. The p-mode oscillations are therefore sensitive to the transition between the helium mantle and hydrogen envelope, whereas g-modes also probe the transition of the He mantle and the convective core. Thus asteroseismological analysis of sdB g-mode pulsators provides an exclusive window into the interior structure of core helium-burning stars.

Quantitative asteroseismological analyses of pulsating sdB stars have been carried out using a forward modeling method [28–31]. The technique uses a neural network of structural sdB models to create theoretical oscillation spectra, which are then compared to the observed frequencies of a given star. The best possible match in the given parameter space holds information about the structural parameters of the star. Pulsational frequencies derived from

 DOI: 10.1051/epjconf/201716004001

light curves of p- and g-mode pulsators using ground-based as well as space-borne (e.g. *CoRoT* and *Kepler*) instruments allowed for detailed asteroseismological analysis. Also, measurements of the g-mode period spacing in *Kepler* light curves of sdB stars [32, 33] enable comparisons with pulsation properties of stellar evolution models (e.g. see [34]).

Using the forward modeling method [e.g. 35, 36] stellar masses, envelope masses, surface gravities and effective temperatures were constrained and agree remarkably well with measurements from other techniques such as light curve modeling of eclipsing binary systems and spectroscopic analyses ([37] Table 3, [38]). As pure structural models, these results are independent of the uncertainties in stellar evolution physics (e.g. convection, nuclear reaction rates or wind mass loss). Hence, a comparison between stellar evolution models and these asteroseismic measurements may help to identify deficiencies in our current understanding.

1.2 Stellar Evolution of Subdwarf B stars

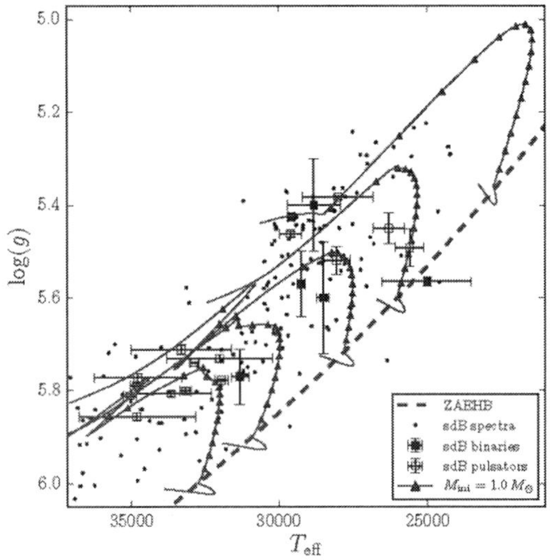

Figure 2: An example of the evolutionary tracks of sdB stars (solid red curves) in the $\log g$–$T_{\rm eff}$ diagram (reproduced from [39]). The evolutionary tracks have a progenitor mass of $M_{\rm ini} = 1.0\,M_\odot$ and sdB masses of $M_{\rm sdB} = 0.4652, 0.4654, 0.4658, 0.4669, 0.4705\,M_\odot$ from bottom to top. The red triangles intersect the lines in intervals of 10^7 years. The dashed line shows the zero age extreme horizontal branch (ZAEHB) for our evolutionary models of $M_{\rm ini} = 1.0\,M_\odot$. The spectroscopic data points (small black dots) for sdB stars [40] agree very well with the open and filled squares with error bars derived from eclipsing binary and asteroseismology analyses, respectively [6].

After stars evolve up the red giant branch and start core helium-burning, losing a substantial amount of their hydrogen envelope in the process, they become horizontal branch stars. Subdwarf B stars on the extreme horizontal branch have hydrogen envelope masses too low to sustain hydrogen shell-burning. It is unsurprising that the first systematic stellar evolution studies relevant to sdB stars targeted horizontal branch stars [e.g. 41–43].

The detailed physics of the envelope mass stripping due to Roche lobe overflow or common envelopes are mostly neglected in one dimensional stellar evolution models. These processes occur on dynamical timescales of $\sim 10^3$ years and are usually mimicked by extreme mass loss by stellar winds on the RGB, until only a tiny hydrogen envelope $M_{\rm env} \lesssim 0.01\,M_\odot$ remains. Very recently, physical arguments for common envelope evolution have been introduced to regulate the extreme mass wind loss in [44].

Similarly, older stellar evolution calculations could not handle the violent helium core flash that occurs in $M \lesssim 2.2\,M_\odot$ stars. Stellar structures at the RGB tip were modified to restart the evolution after the helium flash. The He-core material was artificially enriched by 3%–7% of carbon over a region believed to be plausible for the convective mixing during the He-core-flash (see also [9]). The most widely adopted models of sdB stars were created using this technique [e.g. 7, 29, 43] and show only slight differences from more recent models which follow the evolution through the He-flash [45]. Stellar evolution models of sdB stars [7, 43] have been successful in explaining the general distribution of the population in the $\log g$–$T_{\rm eff}$-diagram and their atmospheric parameters. With the inclusion of physical diffusion processes later studies were further able to self-consistently predict the instability strips for sdB stars [46]. However, tension exists between recent asteroseismological analyses of sdB stellar structures and stellar evolution modeling [34, 39]. As this is also the main focus of this short review, we will expand this discussion in Section 2. In the following paragraphs we highlight some important aspects of sdB stellar evolution models in more depth. .

1.2.1 The He-flash

Modern state-of-the-art stellar evolution codes are capable of evolving stars from the pre-main-sequence through the Helium flash to the ZAEHB [e.g. 45, 47].

However, two- and three-dimensional simulations of the He-flash [48, 49] have shown that the extent of the convective region during the He-flash, as predicted by standard 1D stellar models, is incorrect. Turbulent entrainment on both edges of the convection zone lead to rapid growth of the convective region on a dynamic timescale. As a consequence the main core flash might not be followed by subsequent mini-flashes leading up to core helium-burning.

In the current standard algorithms of convection using Mixing Length Theory (MLT, [50]) turbulent entrainment is not included. The convective region in the standard MLT picture only extends throughout the super-adiabatic region, whereas the convective flows in three dimensional simulations naturally extend into sub-adiabatic regions, where the plumes decelerate ("overshooting"). Additions

DOI: 10.1051/epjconf/201716004001

to the standard algorithms that mimic the physical "overshooting" use prescriptions with a free tunable parameter. Therefore one has to take great caution in interpreting the details of the He-flash calculated by one dimensional stellar models.

1.2.2 The Importance of Physical Diffusion Processes

To explain the p-mode pulsations in sdB stars, Charpinet et al. [20] proposed that radiative levitation creates an over-abundance of iron group elements in the sdB envelope. Partial ionization of these metals can then lead to the driving of pulsation via the κ-mechanism. This was confirmed using static models of sdB stars in diffusive equilibrium [24]. The approach of using static models for the instability calculations was later validated [51], showing that diffusive timescales are short compared to the stellar evolution.

In order to explain the g-mode pulsations, Fontaine et al. [25] argued that the same κ-mechanism was at play. However, the models exhibiting g-mode pulsations were several thousand Kelvin too cool compared to the distribution of observed pulsators. This mismatch was termed the blue-edge problem. The discrepancy could be reduced by using artificially enhanced Fe and Ni abundances in the envelope [26] and the OP [52] instead of the OPAL [53] opacities. In a sequence of pioneering studies, Hu et al. [54–56] solved the diffusion equations for gravitational settling, thermal diffusion, concentration diffusion and radiative levitation in stellar evolution models of sdB stars. These diffusive processes naturally produce the Fe and Ni enhancements. Bloemen et al. [46] finally solved the blue-edge problem by calculating the pulsation properties for stellar models that included the necessary diffusion physics [55].

The inclusion of these diffusive processes is therefore crucial for any study focusing on the atmospheric abundances or pulsation properties of sdB stars.

2 Asteroseismology and Convection in Helium-burning Cores

2.1 A Short Introduction to Convective Mixing in One Dimensional Stellar Evolution

Most stellar evolution codes treat mixing of convective flows as a "diffusive" process. A "diffusion" operator is chosen for mathematical convenience [57], based on mixing length theory (MLT, [e.g. 50, 58]).

The extent of the dynamically unstable regions is often assessed by the Schwarzschild or Ledoux criteria for instability. The Ledoux criterion for convection reads

$$\nabla_{\rm rad} > \nabla_{\rm ad} + \frac{\Phi}{\delta}\nabla_\mu , \qquad (1)$$

where

$$\nabla_{\rm rad} \equiv \left(\frac{\partial \ln T}{\partial \ln p}\right)_{\rm rad}, \quad \nabla_{\rm ad} \equiv \left(\frac{\partial \ln T}{\partial \ln p}\right)_{\rm ad}, \quad \nabla_\mu \equiv \left(\frac{\partial \ln \mu}{\partial \ln p}\right) \qquad (2)$$

and

$$\Phi \equiv \left.\left(\frac{\partial \ln \rho}{\partial \ln \mu}\right)\right|_{P,\mu}, \qquad \delta \equiv -\left.\left(\frac{\partial \ln \rho}{\partial \ln T}\right)\right|_{P,T}. \qquad (3)$$

If the composition term $\Phi/\delta\ \nabla_\mu$ is omitted, one obtains the Schwarzschild criterion for convective instability. Using the Schwarzschild criterion one refers to the unstable region as super-adiabatic ($\nabla_{\rm rad} > \nabla_{\rm ad}$) and to the stable region as sub-adiabatic ($\nabla_{\rm rad} < \nabla_{\rm ad}$).

When the Ledoux criterion is used, composition gradients are able to stabilize regions that would be unstable to the Schwarzschild criterion. A receding convection zone during hydrogen core-burning on the main sequence leaves helium enriched material, stable to mixing if the Ledoux criterion was applied, but unstable to the Schwarzschild criterion. However, the situation is more complicated and double diffusive processes lead to mixing in the seemingly stable region. This is famously called *semiconvective* mixing and is a needed addition to the standard formulation of MLT, whenever the Ledoux criterion is used [e.g. 59].

Another addition to canonical convection is *overshoot* mixing. It refers to the transport of energy and material across the boundary from the dynamically unstable into the stable region. Since MLT is a local theory, the deceleration and turning of the convective flow is not captured and the convective zone only extends to the edge of the super-adiabatic region. To remedy this, overshooting algorithms [e.g. 60, 61] are used to extend the convection region beyond the stable region defined by the instability criterion.

2.2 The Physics of Convection during Core-helium-burning

After the helium flash has lifted the degeneracy of the He-core, core-helium-burning starts. First the triple-alpha process fuses helium into carbon and once a significant carbon abundance has been reached the $^{12}C(\alpha,\gamma)^{16}O$ reaction dominates the energy generation. With the start of core-helium-burning a core convection zone develops.

Three dimensional simulations of convection [48, 62–65] show how plumes of hotter material are accelerated by buoyancy forces in the super-adiabatic region and start to rise up. Once they reach the sub-adiabatic region, buoyancy braking decelerates the plumes until the flow turns and the material descends back down. The plumes transport kinetic energy and thermal energy upwards and mix the convective region. The flow is highly turbulent and the kinetic energy of the flow is converted into internal energy by the physics of the Kolmogorov cascade.

A thin boundary layer develops at the top of the core convection zone. When the convective flow turns, its velocity is perpendicular to the radial coordinate. The material outside the convective zone is not moving and Kelvin-Helmholtz instabilities can develop which slowly entrain material from outside the boundary [65].

The nuclear fusion in the core proceeds to produce carbon and oxygen, which have higher free-free opacities than helium [66]. Therefore the opacity in the convective core slowly rises with the energy production leading to an increase of the super-adiabatic gradient. If helium rich material is entrained, carbon-rich (and later oxygen-rich) material is mixed outward of the convective zone increasing the opacity in the stable region in the process. As a result the boundary region becomes super-adiabatic and the convection zone slowly grows. Secondly, the downward-mixed helium replenishes the nuclear fuel and the phase of core-helium-burning is prolonged.

In one dimensional models, convective instability is determined by the Ledoux or Schwarzschild criterion. Mixing length theory (MLT) predicts the convective velocity locally and calculates the convective energy flux for the energy balance. However, fluxes of the convective velocity across zone boundaries are not included in this framework and therefore deceleration of material (physical overshooting) cannot exist in this picture. Without the addition of "overshoot" prescriptions, an extremely sharp abundance gradient develops at the boundary where convective neutrality ($\nabla_{rad} = \nabla_{ad}$ for the Schwarzschild criterion) is violated. This framework does not allow for entrainment of material at the boundary and as a consequence the convective zone is unable to grow. Historically [66] overshoot prescriptions are used to circumvent this problem. However, they usually introduce a free tunable parameter and therefore the stellar evolution model loses its "predictive" power. Alternatively, concentration diffusion has been shown to soften the abundance gradient and naturally lead to a growth of the convection zone [67]. Yet, the rate of entrainment through diffusion might not necessarily be the same as for convective entrainment. Both approaches do satisfy convective neutrality at the boundary.

This boundary problem is closely related with another phenomenon. It has been shown that the super-adiabatic gradient develops a minimum during core-helium-fusion. [68–71]. Over time the super-adiabatic gradient can decrease until the minimum reaches convective neutrality. Since MLT defines convection to occur in regions with super-adiabatic excess, the definition of the convective zone becomes ambiguous at this point. In consequence standard algorithms will force the convection zone to split. As a solution the material beyond the minimum was mixed with physically motivated but ad-hoc schemes to satisfy convective neutrality in that region. Many similar dedicated algorithms with different approaches and terminology have been developed to achieve this (e.g. "induced semiconvection" [68], "semiconvection" [69], "partial mixing" [72], "maximal overshooting" [34]). We will refer to this process as "induced semiconvection" for the rest of this article.

However, the convection zone does not necessarily split. In some of our models, which include concentration diffusion to enable core growth, the super-adiabatic minimum never decreases to convective neutrality (e.g. see the model with the solid blue line Figure 3 [39]).

Core convection during helium-burning remains a complex problem of stellar evolution. We have briefly mentioned the 3D picture of convection and then discussed core convection during core-helium-burning in one dimensional models. It is time we turned towards observational evidence for guidance and constraints.

2.3 Tension Between 1D Stellar Evolution Models and Asteroseismology

The advent of quantitative asteroseismological analyses in the era of satellite-borne precision photometry opens the possibility to study the deep interior of stellar structures. The results of asteroseismology allow us to contrast stellar evolution models with measurements from real stars and can help to constrain the inadequacies of our current modeling prescriptions. Here we will focus on results regarding core-helium-burning convection, but asteroseismology also yields measures of the internal rotation.

2.3.1 Subdwarf B stars

In contrast with red clump or horizontal branch (HB) stars, sdB stars cannot sustain hydrogen shell-burning nor an outer convective hydrogen envelope. This allows for direct observations of g-mode pulsations in core-helium-burning stars, which makes them unique probes of the core convection zone. Out of the ~15 known g-mode pulsators for which there are sufficient photometric observations for asteroseismology, three have been analyzed using the forward modeling technique described above.

The results directly constrain the extent of the core convection zone as well as the nature of the abundance gradient at the boundary. The analyses for the three pulsators estimate the convection zone to extend out to $M_{cc} = 0.22 \pm 0.01\ M_\odot$ [75], $M_{cc} = 0.28 \pm 0.01\ M_\odot$ [73] and either $M_{cc} = 0.273^{0.008}_{-0.010}\ M_\odot$ or $M_{cc} = 0.225^{+0.011}_{-0.016}\ M_\odot$ [14]. In the last case two equally probable solutions were found. The abundance gradients at the convection zone boundary indicate that all three stars are significantly less than halfway through their He-burning lifetimes, having consumed only about 20%-40% of the helium in their cores.

We conducted a study to test whether we could reproduce the interior structures inferred from asteroseismology using standard algorithms in one dimensional stellar evolution models [39]. We carried out these calculations with the Modules for Experiments in Stellar Astrophysics (MESA, [47, 76]). While our standard model used concentration diffusion to allow for convective core growth (see also [67]), it was not able to reproduce the larger core sizes for these three stars. Only extreme values for additional (exponential) overshoot (as implemented in MESA [47]) were able to bring the extent of the convective core in marginal agreement with the asteroseismological results.

We display the growth of the convective cores for a range of model physics in Figure 3 and contrast them with results of Sweigart et al. [42] using overshoot and induced semiconvection. Our "standard" model shows a monotonically growing convective core that reaches the same extent as the model with induced semiconvection. The convective boundary seems to fluctuate for our models with overshoot, indicating an unstable behavior that might well be

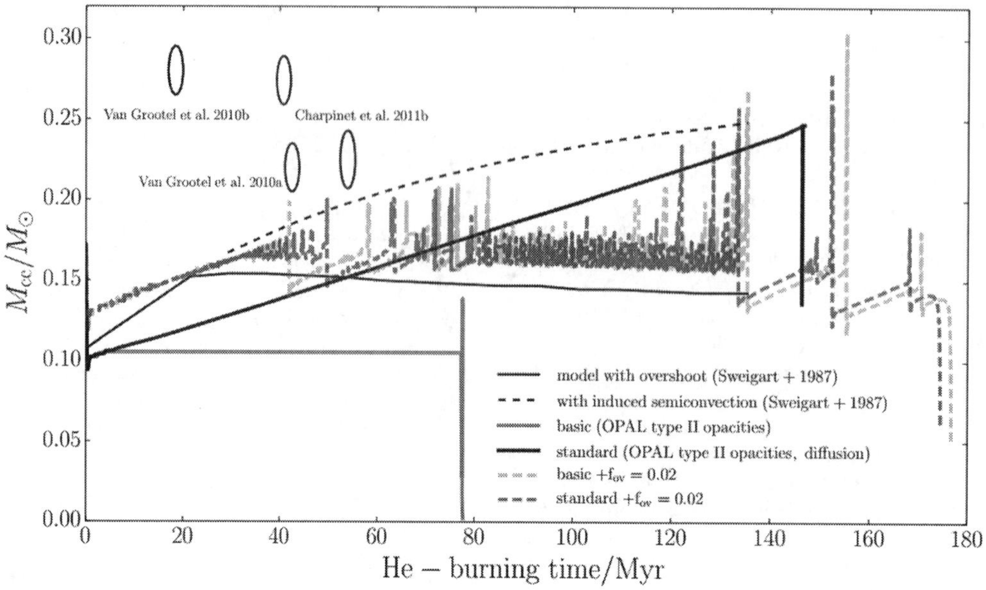

Figure 3: The extent of the convective cores as a function of time for four MESA models [39] (gray and blue curves) and two older models from Sweigart et al. [42] (black curves). The Sweigart models include either overshoot or induced semiconvection. A comparison of the basic and standard MESA models (solid curves) shows the effect of including diffusion in the absence of overshoot. The last two MESA models are the same, except that overshoot is now included. The largest cores result from either overshoot or concentration diffusion to induce core growth. Large values of overshoot do not produce core growth, just larger cores. The rapid fluctuations in the models with overshoot suggest that the mixing algorithm experiences numerical instabilities at the boundary. Finally, we contrast the simulations with asteroseismic results (ovals) from the analysis of three g-mode pulsators [73–75].

numerical in nature. However, none of the convective core sizes match the results from asteroseismology, indicated by the colored error ellipses.

Assumptions inherent in the parameterization of the stellar structures in the forward modeling may also introduce systematic effects. In particular, estimates of the abundance gradient are very sensitive to its assumed shape. While this could introduce uncertainties in the core He-burning lifetime of those three stars, the standard algorithms of convection still seem inadequate to reproduce the inferred stellar structures.

2.3.2 Horizontal Branch and AGB Stars

The period spacing of g-modes is very sensitive to the boundary of the convective helium-burning core, as we discussed above (Section 1.1). In contrast to sdB stars, red clump stars are [6] core He-burning stars which retained their hydrogen envelopes. Light curves of these stars measured by the NASA *Kepler* mission allowed the g-mode spacing to be inferred from mixed modes for hundreds of these stars [77–79].

Bossini et al. [80] constructed stellar evolution models using multiple stellar evolution codes and studied the influence of different convective prescriptions in comparison to the observational data. The goal was to find the convection prescription that would match the AGB luminosity at the AGB bump as well as describe the measured period spacing. While standard models (including induced semiconvection and overshoot) would reproduce the AGB bump luminosity, they could not describe the period spacing. However, models with extreme overshoot reproduced the period spacing but failed to match the AGB bump luminosity. The authors then outlined a candidate model with a moderate overshooting region characterized by adiabatic stratification to fit both constraints, for use in further studies.

Another study conducted by Constantino et al. [34, 81] tested how well a range of stellar models with different convection prescriptions could reproduce the g-mode period spacing as well as the relative numbers of stars on the AGB and RGB in clusters. They demonstrated that the period spacing could only be matched using their new "maximal overshoot" mixing scheme, which produces the largest convective cores possible. Yet, they also stressed that mode trapping can bias the observationally inferred value of the period spacing. If so, the standard model using induced semiconvection would also suffice. Both solutions could explain the cluster counts, if mixing beneath the Schwarzschild boundary during subsequent early-AGB evolution occurred in the "maximal overshoot" models or if models with induced semiconvection could avoid helium core breathing pulses.

DOI: 10.1051/epjconf/201716004001

Neither study provides a clear way forward for the use of convection prescriptions during this phase of stellar evolution. But they do identify important constraints for one dimensional stellar evolution models.

2.3.3 White Dwarfs

Ongoing asteroseismic analyses of the chemical abundance structure of white dwarfs with carbon-oxygen cores (CO-WD) might deliver further clues to understand convection during core-helium-burning (N. Giammichele, private communication). The oxygen abundance profile of a CO-WD is set by the interplay of core-helium-burning and the convection zone as well as the outward moving helium-burning shell and thermal pulses on the AGB. Estimates of the abundance profile in comparison with 1D stellar evolution calculations could provide new constraints on commonly used convection prescriptions.

2.4 On The Issue of Core Breathing Pulses

An interesting phenomenon encountered in low mass stars at the end of the helium-burning lifetime, when the helium abundance is very low, are the so-called "breathing pulses" [82, 83]. During "breathing pulses" the convective zone grows rapidly, mixes new helium fuel down into the core and collapses again. In most cases this effect revives the core and prolongs the helium-burning lifetime considerably. In the late stage of core-helium-burning, the reaction rates depend strongly on the helium abundance. Mixing convective boundary layers with a sharp abundance gradient will lead to a sudden increase in the energy generation rate and can explain the fast expansion of the convective core. Whether this situation occurs in "real" stars is doubtful. Dorman et al. [72] argued that this behavior is exclusively numerical in nature. In the more recent study by Constantino et al. [81], breathing pulses are found in stellar models with overshoot which exhibit a sharp abundance gradient at the convective boundary. The authors show that a monotonically decreasing helium abundance in models with partial mixing, induced semiconvection or maximal overshoot ensures the stability of the convective core. In fact, they point out that non-local treatments of overshooting (see also [84]) seem to avoid avoid breathing pulses naturally [62].

A variety of studies comparing observational constraints to theoretical models [81, 84–86] have led to the conclusion that breathing pulses do not occur in natural environments.

It is the nature of the helium abundance gradient and the rate of entrainment at the convective boundary during core helium-burning that allow for breathing pulses to occur. Their non-existence provides a strong constraint on the prescriptions of convective mixing and the treatment of the convective boundary.

2.5 The $^{12}C(\alpha,\gamma)^{16}O$ Nuclear Rate and Convection

The newest measurements of the $^{12}C(\alpha,\gamma)^{16}O$ nuclear reaction rate (NACRE II, [87]) have reduced the uncertainty on its value. However, in order to reach the conditions for this reaction in low mass stars, the measured rate has to be extrapolated to lower energies, thereby introducing new uncertainties.

Constantino et al. [34] considered the effect of the $^{12}C(\alpha,\gamma)^{16}O$ nuclear reaction rate on the g-mode period spacing, and found that doubling the rate coefficient did not help the stellar models to achieve the larger period spacing observed in red clump stars. Yet, they noted that a higher $^{12}C(\alpha,\gamma)^{16}O$ rate increased the convective core mass.

Asteroseismological studies of white dwarfs [88, 89] inferred the central oxygen abundance to draw conclusions on the effective $^{12}C(\alpha,\gamma)^{16}O$ nuclear reaction rate over the star's lifetime. However, great care has to be taken since diffusion [90] or convection can alter the stellar model's pulsation spectrum and may lead to wrong conclusions.

In fact, core convection during core-helium-burning and the $^{12}C(\alpha,\gamma)^{16}O$ reaction feature a complex interplay [91]. The growth of the convective core depends on the entrainment of more opaque material into the stable region to incite instability. The increase in opacity in turn is due to the carbon and oxygen yields of the reactions of helium-burning. So a higher $^{12}C(\alpha,\gamma)^{16}O$ nuclear rate leads to a higher energy production rate and a larger oxygen yield. Both of these lead to a growing convective core if entrainment or some sort of boundary mixing is included.

On the other hand, the rate of entrainment determines the amount of helium brought into the burning region and therefore influences the rate of the $^{12}C(\alpha,\gamma)^{16}O$ reaction as well. It can be easily shown that the central oxygen abundance depends on the entrainment rate or on the specified amount of "overshoot" at the convective boundary.

These effects complicate studies that like to infer the $^{12}C(\alpha,\gamma)^{16}O$ nuclear reaction rate from asteroseismology by introducing uncertainties that depend on the adopted algorithm of convective entrainment and vice versa.

3 Conclusion

Space-borne precision photometry has stimulated asteroseismological analysis for a variety of stars. Regarding sdB stars much progress has been made in the last fifteen years. It has become clear that the physics of radiative levitation and gravitational settling are needed to reproduce the pulsational instabilities through iron element opacity bumps in their envelopes. Analyses of g-mode pulsations constrained the extent of the convective core, the nature of the abundance gradient as well as internal rotation profiles.

While the instability strips can now be reproduced, difficulties for coherent stellar evolution models still persist. In the binary evolution channels, the prospective sdB star loses most of its hydrogen envelope mass during the common envelope or the stable Roche lobe overflow phase. Current stellar evolution models utilize some form of extreme mass loss prescription to mimic this dynamical event. After the star is stripped of most of its hydrogen envelope, it undergoes the He-flash. Although many recent stellar evolution codes can follow the evolution through this violent event, three dimensional simulations show that the one dimensional picture is not fully correct.

Lastly the standard algorithms employed to simulate convection in stellar evolution models seem to be inadequate during the phase of core-helium-burning. Asteroseismology of g-mode sdB pulsators indicates larger convective cores than can be reproduced in the current framework of mixing algorithms. Recent studies compared the g-mode period spacing found in red clump stars to stellar evolution models. The authors demonstrate that the current, mostly ad-hoc, treatments of overshooting and/or induced semiconvection can reproduce the observations in a few cases. The difficulties can be traced back to the treatment of physical overshooting and the entrainment at the convective boundary. Unfortunately no consistent prescription of core-helium-burning convection using the standard algorithms has emerged yet. This also applies to other stages of stellar evolution. With recent asteroseismological analyses constraining the convective cores during main-sequence evolution [92, 93], a consistent picture might well emerge in the coming years.

Where the asteroseismological results constrain the long term behavior of convection on the stellar structure, three dimensional models of convection offer insight into the dynamics of the physical mechanisms in play. These simulations show the development of a boundary layer and the strong fluctuations of the turbulent convective flow. Physical overshooting and entrainment occur naturally when the moving convective plumes enter the subadiabatic region, are slowed down and turn.

In order to develop a consistent prescription for convection in evolutionary models, one has to undertake the dangerous procedure of integrating over the turbulent fluctuations. One possibility is to use the 3D simulations to provide closure to the Reynolds averaged Navier-Stokes equations. Approximations on the basis of these equations might allow for implementation in a stellar evolutionary code without calibrations to astronomical data [65].

The recent insight into stellar structures from asteroseismology and convective behavior from 3D simulations has highlighted the known inadequacies in our stellar evolution models, but also provided constraints to tackle them. Still, convection, especially in core-helium-burning stars, remains a fundamental problem of modern astrophysics for now.

Acknowledgements

JTS thanks the SOC of "Seismology of the Sun and the Distant Stars 2016" and the Theoretical Astrophysics Program at the University of Arizona for support.

References

[1] U. Heber, A&A **155**, 33 (1986)
[2] R.A. Saffer, P. Bergeron, D. Koester, J. Liebert, ApJ **432**, 351 (1994)
[3] P.F.L. Maxted, U. Heber, T.R. Marsh, R.C. North, MNRAS **326**, 1391 (2001), `astro-ph/0103342`
[4] R. Napiwotzki, C.A. Karl, T. Lisker, U. Heber, N. Christlieb, D. Reimers, G. Nelemans, D. Homeier, Ap&SS **291**, 321 (2004), `astro-ph/0401201`
[5] C.M. Copperwheat, L. Morales-Rueda, T.R. Marsh, P.F.L. Maxted, U. Heber, MNRAS **415**, 1381 (2011), `1103.4745`
[6] G. Fontaine, P. Brassard, S. Charpinet, E.M. Green, S.K. Randall, V. Van Grootel, A&A **539**, A12 (2012)
[7] Z. Han, P. Podsiadlowski, P.F.L. Maxted, T.R. Marsh, N. Ivanova, MNRAS **336**, 449 (2002), `astro-ph/0206130`
[8] Z. Han, P. Podsiadlowski, P.F.L. Maxted, T.R. Marsh, MNRAS **341**, 669 (2003), `astro-ph/0301380`
[9] A.V. Sweigart, ApJ **474**, L23 (1997)
[10] N.L. D'Cruz, B. Dorman, R.T. Rood, R.W. O'Connell, ApJ **466**, 359 (1996), `astro-ph/9511017`
[11] M.M. Miller Bertolami, L.G. Althaus, K. Unglaub, A. Weiss, A&A **491**, 253 (2008), `0808.3580`
[12] R.F. Webbink, ApJ **277**, 355 (1984)
[13] R. Silvotti, S. Schuh, R. Janulis, J.E. Solheim, S. Bernabei, R. Østensen, T.D. Oswalt, I. Bruni, R. Gualandi, A. Bonanno et al., Nature **449**, 189 (2007)
[14] S. Charpinet, G. Fontaine, P. Brassard, E.M. Green, V. Van Grootel, S.K. Randall, R. Silvotti, A.S. Baran, R.H. Østensen, S.D. Kawaler et al., Nature **480**, 496 (2011)
[15] S. Geier, H. Edelmann, U. Heber, L. Morales-Rueda, ApJ **702**, L96 (2009), `0908.1025`
[16] S. Geier, L. Classen, P. Brünner, K. Nagel, V. Schaffenroth, C. Heuser, U. Heber, H. Drechsel, H. Edelmann, C. Koen et al., *Low-mass Stellar and Substellar Companions to sdB Stars*, in *Fifth Meeting on Hot Subdwarf Stars and Related Objects*, edited by D. Kilkenny, C.S. Jeffery, C. Koen (2012), Vol. 452 of *Astronomical Society of the Pacific Conference Series*, p. 153, `1112.2929`
[17] U. Heber, ARA&A **47**, 211 (2009)
[18] U. Heber, PASP **128**, 082001 (2016), `1604.07749`
[19] S. Charpinet, V. Van Grootel, P. Brassard, G. Fontaine, E.M. Green, S.K. Randall, *Asteroseismology of hot B subdwarf stars*, in *European Physical Journal Web of Conferences* (2013), Vol. 43 of *European Physical Journal Web of Conferences*, p. 04005
[20] S. Charpinet, G. Fontaine, P. Brassard, B. Dorman, ApJ **471**, L103 (1996), `astro-ph/9606069`
[21] D. Kilkenny, C. Koen, D. O'Donoghue, R.S. Stobie, MNRAS **285**, 640 (1997)
[22] E.M. Green, G. Fontaine, M.D. Reed, K. Callerame, I.R. Seitenzahl, B.A. White, E.A. Hyde, R. Østensen, O. Cordes, P. Brassard et al., ApJ **583**, L31 (2003)
[23] S. Schuh, J. Huber, S. Dreizler, U. Heber, S.J. O'Toole, E.M. Green, G. Fontaine, A&A **445**, L31 (2006), `astro-ph/0510831`
[24] S. Charpinet, G. Fontaine, P. Brassard, P. Chayer, F.J. Rogers, C.A. Iglesias, B. Dorman, ApJ **483**, L123

(1997)
[25] G. Fontaine, P. Brassard, S. Charpinet, E.M. Green, P. Chayer, M. Billères, S.K. Randall, ApJ **597**, 518 (2003)
[26] C.S. Jeffery, H. Saio, MNRAS **372**, L48 (2006), astro-ph/0607505
[27] P. Chayer, G. Fontaine, M. Fontaine, R. Lamontagne, F. Wesemael, J. Dupuis, U. Heber, R. Napiwotzki, S. Moehler, Ap&SS **291**, 359 (2004)
[28] P. Brassard, G. Fontaine, M. Billères, S. Charpinet, J. Liebert, R.A. Saffer, ApJ **563**, 1013 (2001)
[29] S. Charpinet, G. Fontaine, P. Brassard, B. Dorman, ApJS **131**, 223 (2000)
[30] S. Charpinet, G. Fontaine, P. Brassard, B. Dorman, ApJS **139**, 487 (2002)
[31] S. Charpinet, G. Fontaine, P. Brassard, B. Dorman, ApJS **140**, 469 (2002)
[32] M.D. Reed, A. Baran, A.C. Quint, S.D. Kawaler, S.J. O'Toole, J. Telting, S. Charpinet, C. Rodríguez-López, R.H. Østensen, J.L. Provencal et al., MNRAS **414**, 2885 (2011), 1102.4286
[33] R.H. Østensen, J.H. Telting, M.D. Reed, A.S. Baran, P. Nemeth, F. Kiaeerad, A&A **569**, A15 (2014), 1406.6941
[34] T. Constantino, S.W. Campbell, J. Christensen-Dalsgaard, J.C. Lattanzio, D. Stello, MNRAS **452**, 123 (2015), 1506.01209
[35] S. Charpinet, V. Van Grootel, D. Reese, G. Fontaine, E.M. Green, P. Brassard, P. Chayer, A&A **489**, 377 (2008)
[36] V. Van Grootel, S. Charpinet, G. Fontaine, P. Brassard, E.M. Green, P. Chayer, S.K. Randall, A&A **488**, 685 (2008)
[37] E.M. Green, B. Guvenen, C.J. O'Malley, C.J. O'Connell, B.P. Baringer, A.S. Villareal, T.M. Carleton, G. Fontaine, P. Brassard, S. Charpinet, ApJ **734**, 59 (2011)
[38] V. Van Grootel, S. Charpinet, P. Brassard, G. Fontaine, E.M. Green, A&A **553**, A97 (2013), 1303.3368
[39] J.T. Schindler, E.M. Green, W.D. Arnett, ApJ **806**, 178 (2015), 1410.8204
[40] E.M. Green, G. Fontaine, E.A. Hyde, B.Q. For, P. Chayer, *Systematics of Hot Subdwarfs Obtained from a Large Low Resolution Survey*, in *Hot Subdwarf Stars and Related Objects*, edited by U. Heber, C.S. Jeffery, R. Napiwotzki (2008), Vol. 392 of *Astronomical Society of the Pacific Conference Series*, p. 75
[41] A.V. Sweigart, P.G. Gross, ApJS **32**, 367 (1976)
[42] A.V. Sweigart, ApJS **65**, 95 (1987)
[43] B. Dorman, R.T. Rood, R.W. O'Connell, ApJ **419**, 596 (1993), astro-ph/9311022
[44] H. Xiong, X. Chen, P. Podsiadlowski, Z. Han, ArXiv e-prints (2016), 1608.08739
[45] A. Serenelli, A. Weiss, A&A **442**, 1041 (2005), astro-ph/0507430
[46] S. Bloemen, H. Hu, C. Aerts, M.A. Dupret, R.H. Østensen, P. Degroote, E. Müller-Ringat, T. Rauch, A&A **569**, A123 (2014), 1409.1604
[47] B. Paxton, L. Bildsten, A. Dotter, F. Herwig, P. Lesaffre, F. Timmes, ApJS **192**, 3 (2011), 1009.1622
[48] M. Mocák, E. Müller, A. Weiss, K. Kifonidis, A&A **501**, 659 (2009), 0811.4083
[49] M. Mocák, E. Müller, A. Weiss, K. Kifonidis, A&A **490**, 265 (2008), 0805.1355
[50] E. Böhm-Vitense, ZAp **46**, 108 (1958)
[51] G. Fontaine, E.M. Green, P. Chayer, P. Brassard, S. Charpinet, S.K. Randall, Baltic Astronomy **15**, 211 (2006)
[52] N.R. Badnell, M.A. Bautista, K. Butler, F. Delahaye, C. Mendoza, P. Palmeri, C.J. Zeippen, M.J. Seaton, MNRAS **360**, 458 (2005), astro-ph/0410744
[53] C.A. Iglesias, F.J. Rogers, ApJ **464**, 943 (1996)
[54] H. Hu, G. Nelemans, C. Aerts, M.A. Dupret, A&A **508**, 869 (2009), 0910.5641
[55] H. Hu, C.A. Tout, E. Glebbeek, M.A. Dupret, MNRAS **418**, 195 (2011), 1108.1318
[56] H. Hu, E. Glebbeek, A.A. Thoul, M.A. Dupret, R.J. Stancliffe, G. Nelemans, C. Aerts, A&A **511**, A87 (2010), 0912.4665
[57] P.P. Eggleton, MNRAS **156**, 361 (1972)
[58] J.P. Cox, R.T. Giuli, *Principles of stellar structure* (1968)
[59] N. Langer, M.F. El Eid, K.J. Fricke, A&A **145**, 179 (1985)
[60] J.P. Zahn, A&A **252**, 179 (1991)
[61] F. Herwig, A&A **360**, 952 (2000), astro-ph/0007139
[62] C.A. Meakin, D. Arnett, ApJ **667**, 448 (2007), astro-ph/0611315
[63] M. Viallet, C. Meakin, D. Arnett, M. Mocák, ApJ **769**, 1 (2013), 1212.6365
[64] D. Arnett, C. Meakin, P.A. Young, ApJ **690**, 1715 (2009), 0809.1625
[65] W.D. Arnett, C. Meakin, M. Viallet, S.W. Campbell, J.C. Lattanzio, M. Mocák, ApJ **809**, 30 (2015), 1503.00342
[66] V. Castellani, P. Giannone, A. Renzini, Ap&SS **10**, 340 (1971)
[67] G. Michaud, J. Richer, O. Richard, ApJ **670**, 1178 (2007)
[68] V. Castellani, P. Giannone, A. Renzini, Ap&SS **10**, 355 (1971)
[69] J.W. Robertson, D.J. Faulkner, ApJ **171**, 309 (1972)
[70] B. Paczyński, Acta Astron. **20**, 195 (1970)
[71] P. Demarque, J.G. Mengel, ApJ **171**, 583 (1972)
[72] B. Dorman, R.T. Rood, ApJ **409**, 387 (1993)
[73] V. Van Grootel, S. Charpinet, G. Fontaine, E.M. Green, P. Brassard, A&A **524**, A63 (2010)
[74] S. Charpinet, V. Van Grootel, G. Fontaine, E.M. Green, P. Brassard, S.K. Randall, R. Silvotti, R.H. Østensen, H. Kjeldsen, J. Christensen-Dalsgaard

et al., A&A **530**, A3 (2011)
[75] V. Van Grootel, S. Charpinet, G. Fontaine, P. Brassard, E.M. Green, S.K. Randall, R. Silvotti, R.H. Østensen, H. Kjeldsen, J. Christensen-Dalsgaard et al., ApJ **718**, L97 (2010)
[76] B. Paxton, M. Cantiello, P. Arras, L. Bildsten, E.F. Brown, A. Dotter, C. Mankovich, M.H. Montgomery, D. Stello, F.X. Timmes et al., ApJS **208**, 4 (2013), 1301.0319
[77] B. Mosser, M.J. Goupil, K. Belkacem, E. Michel, D. Stello, J.P. Marques, Y. Elsworth, C. Barban, P.G. Beck, T.R. Bedding et al., A&A **540**, A143 (2012), 1203.0689
[78] B. Mosser, O. Benomar, K. Belkacem, M.J. Goupil, N. Lagarde, E. Michel, Y. Lebreton, D. Stello, M. Vrard, C. Barban et al., A&A **572**, L5 (2014), 1411.1082
[79] M.H. Pinsonneault, Y. Elsworth, C. Epstein, S. Hekker, S. Mészáros, W.J. Chaplin, J.A. Johnson, R.A. García, J. Holtzman, S. Mathur et al., ApJS **215**, 19 (2014), 1410.2503
[80] D. Bossini, A. Miglio, M. Salaris, A. Pietrinferni, J. Montalbán, A. Bressan, A. Noels, S. Cassisi, L. Girardi, P. Marigo, MNRAS **453**, 2290 (2015), 1507.07797
[81] T. Constantino, S.W. Campbell, J.C. Lattanzio, A. van Duijneveldt, MNRAS **456**, 3866 (2016), 1512.04845
[82] A.V. Sweigart, P. Demarque, *Semiconvection and the RR Lyrae Variables*, in *IAU Colloq. 21: Variable Stars in Globular Clusters and in Related Systems*, edited by J.D. Fernie (1973), Vol. 36 of *Astrophysics and Space Science Library*, p. 221
[83] V. Castellani, A. Chieffi, A. Tornambe, L. Pulone, ApJ **296**, 204 (1985)
[84] A. Bressan, G. Bertelli, C. Chiosi, Mem. Soc. Astron. Italiana **57**, 411 (1986)
[85] F. Caputo, A. Chieffi, A. Tornambe, V. Castellani, L. Pulone, ApJ **340**, 241 (1989)
[86] S. Cassisi, V. Castellani, S. Degl'Innocenti, G. Piotto, M. Salaris, A&A **366**, 578 (2001), astro-ph/0112111
[87] Y. Xu, K. Takahashi, S. Goriely, M. Arnould, M. Ohta, H. Utsunomiya, Nuclear Physics A **918**, 61 (2013), 1310.7099
[88] T.S. Metcalfe, M. Salaris, D.E. Winget, ApJ **573**, 803 (2002), astro-ph/0203347
[89] T.S. Metcalfe, ApJ **587**, L43 (2003), astro-ph/0303039
[90] G. Fontaine, P. Brassard, ApJ **581**, L33 (2002)
[91] D. Arnett, *Supernovae and nucleosynthesis. an investigation of the history of matter, from the Big Bang to the present* (1996)
[92] S. Deheuvels, I. Brandão, V. Silva Aguirre, J. Ballot, E. Michel, M.S. Cunha, Y. Lebreton, T. Appourchaux, A&A **589**, A93 (2016), 1603.02332
[93] E. Moravveji, R.H.D. Townsend, C. Aerts, S. Mathis, ApJ **823**, 130 (2016), 1604.02680

 DOI: 10.1051/epjconf/201716004002

Modelling of Red Giant Stars: The state-of-the-art

Santi Cassisi[1,*]

[1] *INAF - Astronomical Observatory of Teramo, Via M. Maggini, sn, 64100 Teramo, Italy*

Abstract. The seismic data obtained by the CoRoT and Kepler space missions have provided inferences of the global and structural properties of thousands of red giants. When compared with stellar model predictions, these inferences can significantly improve our understanding of stellar evolution. We present a brief review of the structure and evolution of red giant stars, devoting some emphasis on the major, still open problems.

1 Introduction

Red giants are cool, highly luminous stars, covering a rather wide domain in mass, age, chemical composition, and evolutionary stage. The three main classes of red giants: Red Giant Branch (RGB) , Red Clump (RC) and Asymptotic Giant Branch (AGB) stars, provide a set of distance and star formation history diagnostics from photometric observations of resolved, old and intermediate-age stellar populations.

The presence of an extended convective envelope can - as in solar-like stars - stochastically excite p- and g-modes of oscillations that have indeed been detected with the CoRoT and Kepler space-based satellites. The use of these asteroseismological data allows robust constraints to be set on the mass and radius of such stars. As a consequence, since the data from the first CoRoT observational runs were analysed, and solar-like oscillations were detected in several thousands of red giant stars (mainly thanks to the Kepler space mission), it has become clear that the newly available observational constraints would allow novel approaches in the study of Galactic stellar populations [1, 2]. This can have a very important impact on our vision of the Galaxy. For instance, if the frequency at maximum power (the so called ν_{max}) of a red giant star, whose parallax is unknown, is derived from asteroseismic observations together with the so-called large frequency separation ($\Delta\nu$), its radius can be estimated and so its distance derived: so oscillating red giants can become new distance indicators. In addition, a spectroscopic analysis of these stars, coupled with these new distance evaluations, could provide a new tool to measure the radial and vertical metallicity gradient in the Galaxy. Moreover, thanks to the estimate of the stellar mass through seismic constraints, it would be possible to characterise the age distributions and age-metallicity relations of the observed populations of giants [3].

Therefore, it appears evident that the impressive amount of seismic data for red giant stars is actually opening the way for 'ensemble seismology' of solar-like stars.

However, as a word of caution, one should recall that the seismic-based age estimates are inherently stellar model dependent, and hence affected by the still-present uncertainties in stellar evolution predictions. So, in order to improve the predictive capability of asteroseismic surveys, it appears mandatory to investigate the lingering uncertainties affecting stellar models for these stars, and to try to improve their accuracy and reliability.

We discuss the main evolutionary and structural properties of red giant stars separately discussing the case of RGB stars and RC ones, and outlining the main improvements that have been recently achieved in the modelling of these stars. We also briefly discuss some non-canonical physical processes - usually ignored in stellar model computations - that we need to properly account for in the next generation of stellar models in order to be able to properly interpret the various observational evidence.

2 Red Giant Branch stars: their evolution up to the RGB Tip

RGB stars are objects with masses lower than ~2.2 $M_\odot$ (the precise value depends on the initial chemical composition), which develop electron degenerate He-cores after the end of central H-burning, surrounded by a thick H-burning shell (thickness of the order of $0.1M_\odot$) and a convective envelope. The envelope temperature gradient is to a large extent adiabatic, apart from the most external layers, where it becomes superadiabatic, and must be treated according to some prescribed convection theory (see Sect. 3.2).

At this stage, the convective envelope progressively deepens and the H-burning shell narrows down to a thickness of the order of 0.001 $M_\odot$[1], while the star crosses the Color-Magnitude diagram (CMD) towards redder colors (i.e. lower $T_{\rm eff}$ and larger radius). There is a large body of literature devoted to the identification of what is (or

*e-mail: cassisi@oa-teramo.inaf.it

[1] This effect is due to the huge dependence on temperature of the *CNO cycle* that is controlling the H-burning process in the shell.

what are, if more than one) the precise physical reason for the expansion of the star to red giant dimensions (a fact that comes out naturally from the integration of the stellar structure equations), but a general consensus has not been yet reached (see, e.g., [4] and references therein).

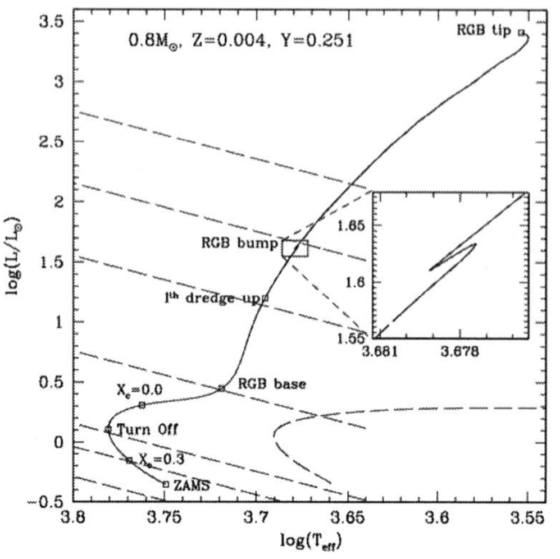

Figure 1: The evolution in the H-R diagram of a low-mass star from the ZAMS up to the RGB tip. The inset shows a enlargement of the RGB bump stage. The long dashed line corresponds to the track of the same mass, computed by setting the value of the mixing length parameter to zero. The thin dashed lines represent loci of constant radius for $R/R_{\odot}$ = 0.6, 0.8, 1.0, 2.0, 5.0, 10 and 20, respectively, from bottom to top.

Due to the steady deepening of the convective region, the lower boundary of the convective envelope enters regions chemically processed (some produced He and the C and N abundances reaching their CN-cycle equilibrium value) by the central H-burning phase; this material is mixed almost instantaneously throughout the convective envelope, thus altering the surface abundances. This phenomenon, called the *first dredge up*, causes an increase of the surface He abundance, and an increase of the N/C ratio with respect to the original values. Convection reaches its maximum extension (in mass) 'near' the base of the RGB, and from this moment on the lower convective boundary slowly recedes towards the surface, due to the steady growth of the He-core which accretes He produced by the H-burning shell.

The evolution along the RGB covers a relatively small range in $T_{\rm eff}$ and a large range of surface luminosity, which increases with time. Due to the growth of the He core mass (M_{core}^{He}), the H-shell will encounter the sharp discontinuity of the H-profile left at the point of maximum extension of the convective envelope. The star reacts to the sudden increase of available fuel by lowering its surface luminosity and slowing down the evolutionary timescale, before starting again to increase its luminosity after the shell has moved past the discontinuity. This occurrence is recorded in the differential luminosity function (LF – number of stars in a given brightness interval as a function of the brightness itself) of old stellar populations as a local peak - the so-called RGB bump - in the differential LF.

When M_{core}^{He} reaches a size of about 0.50 $M_{\odot}$ (the exact value depends weakly on the total mass of the star, being more sensitive to the initial chemical composition), He-ignition occurs in the electron degenerate core, producing the so called He-flash which terminates the RGB phase by removing the electron degeneracy in the core, and drives the star onto its Zero Age Horizontal Branch (ZAHB) location, that marks the start of quiescent central He-burning plus shell H-burning. Figure 1 shows the evolution of a $0.8M_{\odot}$ (Z=0.004, Y=0.251) star from the Zero Age MS (ZAMS) up to the He-flash. The most relevant evolutionary stages are clearly indicated.

3 On the modelling of RGB stars

During these last decades, our knowledge of evolutionary and structural properties of stars is significantly improved. This result has been achieved as a consequence of our improved capability in understanding and describing the physical behavior of matter in the various thermal regimes characteristic of the various stellar mass ranges and evolutionary stages.

This notwithstanding, current generation of stellar models is still affected by significant and, usually, not negligible uncertainties. These uncertainties are related to our poor knowledge of some physical processes occurring in real stars such as, for instance, some nuclear reaction rates, the efficiency of mixing and diffusive processes. These drawbacks of stellar models have to be properly taken into account when comparing theory with observations in order to derive relevant information about the properties of both resolved and unresolved stellar populations, as well as when using stellar models for seismic purposes.

When facing with models for RGB stars, the main issues are those related with their age, effective temperature scale, the brightness of the RGB bump, and mass loss efficiency. In the following we will briefly review the accuracy and reliability of present generation of stellar models concerning these theoretical predictions.

3.1 On the reliability of RGB stellar model age

RGB timescales are of the order of a few percent of the main sequence lifetime, this has the important consequence that the uncertainty on the age of RGB stellar models is largely dominated by the uncertainty on the core H-burning lifetime. When discussing this issue, we need to consider separately the case of low-mass stars, i.e. those that do not develop a convective core during the central H-burning stage from those - the intermediate mass ones - that burn H inside a convective core. The transition mass is around $1.1 - 1.2M_{\odot}$ (the exact value depending on the chemical composition). For stars burning H in a radiative environment, the main sources of uncertainty in the core H-burning lifetime are the nuclear reaction rates and the atomic diffusion efficiency.

DOI: 10.1051/epjconf/201716004002

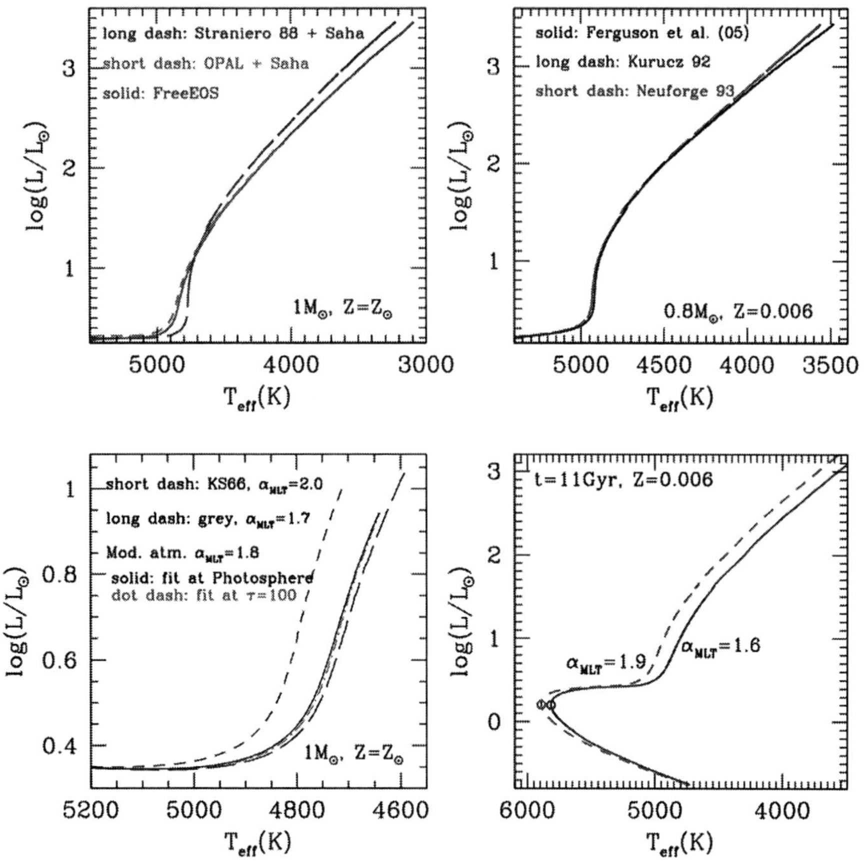

Figure 2: *Upper left panel:* the RGB portion of the evolutionary track of solar metallicity $1M_\odot$ models computed by adopting various EOSs. *Upper right panel:* the RGB loci of $0.8M_\odot$ model for a metallicity Z=0.006, computed by using various prescriptions about the low-T radiative opacities. *Lower left panel:* the sub-giant branch and RGB portion of evolutionary tracks of various SSMs computed with various outer boundary conditions. For each model it has been also listed the corresponding value of the mixing length parameter. *Lower right panel:* two isochrones based on stellar models computed by using the same physical inputs but different values for α_{MLT}.

Concerning the nuclear reaction rates, one of the most important nuclear reaction in stellar astrophysics is the p-capture on ^{14}N nuclei. At odds with the common idea that in low-mass stars the H-burning is always (fully) controlled by the nuclear reactions associated to the *p-p chain* (whose efficiency is indeed very well known after the huge effort devoted to solve the solar neutrino problem), the nuclear reactions related to the *CNO-cycle* can be actually important. In fact, near the end of the Main Sequence stage, due to the paucity of H, the energy supplied by the H-burning becomes insufficient and the star reacts contracting its core in order to produce the requested energy via gravitation. As a consequence, both central temperature and density increase and, when the temperature reaches $\sim 15 \times 10^6 K$, the H-burning process is controlled by the *CNO cycle*, whose efficiency is critically dependent on the $^{14}N(p,\gamma)^{15}O$ reaction rate, because this is the slowest reaction of the whole cycle.

In the past, its rate was uncertain, at least by a factor of 5, because all available laboratory measurements were performed at energies well above the range of interest for astrophysical purposes. The LUNA experiment [5] has significantly improved the low energy measurements, obtaining an estimate which is about a factor of 2 lower than, for instance, the NACRE measurement. When considering an isochrone, the use of this updated reaction rate leads to a brighter and hotter MS turn-off (TO) for a fixed age (see [6, 7] and references therein): the use of these updated isochrones for Galactic globular cluster (GGC) age dating would translate in an increase of the cluster age of ≈ 0.9 Gyr (that is about 10% of the age of old GGCs).

However, this result is mainly due to the change in the interior structure which influences the morphology of the evolutionary tracks, induced by the updated $^{14}N(p,\gamma)^{15}O$ reaction rate. Indeed, when considering the impact on the evolutionary lifetime of individual stellar models, the age

 DOI: 10.1051/epjconf/201716004002

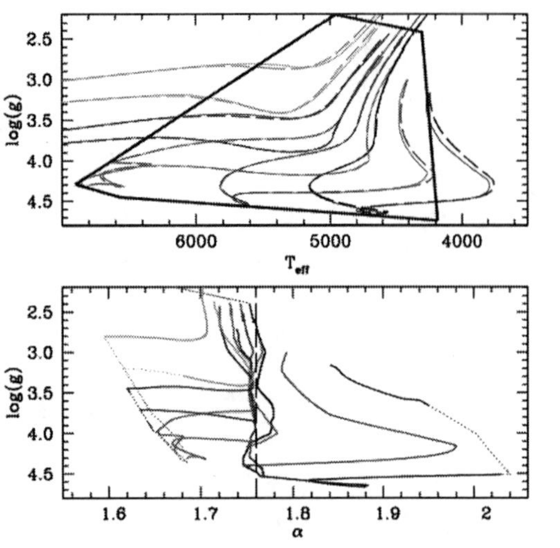

Figure 3: *Upper panel:* tracks for 0.7, 1.0, 1.4, 2.0, 3.0$M_\odot$ stellar models. The region enclosed by the thick black boundary is the g-T_{eff} range covered by the RHD simulations. Thin solid lines denote fully consistent calculations with the RHD calibrated variable mixing length and T(τ) relationships. *Lower panel:* the evolution of the mixing length parameter along each track shown in the upper panel. The dotted portion of each sequence denotes the region where the α_{MLT} values have been extrapolated. Dashed lines in the upper panel display tracks calculated with a constant, solar-calibrated, mixing length equal to 1.76, and the calibrated T(τ) relationship.

changes induced by the new rate amounts to (a negligible) $\approx$ 2%. This means that the lingering uncertainties in the nuclear reaction rates for H-burning processes have practically no impact when using individual stellar models for age dating in the framework of 'ensemble seismology'.

As far as it concerns the impact of atomic diffusion on evolutionary lifetimes, it has been shown that the inclusion of this process in stellar model computations for low-mass stars causes a decrease of the central H-burning lifetime of the order of $\approx$ 0.8 – 1 Gyr (i.e. $\sim$ 10% of the whole core H-burning lifetime).

For stars developing a convective core during the central H-burning stage, several physical mechanisms could extend the size of convective cores beyond the Schwarzschild boundary, such as the overshooting of convective elements in the stable surroundings, and rotational mixings. These processes are still not satisfactorily understood by theory. Their combined effects are often modelled in the standard stellar model computations as a simple extension of the mixed convective core over a distance $d_{ov} = \lambda_{ov} \times H_P$, where H_P is the pressure scale height at the edge of the convective core, and λ_{ov} is a free parameter whose value has to be fixed. It is common to refer to d_{ov} as the overshooting distance, even though overshooting may not be the only mechanism at work. This model is of course simplistic, but its simplicity reflects our current ignorance about the interface between convective and radiative regions. The value of the free parameter λ_{ov} can be fixed by trying to match some empirical constraints such as the CMD morphology around the MS TO in young and intermediate-age star clusters and/or the properties of well studied eclipsing binary systems (see, e.g. [8]).

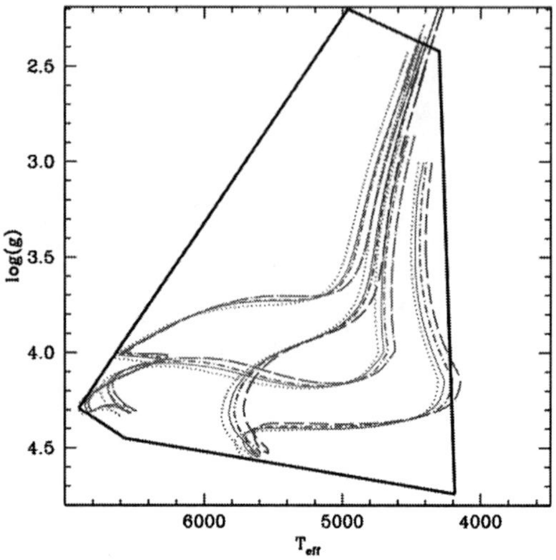

Figure 4: *Lower panel:* tracks for 1.0 and 1.4$M_\odot$ models and various assumptions about the T(τ) relation: thin solid lines correspond to fully consistent calculations with the RHD calibrated variable mixing length and T(τ) relationships; dotted, dash-dotted and dashed lines display models calculated with constant α_{MLT} = 1.76 and the Eddington, KS66 and V81 T(τ), respectively.

When moving deeper inside the star, the pressure scale height steadily increases; this causes a large increase of the size of convective cores in stars whose Schwarzschild convective boundary is fast shrinking, (e.g., for masses below $\sim$ 1.5$M_\odot$) if the overshooting efficiency is kept fixed at a constant fraction of H_P. This occurrence imposes that the value of λ_{ov} has to be decreased to zero for stars with small convective cores. Due to the lack of any physically-grounded prescription for how to deal with this issue, different assumptions about the trend of λ_{OV} with the stellar mass can be adopted. The change of the isochrone morphology can be significant and different choices concerning the core overshoot efficiency in the critical mass range $1.1 \leq M/M_\odot \leq 1.5$ can mimic different isochrone ages. The obvious consequence is that the chosen trend of λ_{OV} with mass, for masses with small convective cores, introduces an additional degree of freedom in stellar evolution models ([8]).

In these last years, a new approach has been envisaged to evaluate the extent of the convective core in central H-burning stars. This method is based on the measurement of the effects on oscillation modes caused by the sharp variations in the mean molecular weight profile at the boundary of the convective core. Preliminary investigations ([9])

 DOI: 10.1051/epjconf/201716004002

have clearly shown that this is a promising avenue. However, only recently seismic surveys of large sample of stars are really exploiting the great potential of asteroseismology to provide tight constrains on the real extension of the convective core (see, e.g., [10] and references therein). As a consequence important improvements in this context are expected in the near future.

3.2 The T_{eff} scale of RGB stars

The main parameters affecting the T_{eff} scale of RGB stars are: the equation of state (EOS), the low-temperature opacity, the efficiency of superadiabatic convection, the outer boundary conditions and the chemical abundances.

EOS: one of the most widely adopted EOS is the OPAL [11] one, whose range of validity does not cover the electron degenerate cores of RGB stars and their cooler, most external layers, below 5000 K. As a consequence, RGB models computed with the OPAL EOS must employ some other EOS to cover the whole thermal stratification of RGB stars. However, there are some exceptions about this as in the case of the models computed by [8, 12, 13]. These sets of stellar models rely on the use of the FreeEOS provided by A. Irwin (2005) which consistently allows the computation of stellar models from the H-burning stage to the He-burning phase.

The upper left panel of fig. 2 shows the RGB of a $1M_{\odot}$ stellar structure computed adopting different assumptions about the EOS: the models based on the OPAL EOS and the FreeEOS are in very good agreement (this comparison is actually meaningful only for T_{eff} larger than about 4500K for the reason discussed before); while there is a significant change in the RGB slope with respect to the model based on the EOS provided by [14] supplemented at the lower temperature by a Saha EOS. On average, there is a difference of about 100K between RGB models based of the two different EOSs.

Low-T opacity: it has been shown by [15] that it is the low-T opacity which mainly determines the T_{eff} location of RGB models, while the high-T one - in particular opacity for temperatures around 10^6 K - enters in the determination of the mass extension of the convective envelope. Present generations of stellar models employ mainly the low-T opacity calculations by [16] which are the most up-to-date computations suitable for stellar modelling, spanning a wide range of chemical compositions. The main improvement of this low-T opacity set with respect to older evaluations is in the treatment of molecular absorption and for a proper treatment of the presence of grains. Although significant improvements are still possible as a consequence of a better treatment of the different molecular opacity sources, we do not expect dramatic changes in the temperature regime where the contribution of atoms and molecules dominate, whilst large variations can be foreseen in the regime ($T < 2000$ K) where grains dominates the interaction between radiation and matter.

Upper right panel of fig. 2 shows a comparison between RGB models computed by adopting the same physical inputs but different low-T opacity evaluations: for effective temperature larger than ~4000 K, the T_{eff} scale of the models computed by using the (nowadays obsolete) opacities provided by [17] and [18] is in very good agreement with that corresponding to the model computed by adopting the [16]'s opacities. However, as soon as the RGB T_{eff} goes below this limit (i.e., when the models approach the RGB tip and/or their metallicity is increased), the opacities by [16] produce progressively cooler models (differences reaching values of the order of 100 K or more), due to the increasing effect of the molecules - in particular H_2O and TiO ones - which contribute substantially to the (increase of the) opacity in this temperature range.

Outer boundary conditions: the procedure commonly adopted for fixing the boundary conditions at the bottom of the atmospheric layers, relies on the integration of the atmosphere by using a functional (semi-empirical or theoretical) ($T(\tau)$) relation between the temperature and the optical depth τ). The effect of using boundary conditions from model atmospheres on RGB stellar models has been extensively investigated by [19]. In lower left panel of fig. 2 it is shown the effects of different $T(\tau)$ relations, namely, the Krishna-Swamy (KS66) solar $T(\tau)$ [20] relationship and the Eddington (grey) one, on a $1M_{\odot}$ solar metallicity model computed by using the same physical inputs and a solar calibrated mixing length parameter. By construction all tracks overlap at the present solar luminosity and effective temperature. On the RGB the track based on the KS66 atmosphere is ~ 100 K hotter than the tracks based on the model atmosphere stratification (in the specific case the MARCS models have been used - see [19] for more details) and the Eddington $T(\tau)$ relationship. It is also important to note that the two tracks obtained matching the MARCS atmospheres to the interior structure at different optical depths: $\tau = 1$ and 100, respectively, do overlap from the Zero Age MS to the RGB. This occurrence is also valid at lower metallicities, and is a proof that the stellar model T_{eff} scale does not depend on the chosen fitting point until a value of τ larger or equal to unity is chosen.

When comparing more metal-poor, low-mass stellar models - based on solar-calibrated α_{MLT} -, the difference in the T_{eff} scale between models based on both types of grey atmospheres and detailed non-grey models appears negligible. It is worth noting that the comparison between theory and observations for the effective temperature scale of approximately solar metallicity RGB stars both in the field and in star clusters, seems to support the T_{eff} scale of stellar models based on the KS66 $T(\tau)$ relation.

Before closing this discussion, we wish to remark that, even if the use of boundary conditions provided by model atmospheres is, in principle, more rigorous, one has also to bear in mind that the convection treatment in the adopted model atmospheres is usually not the same as in the underlying stellar models (i.e., a different mixing length formalism and a different value for the scale height of the convective motion are used).

The treatment of superadiabatic convection: the temperature gradient throughout the bulk of the convective envelope of stellar models can be approximated by the adiabatic value. However, in the outer layers close to the stellar surface the convective gradient becomes strongly superadiabatic. To determine the actual value of the gradient in these outer layers the Mixing Length Theory (MLT, [21]) is almost universally used. The MLT contains four free parameters, whose values affect the predicted T_{eff} of the stellar models. Three parameters are fixed a priori[2] (and define what it is denoted as the MLT *flavour*), and the only one left to be calibrated is α_{MLT}, the ratio of the mixing length to the local pressure scale height H_P, that provides the scale length of the convective motions. As a general rule, an increase of α_{MLT} corresponds to an increase of the convective transport efficiency and an increase of the stellar model T_{eff} .

In stellar evolution calculations the value of α_{MLT} is usually calibrated by reproducing the radius of the Sun at the solar age with a Solar Standard Model (SSM). This solar calibrated α_{MLT} is then kept fixed in all evolutionary calculations of stars of different masses and chemical compositions. The exact numerical value of α_{MLT} varies amongst calculations by different authors because variations of input physics and choices of the outer boundary conditions affect the predicted model T_{eff} values, hence require different α_{MLT} values to match the Sun.

It is clear that, even if a physics input employed in the stellar model computations is not accurate, it is possible to mask this shortcoming - at least from the point of view of the predicted T_{eff} - by simply recalibrating α_{MLT} on the Sun. This guarantees that the models always predict correctly the T_{eff} of at least solar-type stars. However, since the extension of the superadiabatic layers is larger in RGB stars, theoretical RGB models are much more sensitive to α_{MLT} than MS ones. Therefore it is not safe to assume *a priori* that the solar calibrated value of α_{MLT} is also adequate for RGB stars of various metallicities.

A source of concern about *a priori* assumption of a solar α_{MLT} for RGB computations comes from the fact that recent models from various authors, all using a suitably calibrated solar value of α_{MLT}, do not show the same RGB temperatures. A comparison - performed by [23] - of independent sets of RGB stellar models computed with the same initial chemical composition and solar calibrated values of α_{MLT} shows that these models can predict a different T_{eff} scale for the RGB: a realistic estimate of the current uncertainty on this T_{eff} scale is of the order of 200 – 300 K. The reason for this discrepancy must be due to some difference in the input physics which is not compensated by the solar calibration of α_{MLT}. This occurrence clearly points out the fact that one cannot expect the same RGB T_{eff} scale from solar calibrated models that do not employ exactly the same input physics.

The MLT formalism provides only a very simplified description of convection, and there have been several attempts to introduce non-locality in the MLT (see, e.g., [24], and references therein). These 'refinements' are often complex and introduce additional free parameters to be calibrated. The alternative model by [25] includes a spectrum of eddy sizes (rather than the one-sized convective cells of the MLT) and fixes the scale length of the convective motions to the distance to the closest convective boundary. [26] have presented a novel non-local and time-dependent model based on the solution of the Navier-Stokes equations for an incompressible perfect fluid – the so called scale-free convection theory (SFC) –, that does not contain free parameter. A few stellar models based on this SFC have been recently presented by [27]: these preliminary computations seem to shown that, at least for MS stars, the SFC theory yields results very similar to those derived from a solar calibrated MLT regarding the extension of the convective zones, temperature gradients, and energy fluxes, whereas for RGB stars the differences between sets of models based on the two convection theories are significant. Nevertheless, we note that recently the physical basis for this approach for the treatment of the superadiabatic convection has been strongly questioned by [28].

An alternative approach to model the superadiabatic layers of convective envelopes is based on the computation of realistic multidimensional radiation hydrodynamics (RHD) simulations of atmospheres and convective envelopes - where convection emerges from first principles - that cover the range of effective temperatures, surface gravities, and chemical compositions typical of stars with convective outer regions. These simulations have reached nowadays a high level of sophistication ([29]) and for ease of implementation in stellar evolution codes, their results can be used to provide an 'effective hydro-calibration' of α_{MLT}, even though RHD simulations do not confirm the basic MLT picture of columns of convective cells. In this context, [30] produced a grid of convective atmosphere/envelope 3D RHD simulations for the solar chemical composition. The same grid of 3D RHD simulations have been matched ([31]) to 1D hydrostatic equilibrium, spherically symmetric envelope models to calibrate α_{MLT} as function of gravity and effective temperature[3]

Moreover, the same RHD simulations have been employed by [32] to calculate g- and T_{eff}-dependent temperature relations as a function of the τ Rosseland optical depth. The availability ([31]) of a numerical routine to calculate a g- and T_{eff}-dependent RHD-calibrated α_{MLT} and $T(\tau)$ relations and Rosseland opacities consistent with the opacities used in the RHD simulations, enables stellar evolution calculations where boundary conditions, superadiabatic temperature gradient and opacities of the convective envelope are consistent with the RHD simulations[4].

[2] We refer to [22] for a detailed analysis of the impact of choosing different values for these free parameters on stellar model computations. However, as a general rule, these authors have shown that sets of models - based on distinct MLT flavours *but* the same physical inputs - provide consistent results with only minor differences, once the mixing length has been properly calibrated on the Sun. The same sets of models are also able to reproduce the T_{eff} of RGB stars in GGCs.

[3] These RHD simulations provide a value for the α_{MLT} for the Sun equal to 1.76 ± 0.03, not very different from the values obtained with the calibration of SSMs.

[4] We remark that it is particularly important to use both the RHD-calibrated α_{MLT} and $T(\tau)$ relations, because the T_{eff} scale of the stellar models depends on both these inputs as discussed by [23] and [19].

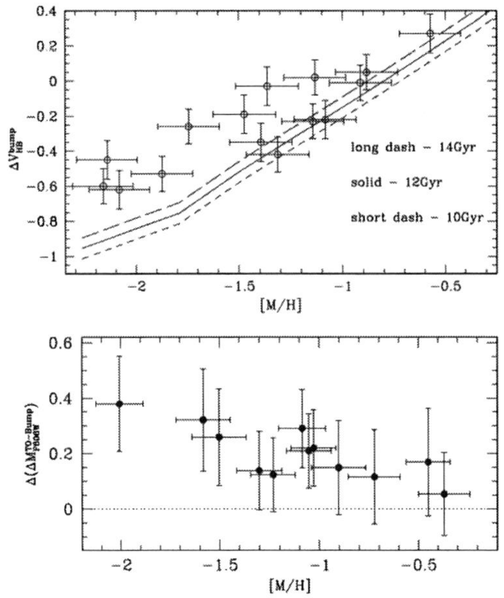

Figure 5: *Upper panel:* comparison between estimates of the brightness difference between the RGB bump and the ZAHB for a sample of Galactic globular clusters, and BaSTI theoretical prescriptions - for the labeled assumptions about the cluster age - as a function of the global metallicity. *Lower panel:* difference between the values of $\Delta M_{F606W_{TO}}^{Bump}$ predicted by models for each individual GC - once its age has been obtained from the absolute TO brightness -, and empirical values, as a function of [M/H].

[33] have been the first to present stellar evolution calculations where this 3D RHD-calibration of α_{MLT} is self-consistently included in a stellar evolution code. Figure 3 shows some results from the quoted investigation: a comparison between fully consistent model calculations with the RHD calibrated variable α_{MLT} and T(τ) relationships, and models based on a constant, solar calibrated α_{MLT} = 1.76 and the same T(τ) relationship. From the point of view of the predicted T_{eff} scale, models calculated with constant RHD calibrated solar mixing length α_{MLT} = 1.76 are very similar to, and often quite indistinguishable from, the models with variable α_{MLT}, maximum differences being at most ~ 30 – 50 K. Indeed, the same analysis has provided a plain evidence that the choice of RHD-calibrated T(τ) relation is more relevant than the use of a variable α_{MLT}, in setting the effective temperature scale of the models as shown in Fig. 4: here it is shown the comparison between models based on the RHD-calibrated T(τ) relation and the T(τ) relationships by Eddington, KS66 and [34] (V81).

3.3 The Red Giant Branch luminosity function

The RGB luminosity function (LF) of GGCs is an important tool to test the chemical stratification inside the stellar envelopes (see [35] and references therein). The RGB LF is a simple straight line on a magnitude-log(N) plane, except in a very narrow magnitude interval corresponding to the RGB bump. This linearity is a direct consequence of the He-core mass - luminosity relation for RGB stars ([35]). The slope of this line allows a major test of the evolutionary rate along the RGB, that is virtually independent of the isochrone age or metallicity. Comparisons between theoretical LFs and empirical ones for GGCs have shown a remarkable good agreement: this represents a plain evidence of the fact that theoretical predictions about the evolutionary rate along RGB are quite reliable.

As previously discussed, the brightness of the RGB bump is related to the location of the H-abundance discontinuity left over by outer convection during the *first dredge up*, in the sense that the deeper the chemical discontinuity is located, the fainter is the bump luminosity. Since any physical inputs and/or numerical assumption adopted in the computations which affects the maximum extension of the convective envelope at the *first dredge up*, strongly affects the bump brightness, a comparison between the predicted RGB bump luminosity and observations in star clusters allows a direct check of how well theoretical models for RGB stars predict the extension of convective regions in the stellar envelope and, then provide a plain evidence of the reliability of the adopted evolutionary framework ([36]).

However, when comparing theory with observations one needs a preliminary estimate of both the cluster metallicity and distance. Current uncertainty in the GGC metallicity scale strongly reduces our capability to constrain the plausibility of the theoretical framework, and for such reason, it has became a common procedure to use simultaneously all available metallicity scales (see [37] for a discussion on this issue). An additional critical issue is related to the knowledge of the cluster distance, whose indetermination could strongly hamper the possibility of a meaningful comparison between theory and observations. In order to overcome this problem, as early suggested by [38], the observed V-magnitude (or filters similar to Johnson V) difference between the RGB bump and the HB at the RR Lyrae instability strip ($\Delta V_{HB}^{Bump} = V_{HB} - V_{bump}$) is usually adopted in order to test the theoretical predictions for the bump brightness. This quantity presents several advantages from the observational point of view and it is empirically well-defined because it does not depend on a previous knowledge of the cluster distance and reddening. However, on the theoretical side, one should keep in mind that such comparison requires the use of a theoretical prediction about the Horizontal Branch brightness which is a parameter still affected by some uncertainty. Nevertheless, empirical estimates about the ΔV_{HB}^{Bump} parameter have been extensively compared with theoretical predictions (see [39] and references therein).

Upper panel of fig. 5 shows a comparison between recent measurements of this parameter for a sample of GGCs and theoretical predictions, and discloses a discrepancy - at the level of ~ 0.20 mag, or possibly larger -, for GCs with [M/H] below ~ −1.5: the predicted ΔV_{HB}^{Bump} being larger than observed. We cannot discriminate whether this is due to too bright theoretial bump luminosities, or underluminous HB models (or a combination of both effects). Due to the strong dependence of ΔV_{HB}^{Bump} parameter on the metallicity, a more quantitative evaluation of the 'real' discrepancy between theory and observation strongly depends on the adopted metallicity scale.

A complementary avenue is offered by the magnitude difference between the TO and RGB bump, $\Delta V_{TO}^{Bump} = V_{TO} - V_{bump}$, which bypasses the HB. Lower panel of fig. 5 shows a comparison between this quantity as measured for a sample of GGCs observed in the F606W filter of the Advanced Camera for Surveys on board of HST and the similar theoretical quantity as estimated once the age of each GC has been accurately derived via the isochrone fitting to the absolute MS TO magnitude obtained using a MS-fitting distance scale (see [40], for a more detailed discussion on the adopted method). This plot shows clearly that the expected $\Delta M_{F606W}{}_{TO}^{Bump}$ values are systematically larger than observed: the mean difference being of the order of 0.20 mag. Given that the observed TO magnitude is by definition matched by the theoretical isochrones to determine the TO age, this discrepancy implies that the theoretical RGB bump luminosity is too bright, e.g., RGB models predict a too shallow maximum depth of the envelope at the *first dredge up*.

On the basis of the most recent updates in the input physics, it does not appear realistic to modify the maximum depth of the convective envelope at the *first dredge up* by reasonable changes in the adopted physics (i.e., in the radiative opacity). The most viable solution of this discrepancy implies some amount of convective overshoot, beyond the Schwarzschild convective boundary, by about $0.25H_P$[5]. The inclusion of atomic diffusion can also help in reducing the discrepancy but it can not completely solve the problem.

New, independent, insights concerning the location of the chemical discontinuity in the envelope of RGB stars can be provided by an accurate analysis of the g-modes in oscillating giants. In fact, recently [41] have investigated the possibility to use the perturbations (also called *glitches*) of the frequencies of high-radial-order g-modes from their asymptotic value, due to sharp variations in the buoyancy frequency induced by local changes in the chemical composition. Although no robust observational results on these buoyancy frequency glitches have been presented so far, one can predict that - once observed - this kind of seismic effects can represent a valuable tool for investigating the inner chemical stratification of cool stars.

In passing, we also note that the RGB LF bump provides other important constraints besides brightness for checking the accuracy of RGB models. More in detail, both the shape and the location of the bump along the RGB LF can be used for investigating on how 'steep' is the H-discontinuity left over by envelope convection at the *first dredge up*. So these features appear, potentially, a useful tool for investigating on the efficiency of non-canonical mixing at the border of the convective envelope ([42]) able to partially smooth the chemical discontinuity. In addition, since the evolutionary rate along the RGB is strongly affected by any change in the chemical profile, it is clear that star counts in the bump region can provide reliable information about the size of the jump in the H profile left over by envelope convection after the *first dredge up* ([37]).

3.4 Mass-loss efficiency along the RGB stage

A major issue in stellar evolution theory is the treatment of the mass loss (ML) during the advanced evolutionary stages. In fact, reliable empirical ML determinations, as well as a comprehensive physical description of the involved processes are so far still lacking; in particular, there is a lack of any empirical law directly calibrated on Population II giants. From a theoretical point of view, our knowledge of the ML timescales, driving mechanisms, dependence on stellar parameters and metallicity is also very poor. The consequence is that there is little theoretical or observational guidance on how to incorporate ML into stellar model computations.

As for the ML efficiency along the RGB, mass loss rates are customarily parametrized in stellar model calculations by means of simple relations like the Reimers formula ([43]). A few other formulae, which are variants of the Reimers one, have been proposed but there is no *a priori* reason for choosing one amongst the different alternatives. These prescriptions are essentially scaling relations between mass loss rates and global stellar parameters like, e.g., surface bolometric luminosity and gravity, effective temperature and/or radius. The zero point of these scaling relations is typically set by a free parameter (η) that needs to be calibrated.

The last decade has witnessed a growing amount of empirical data concerning ML estimates for Pop. II red giants, based on a more direct approach as the detection of outflow motions in the outer regions of the atmospheres (see, e.g., [44]), or the detection of the circumstellar envelopes at larger distances from the stars (see, e.g., [45]). These empirical analyses suggest that the 'actual' ML law should be significantly different (flatter) than the Reimers formula, that seems to be ruled out at the 3σ level. In addition, the RGB ML phenomenon seems to be not a continuous process but an episodic phenomenon, whose efficiency does not appear to be strongly correlated with the metallicity. Recently, [46, 47] applied an alternative approach to estimate the mass lost by RGB stars in the GGC 47 Tuc. These authors determined the rate of diffusion of stars through the cluster core, using a sample of young white dwarfs, and compared the radial distribution of upper MS, RGB and HB stars. As a result, they found that the radial distributions of the various classes of stars are nearly identical, showing that there has been very little time for the young WDs to have diffused through the cluster since their progenitors lost mass. Based on these dynamical consideration, [47] estimated that up to two-thirds of the $\sim 0.4M_\odot$ that 47 Tuc stars are expected to lose between the end of the MS and the beginning of the WD sequence, is shed shortly before the start of the WD cooling, and a typical HB stellar mass of the order of $\sim 0.65M_\odot$

[5]We note that the RGB bump brightness changes at the rate of $0.8 mag/H_P$.

(that corresponds to a mass loss of about 0.25$M_\odot$ along the RGB) is excluded by comparing the radial distribution of HB stars and MS stars.

Needless to say that this result would severely challenge the common idea about the efficiency of mass loss during the RGB stage. Actually, the need for a significant ML during the RGB in order to reproduce the observed cluster HB morphology has been shown by [48]: a detailed comparison between synthetic HB models and data shows that the observed morphology can be reproduced only by accounting for an average mass lost during the RGB equal to $\Delta M \approx 0.23M_\odot$ and an almost negligible Gaussian spread of 0.005$M_\odot$ around this value. When taking into account the possible sources of uncertainty related to the cluster properties such as reddening, chemical composition (also in the context of the multiple population phenomenon) and age, [48] found that the average mass lost during the RGB ranges from a minimum value of $\sim 0.17M_\odot$ (for a GC age of 12.5 Gyr) to a maximum value of $\sim 0.21M_\odot$ (for a GC age of 10.5 Gyr). This analysis suggests the existence of a significant discrepancy between the information coming from cluster dynamics and CMD modelling of the HB. A comparison between the results from these two techniques applied to other GCs is required, to gain more insights about the origin of this apparent disagreement.

An independent approach for measuring the ML efficiency in RGB stars has been adopted by [49]: thanks to the *Kepler* detection of solar-like oscillations in G-K giants in the old open clusters NGC 6791 and 6819, they posed a constraint on the integrated RGB mass-loss efficiency by comparing the average mass of RC stars with that of stars in the faint portion of RGB. As a results they found that very efficient RGB ML can be excluded, and the available data are consistent with ML rates described with a Reimers η parameter smaller than ~ 0.35. These results are interesting, and the seismic method appears a quite viable way to estimate ML efficiency. However, since stellar masses are determined by combining the available semiotics parameters ν_{max} and $\Delta\nu$ with other data (as photometry and distance determinations), more accurate and reliable stellar masses determinations are expected from additional constraints from individual pulsation mode frequencies, spectroscopic $T_{\rm eff}$ determinations and accurate tests of the adopted ν_{max} and $\Delta\nu$ scaling relations.

Before closing this section, we wish to emphasise that an accurate determination of the RGB ML efficiency is a crucial issue also in the context of using stellar models and seismic measurements for Galactic archaelogy ([2]). Indeed, one has to bear in mind that the computation of ages for red giant stars (both RGB stars and RC ones) heavily relies on the knowledge of their masses. In the context of asteroseismologic surveys, the mass of a star is essentially derived by using scaling relations based on the observed $\Delta\nu$ and ν_{max}, and thus the mass is rather independent of the adopted set of stellar models. The same, however, does not hold for ages. Although the mass of a red giant is a useful proxy for age, it is important to distinguish between the initial mass, which sets the evolutionary lifetime of a star,

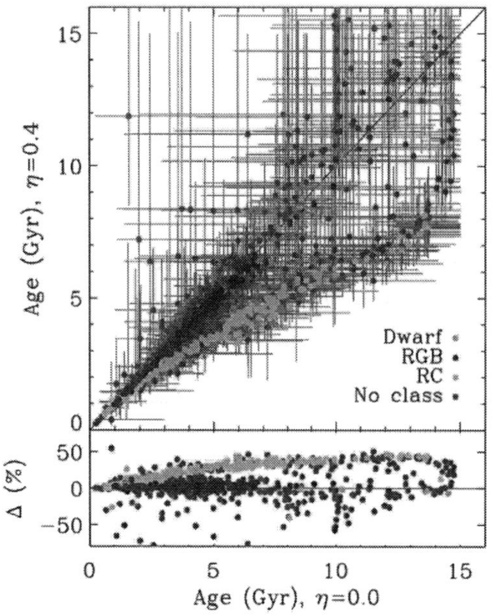

Figure 6: Comparison between ages obtained by alternatively using stellar models accounting for mass-loss (with the parameter η equal to 0.4) and not accounting for ML (from [3]).

and the present-day (i.e., the actual) stellar mass, which is indeed the quantity derived from seismology. Thus, when inferring an age using the actual stellar mass, the value derived will depend on the past history of the star, whether or not significant ML has occurred during its evolution. This issue has been investigated in detail by [3].

Figure 6 shows the comparison of between ages derived with and without mass-loss. Ages of dwarf stars are obviously unaffected by mass-loss, and the same conclusion holds for RGB stars.[6] The effect of ML increases when moving to lower gravities, and it is quite dramatic for stars in the central He-burning stage. Theoretical isochrones including ML return younger ages than those without ML; this can be easily understood since a given mass - seismically inferred - will correspond to a higher initial mass in case of ML, whose evolutionary lifetimes are shorter.

3.5 Additional physical processes

Standard models for red giant stars are non-rotating, spherically symmetric and usually include only convection as element transport mechanism. These models predict that, during the *first dredge up*, the surface chemical composition of low-mass stars is modified due to dredge up into the convective envelope of H-burning processed material.

[6]The distinction between RGB and RC stars is based on the average spacing between mixed modes ([50]); hence a robust identification of RGB stars is possible for $\log(g) \geq 2.6$, i.e. on the lower part of the RGB, where ML turns out to be of little importance in the Reimers formulation. This explains the weak dependence of RGB ages on mass-loss.

DOI: 10.1051/epjconf/201716004002

After this mixing episode, canonical models do not predict any further mixing episode along the RGB. Several spectroscopic observations of metal poor Galactic halo stars provide however compelling evidence of an additional mixing process occurring when RGB stars reach the luminosity of the RGB bump: a sudden drop of the isotopic ratio $^{12}C/^{13}C$, a decrease of the Li and C abundances and an increase of the N abundance is clearly shown by the empirical data. This 'non-canonical' mixing seems to be a universal process as it affects ~ 95% of low-mass stars, regardless of whether they populate the halo field or clusters (see [51] and references therein). The mechanism that has been proposed as a possible source for this mixing episode should be associated with the molecular weight inversion due to the ^{3}He burning through the $^3He(^3He, 2p)^4He$ nuclear reaction, in the outer wing of the H-burning shell, and the associated thermohaline mixing.

In low-mass stars, the main burning mechanism during the MS is the *p-p chain*, that due to its weak dependence on temperature, is efficient also in stellar layers quite far from the star centre. As a consequence, ^{3}He accumulates in a broad zone outside the main energy production region. During the *first dredge up* this ^{3}He is mixed within the convective envelope, with the consequence that during the following RGB evolution, the layers above the H-discontinuity left over by the receding convective envelope at its maximum extension will have a uniform ^{3}He, larger than the initial one. When the shell advances towards the surface during the RGB evolution, in the outer wing above the point of maximum burning efficiency, there is a narrow region where ^{3}He is burnt through the reaction $^3He(^3He, 2p)^4He$. This nuclear reaction is unusual in the sense that 2 nuclei transform into 3 and the mean mass per nucleus decreases from 3 to 2. Because the molecular weight is the mean mass per nucleus, this leads to a small local decrease in the μ gradient. As long as the H-burning shell moves through layers below the H-discontinuity – i.e. the star evolves before the RGB bump – this local negative μ gradient is negligible because the shell is moving in a region with a much larger positive gradient, due to the He profile left over at the end of the MS. However, when the H-burning shell enters the region of uniform chemical profile beyond the H-discontinuity, the local inversion, i.e. negative gradient, in the μ profile, of the order of one part in 10^4, becomes much more important. This situation corresponds to the conditions for thermohaline mixing (see [52] and references therein). This mixing is usually included in stellar evolution codes as a diffusive process that works in the direction to erase the molecular weight inversion, with diffusion coefficient given by [53]

$$D_{th} = C_{th} \frac{K}{c_p\rho} \left(\frac{\varphi}{\delta}\right) \frac{-\nabla_\mu}{(\nabla_{ad} - \nabla)} \quad \text{for } \nabla_\mu < 0 \qquad (1)$$

where ∇_μ is molecular weight gradient, K denotes the thermal conductivity, $C_{th} = \frac{8}{3}\pi^2\alpha^2$ with α a free parameter, $\varphi = (\partial \ln \rho/\partial \ln \mu)_{P,T}$, and $\delta = -(\partial \ln \rho/\partial \ln T)_{P,\mu}$. Evolutionary calculations show that thermohaline mixing extends between the outer wing of the H-burning shell and the inner boundary of the convective envelope, merging with the outer convection in a short time (~ 30 Myr for a model with mass of the order of $1M_\odot$). Therefore, depending on the mixing efficiency, a significant amount of nuclear processed matter in the hotter layers of the H-burning shell can be dredged-up to the surface; an occurrence that allows to reproduce spectroscopic data.

This notwithstanding, there are still several shortcomings in the inclusion of this physical process in stellar model computations. At first, as it has been shown by [51], values of C_{th} ~1000 are required to reproduce the $^{12}C/^{13}C$ abundance pattern in halo RGB stars, but in order to reproduce the same trend for ^{7}Li abundances a quite smaller value is necessary (C_{th} ~200). In addition, hydrodynamical simulations predict a different expression for the diffusion coefficient, that is reasonably reproduced by Eq. 1 with C_{th} ~10 ([54]), i.e. the expected efficiency is much lower than needed to reproduce RGB spectroscopic data. In addition, recently [55] have provided compelling results showing that the surface chemical abundance predictions of stellar models accounting for thermohaline mixing strongly depend on the numerical assumptions on the mesh spacing and time resolution adopted in the evolutionary computations. As a consequence of these results, model predictions about surface chemical abundances should be treated with caution until a more firm assessment on how to manage with thermohaline mixing in stellar model computations is not achieved.

For long time, rotation and the associated rotational mixings have not - in general - been an ingredient of standard stellar models. This occurrence can be attributed largely to the increase in complexity and uncertainty – read free parameters – of models that include rotation, and some estimates that rotation is expected to be at most a (small) perturbation to the structure of low-mass stars. A number of observational data, however, requires mixing mechanisms not present in standard stellar models, and mixing driven by rotation is at least a potential candidate. In addition, nowadays asteroseismic observations are clearly revealing the role played by rotation in the interiors of stars other than the Sun. From the point of view of the inclusion of rotation in model computations, but the huge increase in the complexity of the numerical equations that have to be solved, the main problems are related to our poor knowledge of the interaction between convection and meridional currents, of the efficiency of angular momentum transport and angular momentum loss during the core and shell H-burning stages. The inclusion of these processes in an evolutionary code requires the calibration of several free parameters, that, until a few years, were poorly constrained by standard observations.

In this context, a big improvement has been obtained - and more can be expected in the near future - thanks to seismic observations collected with Kepler satellite which allowed to measure the core rotation rates for hundreds of red giants ([56, 57]). The preliminary comparison between these empirical data and existing sets of models (see [58]) has shown that current RGB models underestimate the angular momentum extraction from the stellar core, i.e. the

 DOI: 10.1051/epjconf/201716004002

amount of torque between the core and the envelope seems to be significantly underestimated.

In order to increase the amount of angular momentum losses from the core, some possible 'candidates' have been identified such as internal gravity waves and large scale magnetic fields. However, a lot of work has to be done before we achieve a firm understanding of the role played by these individual physical processes.

4 On the modelling of Red Clump stars

Concerning the models for RC stars, the main sources of uncertainty are related to the estimation of the rate for the nuclear reaction $^{12}C(\alpha,\gamma)^{16}O$, and of the treatment of convective core boundary.

The $^{12}C(\alpha,\gamma)^{16}O$ reaction is one of the key reactions in stellar evolution. Its rate affects the C/O ratio in the stellar core at the end of the central He-burning stage and, as a consequence, WD cooling times. In addition, during the central He-burning phase, when the abundance of He in the convective core is significantly reduced, α−captures on carbon nuclei becomes strongly competitive with the triple-α reactions in terms of contribution to the nuclear energy budget, with the consequence that the cross-section for this reaction has a strong influence on the core He-burning phase lifetime.

It is known that this reaction has a resonance and a very low cross-section ($\sim 10^{-17}$ barn) at low energies, and the nuclear parameters are difficult to measure experimentally or to predict from theory. In the last decade, the accuracy of this reaction rate has been significantly improved; however its estimate should be still affected by a $\approx 30\%$ uncertainty ([35]).

As discussed by [59], the treatment of convective boundaries during central helium burning is still an open question in stellar evolution calculations, and is handled in different ways by different stellar evolution codes. Recent advances in asteroseismic observations and techniques ([59, 60]) are starting to add very direct observational constraints to the core mixing process during the central He-burning phase, that coupled to theoretical inferences and indications from star counts in GGCs, make a strong case for the core mixed region to be extended beyond the Schwarzschild border.

Questions exist however about the treatment of this mixing beyond the canonical convective border: it is not so clear if a 'standard' overshoot scheme should be preferred to a numerical approach favouriting the formation of an extended, partially mixed region – 'semiconvective' zone - around the canonical convective core. For a detailed discussion on this issue we refer to [35, 60].

The uncertainty in the mixing treatment worsens as the core He-burning phase progresses. If no specific approach is used, the phenomenon of the so-called 'breathing pulses' occurs, that corresponds to a recursive phase of expansion of the convective core occurring when the He abundance in the core becomes quite low, and the feedback from the energy released by the $^{12}C(\alpha,\gamma)^{16}O$ nuclear reaction dominates with respect the triple−α reactions (due to the paucity of He nuclei in the core).

The occurrence of the breathing pulses in real stars has been severely challenged by observations - mainly based on star counts in GGCs -, and their occurrence in models is attributed to the approximation of instantaneous mixing in the convective zones adopted in the model computations. Different methodological approaches have been designed to inhibit the breathing pulses, but they lead to different predictions for the following evolutionary phases (see [35], and references therein).

5 Final remarks

Huge improvements have been made regarding the major physics ingredients entering in stellar models 'cooking' such as opacities, nuclear reaction rates, equation of state, etc. Although - as briefly discussed above - there are still some issues - as the uncertain rates of some specific nuclear reactions - a realistic description of the main input physics needed for stellar computations seems within reach. On the other hand, we still face long-standing problems related to the treatment of mixing in both the stellar interiors and atmosphere. Despite the large efforts devoted to address this problem, we still need to use approximate numerical approaches with some (tunable) free parameter(s). In this context, the role played by rotation and rotationally-induced mixings also has to be fully investigated taking into account the updated constraints coming from seismic measurements on large samples of red giant stars.

We expect that quite important advances towards developing of a comprehensive stellar evolution theory, as well as more realistic stellar models, may come from constraints obtained by current (and future) asteroseismic surveys.

References

[1] A. Miglio, J. Montalbán, F. Baudin, P. Eggenberger, A. Noels, S. Hekker, J. De Ridder, W. Weiss, A. Baglin, A&A **503**, L21 (2009)
[2] L. Casagrande, V. Silva Aguirre, D. Stello, D. Huber, A.M. Serenelli, S. Cassisi, A. Dotter, A.P. Milone, S. Hodgkin, A.F. Marino et al., ApJ **787**, 110 (2014)
[3] L. Casagrande, V. Silva Aguirre, K.J. Schlesinger, D. Stello, D. Huber, A.M. Serenelli, R. Schönrich, S. Cassisi, A. Pietrinferni, S. Hodgkin et al., MNRAS **455**, 987 (2016)
[4] D. Sugimoto, M.Y. Fujimoto, ApJ **538**, 837 (2000)
[5] A. Formicola, G. Imbriani, H. Costantini, C. Angulo, D. Bemmerer, R. Bonetti, C. Broggini, P. Corvisiero, J. Cruz, P. Descouvemont et al., Physics Letters B **591**, 61 (2004)
[6] A. Weiss, A. Serenelli, A. Kitsikis, H. Schlattl, J. Christensen-Dalsgaard, A&A **441**, 1129 (2005)
[7] A. Pietrinferni, S. Cassisi, M. Salaris, A&A **522**, A76 (2010)
[8] A. Pietrinferni, S. Cassisi, M. Salaris, F. Castelli, ApJ **612**, 168 (2004)

[9] V. Silva Aguirre, S. Basu, I.M. Brandão, J. Christensen-Dalsgaard, S. Deheuvels, G. Doğan, T.S. Metcalfe, A.M. Serenelli, J. Ballot, W.J. Chaplin et al., ApJ **769**, 141 (2013)
[10] T. Appourchaux, H.M. Antia, W. Ball, O. Creevey, Y. Lebreton, K. Verma, S. Vorontsov, T.L. Campante, G.R. Davies, P. Gaulme et al., A&A **582**, A25 (2015)
[11] F.J. Rogers, A. Nayfonov, ApJ **576**, 1064 (2002)
[12] D.A. VandenBerg, F.J. Swenson, F.J. Rogers, C.A. Iglesias, D.R. Alexander, ApJ **532**, 430 (2000)
[13] A. Dotter, B. Chaboyer, D. Jevremović, E. Baron, J.W. Ferguson, A. Sarajedini, J. Anderson, AJ **134**, 376 (2007)
[14] O. Straniero, A&AS **76**, 157 (1988)
[15] M. Salaris, A. Chieffi, O. Straniero, ApJ **414**, 580 (1993)
[16] J.W. Ferguson, D.R. Alexander, F. Allard, T. Barman, J.G. Bodnarik, P.H. Hauschildt, A. Heffner-Wong, A. Tamanai, ApJ **623**, 585 (2005)
[17] C. Neuforge, A&A **274**, 818 (1993)
[18] R.L. Kurucz, *Atomic and Diatomic Molecular Opacities for Atmospheres and Envelopes*, in *American Astr. Soc. Meet. Abs. #180* (1992), Vol. 24 of *Bulletin of the American Astronomical Society*, p. 779
[19] D.A. VandenBerg, B. Edvardsson, K. Eriksson, B. Gustafsson, ApJ **675**, 746-763 (2008)
[20] K.S. Krishna Swamy, ApJ **145**, 174 (1966)
[21] E. Böhm-Vitense, Zeitschrift fur Astrophysik **46**, 108 (1958)
[22] M. Salaris, S. Cassisi, A&A **487**, 1075 (2008)
[23] M. Salaris, S. Cassisi, A. Weiss, PASP **114**, 375 (2002)
[24] L. Deng, D.R. Xiong, K.L. Chan, ApJ **643**, 426 (2006)
[25] V.M. Canuto, I. Mazzitelli, ApJ **389**, 724 (1992)
[26] S. Pasetto, C. Chiosi, M. Cropper, E.K. Grebel, MNRAS **445**, 3592 (2014)
[27] S. Pasetto, C. Chiosi, E. Chiosi, M. Cropper, A. Weiss, MNRAS **459**, 3182 (2016)
[28] M.M. Miller Bertolami, M. Viallet, V. Prat, W. Barsukow, A. Weiss, MNRAS **457**, 4441 (2016)
[29] Å. Nordlund, R.F. Stein, M. Asplund, Living Reviews in Solar Physics **6**, 2 (2009)
[30] R. Trampedach, M. Asplund, R. Collet, Å. Nordlund, R.F. Stein, ApJ **769**, 18 (2013)
[31] R. Trampedach, R.F. Stein, J. Christensen-Dalsgaard, Å. Nordlund, M. Asplund, MNRAS **445**, 4366 (2014)
[32] R. Trampedach, R.F. Stein, J. Christensen-Dalsgaard, Å. Nordlund, M. Asplund, MNRAS **442**, 805 (2014)
[33] M. Salaris, S. Cassisi, A&A **577**, A60 (2015)
[34] J.E. Vernazza, E.H. Avrett, R. Loeser, ApJS **45**, 635 (1981)
[35] S. Cassisi, M. Salaris, *Old Stellar Populations: How to Study the Fossil Record of Galaxy Formation* (2013)
[36] S. Cassisi, M. Salaris, MNRAS **285**, 593 (1997), `astro-ph/9702029`
[37] M. Riello, S. Cassisi, G. Piotto, A. Recio-Blanco, F. De Angeli, M. Salaris, A. Pietrinferni, G. Bono, M. Zoccali, A&A **410**, 553 (2003)
[38] F. Fusi Pecci, F.R. Ferraro, D.A. Crocker, R.T. Rood, R. Buonanno, A&A **238**, 95 (1990)
[39] A. Di Cecco, G. Bono, P.B. Stetson, A. Pietrinferni, R. Becucci, S. Cassisi, S. Degl'Innocenti, G. Iannicola, P.G. Prada Moroni, R. Buonanno et al., ApJ **712**, 527 (2010)
[40] S. Cassisi, A. Marín-Franch, M. Salaris, A. Aparicio, M. Monelli, A. Pietrinferni, A&A **527**, A59 (2011)
[41] M.S. Cunha, D. Stello, P.P. Avelino, J. Christensen-Dalsgaard, R.H.D. Townsend, ApJ **805**, 127 (2015)
[42] S. Cassisi, M. Salaris, G. Bono, ApJ **565**, 1231 (2002), `astro-ph/0110247`
[43] D. Reimers, Memoires of the Societe Royale des Sciences de Liege **8**, 369 (1975)
[44] M. Vieytes, P. Mauas, C. Cacciari, L. Origlia, E. Pancino, A&A **526**, A4 (2011)
[45] L. Origlia, F.R. Ferraro, S. Fabbri, F. Fusi Pecci, E. Dalessandro, R.M. Rich, E. Valenti, A&A **564**, A136 (2014)
[46] J. Heyl, H.B. Richer, E. Antolini, R. Goldsbury, J. Kalirai, J. Parada, P.E. Tremblay, ApJ **804**, 53 (2015)
[47] J. Heyl, J. Kalirai, H.B. Richer, P. Marigo, E. Antolini, R. Goldsbury, J. Parada, ApJ **810**, 127 (2015)
[48] M. Salaris, S. Cassisi, A. Pietrinferni, A&A **590**, A64 (2016)
[49] A. Miglio, K. Brogaard, D. Stello, W.J. Chaplin, F. D'Antona, J. Montalbán, S. Basu, A. Bressan, F. Grundahl, M. Pinsonneault et al., MNRAS **419**, 2077 (2012)
[50] T.R. Bedding, B. Mosser, D. Huber, J. Montalbán, P. Beck, J. Christensen-Dalsgaard, Y.P. Elsworth, R.A. García, A. Miglio, D. Stello et al., Nature **471**, 608 (2011)
[51] G.C. Angelou, V. D'Orazi, T.N. Constantino, R.P. Church, R.J. Stancliffe, J.C. Lattanzio, MNRAS **450**, 2423 (2015)
[52] C. Charbonnel, J.P. Zahn, A&A **467**, L15 (2007)
[53] R. Kippenhahn, G. Ruschenplatt, H.C. Thomas, A&A **91**, 175 (1980)
[54] A. Traxler, P. Garaud, S. Stellmach, ApJL **728**, L29 (2011)
[55] J.C. Lattanzio, L. Siess, R.P. Church, G. Angelou, R.J. Stancliffe, C.L. Doherty, T. Stephen, S.W. Campbell, MNRAS **446**, 2673 (2015)
[56] B. Mosser, M.J. Goupil, K. Belkacem, J.P. Marques, P.G. Beck, S. Bloemen, J. De Ridder, C. Barban, S. Deheuvels, Y. Elsworth et al., A&A **548**, A10 (2012)
[57] S. Deheuvels, G. Doğan, M.J. Goupil, T. Appourchaux, O. Benomar, H. Bruntt, T.L. Campante, L. Casagrande, T. Ceillier, G.R. Davies et al., A&A

DOI: 10.1051/epjconf/201716004002

564, A27 (2014)

[58] M. Cantiello, C. Mankovich, L. Bildsten, J. Christensen-Dalsgaard, B. Paxton, ApJ **788**, 93 (2014)

[59] T. Constantino, S.W. Campbell, J. Christensen-Dalsgaard, J.C. Lattanzio, D. Stello, MNRAS **452**, 123 (2015)

[60] D. Bossini, A. Miglio, M. Salaris, A. Pietrinferni, J. Montalbán, A. Bressan, A. Noels, S. Cassisi, L. Girardi, P. Marigo, MNRAS **453**, 2290 (2015)

Nonlinear asteroseismology: insight from amplitude and frequency modulations of oscillation modes in compact pulsators from *Kepler* photometry

Weikai Zong[1,2,⋆], *Stéphane* Charpinet[1,2,⋆], *Gérard* Vauclair[1,2,⋆], *Noemi* Giammichele[1,2,3], and *Valérie* Van Grootel[4]

[1] *CNRS, IRAP, 14 avenue Edouard Belin, F-31400 Toulouse, France*
[2] *Université de Toulouse, UPS-OMP, IRAP, Toulouse F-31400, France*
[3] *Département de Physique, Université de Montréal, CP 6128, Succursale Centre-Ville, Montréal, QC H3C 3J7, Canada*
[4] *Institut d'Astrophysique et de Géophysique, Quartier Agora, Allée du 6 Août 19c, 4000 Liège, Belgium.*

Abstract. Nonlinear mode interactions are difficult to observe from ground-based telescopes as the typical periods of the modulations induced by those nonlinear phenomena are on timescales of weeks, months, even years. The launch of space telescopes, e.g., *Kepler*, has tremendously changed the situation and shredded new light on this research field. We present results from *Kepler* photometry showing evidence that nonlinear interactions between modes occur in the two compact pulsators KIC 8626021, a DB white dwarf, and KIC 10139564, a short period hot B subdwarf. KIC 8626021 and KIC 10139564 had been monitored by *Kepler* in short-cadence for nearly two years and more than three years without interruption, respectively. By analyzing these high-quality photometric data, we found that the modes within the triplets induced by rotation clearly reveal different behaviors: their frequencies and amplitudes may exhibit either periodic or irregular modulations, or remain constant. These various behaviors of the amplitude and of the frequency modulations of the oscillation modes observed in these two stars are in good agreement with those predicted within the amplitude equation formalism in the case of the nonlinear resonant mode coupling mechanism.

1 Introduction

Evolved compact pulsators, white dwarf and hot B subdwarf (sdB) stars, have been well investigated by the technique of asteroseismology. White dwarf stars are the end fates of stellar evolution for 98% of the stars in our Galaxy. They span over quite a large region in the H-R diagram, with effective temperature of $\sim$ 10 000 – 170 000 K and surface gravity of $\sim$ 5.5 – 8.5 dex [1, 2]. SdB stars belong to the extreme horizontal branch in the H-R diagram, with a mass typically of $\sim 0.47 M_{\odot}$, effective temperature of 22 000 – 40 000 K and surface gravity of $\sim$ 5.2 – 6.1 [3]. Observations from ground-based telescopes suggest that pulsation modes in compact pulsating stars may exhibit temporal variations in their amplitudes and/or frequencies, e.g., the GW Virginis variable star PG 0122+200 [4]. However, it is difficult to observe one entire modulation cycle from ground as the modulations are typically much slower than the pulsation periods themselves [5].

The launch of space telescopes, such as *Kepler*, has tremendously changed the situations of uncovering the modulations of amplitude and frequency in pulsating stars. There are 113 candidates of compact stars that had been monitored by *Kepler* during its survey phase, aiming at searching for compact pulsators [6]. In the original *Kepler* mission, the final number of evolved compact pulsators reaches 24, including 6 pulsating white dwarf star and 18 pulsating sdB stars, see., e.g., Ref. [7]. Most of these 24 stars, in particular for sdB pulsators, have been observed by *Kepler* for more than two years without interruption. They are suitable candidates to observe the variability and stability in their oscillation modes.

In this proceeding, we assemble the results of first evidences of amplitude and frequency modulations of multiplet modes in two of these 24 compact stars, a Helium-atmosphere white dwarf, KIC 8626021, and a short-period V 361 Hya stars, KIC 10139654, pointing toward nonlinear mode interactions. These findings suggest that *Kepler* opens a new era to investigate nonlinear asteroseismology.

2 Nonlinear resonant mode couplings

The theoretical explorations of nonlinear asteroseismology, described in the framework of the amplitude equation (AE) formalism, began around 1980's. It concerned the mode resonance in the adiabatic and nonadiabatic cases [8, 9]. A specific case of nonlinear AEs had been then applied to nonradial triplet modes that are induced by stellar rotation [10].

Within nonlinear asteroseismology, the stellar oscillations can be represented by temporal amplitudes as

$$\mathbf{z} = \sum_j \frac{1}{2} A_j(t) e^{i\omega_j t} \mathbf{e}_j + C + O, \quad (1)$$

⋆e-mail: [weikai.zong,stephane.charpinet,gvauclair]@irap.omp.eu where $\mathbf{e}_j$, ω_j, and $A_j(t)$ denote the linear nonadiabatic eigenvectors, eigenfrequency and the temporal amplitude of mode j, respectively, C denotes complex conjugation and O represents 'higher order terms'.

For an $\ell = 1$ triplet, the nonlinear complex nonadiabatic AEs of the three modes are rewritten as a set [10],

$$\frac{dA_-}{dt} = \kappa_- A_- + R_- A_0^2 A_+ \cos(\Phi - \delta_-) - A_-(q_{11}A_-^2 + q_{12}A_0^2 + q_{13}A_+^2), \quad (2a)$$

$$\frac{dA_0}{dt} = \kappa_0 A_0 + R_0 A_0 A_+ A_- \cos(\Phi + \delta_0) - A_0(q_{21}A_-^2 + q_{22}A_0^2 + q_{23}A_+^2), \quad (2b)$$

$$\frac{dA_+}{dt} = \kappa_+ A_+ + R_+ A_0^2 A_- \cos(\Phi - \delta_+) - A_+(q_{31}A_-^2 + q_{32}A_0^2 + q_{33}A_+^2), \quad (2c)$$

$$\frac{d\Phi}{dt} = \delta\omega - 2R_0 A_- A_+ \sin(\Phi + \delta_0) + A_0\Big[R_- \frac{A_+}{A_-}\sin(\Phi - \delta_-) + R_+ \frac{A_-}{A_+}\sin(\Phi - \delta_+)\Big], \quad (2d)$$

where the combination phase $\Phi = \phi_+ + \phi_- - 2\phi_0$, κ_i denote the linear growth rates, q_{ij} and R_i are the coupling coefficients, the phase shifts δ_i with the relation of $r_i = R_i \exp(i\delta_i)$, and the frequency mismatch $\delta\omega = \omega_+ + \omega_- - 2\omega_0$.

The numeric explorations of the above AEs reveal three distinct regimes of amplitude and frequency modulations of the interacting modes [10]:

a) The nonlinear frequency lock regime. The nonlinear couplings force the frequencies to be well equally spaced in the triplet modes, $\delta\omega = 0$. The amplitudes are all constant and typically asymmetric, $A_- \neq A_+$.

b) The intermediate regime. The resonance modes exhibit amplitude and frequency modulations that can be periodic, multi-periodic, irregular, or even chaotic, depending on the real parameters of the star. When the periodic modulations occur, the period of the modulations is about the inverse of the frequency mismatch, $1/\delta\omega$, or $\sim 1/\kappa$.

c) The nonresonant regime. The nonlinear coupling effects are very small and the linear frequencies are very close to the linear results.

In the case of the three-mode resonance condition where $\omega_1 \sim \omega_2 + \omega_3$, the amplitude equation within the adiabatic approximation can be written as,

$$\frac{dA_1}{dt} = \kappa_1 A_1 + \frac{iq}{\omega_1 I_1} A_2 A_3 e^{-i\delta\omega t}, \quad (3a)$$

$$\frac{dA_2}{dt} = \kappa_2 A_2 + \frac{iq}{\omega_2 I_2} A_1 A_3^* e^{-i\delta\omega t}, \quad (3b)$$

$$\frac{dA_3}{dt} = \kappa_2 A_3 + \frac{iq}{\omega_3 I_3} A_1 A_2^* e^{-i\delta\omega t}. \quad (3c)$$

where A_j^* stands for the complex conjugate of A_j, the frequency mismatch is $\delta\omega = \omega_1 - \omega_2 - \omega_3$, term I_j are the mode inertia, the nonlinear coupling coefficient q is very complicated and can be found in [8].

The numerical results suggest that the involved modes can be stable or modulate depending on the type of resonance and parameters such as their linear growth (damping) rates and the frequency mismatch [8].

3 Amplitude and frequency modulations

We have observed clear amplitude and frequency variations in two compact pulsators with photometry from the original *Kepler* mission. As space is limited, we only show here the modulations in the primary triplets in the two compact stars, KIC 8626021, a DB pulsating white dwarf, and KIC 10139564, a short-period pulsating sdB star. One can find more details and interesting mode behaviors in the two recently published papers [11, 12].

KIC 8626021 had been continuously monitored by *Kepler* in short-cadence for nearly two years, from quarter Q10.1 to Q17.2. Figure 1 shows the observed amplitude and frequency variations in the primary triplet near 4310 μHz in that star. The results suggest that both amplitude and frequency have signatures of quasi-periodic patterns which is further confirmed by the best sinusoidal fitting by one pure sine wave. The derived periods are 580 ± 23, 680 ± 10 and 610 ± 43 days for the retrograde mode, central mode and prograde mode, respectively. More interestingly, we also observed that the two side components seem to evolve nearly in phase with similar modulation period of ~ 600 d and nearly in anti-phase with the central component (which exhibits a slightly longer modulation period of ~ 700 d) in this triplet.

KIC 10139564 had been contiguously observed by *Kepler* in short-cadence for more than three years, from quarter Q5.1 to Q17.2. Figure 2 shows the amplitude and frequency modulations of the primary *p*-mode triplet near 5760 μHz in that star. We note that the hyperfine structures of the triplet components are significantly broaden which possibly indicate that the modes are unstable. This is further illustrated by the results shown in the other panels. We observed that the components show clear amplitude and frequency modulations: quasi-periodic frequency variations in all components and quasi-periodic amplitude variations in two components. It is very suggestive that the frequencies of the two side components evolve in anti-phase, which becomes more evident after removing the long-term trend of the frequency variations through quadratic fittings [12].

4 Discussions

The observed periodic behaviors of the triplets discussed here are pointing toward the intermediate regime of nonlinear mode couplings where nonlinear interacting modes may have periodic amplitude and frequency modulations. We discuss the two cases below. More details for different mode behaviors observed in these two targets can be found in Ref. [11, 12], where cases of frequency locked, regular or irregular variations are observed.

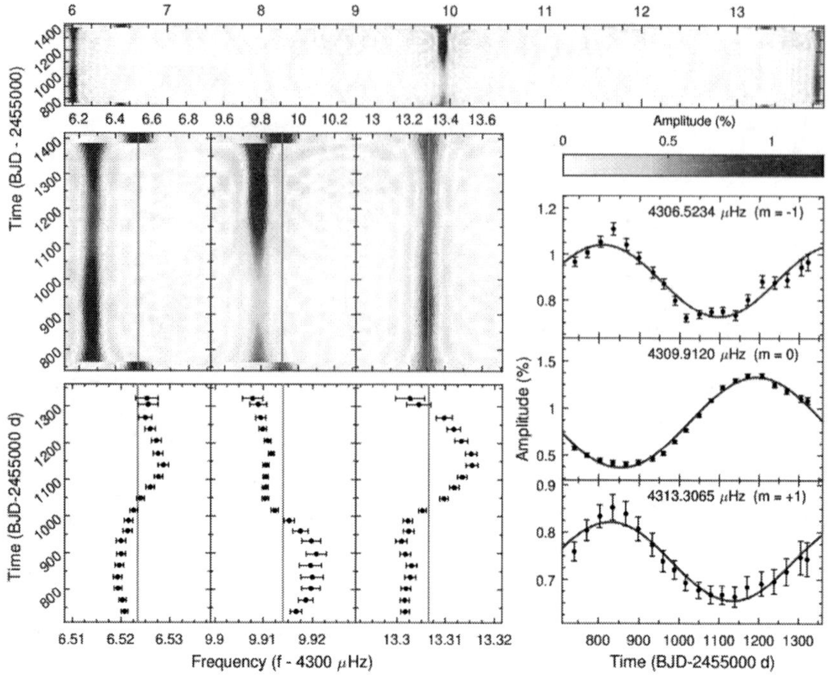

Figure 1: Amplitude and frequency modulations in the triplet near 4310μHz in the DB white dwarf star KIC 8626021. The color panels represent the sliding Lomb-Scargle periodogram. The vertical solid lines are the averaged values of the frequencies. Each data points in the *bottom left* and *lower right* panels are obtained from prewhitening subsets of the photometric data. The red solid curves denote the best sinusoidal fittings of amplitude variations. One can see more details in Ref. [11].

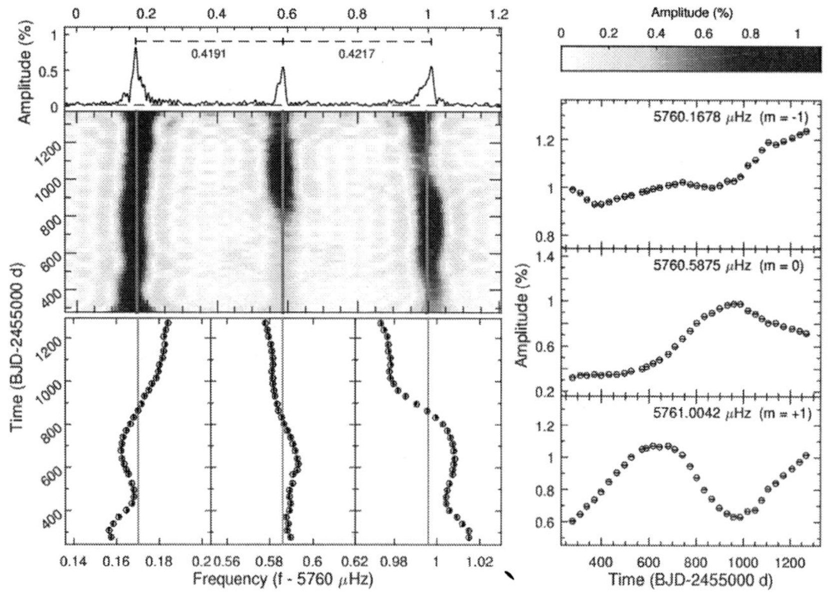

Figure 2: Similar as Fig. 1 but for the triplet modes in the short-period sdB star KIC 10139564 [12], except for the top panel. *Top-left* panel presents the hyperfine structure of the well defined triplet with near symmetric frequency spacing. We note that the error for each measurement is smaller than the symbol itself.

DOI: 10.1051/epjconf/201716004003

For the triplet near 4310 μHz in KIC 8626021, all the components show periodic amplitude and frequency variations on the timescale of $\sim$ 620 d. This value is consistent with the theoretical expectation of $P_m(\text{th}) \sim 1/\delta\omega_t =$ 620 d. This supports the idea that we uncovered the right explanation for the mode behaviors in this triplet. However, as the coupling coefficients in the amplitude equations are unknown, we cannot conduct further quantitative comparisons between theoretical explorations and the observed modulations.

The triplet near 5760 μHz in KIC 10139564 shows quasi-periodic frequency and amplitude modulations that are also linked to the intermediate regime of nonlinear resonant mode couplings. The period of frequency variations in all the three components is about 580 d after removing the long-term trend [11]. This period should correspond to a value of $\sim 0.02\,\mu$Hz, the inverse of the frequency mismatch $\delta\omega$ in this triplet, which can be theoretically met when we consider the rotation up to second order for the *p*-modes [13]. The observed frequency asymmetry $\delta\omega_o$ is only 0.0026 μHz (see Fig. 2). However, we note that the maximum of the observed frequency mismatch of each subset data is about 0.02 μHz. This result suggests that the nonlinear mode interactions possibly force the frequencies to vary around the near exact resonance.

5 Conclusions

The *Kepler* spacecraft provides unprecedented high-quality and uninterrupted photometric data with a time baseline over a couple of years, which is very crucial to investigate behaviors of amplitude and frequency modulations of oscillation modes in pulsating stars. We study in depth the mode behaviors of triplet components in two compact pulsators from *Kepler* photometry, KIC 8626021, a DB white dwarf star, and KIC 10139564, a *p*-mode dominated hot B subdwarf star. We find clear periodic modulations of amplitude and frequency in two triplets in those two stars, that can be associated with the intermediate regime of nonlinear resonant coupling mechanism. The observed timescales of the modulations are consistent with the predictions of nonlinear amplitude equation formalism. We also observed a variety of mode behaviors in many multiplet modes in those two stars and more details can be seen in Ref. [11, 12]. These results are the first evidence of nonlinear mode interactions in pulsating compact stars.

We emphasize that the frequency variations induced by nonlinear effects could jeopardize any attempt to measure the secular rates of the pulsation period changes induced by evolutionary effect since the former one has a much larger influence on the pulsation periods than that from the latter one. To observe reliable change rates of the pulsation periods, one should be extremely careful in correctly extracting the frequencies and remove the nonlinear corrections beforehand.

These intriguing observed amplitude and frequency modulations, induced by nonlinear mode interactions, should bring significant constraints for further progress in the nonlinear stellar pulsation theory, for the development of new methods and techniques to prewhiten observed light curves, for the application of nonlinear resonance to other types of pulsating stars, and for the study of resonant mode behaviors observed from future space missions as well as ground-based telescopes, possibly, leading a new way to the "nonlinear asteroseismology".

References

[1] Fontaine, G., & Brassard, P. 2008, PASP, 120, 1043
[2] Winget, D. E., & Kepler, S. O. 2008, ARA&A, 46, 157
[3] Heber, U. 2016, PASP, 128, 082001
[4] Vauclair, G., Fu, J.-N., Solheim, J.-E., et al. 2011, A&A, 528, A5
[5] Goupil, M. J., Dziembowski, W. A., & Fontaine, G. 1998, Baltic Astronomy, 7, 21
[6] Østensen, R. H., Silvotti, R., Charpinet, S., et al. 2010, MNRAS, 409, 1470
[7] Østensen, R. H., Reed, M. D., Baran, A. S., & Telting, J. H. 2014, A&A, 564, L14
[8] Dziembowski, W. 1982, Acta Astron., 32, 147
[9] Buchler, J. R., & Goupil, M.-J. 1984, ApJ, 279, 394
[10] Buchler, J. R., Goupil, M.-J., & Hansen, C. J. 1997, A&A, 321, 159
[11] Zong, W., Charpinet, S., Vauclair, G., Giammichele, N., & Van Grootel, V. 2016a, A&A, 585, A22
[12] Zong, W., Charpinet, S., & Vauclair, G. 2016b, A&A, 594, A46
[13] Soufi, F., Goupil, M. J., & Dziembowski, W. A. 1998, A&A, 334, 911

EPJ Web of Conferences **160**, 04004 (2017) DOI: 10.1051/epjconf/201716004004
Seismology of the Sun and the Distant Stars II

The K2 RR Lyrae Survey

Róbert Szabó[1,3,⋆], *Katrien* Kolenberg[2,3,4], *László* Molnár[1], *Emese* Plachy[1], *Péter* Klagyivik[1], *Ottó* Hanyecz[1,5], and *the KASC* RR Lyrae and Cepheid Working Group members

[1]*Konkoly Observatory, Research Centre for Astronomy and Earth Sciences, Hungarian Academy of Sciences, H-1121 Budapest, Konkoly Thege Miklós út 15-17, Hungary*
[2]*Harvard-Smithsonian Center for Astrophysics, 60 Garden Street, Cambridge MA 02138, USA*
[3]*Instituut voor Sterrenkunde, KU Leuven, Celestijnenlaan 200D, 3001 Heverlee, Belgium*
[4]*Department of Physics, University of Antwerp, Groenenborgerlaan 171, 2020 Antwerp, Belgium*
[5]*Eötvös Loránd Tudományegyetem, H-1117 Pázmány Péter sétány 1/A, Budapest, Hungary*

Abstract. We have initiated a large survey with K2 to observe thousands of RR Lyrae stars along the Ecliptic. The high photometric precision and the 80-90-day continuous coverage will allow us to investigate the light variation of these galactic structure tracer variable stars with an unprecedented detail. The survey will help us to conduct a thorough statistical study of RR Lyrae pulsation dynamics including old and recently discovered dynamical phenomena, like resonances, non-radial modes, period doubling, and the Blazhko effect. We describe the survey, present the first results and discuss the prospects of combining the K2 RR Lyrae survey, Gaia, and LSST in the context of galactic structure studies.

1 Introduction

With the failure of the second reaction wheel of Kepler, a new mission, dubbed K2 was conceived [1]. The new mission observes ecliptic fields for 80–90 days with a precision which is comparable to the original mission. K2 is a bona fide community endeavour: without a central, pre-determined research plan, exoplanets, stellar (asteroseismic and other variable) sources, extragalactic objects, microlensing events, and even Solar System objects have been proposed by community members. This opened up new possibilities in RR Lyrae research, too, since compared to the original Kepler field, where roughly 50 RR Lyrae stars were found, hundreds or even thousands of RR Lyrae stars can be observed in K2. Members of the Kepler Asteroseismic Science Consortium (KASC) RR Lyrae and Cepheids Working Group initiatied a large survey to ensure that as many RR Lyrae stars will be observed with K2, as possible. This contribution discusses the present status of this undertaking.

2 The K2 RR Lyrae survey

The project has been very successful in proposing and observing RR Lyrae stars with K2, see figure 1 for a graphical representation of observed an planned RR Lyrae stars along the Ecliptic, while figure 2 demonstrates the number of approved/observed RR Lyrae targets [2] and also the success rate of each submitted proposal up to the submission of this contribution. Very briefly we can say that apart from some instrumental problems (shift in the field-of-view, high contamination in the Galactic Bulge) almost all our proposed targets have been accepted and observed. The total number of RR Lyrae targets is over 3400 up to Campaign 13.

The sample covers RR Lyrae in extragalaxies (Sagittarius dwarf galaxy, Leo IV [3]), globular clusters (M80, M4), the Galactic Bulge (Campaign 9), and several Galactic fields along the Ecliptic both in high and low galactic latitudes, the Galactic center and anticenter directions, halo and field stars, as well as prominent halo structures, such as the Sagittarius stream.

The advantage of the K2 RR Lyrae survey is at least threefold. First, it will be a statistical mission. The project will allow us to study the occurrence rate and statistics of newly found dynamical phenomena, like period doubling, resonances, chaos, modulations, nonradial modes and their correlation with age, metallicity, and galactic position. Besides the statistics factor, this extensive survey will also provide rare gems, such as (classical) double-mode stars, stars with anomalous period ratios, rare types of modulation, modulated RRc stars, and so on. Thirdly, the high quality light curves will provide a golden sample to calibrate and test the algorithms and performance of and exploit synergies with upcoming large sky surveys, like Gaia [4] and the Large Synoptic Survey Telescope (LSST) [5].

⋆e-mail: szabo.robert@csfk.mta.hu

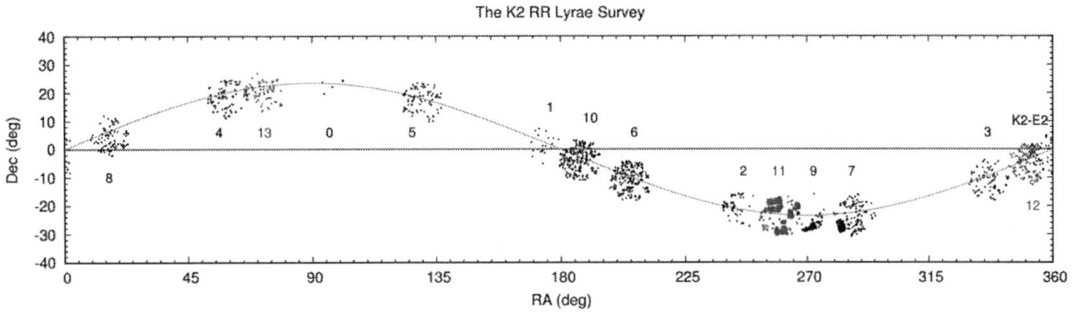

Figure 1: The footprint of the K2 RR Lyrae survey on the sky. Approved (blue) and observed (black) RR Lyrae targets are marked in the K2 campaign fields.

3 Preliminary results

While the K2 RR Lyrae survey is ongoing, some results have already been published or will be published soon. Here we summarize the most important highlights so far:

1. We expect to observe a sizeable sample of RRd stars, that were not present in the original Kepler field, and so far have been observed only with MOST [6] and CoRoT [7].

2. During the 8.9-d long K2 Two-Wheel Concept Engineering Test 33 RR Lyrae stars have been observed [8]. Besides fundamental-mode (RRab) and first overtone (RRc) pulsators, the sample includes the first two double-mode (RRd) stars that Kepler detected. All RRd and non-modulated RRc stars show the additional mode with a period ratio of 0.61 with the first overtone radial mode. As a by-product of the target search and analysis, we identified 165 bona fide double-mode RR Lyrae stars from the Catalina Sky Survey observations throughout the sky, 130 of which are new discoveries.

3. We also discovered a peculiar double-mode (RRd) star, both modes of which are modulated in antiphase. This is hitherto the only modulated RRd star that can be studied with accurate space-borne photometry (Plachy et al. in this proceedings). The results of a detailed investigations will be published in a dedicated paper (Plachy et al. in prep).

4. A strongly modulated (Blazhko) RRc star was found [8]. This is the only modulated first-overtone RR Lyrae ever observed from space, providing a unique window on the mechanism of the Blazhko-modulation in RRc stars.

5. We presented the first observations of extragalactic pulsating stars in the K2 ecliptic survey of the Kepler space telescope. The dwarf spheroidal galaxy Leo IV contains three known RR Lyrae stars. All three were observed by K2 during Campaign 1. These objects are among the faintest targets (Kp 21.5 mag, but see also [9]) the Kepler/K2 missions have observed. We identified one modulated star and another likely Blazhko candidate with periods of 29.8+-0.9 d and more than 80 d, respectively. Therefore we unambiguously detected Blazhko modulation in an extragalaxy beyond the Magellanic Clouds [3]. Two out of three stars are blends with brighter background galaxies in the K2 frames, hence image subtraction was a vital part of the analysis, and will be for future space photometric missions with large pixel sizes (e.g. TESS [10], PLATO [11]), as well.

6. Thanks to high-precision, continuous space-borne as well as ground-based photometry the Petersen-diagram of RR Lyrae stars has started to be populated with exotic and new types of periodicities (see e.g. Molnár et al. in these proceedings). Besides the half-integer frequencies due to period-doubling [12], extra frequencies have been found around the expected location of the second radial overtone frequency [13], around $P_X/P_1 = 0.61$ [14], where P_1 and P_X are the periods of first radial overtone and a presumably nonradial mode which seems to be ubiquitous in RRc and RRd stars [15]. In addition, anomalous first overtone / fundamental mode (P_1/P_0) period ratios were uncovered [16], [17], [18], [19].

All these additional frequencies that have been unveiled in recent years can be studied in exquisite details within the K2 RR Lyrae survey.

4 Conclusions

We introduced the K2 RR Lyrae survey, which will provide excellent means to study the pulsation mechanism and dynamical phenomena of these old, horizontal-branch objects, as well as to conduct galactic structure studies, since many of our targets trace different (extra)galactic components, like globular clusters, halo streams, and dwarf galaxies. The synergies with Gaia and LSST are numerous. Gaia will observe parallaxes and hence luminosities, but the temporal photometric coverage will be inferior compared to K2. LSST will go much deeper and will see much fainter objects, but the sampling rate will not allow the investigation of detailed temporal behavior of the pulsation. Combining the results of these surveys with the K2 sample will offer a unique opportunity to study RR Lyrae stars with a holistic approach. TESS and PLATO will continue and extend that work beyond the Ecliptic fields after K2 comes to an end.

We note that it is not trivial to derive the clean K2 light curves of large-amplitude variables (free of instrumental

DOI: 10.1051/epjconf/201716004004

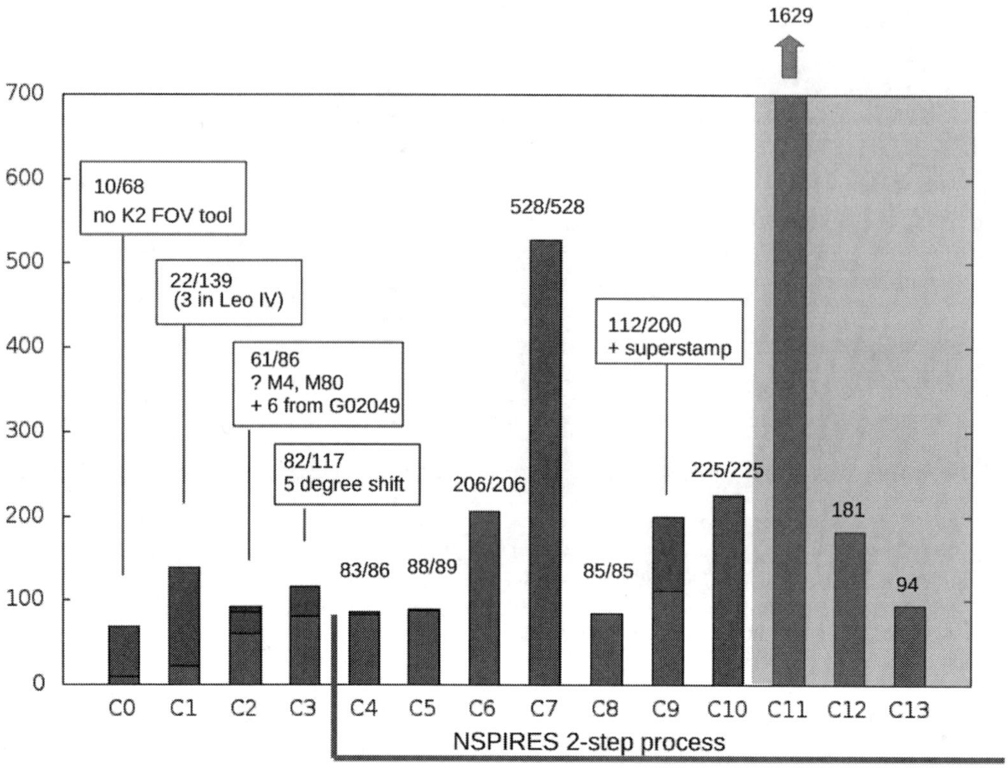

Figure 2: Statistics of the number of proposed (blue) and observed (green) RR Lyrae targets in the K2 campaigns. The K2 RR Lyrae survey has been largely successful in securing most of the known RR Lyrae stars for K2 observations.

effects), see e.g. the contribution of Plachy et al. on the EAP photometric method in these proceedings. Most of the published algorithms are optimized for transit search or the detection of low-amplitude solar-like oscillations. As a service to the community, the corrected K2 RR Lyrae light curves will be publicly available.

Acknowledgements

This project has been supported by the Lendület2009, LP2012-31, and LP2014-17 Young Researchers Programs of the Hungarian Academy of Sciences and by the NKFIH K115709, PD116175, and PD121203 grants of the Hungarian National Research, Development and Innovation Office. The research leading to these results has received funding from the European Community's Seventh Framework Programme (FP7/2007-2013) under grant agreement no. 312844 (SPACEINN), and the ESA PECS Contract No. 4000110889/14/NL/NDe. L. Molnár was supported by the János Bolyai Research Scholarship of the Hungarian Academy of Sciences.

References

[1] Howell, S., Sobeck, C., Haas, M. et al., PASP **126**, 398, (2014)

[2] Plachy, E., Molnár, L., Szabó, R., Kolenberg, K., Bányai, E., CoKon **105**, 19, (2016)

[3] Molnár, L., Pál, A., Plachy, E. et al., ApJ **812**, 2, (2015a)

[4] Gaia Collaboration A&A, in press, arXiv:1609.04153, (2016)

[5] Ivezić, Ž., Tyson, J. A., Abel, B. et al. arXiv:0805.2366, (2008)

[6] Gruberbauer, M., Kolenberg, K., Rowe, J. F. et al., MNRAS **379**, 1498, (2007)

[7] Chadid, M., A&A **540**, 68 (2012)

[8] Molnár, L., Szabó, R., Szabó, Gy. M. et al., MNRAS **452**, 4283, (2015b)

[9] Pál, A., Szabó, R., Szabó, Gy. M. et al., ApJL **804**, L45, (2015)

[10] Ricker, G. R., Winn, J., Vanderspek, R. et al., Journal of Astronomical Telescopes, Instruments, and Systems **1**, id. 014003 (2015)

[11] Rauer, H., Catala, C., Aerts, C. et al: ExpAstr **38**, 249, (2014)

[12] Szabó, R., Kolláth, Z., Molnár, L. et al., MNRAS **409**, 1244, (2010)

DOI: 10.1051/epjconf/201716004004

[13] Benkő, J. M., Kolenberg, K., Szabó, R. et al., MNRAS **409**, 1585, (2010)
[14] Moskalik, P., Smolec, R., Kolenberg, K. et al., MNRAS **447**, 2348, (2015)
[15] Szabó, R., Benkő, J. M., Paparó, M. et al., A&A **570**, A100, (2014)
[16] Molnár, L., Kolláth, Z., Szabó, R. et al., ApJL **757**, L13, (2012)
[17] Soszyński, I., Smolec, R., Dziembowski, W. A. et al., MNRAS **463**, 1332, (2016)
[18] Jurcsik, J., Smitola, P., Hajdu, G., Nuspl, J. ApJ **797**, 3, (2014)
[19] Jurcsik, J., Smitola, P., Hajdu, G. et al., ApJS **219**, 25, (2015)

 DOI: 10.1051/epjconf/201716004005

Rapidly rotating red giants

Charlotte Gehan[1]*, *Benoît* Mosser[1], and *Eric* Michel[1]

[1]*Observatoire de Paris, LESIA, Université Pierre et Marie Curie, Université Paris Diderot, PSL*

Abstract. Stellar oscillations give seismic information on the internal properties of stars. Red giants are targets of interest since they present mixed modes, wich behave as pressure modes in the convective envelope and as gravity modes in the radiative core. Mixed modes thus directly probe red giant cores, and allow in particular the study of their mean core rotation.
The high-quality data obtained by CoRoT and *Kepler* satellites represent an unprecedented perspective to obtain thousands of measurements of red giant core rotation, in order to improve our understanding of stellar physics in deep stellar interiors.
We developed an automated method to obtain such core rotation measurements and validated it for stars on the red giant branch. In this work, we particularly focus on the specific application of this method to red giants having a rapid core rotation. They show complex spectra where it is tricky to disentangle rotational splittings from mixed-mode period spacings. We demonstrate that the method based on the identification of mode crossings is precise and efficient. The determination of the mean core rotation directly derives from the precise measurement of the asymptotic period spacing $\Delta\Pi_1$ and of the frequency at which the crossing of the rotational components is observed.

1 Introduction

Red giant stars are known to exhibit solar-like oscillations, corresponding mostly to pressure modes resulting from acoustic waves stochastically excited by turbulent convection in the external envelope. These low-frequency oscillations are described by the universal red giant oscillation pattern [1]. In combination with effective temperatures, global seismic parameters of pressure modes provide precise estimates of stellar masses and radii which are independent of modelling [2].

However, the space based CoRoT and *Kepler* missions have revealed that red giants also present mixed modes, which behave as pressure modes in the convective envelope and as gravity modes in the radiative interior [3]. Mixed modes thus probe the physical conditions in stellar interiors. In particular, dipole mixed modes allow us to identify the evolutionary stage of evolved stars: it is now possible to distinguish core-helium burning stars from hydrogen-shell burning stars [4]. They moreover provide the measurement of the asymptotic period spacing $\Delta\Pi_1$ [5] which is linked to the size of the core [6]. Their observation also gives access to the differential-rotation profile in red giants [7].

Rotation strongly impacts stellar structure and evolution [8]. The measurement of the mean core rotation of about 300 red giants analyzed by [9] has revealed that a very efficient angular momentum transport process from the core to the envelope is at work in red giants: the mean core rotation slows down as stars evolve along the red giant branch (RGB) while their core is contracting. It is thus crucial to study the evolution of the core rotation for a set of red giants as large as possible in order to reach a deep understanding of the physical mechanisms governing the most inner regions of stars.

The era of massive high-precision photometric data from space initiated by CoRoT and *Kepler* satellites opens the way to thousands of measurements of red giant core rotation. We aim at developing a method allowing an efficient, rapid and automated measurement of red giant core rotation, based on the stretching of the spectra in order to disentangle the complex mixed-mode spectra. In this work we focus in particular on red giants with a rapid core rotation, where the rotational splittings are larger than half the mixed-mode spacing and complicate the interpretation of the spectra.

2 Disentangling red giant spectra in the case of rapidly rotating red giants

When the rotational splitting exceeds half the mixed-mode period spacing, it is very difficult to disentangle the rotational multiplet components from the mixed modes. A part of the difficulty comes from the uneven mixed-mode pattern: pressure dominated mixed-modes are nearly equally spaced in frequency, with a spacing close to

*e-mail: charlotte.gehan@obspm.fr

 DOI: 10.1051/epjconf/201716004005

the large separation $\Delta\nu$, whereas gravity dominated mixed-modes are nearly equally spaced in period, with a spacing close to the gravity mode-period spacing $\Delta\Pi_1$ [5].

In the Astro Fluid 2016 Conference Proceedings edited by EAS Publications Series (in press), we present in detail the method we developed in order to obtain automated measurements of the mean core rotation of red giants, and validate it for stars on the RGB, for which automated measurements are fully consistent with manual measurements [5]. The principle of the method is based on the stretching of frequency spectra, which ensures that mixed modes are regularly spaced. The frequency is changed into

$$d\tau = \frac{d\nu}{\zeta \nu^2}, \tag{1}$$

where τ is the stretched period [10], and ζ is defined by

$$\zeta = \left[1 + \frac{1}{q}\frac{\nu^2 \Delta\Pi_1}{\Delta\nu}\frac{\cos^2\left[\pi\frac{1}{\Delta\Pi_1}\left(\frac{1}{\nu} - \frac{1}{\nu_g}\right)\right]}{\cos^2\left(\pi\frac{\nu-\nu_p}{\Delta\nu}\right)}\right]^{-1}, \tag{2}$$

where q is the coupling parameter of mixed modes, $\Delta\Pi_1$ is the gravity mode period spacing, $\Delta\nu$ is the large separation, ν_g are the pure dipole gravity mode frequencies, ν_p are the pure pressure mode frequencies.

For the pure dipole gravity mode frequencies ν_g we can use the first-order asymptotic expansion [11]

$$\frac{1}{\nu_g} = -(n_g + \varepsilon_g)\,\Delta\Pi_1, \tag{3}$$

where n_g is the gravity radial order usually defined as a negative value, and ε_g is a small but complicated function sensitive to the stratification near the boundary between the radiative core and the convective envelope.

For the pure pressure mode frequencies ν_p we use the universal red giant oscillation pattern [1]

$$\nu_p = \left(n_p + \frac{1}{2} + \varepsilon_p + d_{0\ell} + \frac{\alpha}{2}[n_p - n_{max}]^2\right)\Delta\nu, \tag{4}$$

where n_p is the pressure radial order, ε_p is the pressure mode offset, $d_{0\ell}$ is the small separation, α represents the curvature of the oscillation pattern, and $n_{max} = \nu_{max}/\Delta\nu - \varepsilon_p$ is the non-integer order at the frequency ν_{max} of maximum oscillation signal.

In the stretched spectra, rotation induces a small departure from an evenly spaced pattern. The stretched period spacing between rotational multiplet components with the same azimuthal order m expresses [10]

$$\Delta\tau_m = \Delta\Pi_1(1 + m x_{rot}) \tag{5}$$

It presents a small departure from $\Delta\Pi_1$, expressed by

$$x_{rot} = 2\,\zeta\,\frac{\delta\nu_{rot}}{\nu}, \tag{6}$$

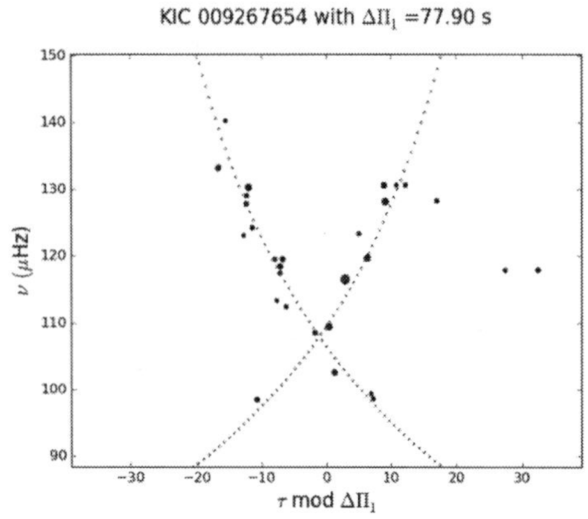

Figure 1: Échelle diagram for KIC 9267654 star with $\Delta\Pi_1$ = 77.9 s. The rotational multiplet component corresponding to the azimuthal order m = +1 is identified in red, and the rotational multiplet component corresponding to the azimuthal order m = − 1 is identified in green.

where $\delta\nu_{rot}$ is the rotational splitting and ν_{max} is the frequency of maximum oscillation signal.

It is then possible to build échelle diagrams based on this stretched period, where the different components of the rotational multiplet draw ridges and are disentangled. Hence, it is easy to obtain measurements of the mean core rotation.

In the present work, we examine the case of stars with a large rotational splitting, which complicates the frequency spectra pattern. In échelle diagrams representing ν as a function of τ modulo $\Delta\Pi_1$, a crossing of the different rotational multiplet components occurs when the rotational splitting is a multiple of half the mixed-mode frequency spacing. Figure 1 represents the échelle diagram of the rapid rotator KIC 9267654. This star is seen equator-on so only the two rotational multiplet components with the azimuthal orders $m = \pm 1$ are visible. We clearly identify a crossing of the components at the frequency ν = 108 μHz.

3 Crossing of rotational multiplet components

We aim at deriving a relation between the rotational splitting and the mixed-mode spacing which expresses the conditions leading to a crossing of the different multiplet components. The condition to have a crossing of the multiplet components is

$$\delta\nu_{rot} = k\,\frac{\delta\nu}{2}, \tag{7}$$

where $\delta\nu$ is the mixed-mode frequency spacing, and k is a positive integer.

The rotational splitting expresses [12]

$$\delta\nu_{rot} = \zeta\,\delta\nu_{rot,core} + (1-\zeta)\,\delta\nu_{rot,env}, \tag{8}$$

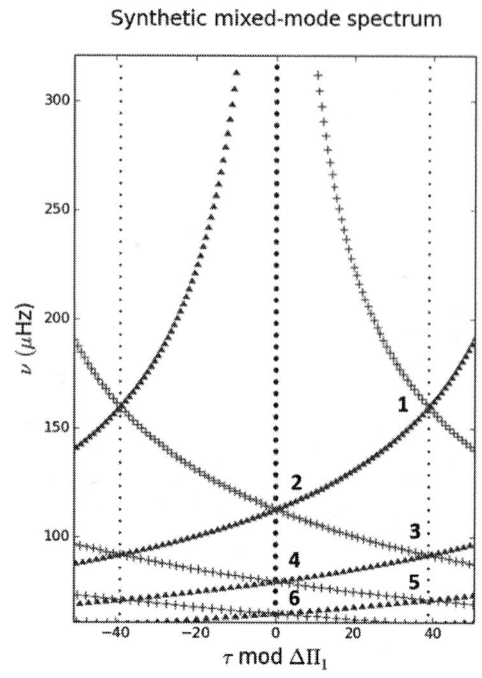

Figure 2: Échelle diagram obtained with Equations (5) and (6) for a synthetic mixed-mode spectrum with $\Delta\Pi_1 = 77.9$ s and $\delta\nu_{\rm rot} = 1\ \mu$Hz. The rotational multiplet components corresponding to the azimuthal orders $m = \{-1, 0, 1\}$ are respectively represented by green triangles, large blue dots and red crosses. Vertical black dotted lines represent the $\tau = (\pm\Delta\Pi_1/2) \bmod \Delta\Pi_1$ values. The crossing orders are identified by black numbers.

where $\delta\nu_{\rm rot,core}$ and $\delta\nu_{\rm rot,env}$ are the rotational splittings that represent respectively the mean rotation of the radiative core and the mean rotation of the convective envelope.

Red giants have a radiative core rotating much faster than the envelope. When the surface rotation can be neglected, we can rewrite Equation (8) as

$$\delta\nu_{\rm rot} = \zeta\,\delta\nu_{\rm rot,core}, \tag{9}$$

which means that the rotational splittings associated to the core rotation are modulated along the oscillation spectra by the ζ function.

The period spacing between consecutive mixed-modes expresses [10]

$$\Delta P = \zeta\,\Delta\Pi_1. \tag{10}$$

The relation between P and ν expresses

$$\frac{{\rm d}P}{P} = -\frac{{\rm d}\nu}{\nu}. \tag{11}$$

Hence the frequency spacing $\delta\nu$ between consecutive mixed-modes expresses

$$\delta\nu = \nu^2\,\Delta P. \tag{12}$$

Thus we can rewrite Equation (7) with Equation (9) and Equation (12) to obtain the condition to have a crossing of the rotational multiplet components. This crossing at frequency $\nu_{\rm k}$ is linked to the global seismic properties $\delta\nu_{\rm rot,core}$ and $\Delta\Pi_1$ by

$$\zeta\delta\nu_{\rm rot,core} = k\,\frac{\nu_{\rm k}^2\,\zeta\,\Delta\Pi_1}{2}, \tag{13}$$

where the $\nu_{\rm k}$ are the frequencies where a crossing of the rotational multiplet components occurs at the order k. This relation simplifies into

$$\delta\nu_{\rm rot,core} = k\,\frac{\nu_{\rm k}^2\,\Delta\Pi_1}{2}. \tag{14}$$

We note that the ζ function is not involved in the measurement of the rotational splitting. The $\Delta\Pi_1$ parameter has been measured for about 5000 stars [13]. We just need to identify in échelle diagrams one of the frequencies $\nu_{\rm k}$ corresponding to a crossing of the multiplet components and the order of the crossing k in order to obtain a precise estimate of the core rotation of these rapidly rotating red giants.

4 Measurement of the rotational splitting

The next step consists in identifying the order k of the crossing of the rotational multiplet components.

4.1 Crossings of the rotational multiplet components in a synthetic spectrum

Figure 2 represents an échelle diagram for a synthetic mixed-mode spectrum based on Equations (5) and (6). We note that several crossings of the multiplet components occur at different frequencies. The order of the crossing decreases with frequency. At high frequency, the multiplet components are almost parallel and no crossing is possible. We note that the frequency spacing of the crossings increases with frequency. The parity of the order k modifies the pattern drawn by the multiplet components in the échelle diagram. When k is an even number the three components corresponding to the azimuthal order values $m = \{-1, 0, 1\}$ overlap at $\tau \bmod \Delta\Pi_1 = 0$, while when k is an odd number only the $m = \pm 1$ components overlap at $\tau = (\pm\Delta\Pi_1/2) \bmod \Delta\Pi_1$.

Usually, the number of crossings in the range where mixed modes are excited is low and only one crossing is visible, as it is the case for KIC 9267654 star. The existence or the absence of possible crossings at $\tau = (\pm\Delta\Pi_1/2) \bmod \Delta\Pi_1$ allows the identification of the crossing order k.

4.2 Identification of the order of the observed crossing

The identification of the order k of the crossing of the rotational multiplet components is achieved by computing the

 DOI: 10.1051/epjconf/201716004005

Table 1: Determination of the order k of the crossing

k	ν_{k-1} (μHz)	ν_k (μHz)	ν_{k+1} (μHz)	$\delta\nu_{rot,core}$ (nHz)
1	-	108.0	76.4	454
2	152.7	108.0	88.2	909
3	132.3	108.0	93.5	1363
4	124.7	108.0	96.6	1817

Order k of the observed crossing, frequencies $\{\nu_{k-1}, \nu_k, \nu_{k+1}\}$ where a crossing occurs and associated $\delta\nu_{rot,core}$ values for KIC 9267654 star having $\Delta\Pi_1 = 77.9$ s.

Table 2: Measurement of rotational splittings

KIC	k	$\Delta\Pi_1$ (s)	ν_{k-1} (μHz)	ν_k (μHz)	ν_{k+1} (μHz)	$\delta\nu_{rot,core}$ (nHz)
5380775	3	78.20	91.9	75.0	65.0	660
6949009	2	72.79	113.1	80.0	65.3	466
7257175	2	72.60	134.4	95.0	77.6	655
7267119	3	73.90	98.0	80.0	69.3	709
9227589	4	76.00	207.8	180.0	161.0	4925
10387370	3	70.80	89.4	73.0	63.2	566

frequencies where crossings occur. These frequencies are given by

$$\nu_k = \sqrt{\frac{2\,\delta\nu_{rot,core}}{k\,\Delta\Pi_1}}. \qquad (15)$$

In order to correctly span the possibilites, the different possible k values are explored for a fixed $\delta\nu_{rot,core}$ value obtained via Equation (14). Then the frequencies ν_k which match the observations provide the value of the order k. An example is given in Table 1 for KIC 9267654 star. For this star, the order of the crossing is most likely $k = 1$ and the corresponding rotational slitting value $\delta\nu_{rot,core} = 454$ nHz, since no crossing occurs for frequencies greater than 108 μHz.

Table 2 gives the measured rotational splitings for other stars presenting overlaping rotational multiplet components. We note that the identication of the frequency where multiplet components overlap leads to the measurement of high core rotation rates as large as 5 μHz, as it is the case for KIC 9227589.

5 Uncertainties

The uncertainty on $\delta\nu_{rot}$ expresses

$$\frac{\delta(\delta\nu_{rot,core})}{\delta\nu_{rot,core}} = \sqrt{\left[\frac{\delta(\Delta\Pi_1)}{\Delta\Pi_1}\right]^2 + 4\left[\frac{\delta\nu_k}{\nu_k}\right]^2} \qquad (16)$$

where $\delta(\Delta\Pi_1)$ and $\delta\nu_k$ are respectively the uncertainties on $\Delta\Pi_1$ and on ν_k.

Typical values for the relative uncertainties are $\delta(\Delta\Pi_1)/\Delta\Pi_1 \simeq 1\%$ and $\delta\nu_k/\nu_k \simeq 1\%$. Thus the relative uncertainty on $\delta\nu_{rot,core}$ is $\delta(\delta\nu_{rot,core})/\delta\nu_{rot,core} \simeq 3\%$.

6 Conclusions

By correcting frequency spectra into stretched period spectra where mixed modes are regularly spaced, we were able to disentangle intricated red giant spectra where the rotational splitting exceeds half the mixed-mode period spacing. These large rotational splittings are the signature of large core rotation values, which induce crossings of the rotational multiplet components in stretched period échelle diagrams. This method succeeds in measuring precise high core rotation rates with the knowledge of the gravity mode period spacing $\Delta\Pi_1$, by identifying the frequency corresponding to a crossing of the multiplet components. The relative uncertainty obtained on the rotational splitting measurement is around 3%.

References

[1] B. Mosser, E. Michel, K. Belkacem et al., A&A **550**, A126 (2013)
[2] T. Kallinger, B. Mosser, S. Hekker et al., A&A **522**, A1 (2010)
[3] P.G. Beck, T.R. Bedding, B. Mosser et al., Science **332**, 205 (2011)
[4] B. Mosser, O. Benomar, K. Belkacem et al., A&A **572**, L5 (2014)
[5] B. Mosser, M.J. Goupil, K. Belkacem et al., A&A **540**, A143 (2012)
[6] J. Montalbán, A. Miglio, A. Noels et al., ApJ **766**, 118 (2013)
[7] P.G. Beck, J. Montalban, T. Kallinger et al., Nature **481**, 55 (2012)
[8] N. Lagarde, T. Decressin, C. Charbonnel et al., A&A **543**, A108 (2012)
[9] B. Mosser, M.J. Goupil, K. Belkacem et al., A& A **548**, A10 (2012)
[10] B. Mosser, M. Vrard, K. Belkacem et al., A&A **584**, A50 (2015)
[11] M. Tassoul, ApJS **43**, 469 (1980)
[12] M.J. Goupil, B. Mosser, J.P. Marques et al., A&A **549**, A75 (2013)
[13] M. Vrard, B. Mosser, R. Samadi et al., A&A **588**, A87 (2016)

DOI: 10.1051/epjconf/201716004006

Evolutionary states of red-giant stars from grid-based modelling

Saskia Hekker[1,2,⋆], *Yvonne* Elsworth[3,2], *Sarbani* Basu[4], and *Earl* Bellinger[1,2,4]

[1] *Max Planck Institute for Solar System Research, Göttingen, Germany*
[2] *Stellar Astrophysics Centre, Aarhus, Denmark*
[3] *University of Birmingham, Birmingham, United Kingdom*
[4] *Yale University, New Haven, United States of America*

Abstract. From its surface properties it can be difficult to determine whether a red-giant star is in its helium-core-burning phase or only burning hydrogen in a shell around an inert helium core. Stars in either of these stages can have similar effective temperatures, radii and hence luminosities, i.e. they can be located at the same position in the Hertzsprung-Russell diagram. Asteroseismology – the study of the internal structure of stars through their global oscillations – can provide the necessary additional constraints to determine the evolutionary states of red-giant stars. Here, we present a method that uses grid-based modelling based on global asteroseismic properties (ν_{max}, frequency of maximum oscillation power; and $\Delta\nu$, frequency spacing between modes of the same degree and consecutive radial orders) as well as effective temperature and metallicity to determine the evolutionary phases. This method is applicable even to timeseries data of limited length, although with a small fraction of miss-classifications.

1 Introduction

The long near-uninterrupted photometric timeseries of *Kepler* data have unambiguously shown the presence of mixed pressure-gravity oscillation modes. These modes provide information about the stellar structure of red giants and from that allow to identify the evolutionary state of a red giant [1, 12], i.e. whether the star is on the red-giant branch (RGB) or in the He-core burning phase. However, resolving individual mixed modes is not always feasible in shorter timeseries data such as those of the K2 [7] and TESS [16] missions. Here we present a method that does not rely on the detection of the individual mixed modes, but uses grid-based modelling (GBM).

GBM is the technique in which a set of observed parameters of a star are compared to these parameters obtained from a grid of stellar models to infer in this way the most likely stellar parameters of the observed star. In asteroseismic GBM of red giants the observed parameters are generally ν_{max} (frequency of maximum oscillation power), $\Delta\nu$ (frequency spacing between modes of the same degree and consecutive radial orders), effective temperature and metallicity.

Within the asteroseismology community there is a general thought that for GBM applied to red-giant stars it is important to know a priori the evolutionary phase, i.e., whether a star is in the hydrogen-shell burning red-giant branch (RGB) phase or in the He-core burning phase. This knowledge is important to avoid biases in the resulting stellar parameters, i.e., mass, radius, age etc. Here, we follow this reasoning in the opposite direction and seek to answer the question whether it is possible to determine the evolutionary phases of red-giant stars from grid-based modelling using as input $\Delta\nu$, ν_{max}, T_{eff} and [Fe/H]. In other words, we want to determine whether or not it is possible to find loci in this four dimensional space where RGB or He-core burning stars are uniquely defined, thereby allowing us to provide a reliable identification of the evolutionary phase.

An earlier version of the method presented here has been applied together with the methods by [4, 9, 12] to the APOKASC stars [14, see also the contributions by Marc Pinsonnealt and Yvonne Elsworth in these proceedings] to provide consolidated evolutionary phases.

⋆e-mail: Hekker@mps.mpg.de

2 Methodology

By using GBM to find the evolutionary phase we are searching in a four-dimensional ($\Delta\nu$, ν_{max}, T_{eff}, [Fe/H]) space for loci that uniquely define the evolutionary phase. To do so we have used a grid of BASTI models [13] to which we applied a Support Vector Machine (SVM). We did this in the four-dimensional space described above as well as in a three-dimensional space where we used the ratio $\Delta\nu/\nu_{max}$ together with T_{eff}, [Fe/H]. The distribution of the models in this reduced parameter space is shown in figure 1. More details of the models and the SVM are discussed below.

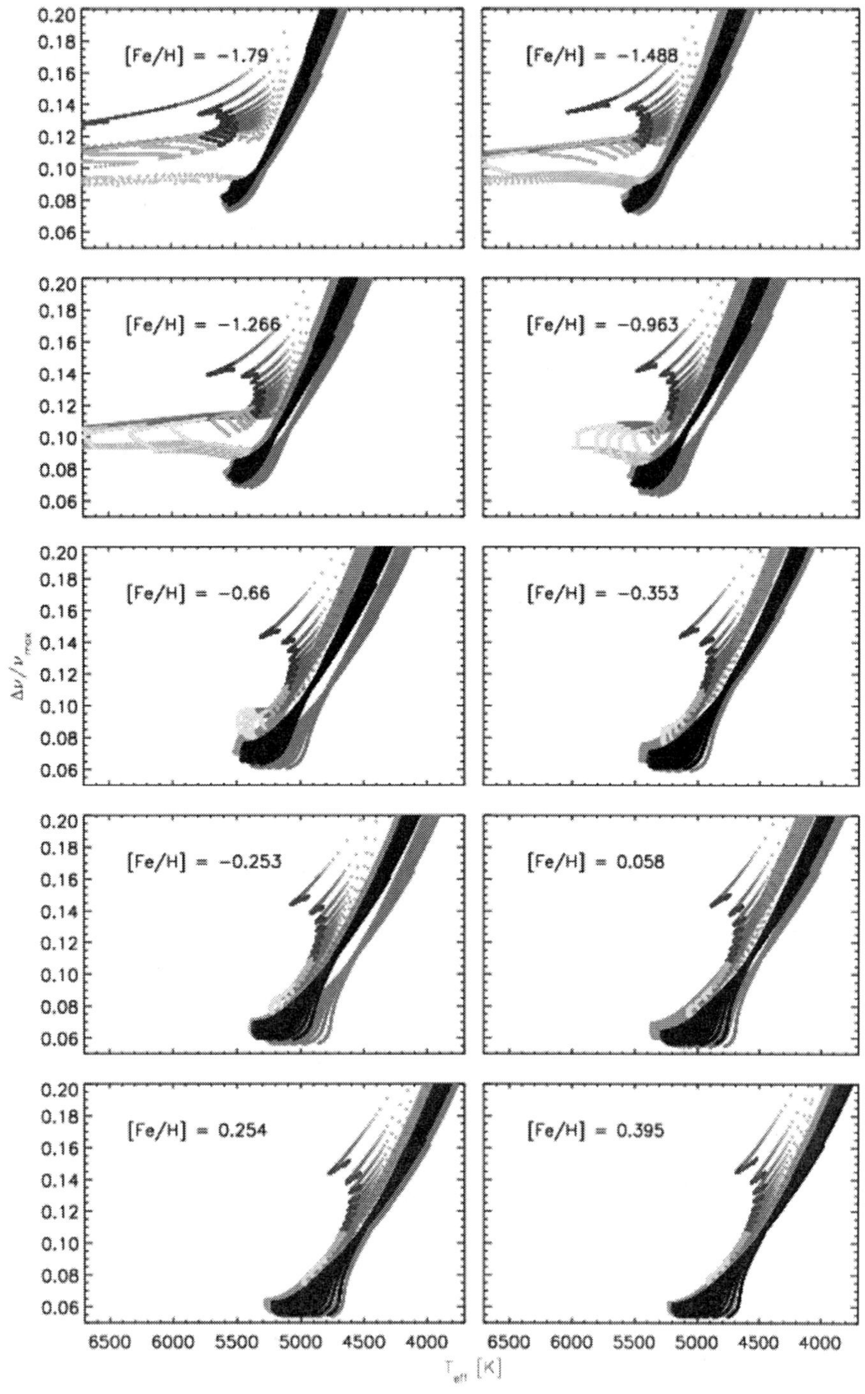

Figure 1: $\Delta\nu/\nu_{\max}$ as a function of $T_{\rm eff}$ for different [Fe/H] values as indicated in the legend of each panel for BASTI models. Models of RGB stars are indicated in black. He-burning stars with He core mass fraction > 0.1 are indicated in red, orange and yellow for models with masses < 1.5 $M_{\odot}$, masses between 1.5 and 2.0 $M_{\odot}$ and masses 2.0 and 2.5 $M_{\odot}$, respectively. He-core burning stars with He core mass fraction between 0.0 and 0.1 are shown in gray. The models in light blue (pink) indicate RGB models with lower (higher) metallicities in the grid. In this figure we have defined RGB stars as those stars that lie between the base and the tip of the red-giant branch as defined by evolutionary key points 5 and 8 in the BaSTI documentation.

2.1 BaSTI models

The models are from the canonical BaSTI grid[1] [13] using an updated version of the code described by [2] and [17]. The grid spans masses from 0.7 $M_\odot$ to 2.5 $M_\odot$ in steps of 0.1 $M_\odot$ and metallicities of Z = 0.0003, 0.0006, 0.001, 0.002, 0.004, 0.008, 0.01, 0.0198, 0.03 and 0.04.

To compute $\Delta\nu$ and $\nu_{\rm max}$ for the models we use the scaling relations [11]:

$$\Delta\nu \propto \sqrt{\frac{M}{R^3}} \tag{1}$$

$$\nu_{\rm max} \propto \frac{M}{R^2\sqrt{T_{\rm eff}}} \tag{2}$$

for which we used the following values as references: $\nu_{\rm max,ref} = \nu_{\rm max,\odot} = 3090\mu$Hz [8], and $\Delta\nu_{\rm ref}$ computed by [6] to mitigate the known temperature and metallicity dependence.

2.2 Support Vector Machine

To find loci that uniquely define the evolutionary phases of red giants we train a support vector machine (SVM) for classification. SVMs are supervised learning models that are based on the concept of decision planes that define decision boundaries, with support vectors being the training data that are closest to the decision boundary. A decision plane is a plane that separates a set of objects having different class memberships. The basic idea behind SVM is to map the original objects using a set of mathematical functions, known as kernels, from the real space to a high-dimensional feature space in which classes can be separated. In this case we used a Gaussian radial basis function as a kernel:

$$\phi(r) = e^{-(\epsilon r)^2} \tag{3}$$

with $r = \|x_j - x_i\|$ being the distance between the input values of x_j and x_i, and ϵ is an adjustable parameter.

In addition to the kernels one can choose the type of SVM. We used a C-type SVM in which training involves minimisation of the following error function:

$$\frac{1}{2}w^T w + c\Sigma_{i=1}^{N}\xi_i \tag{4}$$

subject to the constraints:

$$y_i(w^T\phi(x_i) + b) \geq 1 - \xi_i \tag{5}$$

in which $\xi_i \geq 0$, $i = 1, ..., N$, c is the capacity constant, w is the vector of coefficients, b is a constant, and ξ_i represents parameters for handling non-separable input data, i.e. training data that cannot be classified correctly. The index i labels the N training cases.

A SVM searches for a hyperplane with the largest minimum margin separating the two classified groups. At the same time a SVM searches for a hyperplane that correctly separates as many instances as possible. The problem is that these requirements cannot always be accomplished at the same time. The capacity constant determines the balance between these requirements, with a low value of c favouring towards a larger minimum margin and a high value of c favouring the largest number of correct classifications in the training data.

For more details about SVM we refer to the original SVM paper [18] and the paper that made them computationally viable [3]. Additionally, both radial basis functions and SVM are covered in detail in [5].

3 Results

For our current set-up we used the R [15] library kernlab [10] to train a C-type classification SVM, with $c = 1$ and the a Gaussian radial base kernel as described above. In this set-up we trained models with masses in the range 0.75 $M_\odot \leq M \leq$ 1.75 $M_\odot$ and used as input parameters $T_{\rm eff}$, $\Delta\nu/\nu_{\rm max}$ and [Fe/H] (see Figure 1). For these models we can show that the RGB and He-core burning stars are indeed uniquely defined, i.e., we find a training error of 0 in the ideal case without uncertainties. We repeated the same experiment using a four-dimensional parameters space with $\Delta\nu$, $\nu_{\rm max}$, $T_{\rm eff}$, [Fe/H] as input data. In this case we could not reduce the training error to 0. This indicates that the four-dimensional input space is too complex to find a decision boundary that correctly separates all input data, while this is possible in the reduced three-dimensional parameter space.

4 Prospects

The determination of the evolutionary phase of red-giant stars based on grid-based modelling seems very promising. A preliminary estimate for the success rate of this method is on the order of 80 to 90%. As this method is applicable to relatively short asteroseismic timeseries, i.e., 30 to 80 days, as obtained with the TESS and K2 missions for which only $\Delta\nu$ and $\nu_{\rm max}$ can be extracted, this can be of significant added value.

The preliminary analysis presented here will need to undergo further tests to investigate the importance of observational uncertainties on the classification as well as the impact of different stellar evolution models. Results of the full series of tests and validations as well as success rate of this method will be published in a forthcoming paper.

Acknowledgements

The research leading to the presented results has received funding from the European Research Council under the European Community's Seventh Framework Programme (FP7/2007-2013) / ERC grant agreement no 338251 (StellarAges). YE acknowledge the support of the UK Science and Technology Facilities Council (STFC). SB acknowledges support from NSF grant AST-1514676 and NASA grant NNX16AI09G. Funding for the Stellar Astrophysics Centre (SAC) is provided by The Danish National Research Foundation (Grant agreement no.: DNRF106). This work has made use of BaSTI web tools.

[1] http://albione.oa-teramo.inaf.it/

References

[1] Bedding, T. R., Mosser, B., Huber, D., et al. 2011, Nature,, 471, 608

[2] Cassisi, S., & Salaris, M. 1997, MNRAS,, 285, 593

[3] Cortes, C, Vapnik, V, 1995, Machine learning 20.3, 273

[4] Elsworth, Y, Hekker, S, Basu, S, Davies, G. R. submitted to MNRAS,

[5] Friedman, J, Hastie, T, Tibshirani, R, 2001, The Elements of Statistical Learning, Vol. 1. Springer, Berlin

[6] Guggenberger, E., Hekker, S., Basu, S., & Bellinger, E. 2016, MNRAS,, 460, 4277

[7] Haas, M. R., Barclay, T., Batalha, N. M., et al. 2014, American Astronomical Society Meeting Abstracts #223, 223, 228.01

[8] Hekker, S., Elsworth, Y., Mosser, B., et al. 2013, A&A,, 556, A59

[9] Kallinger, T., Hekker, S., Mosser, B., et al. 2012, A&A,, 541, A51

[10] Karatzoglou, A, Smola, A, Hornik, K, Zeileis, A, 2004, Journal of Statistical Software, 11, 1

[11] Kjeldsen, H., & Bedding, T. R. 1995, A&A,, 293, 87

[12] Mosser, B., Benomar, O., Belkacem, K., et al. 2014, A&A,, 572, L5

[13] Pietrinferni, A., Cassisi, S., Salaris, M., & Castelli, F. 2004, ApJ,, 612, 168

[14] Pinsonneault, M. H., Elsworth, Y., Epstein, C., et al. 2014, ApJS,, 215, 19

[15] R Core Team, 2014, R: A Language and Environment for Statistical Computing, R foundation for Statistical Computing

[16] Ricker, G. R., Winn, J. N., Vanderspek, R., et al. 2015, Journal of Astronomical Telescopes, Instruments, and Systems, 1, 014003

[17] Salaris, M., & Cassisi, S. 1998, MNRAS,, 298, 166

[18] Vapnik, V, Chervonenkis, A, 1964, Automation and remote control 25.1

 DOI: 10.1051/epjconf/201716004007

An Analysis of Pulsating Subdwarf B Star EPIC 203948264 Observed During Campaign 2 of *K2*

Laura Ketzer[1,*] and *Mike* Reed[1,**]

[1] *Department of Physics, Astronomy and Materials Science, Missouri State University, 901 S. National, Springfield, MO 65897, USA*

Abstract. We present a preliminary analysis of the newly-discovered pulsating subdwarf B (sdB) star EPIC 203948264. The target was observed for 83 days in short cadence mode during Campaign 2 of *K2*, the two-gyro mission of the *Kepler* space telescope. A time-series analysis of the data revealed 22 independent pulsation frequencies in the g−mode region ranging from 100 to 600 μHz (0.5 to 2.8 hours). The main method we use to identify pulsation modes is asymptotic period spacing, and we were able to assign all but one of the pulsations to either $\ell = 1$ or $\ell = 2$. The average period spacings of both sequences are 261.34 ± 0.78 s and 151.18 ± 0.34 s, respectively. The pulsation amplitudes range from 0.77 ppt down to the detection limit at 0.212 ppt, and are not stable over the duration of the campaign. We detected one possible low-amplitude, $\ell = 2$, rotationally split multiplet, which allowed us to constrain the rotation period to 46 days or longer. This makes EPIC 203948264 another slowly rotating sdB star.

1 Introduction

The K2 mission is the extension of the very successful *Kepler* mission. With only two remaining gyros, the spacecraft uses solar radiation pressure and regular thruster firings to ensure accurate pointing to within a pixel. Observing fields are restricted to lie along the ecliptic plane, where the spacecraft can observe them for around ~ 80 days at a time [1].

Despite some limitations of the *K2* mission, the unprecedented photometric data obtained by *Kepler* has enabled asteroseismologists to study pulsating subdwarf B stars (sdB) in great detail. These stars are located on the extreme horizontal branch and consist of helium-rich cores that are surrounded by thin hydrogen envelopes. The processes for stripping off the envelopes are not well understood, but a likely mechanism is binary interaction [2]. After leaving the horizontal branch, their hydrogen shells are too thin to sustain shell fusion and they proceed directly to the white dwarf cooling track. Subdwarf B stars typically have masses around $0.47 M_{\odot}$, and effective temperatures ranging from 20,000 to 40,000 K [3].

Many sdB stars have been found to exhibit stellar pulsations. The V361 Hya class are short-period, pressure(p)-mode pulsators with periods of a few minutes while the V1093 Her class are longer period, gravity (g)-mode pulsators with periods of one to a few hours. The V361 Hya stars tend to be hotter, the V1093 Her cooler, and hybrid pulsators, with both types of pulsations have intermediate temperatures. For a pre-Kepler pulsation review, see Østensen[4].

EPIC 203948264 was observed during Campaign 2 as part of our Guest Observer proposal. We obtained spectroscopy which indicated EPIC 203948264 to be in the g−mode instability region and these K2 observations indicate it to be a pulsator. Our goal for EPIC 203948264 was to identify the modes of the detected pulsations using asymptotic period spacings and rotationally induced frequency multiplets. Reed et al. discovered that sdB stars exhibit asymptotic period spacings in the g−mode region, with the $\ell = 1$ sequence equally spaced by about 250 s [5]. In addition, rotationally split multiplets have been used to identify the ℓ and m modes of pulsations [e.g. 6, 7]. Stellar rotation lifts the frequency degeneracy and gives rise to frequency multiplets whose splittings follow the relation $\Delta\nu = \Delta m\Omega(1 - C_{n,\ell})$. $\Delta\nu$ represents the shift from the central frequency ($m = 0$), Ω is the rotation frequency of the star, and the Ledoux constant for g−modes is given by $C_{n,\ell} \approx 1/[\ell(\ell + 1)]$ [8]. We can determine the Ledoux constant from period spacing and if we can observe and measure the splitting of multiplets, we can calculate the rotation period of the star.

2 The Data

Kepler observed EPIC 203948264 during Campaign 2 of *K2* in short cadence (SC) mode with an integration time of 58.85 seconds. The Kepler Science office does not

*e-mail: ketzer.laura@gmail.com
**e-mail: MikeReed@missouristate.edu

DOI: 10.1051/epjconf/201716004007

provide processed SC lightcurves and so we have developed our own custom processing pipeline to extract a lightcurve from the pixel files. Our custom method uses centroiding to determine the stellar center which we follow with round apertures of multiple sizes. We choose the aperture which optimizes the signal-to-noise. Similar to the *Kepler*–developed PyKE software [9], we use pixel–brightness correlations to remove flux variations caused by spacecraft-induced motion from the drift and thruster firings. We compare our custom processed data to PyKE extraced data in oder to search for low-level pulsations that are consistent in both methods.

3 The Analysis

Due to pointing adjustments in the beginning of Campaign 2, we processed only 79 days of the 83-day-long observing campaign. The 1/T temporal resolution of the data is 0.147 μHz, and we calculated a standard deviation of the Fourier spectrum to be 0.049 ppt. To make it statistically unlikely for peaks to be generated by noise, we chose a detection limit of 4.3σ (0.212 ppt) [10]. Based on simulated data, we chose a second detection threshold of 5.2σ (0.256 ppt) [11].

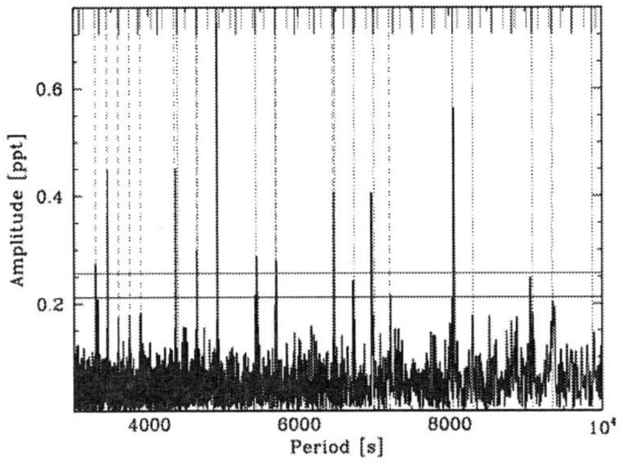

Figure 1: Period transform highlighting the evenly spaced periods in the g–mode region. The short green (blue) lines at the top indicate the asympic $\ell = 1$ (2) sequence. The dashed lines indicate which periods match the asympic sequences. The horizontal blue (red) line represents the 4.3σ (5.2σ) detection limit.

We plotted the Fourier spectrum and selected all the peaks above the detection threshold. Figure 1 shows a period transform of the 22 detected pulsations. Their periods range from 1660 to nearly 9900 s (101μ to 602μ Hz). Asymptotic period spacings have been detected in all Kepler-observed sdB g–mode pulsators [5] and they all showed similar spacings of around 250 s for the $\ell = 1$ sequence and 145 s for $\ell = 2$. Once we had our list of detected periods, we started testing for evenly spaced periods using a Kolmogorov-Smirnov (KS) test. This test has been successfully used in sdB stars to indicate common spacings in the pulsation periods. The result of the KS test revealed two common spacings: 260 s and 150 s. We pick out periods with these separations, assign relative overtones, n, and then do a linear regression fit. The solutions, 261.34 ± 0.78 for the $\ell = 1$ and 151.18 ± 0.34 s for the $\ell = 2$ sequence agree with asymptotic theory, $\prod_{\ell=2} = \prod_{\ell=1} / \sqrt{3}$ [8], and are in agreement with other *Kepler*–observed g–mode pulsators [5]. Using asymptotic period spacings we identified all but one of the pulsations as $\ell = 1$ or $\ell = 2$, with four of them matching both sequences. The unidentified period was the shortest one, which likely does not match the asymptotic criterion of $n \gg \ell$. We also searched for frequency multiplets with the goal of determining the rotation period of the star and identifying ℓ and m modes. In the case of EPIC 203948264, we were only able to find one possible multiplet, an $\ell = 2$ triplet with a splitting of 0.210 ± 0.004μHz.

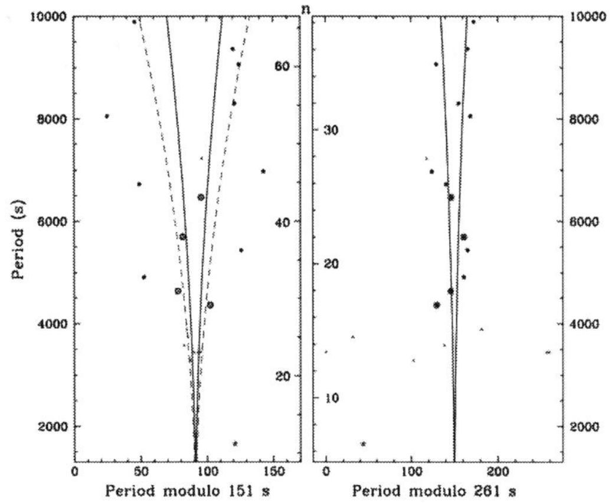

Figure 2: Échelle diagram for the $\ell = 2$ (left) and $\ell = 1$ (right) g-modes. Periods found to be $\ell = 1$ ($\ell = 2$) are shown in black (red). Periods that fit both sequences are displayed in both colors, and the unidentified period is shown as a blue star. The curved lines show how frequency splittings translate into period space.

From *Kepler* data, we know that sdB pulsators have amplitudes which can change substantially and show variability in frequency with time. Therefore we use sliding Fourier transforms (SFTs) to investigate the behavior of pulsations in the time domain. For EPIC 203948264, we used data spanning eight days with one day steps, and these are show n in Figure 3 for all regions of variability.

4 Conclusions

From 79 days of K2 data we were able to extract 22 individual pulsations and one possible multiplet. Asymptotic period spacing was used to identify all but one of the pulsations with low-degree ℓ modes. The period transform in Figure 1 shows the pulsations that fall on the $\ell = 1$ and/or $\ell = 2$ sequence. Upon closer inspection one can see that some of the peaks deviate from the sequence lines. However, the deviations are small which implies that the compact, high gravity sdB stars are less stratified

DOI: 10.1051/epjconf/201716004007

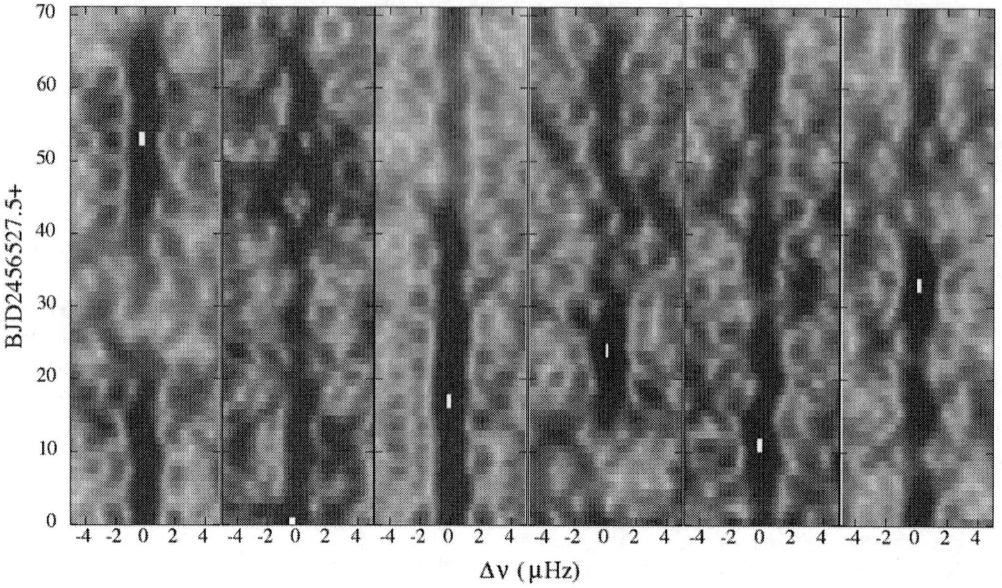

Figure 3: Series of sliding Fourier transforms highlighting the amplitude and frequency variations of six pulsations over time. Note: color indicates the amplitude in terms of σ, but the windows do not follow the same scale.

and more homogeneous than originally considered. The échelle digram (Fig. 2) makes these slight deviation more apparent. The $\ell = 1$ sequence in the right panel of the plot shows a clear "hook" feature, which has been observed in some other sdB pulsators [12, 13]. Models have also shown this feature [14] and it is attributed to a mode trapping feature dependent upon evolution and envelope mass. It will be interesting to calculate échelle digrams for other *Kepler*–observed sdB stars, do an ensemble analysis to correlate this "hook" feature with spectroscopic quantities and to supply additional constraints for models of sdB interiors.

We have identified the only possible multiplet as $\ell = 2$ via asymptotic period spacings. Multiplets are usually related to stellar rotation which lifts the azimuthal degeneracy and gives rise to separations in frequency. If we observe a triplet with $\Delta m = 1$, the rotation period would be 45.9 ± 0.8 days, and if we observe only the outside members of a quintuplet, $\Delta m = 2$, the rotation period would be 91.8 ± 1.5 days. Both rotation periods agree with values measured in other apparently single sdBV stars [15]. An explanation why we do not see more multiplets might be the slow rotation period of the star. Even at the shorter rotation period of about 46 days, $\ell = 1$ multiplets would have a splitting of 0.13μHz, which makes them unresolved from 79 days of data.

Follow-up spectroscopy places EPIC 203948264 in the middle of the g–mode instability region (Telting, private communication) and we find it to be a typical Kepler-observed sdB pulsator in that we easily assigned pulsation modes to the observations. Such constraints will be extremely useful producing models which match the structure defined by the features within the asymptotic sequence. So we add EPIC 203948264 to our list of "solved" sdB pulsators. We note it is one of the few sdB stars showing the "hook" feature (only in the $\ell = 1$ sequence) and our temporal analysis of the pulsations show them to be variable in amplitude, though their frequencies look to be consistent, within the errors. These K2 observations will be invaluable in our understanding of these stars, which represent the cores of all horizontal branch stars. Kepler has proven to be the *only* instrument which provides mode identifications and only through these observations will we be able to obtain an understanding of helium-fusing stars.

Acknowledgements

Funding for this research was provided by the National Science Foundation grant#1312869. Any opinions, findings, and conclusions or recommendations expressed in this material are those of the author(s) and do not necessarily reflect the views of the National Science Foundation. These observations were obtained as part of our K2 Guest Observer proposal. This paper includes data collected by the Kepler mission. Funding for the Kepler mission is provided by the NASA Science Mission directorate. Data presented in this paper were obtained from the Mikulski Archive for Space Telescopes (MAST). STScI is operated by the Association of Universities for Research in Astronomy, Inc., under NASA contract NAS5-26555. Support for MAST for non-HST data is provided by the NASA Office of Space Science via grant NNX13AC07G and by other grants and contracts. This research has used the services of ASTROSERVER.ORG.

References

[1] S.B. Howell, C. Sobeck, M. Haas, M. Still, T. Barclay, F. Mullally, J. Troeltzsch, S. Aigrain, S.T. Bryson, D. Caldwell et al., PASP, **126**, 398 (2014), 1402.5163

[2] Z. Han, P. Podsiadlowski, P.F.L. Maxted, T.R. Marsh, MNRAS, **341**, 669 (2003), arXiv:astro-ph/0301380

[3] U. Heber, PASP, **128**, 082001 (2016), 1604.07749

[4] R.H. Østensen, Astronomische Nachrichten **331**, 1026 (2010), 1010.3214

[5] M.D. Reed, A.S. Baran, A.C. Quint, S.D. Kawaler, S.J. O'Toole, J. Telting, et al., MNRAS **414**, 2885 (2011), 1102.4286

[6] A.S. Baran, M.D. Reed, D. Stello, R.H. Østensen, J.H. Telting, E. Pakštienė, S.J. O'Toole, R. Silvotti, P. Degroote, S. Bloemen et al., MNRAS **424**, 2686 (2012), 1206.3841

[7] A.S. Baran, ActaAstron **62**, 179 (2012)

[8] C. Aerts, J. Christensen-Dalsgaard, D.W. Kurtz, *Asteroseismology* (Springer, 2010)

[9] M. Still, T. Barclay, *PyKE: Reduction and analysis of Kepler Simple Aperture Photometry data*, Astrophysics Source Code Library (2012), 1208.004

[10] P.R. Bevington, D.K. Robinson, *Data reduction and error analysis for the physical sciences* (McGraw-Hill, 2003)

[11] A.S. Baran, C. Koen, B. Pokrzywka, MNRAS **448**, L16 (2015), 1412.2753

[12] A.S. Baran, A. Winans, ActaAstron **62**, 343 (2012)

[13] A.S. Baran, J.H. Telting, P. Németh, S. Bachulski, J. Krzesiński, A&A, **573**, A52 (2015)

[14] S. Charpinet, P. Brassard, V. Van Grootel, G. Fontaine, *On Interpreting g-Mode Period Spacings in sdB Stars*, in *6th Meeting on Hot Subdwarf Stars and Related Objects*, edited by V. van Grootel, E. Green, G. Fontaine, S. Charpinet (2014), Vol. 481 of *Astronomical Society of the Pacific Conference Series*, p. 179

[15] M.D. Reed, H. Foster, J.H. Telting, R.H. Østensen, L.H. Farris, R. Oreiro, A.S. Baran, MNRAS **440**, 3809 (2014)

EPJ Web of Conferences **160**, 04008 (2017)
Seismology of the Sun and the Distant Stars II
DOI: 10.1051/epjconf/201716004008

The additional-mode garden of RR Lyrae stars

László Molnár[1,*], *Emese* Plachy[1], *Péter* Klagyivik[1], *Áron L.* Juhász[1,2], *Róbert* Szabó[1], *Zachary* D'Alessandro[3], *Benjamin* Kratz[3], *Justin* Ortega[3], and *Shashi* Kanbur[3]

[1] *Konkoly Observatory, MTA CSFK, Konkoly Thege Miklós út 15-17, H-1121 Budapest, Hungary*
[2] *Eötvös Loránd University, Pázmány Péter sétány 1/a, H-1117 Budapest, Hungary*
[3] *State University New York, 7060 Route 104, Oswego, NY 13126-3599, USA*

Abstract. Space-based photometric missions revealed a surprising abundance of millimagnitude-level additional modes in RR Lyrae stars. The modes that appear in the modulated fundamental-mode (RRab) stars can be ordered into four major categories. Here we present the distribution of these groups in the Petersen diagram, and discuss their characteristics and connections to additional modes observed in other RR Lyrae stars.

1 Introduction

RR Lyrae stars have been classified into three main categories based on the pulsation modes present in their light variations. These classical Bailey classes are designated as RRab, RRc, and RRd for the fundamental-mode, first-overtone and double-mode stars, respectively. (A fourth, second-overtone group is claimed to exist in the Magellanic Clouds, but these RRe stars are yet to be found in the Milky Way [1].) However, the discoveries of the past decade, mainly from the OGLE survey and the photometric space missions, complicated this simple picture. The RRc population, for example, has been extensively studied, and a new family of modes called 0.61-type, or f_X modes was discovered. These modes are strongly connected to the first radial overtone, and can be detected in both first-overtone and double-mode RR Lyrae and Cepheid stars. These new results strongly suggest that RRc and RRd stars should be considered as stars dominated by one or two radial modes, but are not, in fact, single- and double-mode stars.

Also, it seems that at least four main groups can be recognized, by splitting the RRab group into two subgroups: the truly single-mode, and the modulated-multimode types [2]. According to the results of *Kepler* and *CoRoT*, non-Blazhko RRab stars seem to be pure radial pulsators with no indications of other modes down to sub-mmag levels. In contrast, modulated RRab stars exhibit a variety of low-amplitude additional modes (see, e.g., [3]). However, the distribution of these modes was not investigated yet in detail.

In this paper we present our first results from a systematic survey of additional modes in RRab stars, based mostly on the early campaigns of the K2 mission of the *Kepler* space telescope. Here we only discuss the new landscape and identify the main groups of the new modes. Detailed findings, including the amplitudes and tabulated values will be presented in a forthcoming paper. Another restriction is that here we only focus on modes that appear at frequencies between f_0 and $2f_0$, e.g., we do not include ones that could be g-modes at low frequencies or higher-order p-mode overtones [4, 5]. The only exception is period doubling: the strong 9:2 resonance that destabilises the limit cycle of the fundamental mode and causes it to bifurcate, occurs between the fundamental mode and the 9th radial overtone [6]. However, because of the resonance, the mode itself is usually not directly observable, but the strongest signature of the bifurcation, the half-integer peak at $\approx 3/2f_0$, appears within the search range of our investigation.

*e-mail: molnar.laszlo@csfk.mta.hu

2 Additional modes in RRab stars

The first additional modes in RRab stars were discovered by the observations of the *CoRoT* space telescope [7, 8]. In parallel, the first observations of the *Kepler* space telescope also revealed peaks in multiple stars at a a level of 1-10 mmag in the Fourier spectra [9]. These first detections are summarised in the Petersen diagram in figure 1. The small grey and black dots represent the classical RRd stars in the OGLE-III sample (Magellanic Clouds and the bulge) and the Galactic field, respectively [1, 10, 11]. Two groups can be readily identified in the Petersen diagram. One is at $P/P_0 = 2/3$, or at the inverse of the half-integer frequency peak: these are the period-doubled (PD) stars. The other group appears around $P/P_0 \approx 0.59$. This value agrees with the expected range of the second radial overtone ($P_2/P_0 = 0.55\ldots0.62$), hence it is commonly labeled as the f_2 group, even though the current one-dimensional, non-linear hydrodynamical models cannot generate second-overtone pulsations to confirm the nature of this group. The rest of the modes spread between

DOI: 10.1051/epjconf/201716004008

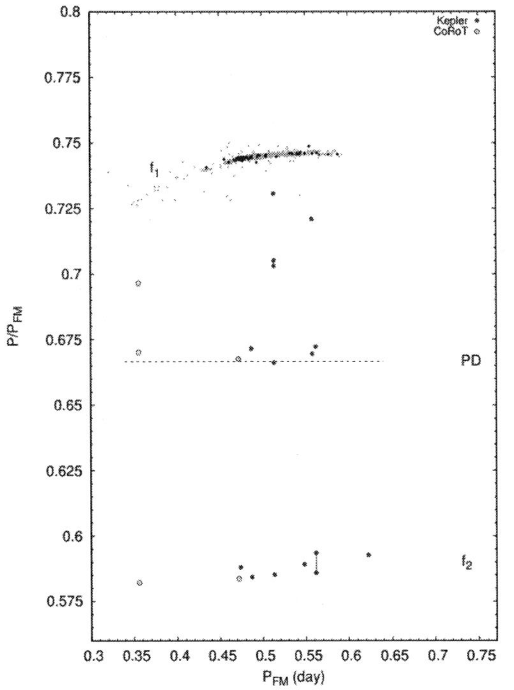

Figure 1: Petersen diagram of the first additional modes detected in fundamental-mode stars, compared to the distribution of classical double-mode stars (grey and black). The dashed line marks the 2/3 ratio where the half-integer peaks of period doubling are expected.

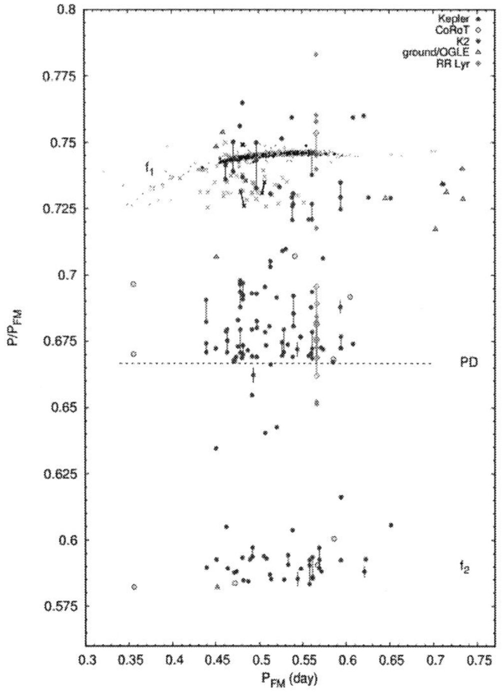

Figure 2: Petersen diagram of RRab stars, with the K2-E, K2 campaigns 0–3 and 6, and other data sets included. See the text for more details.

the half-integer ones and the position of the first overtone, labeled as f_1.

2.1 The current sample of additional modes

During the six years that passed since the first additional modes described in V1127 Aql, the sample has expanded greatly, mostly due to the efforts of the K2 RR Lyrae Survey (Szabó et al., in these proceedings). One important result came from the star RR Lyrae itself. The continuous *Kepler* light curve of the star revealed that not only a strongly variable period doubling, but another mode is also present in the star that could be the first radial overtone [12]. A reanalysis of the *Kepler* quarters 5, 6, 13, and 14 uncovered a complicated structure of peaks at the regions of both the half-integer peak and the first overtone. These are marked with golden diamonds in figure 2. Here the empty diamonds mark peaks that are separated approximately by the modulation frequency of RR Lyr, suggesting that these modes also experience the Blazhko effect to at least some extent.

Additional modes were identified in a few more stars that were observed by *CoRoT*, the Konkoly Blazhko Survey, OGLE-IV and from Antarctica [5, 13–16], but most of the new detections originate from the K2 mission. This is due to the large field-of-view and aperture of *Kepler*, the step-and-stare approach of the mission, and the plethora of newly discovered RR Lyrae stars that can be targeted, discovered mainly by the LINEAR and Catalina Sky Surveys [17, 18]. For this initial study, we analysed the frequency contents of 307 RRab stars observed in K2 Campaigns 0-3 and 6, and included the results from the 27 stars detected in the K2-E2 engineering run [4]. The K2 observations were reduced with the Extended Aperture Photometry method (Plachy et al., these proceedings), and the pulsation frequencies were determined with the Period04 software [19]. Overall, we detected additional modes in 40 stars, e.g., 12% of the K2 sample.

The investigation of the K2 data revealed interesting patterns in the occurrence of these additional modes. In most cases, only modes belonging to one or two main groups could be identified. A rare example with all three types (f_1, f_2, PD) present in the star simultaneously, is shown in the lower panel of figure 3. Another feature we observed in many stars is when multiple peaks are present, the strongest ones in the groups are not necessary the central peaks, and/or the strongest peaks are offset from the expected, canonical values of period doubling and/or the first overtone. The upper panel of figure 3 clearly shows these signatures. The f_1 group is centered at the canonical value, but the highest-amplitude one is the one of highest frequency; whereas the PD group is shifted from the 3/2 f_0 value, but it is symmetric in terms of amplitudes. These

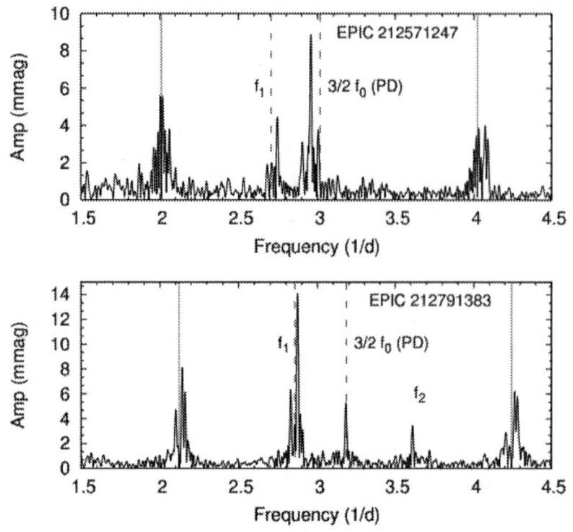

Figure 3: Two examples from Campaign 6. Upper panel: note the offset from canonical values (red dashed lines). Lower panel: a rare example with all three types of additional modes present. Pulsation peaks (blue lines) have been prewhitened.

details about the RRab additional modes are currently not understood.

For completeness, we also included the new, anomalous RRd stars in figure 2 with small crosses: grey crosses are OGLE-IV stars that feature either peculiar period ratios or modulations or both [20], whereas black crosses are the Blazhko-RRd stars discovered in the globular cluster M3 [21].

2.2 Distribution of the modes by period ratios

The main aim of the study was to identify sub-structures in the distribution of the additional modes, just like the three distinct groups found in the Petersen diagram of RRc stars [22]. As figure 2 illustrates, this was not fully realised. The f_2 group remained a broad but distinct feature at the bottom of the Petersen diagram. Upwards, a few points show up in the previously empty range between P/P_0 = 0.60...0.66.

The most dramatic change can be seen at and above the expected position of the peaks of period doubling. Here, in the middle of the figure, we no longer see a definite ridge along the $P/P_0 = 2/3$ line. The points spread upwards, to $P/P_0 = 0.70$. We suspect that in several cases, the temporal variations of either the additional mode or the fundamental mode, or both, give rise to multiple peaks. Some obvious cases are linked with lines in the figure. However, this does not explain the occurrence of all peaks. It seems that bona-fide period doubling (resonant interaction with the 9th overtone) may suffer confusion with other modes that are excited in the P/P_0 = 0.66...0.70 range.

Finally, by now, multiple stars can be identified in which the additional modes overlap with region of the first overtone in RRd stars, especially with the ones with anomalous period ratios. The existence of this group

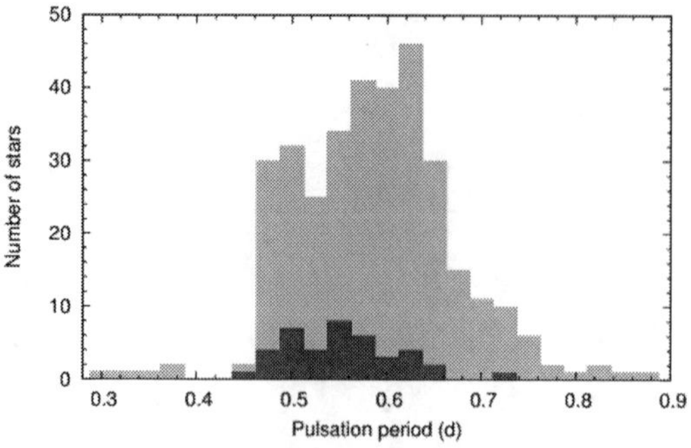

Figure 4: In grey: the period distribution of all known RRab stars in the K2 sample investigated here: In blue: the distribution of stars where addtional modes were identified.

can be described by pure radial pulsations: hydrodynamic models showed that once the period-doubling bifurcation occurs, the new limit cycle can become unstable against the first overtone. These triple-mode models are unlike RRd stars: they are still dominated by the fundamental mode, and may occur over a broader parameter regime, and the interaction between the modes can lead to the appearance of resonant and chaotic states [23].

2.3 Distribution by pulsation periods

Interestingly, all but one RRab stars with additional modes from the K2 sample are restricted to the period range of P_0 = 0.44...0.66 d which is only a subset of the accessible period range for these stars, as shown in figure 4. At longer periods only a single K2 star and a few bulge stars display modes that fall approximately into the f_1 group. This cut-off could be plausibly explained by the occurrence of the Blazhko effect itself. A recent study found that most modulated stars have pulsation periods below 0.7 d, and their mean period is 0.54 d [24]. This is in perfect agreement with the distribution of the K2 stars with additional modes, even though the distribution of all RRab stars peaks at about 0.6 d.

However, modulated stars, although in lower numbers, extend to periods even shorter than 0.4 d. One prime example is V1127 Aql itself, the single occupant of the short-period end of figures 1–2. This lack of stars can be attributed the small numbers of short-period RRab stars observed by K2 in general so far: a larger sample could uncover similar stars from the K2 sample as well.

The long-period OGLE stars were found in a study that specifically targeted bulge RRab stars with $P_0 > 0.6$ d. Although this criterion strongly limits the bulge sample, one would still expect an increase in the number of stars towards the cut-off. Yet, this not the case, hinting that the distribution of additional modes might be different for field and bulge RR Lyrae stars.

3 Conclusions

New observations, especially from the K2 RR Lyrae Survey, have uncovered a plethora of low-amplitude additional modes in fundamental-mode RR Lyrae stars. These can be grouped into three broad categories established in recent years (namely, f_1, f_2, and PD), but our efforts to detect further structures in the Petersen diagram of these modes have not been successful so far. We suspect that multiple radial and non-radial modes are behind these groups and that temporal variation of the modes (either intrinsic or connected to the Blazhko effect) further complicate the identification of the modes. A more detailed study to incorporate more K2 campaigns and examine the amplitudes of the frequency peaks is currently under way.

Acknowledgements

Z.D., B.K., and J.O. wishes to acknowledge the hospitality of the Konkoly Observatory during their stay. This project has been supported by the NKFIH K-115709, PD-116175, and PD-121203 grants of the Hungarian National Research, Development and Innovation Office, and by the Lendület LP2014-17 Program of the Hungarian Academy of Sciences. The research leading to these results has received funding from the ESA PECS Contract No. 4000110889/14/NL/NDe, and from the European Community's Seventh Framework Programme (FP7/2007-2013) under grant agreement no. 312844 (SPACEINN). L.M. was supported by the János Bolyai Research Scholarship of the Hungarian Academy of Sciences.

References

[1] I. Soszyński, A. Udalski, M.K. Szymański, M. Kubiak, G. Pietrzyński, Ł. Wyrzykowski, O. Szewczyk, K. Ulaczyk, R. Poleski, AcA **59**, 1 (2009), `0903.2482`

[2] L. Molnár, Comm. Konkoly Obs. **105**, 11 (2016), `1602.00842`

[3] J.M. Benkő, R. Szabó, ApJL **809**, L19 (2015), `1507.06814`

[4] L. Molnár, R. Szabó, P.A. Moskalik, J.M. Nemec, E. Guggenberger, R. Smolec, R. Poleski, E. Plachy, K. Kolenberg, Z. Kolláth, MNRAS **452**, 4283 (2015), `1507.04714`

[5] Á. Sódor, G. Hajdu, J. Jurcsik, B. Szeidl, K. Posztobányi, Z. Hurta, B. Belucz, E. Kun, MNRAS **427**, 1517 (2012), `1208.0139`

[6] Z. Kolláth, L. Molnár, R. Szabó, MNRAS **414**, 1111 (2011), `1102.0157`

[7] M. Chadid, J.M. Benkő, R. Szabó, M. Paparó, E. Chapellier, K. Kolenberg, E. Poretti, G. Bono, J.F. Le Borgne, H. Trinquet et al., A&A **510**, A39 (2010)

[8] E. Guggenberger, K. Kolenberg, E. Chapellier, E. Poretti, R. Szabó, J.M. Benkő, M. Paparó, MNRAS **415**, 1577 (2011), `1104.1726`

[9] J.M. Benkő, K. Kolenberg, R. Szabó, D.W. Kurtz, S. Bryson, J. Bregman, M. Still, R. Smolec, J. Nuspl, J.M. Nemec et al., MNRAS **409**, 1585 (2010), `1007.3928`

[10] I. Soszyński, W.A. Dziembowski, A. Udalski, R. Poleski, M.K. Szymański, M. Kubiak, G. Pietrzyński, Ł. Wyrzykowski, K. Ulaczyk, S. Kozłowski et al., AcA **61**, 1 (2011), `1105.6126`

[11] I. Soszyński, A. Udalski, M.K. Szymański, J. Kubiak, G. Pietrzyński, Ł. Wyrzykowski, K. Ulaczyk, R. Poleski, AcA **60**, 165 (2010), `1009.0528`

[12] L. Molnár, Z. Kolláth, R. Szabó, S. Bryson, K. Kolenberg, F. Mullally, S.E. Thompson, ApJL **757**, L13 (2012), `1208.4908`

[13] R. Szabó, J.M. Benkő, M. Paparó, E. Chapellier, E. Poretti, A. Baglin, W.W. Weiss, K. Kolenberg, E. Guggenberger, J.F. Le Borgne, A&A **570**, A100 (2014), `1408.0653`

[14] R. Smolec, Z. Prudil, M. Skarka, K. Bakowska, MNRAS **461**, 2934 (2016), `1606.06928`

[15] C. Zalian, M. Chadid, J. Vernin, L. Abe, A. Agabi, MNRAS **456**, 192 (2016)

[16] J.M. Benkő, R. Szabó, A. Derekas, Á. Sódor, MNRAS (2016), `1608.06418`

[17] A.J. Drake, M.J. Graham, S.G. Djorgovski, M. Catelan, A.A. Mahabal, G. Torrealba, D. García-Álvarez, C. Donalek, J.L. Prieto, R. Williams et al., ApJS **213**, 9 (2014), `1405.4290`

[18] B. Sesar, Ž. Ivezić, J.S. Stuart, D.M. Morgan, A.C. Becker, S. Sharma, L. Palaversa, M. Jurić, P. Wozniak, H. Oluseyi, AJ **146**, 21 (2013), `1305.2160`

[19] P. Lenz, M. Breger, CoAst **146**, 53 (2005)

[20] I. Soszyński, A. Udalski, M.K. Szymański, Ł. Wyrzykowski, K. Ulaczyk, R. Poleski, P. Pietrukowicz, S. Kozłowski, D.M. Skowron, J. Skowron et al., AcA **66**, 131 (2016), `1606.02727`

[21] J. Jurcsik, P. Smitola, G. Hajdu, J. Nuspl, ApJL **797**, L3 (2014), `1411.1987`

[22] H. Netzel, R. Smolec, P. Moskalik, MNRAS **453**, 2022 (2015), `1507.08414`

[23] E. Plachy, Z. Kolláth, L. Molnár, MNRAS **433**, 3590 (2013), `1306.1526`

[24] M. Skarka, J. Liška, R.F. Auer, Z. Prudil, A. Juráňová, Á. Sódor, A&A **592**, A144 (2016), `1606.09114`

DOI: 10.1051/epjconf/201716004009

Extended Aperture Photometry of K2 RR Lyrae stars

Emese Plachy[1,*], *Péter* Klagyivik[1], *László* Molnár[1], *Ádám* Sódor[1] and *Róbert* Szabó[1]
[1]*Konkoly Observatory, MTA CSFK, Konkoly Thege Miklós út 15-17, H-1121 Budapest, Hungary*

Abstract. We present the method of the Extended Aperture Photometry (EAP) that we applied on K2 RR Lyrae stars. Our aim is to minimize the instrumental variations of attitude control maneuvers by using apertures that cover the positional changes in the field of view thus contain the stars during the whole observation. We present example light curves that we compared to the light curves from the K2 Systematics Correction (K2SC) pipeline applied on the automated Single Aperture Photometry (SAP) and on the Pre-search Data Conditioning Simple Aperture Photometry (PDCSAP) data.

1 The need for custom photometric pipelines

K2 observations suffer from various instrumental effects. The most serious one is the variation in spacecraft attitude that is corrected on a 6-hour timescale. Various methods have been developed to tackle these issues but neither is really optimal for RR Lyrae stars. Main characteristics of RR Lyrae photometry are the ~6-12 hour periods (in the range of the correction frequency) and sharp features in the light curve (false positives for outliers). Also, we are interested in the full-amplitude variation, and not in a detrended (systematics-corrected) light curve that other pipelines offer. The KepSFF pipeline [5], for example, is confused by the pulsation, and detrended light curves are of lower quality than the raw fluxes, in several cases. The K2SC pipeline [1] removes the position-dependent systematics from the SAP and PDCSAP light curves. PDCSAP data are already free from systematic trends removed by the PDC (Pre-search Data Conditioning) method [4].

Here we compare our solution for RR Lyrae stars with the outputs of K2SC pipeline. In many cases, EAP photometry retains the full amplitude much better than SAP/PDCSAP-based light curves, but it is sensitive to contamination from blended sources.

2 A simple way: Extended Aperture Photometry (EAP)

We created new apertures for every star. The apertures were defined to contain the movements of the stars within the target pixel masks, but to avoid contamination from nearby stars whenever possible. Although this method produces a very slightly increased noise from the background pixels, in many cases it also preserves the pulsation amplitudes throughout the campaign much better than the SAP and PDCSAP fluxes.

After the photometry, we removed all points with SAP_QUALITY flags larger than 0, except for C2, where we also kept the large number of points with flags value of 2^{15} (16384).

We then ran an automated Fourier analysis script on the light curves to construct an initial fit for sigma clipping [2]. The fit was subtracted from the data, all >3 sigma outliers from the residuals were removed, then the fit was added back.

3 Examples

Figure 1 displays four examples of RR Lyrae stars from Campaign 4.

3.1 EPIC 210438688 (Kp = 17.387 mag)

The aperture used by the Kepler pipeline is clearly undersized. The SAP data indicates that the PSF periodically moved out of the SAP aperture during the second half of the campaign, leading to significant flux loss. The PDCSAP and the K2SC-corrected light curves recovered some of the intrinsic variation, but at the cost of increased scatter. In contrast, our EAP photometry collected the flux troughout the campaign, and provided us with an improved light curve for this first-overtone RR Lyrae star.

3.2 EPIC 210600482 (Kp = 17.636 mag)

The SAP/PDCSAP fluxes experienced flux loss in the case of this faint RRd star too. The K2SC correction recovered most of the variation, but the last third of the data has larger scatter, and also changes in the average flux level. The EAP light curve nicely follows the double-mode variation along the campaign.

*e-mail: eplachy@konkoly.hu

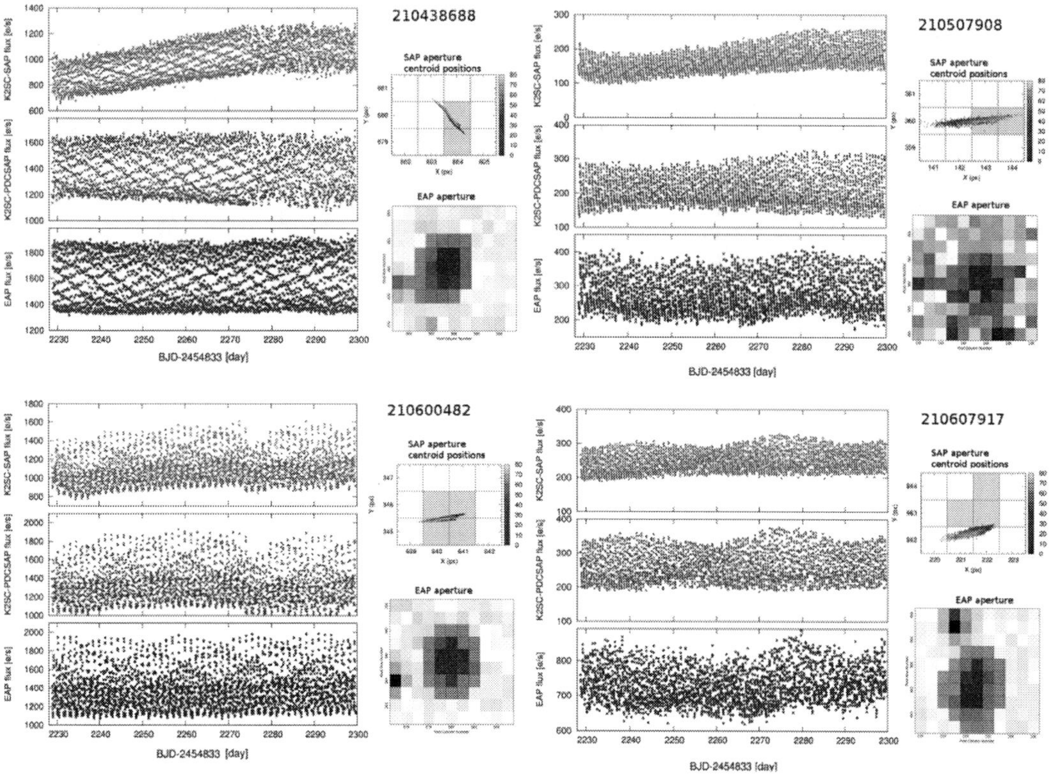

Figure 1: Different photometric solutions for EPIC 210438688, EPIC 210600482, EPIC 210507908, EPIC 210607917 are presented. The upper panels (grey) show the K2SC-corrected SAP light curves. Middle panels (blue) show the K2SC and PDC corrected light curves. Lower panels display the EAP photometry. The applied SAP aperture with the PSF centroid positions are on the small panels on the right side of the light curve panels. Colour bars indicate the time evolution of the position in days. The green pixels on the lower small panels show the aperture used in the EAP photometry.

3.3 EPIC 210507908 ($Kp = 19.085$ mag)

This star suffers from contamination from nearby sources. Interestingly, both the original PDCSAP and the systematics-corrected one shows an apparent amplitude change that follows the average flux changes of the SAP flux. The EAP flux shows small changes in the average brightness (implying contamination), but no such amplitude variation. We noticed this behaviour in other stars too. This effect could mimic the Blazhko modulation in stars where it is not present.

3.4 EPIC 210607917 ($Kp = 18.898$ mag)

Both the SAP and PDCSAP fluxes suggest the presence of amplitude modulation (Blazhko effect) in this RRab star. The EAP light curve shows much higher flux and strong contamination from a blended source, but no amplitude variation. Here the photocenter tracks the brighter, non-RR Lyrae star.

4 Conclusions

Our study revealed a distortion effect of the PDC pipeline on K2 RR Lyrae light curves. When the star moves in and out from the undersized mask a significant amplitude change occurs. After removing the systematics by extracting the Cotrending Basis Vectors, the amplitude variation becomes more prominent and mimics the Blazhko effect. This is very typical when contaminations play a role. On the other hand EAP light curves contain all the fluxes from the blended stars and their variability may be hard to distinguish from the RR Lyrae pulsation. Taken all these together we believe that EAP gives better light curves solutions for RR Lyrae stars than other available methods. However, we note that blended stars require careful analysis.

Acknowledgements

This work has used K2 targets selected and proposed by the RR Lyrae and Cepheid Working Group of the Kepler Asteroseismic Science Consortium (proposal number GO4069). Funding for the Kepler and K2 missions is provided by the NASA

Science Mission directorate. This project has been supported by the LP2014-17 Program of the Hungarian Academy of Sciences, and by the NKFIH K-115709, PD-116175 and PD-121203 grants of the Hungarian National Research, Development and Innovation Office. L.M. and Á.S. were supported by the János Bolyai Research Scholarship of the Hungarian Academy of Sciences. The research leading to these results has received funding from the European Community's Seventh Framework Programme (FP7/2007-2013) under grant agreement no. 312844 (SPACEINN). The authors also acknowledge support from to ESA PECS Contract No. 4000110889/14/NL/NDe. This work made use of PyKE [3], a software package for the reduction and analysis of Kepler data.

References

[1] Aigrain S. et al., MNRAS **447**, 2880 (2015)
[2] Sódor Á., Konkoly Observatory Occasional Technical Notes **15**, 1 (2012)
[3] Still M., Barclay T., Astrophysics Source Code Library, record ascl:1208.004 (2012)
[4] Twicken, J. D. et al., SPIE **7740**, 1 (2010)
[5] Vanderburg A., Johnson J. A., PASP **126**, 948 (2014)

EPJ Web of Conferences **160**, 04010 (2017)
Seismology of the Sun and the Distant Stars II DOI: 10.1051/epjconf/201716004010

A modulated RRd star observed by K2

Emese Plachy[1,★], *Péter* Klagyivik[1], *László* Molnár[1] and *Róbert* Szabó[1]

[1]*Konkoly Observatory, MTA CSFK, Konkoly Thege Miklós út 15-17, H-1121 Budapest, Hungary*

Abstract. We report the analysis of the double-mode RR Lyrae star EPIC 205209951, the first modulated RRd star observed from space. The amplitude and phase modulation are present in both modes.

1 Introduction

Modulated (Blazhko) RRd stars have been recently discovered by the OGLE (Optical Gravitational Lensing Experiment) Survey in the Galactic Bulge [4, 5] and in the Magellanic Clouds [3]. These stars belong to the "anomalous RRd variables" that have different period and amplitude ratios than typical double-mode RR Lyraes and the fundamental mode often dominates in their pulsation. The investigation of RR Lyrae stars in the globular cluster M3 also revealed Blazhko RRd stars with similar characteristics [6]. Here we present EPIC 205209951, the first modulated RRd star observed in a space mission. This star was previously known as a strongly modulated RRab star: data from ground-based surveys revealed only the modulated fundamental mode that reaches higher amplitudes than the first overtone.

2 Data and analysis

EPIC 205209951 was observed with 1-minute sampling mode in Campaign 2 of the K2 mission. We used the PYKE Software [2] and the Extended Aperture Photometry (EAP, Plachy et al. these proceedings) to derive brightness data from the target pixel files. The main idea of the method is to determine the optimal aperture that contains the star with the movements due to the attitude control manoeuvre corrections but avoid nearby stars.

The EAP light curve of EPIC 205209951 is displayed in figure 1. The figures shows how the changing amplitudes shape the light curve. The middle part resembles a classical RRd star, but elsewhere it is dominated by the fundamental mode and looks very different (see figure 2).

Period04 software [1] was used to calculate the Fourier transform and the time variation of the pulsation modes. The fundamental mode and the first overtone frequencies were determined to be f_0=2.124311 c/d and f_1=2.86714 c/d. The fundamental mode has a higher average amplitude than the first overtone. While in most RRd stars the amplitude ratio is the opposite, this is typical for the anomalous RRd stars. The ~0.741 period ratio is also slightly outside the canonical 0.742–0.748 range for the double-mode RR Lyrae period ratios.

The temporal variations of the A_0, A_1, ϕ_0 and ϕ_1 parameters indicate that both modes are modulated, and likely not independently. While the modulation amplitudes show clear anticorrelation, the phases are nearly correlated except in the first twenty days of the observations, as it is seen in figure 3.

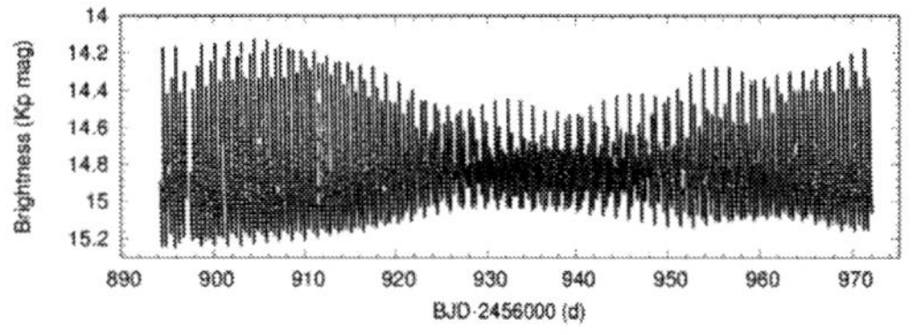

Figure 1: The short cadence light curve of EPIC 205209951 obtained with EAP photometry.

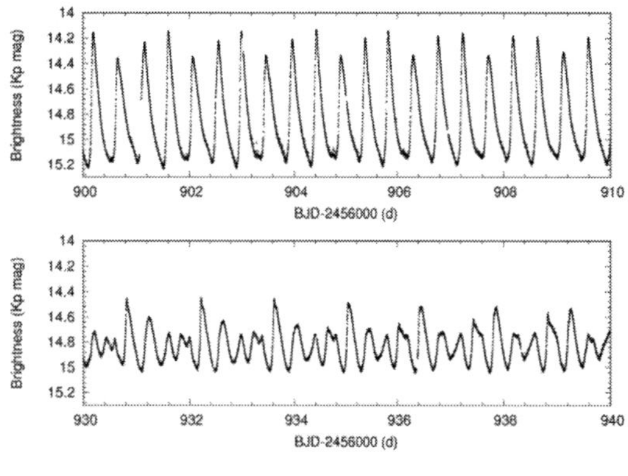

Figure 2: Different sections of the light curve.

★e-mail: eplachy@konkoly.hu

3 Conclusions

EPIC 205209951 has all the characteristics that anomalous RRd stars do: the unusual period ratio, the dominance of the fundamental mode in the amplitude and the modulation. Thanks to the high-precision data provided by *Kepler*, a new member of this strange group can be examined in great detail. The preparation of a detailed investigation of the light curve characteristics and frequency content of EPIC 205209951 is in progress (Plachy et al. in prep).

Acknowledgements

This work has used K2 targets selected and proposed by the RR Lyrae and Cepheid Working Group of the Kepler Asteroseismic Science Consortium (proposal number GO4069). Funding for the Kepler and K2 missions is provided by the NASA Science Mission directorate. This project has been supported by the LP2014-17 Program of the Hungarian Academy of Sciences, and by the NKFIH K-115709, PD-116175 and PD-121203 grants of the Hungarian National Research, Development and Innovation Office. L.M. was supported by the János Bolyai Research Scholarship of the Hungarian Academy of Sciences. The research leading to these results has received funding from the European Community's Seventh Framework Programme (FP7/2007-2013) under grant agreement no. 312844 (SPACEINN). The authors also acknowledge support from to ESA PECS Contract No. 4000110889/14/NL/NDe. This work made use of PyKE [2], a software package for the reduction and analysis of Kepler data.

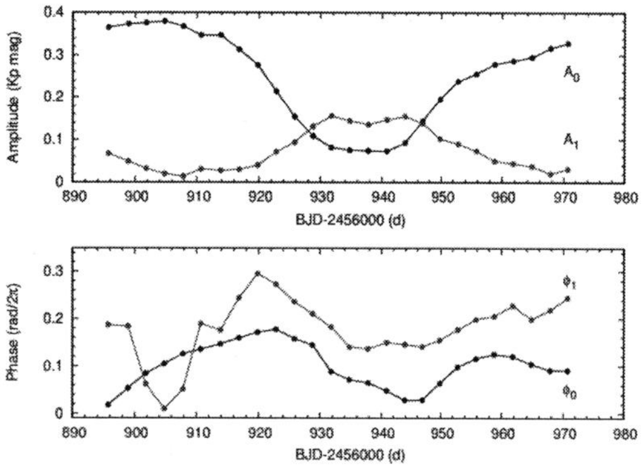

Figure 3: The temporal variation of the amplitudes and phases of the different the fundamental mode (black) and the first-overtone mode (blue) calculated with the Period04 software using 3 day bins.

References

[1] Lenz, P., Breger, M., Communications in Asteroseismology **146**, 53 (2005)
[2] Still M., Barclay T., Astrophysics Source Code Library, record ascl:1208.004 (2012)
[3] Soszyński I. et al., MNRAS **463**, 1332 (2016)
[4] Smolec, R. et al., MNRAS **447**, 3756(2015)
[5] Soszyński I. et al., AcA **64**, 177 (2014)
[6] Jurcsik, J. et al., ApJ **797**, 3 (2014)

EPJ Web of Conferences **160**, 04011 (2017) DOI: 10.1051/epjconf/201716004011
Seismology of the Sun and the Distant Stars II

RR Lyrae Stars in M4

Charles A Kuehn[1,⋆], *Pawel* Moskalik[2], and *Jason A.* Drury[3,4]

[1] *University of Northern Colorado, United States*
[2] *Copernicus Astronomical Center, Poland*
[3] *The University of Sydney, Australia*
[4] *Stellar Astrophysics Center (SAC) - Aarhus University, Denmark*

Abstract. Observations by Kepler/K2 have revolutionized the study of RR Lyrae stars by allowing the detection of new phenomna, such as low amplitude additional modes and period doubling, which had not previously been seen from the ground. During campaign 2, K2 observed the globular cluster M4, providiing the first opportunity to study a sizeable group of RR Lyrae stars that belong to a single population; the other RR Lyrae stars that have been observed from space are field stars in the galactic halo and thus belong to an assortment of populations. In this poster we present the results of our study of the RR Lyrae variables in M4 from K2 photometry. We have identified additional, low amplitude pulsation modes in both observed RRc stars. In 3 RRab stars we have found the Blazhko effect with periods of 16.6d, 22.4d, and 44.5d.

1 Introduction

The globular cluster M4 was observed during K2's Campaign 2, making it the first globular cluster, along with M80 observed during the same campaign, to be observed by Kepler/K2. These globular clusters represent the first opportunity to use Kepler/K2 to study single populations of RR Lyrae stars. While many RR Lyrae had perviously been observed by Kepler/K2, these were all field stars in the Milky Way halo and thus came from a variety of different populations.

Only a portion of M4 fell on silicon during Campaign 2 with most of the cluster falling in the gap between CCDs. A large superstamp measuring approximately $300x150$ pixels ($\approx 20x \approx 10$ arcminutes), Figure 1, was obtained in long cadence mode. These superstamps are similar to those obtained for the open clusters NGC 6791 and NGC 6819 in the original Kepler mission.

2 Photometry

Studying star clusters with Kepler/K2 is a challenge due to the crowded nature of the clusters, especially their centers, and the large pixel size scale (3.98 arc seconds on a side) of the telescope. This results in a large number of the stars in M4 being at least partially blended with the degree of blending increasing toward the cluster center. M4 is one of the two nearest globular clusters and thus will appear more spread out on the sky, making it the best chance for us to study the stars in a globular cluster with K2. In order to obtain photometry for these blended stars, we performed image subtraction using Wojtek Pych's DIAPL2 package[1]; DIAPL2 is an improved version of the DIAPL package [1].

Despite the best efforts to minimize the drift of the telescope, K2 still suffers from a slow drift that necessitates thruster firing to repoint the telescope approximately every six hours. This drift and repointing presents complications for photometry with K2 compared to the original Kepler mission. The differing pixel sensitivies result in an additional signal as the stars drift across the pixels. Figure 2 shows some of the systematic effects seen in the lightcurves of stars in M4 including a general increase in the differential flux across the entire campagn (left panel) which is due to a change in the sky brightness. The effects of the drift and repointing every 6 hours can be seen in the right panel.

The large pulsational amplitudes of RR Lyrae stars meant that these systematic effects had little impact on our ability to identify the main pulsation frequencies in these stars though they likely reduced our ability to find all low amplitude frequencies. We attempted to remove the systematics from the lightcurves using a Principle Component Analysis (PCA) code that we adapted from one originally written by Tom Barclay and Daniel Huber for performing photometry on K2 postage stamps. This produced mixed results but did not lead to the reliable detection of any additional frequencies.

3 RR Lyrae Variables

3.1 RRab Stars

Nine RRab stars in M4 fell on silicon during the K2 observations. We detected the Blazhko effect in three of those

⋆e-mail: charles.kuehn@unco.edu

[1] http://users.camk.edu.pl/pych/DIAPL/

DOI: 10.1051/epjconf/201716004011

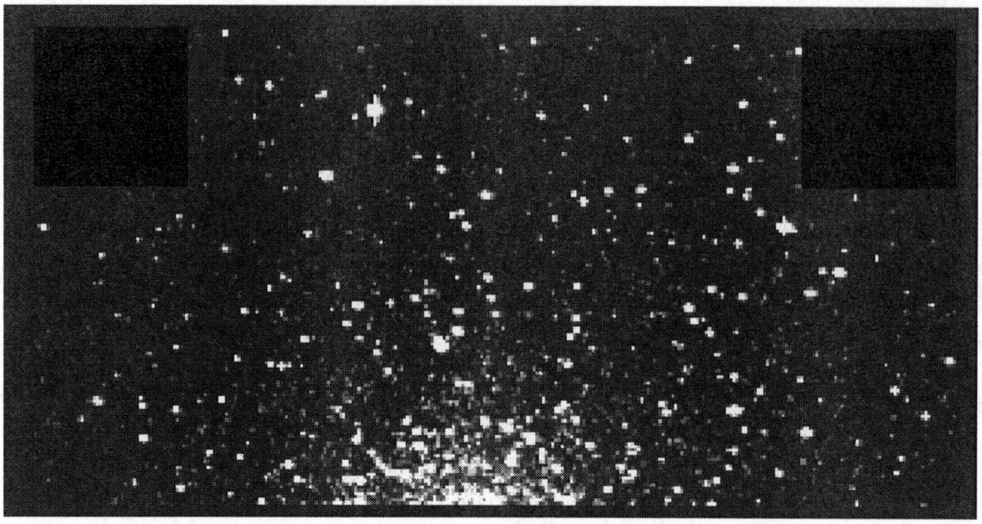

Figure 1: Superstamp of M4 observed by K2.

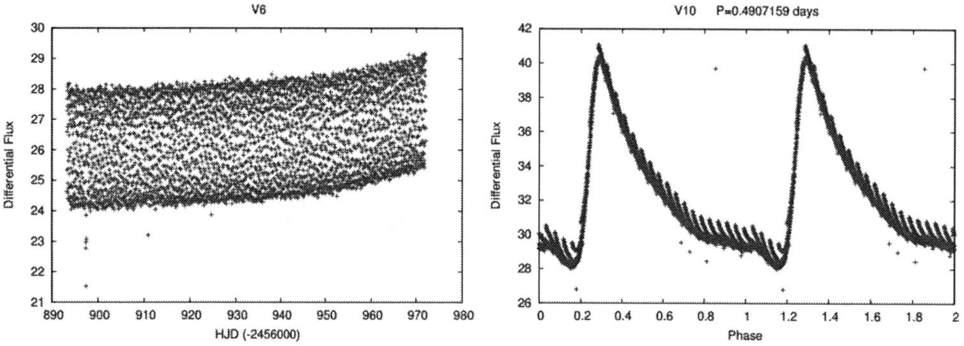

Figure 2: (Left panel) Lightcurve of the RRc star V6. A general increase in the differential flux can be seen over the course of the observing period. (Right panel) Phase-folded lightcurve of RRab star V10; the effects of K2's drifting and repointing every 6 hours is clearly visible. Similar systematics are seen in all of the M4 lightcurves.

stars (V19, V29, V35). V29 and V35 had previously been identified as candidate Blazhko stars [2] while V19 had not previously been considered a Blazhko star. Figure 3 shows the amplitude spectra for the three Blazhko RR Lyrae stars; the measured fluxed were converted to magnitudes prior to the frequency analysis being performed. The fundamental radial mode and its harmonics have been subtracted making the characteristic Blazhko sidepeaks visible. We found Blazhko periods of 16.554 days for V19, 22.419 days for V29, and 44.514 days for V35; this is the first time that Blazhko periods have been determined for these stars.

3.2 RRc Stars

Two RRc stars fell on silicon and both showed additional modes. Figure 4 shows the amplitude spectrum for V42; the first radial overtone and its harmonics have been removed, their locations are marked by dotted lines. The thruster firing frequency at 4.08 cycles per days and its harmonics (8.16 d^{-1}, 12.24 d^{-1}) can be easily seen in amplitude spectrum. We detected three secondary oscillation frequencies with frequencies of f_{2a} = 5.218d^{-1}, f_{2b} = 5.298d^{-1}, and f_{2c} = 5.373d^{-1}; this triplet of peaks corresponds to nonradial modes (see however [3]). The period ratios between these modes and the first radial overtone are 0.631, 0.622, and 0.613, and fall perfectly well on the three sequences in the Petersen diagram established by [4]. Nonradial modes with similar period ratios have been found in many RRc stars but excitation of all three modes in a single star is rare. The three peaks around 8.6d^{-1} are combination peaks of the first radial overtone and the triplet of nonradial modes. We also see subharmonics of the nonradial modes (two peaks at around 8.0d^{-1} and two weak peaks at around 2.6d^{-1}) and a combination peak involving one of the subharmonics (11.345d^{-1}). Table 1 shows the detected frequencies for

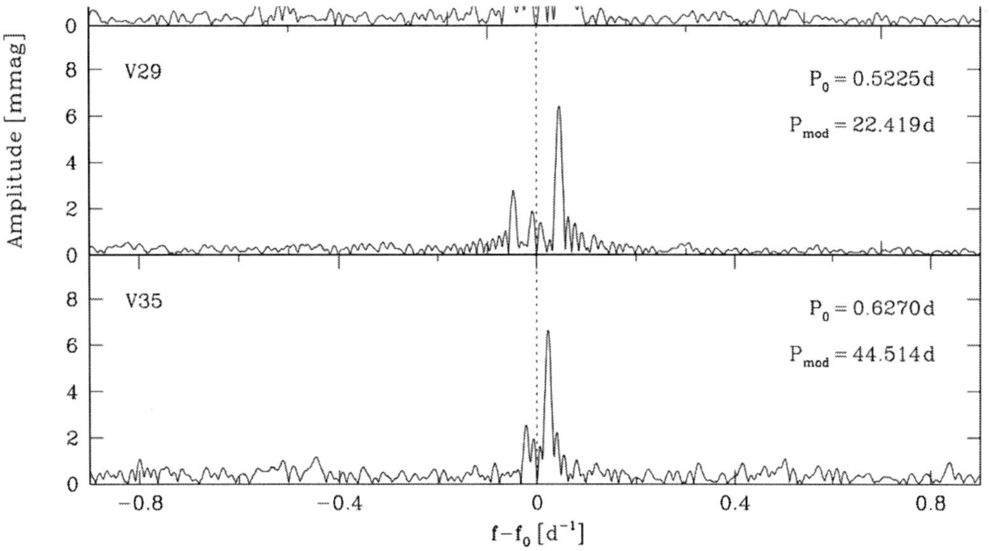

Figure 3: Amplitude spectra for the three RRab stars in M4 observed by K2 that show the Blazhko effect. The fundamental radial mode has been removed and its location is marked with a dotted line. The characteristic Blazhko sidepeaks are visible.

Table 1

V42		V6	
f (d^{-1})	Identification	f (d^{-1})	Identification
3.293	f_1	3.120	f_1
5.218	f_{2a}	2.474	f_2
5.298	f_{2b}	4.952	$2f_2$
5.373	f_{2c}	7.415	$3f_2$
8.511	$f_{2a}+f_1$	8.072	$2f_2+f_1$
8.590	$f_{2b}+f_1$	8.714	f_2+2f_1
8.665	$f_{2c}+f_1$	10.535	$3f_2+f_1$
11.803	$f_{2a}+2f_1$		
11.958	$f_{2c}+2f_1$		
2.600	$\approx 0.5f_{2a}$		
2.687	$\approx 0.5f_{2c}$		
7.980	$\approx 1.5f_{2b}$		
8.052	$\approx 1.5f_{2c}$		
11.345	$\approx 1.5f_{2c}+f_1$		

V42 and identifies the combination frequencies.

Figure 5 shows the amplitude spectrum for V6; the first radial overtone and its harmonics have been subtracted and their locations are marked by dotted lines. As was the case with V42 the thruster firing frequency and its harmonics are clearly seen in the amplitude spectrum. A second frequency at 2.47 d^{-1} and its harmonics are clearly detected. The period ratio of $P_1/P_2 = 0.7928$ is tricky to explain. This ratio would suggest that P_1 and P_2 correspond to the second and first radial overtones, respectively. This scenario is contradicted by the shape of the lightcurve of the dominant mode (P_1) which is very typical for first overtone pulsators. Thus, P_2 must be interpreted as corresponding to a nonradial mode. Alternateively, variability with period P_2 could be contamination from another star but the presence of combination peaks with the dominant mode at frequencies $f_1 + 2f_2$, $f_1 + 3f_2$, and $2f_1 + f_2$ makes this explanation less likely. Table 1 shows the detected frequencies for V42 and identifies the combination frequencies.

References

[1] Woźniak, P.R., AcA **50**, (2000) 421
[2] Stetson, P.B., et al., PASP **126**, (2014) 521
[3] Dziembowski, W.A., CoKon **105**, (2016) 23
[4] Netzel, H. et al., MNRAS **453**, (2015) 2022

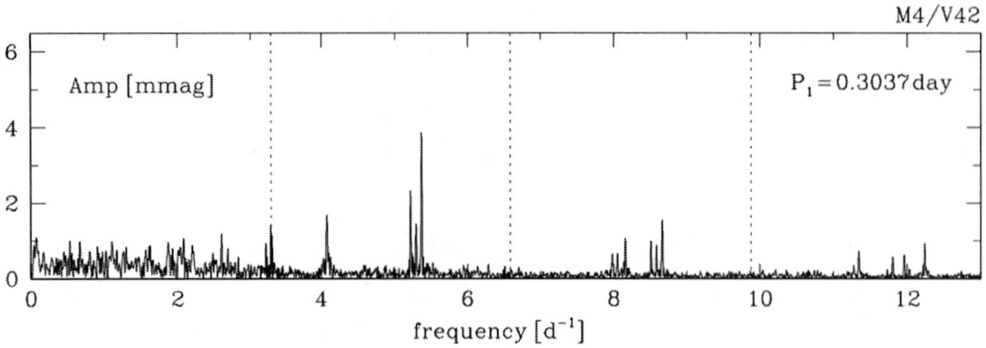

Figure 4: Amplitude spectra for the RRc star V42. The first radial overtone and its harmonics have been removed but their locations are marked with dotted lines. The thruster firing frequency ($4.08d^{-1}$) and its harmonics ($8.16d^{-1}$, $12.24d^{-1}$) are visible. The triplet of peaks around $5.3d^{-1}$ correspond to nonradial modes. Combination peaks, subharmonics of the nonradial modes, and a combination peak involving one of the subharmonics are also seen.

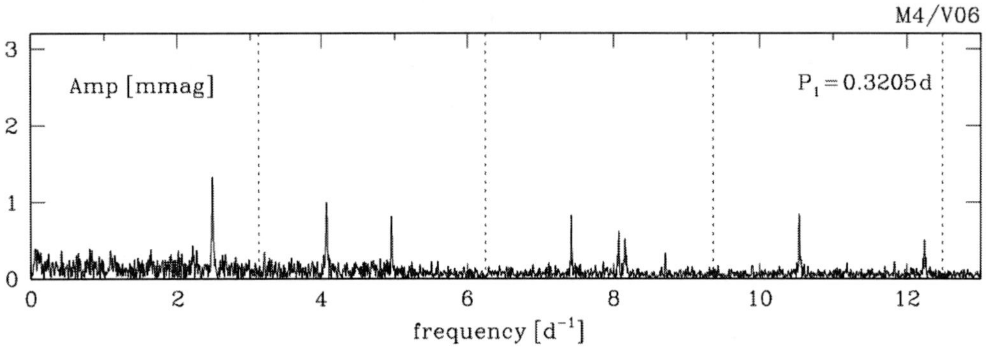

Figure 5: Amplitude spectra for the RRc star V6. The location of the first radial overtone and its harmonics are marked with dotted lines. As was the case in Figure 4 the thruster firing frequency and its harmonics are visible. A secondary frequency at $2.47d^{-1}$ and its harmonics ($4.95d^{-1}$, $7.42d^{-1}$) are detected as are several combination peaks of the dominant and secondary mode.

EPJ Web of Conferences **160**, 04012 (2017) DOI: 10.1051/epjconf/201716004012
Seismology of the Sun and the Distant Stars II

Radial mode widths in red giant stars spectra observed by *Kepler*

Mathieu Vrard[1,2,⋆], *Benoît* Mosser[2], and *Caroline* Barban[2]

[1] *Instituto de Astrofísica e Ciências do Espaço, Universidade do Porto, CAUP, Rua das Estrelas, 4150-762 Porto, Portugal*
[2] *LESIA, CNRS, PSL Research University, Université Pierre et Marie Curie, Université Denis Diderot, Observatoire de Paris, 92195 Meudon cedex, France*

Abstract. The *Kepler* space mission has observed many solar-like pulsators, and helped to decipher their fundamental parameters (e.g: mass, radius, rotation). Most of the achievements recently obtained in that domain result from the analysis of the mode frequencies. However, unique information on non-adiabatic physics derives from the height and width of the modes. In this study, we aim at measuring the mode widths of the pressure modes in thousands of *Kepler* red giants and to analyze their variations in function of stellar parameters. To achieve that, we used a peakbagging technique on the star radial modes. The results show a relation between the radial mode linewidth and the effective temperature of the star as theoretically predicted. We also unveil a clear dependence with mass and stellar evolution for the radial mode width. This means that the mode damping depends on the evolutionary status of the stars.

1 Introduction

Thanks to the CoRoT [2] and *Kepler* [8] missions, solar-like oscillations have been detected in thousands of red giants, which enable the precise measurement of their mass and radius [15]. Their oscillation spectra exhibit pressure modes as well as mixed modes [10], allowing the characterization of their internal structure by the measurement of the mode frequencies. Observations enable us to distinguish the stars that have ignited their helium in their core (clump stars) from the ones that burn their hydrogen on a shell (RGB stars) [e.g. 6, 19]. These data also bring constraints on the physical processes occurring in the star interiors. For example, the measurement of the mode width gives access to the mode lifetime. Therefore it allows us to obtain informations on the damping mechanism [e.g. 13]. The previous studies conducted on this subject [4, 18] have shown that radial mode widths in red giants behave as in main-sequence stars. However, slight differences arise in the scaling relations between the mode damping and the effective temperature. A dependence with the evolutionary state is also observed. The aim of this work is to confirm and extend the previous results using the most recent data release from the *Kepler* satellite.

2 Data analysis method

2.1 Data set

Long-cadence data from *Kepler* up to the quarter Q17 corresponding to 44 months of photometric observations were used. The red giant sample was selected based on the precise determination of the star seismic parameters and their evolutionary states performed in [22]. It corresponds to a sample of more than 4700 stars.

2.2 Measurement of the star radial mode linewidths

The red giant oscillation spectra exhibit pressure modes and mixed modes [e.g. 5], the latter resulting from the coupling of the acoustic waves propagating inside the stellar envelop and the gravity waves propagating in their core [e.g. 20]. Mixed modes present long lifetime and rapid variations of the mode width which made the characterization of the mode damping difficult [e.g. 9]. As a consequence, we focus our study on radial modes only since they are pure pressure modes.

For each of these stars, a first estimate of the global seismic parameters (the large separation $\Delta\nu$ and the frequency of maximum oscillation $\nu_{\max}$) has been obtained with the envelope autocorrelation function [16]. The $\Delta\nu$ values are then refined by the use of the universal pattern [17], which also helps locating the position of the radial modes. The next step was to automatically identify the modes present near the first estimate of the radial mode frequency position given by the universal pattern. In order to do that, we smoothed the power spectra of the light curves. We then locate the nearby local maximas in the smoothed spectrum within a range of frequencies equivalent to a tenth of the large frequency separation around the estimated frequency position of the radial modes. These local peaks are considered to be significant when their heights are above a threshold value corresponding to the rejection of the pure noise hypothesis with a confidence

⋆e-mail: mathieu.vrard@astro.up.pt

DOI: 10.1051/epjconf/201716004012

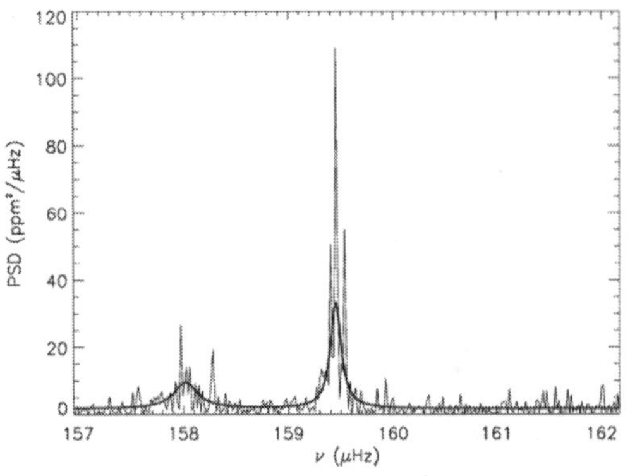

Figure 1: Power density spectrum as a function of frequency for the star KIC3758505 (in black). The fit based on Lorentzian is shown with the solid green line. The peak around 159.5 μHz is a radial mode and the one on the left is the nearby $\ell = 2$ mode.

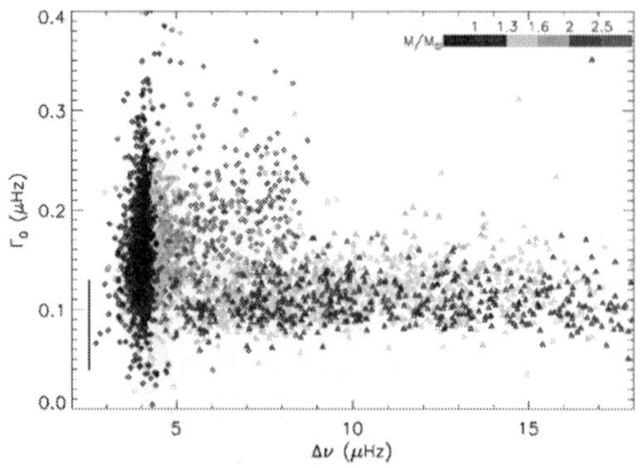

Figure 2: Radial mode width (Γ_0) as a function of the large separation ($\Delta\nu$). The color code indicates the star masses. The diamonds indicate clump stars and the triangles RGB stars. The black line corresponds to the mean value of Γ_0 uncertainties.

level of 99.9%. Among those modes, the doublet $\ell = 0$ and $\ell = 2$ can be easily identified by being the modes with the highest height and mode width value.

After that, we fit the radial modes with Lorentzians as well as the nearby modes (like the quadrupolar modes, see Fig. 1) following the MLE technique described in [3] and [21]. The fit is composed of a linear background component and several Lorentzian profiles, one for each of the modes. The fit being local, a complicate modelling of the background is not necessary. This allows us to extract the frequency, height and width of the radial modes.

Individual mode widths (Γ_0) were computed following [1] for each of the stars. A mean value of Γ_0 is derived from the weighted average of the width of the three radial modes closest to ν_{max}. This way of computing Γ_0 proved to be less sensitive to systematic effects. The individual linewidths are indeed expected to be nearly uniform around ν_{max} as shown by [13]. We also used the seismic scaling relations in order to have an estimate of the stellar masses [15].

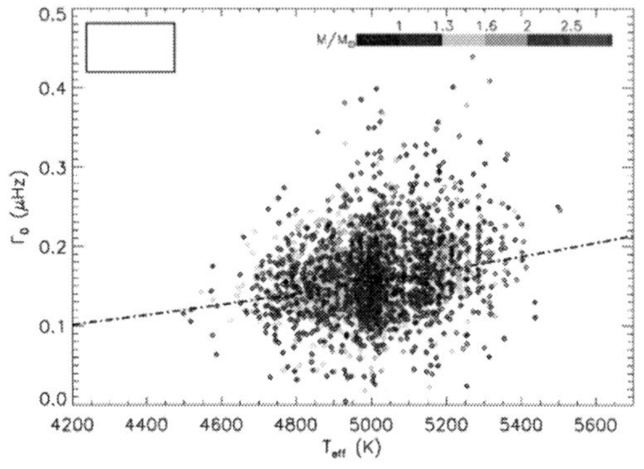

Figure 3: Radial mode width (Γ_0) as a function of the stellar effective temperature for clump stars. The color code indicates the stellar masses. The black square in the upper left corner corresponds to the mean value of the uncertainties. The black dashed-dotted line represents the width-temperature relation measured for clump stars.

3 Results

3.1 Evidence of a mass dependence

We obtained Γ_0 for more than 4700 stars. We first chose to plot the variation of Γ_0 with the global seismic parameter $\Delta\nu$ (Fig. 2). It allows us to make appear clear variations of Γ_0 in function of the stellar parameters. We could note that, for stars with similar masses, the radial mode widths in the clump are significantly higher than on the RGB. We also put into light a slight but clear mass dependence, regardless of the evolutionary status. This phenomenon is particularly obvious for helium burning stars. The values of Γ_0 for clump stars show indeed a clear dependence on $\Delta\nu$ especially in the secondary clump which is a consequence of the mass dependence. Mode widths are directly related to the mode damping. These observational results will be used in a further step for assessing the modes damping in red giants.

3.2 The Γ_0-T_{eff} relation

According to theoretical predictions, the star linewidths should be highly dependent on the stellar effective temperature [7]. However, previous studies have not seen this dependence for red giant stars [4] contrary to subgiants and main-sequence stars. In order to test the theoretical expectations we used the measured effective temperature described in [14] for the *Kepler* targets. Since we put into light a clear difference of behavior for Γ_0 in function of the stellar evolutionary states, we separated the clump and RGB stars in the following analyzes.

The results for clump stars are shown on Fig. 3, the one for RGB stars on Fig. 4. We found a dependence between

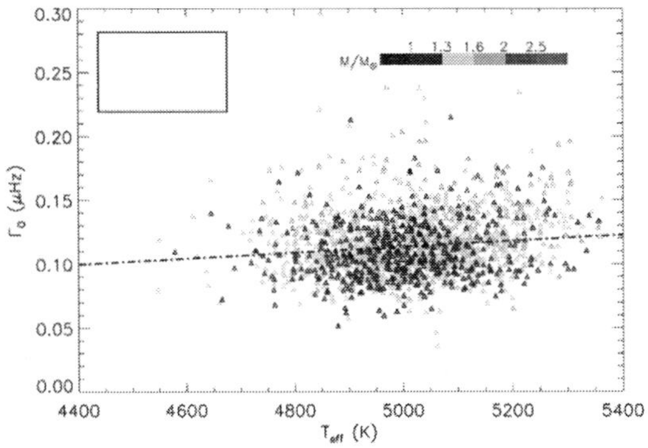

Figure 4: Radial mode width (Γ_0) as a function of the stellar effective temperature for RGB stars. The color code indicates the stellar masses. The black square in the upper left corner corresponds to the mean value of the uncertainties. The black dashed-dotted line represents the width-temperature relation measured in this work for RGB stars.

the two parameters corresponding to $\Gamma_0 \propto T_{\rm eff}^{2.43\pm0.21}$ for clump stars and $\Gamma_0 \propto T_{\rm eff}^{1.02\pm0.22}$ for RGB stars which are non negligible in both cases. For clump stars, the fit can be affected by the presence of high-mass stars with high $T_{\rm eff}$ but it is not the case for RGB stars. On the RGB, we note a double dependence: first, the radial mode width decreases when $T_{\rm eff}$ decreases; second, at fixed $T_{\rm eff}$, the larger the mass, the larger Γ_0. Previous observations did not report a similar $T_{\rm eff}$ dependence [4]. On the contrary, this kind of $T_{\rm eff}$ dependence was firmly established for main-sequence and subgiant stars [1, 4]. However, we find a power law with a small exponent, whereas a high value is observed for the other solar-like pulsators, so that our results are far from the ones observed for main-sequence stars.

Finally, the observed relationship significantly differs from the theoretical expectations. The small exponent we find for the power law does not match the high value predicted by the modelling. It follows that our results are far from the expected dependence [7]. On the one side, our sample benefits from a large number of red giants and represents the largest set of data considered up to now for addressing the radial mode widths; as a result, we may consider that our results benefit from a high statistical reliability. On the other side, the data set covers a limited range in effective temperature, so that any possible bias on these $T_{\rm eff}$ may severely affect the fit.

4 Conclusions

In this study, we measured the radial mode width in thousands of red giant stars. Our results emphasize a relationship between the radial mode width and the star effective temperature, which can be modelled with a power law. The exponent is however much below the value observed for subgiants and main-sequence stars. A difference of behavior in function of the stellar evolutionary states was also unveiled, the clump stars having a much higher radial mode width value than RGB stars. We, then, put into light a new mass dependence for the radial mode width present throughout the star evolution. The mode width being directly related to the mode damping, this offers new constraints on this parameter. Theoretical studies about mode damping and excitation suggest that the discovered mass dependence corresponds to theoretical expectations [11, 12]. However, further theoretical studies are needed to fully exploit the results. Meanwhile, the data analysis will continue, especially on the mode height thus on the oscillation excitation mechanism.

Acknowledgements

MV acknowledges funding by the Portuguese Science foundation through the grant with reference CIAAUP-03/2016-BPD, in the context of the project FIS/04434, co-funded by FEDER through the program COMPETE.

References

[1] Appourchaux T. et al., A&A **537**, A134 (2012)
[2] Baglin A. et al., ESA Special Publication **1306**, 33 (2006)
[3] Barban C. et al., Astronomische Nachrichten **331**, 1016 (2010)
[4] Baudin F. et al., A&A **529**, A84 (2011)
[5] Bedding T. et al., ApJ **713**, L176-L181 (2010)
[6] Bedding T. et al., Nature **471**, 608-611 (2011)
[7] Belkacem K. et al., A&A **540**, L7 (2012)
[8] Borucki W. J. et al., Science **327**, 977 (2010)
[9] Corsaro E. et al., A&A **579**, A83 (2015)
[10] De Ridder J. et al., Nature **459**, 398-400 (2009)
[11] Dupret M.-A. et al., A&A **506**, 57-67 (2009)
[12] Grosjean M. et al., A&A **572**, A11 (2014)
[13] Houdek G. et al., A&A **351**, 582-596 (1999)
[14] Huber D. et al., ApJ **211**, 2 (2014)
[15] Kallinger T. et al., A&A **522**, A1 (2010)
[16] Mosser B. & Appourchaux T., A&A **508**, 877-887 (2009)
[17] Mosser B. et al., A&A **525**, L9 (2011)
[18] Mosser B. et al., A&A **537**, A30 (2012)
[19] Mosser B. et al., A&A **540**, A143 (2012)
[20] Unno W. et al., *Nonradial oscillations of stars* (University of Tokyo Press, Tokyo, 1989) etc
[21] Vrard M. et al., A&A **579**, A84 (2015)
[22] Vrard M. et al., A&A **588**, A87 (2016)

 DOI: 10.1051/epjconf/201716005001

Probing the Deep End of the Milky Way with New Oscillating *Kepler* Giants

Savita Mathur[1,⋆], *Rafael A.* García[2], *Daniel* Huber[3,4,5], *Clara* Regulo[6,7], *Dennis* Stello[3,5], *Paul G.* Beck[2], *Kenza* Houmani[2], and *David* Salabert[2]

[1] *Center for Extrasolar Planetary Systems, Space Science Institute, 4750 Walnut street Suite#205, Boulder, CO 80301, USA*
[2] *Laboratoire AIM, CEA/DRF-CNRS-Université Paris Diderot; IRFU/SAp, Centre de Saclay, 91191 Gif-sur-Yvette Cedex, France*
[3] *Sydney Institute for Astronomy (SIfA), School of Physics, University of Sydney, NSW 2006, Australia*
[4] *SETI Institute, 189 Bernardo Avenue, Mountain View, CA 94043, USA*
[5] *Stellar Astrophysics Centre, Department of Physics and Astronomy, Aarhus University, Ny Munkegade 120, DK-8000 Aarhus C, Denmark*
[6] *Universidad de La Laguna, Dpto de Astrofísica, 38206, Tenerife, Spain*
[7] *Instituto de Astrofísica de Canarias, 38205, La Laguna, Tenerife, Spain*

Abstract. The *Kepler* mission has been a success in both exoplanet search and stellar physics studies. Red giants have actually been quite a highlight in the *Kepler* scene. The *Kepler* long and almost continuous four-year observations allowed us to detect oscillations in more than 15,000 red giants targeted by the mission. However by looking at the power spectra of 45,000 stars classified as dwarfs according to the Q1-16 *Kepler* star properties catalog, we detected red-giant like oscillations in 850 stars. Even though this is a small addition to the known red-giant sample, these misclassified stars represent a goldmine for galactic archeology studies. Indeed they happen to be fainter (down to Kp∼16) and more distant (d>10kpc) than the known red giants, opening the possibility to probe unknown regions of our Galaxy. The faintness of these red giants with detected oscillations is very promising for detecting acoustic modes in red giants observed with K2 and TESS. In this talk, I will present this new sample of red giants with their revised stellar parameters derived from asteroseismology. Then I will discuss about the distribution of their masses, distances, and evolutionary states compared to the previously known sample of red giants.

1 Introduction

The *Kepler* mission [1] has been observing more than 197,000 stars for almost 4 years [2]. While the mission has been a success for its main goal in terms of looking for extrasolar planets and characterizing them (for habitability, occurrence rates etc.) [e.g. 3–5] the excellent quality of the photometric data has enabled us to make tremendous progress on the understanding of stellar evolution and dynamics [6–13]. In particular, the mission showed the power of asteroseismology to better constrain the fundamental properties of planet host stars (such as radius and age) needed to characterize planetary systems but also to probe the structure and rotation in deeper layers of stars.

Some additional studies were also done for binary stars whether to understand their interaction [e.g. 14, 15] or to test scaling relations used in asteroseismology [e.g. 16].

More recently, photometric data from CoRoT [Convection, Rotation, and Transits 17] and *Kepler* have been combined to large spectroscopic surveys such as APOGEE [Apache Point Observatory Galactic Evolution Experiment 18] leading to interesting discoveries on chemical abundances evolution of the Galaxy [19, 20], distances of stars [21].

⋆e-mail: smathur@spacescience.org

Here we present the discovery of additional red giants in the *Kepler* field that were classified as cool dwarfs in the latest *Kepler* star properties catalog of [22]. For more details this work was presented by [23].

2 Looking at dwarfs

We selected the sample of cool dwarfs based on *Kepler* star properties catalog available then [22] with the following criteria on their gravities and temperatures: $\log g > 4.2$ and $T_{\rm eff} < 6125K$ as shown in Figure 1 by the box overlaid on the Hertzprung-Russell Diagram from that catalog. This sample was constituted of approximately 45,000 dwarfs. The first purpose of this selection was to study the photon noise of the *Kepler* mission as done by [24] but using the 17 quarters available to see the temporal variation of the instrumental noise.

Figure 2 shows an example of the power spectrum of one of the stars (KIC 3216802) of this sample that had the following parameters according to the [22] catalog:

- *Kepler* magnitude, Kp=15.9
- $T_{\rm eff} = 4745 \pm 126$ K
- $\log g = 4.8$ dex
- M∼0.5 $M_\odot$

 DOI: 10.1051/epjconf/201716005001

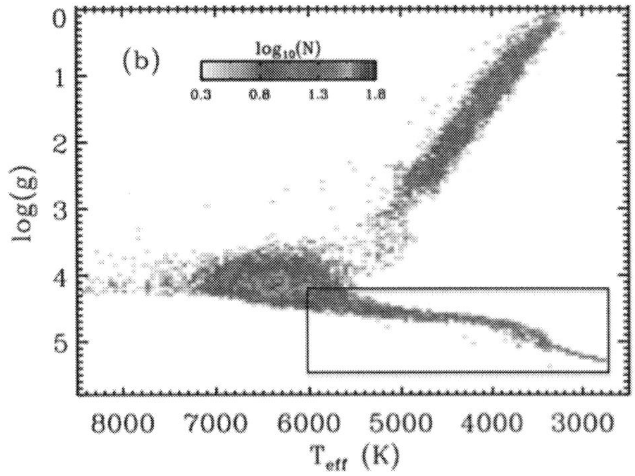

Figure 1: Hertzprung-Russell Diagram from the *Kepler* star properties catalog of [22] showing the number density of stars. The box represents the dwarfs selected based on the criteria given in the text for which we visually inspected the power spectra.

- R~0.48 $R_\odot$

The power spectrum clearly shows an excess power around 25 μHz, which is the region of acoustic modes corresponding to a red giant. The global seismic parameters of this power spectrum suggest that this star has a surface gravity of ~2.38 dex, a mass of $0.73 \pm 0.3 M_\odot$ and a radius of 9.1$R_\odot$. Given this discovery we visually checked the power spectra of the full sample of stars we had selected. We also added some hotter stars that had been observed in short cadence during the survey phase. This led to more than 1000 candidate dwarfs that could be red giants.

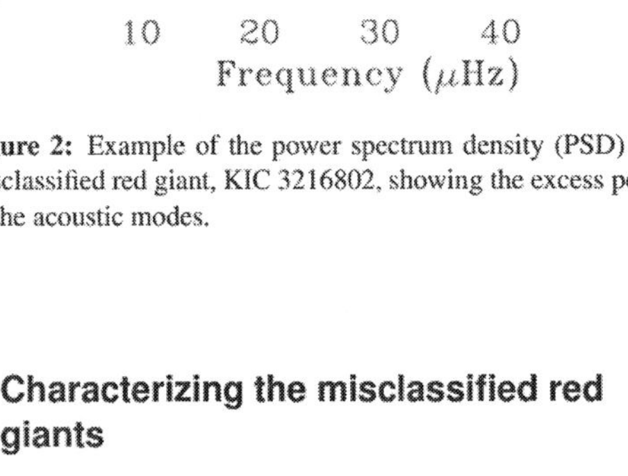

Figure 2: Example of the power spectrum density (PSD) of a misclassified red giant, KIC 3216802, showing the excess power of the acoustic modes.

3 Characterizing the misclassified red giants

We first calibrated the light curves following the KADACS pipeline as described in [25] and then filled gaps shorter than 20 days with the inpainting technique [26, 27]. We then performed an asteroseismic analysis of the light curves and power spectrum using two different pipelines, A2Z [28] and SYD [29] in order to measure the two global acoustic mode parameters: the mean large frequency separation, $\Delta\nu$ and the frequency of maximum power, $\nu_{\rm max}$.

For the A2Z pipeline, we actually used an improved version called A2Z+ that leads a more precise value of $\Delta\nu$ by applying the auto-correlation and masking the ℓ=1 modes. Moreover, for stars with modes above 200 μ Hz, they could result from modes that are the reflection of modes above the Nyquist frequency [30] so we did a specific analysis to check whether it was the case or not.

The results from the different pipelines agreed within 1σ. We re-derived the effective temperatures and distances of the misclassified red giants with an isochrone fitting code using broadband photometry, the asteroseismic observables, and a grid of Parsec isochrones [31].

4 Checking for pollution

We then checked for possible pollution that could lead to the detection of acoustic modes in the red-giant regime. First we looked at the amplitude of the modes. We know there is a tight relation between the maximum amplitude of the modes, $A_{\rm max}$ and the frequency of maximum power, $\nu_{\rm max}$, as showed in [32]. Stars with mode amplitudes that are lower than the scaling relation derived from a known sample of red giants observed by *Kepler* could result from a blending effect diluting the light of the main target. We then checked the J-band images of all the stars and found 36 stars out of the 50 stars with low $A_{\rm max}$ could have a nearby bright star that could pollute the light curves.

There could also be a nearby red giant that was previously known that could pollute a nearby target. To check for such a phenomena, we compared the power spectra of our red-giant candidates with the power spectra of known red giants located within a radius of 1 arcmin. Figure 3 shows the superposition of the a red-giant candidate KIC 3103693, superimposed on the power spectrum of a known red giant KIC3103685. We clearly see that in addition of having the same $\nu_{\max}$ the shape of the p-mode bell is exactly the same. This means that the misclassified candidate is actually coming from pollution so we discarded all the stars where the power spectra are as similar as in Figure 3.

Finally, we also looked at the crowding, which is the measurement of the target flux compared to the total flux in the optimal aperture as defined by the NASA data. In this definition, a crowding of 1 means that the flux belongs to the main target while a crowding of 0 corresponds to a completely polluted target. Stars with a crowding smaller than 75% were flagged in the catalog of misclassified stars.

After these checks for possible blends or pollution, we end up with a sample of misclassified red giants of 854 stars.

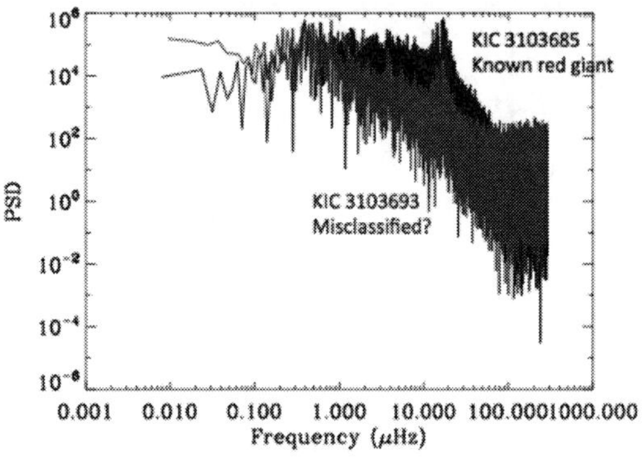

Figure 3: Illustration of pollution of a red giant candidate. KIC 3103685 power spectrum (black) from the oscillations of a known red giant, KIC 3103693 (blue), located within 1 arcmin of the main target.

5 Conclusions

We estimated the surface gravities, masses and radii of these stars with scaling relations based on the Sun [e.g. 33, 34] using the new values of effective temperatures. Figure 4 shows the HR Diagram of the misclassified red giants with their original parameters (blue circles) and the revised ones (red ones).

We also measured the period spacings of the mixed modes [35, 36] for a subsample of 280 stars allowing us to determine whether the stars are on the Red Giant Branch or Red Clump stars.

The misclassified red giants are in general fainter than the previously known sample from the *Kepler* mission, peaking at a magnitude 16 instead o 14. This proves that asteroseismology can be applied to such faint stars. which is promising for missions such as K2 and TESS.

We also found that misclassified stars are less massive and more distant than the known sample of red giants with distances larger than 10kpc. This sample of stars will thus allow us to probe in more details more distant regions of the Galaxy.

Figure 5 shows the distribution of the height of the stars compared to the galactic plane. The comparison of our sample with a a distribution from a synthetic population of halo stars from the Galaxia model [37] suggests that around 45 red giants of our sample could be halo stars.

A sample of around 30 stars have already been observed by the APOGEE survey and the spectroscopic analysis agrees with the new seismic classification provided here. The full sample of stars will be studied more thoroughly with the new spectroscopic data for galacto-archeology investigations.

References

[1] W.J. Borucki, D. Koch, G. Basri, N. Batalha, T. Brown, D. Caldwell, J. Caldwell, J. Christensen-Dalsgaard, W.D. Cochran, E. DeVore et al., Science **327**, 977 (2010)

[2] S. Mathur, D. Huber, N.M. Batalha, D.R. Ciardi, F.A. Bastien, A. Bieryla, L.A. Buchhave, W.D. Cochran, M. Endl, G.A. Esquerdo et al., ArXiv e-prints 1609.04128 (2016), 1609.04128

[3] W.J. Borucki, D.G. Koch, N. Batalha, S.T. Bryson, J. Rowe, F. Fressin, G. Torres, D.A. Caldwell, J. Christensen-Dalsgaard, W.D. Cochran et al., ApJ, **745**, 120 (2012), 1112.1640

[4] C.J. Burke, J.L. Christiansen, F. Mullally, S. Seader, D. Huber, J.F. Rowe, J.L. Coughlin, S.E. Thompson, J. Catanzarite, B.D. Clarke et al., ApJ, **809**, 8 (2015), 1506.04175

[5] S.R. Kane, M.L. Hill, J.F. Kasting, R.K. Kopparapu, E.V. Quintana, T. Barclay, N.M. Batalha, W.J. Borucki, D.R. Ciardi, N. Haghighipour et al., ArXiv e-prints (2016), 1608.00620

[6] T.R. Bedding, B. Mosser, D. Huber, J. Montalbán, P. Beck, J. Christensen-Dalsgaard, Y.P. Elsworth, R.A. García, A. Miglio, D. Stello et al., Nature, **471**, 608 (2011), 1103.5805

[7] P.G. Beck, J. Montalban, T. Kallinger, J. De Ridder, C. Aerts, R.A. García, S. Hekker, M.A. Dupret, B. Mosser, P. Eggenberger et al., Nature, **481**, 55 (2012), 1112.2825

[8] R.A. García, T. Ceillier, D. Salabert, S. Mathur, J.L. van Saders, M. Pinsonneault, J. Ballot, P.G. Beck, S. Bloemen, T.L. Campante et al., A&A, **572**, A34 (2014), 1403.7155

[9] R.A. García, F. Pérez Hernández, O. Benomar, V. Silva Aguirre, J. Ballot, G.R. Davies, G. Doğan, D. Stello, J. Christensen-Dalsgaard, G. Houdek et al., A&A, **563**, A84 (2014), 1311.6990

[10] S. Mathur, R.A. García, J. Ballot, T. Ceillier, D. Salabert, T.S. Metcalfe, C. Régulo, A. Jiménez, S. Bloemen, A&A, **562**, A124 (2014), 1312.6997

[11] J. Fuller, M. Cantiello, D. Stello, R.A. Garcia, L. Bildsten, Science **350**, 423 (2015), 1510.06960

[12] T. Ceillier, J. van Saders, R.A. García, T.S. Metcalfe, O. Creevey, S. Mathis, S. Mathur, M.H. Pin-

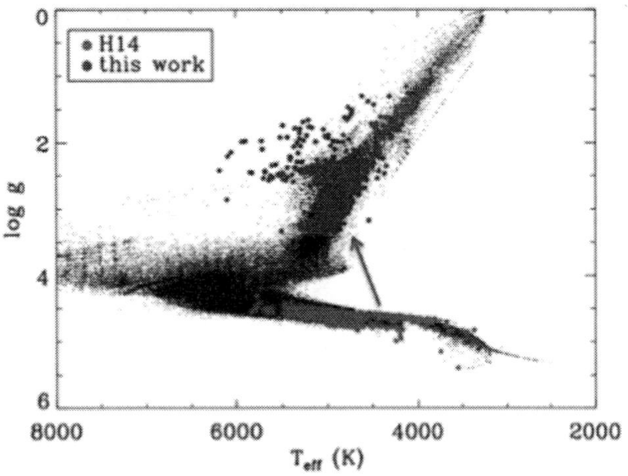

Figure 4: HR Diagram of the full sample of *Kepler* targets where we superimposed the stars studied in this work with their original parameters (red circles) and with their updated parameters (blue circles). Adapted from [23].

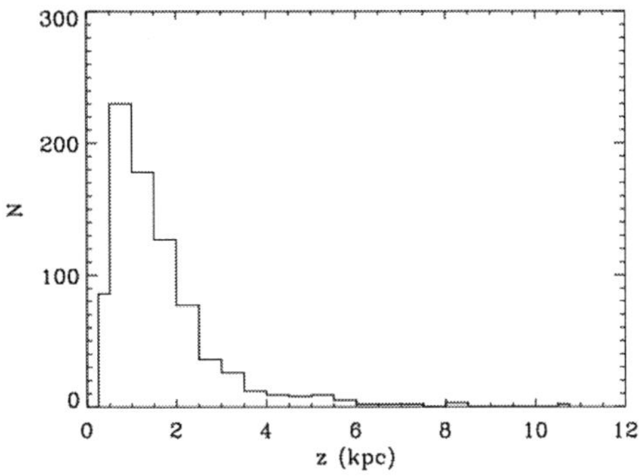

Figure 5: Distribution of the heights of the 854 misclassified red giants (from [23]).

sonneault, D. Salabert, J. Tayar, MNRAS, **456**, 119 (2016), 1510.09023

[13] J.L. van Saders, T. Ceillier, T.S. Metcalfe, V.S. Aguirre, M.H. Pinsonneault, R.A. García, S. Mathur, G.R. Davies, Nature, **529**, 181 (2016), 1601.02631

[14] P.G. Beck, K. Hambleton, J. Vos, T. Kallinger, S. Bloemen, A. Tkachenko, R.A. García, R.H. Østensen, C. Aerts, D.W. Kurtz et al., A&A, **564**, A36 (2014), 1312.4500

[15] A. Prsa, K.E. Conroy, M. Horvat, H. Pablo, A. Kochoska, S. Bloemen, J. Nemravova, J. Giammarco, K.M. Hambleton, P. Degroote, ArXiv e-prints (2016), 1609.08135

[16] P. Gaulme, J. McKeever, J. Jackiewicz, M.L. Rawls, E. Corsaro, B. Mosser, J. Southworth, S. Mahadevan, C. Bender, R. Deshpande, ArXiv e-prints (2016), 1609.06645

[17] A. Baglin, M. Auvergne, L. Boisnard, T. Lam-Trong, P. Barge, C. Catala, M. Deleuil, E. Michel, W. Weiss, *CoRoT: a high precision photometer for stellar evolution and exoplanet finding*, in *36th COSPAR Scientific Assembly* (2006), Vol. 36 of *COSPAR, Plenary Meeting*, p. 3749

[18] S.R. Majewski, APOGEE Team, APOGEE-2 Team, Astronomische Nachrichten **337**, 863 (2016), 1509.05420

[19] M. Martig, H.W. Rix, V. Silva Aguirre, S. Hekker, B. Mosser, Y. Elsworth, J. Bovy, D. Stello, F. Anders, R.A. García et al., MNRAS, **451**, 2230 (2015), 1412.3453

[20] F. Anders, C. Chiappini, I. Minchev, A. Miglio, J. Montalbán, B. Mosser, T.S. Rodrigues, B.X. Santiago, F. Baudin, T.C. Beers et al., ArXiv e-prints (2016), 1608.04951

[21] T.S. Rodrigues, L. Girardi, A. Miglio, D. Bossini, J. Bovy, C. Epstein, M.H. Pinsonneault, D. Stello, G. Zasowski, C.A. Prieto et al., MNRAS, **445**, 2758 (2014), 1410.1350

[22] D. Huber, V. Silva Aguirre, J.M. Matthews, M.H. Pinsonneault, E. Gaidos, R.A. García, S. Hekker, S. Mathur, B. Mosser, G. Torres et al., ApJS, **211**, 2 (2014), 1312.0662

[23] S. Mathur, R.A. García, D. Huber, C. Regulo, D. Stello, P.G. Beck, K. Houmani, D. Salabert, ApJ, **827**, 50 (2016), 1605.03675

[24] J.M. Jenkins, D.A. Caldwell, H. Chandrasekaran, J.D. Twicken, S.T. Bryson, E.V. Quintana, B.D. Clarke, J. Li, C. Allen, P. Tenenbaum et al., ApJL, **713**, L120 (2010), 1001.0256

[25] R.A. García, S. Hekker, D. Stello, J. Gutiérrez-Soto, R. Handberg, D. Huber, C. Karoff, K. Uytterhoeven, T. Appourchaux, W.J. Chaplin et al., MNRAS, **414**, L6 (2011), 1103.0382

[26] R.A. García, S. Mathur, S. Pires, C. Régulo, B. Bellamy, P.L. Pallé, J. Ballot, S. Barceló Forteza, P.G. Beck, T.R. Bedding et al., A&A, **568**, A10 (2014), 1405.5374

[27] S. Pires, S. Mathur, R.A. García, J. Ballot, D. Stello, K. Sato, A&A, **574**, A18 (2015), 1410.6088

[28] S. Mathur, R.A. García, C. Régulo, O.L. Creevey, J. Ballot, D. Salabert, T. Arentoft, P. Quirion, W.J. Chaplin, H. Kjeldsen, A&A, **511**, A46 (2010), 0912.3367

[29] D. Huber, D. Stello, T.R. Bedding, W.J. Chaplin, T. Arentoft, P. Quirion, H. Kjeldsen, Communications in Asteroseismology **160**, 74 (2009), 0910.2764

[30] W.J. Chaplin, Y. Elsworth, G.R. Davies, T.L. Campante, R. Handberg, A. Miglio, S. Basu, MNRAS, **445**, 946 (2014), 1409.0696

[31] A. Bressan, P. Marigo, L. Girardi, B. Salasnich, C. Dal Cero, S. Rubele, A. Nanni, MNRAS, **427**, 127 (2012), 1208.4498

[32] D. Huber, T.R. Bedding, D. Stello, S. Hekker, S. Mathur, B. Mosser, G.A. Verner, A. Bonanno, D.L. Buzasi, T.L. Campante et al., ApJ, **743**, 143 (2011), 1109.3460

[33] H. Kjeldsen, T.R. Bedding, A&A, **293**, 87 (1995), arXiv:astro-ph/9403015

[34] H. Kjeldsen, T.R. Bedding, A&A, **529**, L8 (2011), 1104.1659

[35] P.G. Beck, T.R. Bedding, B. Mosser, D. Stello, R.A. Garcia, T. Kallinger, S. Hekker, Y. Elsworth, S. Frandsen, F. Carrier et al., Science **332**, 205 (2011)

[36] B. Mosser, C. Barban, J. Montalbán, P.G. Beck, A. Miglio, K. Belkacem, M.J. Goupil, S. Hekker, J. De Ridder, M.A. Dupret et al., A&A, **532**, A86 (2011), 1105.6113

[37] S. Sharma, J. Bland-Hawthorn, K.V. Johnston, J. Binney, ApJ, **730**, 3 (2011), 1101.3561

EPJ Web of Conferences **160**, 05002 (2017) DOI: 10.1051/epjconf/201716005002
Seismology of the Sun and the Distant Stars II

Formation history of open clusters constrained by detailed asteroseismology of red giant stars observed by *Kepler*

Enrico Corsaro[1,2,3,⋆], *Yueh-Ning* Lee[1], *Rafael A.* García[1], *Patrick* Hennebelle[1], *Savita* Mathur[4], *Paul G.* Beck[1], *Stephane* Mathis[1], *Dennis* Stello[5,6], and *Jérôme* Bouvier[7]

[1] *Laboratoire AIM Paris-Saclay, CEA/DRF — CNRS — Université Paris Diderot, IRFU/SAp Centre de Saclay, F-91191 Gif-sur-Yvette Cedex, France*
[2] *Instituto de Astrofísica de Canarias, E-38200 La Laguna, Tenerife, Spain*
[3] *Departamento de Astrofísica, Universidad de La Laguna, E-38205 La Laguna, Tenerife, Spain*
[4] *Space Science Institute, 4750 Walnut street Suite 205, Boulder, CO 80301, USA*
[5] *Sydney Institute for Astronomy (SIfA), School of Physics, University of Sydney, Sydney, New South Wales 2006, Australia*
[6] *Stellar Astrophysics Centre, Department of Physics and Astronomy, Aarhus University, Ny Munkegade 120, DK-8000 Aarhus C, Denmark*
[7] *Université Grenoble Alpes, IPAG, F-38000 Grenoble, France; CNRS, IPAG, F-38000 Grenoble, France*

Abstract. Stars originate by the gravitational collapse of a turbulent molecular cloud of a diffuse medium, and are often observed to form clusters. Stellar clusters therefore play an important role in our understanding of star formation and of the dynamical processes at play. However, investigating the cluster formation is difficult because the density of the molecular cloud undergoes a change of many orders of magnitude. Hierarchical-step approaches to decompose the problem into different stages are therefore required, as well as reliable assumptions on the initial conditions in the clouds. We report for the first time the use of the full potential of NASA *Kepler* asteroseismic observations coupled with 3D numerical simulations, to put strong constraints on the early formation stages of open clusters. Thanks to a Bayesian peak bagging analysis of about 50 red giant members of NGC 6791 and NGC 6819, the two most populated open clusters observed in the nominal *Kepler* mission, we derive a complete set of detailed oscillation mode properties for each star, with thousands of oscillation modes characterized. We therefore show how these asteroseismic properties lead us to a discovery about the rotation history of stellar clusters. Finally, our observational findings will be compared with hydrodynamical simulations for stellar cluster formation to constrain the physical processes of turbulence, rotation, and magnetic fields that are in action during the collapse of the progenitor cloud into a proto-cluster.

1 Introduction

Most of the star formation in the Galaxy is occurring within massive giant molecular clouds [1]. Giant molecular clouds are expected to form hundred of thousands of stars that can originate stellar clusters if the system is able to remain gravitationally bound well beyond its formation phase. This shows that investigating the early dynamical evolution of the molecular clouds is crucial for understanding the star and planet formation processes at different epochs and by taking into account the diversity of environments in the Galaxy [2, 3]. However, the star forming regions are obscured by dust, hence direct observations are limited to the infrared and radio bands. Stellar clusters, which are born embedded to giant molecular clouds, therefore represent important benchmarks to test stellar and planetary formation, and stellar evolution theories. Open clusters in particular generally contain a small amount or no interstellar medium, which makes them well suitable for studies involving a broad range of wavelengths other than the infrared and radio ones. In addition, conversely to globular clusters, stars in open clusters are sparse enough, with a density of at least 1 $M_\odot$ pc^{-3}, thus providing the opportunity to perform detailed analyses on individual stellar members. The great advantage of this approach is that the stars in a cluster — as opposed to field stars that often originate from the dissolution of small stellar systems — can preserve the signature of the initial conditions of the progenitor cloud.

Little is known on the role of angular momentum on the early stage formation of stars in open clusters. It is believed that molecular clouds satisfying the Jeans instability undergo a gravitational collapse in which the internal motions are strongly influenced by turbulence [4]. This implies that the angular momentum from the progenitor gas cloud cannot leave any significant imprint of their action in the resulting stars of the cluster. From a theoretical point of view, if the angular momentum vector of the cloud is transferred to the individual pre-stellar cores then rotation and possibly magnetic fields could be responsible of guiding the gravitational accretion of mass along preferential

⋆e-mail: enrico.corsaro@cea.fr

directions in space. For understanding this aspect, reliable measurements of the space orientation of the stellar-spin axis are required. Previous analyses conducted on young open clusters support no evidence of stellar-spin alignment [5], but they relied on a combination of measurements for stellar rotational periods from light curves, projected equatorial velocities from spectroscopy, and stellar radii based on the cluster distance. As a result, the adopted methodology could be prone to significant systematics, and it is not applicable for distant and old populations of stars, whereas the required primary information will lack.

Asteroseismology, the study of stellar oscillations, has proven to be a powerful tool to extract direct information on the stellar interiors, and in particular on the stellar angular momentum vector, thus allowing for accurate and detailed results, especially in the case of red giant stars [6]. Red giants are low- and intermediate-mass stars that have evolved off the main sequence to settle in either a hydrogen-shell-burning or helium-core-burning phase of the stellar evolution. Many red giants exhibit stellar oscillations that can be analyzed through a Fourier spectrum of the light curve. Such a Fourier spectrum contains a regular pattern of tens and sometimes more than a hundred of radial and non-radial stellar oscillation modes [7]. Each oscillation mode can be identified by means of an angular degree l and an azimuthal number m, the latter related to the degeneracy lifted by the stellar rotation to form a multiplet of $(2l + 1)$ components [8, 9]. The dipolar ($l = 1$) mixed modes [10] are the most suited for measuring the orientation of the spin axis.

2 Data analysis & results

Here we investigate 48 oscillating red giant members of the open clusters NGC 6791 and NGC 6819 [11–13], with masses spanning from 1.1 to 1.7 $M_\odot$ [14]. Both clusters are old, with NGC 6791 one of the oldest known in our Galaxy [15]. Despite their distance (on galactic scales up to ~4 kpc), thanks to the high luminosity of the red giants, the individual stellar members could be monitored by the photometric space mission NASA *Kepler* for a period of about 4 years. By performing detailed asteroseismic analyses on each star [16–20], we obtain a total of about 380 rotationally split dipolar mixed modes — identified from a set of more than 3900 oscillation modes — and we use them to measure the spin inclination angles.

The analysis of the rotationally split dipolar mixed modes shows that about 70% of the stars in each cluster has a strong preferential spin orientation, with inclination angles between 20° and 30°. This is contrary to what would be expected by assuming an uniform 3D distribution of spin vectors, where most of the inclination angles would then be found close to 90°. The spatial distribution of the stars also guarantees that they adequately sample the observed spherical morphology of the two stellar clusters.

Existing N-body simulations for old open clusters [21] show that the stellar angular momentum can influence the evolution of colliding stars and of the orbital configurations of multiple stellar systems, e.g. the orbital eccentricities and periods of binaries. In the case of single stars we instead observe only a spin down over time [22]. Because of the larger average distances among the star members of an open cluster, the N-body interactions cannot produce any appreciable effect on redistributing stellar spin inclinations and yielding a spin alignment even for timescales of several Gyr. As a consequence, the spin alignment that is observed in the stars of our sample should have a different origin, linked to the formation epoch of the cluster.

We have performed 3D hydrodynamical simulations that reproduce the collapse of a dense molecular cloud forming a proto-cluster [23]. The simulations show that when we consider a cloud dominated by turbulent kinetic energy, E_{tur}, over that of global rotation, E_{rot} (yielding the condition $E_{\text{rot}}/E_{\text{tur}} < 1$), no significant stellar spin alignment is produced in the final pre-stellar cores and the distribution of spin inclinations resembles that of a uniform orientation in 3D space. The scenario becomes comparable to that of our observations when $E_{\text{rot}}/E_{\text{tur}} \simeq 1$ instead. In addition, when considering masses below 0.7 $M_\odot$, even an energy equipartition between rotational and turbulent kinetic energy is not able to produce a significant spin alignment in the resulting pre-stellar cores.

3 Conclusion

Our results indicate that a mass above 0.7 $M_\odot$, accreted into a pre-stellar core from the progenitor molecular cloud, would be required in order to allow the cloud's average angular momentum to be transferred from the scale of the molecular cloud to those of individual stars. At least 50 % of the cloud's rotational kinetic energy can be responsible of efficiently aligning the spin axes within the stellar members of a cluster. Therefore, the two open clusters NGC 6791 and NGC 6819 are likely to be originating from a formation process that involved a dense collapsing molecular cloud with a strong rotational energy component. Through the measurement of stellar spin inclinations for stars with masses comparable to that of our Sun and belonging to open clusters we can obtain information on the initial energy budget of the progenitor molecular cloud, its global rotation, and the efficiency by which the cloud's averaged angular momentum is transferred to the individual stellar members of the clusters. This will lead to the exploration and reconstruction of the dynamical evolution of galactic star forming regions that have originated clusters with ages up to several Gyr.

Acknowledgements

E.C. is funded by the European Community's Seventh Framework Programme (FP7/2007-2013) under grant agreement n°312844 (SPACEINN). Y.-N.L. and P.H. acknowledge funding by the European Research Council under the European Community's Seventh Framework Programme (FP7/2007-2013 Grant Agreement no. 306483) and the HPC resources of CINES under the allocation x2014047023 made by GENCI (Grand Equipement National de Calcul Intensif). R.A.G. received funding from the CNES GOLF and PLATO grants at CEA. R.A.G. and P.G.B. received funding from the ANR (Agence Nationale de

DOI: 10.1051/epjconf/201716005002

la Recherche, France) program IDEE (n° ANR-12-BS05-0008) "Interaction Des Étoiles et des Exoplanètes". Sa.M. acknowledges support from the NASA grant NNX12AE17G. St.M. acknowledges funding by the European Research Council through ERC grant SPIRE 647383. D.S. is the recipient of an Australian Research Council Future Fellowship (project number FT140100147). J.B. acknowledges financial support from grant ANR 2011 Blanc SIMI5-6 020 "Toupies: Towards understanding the spin evolution of stars". This work has received funding from the CNES grants at CEA.

References

[1] Lee, E. J., Murray, N., & Rahman, M. 2012, ApJ, 752, 146
[2] Lada, C. J. & Lada, E. A. 2003, ARA&A, 41, 57
[3] Longmore, S. N., Kruijssen, J. M. D., Bastian, N., et al. 2014, Protostars and Planets VI, 291
[4] Shu, F. H., Adams, F. C., & Lizano, S. 1987, ARA&A, 25, 23
[5] Jackson, R. J. & Jeffries, R. D. 2010, MNRAS, 402, 1380
[6] Huber, D., Carter, J. A., Barbieri, M., et al. 2013, Science, 342, 331
[7] Bedding, T. R., Mosser, B., Huber, D., et al. 2011, Nature, 471, 608
[8] Beck, P. G., Montalban, J., Kallinger, T., et al. 2012, Nature, 481, 55
[9] Gizon, L. & Solanki, S. K. 2003, ApJ, 589, 1009
[10] Beck, P. G., Bedding, T. R., Mosser, B., et al. 2011, Science, 332, 205
[11] Basu, S., Grundahl, F., Stello, D., et al. 2011, ApJ, 729, L10
[12] Corsaro, E., Stello, D., Huber, D., et al. 2012, ApJ, 757, 190
[13] Stello, D., Meibom, S., Gilliland, R. L., et al. 2011, ApJ, 739, 13
[14] Miglio, A., Brogaard, K., Stello, D., et al. 2012, MNRAS, 419, 2077
[15] Brogaard, K., VandenBerg, D. A., Bruntt, H., et al. 2012, A&A, 543, A106
[16] García, R. A., Hekker, S., Stello, D., et al. 2011, MNRAS, 414, L6
[17] García, R. A., Mathur, S., Pires, S., et al. 2014, A&A, 568, A10
[18] Mathur, S., García, R. A., Régulo, C., et al. 2010, A&A, 511, A46
[19] Corsaro, E. & De Ridder, J. 2014, A&A, 571, A71
[20] Corsaro, E., De Ridder, J., & García, R. A. 2015, A&A, 579, A83
[21] Geller, A. M., Hurley, J. R., & Mathieu, R. D. 2013, AJ, 145, 8
[22] Meibom, S., Barnes, S. A., Platais, I., et al. 2015, Nature, 517, 589
[23] Lee, Y.-N. & Hennebelle, P. 2016, A&A, 591, A30

EPJ Web of Conferences **160**, 05003 (2017) DOI: 10.1051/epjconf/201716005003
Seismology of the Sun and the Distant Stars II

Stellar Parameters in an Instant with Machine Learning

Application to *Kepler* LEGACY Targets

Earl P. Bellinger[1,2,3,4,⋆], *George C.* Angelou[1,3], *Saskia* Hekker[1,3], *Sarbani* Basu[2], *Warrick H.* Ball[5], and *Elisabeth* Guggenberger[1,3]

[1] *Max-Planck-Institut für Sonnensystemforschung, Göttingen, Germany*
[2] *Department of Astronomy, Yale University, New Haven, CT, USA*
[3] *Stellar Astrophysics Centre, Department of Physics and Astronomy, Aarhus University, Denmark*
[4] *Institut für Informatik, Georg-August-Universität Göttingen, Germany*
[5] *Institut für Astrophysik, Georg-August-Universität Göttingen, Germany*

Abstract. With the advent of dedicated photometric space missions, the ability to rapidly process huge catalogues of stars has become paramount. Bellinger and Angelou et al. [1] recently introduced a new method based on machine learning for inferring the stellar parameters of main-sequence stars exhibiting solar-like oscillations. The method makes precise predictions that are consistent with other methods, but with the advantages of being able to explore many more parameters while costing practically no time. Here we apply the method to 52 so-called "LEGACY" main-sequence stars observed by the *Kepler* space mission. For each star, we present estimates and uncertainties of mass, age, radius, luminosity, core hydrogen abundance, surface helium abundance, surface gravity, initial helium abundance, and initial metallicity as well as estimates of their evolutionary model parameters of mixing length, overshooting coefficient, and diffusion multiplication factor. We obtain median uncertainties in stellar age, mass, and radius of 14.8%, 3.6%, and 1.7%, respectively. The source code for all analyses and for all figures appearing in this manuscript can be found electronically at https://github.com/earlbellinger/asteroseismology [2].

1 Introduction

The *Kepler* seismic LEGACY sample data represents the best-quality observations of cool dwarf stars obtained during the nominal and extended 4-year mission of the *Kepler* spacecraft. These stars, thought to be of a similar evolutionary stage as our Sun, serve as an excellent testbed for theories of stellar structure and evolution. In Bellinger and Angelou et al. (hereinafter Paper 1), we introduced a method for determining the current structural parameters and evolutionary model parameters of main-sequence stars from asteroseismic observations. Here we apply that method to the LEGACY sample and present estimates of their parameters.

2 Data

The *Kepler* seismic LEGACY sample data were obtained from Lund et al. [3]. These data include individual frequencies, effective temperatures, frequencies of maximum oscillation power, and metallicities of 66 stars. Although none of the LEGACY stars show mixed modes, which would be an indication that the core hydrogen burning evolutionary phase has ceased, there is no way *a priori* to determine the evolutionary status of a star. Some of these stars may have already depleted their supply of core hydrogen and begun hydrogen shell burning. As our method is currently restricted to stars on the main sequence, i.e. stars with a fractional core hydrogen abundance $X_c \geq 10^{-3}$, we wish to only apply the method to stars that are still in this phase. Therefore, in order to be confident in our estimates, we adopt a very conservative inclusion criterion and do not present estimates for any stars with any part of their estimated core hydrogen distribution having $X_c \leq 10^{-2}$. In order to perform a selection with this criterion, we first ran our algorithm on all 66 stars. Then, for each star, we checked whether any of the 10 000 samples we obtained from the posterior X_c distribution were smaller than that cutoff, and excluded the ones that were. Of the original set, 52 stars remain. The stars are visualized in an asteroseismic frequency separation diagram in Figure 1.

3 Results

In Table 1 we present the means and standard deviations of current and initial parameters for 52 stars of the *Kepler* LEGACY sample as inferred via the machine learning method presented in Paper 1. Figure 2 further shows the cumulative distributions of uncertainties for each of these parameters. Nearly all of the masses are estimated to better than 5% accuracy, with an overall average uncertainty of

⋆e-mail: earl.bellinger@yale.edu

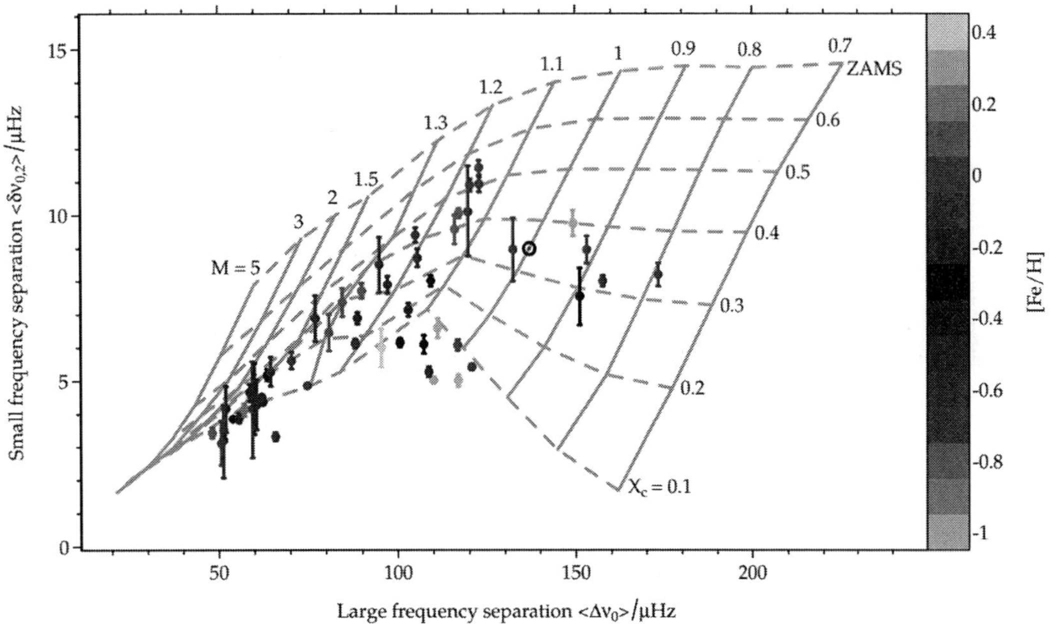

Figure 1: Small frequency separations against large frequency separations with [Fe/H] indicated by color for 52 main-sequence *Kepler* LEGACY stars overplotted on top of evolutionary models varied in mass with solar-calibrated mixing length and abundances generated using MESA [4] with frequencies calculated using GYRE [5]. If all stars had the solar abundances and solar mixing length, it would suffice to look up their mass and core-hydrogen abundance in this diagram. Since they do not, a more sophisticated approach is required; here we employ the method introduced in Paper 1 for this task.

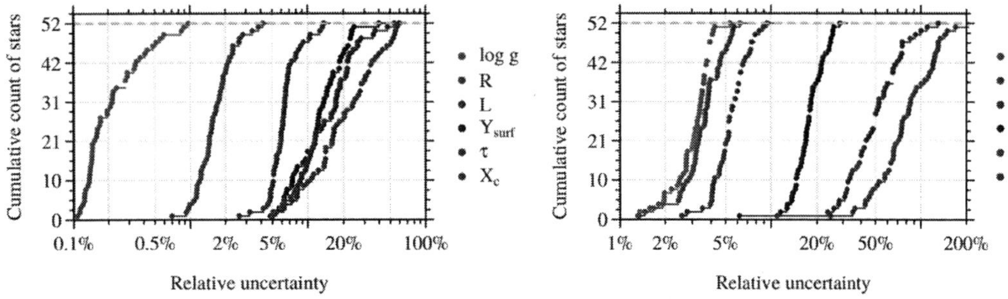

Figure 2: Cumulative distribution functions showing the relative uncertainties in estimated current parameters (left) and initial parameters (right) for 52 main-sequence *Kepler* LEGACY stars.

3.6%. The star with the best-constrained mass and age is KIC 8760414, an old star of 10.56 Gyr that is less massive but larger than our Sun with a mass uncertainty of 1.34% and an age uncertainty of only 5%. The star KIC 9139151 also has a very well-constrained age: a young star of 1.85 Gyr, its age is estimated with an uncertainty of just 210 million years.

Surface gravity estimates are nearly an order of magnitude more precise than any other quantity, with all stars in the sample measured to better than 1% uncertainty. In second place are radius estimates, then followed by initial helium Y_0. At first glance, this may seem surprising. However, our grid imposes a uniform prior in Y_0 spanning from $a = 0.22$ to $b = 0.34$. Thus, the largest uncertainty that would be theoretically possible is

$$\max\left(\frac{\sigma^2(Y_0)}{Y_0}\right) = \frac{|b-a|}{|a|} \cdot 100 = 54.51\%. \tag{1}$$

Hence, the "small" relative uncertainties on Y_0 are actually unsurprising despite the intrinsic uncertainty in retrodicting this quantity (c.f. Paper 1 §2.3.3). Contrast this to stel-

Table 1: Means and standard deviations for current age τ, core hydrogen abundance X_c, surface gravity $\log g$, luminosity L, radius R, surface helium abundance Y_{surf}; and initial conditions of mass M, initial helium Y_0, initial metallicity Z_0, mixing length parameter α_{MLT}, overshooting coefficient α_{ov}, and diffusion multiplication factor D of the *Kepler* Legacy data set inferred via machine learning.

KIC	τ/Gyr	X_c	$\log g$	$L/L_\odot$	$R/R_\odot$	Y_{surf}	$M/M_\odot$	Y_0	Z_0	α_{MLT}	α_{ov}	D
1435467	2.10 ± 0.27	0.328 ± 0.037	4.133 ± 0.011	4.22 ± 0.23	1.705 ± 0.032	0.183 ± 0.038	1.439 ± 0.044	0.2645 ± 0.0089	0.0262 ± 0.0038	1.798 ± 0.091	0.095 ± 0.072	2.7 ± 2.7
3427720	2.8 ± 1.6	0.44 ± 0.13	4.376 ± 0.018	1.56 ± 0.12	1.123 ± 0.026	0.223 ± 0.031	1.090 ± 0.056	0.280 ± 0.011	0.0193 ± 0.0038	1.87 ± 0.14	0.27 ± 0.17	15 ± 11
3456181	3.27 ± 0.60	0.091 ± 0.050	3.940 ± 0.022	6.36 ± 0.49	2.056 ± 0.060	0.254 ± 0.029	1.342 ± 0.084	0.277 ± 0.011	0.0147 ± 0.0031	1.95 ± 0.11	0.42 ± 0.15	1.3 ± 1.8
3632418	3.80 ± 0.71	0.102 ± 0.058	4.015 ± 0.024	4.62 ± 0.32	1.857 ± 0.049	0.222 ± 0.037	1.298 ± 0.071	0.2718 ± 0.0098	0.0179 ± 0.0039	1.91 ± 0.13	0.32 ± 0.15	3.5 ± 3.6
3735871	1.51 ± 0.33	0.515 ± 0.033	4.3997 ± 0.0080	1.571 ± 0.086	1.119 ± 0.011	0.216 ± 0.031	1.145 ± 0.035	0.271 ± 0.011	0.0208 ± 0.0036	1.95 ± 0.15	0.097 ± 0.064	17 ± 12
5184732	3.86 ± 0.73	0.246 ± 0.078	4.266 ± 0.012	2.01 ± 0.13	1.374 ± 0.020	0.254 ± 0.029	1.271 ± 0.047	0.274 ± 0.011	0.0403 ± 0.0068	1.85 ± 0.16	0.050 ± 0.045	2.1 ± 3.5
5773345	2.56 ± 0.22	0.198 ± 0.030	4.029 ± 0.013	4.98 ± 0.30	1.964 ± 0.030	0.239 ± 0.025	1.496 ± 0.023	0.2728 ± 0.0095	0.0329 ± 0.0037	1.742 ± 0.064	0.180 ± 0.091	2.1 ± 1.8
6116048	5.96 ± 0.43	0.080 ± 0.030	4.2776 ± 0.0051	1.861 ± 0.087	1.232 ± 0.013	0.220 ± 0.020	1.048 ± 0.028	0.2590 ± 0.0064	0.0130 ± 0.0018	1.805 ± 0.074	0.060 ± 0.018	2.2 ± 1.4
6225718	2.23 ± 0.38	0.390 ± 0.053	4.3203 ± 0.0058	2.25 ± 0.12	1.252 ± 0.017	0.218 ± 0.033	1.194 ± 0.039	0.2730 ± 0.0086	0.0191 ± 0.0033	1.97 ± 0.13	0.096 ± 0.070	6.0 ± 5.2
6508366	3.27 ± 0.58	0.105 ± 0.053	3.938 ± 0.034	6.33 ± 0.60	2.087 ± 0.081	0.256 ± 0.026	1.371 ± 0.079	0.279 ± 0.010	0.0179 ± 0.0040	1.99 ± 0.12	0.46 ± 0.14	2.7 ± 3.2
6603624	7.65 ± 0.62	0.063 ± 0.016	4.3355 ± 0.0071	1.301 ± 0.074	1.182 ± 0.015	0.255 ± 0.012	1.104 ± 0.040	0.2609 ± 0.0098	0.0327 ± 0.0061	1.87 ± 0.11	0.046 ± 0.024	0.30 ± 0.43
6679371	2.54 ± 0.34	0.109 ± 0.053	3.935 ± 0.019	7.38 ± 0.51	2.159 ± 0.052	0.271 ± 0.017	1.461 ± 0.056	0.280 ± 0.012	0.0189 ± 0.0029	1.98 ± 0.11	0.37 ± 0.14	0.9 ± 1.2
7103006	3.01 ± 0.63	0.181 ± 0.098	4.038 ± 0.041	5.12 ± 0.54	1.868 ± 0.082	0.243 ± 0.033	1.379 ± 0.063	0.2782 ± 0.0070	0.0228 ± 0.0050	1.94 ± 0.13	0.33 ± 0.15	4.3 ± 5.3
7106245	5.64 ± 0.92	0.143 ± 0.032	4.3081 ± 0.0091	1.478 ± 0.065	1.090 ± 0.014	0.099 ± 0.040	0.879 ± 0.021	0.2821 ± 0.0098	0.0054 ± 0.0012	1.97 ± 0.13	0.061 ± 0.039	12.1 ± 4.3
7510397	3.66 ± 0.34	0.087 ± 0.015	4.031 ± 0.0040	4.36 ± 0.28	1.822 ± 0.038	0.192 ± 0.022	1.302 ± 0.058	0.2601 ± 0.0048	0.0157 ± 0.0026	1.893 ± 0.094	0.225 ± 0.068	3.0 ± 1.3
7871531	9.1 ± 2.0	0.285 ± 0.070	4.47 ± 0.012	0.675 ± 0.052	0.890 ± 0.017	0.211 ± 0.027	0.852 ± 0.033	0.2645 ± 0.0093	0.0137 ± 0.0023	1.881 ± 0.079	0.19 ± 0.15	6.2 ± 4.6
7940546	3.47 ± 0.43	0.068 ± 0.031	4.0075 ± 0.0088	4.91 ± 0.32	1.881 ± 0.036	0.215 ± 0.027	1.311 ± 0.058	0.2655 ± 0.0072	0.0160 ± 0.0027	1.79 ± 0.10	0.197 ± 0.080	1.9 ± 1.4
7970740	10.41 ± 0.98	0.299 ± 0.023	4.5409 ± 0.0052	0.443 ± 0.022	0.7770 ± 0.0048	0.181 ± 0.027	0.764 ± 0.011	0.2485 ± 0.0050	0.0088 ± 0.0014	2.015 ± 0.098	0.039 ± 0.030	4.8 ± 2.4
8006161	5.1 ± 1.1	0.412 ± 0.037	4.4862 ± 0.0088	0.748 ± 0.046	0.953 ± 0.016	0.262 ± 0.016	1.014 ± 0.031	0.2823 ± 0.0087	0.0387 ± 0.0078	2.055 ± 0.096	0.25 ± 0.13	3.1 ± 2.5
8150065	2.83 ± 0.32	0.304 ± 0.037	4.2363 ± 0.0059	2.66 ± 0.18	1.403 ± 0.024	0.177 ± 0.043	1.235 ± 0.042	0.2680 ± 0.0086	0.0202 ± 0.0035	1.88 ± 0.12	0.061 ± 0.026	6.7 ± 4.2
8179536	1.92 ± 0.90	0.43 ± 0.12	4.275 ± 0.017	2.76 ± 0.17	1.358 ± 0.025	0.220 ± 0.042	1.265 ± 0.069	0.276 ± 0.011	0.0216 ± 0.0049	1.88 ± 0.17	0.15 ± 0.13	8.9 ± 9.9
8228742	3.73 ± 0.31	0.049 ± 0.016	4.0231 ± 0.0043	4.37 ± 0.22	1.850 ± 0.031	0.222 ± 0.018	1.317 ± 0.051	0.2637 ± 0.0053	0.0192 ± 0.0042	1.834 ± 0.078	0.195 ± 0.063	1.83 ± 0.75
8379927	1.43 ± 0.35	0.523 ± 0.027	4.3885 ± 0.0091	1.56 ± 0.13	1.126 ± 0.016	0.196 ± 0.035	1.131 ± 0.050	0.277 ± 0.015	0.0205 ± 0.0047	1.90 ± 0.15	0.133 ± 0.089	23 ± 14
8394589	3.62 ± 0.40	0.272 ± 0.038	4.3255 ± 0.0057	1.842 ± 0.089	1.184 ± 0.013	0.193 ± 0.030	1.081 ± 0.028	0.2625 ± 0.0064	0.0130 ± 0.0019	1.849 ± 0.098	0.056 ± 0.024	6.1 ± 3.4
8424992	8.98 ± 0.52	0.096 ± 0.011	4.3643 ± 0.0050	1.081 ± 0.049	1.052 ± 0.018	0.234 ± 0.018	0.932 ± 0.031	0.2677 ± 0.0054	0.0153 ± 0.0031	1.781 ± 0.066	0.037 ± 0.013	1.6 ± 1.1
8760414	10.56 ± 0.54	0.0553 ± 0.0071	4.3194 ± 0.0063	1.191 ± 0.037	1.0373 ± 0.0086	0.141 ± 0.023	0.816 ± 0.011	0.2599 ± 0.0042	0.00499 ± 0.00065	1.888 ± 0.046	0.115 ± 0.032	7.6 ± 2.0
9025370	4.1 ± 1.5	0.374 ± 0.075	4.42 ± 0.015	0.76 ± 0.11	1.009 ± 0.024	0.147 ± 0.037	0.972 ± 0.047	0.2722 ± 0.0092	0.0287 ± 0.0072	1.879 ± 0.087	0.177 ± 0.099	30 ± 13
9098294	8.75 ± 0.56	0.059 ± 0.016	4.3139 ± 0.0049	1.355 ± 0.071	1.128 ± 0.016	0.236 ± 0.015	0.955 ± 0.029	0.2630 ± 0.0060	0.0132 ± 0.0023	1.760 ± 0.071	0.106 ± 0.035	1.32 ± 0.86
9139151	1.85 ± 0.21	0.451 ± 0.032	4.3826 ± 0.0054	1.93 ± 0.10	1.167 ± 0.018	0.260 ± 0.017	1.199 ± 0.042	0.2726 ± 0.0099	0.0225 ± 0.0042	2.08 ± 0.11	0.084 ± 0.044	1.6 ± 2.1
9139163	1.94 ± 0.50	0.408 ± 0.067	4.199 ± 0.014	3.71 ± 0.21	1.566 ± 0.025	0.224 ± 0.029	1.415 ± 0.053	0.2754 ± 0.0085	0.0307 ± 0.0052	2.00 ± 0.15	0.25 ± 0.15	3.9 ± 4.3
9206432	1.75 ± 0.39	0.358 ± 0.074	4.2144 ± 0.0089	3.69 ± 0.20	1.496 ± 0.024	0.274 ± 0.021	1.339 ± 0.059	0.291 ± 0.012	0.0226 ± 0.0042	1.89 ± 0.18	0.033 ± 0.035	0.48 ± 0.92
9353712	2.98 ± 0.38	0.117 ± 0.033	3.9332 ± 0.0052	6.34 ± 0.38	2.121 ± 0.030	0.257 ± 0.027	1.406 ± 0.045	0.282 ± 0.010	0.0190 ± 0.0031	1.937 ± 0.092	0.43 ± 0.13	1.12 ± 0.93
9410862	7.17 ± 0.84	0.072 ± 0.029	4.3060 ± 0.0059	1.62 ± 0.094	1.148 ± 0.019	0.235 ± 0.016	0.971 ± 0.034	0.2641 ± 0.0075	0.0103 ± 0.0018	1.817 ± 0.095	0.087 ± 0.047	1.6 ± 1.3
9414417	3.66 ± 0.79	0.129 ± 0.078	4.017 ± 0.034	4.81 ± 0.39	1.856 ± 0.063	0.224 ± 0.042	1.301 ± 0.074	0.2740 ± 0.0093	0.0179 ± 0.0049	1.92 ± 0.12	0.34 ± 0.15	4.4 ± 5.7
9812850	2.46 ± 0.43	0.263 ± 0.065	4.095 ± 0.021	4.53 ± 0.31	1.765 ± 0.045	0.185 ± 0.041	1.412 ± 0.054	0.266 ± 0.011	0.0235 ± 0.0045	1.83 ± 0.11	0.14 ± 0.10	4.5 ± 5.7
9955598	7.2 ± 1.2	0.363 ± 0.039	4.4916 ± 0.0077	0.662 ± 0.041	0.902 ± 0.013	0.246 ± 0.020	0.920 ± 0.029	0.270 ± 0.010	0.0223 ± 0.0038	1.940 ± 0.092	0.22 ± 0.14	3.0 ± 2.2
9965715	3.49 ± 0.59	0.281 ± 0.049	4.269 ± 0.012	1.96 ± 0.18	1.267 ± 0.028	0.140 ± 0.033	1.068 ± 0.048	0.2814 ± 0.0086	0.0154 ± 0.0036	1.709 ± 0.073	0.080 ± 0.036	15.8 ± 7.5
10068307	3.91 ± 0.29	0.071 ± 0.014	3.9423 ± 0.0054	5.24 ± 0.29	2.012 ± 0.020	0.225 ± 0.027	1.287 ± 0.028	0.2720 ± 0.0035	0.0145 ± 0.0018	1.900 ± 0.066	0.431 ± 0.048	2.5 ± 1.8
10079226	2.21 ± 0.62	0.447 ± 0.065	4.3701 ± 0.0097	1.57 ± 0.10	1.174 ± 0.022	0.217 ± 0.033	1.177 ± 0.046	0.271 ± 0.011	0.0284 ± 0.0050	1.90 ± 0.15	0.13 ± 0.11	14 ± 12
10162436	3.43 ± 0.29	0.085 ± 0.030	3.9788 ± 0.0072	5.07 ± 0.29	1.970 ± 0.028	0.202 ± 0.026	1.345 ± 0.042	0.2696 ± 0.0047	0.0182 ± 0.0025	1.896 ± 0.076	0.280 ± 0.096	3.7 ± 1.8
10454113	1.36 ± 0.32	0.490 ± 0.030	4.3257 ± 0.0064	2.11 ± 0.12	1.263 ± 0.015	0.163 ± 0.035	1.229 ± 0.029	0.2786 ± 0.0094	0.0251 ± 0.0038	2.00 ± 0.12	0.098 ± 0.066	25 ± 11
10644253	1.06 ± 0.28	0.562 ± 0.034	4.4029 ± 0.0070	1.525 ± 0.089	1.123 ± 0.016	0.234 ± 0.026	1.163 ± 0.044	0.277 ± 0.011	0.0239 ± 0.0038	1.85 ± 0.13	0.093 ± 0.055	16 ± 12
10730618	4.16 ± 0.80	0.171 ± 0.024	4.0547 ± 0.0087	4.01 ± 0.52	1.760 ± 0.037	0.206 ± 0.050	1.275 ± 0.071	0.2764 ± 0.0077	0.0199 ± 0.0053	1.994 ± 0.099	0.42 ± 0.11	5.8 ± 4.7
10963065	4.35 ± 0.46	0.158 ± 0.044	4.2929 ± 0.0062	1.998 ± 0.099	1.234 ± 0.014	0.224 ± 0.026	1.089 ± 0.030	0.2633 ± 0.0076	0.0136 ± 0.0024	1.81 ± 0.11	0.052 ± 0.023	2.4 ± 1.8
11081729	1.69 ± 0.27	0.428 ± 0.033	4.2537 ± 0.0051	3.46 ± 0.18	1.446 ± 0.019	0.247 ± 0.020	1.368 ± 0.044	0.2739 ± 0.0081	0.0249 ± 0.0044	2.11 ± 0.14	0.118 ± 0.059	1.8 ± 2.5
11253226	1.75 ± 0.60	0.456 ± 0.066	4.19 ± 0.020	4.34 ± 0.28	1.567 ± 0.028	0.224 ± 0.040	1.388 ± 0.069	0.2735 ± 0.0092	0.0214 ± 0.0054	1.90 ± 0.11	0.35 ± 0.17	3.7 ± 4.8
11772920	9.08 ± 0.92	0.305 ± 0.014	4.4958 ± 0.0064	0.497 ± 0.067	0.8635 ± 0.0095	0.202 ± 0.048	0.851 ± 0.026	0.265 ± 0.011	0.0208 ± 0.0058	1.92 ± 0.10	0.124 ± 0.079	6.2 ± 5.6
12009504	3.64 ± 0.32	0.181 ± 0.029	4.2147 ± 0.0053	2.72 ± 0.13	1.424 ± 0.017	0.213 ± 0.028	1.212 ± 0.037	0.2649 ± 0.0080	0.0177 ± 0.0024	1.90 ± 0.11	0.047 ± 0.022	2.2 ± 1.5
12069127	2.71 ± 0.30	0.106 ± 0.048	3.9052 ± 0.0071	7.03 ± 0.41	2.232 ± 0.026	0.278 ± 0.018	1.460 ± 0.037	0.292 ± 0.010	0.0234 ± 0.0035	1.867 ± 0.090	0.41 ± 0.15	0.57 ± 0.57
12069449	6.89 ± 0.36	0.154 ± 0.027	4.3547 ± 0.0064	1.234 ± 0.038	1.115 ± 0.012	0.244 ± 0.013	1.024 ± 0.015	0.2675 ± 0.0049	0.0211 ± 0.0014	1.791 ± 0.048	0.103 ± 0.033	1.34 ± 0.73
12258514	4.25 ± 0.34	0.0677 ± 0.0097	4.1225 ± 0.0050	3.06 ± 0.15	1.618 ± 0.021	0.217 ± 0.027	1.264 ± 0.043	0.2646 ± 0.0073	0.0221 ± 0.0032	1.85 ± 0.10	0.100 ± 0.024	2.7 ± 1.9
12317678	2.73 ± 0.28	0.094 ± 0.035	4.0494 ± 0.0052	5.39 ± 0.34	1.781 ± 0.029	0.262 ± 0.019	1.297 ± 0.047	0.2767 ± 0.0087	0.0114 ± 0.0015	1.85 ± 0.11	0.155 ± 0.091	0.20 ± 0.23
Sun	4.62 ± 0.17	0.355 ± 0.012	4.4393 ± 0.0042	1.015 ± 0.043	1.0004 ± 0.0064	0.2458 ± 0.0077	1.0022 ± 0.0095	0.2665 ± 0.0035	0.0182 ± 0.0011	1.819 ± 0.036	0.072 ± 0.023	2.14 ± 0.72

Note. The values obtained from degraded solar data (c.f. Paper 1 §3.2) predicted on these quantities are shown for reference.

Table 2: Maximum theoretically possible uncertainties, average uncertainties attained, and truncated average explained variance of the twelve estimated quantities for the 52 *Kepler* LEGACY stars.

Quantity	σ^2_{max}	σ^2_{mean}	$V^{\text{trunc}}_{\text{e, mean}}$
τ	∞	16.76%	0.710
X_c	32 612 260%	23.98%	0.653
$\log g$	56.42%	0.26%	0.932
L	71 054.89%	6.59%	0.818
R	919.52%	1.74%	0.906
Y_{surf}	∞	13.58%	0.739
M	128.48%	3.67%	0.830
Y_0	54.51%	3.14%	0.759
Z_0	999 060%	18.19%	0.698
α_{MLT}	66.65%	5.60%	0.710
α_{ov}	∞	53.29%	0.483
D	∞	86.86%	0.340

Note. Quantities that can take on a value of zero, such as a ZAMS star with $\tau = 0$, have $\sigma^2_{\text{max}} = \infty$.

lar mass, whose uncertainties are comparable, but whose maximum possible uncertainty is 128%.

In Table 2, we show these maximum theoretically possible uncertainties for all twelve quantities that we estimate. We compare them with the actual average uncertainties obtained across the 52 stars analyzed here. We also calculate a truncated average explained variance score

$$V^{\text{trunc}}_{\text{e, mean}} = 1 - \frac{\sqrt{\sigma^2_{\text{mean}}}}{\sqrt{\min\left(\sigma^2_{\text{max}}, 200\right)}} \tag{2}$$

(c.f. Paper 1 equation 8), which gives an indication of how well the predictions compare with a random guess, with a score of zero being no better and a score of one being much better. We truncate at 200 because quantities that can take on a value of zero otherwise vacuously give $V_{\text{e, mean}} = 1$.

Based on these scores, the most well-constrained parameters are $\log g$, R, and M. The parameters that are hardest to constrain are the α_{ov} and D. All of the stars have $\sigma^2(D)/D > 0.2$ and nearly a third of them are more than 100% uncertain. This highlights several aspects. First, since D can take on a value of zero, an infinite relative uncertainty is possible. Second, D is highly degenerate with the parameters controlling the initial chemical composition. These uncertainties may merely represent that degeneracy. Third, there are mixing processes that are not correctly accounted for in one-dimensional stellar modelling. Extreme values of D may therefore be compensating for those processes. Finally, there may be seismic diagnostics, e.g. glitch analysis, that would be able to better constrain diffusion, but are absent from the present analysis.

4 Conclusions

In this paper, we applied machine learning techniques to estimate structural and evolutionary parameters of main-sequence stars. We achieved extremely precise estimates of stellar mass and radius using asteroseismologic diagnostics that are competitive with orbital modelling and even direct interferometric measurements. Hence, these estimates represent one of the largest and most precise collections of main-sequence stellar parameters.

There are other modelling efforts of this LEGACY sample. Silva Aguirra et al. [6] applied seven different techniques based on iterative optimization to estimate the parameters of these stars. Although we have shown in Paper 1 that the results are in good agreement, the philosophy of our approach is fundamentally different from those seven. Those approaches are based on various strategies for searching through grids of models in order to optimize a goodness-of-fit criterion. Our approach, which is based on classification and regression trees (CART), works without searching and essentially without the tuning of hyperparameters. Moreover, our approach enables estimation of many more stellar parameters, such as the initial helium abundance, mixing length parameter, overshooting coefficient, and diffusion multiplication factor, which would be too computationally expensive to vary with search-based methods, while still only taking seconds per star.

We have omitted several stars due to their proximity to the end of the main sequence. We are currently working on extending this method to more evolved stellar types, and we are soon to release a follow-up paper analyzing these omitted stars as well as more evolved ones.

Acknowledgements

The research leading to the presented results has received funding from the European Research Council under the European Community's Seventh Framework Programme (FP7/2007-2013) / ERC grant agreement no 338251 (StellarAges). This research was undertaken in the context of the International Max Planck Research School for Solar System Research. S.B. acknowledges partial support from NSF grant AST-1514676 and NASA grant NNX13AE70G. W.H.B. acknowledges research funding by Deutsche Forschungsgemeinschaft (DFG) under grant SFB 963/1 "Astrophysical flow instabilities and turbulence" (Project A18).

Software

Analysis was performed with R 3.2.3 [7] and the R libraries magicaxis 1.9.4 [8], parallelMap 1.3 [9], and matrixStats 0.50.1 [10].

References

[1] Bellinger, E. P. and Angelou, G. C. et al., ApJ **828**, (2016) 20
[2] Bellinger, E. P., asteroseismology, DOI: 10.5281/zenodo.160399
[3] Lund, M. L. et al., submitted (2016)
[4] Paxton, B., Bildsten, L., Dotter, A., et al., ApJSS **192**, (2011) 3
[5] Townsend, R. H. D. and S. A. Teitler, MNRAS **435** (2013) 4
[6] Silva Aguirra, V. et al., submitted (2016)

DOI: 10.1051/epjconf/201716005003

[7] R Core Team, R 3.2.3 (2014)
[8] Robotham, A., magicaxis 1.9.4 (2015)
[9] Bischl, B. and Lang, M., parallelMap 1.3 (2015)
[10] Bengtsson, H., matrixStats 0.14.2 (2015)

New Insights of High-precision Asteroseismology: Acoustic Radius and χ^2-matching Method for Solar-like Oscillator KIC 6225718

Tao Wu[1,2,*] and *Yan* Li[1,2,**]

[1] *Yunnan Observatories, Chinese Academy of Sciences, P.O. Box 110, Kunming 650216, China*
[2] *Key Laboratory for Structure and Evolution of Celestial Objects, Chinese Academy of Sciences, P.O. Box 110, Kunming 650216, China*

Abstract. Asteroseismology is a powerful tool for probing stellar interiors and determining stellar fundamental parameters. In the present work, we adopt the χ^2-minimization method but only use the observed high-precision seismic observations (i.e., oscillation frequencies) to constrain theoretical models for analyzing solar-like oscillator KIC 6225718. Finally, we find the acoustic radius τ_0 is the only global parameter that can be accurately measured by the χ^2-matching method between observed frequencies and theoretical model calculations for a pure p-mode oscillation star. We obtain $\tau_0 = 4601.5^{+4.4}_{-8.3}$ seconds for KIC 6225718. It leads that the mass and radius of the CMMs are degenerate with each other. In addition, we find that the distribution range of acoustic radius is slightly enlarged by some extreme cases, which posses both a larger mass and a higher (or lower) metal abundance, at the lower acoustic radius end.

1 Introduction

Asteroseismology is a useful tool, which is usually used to determine stellar fundamental parameters, such as stellar mass, radius, surface gravity, and mean density. It is as well as used to probe the interiors of stars, such as constraining the convective overshooting of stellar convective core [1, 2] and distinguishing the status of stellar core on the red giant branches – H-shell burning stars whose center is an radiative electric degenerate He-core or He-core burning stars whose center is a convective core [3, 4].

Thanks to the space-based missions, such as *CoRoT* [5], *Kepler* [6, 7], and the following *K2* [8, 9], more and more unexpected high-precision observations have been obtained. It makes the asteroseismic researches enter the age of high-precision. In the next few years, the observation missions, such as *TESS* [10] and *PLATO 2.0* [11], will support more and more high-precision observations for the researches of asteroseismology. They will lead to the research of asteroseismology reach a new age.

In order to determine the stellar fundamental parameters and/or probe the interiors of stars, comparisons between observations and model calculations are usually used to decide the best-fitting model whose parameters and interiors will be used to characterize those of the observations [1, 2, 12–16]. The goodness of the comparison is characterized by a χ^2_x-minimization method, which is defined as [17, 18]:

$$\chi^2_x = \frac{1}{N}\sum_{i=1}^{N}\frac{\left(x_i^{\rm obs} - x_i^{\rm mod}\right)^2}{\sigma^2_{x_i^{\rm obs}} + \sigma^2_{x_i^{\rm mod}}}, \qquad (1)$$

where superscripts "obs" and "mod" denote the observations and model calculations, respectively. σ and N are the uncertainty of the quantity x and the numbers of available quantity x, respectively. It is turn to

$$\chi^2_x = \frac{1}{N}\sum_{i=1}^{N}\left(\frac{x_i^{\rm obs} - x_i^{\rm mod}}{\sigma_{x_i^{\rm obs}}}\right)^2, \qquad (2)$$

when $\sigma_{x_i^{\rm mod}} << \sigma_{x_i^{\rm obs}}$. Obviously, for the value of χ^2_x, the smaller, the better. The theoretical model is closer to the observations.

In the present work, we will try to only use those high-precision seismic observations (i.e., oscillation frequencies ν_{nl}) to analyze the theoretical calculation models. However, what can the χ^2_ν analysis tell us about the physical information of the investigated targets? Or, in other words, what physical parameters of the investigated target can be determined best with such an analysis method? In the following, we will try to answer such question via analyzing the theoretical calculation models and the oscillation frequencies of the solar-like oscillator *KIC 6225718*.

KIC 6225718

KIC 6225718 is also named HIP (HIC) 97527 and HD 187637. It is a low-mass main-sequence star [19], whose

* e-mail: wutao@ynao.ac.cn
** e-mail: ly@ynao.ac.cn

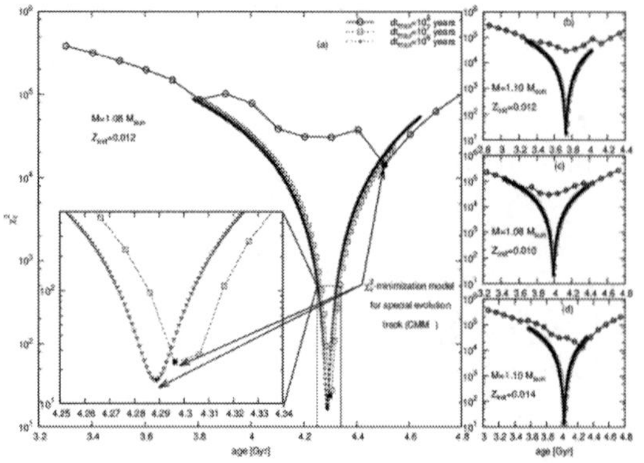

Figure 1: χ^2_ν as a function of stellar age $\tau_{\rm age}$. Different panels show different masses and initial metal abundances. They are symbolled at the figure with text. In panels, different point styles represent different time step. They are about 10^8, 10^7, and 10^6 years, respectively. In panel (a), filled points denote the χ^2_ν-minimization models (CMMs) [18].

mass ranges from 1.08 to 1.37 $M_\odot$. In addition, its metal abundance [Fe/H] ranges from −0.24 to −0.10 dex [18]. Its mass M and metal abundance [Fe/H] are 1.03 − 1.52 $M_\odot$ and −0.38 − +0.0.02 dex, respectively, when their observational uncertainties are considered [18]. Based on the observations of *Kepler* (there are 18 long-cadence (about 30 minutes exposure time) quarters Q0-Q17 and 13 short-cadence (about 1 minute exposure time) quarters Q0 and Q6-Q17) during more than 4 years, the obvious works analyzed parts of the light curves and extracted the frequency of maximum oscillation power $\nu_{\rm max}$ = 2301 − 2351 μHz and the large frequency separation $\Delta\nu \simeq 105.8\,\mu$Hz [1, 20–23]. In the present work, we use the 33 individual frequencies and their uncertainties, which is from the work of Tian et al. (2014) [1].

Modeling

We use MESA (v6208) evolutionary code to calculate the theoretical models and their oscillation information (i.e., oscillation frequencies) [24–26]. For the stellar mass M and the initial heavy element mass fraction $Z_{\rm init}$, they are M = 1.00 − 1.26 $M_\odot$ with a step of 0.02 $M_\odot$ and $Z_{\rm init}$ = 0.010 − 0.036 with a step of 0.002, respectively. For more descriptions of the physical input and other initial input parameters of the model calculations, please refer to the work of Wu & Li (2016) [18]. In order to void the influence due to the outer boundary of stellar model for oscillation frequencies, the outer boundary of stellar model is fixed at optical depth $\tau = 10^{-3}$ in our model calculations.

2 Results and Discussions

In calculating stellar theoretical models, the time steps (or time resolutions) will affect the precision of model. Figure

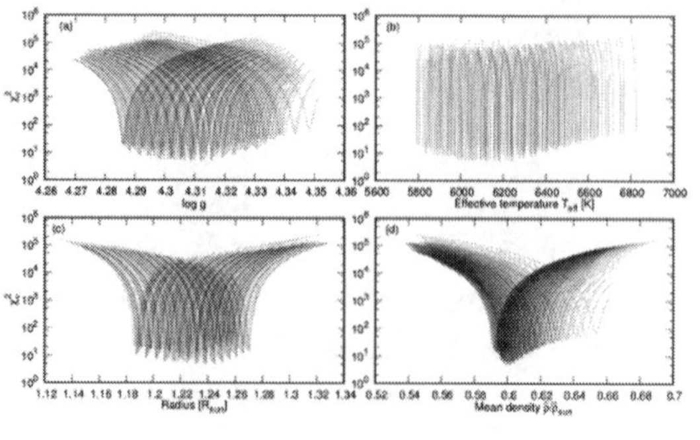

Figure 2: χ^2_ν as a function of surface gravity $\log g$–panel (a), effective temperature $T_{\rm eff}$–panel (b), radius R–panel (c), and of mean density $\bar{\rho}$–panel (d), respectively [18].

1 shows four different evolutionary tracks. For every evolutionary tracks, there are three different time steps. They are 10^8, 10^7, and 10^6 years, respectively. It can be seen from Figure 1 that the stellar evolutions whose time step is 10^8 years are fully different from those of 10^7 years time step. For the time steps of 10^7 and 10^6 years, except for the latter has larger time resolution (i.e., smaller time step) to precisely decide the χ^2-minimization model (CMM) for every evolutionary tracks, they are almost the same. In addition, it can be found from Figure 1 that the evolutions need larger time resolution (i.e., smaller time step), when the theoretical models are closer to the real situation of investigated target.

All of the calculated χ^2_ν are shown in Figures 2 and 3. In the two figures, every lines represent an evolutionary tracks. In Figure 2, the χ^2_ν as a function of surface gravity $\log g$, effective temperature $T_{\rm eff}$, radius R, and of mean density $\bar{\rho}$ of stars, respectively. It can be seen from Figure 2 that the profiles of χ^2_ν are in a state of disorder not only with $\bar{\rho}$ but also with $\log g$, $T_{\rm eff}$, and R. Despite they are related with the frequency of the maximum oscillation power $\nu_{\rm max}$ and the large frequency separation $\Delta\nu$ with the following relations [27–29]:

$$\frac{\nu_{\rm max}}{\nu_{\rm max,\odot}} = \frac{M/M_\odot}{(R/R_\odot)^2\sqrt{T_{\rm eff}/T_{\rm eff,\odot}}} = \frac{g/g_\odot}{\sqrt{T_{\rm eff}/T_{\rm eff,\odot}}}, \quad (3)$$

and

$$\frac{\Delta\nu}{\Delta\nu_\odot} = \sqrt{\frac{M/M_\odot}{(R/R_\odot)^3}} = \sqrt{\bar{\rho}/\bar{\rho}_\odot}. \quad (4)$$

In Figure 3, however, the profiles of $\chi^2_\nu(\tau_0)$ are convergence on the acoustic radius τ_0. It is defined as [30, 31]:

$$\tau_0 = \int_0^R \frac{dr}{c_s}, \quad (5)$$

where c_s is the adiabatic sound speed. The $\chi^2_\nu(\tau_0)$ curves are close each other, even overlap. All of the calculations construct a perfect "V" shape. All the CMMs almost point the same point on acoustic radius τ_0. It is about 4600 seconds. From this, we conclude that the acoustic radius τ_0 is the global model parameter that can be determined best

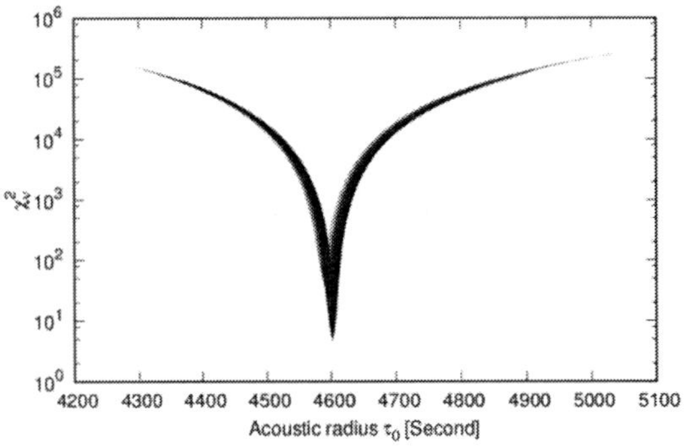

Figure 3: χ^2_ν as a function of stellar acoustic radius τ_0 [18].

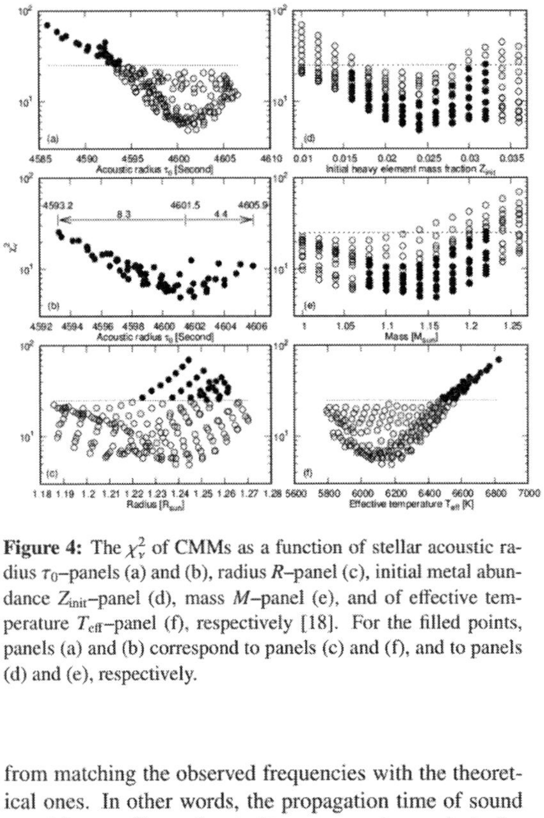

Figure 4: The χ^2_ν of CMMs as a function of stellar acoustic radius τ_0–panels (a) and (b), radius R–panel (c), initial metal abundance $Z_{\rm init}$–panel (d), mass M–panel (e), and of effective temperature $T_{\rm eff}$–panel (f), respectively [18]. For the filled points, panels (a) and (b) correspond to panels (c) and (f), and to panels (d) and (e), respectively.

from matching the observed frequencies with the theoretical ones. In other words, the propagation time of sound speed from stellar surface to its center can be precisely determined without any other constraint.

The distributions of CMMs ($\chi^2_{\nu,\rm CMM}$) are shown in Figure 4 against the acoustic radius τ_0, radius R, mass M, initial abundance $Z_{\rm init}$, and effective temperature $T_{\rm eff}$. It can be seen from Figure 4(a) that the distribution of CMMs ($\chi^2_{\nu,\rm CMM}(\tau_0)$) looks like a ladle. They are divided into two parts with the dashed line of $\chi^2_{\nu,\rm CMM} = 25$. Most of the CMMs are located in the bowl-shape part ($\chi^2_{\nu,\rm CMM} < 25$) and the rest form the handle ($\chi^2_{\nu,\rm CMM} > 25$). It can be seen from Figures 4(d)-(e) that the minimum of $\chi^2_{\nu,\rm CMM}$ is realized by stellar models with $Z_{\rm init} = 0.024$, and $M = 1.14\ M_\odot$ and $1.16\ M_\odot$. The CMMs whose $1.08 \leq M \leq 1.22\ M_\odot$, and $0.016 \leq Z_{\rm init} \leq 0.032$ occupy the center part

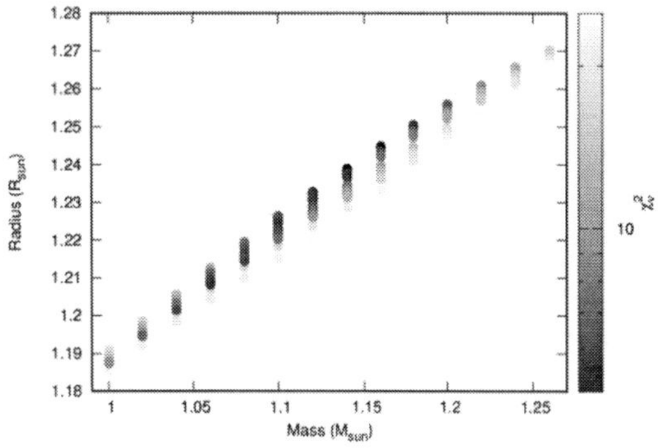

Figure 5: The radius R of CMMs as a function of their masses M.

in Figures 4(d)-(e). They from the lower edge of the ladle (see Figure 4(b)). It can be found from Figure 4(b) that the acoustic radius of KIC 6225718 can be determined as $\tau_0 = 4601.5^{+4.4}_{-8.3}$ seconds.

It can be found from Figures 4(a), (d)-(e) that the outlying part (handle part) has larger mass M and smaller or higher metal abundance $Z_{\rm init}$ comparing with the normal part (bowl-shape part). The outlying part has smaller acoustic radius τ_0 (see Figure 4(a)) than the normal part. It is due to the increase of stellar radius R (see Figure 4(c)) can not offset the increase of adiabatic sound speed $c_s(r)$, i.e., temperature $T(r)$ (see Figure 4(f)). From this, we conclude that the larger stellar mass and higher or lower metal abundance slightly affect the distribution of $\chi^2_{\nu,\rm CMM}(\tau_0)$. It leads that the CMMs have smaller acoustic radius.

The distribution of $R(M)$ for all of the calculated CMMs are shown in Figure 5. It can be seen from Figure 5 that those CMMs are located in a narrow band. They are close each other and almost can not be clearly distinguished. In other words, the stellar mass is degenerate with its radius for those CMMs. In the theory of stellar oscillation, the acoustic radius τ_0 is related with the stellar mass M and radius R with the relation of $\tau_0 \sim (M/R^3)^{-1/2}$ [30, 31]. While, the above analysis indicates that, for a given star, the acoustic radius of CMMs can be roughly referred as a constant. Therefore, we obtain the relation: $(M/R^3)^{-1/2} \simeq const.$. The stellar mass of CMMs vary with its radius $M \sim R^3$.

3 Conclusions

In the present work, we only use the high-precision asteroseismic observations (i.e., oscillation frequencies) to constrain the theoretical models with the method of χ^2_ν-matching to analyze the solar-like oscillator KIC 6225718. On the other hands, we consider the influence of the model precision for deciding the CMM and set its time step as 10^6 years. Finally, we find that the acoustic radius τ_0 is the global parameters that can be determined best from matching the observed frequencies with the theoretical ones via the method of χ^2_ν. It is about $\tau_0 = 4601.5^{+4.4}_{-8.3}$ seconds. The calculations suggest that the distribution of acoustic radius is slightly affected by larger stellar mass, and higher or lower metal abundance at the direction of smaller values. In addition, the approximate same τ_0 leads that the mass and radius of CMMs are degenerate with each other.

Acknowledgements

This work is cosponsored by the NSFC of China (grant Nos. 11333006, 11503076, and 11521303), and by the Chinese Academy of Sciences (grant No. XDB09010202). The authors express their sincere thanks to NASA and the Kepler team for allowing them to work with and analyze the Kepler data making this work possible and also gratefully acknowledge the computing time granted by the Yunnan Observatories.

References

[1] Tian, Z. J., et al., MNRAS, 445, 2999 (2014)
[2] Moravveji, E., et al., A&A, 580, A27 (2015)
[3] Bedding, T. R., et al., Nature, 471, 608 (2011)
[4] Mosser, B., et al., A&A, 572, 5 (2014)
[5] Baglin, A., et al., in COSPAR, Plenary Meeting, Vol. 36, 36th COSPAR Scientific Assembly, 3749 (2006)
[6] Gilliland, R. L., et al., PASP, 122, 131 (2010)
[7] Koch, D. G., et al., ApJL, 713, L79 (2010)
[8] Haas, M. R., et al., American Astronomical Society Meeting Abstracts #223, 223, #228.01 (2014)
[9] Howell, S. B., et al., PASP, 126, 398 (2014)
[10] Ricker, G. R., et al., Bulletin of the American Astronomical Society, 41, 403.01 (2009)
[11] Catala, C., et al., Pathways Towards Habitable Planets, 430, 260 (2010)
[12] Guenther, D. B., & Brown, K. I. T., ApJ, 600, 419 (2004)
[13] Lebreton, Y., Progress in Solar/Stellar Physics with Helio- and Asteroseismology, 462, 469 (2012)
[14] Lebreton, Y., & Goupil, M. J., A&A, 569, A21 (2014)
[15] Roxburgh, I. W., & Vorontsov, S. V., A&A, 560, A2 (2013)
[16] Roxburgh, I. W., A&A, 585, A63 (2016)
[17] Eggenberger, P., et al., A&A, 417, 235 (2004)
[18] Wu, T., & Li, Y., ApJL, 818, L13 (2016)
[19] Boyajian, T. S., et al., ApJ, 771, 40 (2013)
[20] Bruntt, H., et al., MNRAS, 423, 122 (2012)
[21] Chaplin, W. J., et al., ApJS, 210, 1 (2014)
[22] Huber, D., et al., ApJ, 760, 32 (2012)
[23] Silva Aguirre, V., et al., ApJ, 757, 99 (2012)
[24] Paxton, B., et al., ApJS, 192, 3 (2011)
[25] Paxton, B., et al., ApJS, 208, 4 (2013)
[26] Christensen-Dalsgaard, J., Ap&SS, 316, 113 (2008)
[27] Brown, T. M., et al., ApJ, 368, 599 (1991)
[28] Belkacem, K., et al., A&A, 530, A142 (2011)
[29] Kjeldsen, H., & Bedding, T. R., A&A, 293, 87 (1995)
[30] Christensen-Dalsgaard, J., Lecture Notes on Stellar Oscillations, Aarhus University, (2003)
[31] Aerts, C., Christensen-Dalsgaard, J., & Kurtz, D. W., Asteroseismology (Astronomy and Astrophysics Library; Dordrecht: Springer Science+Business Media B.V.), 207 (2010)

DOI: 10.1051/epjconf/201716005005

Isochrones of M67 with an Expanded Set of Parameters

Lucas Viani[1,*] and *Sarbani* Basu[1]

[1] *Department of Astronomy, Yale University, P.O. Box 208101, New Haven, CT 06520-8101, USA*

Abstract. We create isochrones of M67 using the Yale Rotating Stellar Evolution Code. In addition to metallicity, parameters that are traditionally held fixed, such as the mixing length parameter and initial helium abundance, also vary. The amount of convective overshoot is also changed in different sets of isochrones. Models are constructed both with and without diffusion. From the resulting isochrones that fit the cluster, the age range is between 3.6 and 4.8 Gyr and the distance is between 755 and 868 pc. We also confirm Michaud et al. (2004) claim that M67 can be fit without overshoot if diffusion is included.

1 Introduction

A better understanding of stellar properties and ages is crucial to many fields of astronomy. One widely used method of studying stellar clusters is isochrone fitting. Since stars in a stellar cluster formed around the same time and under the same physical conditions, they have a very small age and metallicity distribution. As a result, the cluster properties (for example age, distance, reddening, and metallicity) can be estimated through the fitting of isochrones. The properties of the isochrone that best fit the cluster's CMD can be used to estimate the properties of the cluster itself. While stellar ages and other properties determined through isochrone fitting are not as precise as asteroseismic methods, obtaining a basic age and property estimate (i.e. mass) of stars helps us later know what frequencies to expect. Here we will generate and fit isochrones to the M67 cluster data, allowing for a wider range of free parameters in our isochrone construction than is typically used.

The M67 cluster (NGC 2682) is the nearest open cluster that is around the solar age. It is estimated to be between 800 and 900 pc away (e.g. as in Sarajendini et al., 2009) at an age of around 4 Gyr (Richer et al., 1998). The cluster has a metallicity around solar, with [Fe/H] = 0.00 ± 0.06 (Heiter et al., 2014) and a reddening of $E(B-V) = 0.041 \pm 0.004$ (Taylor 2007). A color magnitude diagram (CMD) of M67 can be seen in Figure 1 (data from Geller et al., 2015). It is of importance to note that the M67 CMD has a distinctive "hook" feature below the main sequence turn-off. Any isochrone produced must also have this feature in order to be considered a reasonable fit to the cluster.

While many resources currently exist for the quick generation of a desired set of isochrones, such as the YY isochrones (Yi et al., 2001), PARSEC isochones (Bressan et al., 2012), and BaSTI isochrones (Pietrinferni et al., 2004), these methods are limited in the number of parameters that the user can fine-tune and control. For example, the mixing length, initial helium abundance, and the inclusion and amount of diffusion or convective overshoot. As a result, we generated our own set of isochrones through the creation of stellar models, allowing these parameters which are typically held fixed to vary.

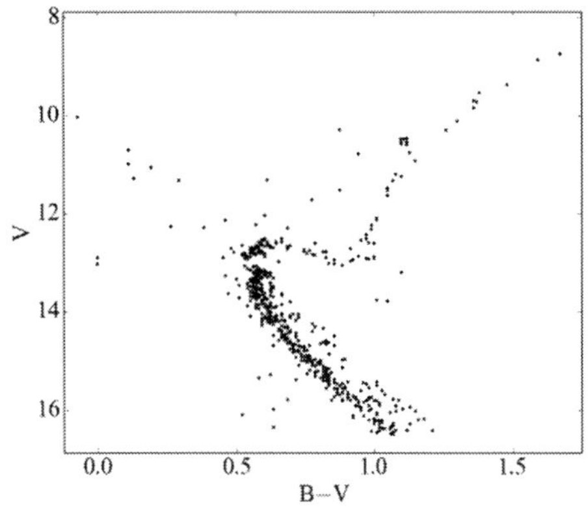

Figure 1: Color magnitude diagram for M67, data from Geller et al. (2015). Note the hook-like feature below the main sequence turn-off.

The difficulties in accurately modeling convection led to the use of the "mixing length theory" approximation (Böhm-Vitense 1958). Here the mean-free path that a convective element travels is determined by αH_P where H_P is the pressure scale height and α is the mixing length parameter. Traditionally, the value of α is determined by creating a solar model. To do this, a relation between metallicity, Z, and Y_0 is assumed. Then the value of α is adjusted until the model has a luminosity of 1 $L_\odot$ and radius 1 $R_\odot$ at the

*e-mail: lucas.viani@yale.edu

DOI: 10.1051/epjconf/201716005005

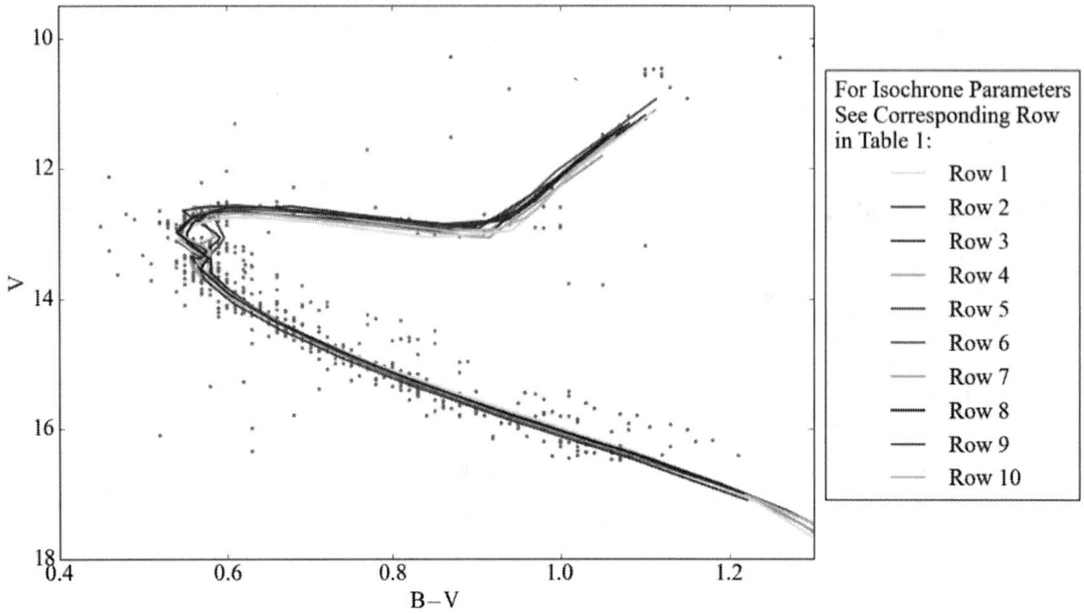

Figure 2: The isochrones that fit the M67 CMD. To be a reasonable fit isochrones had to fit the main sequence, turn-off, giant branch, and reproduce the hook-like feature in the CMD. There were 10 such isochrones that fit the M67 CMD. The legend refers to rows in Table 1 where the parameters of the isochrones are listed.

solar age. Typically isochrones are then created with the values of α and Y_0 held fixed. This investigation however will allow both α and Y_0 to be free parameters.

It is also of interest to examine the need for convective overshoot or diffusion in M67. Michaud et al. (2004) shows that the M67 CMD can be fitted with isochrones of models without overshoot if diffusion is included. Additionally, from spectroscopic results Önehag et al. (2014) claims M67 must have diffusion. Isochrones with varying degrees of diffusion and overshoot will be created for M67.

This article is organized as follows. Section 2 discusses the creation of the stellar models and isochrones, Section 3 describes the isochrone fitting to the CMD as well as the results, and Section 4 summarizes the best-fit parameters and provides concluding remarks.

2 Stellar Models

The stellar models were all created using the Yale Rotating Evolutionary Code (YREC) (Demarque et al., 2008). The OPAL equation of state (Rogers & Nayfonov 2002) was used along with the OPAL high temperature opacities (Iglesias & Rogers 1996) and low temperature opacities (Ferguson et al., 2005). The Adelberger et al. (1998) nuclear reaction rates were used, with the Formicola et al. (2004) value for the $^{14}N(p,\gamma)^{15}O$ reaction. In all cases the solar value of $Z_\odot/X_\odot = 0.023$ was used (Grevesse & Sauval 1998) and models were created using the Eddington T-τ relation in the atmosphere.

The models created spanned a wide range of parameter space. Models ranged in mass from M = 0.65 to 1.50 $M_\odot$, metallicity of [Fe/H] = −0.1, 0.0, and 0.1, initial He abundance from $Y_0 = 0.248$ to 0.320, convective overshoot of 0.0, 0.2, and mass-dependent (from Demarque et al., 2004), and mixing lengths of $\alpha = 1.5$ to 2.0. Additionally, models were created both with and without diffusion. The mass based overshoot, from Demarque et al. (2004), gives the following overshoot coefficient (Λ_{os}) values:
$\Lambda_{os} = 0.00$ for $M < M_{crit}$,
$\Lambda_{os} = 0.05$ for $M = M_{crit}$,
$\Lambda_{os} = 0.10$ for $M = M_{crit} + 0.1$,
$\Lambda_{os} = 0.15$ for $M = M_{crit} + 0.2$,
$\Lambda_{os} = 0.20$ for $M > M_{crit} + 0.2$,
where M and M_{crit} are in solar units. The value of M_{crit} depends on metallicity, as well as the composition of the star in terms of the X, Y, and Z abundance ratios. For this set of models $M_{crit} = 1.2\ M_\odot$.

For the models with diffusion, the diffusion coefficient was set as 1 for masses below 1.25 $M_\odot$ and for masses larger than 1.25 $M_\odot$ was given by

$$\text{diff coefficient} = \exp\left(\frac{-(M-1.25)^2}{2(0.085)^2}\right) \quad (1)$$

In this way the value of the diffusion coefficient smoothly decreased for stars more massive than 1.25 $M_\odot$.

 DOI: 10.1051/epjconf/201716005005

Table 1: The model physics and the best fit values for the isochrones that were reasonable fits to the M67 CMD. These are the isochrone parameter combinations which were able to reproduce the M67 main sequence, turn-off, giant branch, and "hook" feature.

Model Physics						Best Fit Values		
Row Reference Number	**Diffusion**	**Overshoot**	**[Fe/H]**	α	Y_0	**Age (Gyr)**	**Reddening** $E(B-V)$	**Distance Modulus**
1	Yes	None	0.1	2.0266	0.267	4.8	0.000	9.616
2	Yes	None	0.1	2.0266	0.320	4.2	0.022	9.495
3	Yes	None	−0.1	2.0266	0.320	3.7	0.096	9.398
4	Yes	0.2	0.1	2.0266	0.248	3.9	0.015	9.693
5	Yes	0.2	0.0	2.0266	0.320	3.6	0.062	9.527
6	Yes	0.2	−0.1	2.0266	0.320	3.8	0.087	9.390
7	Yes	Mass-Based	0.0	2.0266	0.248	4.4	0.035	9.608
8	Yes	Mass-Based	0.0	2.0266	0.320	3.9	0.059	9.443
9	Yes	Mass-Based	−0.1	2.0266	0.248	4.1	0.071	9.520
10	Yes	Mass-Based	−0.1	2.0266	0.267	3.9	0.085	9.513

2.1 Results

2.2 Fitting Isochrones to the Cluster CMD

For each set of parameters, isochrones for M67 were created at different ages. The best fit age, distance modulus, and reddening value for each set of parameters was then determined using 2-dimensional χ^2 fitting between the points of the M67 CMD (Geller et al., 2015) and the isochrone. Only M67 stars with radial velocity membership probability of 90% or greater were used (from Geller et al., 2015). Allowing the distance modulus, reddening, and age to be free parameters, a χ^2 value for both the magnitude, V, and the color, $B-V$, was determined for each star in M67. These χ^2_V and $\chi^2_{(B-V)}$ values were summed to create the parameter χ^2_{total}. The age, reddening, and distance modulus combination that produced the lowest χ^2_{total} was determined to be the best-fit isochrone for that model parameter set.

For each combination of model parameters, the best-fit isochrone was then visually inspected to determine if the isochrone was indeed a reasonable or realistic fit to the M67 cluster. In other words, that the isochrone fit the main sequence, turn off, giant branch, and reproduce the "hook" in the CMD. There were 10 best-fit isochrones that were reasonable fits to the M67 CMD. The properties of these isochrones are listed in Table 1 and they are plotted along with the M67 CMD in Figure 2.

2.3 Effects of Model Physics on the M67 "Hook"

From the generated isochrones we can also investigate the parameters which give rise to the hook-like feature in the M67 CMD. As expected, isochrones without any diffusion and without any convective overshoot do not reproduce the hook. Models with convective overshoot do reproduce the hook in the CMD. However, we see that overshoot is not required to reproduce the hook if diffusion is included in the models. This confirms Michaud et al. (2014) claim that convective overshoot is not needed to model M67 if diffusion is implemented. A comparison of different isochrone properties and their effects on creating a hook-feature can be seen in Figure 3. In Figure 3 the isochrones are identical except with respect to their convective overshoot and diffusion amounts. The blue isochrone has neither overshoot or diffusion and fails to reproduce the hook feature. The green isochrone has no diffusion but does include convective overshoot and as a result has a hook in the CMD. The red isochrone does not have convective overshoot but does implement diffusion and also is able to reproduce this hook-like feature.

3 Conclusions

From the isochrones that fit the cluster, as seen in Table 1, we can get an age, distance, and reddening range. This gives an age of between 3.6 and 4.8 Gyr, a distance between 755 and 868 pc, and a reddening between $E(B-V) = 0.000$ and 0.096. These ranges all agree well with the M67 parameter values from literature discussed in Section 1.

The best-fit isochrone has an age of 4.4 Gyr, distance of 834.9 pc, metallicity of [Fe/H] = 0.0, and reddening of $E(B-V) = 0.035$. The physics of the best fit isochrone is with diffusion included, a mass-based overshoot, a mixing length of $\alpha = 2.0266$, and an initial He abundance of $Y_0 = 0.248$.

We have also demonstrated the importance of allowing parameters that are generally held constant to vary and the effect that convective overshoot and diffusion can have on isochrone behavior. These results provide a good basis for parameter estimates which can aid in future asteroseismic studies of the cluster, for example helping to estimate expected frequencies of target stars.

Acknowledgments

This work has been supported by NSF grant AST-1514676 and NASA grant NNX16AI09G to SB.

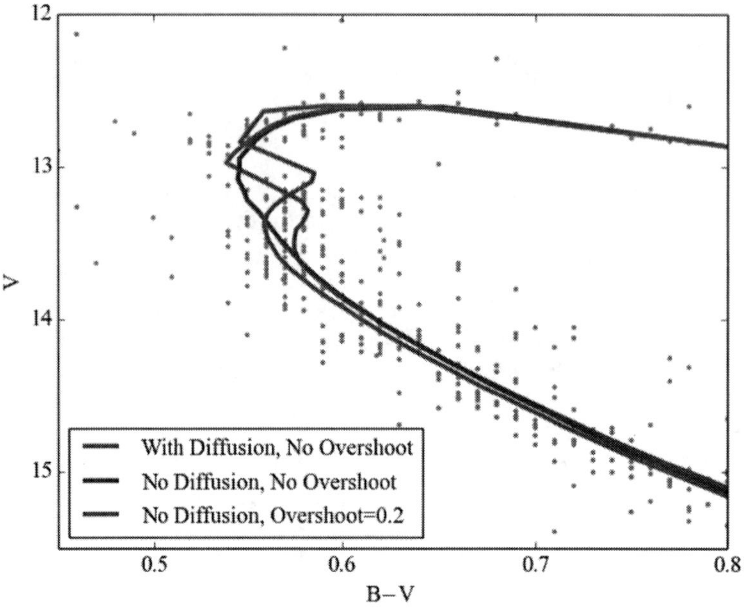

Figure 3: A closer look at reproducing the M67 hook-like CMD feature. The red isochrone includes diffusion but no overshoot, the blue isochrone has no diffusion and no overshoot, and the green isochrone has no diffusion but does have convective overshoot. All other parameters of the isochrones are identical. As can be seen, both the presence of overshoot or the presence of diffusion can reproduce this hook-like feature while an isochrone with no diffusion and no overshoot cannot.

References

[1] Adelberger, E. G., Austin, S. M., Bahcall, J. N., et al., Reviews of Modern Physics **70**, 1265 (1998)

[2] Böhm-Vitense, E., ZAp **46**, 108 (1958)

[3] Bressan, A., Marigo, P., Girardi, L., et al., MNRAS **427**, 127 (2012)

[4] Demarque, P., Guenther, D. B., Li, L. H., Mazumdar, A., & Straka, C.W., Ap&SS **316**, 31 (2008)

[5] Demarque, P., Woo, J.-H., Kim, Y.-C., & Yi, S. K., ApJS **155**, 667 (2004)

[6] Ferguson, J. W., Alexander, D. R., Allard, F., et al., ApJ **623**, 585 (2005)

[7] Formicola, A., Imbriani, G., Costantini, H., et al., Physics Letters B **591**, 61 (2004)

[8] Geller, A. M., Latham, D. W., & Mathieu, R. D., AJ **150**, 97 (2015)

[9] Grevesse, N. & Sauval, A. J., Space Sci. Rev. **85**, 161 (1998)

[10] Heiter, U., Soubiran, C., Netopil, M., & Paunzen, E., A&A **561**, A93 (2014)

[11] Iglesias, C. A. & Rogers, F. J., ApJ **464**, 943 (1996)

[12] Michaud, G., Richard, O., Richer, J., & VandenBerg, D. A., ApJ **606**, 452 (2004)

[13] Önehag, A., Gustafsson, B., & Korn, A., A&A **562**, A102 (2014)

[14] Pietrinferni, A., Cassisi, S., Salaris, M., & Castelli, F., ApJ **612**, 168 (2004)

[15] Richer, H. B., Fahlman, G. G., Rosvick, J., & Ibata, R., ApJL **504**, L91 (1998)

[16] Rogers, F. J. & Nayfonov, A., ApJ **576**, 1064 (2002)

[17] Sarajedini, A., Dotter, A., & Kirkpatrick, A., ApJ **698**, 1872 (2009)

[18] Taylor, B. J. AJ **133**, 370 (2007)

[19] Yi, S., Demarque, P., Kim, Y.-C., et al., ApJS **136**, 417 (2001)

 DOI: 10.1051/epjconf/201716005006

Galactic Archaeology with TESS: Prospects for Testing the Star Formation History in the Solar Neighbourhood

Alexandra Thomas[1], *Emma* Stevenson[1], *Fabian W. R.* Gittins[1], *Andrea* Miglio[1,2,*], *Guy* Davies[1,2], *Léo* Girardi[3], *Tiago L.* Campante[1,2], and *Mathew* Schofield[1,2]

[1] *School of Physics and Astronomy, University of Birmingham, Edgbaston, Birmingham, B15 2TT, UK*
[2] *Stellar Astrophysics Centre, Department of Physics and Astronomy, Aarhus University, Ny Munkegade 120, DK-8000 Aarhus C, Denmark*
[3] *Osservatorio Astronomico di Padova, Vicolo dell'Osservatorio 5, I-35122 Padova, Italy*

Abstract.
A period of quenching between the formation of the thick and thin disks of the Milky Way has been recently proposed to explain the observed age-[α/Fe] distribution of stars in the solar neighbourhood. However, robust constraints on stellar ages are currently available for only a limited number of stars. The all-sky survey TESS (Transiting Exoplanet Survey Satellite) will observe the brightest stars in the sky and thus can be used to investigate the age distributions of stars in these components of the Galaxy via asteroseismology, where previously this has been difficult using other techniques. The aim of this preliminary study was to determine whether TESS will be able to provide evidence for quenching periods during the star formation history of the Milky Way. Using a population synthesis code, we produced populations based on various stellar formation history models and limited the analysis to red-giant-branch stars. We investigated the mass-Galactic-disk-height distributions, where stellar mass was used as an age proxy, to test for whether periods of quenching can be observed by TESS. We found that even with the addition of 15% noise to the inferred masses, it will be possible for TESS to find evidence for/against quenching periods suggested in the literature (e.g. between 7 and 9 Gyr ago), therefore providing stringent constraints on the formation and evolution of the Milky Way.

1 Introduction

Uncertainties in the age distribution of stars in the solar neighbourhood have thus far limited the determination of the star formation history (SFH) of the local Galactic disk [1], one of the key constraints to test competing models of formation and evolution of the Milky Way. It is hoped that this barrier may be overcome with the reception of all-sky asteroseismic data.

In this study we investigated whether the TESS (Transiting Exoplanet Survey Satellite) mission [2] will be able to make a substantial contribution to the field of Galactic archaeology by providing evidence for quenching periods during the formation of the Galaxy. We used the TRILEGAL code (TRIdimensional modeL of thE GALaxy) [3] to generate a set of synthetic stellar populations based on various SFH models, and considered only red-giant-branch (RGB) stars which would display oscillations detectable by TESS.

The layout of this conference proceedings is as follows. We begin in Sect. 2 by describing the nature of the TRILEGAL population software and the code used to test simulated stars for detectable solar-like oscillations by TESS. In Sect. 3 we outline the SFH models used when generating stellar populations. These are based around ideas of quenching of star formation rate (SFR) between formation of the thin and thick disks from [4]. We then proceed in Sect. 4 to examine the results from analysis of age/mass distributions of the synthetic populations. We conclude in Sect. 5 with a discussion of whether the periods of quenching in the formation of the Milky Way explored in this study could be determined by analysing TESS data.

2 Stellar Population Synthesis

The Monte Carlo population synthesis code, TRILEGAL [3], was used to create all-sky populations of the Milky Way. The sky was split into 1280 tiles of equal areas for our first simulation and cylindrical symmetry was assumed. For subsequent populations, half of the sky was simulated for computational efficiency. For each tile, an extinction was calculated at infinity using a model provided by [5]. It was assumed that all stars could be considered to be resolved by TESS. This assumption would not necessarily be valid for those tiles in close proximity to the Galactic centre, and thus regions of the sky below 9 deg from the centre of the Galactic disk were omitted during analysis.

*e-mail: miglioa@bison.ph.bham.ac.uk

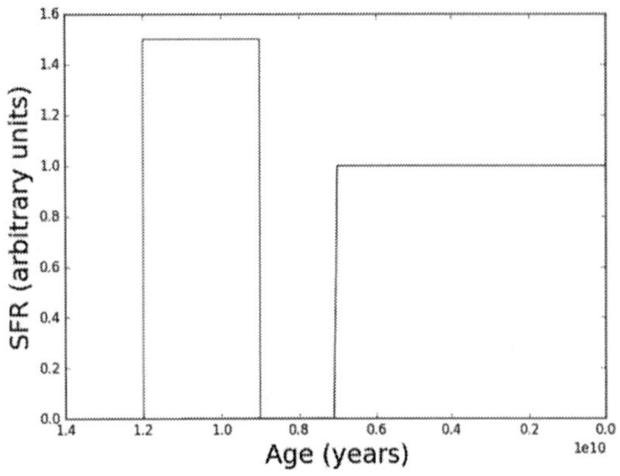

(a) Model 1

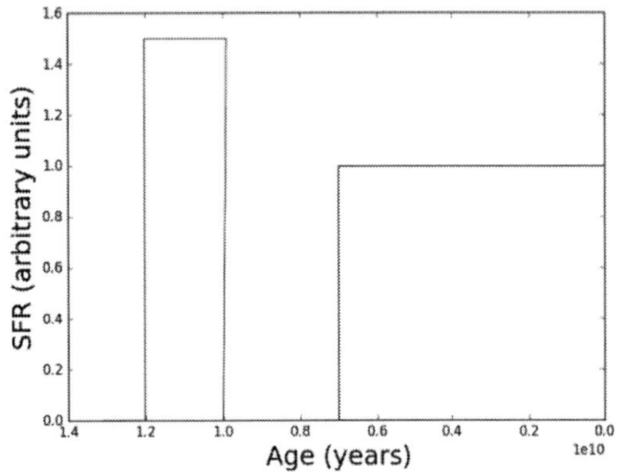

(b) Model 2

Figure 1: Star-formation-rate models of the thick- and thin-disk components, used as a basis for 2 simulated populations using TRILEGAL. Thick-disk formation rate is represented in red, thin-disk SFR in blue.

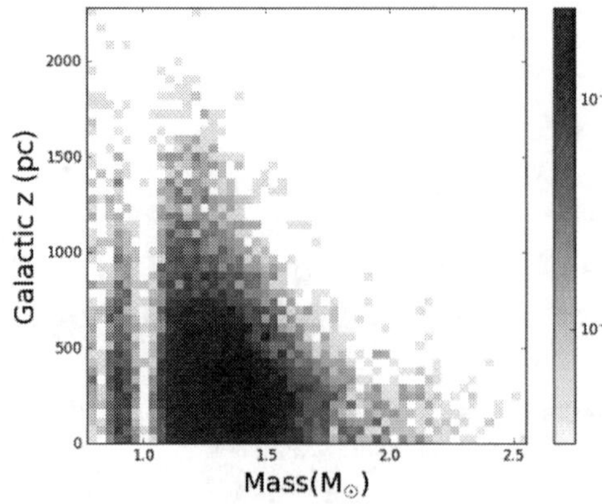

(a) Model 1

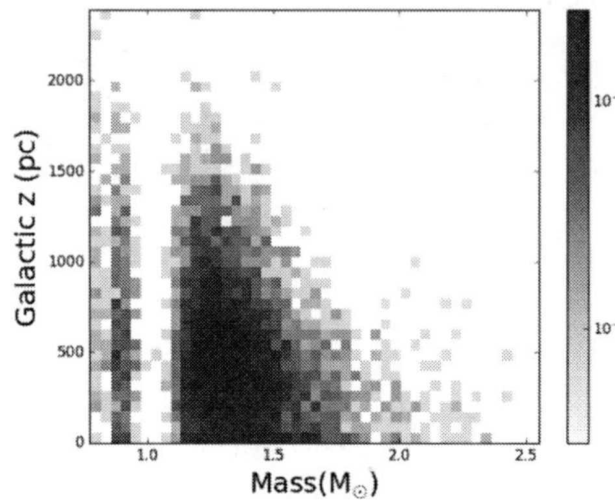

(b) Model 2

Figure 2: Two dimensional histogram distribution of RGB star mass (proxy for age) with disk height, z. Each distribution of synthesised stars was generated from the corresponding SFR models shown in Fig. 1.

Age is possibly one of the most distinguishing properties between the thin and thick disks. TRILEGAL models four components of the Galaxy separately: thin disk, thick disk, halo and bulge. Each component of the Galaxy is populated by stars generated from the same initial mass function (IMF) but with a different SFR and age-metallicity relation (AMR). For the purpose of this study, SFH of the bulge and halo were unchanged for all simulated populations. Models for the SFH of the two disk components are outlined in Sect. 3.

Stars within the simulated populations were limited to red giants with detectable oscillations by TESS, as determined by the detection-probability algorithm described in [6]. For this investigation we focused on RGB stars since they are expected to follow a well-defined age-mass relation, and so the ages of these stars can be determined thanks to asteroseismic constraints on their masses.

3 Star-Formation-Rate Models

The two disk components, which dominate the solar neighbourhood, were modelled to have distinct star-formation histories. Models were designed to test for the length of quenching period and durations of thin and thick disk formation.

Model 1 was based on ideas of quenching from [4], where a cessation in the SFR between thick and thin disk formation is present. This mechanism can be used to explain the observed age-[α/Fe] abundance distribution [4] and is a phenomenon likely to depend on the mass of the galaxy [7]. Using age ranges suggested by [8], [9],[10] the thick-disk formation was chosen to extend from 12 to 9 Gyr ago and the thin disk from 7 Gyr ago to the present epoch. A quenching of duration 2 Gyr was used. The model also assumed a higher SFR during thick disk formation in comparison to the thin disk. This model can be seen in Fig 1a.

DOI: 10.1051/epjconf/201716005006

Model 2, shown in Fig. 1b, was based on the same ideas of quenching but used to study the consequences of a longer period of cessation since a range of quenching durations was found in literature [8]. This model used a quenching gap of 3 Gyr extending from 10 to 7 Gyr ago.

4 Results

4.1 Age-Mass Distributions

In order to reconstruct the local star-formation history of the Milky Way, the age distributions of stars in the solar vicinity were studied. Since the stellar populations of the Galactic components are primarily discernible with height from the Galactic plane, z, we expect the age-z distributions would differ for each SFR model. These are displayed in Fig. 2 and are associated to the SFR models in Fig. 1.

Red-giant-branch stars were chosen since these have a well-defined age proxy using the initial mass of the red giant's progenitor [e.g. see 11] and so the ages of these stars can be determined by coupling asteroseismic constraints from TESS with photospheric properties from photometric/spectroscopic surveys [12, 13]. Distances to stars may be determined either by seismic constraints or by dedicated astrometric missions, such as Gaia.

From Fig. 2 it can be seen that the distinct stages for formation of the thin and thick disk are separated by a thin gap in mass. It is obviously easier to discriminate models with longer quenching periods since the separation between these components is larger. For stars with $M \gtrsim 1.1\,M_\odot$ the vertical structure of the mass distribution is determined by the assumption that the geometrical vertical scale height of the thin disk increases with age.

4.2 Kolmogorov-Smirnov Testing

A Kolmogorov-Smirnov (K-S) test was conducted to determine whether TESS will be able to distinguish between populations modelled with varying duration of quenching period. Gaussian noise of 15% [13] was added to the simulated mass data of models 1 and 2 by running 1000 Monte Carlo simulations (see Fig. 3). The null hypothesis, that the data both with and without added noise came from the same distribution, was rejected. Models with different quenching lengths were found to be discernible, and only after 30% noise was added were models no longer distinguishable. After the addition of 25% noise model 2 showed quenching where model 1 did not. Since the expected error on asteroseismic mass is less than 30% [13, 14], we can conclude that it should be possible for TESS to distinguish between models 1 and 2, and therefore to provide evidence for or against quenching.

5 Conclusions

The aim of this preliminary study was to investigate whether the all-sky asteroseismic mission TESS could significantly contribute to studies of Galactic archaeology in the solar neighbourhood. This was conducted by proposing two SFH models for the thin and thick disk which can be compared to data from TESS once observations have been made. The models used in this investigation tested for quenching periods during Galaxy formation and relative SFR of the disk components.

The analysis was restricted to RGB stars since they have a well understood age-mass relation. Since the Sun is very close to the Galactic mid-plane, half sky populations were simulated. Comparisons were made between distributions of stellar mass (proxy for age) with Galactic height for each model. A K-S test revealed that with the addition of 15% noise on the asteroseismically determined mass, we will be able to discriminate between star-formation quenching periods proposed in the literature.

We have concluded that it will be possible for TESS to distinguish between models which include different lengths of quenching periods and therefore set an upper limit on the duration of a detectable quenching episode. The mission should be able to provide evidence for or against this scenario. It will be more difficult to determine relative SFR of the thin and thick disks, unless independent information (e.g. chemical abundances, kinematics) is considered.

Acknowledgements

This work was done as part of a third-year undergraduate project at the School of Physics and Astronomy, University of Birmingham. AM, GRD, TC, and MS acknowledge the support of the UK Science and Technology Facilities Council (STFC). Funding for the Stellar Astrophysics Centre is provided by The Danish National Research Foundation (Grant agreement no.: DNRF106). LG acknowledges the support from the PRIN INAF 2014 – CRA 1.05.01.94.05.

References

[1] K. Freeman, *Astrophysics and Space Science Proceedings* (Springer, 2011)

[2] G.R. Ricker, J.N. Winn, R. Vanderspek, D.W. Latham, G.Á. Bakos, J.L. Bean, Z.K. Berta-Thompson, T.M. Brown, L. Buchhave, N.R. Butler et al., Journal of Astronomical Telescopes, Instruments, and Systems **1**, 014003 (2015)

[3] L. Girardi, M.A.T. Groenewegen, E. Hatzimi-naoglou, L. da Costa, Astronomy and Astrophysics **436**, 895 (2005)

[4] O.N. Snaith, M. Haywood, P. Di Matteo, M.D. Lehnert, F. Combes, D. Katz, A. Gómez, The Astrophysical Journal, Letters **781**, L31 (2014)

[5] R. Drimmel, A. Cabrera-Lavers, M. López-Corredoira, Astronomy and Astrophysics **409**, 205 (2003)

[6] T.L. Campante, M. Schofield, J.S. Kuszlewicz, L. Bouma, W.J. Chaplin, D. Huber, J. Christensen-Dalsgaard, H. Kjeldsen, D. Bossini, T.S.H. North et al., The Astrophysical Journal **830**, 138 (2016)

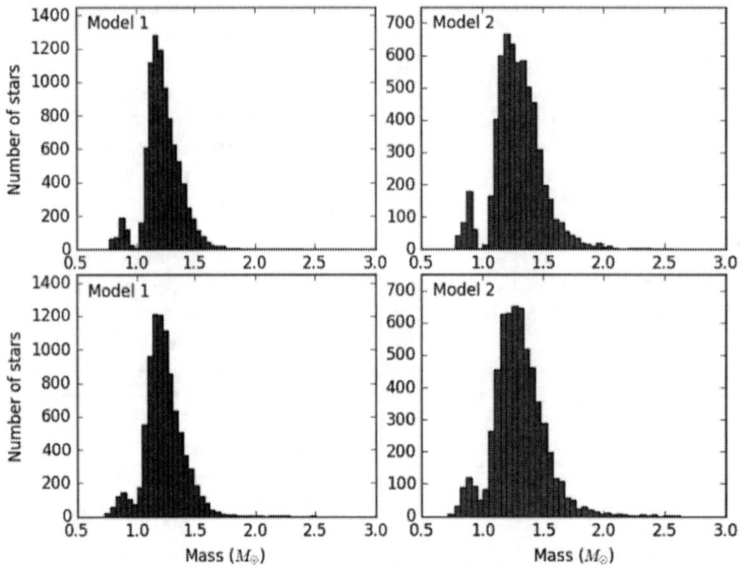

Figure 3: Histogram of mass distributions for models 1 (left) and 2 (right). Top row: No noise added. Bottom row: 15% noise added.

[7] A. Cimatti, E. Daddi, A. Renzini, Astronomy and Astrophysics **453**, L29 (2006)

[8] M. Haywood, M.D. Lehnert, P. Di Matteo, O. Snaith, M. Schultheis, D. Katz, A. Gómez, Astronomy and Astrophysics **589**, A66 (2016)

[9] J. Bernkopf, K. Fuhrmann, Monthly Notices of the Royal Astronomical Society **369**, 673 (2006)

[10] K. Fuhrmann, Monthly Notices of the Royal Astronomical Society **414**, 2893 (2011)

[11] A.W. R. Kippenhahn, *Stellar Structure and Evolution*, 1st edn. (Springer, New York, 1990)

[12] A. Miglio, C. Chiappini, T. Morel, M. Barbieri, W.J. Chaplin, L. Girardi, J. Montalbán, M. Valentini, B. Mosser, F. Baudin et al., Monthly Notices of the Royal Astronomical Society **429**, 423 (2013)

[13] G.R. Davies, A. Miglio, Astronomische Nachrichten **337**, 774 (2016)

[14] W.J. Chaplin, A. Miglio, Annual Review of Astronomy and Astrophysics **51**, 353 (2013)

EPJ Web of Conferences **160**, 05007 (2017)
Seismology of the Sun and the Distant Stars II
DOI: 10.1051/epjconf/201716005007

Dynamical Asteroseismology: towards improving the theories of stellar structure and (tidal) evolution

Andrew Tkachenko[1,*]

[1] *Instituut voor Sterrenkunde, KU Leuven, Celestijnenlaan 200D, B-3001 Leuven, Belgium*

Abstract. The potential of the dynamical asteroseismology, the research area that builds upon the synergies between the asteroseismology and binary stars research fields, is discussed in this manuscript. We touch upon the following topics: i) the mass discrepancy observed in intermediate- to high-mass main-sequence and evolved binaries as well as in the low mass systems that are still in the pre-main sequence phase of their evolution; ii) the rotationally induced mixing in high-mass stars, in particular how the most recent theoretical predictions and spectroscopic findings compare to the results of asteroseismic investigations; iii) internal gravity waves and their potential role in the evolution of binary star systems and surface nitrogen enrichment in high-mass stars; iv) the tidal evolution theory, in particular how its predictions of spin-orbit synchronisation and orbital circularisation compare to the present-day high-quality observations; v) the tidally-induced pulsations and their role in the angular momentum transport within binary star systems; vi) the scaling relations between fundamental and seismic properties of stars across the entire HR-diagram.

1 Introduction

Asteroseismology is a study of deep interiors of stars through interpretation of frequencies, amplitudes, and phases of their oscillation modes. Similar to the seismology of our own Earth, one makes use of the fact that different pulsation modes probe different layers inside a star, providing information about the physical conditions in the stellar interiors. Depending on their position in the Hertzspung-Russel (HR) diagram, stars are found to pulsate in acoustic (p-), gravity (g-), or mixed modes. These latter oscillations have the properties of g-modes near the stellar core while posses the p-mode character in the outer regions of a star. Pressure and gravity are the restoring forces for the acoustic and gravity waves, respectively. Pulsating stars occupy many different places in the HR-diagram making the assteroseismic studies possible for a variety of stellar objects in a wide range of stellar masses and at different evolutionary stages (see Fig. 1). A discussion of different types of pulsating stars as well as the wave excitation mechanisms is beyond the scope of this paper. We thus refer the interested reader to [1] for more details.

Asteroseismology has been revolutionized with the launch of space missions such as MOST [36], CoRoT [5], and Kepler [9], which delivered photometric data of unprecedented quality and allowed detailed studies of stellar objects across the whole HR-diagram. An observational probe of the near core regions, as well as the measurements of the internal rotation have recently become available for main-sequence (MS) stars of spectral types B to F [e.g., 22, 26, 33, 35]. Space-based data provided us with new insights into stellar objects at advanced evolutionary stages (e.g., RR Lyrae stars - [25]; subdwarf B-type stars - [8]), revolutionized our understanding of evolved red-giant stars exhibiting stochastically excited solar-like oscillations [e.g., 7], and significantly improved our understanding of stars during early, pre-main-sequence (pre-MS) stages of their evolution [43].

Despite the huge success of asteroseismology in the past decade, it still remains a model dependent research field, with the model dependency being significantly larger for intermediate- and high-mass stars than for sun-like stars. Insufficient quality of observations is no longer a limiting factor for a new series of progressions in asteroseismology, but the shortcomings in the current theories of stellar structure and evolution become a real bottleneck. Thanks to the fact that many pulsating stars are found in binary systems across the whole HR-diagram, binary stars - systems consisting of two stellar objects that periodically revolve around their common center of mass - can provide us with additional, model-independent observational constraints in terms of the dynamical masses and radii of stars. Nowadays, we talk about dynamical measurements of fundamental properties of stars in large mass and age ranges to precisions approaching 1% [e.g., 15, 31]. This way, asteroseismic investigations can be confronted/supplied with model-independent fundamental properties of stars, providing a valuable test of the current theories of stellar structure and evolution. In addition, pulsating stars in binary systems allow us to test and further develop theories of tidal evolution and angular momentum transport - both theories can be observationally tested and further refined thanks to asteroseismology delivering pre-

* e-mail: Andrew.Tkachenko@ster.kuleuven.be

DOI: 10.1051/epjconf/201716005007

cise observational constraints on the internal structure of stars. Dynamical tides are known to be effective in resonant excitation of stellar pulsations but magnetic fields in binaries are expected to strongly affect the transfer of mass and angular momentum as well as the global properties of stellar oscillations. The interplay between stellar magnetic fields, binarity, and oscillations has yet to be investigated in detail, and will provide even further insight into the tidal evolution theory. A few applications of the dynamical asteroseismology (the field that builds upon the synergies between asteroseismology and binary star research areas) as well as its potential impact on the stellar astrophysics as a whole are discussed in this manuscript.

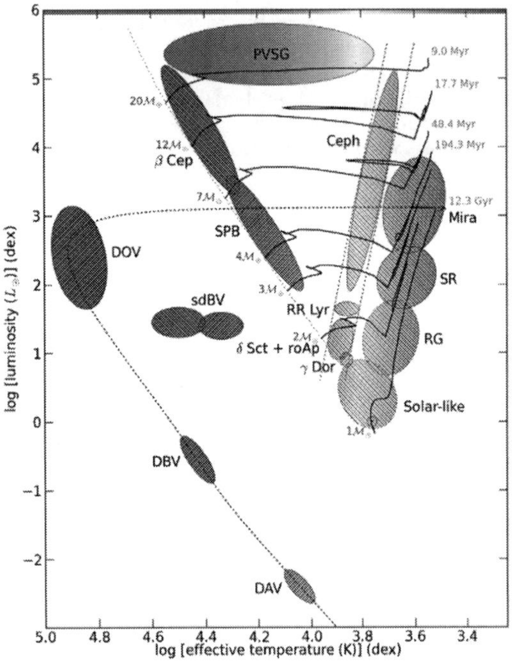

Figure 1: Pulsation HR-diagram showing different types of stars for which asteroseismic studies are possible. Colour coding (approximately) resembles temperature scale, with orange/blue indicating coolest/hottest stars. A few evolutionary tracks are overplotted to give a hint on the range of stellar masses. Image credit: Dr. P. I. Pápics.

2 Mass discrepancy

One of the problems that is currently pending a solution in the research field of massive binary stars is the mass discrepancy problem. In its original formulation, the term mass discrepancy refers to the disagreement between the spectroscopic and evolutionary masses of the star [17]. The evolutionary mass is determined by fitting evolutionary tracks to a position of the star in the HR-diagram, whereas the spectroscopic mass is computed from the spectroscopically derived value of the surface gravity and from the radius of the star. This original definition of the mass discrepancy has nothing to do specifically with binary systems but refers to massive stars in general. We keep on using the term mass discrepancy in application to binary systems throughout the manuscript, and warn the reader that we refer to the discrepancy between the evolutionary masses and those inferred from binary dynamics (dynamical masses).

The V380 Cyg binary system is probably one of the best illustrations of the mass discrepancy observed in massive binary stars. The system was originally investigated by [16] based on ground-based multicolour photometry and high-resolution spectroscopy. One of the main results obtained in the study is depicted in the left panel of Fig. 2. In particular, the authors found that the position of the more evolved primary component in the Kiel diagram can only be explained when a large amount of extra mixing in terms of the overextension of the convective core (core overshoot α) is included in the models. The system has been recently revisited by [31] based on high-quality space-based *Kepler* photometry and ground-based, high-resolution HERMES [27] spectroscopy. The authors found that the evolutionary tracks computed for the dynamical masses of the stars fail to explain their positions in the Kiel diagram (see Fig. 2, right panel, solid lines), similar to the result obtained by [16] for the primary component. Either the mass/metallicity has to be changed to higher/lower value or an extra mixing had to be included in the models in terms of the core overshoot or rotational mixing to fit the components' positions in the Kiel diagram (dashed and dotted lines in the right panel of Fig. 2). An excess of mass, or the mass discrepancy, that was obtained by [31] amounts to about 10% which transforms into the age uncertainty of about 50%. Since the mixing, mass, and metallicity correlate in the models and have similar effect on the tracks by shifting them towards higher temperature and extending towards lower gravities in the Kiel diagram, the problem can also be referred to just as "discrepant models". The term mass discrepancy is essentially used to quantify the level to which the models fail to explain high-quality observations of eclipsing binary stars.

The fact of evolutionary models lacking a certain amount of extra mixing is also asteroseismology proved. As it has been shown by [2], an extra mixing needs to be introduced in the terms of the core overshoot into the evolutionary models to match them with the observed seismic properties of binary components. However, it is barely the case for the majority of single stars of similar mass, where about the standard value of the core overshoot $\alpha = 0.2$ is preferred, probably because the mass of a star can be way more relaxed in the modelling compared to the case of binary systems.

The same story seems to be repeating at early (pre-MS) stages of stellar evolution and at a lower mass end (below one solar mass). Recent studies of low-mass pre-MS stars based on K2 photometry [e.g., 11, 12, 20] show that the mass discrepancy is relevant for these objects as well and is even more pronounced (above 40%) than in the case of high-mass main-sequence and giant stars. All three studies

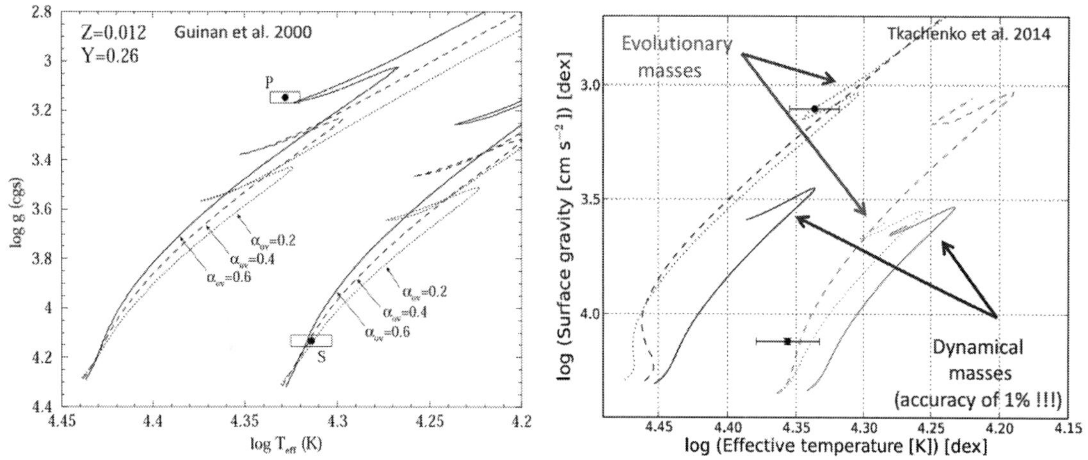

Figure 2: Position of both stellar components of the V380 Cyg system in the T_{eff}-log g Kiel diagram as obtained by [16] (left) and by [31] (right). The more evolved primary component has lower log g value and thus appears on top in both panels. The lines of different styles in the left panel represent evolutionary tracks computed assuming different values of the overshoot parameter (the exact values are indicated in the plot in the units of pressure scale height) and the (dynamical) masses obtained from the binary modelling. The solid lines in the right panel refer to the tracks computed for the dynamical masses of the stars and assuming standard value of overshoot of $\alpha = 0.2$. The dashed and dotted lines represent the best fit evolutionary tracks obtained by including extra mixing into the models in terms of the overshoot and rotational mixing. See text for a discussion.

arrive to a similar conclusion: the models are not fully adequate as they generally tend to over-predict temperatures which is equal to under-predicting the masses. The problem seems to be mass-dependent and is more pronounced towards higher masses. The authors also found that different sets of evolutionary models are often inconsistent with each other. Moreover, the models are not internally consistent, which is nicely illustrated in Fig. 3. The plots show the positions of both stellar components of the UScoCTIO 5 binary system in the HR-diagram (top) and in the mass-radius diagram (bottom). The age obtained by fitting the isochrones to the stellar positions in the HR-diagram is a factor of two different from the one obtained by performing a similar fit in the observational mass-radius diagram. The authors find it very probable that the models fail to correctly predict the temperatures, possibly because they do not fully encompass the physics of energy transport (via convection and/or missing opacities). There is also a hint that incorporating the magnetic fields into the models might help to partially resolve the discrepancies. Magnetic fields inhibit the convection which results in decreasing the emergent flux and thus provides a slower contraction of a star during its pre-MS phase of evolution.

As it has been mentioned before, the mass discrepancy observed in binary stars is a manifestation of various shortcomings in the current theories/models of stellar structure and evolution. Thus, by studying in detail and resolving the mass discrepancy problem will help us to reveal the shortcomings in the theories and improve upon them. Observations wise, this requires an access to high-quality (space-based) photometric data which have to be complemented with the ground-based, high-resolution spectroscopy. Unfortunately, neither pre-MS nor massive stars have been represented in sufficient quantity in the field of view of the original *Kepler* mission, and only a few of them have been observed by CoRoT. However, many of these objects will be observed by the K2 and TESS missions, for the first time allowing us to approach the problem from statistical point of view, by analysing and charactiresing large star samples in a wide range of stellar masses.

3 Rotationally induced mixing in high-mass stars

The rotationally-induced mixing in high-mass stars is yet another topic that deserves a specific attention and is thus discussed here. The current theory predicts a significant enhancement of the nitrogen abundance at the surface of rapidly rotating stars, where the elements are transported by the rotational mixing to. [18] presented an observational confirmation of these theoretical predictions by finding a correlation between the surface nitrogen abundance and the projected rotational velocity from the analysis of a sample of 135 B-type stars in the Large Magellanic Cloud (see top panel of Fig. 4). However, the authors also reported on the detection of two groups of stars that contradict the theoretical predictions for the surface nitrogen enhancement due to the rotational mixing. These groups are termed "Group1" and "Group2" in the top panel of Fig. 4 and represent rapidly rotating stars with surface nitrogen abundances consistent with the LMC baseline composition and slowly rotating stars with enhanced surface nitrogen abundance, respectively. Assuming the theoretical predictions of the surface nitrogen enrichment are correct, these two groups of stars are of particular interest as there

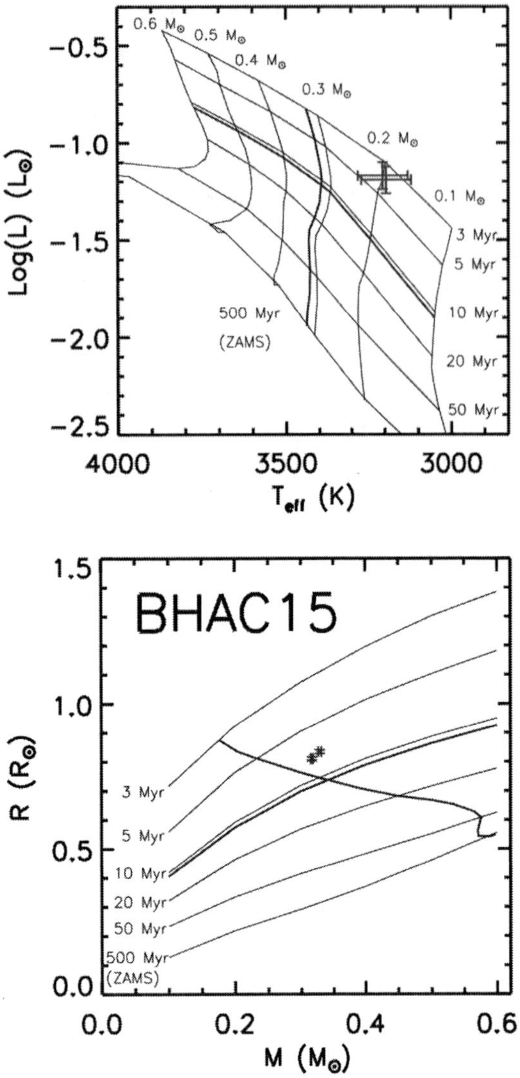

Figure 3: Positions of both stellar components of the UScoCTIO 5 binary system in the HR-diagram (top) and in the mass-radius diagram (bottom). The isochronal and isomass sequences are those predicted by the BHAC 15 models [6]. The figure is taken from [20].

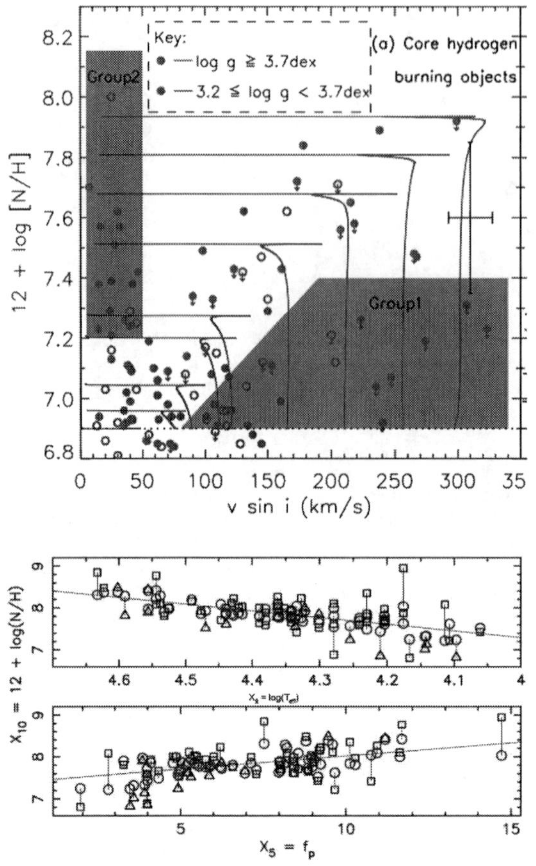

Figure 4: Top: Surface nitrogen abundance (12 + log [N/H]) versus the projected rotational velocity ($v \sin i$) diagram for 135 B-type stars in the Large Magellanic Cloud. The typical error bars are indicated in the plot. Figure is taken from [18]. **Bottom:** Correlation of the surface nitrogen abundance with the effective temperature of the star and the frequency of its dominant acoustic oscillation mode. Figure has been adopted from [3].

must be some additional mechanisms involved (magnetic fields, binarity, etc.) that "pull out" the stars of the predicted correlation.

Although the correlation found by [18] indeed seems to be in line with the present theoretical predictions, neither of the two agree with the recent asteroseismic findings. In particular, [3] presented a detailed analysis of a sample of 70 B-type stars using multivariate statistical analysis techniques. Both the surface nitrogen abundance and the rotational frequency of the star has been included as the variables in the statistical analysis and, contrary to [18], the authors found that rotation has zero predictive power for the surface nitrogen abundance. Instead, the latter was found to correlate with the effective temperature of the star as well as with the frequency of its dominant acoustic oscillation mode (see bottom panel of Fig. 4). Thus, the largely observational asteroseismic analysis appears to be in contradiction with both the spectroscopic findings and the theoretical predictions as for the surface nitrogen enrichment driven by the rotational mixing in B-type stars.

Pulsating B-type stars in (double-lined spectroscopic eclipsing) binaries can help to unravel the ambiguity with the surface nitrogen abundance enrichment and its physical cause. Chemical composition analysis of B-type stellar components in binaries benefits greatly from the precisely determined values of the surface gravity (from dynamical mass and radius) and the effective temperature (net result of the removed degeneracy between $\log g$ and T_{eff}), allowing to reach a much higher precision in the individual

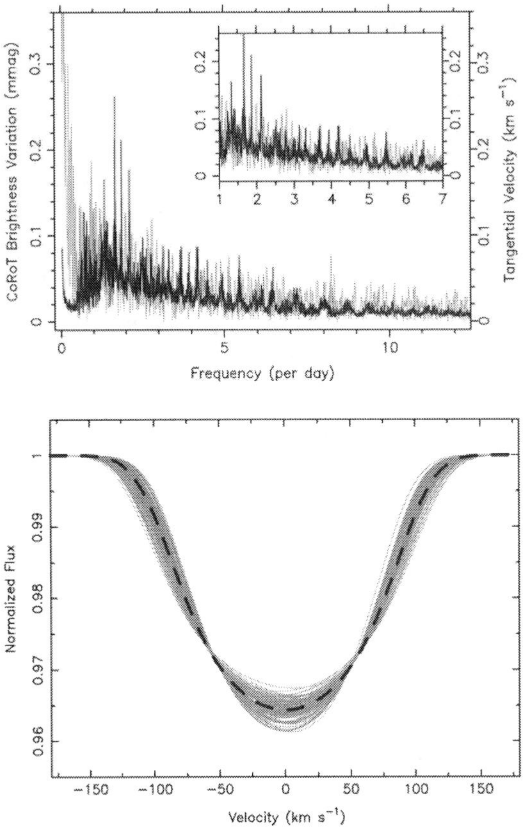

Figure 5: Top: Comparison of the amplitude spectrum constructed based on the 2D simulations of the IGWs (red) with the amplitude spectrum of HD 46150 observed by CoRoT (light grey). **Bottom:** Simulations of the line profile variations (light grey) caused by the IGWs in comparison with the line profile shape due to the rotational broadening only (red). Both plots has been adopted from [4].

abundance measurements than for single stars in the same mass/spectral type range. The results of such an analysis could be confronted with those obtained by [18] and [3] with the ultimate goal to 1) check whether any of the announced correlations are also applicable to binary stars, and 2) see whether the (possible) differences in the results can be attributed to the effect of the tides in binary systems.

4 Internal gravity waves

Another interesting phenomenon that is worth touching upon here is the Internal Gravity Waves (IGWs). The IGWs get excited in a stratified medium where the density decreases continuously or discontinuously with depth/height. In stars, these waves are generated at the boundary between the convective and radiative regions by convective motions. The IGWs has been studied theoretically and numerically [e.g., 21, 24, 28, 29, 32] quite extensively but essentially remained observationally undetectable until very recently. A quantitative comparison of the amplitude spectra constructed based on 2D simulations of IGWs with those observed by CoRoT have been performed by [4] for three O-type stars. Fig. 5 (top panel) illustrates a good overall agreement between the observed and simulated shapes of the Fourier spectrum of HD 46150, which is also representative of the other two stars studied, HD 46223 and HD 46966. The bottom panel of Fig. 5 illustrates the simulated line profile variations due to the surface velocity field generated by the IGWs. The variability is quite remarkable and appears to be consistent with additional line broadening observed in OB stars due to time-dependent macroturbulent velocity field.

The IGWs are also believed to be very efficient in transporting angular momentum and chemical species inside stars. Having said that, it is interesting to come back to the study performed by [18] for 135 B-type stars in the Large Magellanic Could, namely to the two groups of stars that were found to contradict the theory of rotationally induced mixing (see top panel of Fig. 4). The question to raise would be: could the existence of those two groups of stars be explained within the framework of IGWs? At first sight, the answer is probably "yes", assuming the IGWs can bring the angular momentum to the outer layers of a star as well as to extract it from them. The "Group1" stars (rapidly rotating stars with no surface nitrogen enrichment) could then be seen as intrinsically slow rotators in which the rotationally induced mixing is inefficient to bring nitrogen to the surface but whose outer layers are spun up by the IGWs leading to the detection of large $\upsilon \sin i$ values. And the other way around, the "Group2" stars would then be intrinsically rapid rotators whose outer layers got spun down due to the angular momentum extracted from them by the IGWs. However, this hypothesis still contradicts the asteroseismic findings by [2] who found no correlation between the surface nitrogen abundance and the rotational frequency/projected rotational velocity of the star based on the multivariate statistical analysis of a sample of 70 B-type stars. Nevertheless, the IGWs can still be (and probably are) largely involved in the transport of the chemical elements inside stars. Thus, it would be worth looking for a way to quantify the effect of the IGWs in massive stars and to include the corresponding parameter into the multivariate analysis similar to that performed by [2]. Should any correlation between the parameter describing the effect of the IGWs and the surface nitrogen abundance be found, this would point to the fact that the effect of the IGWs cannot be ignored when addressing the problem of the surface nitrogen enrichment in high-mass stars but must be taken into account along with the rotationally induced mixing.

Since the IGWs are probably generated in the majority of stars and are likely to play a major role in the transport of angular momentum inside them, it is also expected that these waves play a major role in the evolution of binary stars. The connection between the IGWs and the evolution of binary star systems in terms of circularisation of the orbit and synchronisation of the stellar components' spins are briefly touched upon in the next section.

5 Binary stars and tidal evolution theory

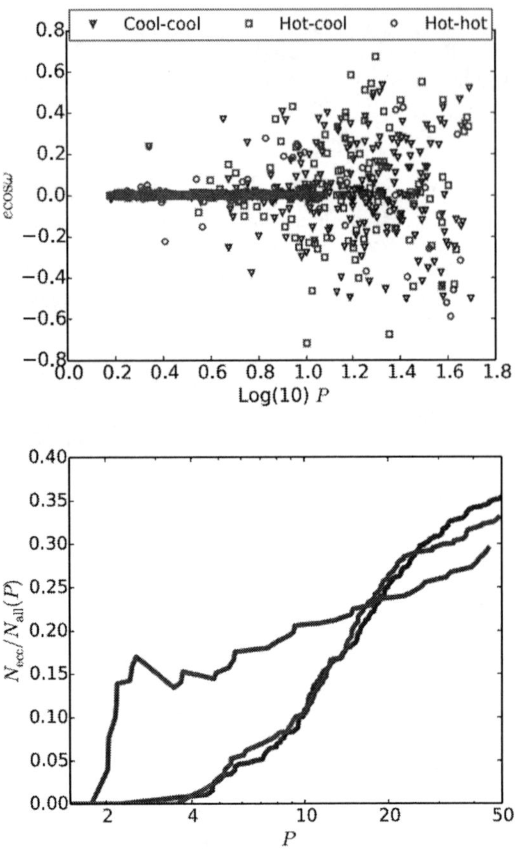

Figure 6: Top: A sample of some 900 *Kepler* binary systems on the $e \cos \omega$–$\log P$ diagram studied by [34]. The colours represent different system configurations in terms of the masses/spectral types of the individual stellar components: cool-cool (blue), hot-cool (green), and hot-hot (red). **Bottom:** The cumulative fraction of eccentric systems as a function of the orbital period. The meaning of the colours is the same as in the top panel. Both plots have been adopted from [34].

The theory of tidal evolution describes the path that a binary star follows starting from its original configuration to the stage of lowest possible kinetic energy that is consistent with the conservation of angular momentum in the system. The lowest energy state is characterised by the alignment of the rotational axes of both stellar components with the normal to the orbital plane (also called spin-orbit alignment), circular orbit, and rotational periods of the stars equal to the orbital period of the system (also called spin-orbit synchronisation). Keeping it simple, the general pattern of the evolutionary path is as follows: in eccentric (close) binaries with individual components having non-synchronized spins, the tidal forces will raise differential motions in the stars, which will be dissipated into heat that is ultimately radiated into space. The excess energy comes at the expense of either the orbital energy or the spin energies of the stars, and a binary system tends to evolve into the state characterised by a circular orbit with the two stars having synchronised spins. Turbulent viscosity is one of the main dissipation mechanisms in stars with (large) outer convective zones [39–41], whereas damping of the resonances of the free pulsation g-modes by radiative dissipation is believed to be the dominant process in the tidal evolution for stars without (large) outer convective zones [42].

There is a number of issues with the current theoretical predictions concerning the synchronisation time scales. In particular, there are many examples in the literature where the individual stellar components exhibit either sub- or super-synchronous rotation while the theory predicts full synchronisation with the orbit. The situation is complicated by the fact that calculation of the spin-orbit synchronisation rates requires deep understanding of angular momentum transport inside individual stellar components. This is because the tidal torque that drives binary orbital evolution and synchronisation in particular is mostly applied to the outermost layers of the star, thus the synchronisation must be occurring from the stellar surface inwards. Looking at the effect of the IGWs in the context of spin-orbit synchronisation is particularly interesting as these waves are very efficient in transporting angular momentum inside stars, as it has been mentioned above already. In fact, the IGWs may just be "disturbing" the individual stellar components of binary systems from their synchronous rotation by spinning up or down their outermost layers. This would make the stars to appear observationally non-synchronised whereas they are in fact temporary pulled out of the spin-orbit synchronisation state by the IGWs transporting/redistributing angular momentum inside the stars. This hypothesis requires a further investigation which implies the observational detection of the IGWs in a wide range of stellar masses as well as the incorporation of the effect of these waves into the tidal evolution theory.

Contrary to the effect of the spin-orbit synchronisation, the current theory of tidal evolution performs reasonably well when it comes to the prediction of the orbital circularisation time scales. Recently, [34] investigated a sample of some 900 eclipsing binary stars observed by the original *Kepler* mission with the ultimate goal to confront the observations with the theoretical predictions concerning the circularisation time scales. Because of the lack of follow-up spectroscopic data, the parameter $e \cos \omega$ was used as a proxy for the eccentricity of the system. The sample was limited to stars with effective temperatures below 10 000 K, and three groups of binaries were defined according to the effective temperatures of the individual stellar components. These groups are: 1) cool-cool, 2) cool-hot, and 3) hot-hot binaries, with hot being a star with the effective temperature above 6 250 K (see top panel of Fig. 6). All stars were assumed to be on the main-sequence, so that the mass-radius-age relations could be used to estimate parameters necessary for the calculation of the circularisation time scales. The authors came to the conclusion that the theoretical predictions are in most cases in line with the observations, in particular for bi-

DOI: 10.1051/epjconf/201716005007

naries containing at least one cool companion, although a few outliers were detected. Some cut-off periods were proposed beyond which the circularisation occurs at almost 100% rate, and it is worth noting that binaries with one cool and one hot component behave likewise the systems containing two cool stars (see bottom panel of Fig. 6). This result points to the fact that the tidal dissipation plays a major role in the observed distribution of the eccentricities for two groups of binaries with at least one cool component, whereas age is the dominant contributor for binaries formed of two hot stars.

Although very interesting and valuable by itself, the study of [34] can be extended/improved on several aspects. First, this concerns including more massive and hotter stars into the sample so that it also covers the entire spectral type B and thus binaries with more extreme mass ratios. Secondly, it would be worth including the spin-orbit synchronisation aspect to quantify by how much the models fail to predict the corresponding time scales and to identify the mass-temperature regime that perhaps deserves a specific attention. Last but not least, it would be very beneficial for this type of the analysis to replace the many assumptions made with the observational constraints: replace the $e \cos \omega$ parameter by the eccentricity itself, feed the statistical analysis with the masses, radii, ages, etc. derived from binary dynamics and/or asteroseismology instead of assuming all stars being on the main-sequence and using the corresponding mass-radius-age relations, etc. The K2, TESS, and PLATO missions will provide a significant contribution to the problem as they cover a larger parameter space (in terms of stellar masses and evolutionary stages) than the original *Kepler* mission did. The TESS and PLATO missions will also target brighter stars which implies high-quality ground-based spectroscopic data can be obtained providing some of the missing observational constraints.

Despite the fact that the tidal evolution theory performs quiet well concerning the circularisation time scale predictions, the objects showing a remarkable disagreement between the theory and the observations are reported in the literature from time to time. The NGC 6791 turn-off binary system V568 Lyr is one of those objects where the observational measurements obtained from a combination of *Kepler* space-based photometry and ground-based high-resolution spectroscopy significantly disagree with the predictions of the tidal evolution theory. In particular, [38] found the system to be about 7.5 Gyr old and to have a circular orbit. Contrary to these findings, the tidal evolution theory predicts a significant eccentricity for the system and the tidal friction would need to be increased by a factor of $\sim$100 in the models to make them consistent with the observations. This example clearly illustrates that there are cases where the theory also fails to predict the circularisation time scales, and such "outliers" are worth looking at to reveal the shortcomings in and improve upon the current theory of tidal evolution.

6 Tidally-induced pulsations

Tidally-induced pulsations in binary systems represent yet another physical mechanism that emphasises the importance of synergies between asteroseismology and binary star research fields. These are the gravity-mode free oscillations of stars that get resonantly driven by the dynamic tide in a binary system, typically during the periastron passage of the binary components where the tidal forces are the most efficient. As briefly discussed above, the tidally-induced oscillations play a major role in the evolution of intermediate- to high-mass binary system as their radiative damping serves as the main energy dissipation mechanism in those objects. However, recent studies of intermediate-mass binary systems observed by *Kepler* clearly showed that some of the resonantly driven modes are not damped in the radiative regions of stars but appear to have quite stable and large amplitudes so that they get observed at the stellar surfaces. The KOI-54 system discovered and observationally studied in detail by [37] is probably one of the most remarkable examples of pulsating binary stars where large fraction of the observed oscillation modes are resonantly driven by the tides. A distinct property of the tidally-induced pulsations is that they are observationally detected as exact harmonics of the orbital frequency in the Fourier spectrum, and the KOI-54 system exhibits the 90th and 91st harmonics as its largest amplitude oscillation modes. Several challenges have been raised by [37] concerning the pulsation properties of the KOI-54 system: 1) is it just one or both stars that are pulsating in the system?, 2) why do the 90th and 91st harmonics have the largest amplitudes?, and 3) what is the exact mechanism that feeds the observed pulsation modes and keeps them so stable in time, and why are not they damped in the radiative regions of a star? [13] investigated the KOI-54 system within the extended theory of tidal evolution with the ultimate goal to explain the observed oscillation properties of the star. The authors computed the predicted flux variations due the geometric and temperature effects of the modes and compared them to the observations of the two dominant modes in the system. Both observed oscillation modes could be explained by resonantly locked quadruple modes. The stability of these two modes over time is achieved through their resonant locking driven by the dynamical tides in the system and the intrinsic stellar spin down. The authors also show that the resonant locking provides an efficient way of transporting angular momentum from the orbit to the individual stellar components which implies that the KOI-54 system is going to evolve rapidly to zero eccentricity from the actual value of e=0.83. Thus, the tidally-induced pulsations represent an efficient way of transporting angular momentum within binary star systems and the space missions like *Kepler*, TESS, and PLATO provide us with the required high-quality observations to test and improve upon the current theories of tidal evolution.

7 Scaling relations

The scaling relations is the last topic that is discussed in this manuscript in the context of synergies between as-

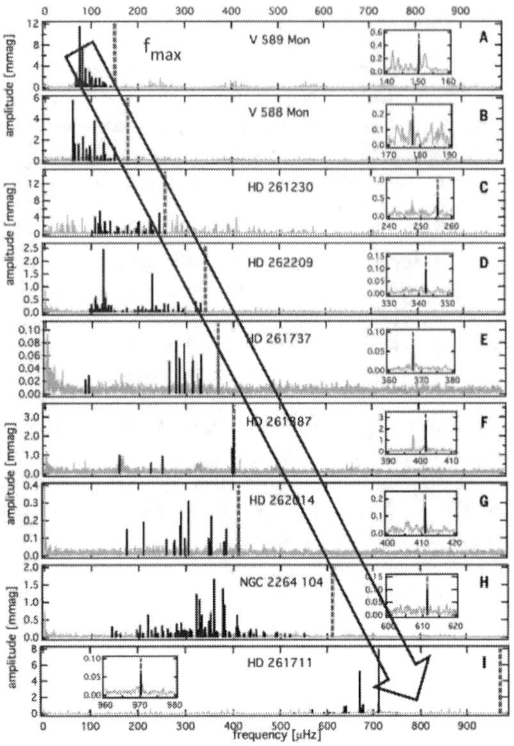

Figure 7: Correlation between the maximum pulsation frequency of the p-mode observed in a star and its relative age. The red arrow (schematically) shows how the pulsation properties of a star change as its moving towards the Zero Age Main Sequence (ZAMS). The plot has been adopted from [43].

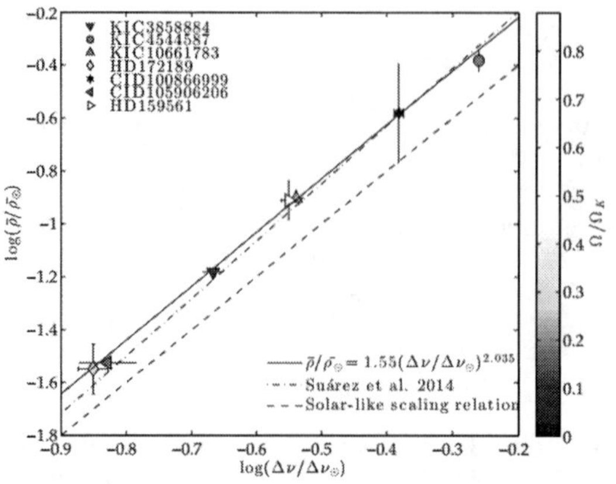

Figure 8: A relation between the mean density of the star and its large frequency separation in the low radial order regime for seven binary systems with main-sequence δ Sct type pulsating components. The solid line indicates the obtained empirical relation for MS δ Sct pulsators, the dashed line refers to the solar-like scaling relation, and the dashed-dotted line stands for the theoretically predicted relation for low radial order p-mode oscillations. The figure has been adopted from [14].

teroseismology and binary stars. The scaling relations in asteroseismology stand for the relations between the fundamental properties of stars (masses and radii) and their seismic characteristics like large frequency separation $\Delta\nu$ and frequency of the maximum power $\nu_{\rm max}$. The relations hold for stars with extended convective envelopes that excite their pulsation modes stochastically and read

$$\begin{aligned} \frac{M}{M_\odot} &\simeq \left(\frac{\nu_{\rm max}}{\nu_{\rm max,\odot}}\right)^3 \left(\frac{\Delta\nu}{\Delta\nu_\odot}\right)^{-4} \left(\frac{T_{\rm eff}}{T_{\rm eff,\odot}}\right)^{3/2} \\ \frac{R}{R_\odot} &\simeq \left(\frac{\nu_{\rm max}}{\nu_{\rm max,\odot}}\right) \left(\frac{\Delta\nu}{\Delta\nu_\odot}\right)^{-2} \left(\frac{T_{\rm eff}}{T_{\rm eff,\odot}}\right)^{1/2} \end{aligned} \quad (1)$$

Large frequency separation $\Delta\nu$ refers to the frequency spacing between the adjacent modes of stellar oscillation having the same degree l (the total number of the surface nodal lines where no movement of the material occurs during the pulsation cycle) but different radial order n (the total number of nodal lines inside the star in the radial direction). The relations were theoretically defined by [19] and are widely used nowadays to determine fundamental properties of stars pulsating in solar like oscillations. The relations assume that the properties of the star in question are close to those of the Sun which is often not the case (especially when the relations are applied to red giants). For example, [23] pointed out that masses obtained from the scaling relations are on average accurate to about 10%, which results in about 30% uncertainty on stellar age. [10] come to a similar conclusion, emphasizing that these uncertainties are due to the combined effects of models and adopted analysis methodology. The authors explicitly mention the need of more independent tests of the scaling relations, in particular tests of the estimated masses. The most obvious and precise way of tuning the scaling relations is to use the dynamical asteroseismology, which involves the study of solar-like pulsators in binary systems, where the fundamental, model-independent properties of stars inferred from binary dynamics can be confronted to the values obtained by means of the asteroseismic analysis. This field is currently being explored, in particular there was a talk given by Jean McKeever (see the corresponding contribution in these proceedings) in the session S10: Seismology: Stars Beyond the Main Sequence. The conclusions are essentially the same as those presented by [10, 23]: the relations perform quite well for the radius estimates whereas the masses are uncertain to about 10% level.

The idea of scaling relations can actually be extended towards the stars exhibiting heat-driven oscillations. However, instead of introducing theoretical relations, the synergies between asteroseismology and binary stars can be fully exploited to establish empirical relations between the fundamental properties of stars (obtained from binary dynamics) and their oscillation characteristics. For example, a clear relation between seismic properties and the age of a star has been found by [43] for pre-MS δ Sct type pulsators (see Fig. 7). The relation has been determined form the analysis of a sample of 9 pre-MS single stars and does not presently have a defined zero point, i.e. it allows to derive a relative age of a young star from its observed p-mode pulsation properties. Detecting a δ Sct type pre-MS pulsator in a binary system would allow to put this relation on the absolute scale, which will certainly become possible in the near future with the K2, TESS, and PLATO observations.

[14] have studied seven binary systems containing at least one main-sequence δ Sct type pulsator. A clear relation between the mean density of the star (computed from the dynamical mass and radius) and the analog of the large frequency separation from solar case but in low radial order regime has been detected (see Fig. 8). The authors also found the empirical relation to be in a reasonable agreement with the theoretical relation computed for p-mode oscillations in the low radial order regime.

The studies by [43] and [14] clearly demonstrate the potential of the dynamical asteroseismology in establishing empirical scaling relations between fundamental and seismic properties of stars exhibiting heat-driven oscillations. This opens a window for the determination of precise fundamental parameters of pulsating stars in wide range of stellar masses and evolutionary stages, thus providing critical observational input for testing and if necessary refining the current theories of stellar structure and evolution.

8 Summary

This paper touches upon just a few of many topics in stellar astrophysics that naturally arise from combining the research fields of asteroseismology and binary stars. The major points that have been discussed in the manuscript are summarized here:

- Many of the intermediate- to high-mass binary stars are found to exhibit the mass discrepancy problem which points to the shortcomings in the current theories of stellar structure and evolution. In the majority of cases, the inclusion of an extra mixing into the models in terms of the convective core overshoot allows to minimize the discrepancy between the dynamical and evolutionary masses of stars, but the high-overshoot values appear to be inconsistent with those seismically derived for single stars of similar masses and spectral type. Combining the fields of asteroseismology and binary stars will help us to reveal the shortcomings in and improve upon the current theories of stellar structure and evolution.
- The recent asteroseismic studies of B-type stars do not support the theoretical predictions and some of the spectroscopic findings concerning the rotationally induced mixing of chemical elements inside intermediate- to high-mass stars. Studying (pulsating) binary stars in similar mass/spectral type range would help to 1) significantly improve upon the accuracy of spectroscopically measured individual abundances, 2) explore whether there is a link between binarity and surface nitrogen enrichment (tidal effects, mergers, etc.), and 3) investigate a possible connection with the internal gravity waves excited in stellar interiors.
- Internal gravity waves are very efficient in transporting angular momentum inside stars. Moreover, these waves are also believed to play a major role in transporting chemical elements, thus might be an important physical ingredient in the surface nitrogen enrichment observed in B-type stars. Although the IGWs has been extensively studied theoretically in the past and both 2D and 3D numerical simulations are currently performed, the observational evidences of these waves are yet to be searched for and the effect is to be quantified using both space-based photometry and ground-based high-resolution spectroscopy. The precise observational constraints will help us to improve upon the existing theories and to better understand the role of the IGWs in the evolution of single stars and multiple systems.
- The present-day theory of tidal evolution seems to be performing reasonably well when it comes to the prediction of the circularisation time scales in binary stars. However, the theory is still likely to be missing some important physical ingredients as it is often found to disagree with the observations of synchronisation rates of the individual stellar components. The high-quality space-based photometric observations delivered by *Kepler*/K2, TESS, and PLATO and complemented with high-resolution ground-based spectroscopy open a window for studying new physical effects (tidally-induced pulsations, IGWs, etc.) and incorporating them into the tidal evolution theory.
- Radiative damping of the resonantly driven oscillation modes is the main energy dissipation mechanism that drives the evolution of intermediate- to high-mass binary stars. However, the recent *Kepler* observations also show that some of those resonant modes get high and stable amplitudes instead of being damped. For the first time, we get the opportunity to confront the existing theories with the high-quality observations and to learn how the resonantly locked oscillation modes impact/change the evolutionary path of the individual binary components as well as the evolution of the system as a whole.
- The dynamical asteroseismology is a perfect tool for testing and if necessary refining the asteroseismic scaling relations for solar-like pulsators. It also holds a great potential for establishing empirical relations for stars of different masses and evolutionary stages that exhibit heat-driven oscillations.

Acknowledgments

The research leading to these results has received funding from the European Research Council (ERC) under the European Union's Horizon 2020 research and innovation programme (grant agreement N°670519: MAMSIE).

References

[1] Aerts, C., Christensen-Dalsgaard, J., & Kurtz, D. W., *Asteroseismology*, (Astronomy and Astro-

physics Library. ISBN 978-1-4020-5178-4. Springer Science+Business Media B.V., 2010)

[2] Aerts, C., EAS Publications Series, **64**, 323 (2013)

[3] Aerts, C., Molenberghs, G., Kenward, M. G., & Neiner, C., ApJ, **781**, 88 (2014)

[4] Aerts, C., & Rogers, T. M., ApJL, **806**, L33 (2015)

[5] Auvergne, M., Bodin, P., Boisnard, L., et al., A&A, **506**, 411 (2009)

[6] Baraffe, I., Homeier, D., Allard, F., & Chabrier, G., A&A, **577**, A42 (2015)

[7] Bedding, T. R., Mosser, B., Huber, D., et al., Nature, **471**, 608 (2011)

[8] Bloemen, S., Hu, H., Aerts, C., et al., A&A, **569**, A123 (2014)

[9] Borucki, W. J., Koch, D., Basri, G., et al., Science, **327**, 977 (2010)

[10] Chaplin, W. J., Basu, S., Huber, D., et al., ApJS, **210**, 1 (2014)

[11] David, T. J., Conroy, K. E., Hillenbrand, L. A., et al., AJ, **151**, 112 (2016a)

[12] David, T. J., Hillenbrand, L. A., Cody, A. M., Carpenter, J. M., & Howard, A. W., ApJ, **816**, 21 (2016b)

[13] Fuller, J., & Lai, D., MNRAS, **420**, 3126 (2012)

[14] García Hernández, A., Martín-Ruiz, S., Monteiro, M. J. P. F. G., et al., ApJL, **811**, L29 (2015)

[15] Gillen, E., Aigrain, S., McQuillan, A., et al., A&A, **562**, AA50 (2014)

[16] Guinan, E. F., Ribas, I., Fitzpatrick, E. L., et al., ApJ, **544**, 409-422 (2000)

[17] Herrero, A., Kudritzki, R. P., Vilchez, J. M., et al., A&A, **261**, 209 (1992)

[18] Hunter, I., Brott, I., Lennon, D. J., et al., ApJL, **676**, L29 (2008)

[19] Kjeldsen, H., & Bedding, T. R., A&A, **293**, 87 (1995)

[20] Kraus, A. L., Cody, A. M., Covey, K. R., et al., American Astronomical Society Meeting Abstracts, **227**, 236.12 (2016)

[21] Kumar, P., Talon, S., & Zahn, J., ApJ, **520**, 859 (1999)

[22] Kurtz, D. W., Saio, H., Takata, M., et al., MNRAS, **444**, 102 (2014)

[23] Miglio, A., Chiappini, C., Morel, T., et al., MNRAS, **429**, 423 (2013)

[24] Montalban, J., & Schatzman, E., A&A, **354**, 943 (2000)

[25] Nemec, J. M., Cohen, J. G., Ripepi, V., et al., ApJ, **773**, 181 (2013)

[26] Pápics, P. I., Tkachenko, A., Aerts, C., et al., ApJL, **803**, L25 (2015)

[27] Raskin, G., van Winckel, H., Hensberge, H., et al., A&A, **526**, A69 (2011)

[28] Rogers, T. M., & MacGregor, K. B., MNRAS, **410**, 946 (2011)

[29] Rogers, T. M., Lin, D. N. C., McElwaine, J. N., & Lau, H. H. B., ApJ, **772**, 21 (2013)

[30] Saio, H., Kurtz, D. W., Takata, M., et al., MNRAS, **447**, 3264 (2015)

[31] Tkachenko, A., Degroote, P., Aerts, C., et al., MNRAS, **438**, 3093 (2014)

[32] Townsend, A. A., JFM, **24**, 307 (1966)

[33] Triana, S. A., Moravveji, E., Pápics, P. I., et al., ApJ, **810**, 16 (2015)

[34] Van Eylen, V., Winn, J. N., & Albrecht, S., ApJ, **824**, 15 (2016)

[35] Van Reeth, T., Tkachenko, A., & Aerts, C., A&A, in press (2016, arXiv:1607.00820)

[36] Walker, G., Matthews, J., Kuschnig, R., et al., PASP, **115**, 1023 (2003)

[37] Welsh, W. F., Orosz, J. A., Aerts, C., et al., ApJS, **197**, 4 (2011)

[38] Yakut, K., Eggleton, P. P., Kalomeni, B., Tout, C. A., & Eldridge, J. J., MNRAS, **453**, 2937 (2015)

[39] Zahn, J.-P., Ann. Astrophysics, **29**, 313 (1966a)

[40] Zahn, J.-P., Ann. Astrophysics, **29**, 489 (1966b)

[41] Zahn, J.-P., Ann. Astrophysics, **29**, 565 (1966c)

[42] Zahn, J.-P. 1975, A&A, **41**, 329-344

[43] Zwintz, K., Fossati, L., Ryabchikova, T., et al., Science, **345**, 550 (2014)

DOI: 10.1051/epjconf/201716005008

Constraining stellar physics from red-giant stars in binaries – stellar rotation, mixing processes and stellar activity

P. G. Beck[1], T. Kallinger[2], K. Pavlovski[3], A. Palacios[4], A. Tkachenko[5], R. A. García[1], S. Mathis[1], E. Corsaro[1], C. Johnston[5], B. Mosser[6], T. Ceillier[1], J.-D. do Nascimento Jr.[7,8], and G. Raskin[5]

[1]*Laboratoire AIM Paris-Saclay, CEA/DRF — CNRS — Université Paris Diderot, IRFU/SAp Centre de Saclay, F-91191 Gif-sur-Yvette Cedex, France*
[2]*Institut für Astronomie der Universität Wien, Türkenschanzstr. 17, 1180 Wien, Austria*
[3]*Department of Physics, Faculty of Science, University of Zagreb, Croatia*
[4]*LUPM, Université Montpellier II, Place Eugéne Bataillon cc -0072, F-34095, Montpellier cedex 5, France*
[5]*Instituut voor Sterrenkunde, KU Leuven, 3001 Leuven, Belgium*
[6]*LESIA, CNRS, Université Pierre et Marie Curie, Université Denis Diderot, Observatoire de Paris, 92195 Meudon cedex, France*
[7]*Harvard-Smithsonian Center for Astrophysics, Cambridge, MA 02138, USA*
[8]*Departamento de Física Teórica e Experimental, Universidade Federal do Rio*

Abstract. The unparalleled photometric data obtained by NASA's *Kepler* Space Telescope has led to an improved understanding of stellar structure and evolution - in particular for solar-like oscillators in this context. Binary stars are fascinating objects. Because they were formed together, binary systems provide a set of two stars with very well constrained parameters. Those can be used to study properties and physical processes, such as the stellar rotation, dynamics and rotational mixing of elements and allows us to learn from the differences we find between the two components. In this work, we discussed a detailed study of the binary system KIC 9163796, discovered through *Kepler* photometry. The ground-based follow-up spectroscopy showed that this system is a double-lined spectroscopic binary, with a mass ratio close to unity. However, the fundamental parameters of the components of this system as well as their lithium abundances differ substantially. *Kepler* photometry of this system allows to perform a detailed seismic analysis as well as to derive the orbital period and the surface rotation rate of the primary component of the system. Indications of the seismic signature of the secondary are found. The differing parameters are best explained with both components located in the early and the late phase of the first dredge up at the bottom of the red-giant branch. Observed lithium abundances in both components are in good agreement with prediction of stellar models including rotational mixing. By combining observations and theory, a comprehensive picture of the system can be drawn.

1 Introduction

Asteroseismology is a powerful tool to study stellar structure and dynamics as well as their evolution [1]. The NASA *Kepler* mission [2] delivered unprecedented, high-precision photometric data, enabling us to perform detailed asteroseismic analysis for thousands of stars [e.g. 3–7]. Seismology provides important inferences on the internal structure of stars such as the evolutionary state [e.g. 8–11], stellar rotation [e.g. 12–17], or the characterisation of members of open clusters [18–21]. Also distances can be estimated through asteroseismology [e.g. 22] and now can be compared to distances from trigonometric parallaxes, provided by GAIA [e.g. 23]. Such wealth of information allows us to put strong constraints on the interpretation of observations of stars and view the star in a wider context. Asteroseismology unfolds its full potential of precision when combined with fundamental parameter from spectroscopy. This allows an accurate determination of the mass and radius of the star [7, 24]. A particular treasure trove for stellar astrophysics are oscillating stars in binary systems. For such stars we have the unique possibility to constrain radius and mass with two different independent methods [e.g. 26, 27]. Indeed, comparing masses from binary solutions and stars in cluster, evidence was found that the seismic scaling relations [e.g. 7] are systematically overestimating the stellar mass by about 15% [21, 25].

Binary systems are rewarding to study, since many parameters are well constrained (e.g. mass, radius, metallicity). Unless a binary system was created through a rare event of capturing, both components were born together, they have the same age and metallicity. For stars in binary systems, the only differences between the two components are the difference in mass and eventually from the surface rotation if the system is synchronised. The difference in mass then translates into different speeds of stellar evolution [e.g. 28]. In case of a double-lined spectroscopic binary system (SP2), this parameters can precisely be determined from the ratio of the radial velocity amplitudes, even for unknown inclinations [26]. As we will demon-

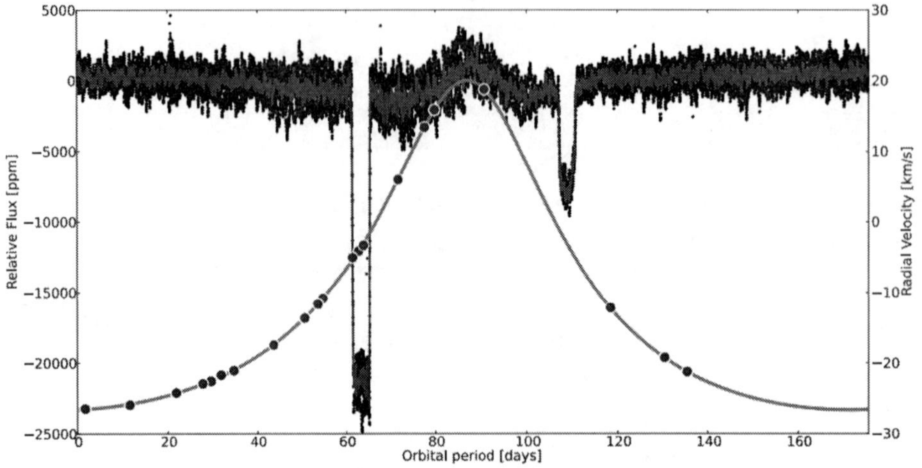

Figure 1: Combined phase diagram for the eclipsing system KIC 9540226 observed using space photometry from the *Kepler* satellite and radial velocities from ground-based spectroscopic observations with the Hermes spectrograph. The *Kepler* photometry is presented on the left vertical axis. Black data points are the original measurements, while the red line shows the averaged light curve, which was determined by rebinning the phase diagrams into 30 minutes bins. The deviation from the averaged light curve is caused by solar-like oscillations. The radial velocities are shown as solid, blue dots (right vertical axis). The solid blue line represents the orbital model of a Keplerian orbit. (Figure from [44])

strate later in this proceedings paper, the SB2 feature, actually allows to measure mass ratio between the two components with an accuracy of about 1%. Using SB2 thus allows a model independent approach. If one or both components show stellar oscillations, seismic techniques can be applied to further study the stellar structure of the oscillating component in detail. With such a well constrained set of parameters, one can draw a comprehensive picture of the full system and its components.

By now about 3000 stars, i.e. ~1.3% of all stars observed by *Kepler*, are found to be binaries [29, and references therein]. About 40 systems with an oscillating red-giant component are known and characterised whereby some studies focused on the eclipsing systems [30, 31], while another presented a catalogue of red giants in binary systems with tidally induced flux modulation [32]. A set of additional candidates of red giants in binaries was proposed by [33]. Finally, binaries with solar-like oscillators on the main-sequence were found [e.g. 34–36].

Several binaries with a red-giant component have been analysed in detail using *Kepler* photometry and ground-based spectroscopy: KIC 8410637 [27, 37, 38], KIC 5006817 [32], KIC 9246715 [39], KIC 5640750 [38], and KIC 9540226 [38]. With this work, we added KIC 9163796 [40] (also nicknamed *Asterix & Obelix*) to the sample of well studied binary systems.

In this paper, we first discuss in Section 2 the different types of observational features that can be used to detect binaries and highlight the benefit of ground-based spectroscopy. The *Kepler* photometry and Hermes spectroscopy of the system of KIC 9163796 is described in detail in Section 3. The seismology of the primary and secondary component of KIC 9163796 are discussed in Section 4. The indicators of stellar activity are described in Section 5 and the measured and predicted abundances of lithium are compared in Section 6. The findings are summarised in Section 7.

2 Observations of Binaries

2.1 *Kepler* photometry

Given the unprecedented quality of the photometric data provided by the *Kepler* space mission, binaries can be detected by searching for four different features. First, there are the classical eclipsing binary systems, which can be detected through the periodic eclipses. A second class of binaries are the ellipsoidal variables, in which the flux modulation originates from tidal interaction [e.g. 41–43]. A special case of these ellipsoidal variables are the eccentric cases, in which the flux modulation only occurs during the periastron passage. This leads to a periodic modulation of the observed object brightness. The characteristic shape of the light curve at periastron triggered the colloquial but controversial nick name of *'heartbeat stars'*. This term, coined by [43], was debated during this conference. Physically, eccentric eclipsing binaries and heartbeat stars are the same and thanks to the photometric quality of *Kepler* even combinations of eclipses and heartbeat events are found [e.g. KIC 2697935, KIC 9540226, KIC 10614012; in 32, 44, see also Figure 1]. The only difference between the two sets is that eclipses allow a better determination of the inclination, though are restricted to a very narrow degree ranges, while ellipsoidal variables can be found at nearly all inclination. Therefore, separating heartbeat stars from eclipsing binaries is unpractical and unnecessarily narrowing down sample sizes. The third and fourth way of detecting binaries are not found in the light curves, but in the Fourier space. On one hand, it was shown that oscillating stars in binaries can be identified from the phase modulation of coherent modes [45, 85]. However, the stochastic nature of solar-like oscillations and the granulation background make this approach impracticable for solar-like and red-giant stars. Still binaries can be detected from the oscillation spectrum, if both components are visible. These seismic binaries (hereafter in analogy to the double-lined spectroscopic binaries, SB2, also referred to as PB2), however can only be found in a very narrow mass range and have smaller amplitudes due to photometric dilution [47, 48].

On the other hand, an additional phenomenon that in principle is present in binaries is the so-called *Doppler beaming* [49, 50]. This effect originates from the varying radial velocity in the binary, periodically shifting the stars' spectral energy distribution with respect to *Kepler's* photometric pass band and an alternated arrival rate of photons. In a single-lined binary (SB1), a net increase or decrease would be observed if the primary component moves toward or away from the observer, respectively. While this effect was predicted in 2003 [49, 50], its existence could only be shown from high-quality photometry provided by the CoRoT and *Kepler* space missions for binary systems with large radial-velocity amplitudes [e.g. 51]. Due to the combination of the red-biased spectral energy distribution of red-giants and the pass band of *Kepler*, Doppler beaming should also be observable for binaries with red-giant components. In particular, for the low-luminosity red-giant star KIC 5006817, models suggest that the net amplitude of this effect over the full orbital cycle should be about 300 ppm [32]. Therefore, this effect is influencing the light curve and needs to be taken into account for its modelling. However, finding a small effect in binaries with periods compatible or longer than the length of a *Kepler* data quarter (90 days) is challenging, as it is nearly impossible to distinguish it from instrumental noise and therefore will be filtered out in the data calibration.

2.2 Ground-based spectroscopy

In addition to the photometric or seismic detection of binaries, the analysis of multiple star systems requires the combination of space photometry with high-resolution time-resolved spectroscopy. The most fundamental task of ground-based follow-up observations is to obtain an independent confirmation of the binary system itself. **This is particularly important for systems where the contamination through neighbouring field stars or some combination of activity or instrumental systematic effects could mimic the signature of stellar binarity.** Only periodic radial velocity (RV) variations can confirm such systems. Having all phases of the orbit well covered with observations further allows to derive the orbital elements. Although it is in principle possible to derive all orbital parameters from the average light curve, it is beneficial, especially in case of non-eclipsing heartbeat stars, to narrow down the used parameters space by including the orbital parameters from spectroscopy. This is particularly true for ellipsoidal variables.

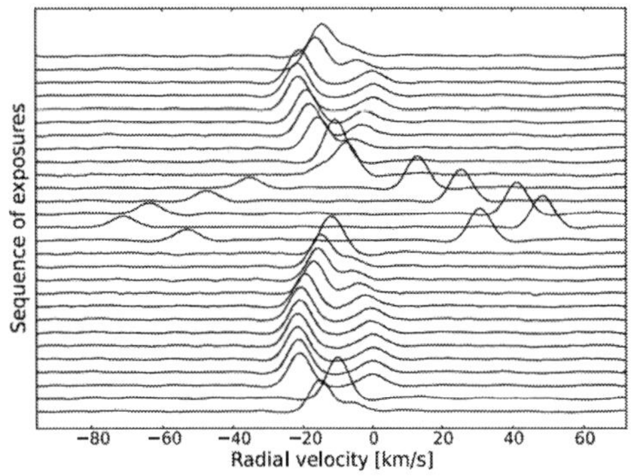

Figure 2: Time series of cross-correlation profiles of individual spectra of KIC 9163796 with a red-giant-star template. The higher and the smaller peak corresponds to the average line profile of the primary and secondary component in the system.

In the optimal case, a binary system is found to be a double-line binary. In such a system the stellar spectra of both components are moving in anti-phase to each other. The RV amplitudes of such systems provide directly the mass ratio. In addition, having the information on the inclination allows us to obtain the dynamical masses of the system [26].

Furthermore, using all the spectra obtained to measure the RV of the system, allows to substantially increase the signal-to-noise (S/N) in the combined spectrum. Such spectrum can be used to perform a detailed spectroscopic analysis of one or both components of the system. Especially for SB2s', several methods have been developed to search and extract the individual stellar spectra from the composite spectrum, [e.g. TodCor or FDBinary; 52, 53, respectively]. Recent advances in the analysis techniques and data quality have pushed the limit for a positive detection below 1 percent of contribution of the secondary or even tertiary component to the total flux of the system [e.g. 54, 55]. Even the non-detection of the secondary is a valuable information, as it sets the upper limit for luminosity of the secondary [e.g. 32].

3 The red-giant/red-giant SP2 system KIC 9163796

In a previous study [32], it was shown that KIC 9163796 is a non-eclipsing binary system, though it exhibits clear ellipsoidal flux modulation every 121.3 days during periastron (see Figure 3). The analysis of radial velocities from ground-based spectroscopy of this system confirmed that KIC 9163796 is an eccentric binary system ($e \simeq 0.7$). The primary component shows an oscillation power excess at about 160 μHz. A visual inspection of the cross-

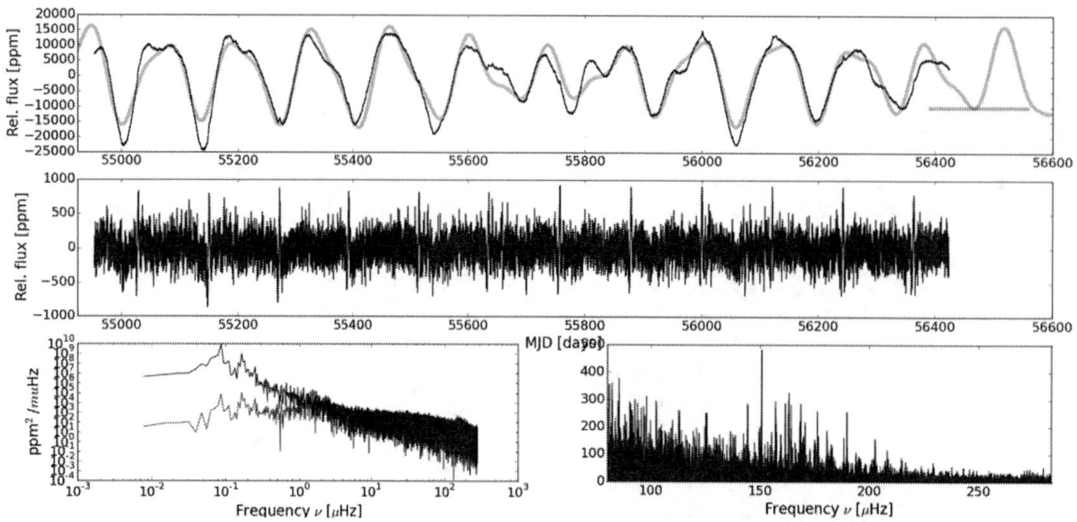

Figure 3: Light curve of KIC 9163796. The top and lower panel show the full light curve (Q0-Q17), smoothed with a filter of 150 and 4 days, respectively. The synthetic light curve, depicted as solid light grey line is from the four significant frequencies found through a prewhitening procedure. The time base of the spectroscopic monitoring is indicated through the horizontal grey bar. The bottom left and right panel show the power spectral density in logarithmic and linear scalings, whereby the red and black curves represent the spectra originating from the inpainted *Kepler* data set with a smoothing of 4 and 150 days, respectively.

correlation profiles of the time series revealed that this system is an impressive case of an SB2 system [44]. The time series of the individual responses of the cross correlation of the stellar spectra with a red-giant template is depicted in Figure 2. In this section, we are summarising the results of our study, dedicated to obtain a comprehensive and complete picture of KIC 9163796 [36].

3.1 Long period flux modulation

A careful rereduction of the full available *Kepler* data set (Q0-Q17), as described in [56, 57], reveals that this system exhibits strong flux modulations (see top panel of Figure 3). Among the red-giant stars observed with *Kepler*, photometric rotational variability is found only for a small fraction of stars. In a very recent study, only ~300 red-giant stars were found to exhibit surface modulation in the range of a few to several hundred days [58]. Applying optimised techniques [59, 60] to the light curve of this system, such as the wavelet analysis depicted in Figure 4, we find the period of the modulation to be ~130 d.

The photometric period, likely originating from stellar rotation of the primary component is close to the orbital period of ~120 d. By comparing with values of solar and anti-solar differentially surface rotation reported in the literature [e.g. 61, 62] we cannot reproduce this discrepancy of 10 d. Therefore, rotation of the primary component and the orbital motion are currently not synchronised. One might speculate that the system was in a synchronised state on the main sequence, but the rapid changes of the stellar structure and radius in particular could have lifted this state or that the period of the system is to long and the system too wide to reach a perfect synchronisation (see [63] for a detailed discussion of the tidal evolution of binaries containing red-giant stars).

3.2 Spectroscopic study & fundamental parameters

Since the identification of KIC 9163796 as a binary, this system was monitored with the HERMES spectrograph [64, 65], mounted to the 1.2 m MERCATOR telescope on La Palma, Canary islands in 2013 [32] and 2015 [40]. The cross-correlation response function of the individual spectra, depicted in Figure 2, show that this system is an SB2 with strong contributions of both components. For this analysis of the SB2 feature, we applied spectral disentangling in the Fourier domain [66] with the FDBINARY code [53] to search for the optimal set of orbital elements. Experience has shown that the best region in the spectrum to derive a consistent orbital solution is located at 520 nm. This range includes many prominent absorption lines and enough regions to determine the pseudo continuum which is crucial for the normalisation of the spectra. Because the signature of the secondary is clearly visible in the spectrum, the initial set of parameters is straightforward to define and the orbital period is well defined from the analysis of the *Kepler* light curve. For details of the full spectral disentangling we refer the reader to [40].

The RV semi amplitudes are $K_1 = 35.25 \pm 0.12$ km/s and $K_2 = 35.77 \pm 0.13$ km/s for the primary and secondary, respectively. These radial velocity amplitudes translate into a mass ratio, $q = M_1/M_2 = 1.015 \pm 0.005$. Therefore, the SB2-nature of this system allows us to measure mass ratio between the two components much more precisely than what could be achieved through seismology.

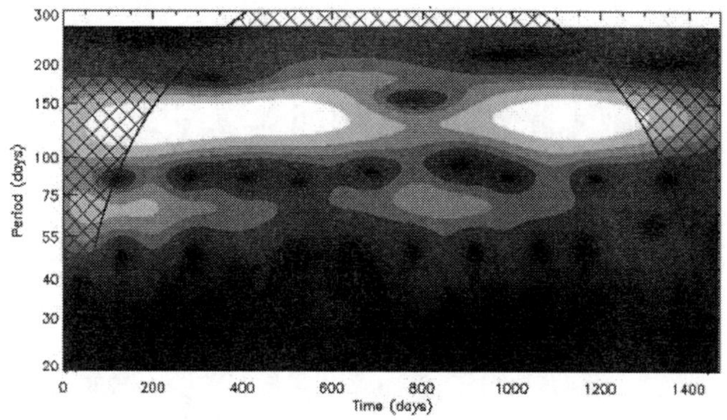

Figure 4: Analysis of the long periodic light curve modulation of KIC9163796 in the period range between 20 and 200 days. The left panel depicts the wavelet analysis of the full time series in period space. Period regions in the wavelet for which the window function is larger than the covered time base are indicated. The right panel shows the average signal over the full time base.

The cross-correlation response functions of the individual spectra (Figure 2) show that average-absorption line profiles of the primary and secondary components are equal in hight in the composite spectrum. This indicates that the primary contributes substantially more flux to the integrated brightness of the system than the secondary. This indicates that the fundamental parameters of both stars are quite different for these stars with a mass ratio close to unity.

The final orbital solution was used to extract a 20 nm wide spectral segment around the H_β line as well as the Mg I triplet which were subsequently used to derive the spectroscopic analysis with the *Grid Search in Stellar Parameters* (GSSP[1]) software package by [67]. In such an analysis the contribution of both components to the total flux need to be taken into account and correctly normalised to the individual continuum flux. Although the mass ratio derived from the solution of spectral disentangling indicates that the masses of the two stars is very similar the spectroscopic analysis finds that both components differ substantially in terms of the effective temperature with a difference of 600 K. Consequently, the primary is substantially more luminous than the secondary and contributes to ~60% of the total flux. The spectroscopic analysis shows that both components share the same value of the stellar metallicity, $[M/H] \simeq -0.37$ dex, which is coherent with the fact that both stars were born in the same cloud.

The main difference between the two components, besides the effective temperature was found in the abundance of lithium (Li). The orbital solution was used to extract a 5 nm wide segment around the lithium resonance doublet at 670.78 nm. While the primary component shows 1.3 dex, the secondary exhibits a stronger lithium line, leading to 2.6 dex.

[1]The GSSP package is available for download at https://fys.kuleuven.be/ster/meetings/binary-2015/gssp-software-package.

4 Asteroseismic analysis of KIC 9163796

From the preceding study of this system it was known that the oscillation amplitude of the system is very small. The power spectral density (PSD) is shown in the bottom panels of Figure 3. In the light of the spectroscopic results this is now easy to understand due to the effects of photometric dilution [e.g. 47]. Further challenges for the analysis of the PSD are introduced by the regular gaps, originating from on-board manoeuvres of the *Kepler* satellite which complicate the spectral-window function [56, 57]. The effect is depicted through the difference of the original and inpainted *Kepler* data as red and black curves, respectively in the bottom panels of Figure 3. These alias frequencies are dominating the high-frequency regime of the PSD and if not corrected for, the low-amplitude oscillation signal is hidden by the window function. Such alias frequencies are reduced by applying inpainting techniques described in [56, 57].

Following the standard approach of fitting several components seen in the PSD of a red giant, (for a summary we refer the reader to [3, 70]), we obtain the normalised PSD of KIC 9163796 shown in the top spectrum of Figure 5. From scaling relations [e.g. 3, 5], we find an estimate of mass and radius of 1.4 $M_\odot$ and 5.4 $R_\odot$, respectively.

The unambiguous identification of the excited nonradial modes was found to be challenging due to the low S/N in the spectrum. In the case of the binary KIC 9246715 [39], two overlapping power excesses were found, further complicating the PSD. For that particular binary, both components are found to be located in the secondary clump and therefore are already in the stable phase of helium core-burning. The frequency of the power excess and the period spacing of consecutive dipole modes [8, 68, 69] indicate that the more massive primary of KIC 9163796 is a low-luminosity red-giant star in the hydrogen-shell burning phase. In this phase, at the bottom of the red-giant branch, changes of the radius and luminosity are happening on much shorter time scales and there-

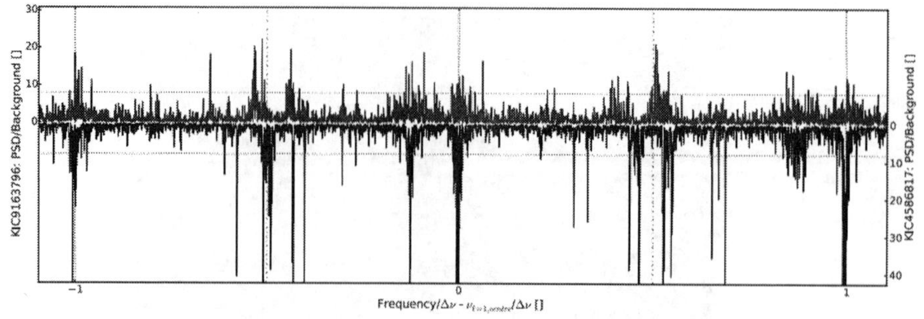

Figure 5: Normalised power spectral density of KIC 9163796 (blue, top spectrum) and KIC 4586817 (black, bottom spectrum). The background contribution has been removed by division of the PSD through the background model. The frequency was divided by the respective large separation and centred on the central radial mode. For KIC 9163796, the light grey spectrum depicts the dilution factor corrected spectrum. The frequencies of the $\ell = 0$ and 1 modes are marked by magenta dotted and red dash dotted vertical lines. The dotted horizontal lines mark the traditional significance threshold of 8 times the background signal.

fore, we expected the power excesses of the two stars to be well separated. To be sure that all peaks belong only to the primary component and to guide the eye for the mode identification, the PSD of a single red-giant star is shown. This comparison of KIC 9163796 to KIC 4586817, depicted in Figure 5, shows a very good resemblance and was guiding the eye in the initial mode identification.

From the mass ratio between the two components and the fundamental parameters of the secondary of this system, it is expected that also the less massive star should be found to oscillate. However, the secondary power excess would be hard to detect due to small amplitudes and the effects of photometric dilution. Based on the Bayesian evidence we find that a model with two Gaussian components significantly better fits the observed PSD than a model with only one Gaussian. The second power excess is located at ~215 μHz. This value is in qualitative good agreement with the estimated position of the secondary power excess, if scaled from the output parameters of the spectral disentangling analysis. Unfortunately, the mode visibility of the secondary is too low to determine a reliable estimate for the large frequency separation $\Delta\nu$.

The PSD of KIC 9163796 also shows the rotational splitting of dipole mixed modes (see Figure 5). By using the Diamonds *(high-DImensional And multi-MOdal NesteD Sampling)* code [71], precise frequency values of these non-radial modes were extracted and the rotational splitting computed. Following the standard forward-modelling approach of using a two-zone model of assuming a rigidly rotating core and envelope, we used a representative model, calculated with the *Yale Stelar Evolution Code, YREC* [72, 73]. The rotation rates of both zones was optimised until a good agreement with the set of measured rotational splittings was reached. The seismic result is a surface rotation period of 81 nHz, which is in good agreement with the main period of the surface brightness modulation of 79 nHz. In the primary of this system the core rotates about 6.9 times faster than the surface. This is in the lower range of observed rotational gradients between the core and the envelope of 10-30 times [12, 14, 15, 17, 74], but not exceptionally low. This differs from the quasi-rigid rotation, which was found in red giants, located in the secondary clump [16].

5 Stellar activity

The light curve, depicted in Figure 3, indicates that at least one star must show strong stellar activity. The existing observational material allows us to estimate the stellar activity in two different ways. The classical way of measuring activity is the $\mathcal{S}$-index, quantifying the emission line at the core of the Ca H&K lines in the near Ultra-Violet. The spectra of KIC 9163796 show a very strong emission, compared to other red giants and solar-analogues, observed with the Hermes spectrograph [32, 55, 78–80]. However, it is very challenging to extract a reliable $\mathcal{S}$-index for a star in a SB2 system with a bright secondary component.

Another proxy is the linewidth of the oscillation modes. Several studies have presented indications that stellar activity also reduces the amplitudes of solar-like oscillations and increases the damping [e.g. 31, 75–77]. Because the full width at half maximum is independent of photometric dilution through the other stellar component, it can be directly compared to measured mode in other stars. Using the sample of [81] as a benchmark, we see that the full width at half maximum of dipole modes in KIC 9163796 is about four times broader than those in normal active stars on average. This can only be explained through strong mode damping.

6 Stellar modelling

KIC 9163796 provides a set of stars, well constrained through photometry, spectroscopy, and seismology. Therefore, this binary system is a rewarding target for modelling.

Both components were modelled with the *Toulouse Geneva Evolutionary Code* [TGEC, 82, 83]. The first result is that the difference in fundamental parameters is in

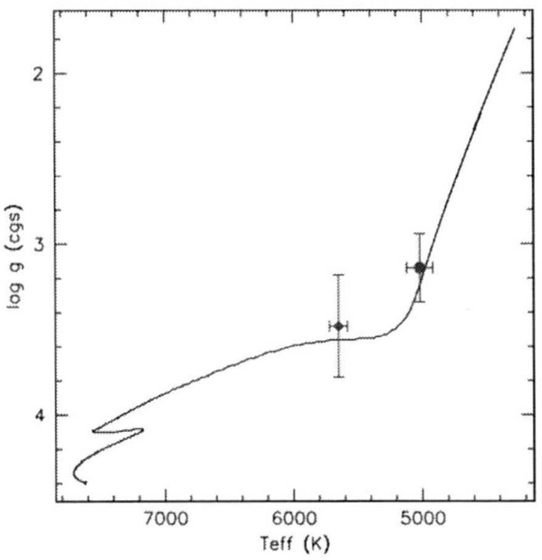

Figure 6: Evolutionary track for a representative model of both components of KIC 9163796. The current position of the primary and secondary are indicated through the blue and red marker, respectively.

agreement with the position of the primary and secondary being in the late and early phase of the first dredge-up event (FDU) at the bottom of the red-giant branch as illustrated in Figure 6. Therefore, both components are framing one of the most important events on internal mixing in stellar evolution.

In a second step, these models were used to calculate and compare the lithium abundance with various scenarios of rotational histories. The best agreement was found with a (quasi) rigidly rotating progenitor. This is in agreement with findings of modern asteroseismology [e.g. 84, 85]

7 Conclusions

KIC 9163796 is an intriguing binary system. Through a multi-technique approach we arrived to the conclusion that both the primary and secondary are oscillating and are located in the late and early phase of first dredge-up event. This is astonishing, because both stars have a very similar mass. Therefore, this system is an impressive example of the how sensitive is the stellar evolution to small differences in mass. An extensive discussion of the results is provided in the main paper [40].

Acknowledgements

We acknowledge the work of the team behind *Kepler*. Funding for the *Kepler* Mission is provided by NASA's Science Mission Directorate. We thank the technical team as well as the observers of the HERMES spectrograph and Mercator Telescope, operated on the island of La Palma by the Flemish Community, at the Spanish Observatorio del Roque de los Muchachos of the Instituto de Astrofísica de Canarias. PGB, RAG and BM acknowledge the ANR (Agence Nationale de la Recherche, France) program IDEE (n° ANR-12-BS05-0008) "Interaction Des Etoiles et des Exoplanetes". PGB, RAG, StM and BM also received funding from the CNES grants at CEA. KP was supported by the Croatian Science Foundation grant 2014-09-8656. E.C. is funded by the European Community's Seventh Framework Programme (FP7/2007-2013) under grant agreement N°312844 (SPACEINN). StM acknowledges support by the ERC through ERC SPIRE grant No. 647383. The research leading to these results has received funding from the European Community's Seventh Framework Programme ([FP7/2007-2013]) under grant agreement No. 312844 (SPACEINN) and under grant agreement No. 269194 (IRSES/ASK).

References

[1] Aerts, C., Christensen-Dalsgaard, J. and Kurtz, D. W., 2010, Asteroseismology, Astronomy and Astrophysics Library. ISBN 978-1-4020-5178-4

[2] Borucki, W. J., Koch, D., Basri, G., et al., 2010, Science, 327, 977

[3] Kallinger, T., Mosser, B., Hekker, S., et al., 2010, A&A, 522, A1

[4] Hekker, S., Elsworth, Y., De Ridder, J., et al., 2011, A&A, 525, A131

[5] Chaplin, W. J., Kjeldsen, H., Christensen-Dalsgaard, J., et al., 2011, Science, 332, 213

[6] Huber, D., Bryson, S. T., Haas, M. R., et al., 2016, ApJS, 224, 2

[7] Chaplin, W. J., Basu, S., Huber, D., et al., 2014, ApJS, 2010, 1

[8] Bedding, T. R., Mosser, B., Huber, D., et al., 2011, Nature, 471, 608

[9] Mosser, B., Vrard, M., Belkacem K., et al., 2015, A&A, 584, A50

[10] Vrard, M., Mosser, B., Barban, C., et al., 2015, A&A, 579, A84

[11] Kallinger, T., Hekker, S., Garcia, R. A., et al., 2016, Science Astronomy, 2, 250654

[12] Beck, P. G., Montalban, J., Kallinger, et al. T., 2012, Nature, 481, 55

[13] Mosser, B., Goupil, M. J., Belkacem, K., et al. 2012, A&A, 548, A10

[14] Deheuvels, S., García, R. A., Chaplin, W. J., et al. 2012, ApJ, 756, 19

[15] Deheuvels, S., Dogan, G., Goupil, M. J., et al. 2014, A&A, 564, A27

[16] Deheuvels, S., Ballot, J., Beck, P. G., et al., 2015, A&A, 580, A96

[17] Di Mauro, M. P., Ventura, R., Cardini, D., et al., 2016, ApJ, 817, 65

[18] Miglio, A., Brogaard, K., & Stello, D. 2012, MNRAS, 419, 2077

[19] Corsaro, Enrico; Stello, Dennis; Huber, Daniel; 2012, ApJ, 757, 190C

[20] Brogaard, K., Jessen-Hansen, J., Handberg, R., et al., 2016, AN, 337, 793

[21] Epstein, C. R., Elsworth, Y. P., Johnson, J. A., et al., 2014, ApJ, 785L, 28
[22] Mathur, S., Garcia, R. A., Huber, D., et al., 2016, ApJ, 827, 50
[23] De Ridder, J., Molenberghs, G., Eyer, L., and Aerts, C., 2016, arXiv: 1609.08945 (A&A in press)
[24] Lebreton, Y., & Goupil, M. J., 2014, A&A, 569, A21
[25] Gaulme, P., McKeever, J., Jackiewicz, J., et al., 2016, arXiv:1609.06645
[26] Torres, G., Andersen, J., Giménez, A., 2010, A&ARv, 18, 67
[27] Frandsen, S., Lehmann, H., Hekker, S., et al., 2013, A&A, 556, A138
[28] Kippenhahn, R., Weigert, A., & Weiss, A. 2012, 'Stellar Structure and Evolution', ISBN 978-3-642-30255-8. Springer-Verlag Berlin Heidelberg
[29] Kirk, B., Conroy, K., Prša, A., 2016, AJ, 151, 68
[30] Gaulme, P. , McKeever, J. , Rawls, et al., 2013, ApJ, 767, 82
[31] Gaulme, P., Jackiewicz, J., Appourchaux, T. & Mosser, B., 2014, ApJ, 785, 5
[32] Beck, P. G., Hambleton, K., Vos, J. et al. 2013, A&A, 564A, 36
[33] Beck, P. G., Hambleton, K., Vos, J., et al., 2015, EPJWC, 10106004
[34] Appourchaux, T., Antia, H. M., Ball, W., et al. 2015A&A, 582, A25
[35] White, T. R., Benomar, O., Silva Aguirre, V., et al., 2016arXiv160909581W
[36] Beck,P. G., doNascimento Jr., J.-D., Duarte, T., et al., 2016, A&A, submitted
[37] Hekker, S., Debosscher, J., Huber, D., et al., 2010, ApJ, 713L, 187
[38] Themeßl, N., Hekker, S., Southworth, J., et al., 2016, MNRAS, submitted
[39] Rawls, M. L., Gaulme, P., McKeever, J., et al., 2016, ApJ, 818, 108
[40] Beck, P. G., Kallinger, T., Pavlovski, K., et al., 2016, A&A, submitted
[41] Welsh, W. F., Orosz, J. A., Aerts, C., et al. 2011, ApJS, 197, 4.
[42] Fuller, J. & Lai, D. 2012, MNRAS, 420, 3126
[43] Thompson, S. E., Everett, M., Mullally, et al., 2012, ApJ, 753, 86
[44] Beck, P. G., 2013, PhD thesis, University of Leuven, Belgium, ISBN 978-90-8649-677-8
[45] Murphy, S. J., Shibahashi, H., & Bedding, T. R., 2016, MNRAS, 461, 4215
[46] Shibahashi, H., & Kurtz, D. W., 2012, MNRAS, 422, 738
[47] Miglio, A., Chaplin, W. J., Farmer, R., et al., 2014, ApJ, 784L 3
[48] Johnston, C., Beck, P. G., Prsa, A., et al., in prep.
[49] Loeb, A. and Gaudi, B. S., 2003, ApJ, 588L, 117
[50] Zucker, S., Mazeh, T., & Alexander, T., 2007, ApJ, 670, 1326
[51] Bloemen, S., Marsh, T. R., Østensen, R. H., et al., 2011, MNRAS, 410, 1787
[52] Mazeh, T., and Zucker, S., 1994, Ap&SS, 212, 349
[53] Ilijic, S., Hensberge, H., Pavlovski, K., and Freyhammer, L. M. 2004, in ASPCS, Vol. 318, 111
[54] Kolbas, V., Pavlovski, K., Southworth, J., et al., 2015, MNRAS, 451, 4150
[55] Beck, P. G., Pavlovski, K., Torres, G., et al., in prep.
[56] García, R. A., Hekker, S., Stello, D., et al. 2011, MNRAS, 414L, 6
[57] García, R. A., Mathur, S., Pires, S., et al., 2014, A&A, 568, A10
[58] Ceillier, T., Tayar, J., Mathur, S., et al., A&A, subm.
[59] García, R. A.; Ceillier, T.; Salabert, D., et al. 2014, A&A, 572, A34
[60] Ceillier, T., van Saders, J., García, R. A., et al. 2016, MNRAS, 456, 119
[61] Kővári, Z., Kriskovics, L., Künstler, A., et al., 2015, A&A, 573, A98
[62] Künstler, A., Carroll, T. A., & Strassmeier, K. G. 2015, A&A, 578, A101
[63] Verbunt, F., & Phinney, E. S. 1995, A&A, 296, 709
[64] Raskin, G., van Winckel, H., Hensberge, H., et al., 2011, A&A, 526, A69
[65] Raskin, G., 2011, PhD thesis, University of Leuven, Belgium, ISBN 978-94-6018-431-4
[66] Hadrava, P. 1995, AApS, 114, 393
[67] Tkachenko, A. 2015, A&A, 581, A129
[68] Beck, P. G., Bedding, T. R., Mosser, B., et al. 2011, Science, 332, 205
[69] Mosser, B., Barban, C., Montalban, J., et al. 2011, A&A, 532, A86
[70] Buysschaert, B., Beck, P. G., Corsaro, E., et al., 2016, A&A, 588, A82
[71] Corsaro, E. and De Ridder, J., 2014, A&A, 571, A71
[72] Demarque, P., Guenther, D. B., Li, L. H., Mazumdar, A., & Straka, C. W. 2008, ApSS, 316, 31
[73] Guenther, D. B. 1994, ApJ, 422, 400
[74] Goupil, M. J., Mosser, B., Marques, J. P., et al. 2013, A&A, 549, A75
[75] Mosser, B., Michel, E., Appourchaux, T., et al. 2009, A&A, 506, 33
[76] García, R. A., Mathur, S., Salabert, D., et al. 2010, Science, 329, 1032
[77] Bonanno, A., Corsaro, E., & Karoff, C. 2014, A&A, 571, A35
[78] Beck, P. G., Kambe, E., Hillen, M., et al. 2015, A&A, 573, A138
[79] Beck, P. G., Allende Prieto, C., Van Reeth, T., et al. 2016, A&A, 589, A27
[80] Salabert, D., García, R. A., Beck P. G. et al. 2016, A&A accepted, arXiv: 1608.01489
[81] Corsaro, E., De Ridder, J., & García, R. A. 2015, A&A, 579, A83
[82] Hui-Bon-Hoa, A. 2008, Ap&SS, 316, 55
[83] Théado, S., Alecian, G., LeBlanc, F., & Vauclair, S., 2012, A&A, 546, A100

DOI: 10.1051/epjconf/201716005008

[84] Benomar, O., Belkacem, K., Bedding, T. R., et al., 2014, ApJ, 781L, 29

[85] Kurtz, D. W., Saio, H., Takata, M., et al., 2014, MNRAS, 444, 102

DOI: 10.1051/epjconf/201716005009

Presence of mixed modes in red giants in binary systems

Nathalie Themeßl[1,2,3,★], *Saskia* Hekker[1,2], and *Yvonne* Elsworth[4,2]

[1] *Max Planck Institute for Solar System Research, Justus-von-Liebig-Weg 3, 37077 Göttingen, DE*
[2] *Stellar Astrophysics Centre, Department of Physics and Astronomy, Aarhus University, Ny Munkegade 120, 8000 Aarhus C, DK*
[3] *Institute for Astrophysics, Georg-August University Göttingen, Friedrich-Hund-Platz 1, 37077 Göttingen, DE*
[4] *School of Physics and Astronomy, U. Birmingham, Edgbaston, Birmingham B15 2TT, UK*

Abstract. The frequencies of oscillation modes in stars contain valueable information about the stellar properties. In red giants the frequency spectrum also contains mixed modes, with both pressure (p) and gravity (g) as restoring force, which are key to understanding the physical conditions in the stellar core. We observe a high fraction of red giants in binary systems, for which g-dominated mixed modes are not pronounced. This trend leads us to investigate whether this is specific for binary systems or a more general feature. We do so by comparing the fraction of stars with only p-dominated mixed modes in binaries and in a larger set of stars from the APOKASC sample. We find only p-dominated mixed modes in about 50% of red giants in detached eclipsing binaries compared to about 4% in the large sample. This could indicate that this phenomenon is tightly related to binarity and that the binary fraction in the APOKASC sample is about 8%.

1 Introduction

In its nominal mission the *Kepler* space telescope [1] provided long-term photometric data for more than 100 000 stars, among which many pulsating red giants were found. Red giants are evolved stars showing solar-like oscillations which are stochastically excited and intrinsically damped by the near surface convection. Their oscillation spectrum comprises several overtones of radial order (n) and spherical degree (ℓ) modes. The most significant peaks form a well-defined pattern which reveals the structure of the radial ($\ell = 0$) and non-radial ($\ell = 1, 2$) modes [2, 3]. Radial modes are pure pressure modes that are mostly sensitive to the outer layers of the star. In contrast, non-radial modes have a mixed nature with gravity as the restoring force in the deep interior and pressure as the restoring force close to the stellar surface. The dipole ($\ell = 1$) modes consist of one central most p-dominated mode, which is surrounded by several mixed modes that have a more g-dominated character. In a significant number of red giants in detached binary systems these g-dominated mixed modes are missing. In close binary systems, tidal interactions can impact upon the oscillation spectra of the individual stars in the form of mode damping or even complete mode suppression [4]. This interplay between binary components is not expected in detached systems, where stars are assumed to evolve separately. In our study, we investigate if the lack of pronounced g-dominated mixed modes is mainly observed in red giants that belong to binary or triple-star systems or if it is also a common phenomenon among field red giants.

★e-mail: themessl@mps.mpg.de

2 Data

2.1 Binary and triple-star systems with red-giant components

We used *Kepler* long-cadence (~ 29.4 min) data for a sample of red giants that were confirmed to be components of binary or triple-star systems [4–7]. For this study, we adopted the unweighted power density spectra of the corrected and concatenated timeseries which are provided by KASOC [8].

2.2 APOKASC red giants

For comparison, we chose 6604 red giants from the APOKASC [9] sample, which contains many astrophysically interesting targets whose stellar properties are available from asteroseismology and complementary spectroscopy. We adopted the corrected lightcurves [8] that are available in the APOKASC data bundle to calculate the power density spectra of the red-giant stars. These data are the same as used by Elsworth et al. (submitted).

3 Method

In this study, we attempt to classify the red giants into three different categories: (a) stars where g-dominated mixed modes are clearly present, (b) stars that show only p-dominated dipole modes, and (c) stars exhibiting non-radial mode suppression. See also Figure 1 for an example of each of these classes.

For the characterization of the red giants, we performed a visual inspection of the dipole $\ell = 1$ modes in the oscillation region close to the frequency of maximum oscillation power ($\nu_{\max}$). Thus, we restricted the power density spectrum of each star to $\pm 2.5\,\Delta\nu$ around $\nu_{\max}$, where $\Delta\nu$ is the large frequency separation between modes of the same spherical degree and consecutive radial orders. Based on this approach, we inspected 18 red giants in binary and triple-star systems and we extended this classification to a larger set of stars from the APOKASC sample. The latter contains red giants that cover different evolutionary stages. We were not able to categorize 284 of these stars either due to very low signal-to-noise spectra, oscillation regions that were too close to the Nyquist frequency or the absence of solar-like oscillations. Table 1 presents the preliminary results with the percentage of red giants according to the three different types that we defined.

4 Discussions

In this preliminary study, we found a high percentage ($\sim$ 50%) of red giants in known detached binary systems that show mainly p-dominated dipole modes. To investigate the significance of this result, we also examined a large number of red giants from the APOKASC sample. Elsworth et al. (submitted) detected distinct mixed dipole modes in a large fraction of these stars. We found that the remaining small fraction of stars show either p-dominated or suppressed dipole modes. Most interestingly, the fraction of stars exhibiting p-dominated dipole modes is much smaller than what we observed for the binaries, where about half of the stars showed a tendency towards this category. In contrast, the fraction of suppressed dipole modes is of the same order of magnitude for both samples. As we use a statistically insignificant number of known binaries, we are not able to draw any firm conclusions from this. We can however speculate that the binary nature does not seem to be the cause for mode suppression. Yet binarity appears to have some influence on the g-dominated mixed modes, which we will investigate further. If we take this one step further, saying that the presence of mainly p-dominated mixed modes only appears in about half the binaries, this would mean that the binary fraction in the APOKASC sample would be $\sim$ 8%.

Acknowledgements

NT and SH have received funding from the European Research Council under the European Communitys Seventh Framework Programme (FP7/2007-2013) / ERC grant agreement no 338251 (StellarAges). NT also acknowledges travel support from the EC Project SPACEINN (FP7-SPACE-2012-312844). YE acknowledges the support of the UK Science and Technology Facilities Council (STFC). Funding for the Stellar Astrophysics Centre (SAC) is provided by The Danish National Research Foundation (Grant agreement no.: DNRF106).

References

[1] W.J. Borucki, D. Koch, G. Basri, N. Batalha, T. Brown, D. Caldwell, J. Caldwell, J. Christensen-Dalsgaard, W.D. Cochran, E. DeVore et al., Science **327**, 977 (2010)

[2] M. Tassoul, The Astrophysical Journal Supplement **43**, 469 (1980)

[3] B. Mosser, K. Belkacem, M.J. Goupil, E. Michel, Y. Elsworth, C. Barban, T. Kallinger, S. Hekker, J. De Ridder, R. Samadi et al., Astronomy & Astrophysics **525**, L9 (2011), 1011.1928

[4] P. Gaulme, J. McKeever, M.L. Rawls, J. Jackiewicz, B. Mosser, J.A. Guzik, The Astrophysical Journal **767**, 82 (2013), 1303.1197

[5] P.G. Beck, K. Hambleton, J. Vos, T. Kallinger, S. Bloemen, A. Tkachenko, R.A. García, R.H. Østensen, C. Aerts, D.W. Kurtz et al., Astronomy & Astrophysics **564**, A36 (2014), 1312.4500

[6] P. Gaulme, J. Jackiewicz, T. Appourchaux, B. Mosser, The Astrophysical Journal **785**, 5 (2014), 1402.3027

[7] P. Gaulme, J. McKeever, J. Jackiewicz, M.L. Rawls, E. Corsaro, B. Mosser, J. Southworth, S. Mahadevan, C. Bender, R. Deshpande, ArXiv e-prints (2016), 1609.06645

[8] R. Handberg, M.N. Lund, Monthly Notices of the Royal Astronomical Society **445**, 2698 (2014), 1409.1366

[9] M.H. Pinsonneault, Y. Elsworth, C. Epstein, S. Hekker, S. Mészáros, W.J. Chaplin, J.A. Johnson, R.A. García, J. Holtzman, S. Mathur et al., The Astrophysical Journal Supplement **215**, 19 (2014), 1410.2503

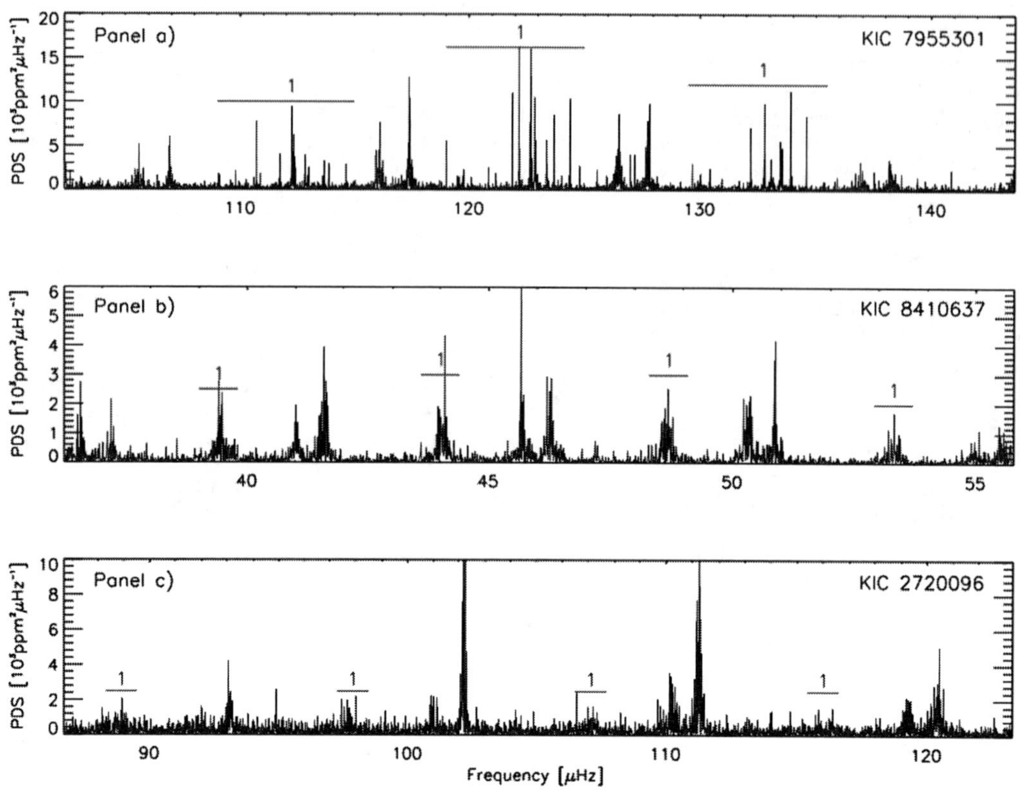

Figure 1: The frequency range of oscillations around the frequency of maximum oscillation power. Panel a) shows KIC 7955301 which exhibits clear mixed modes. Panel b) shows KIC 8410637 where only the p-dominated non-radial modes are distinct. Panel c) shows KIC 2720096, a red giant with suppressed dipole modes.

Table 1: The percentage of stars in the different categories.

	Number of stars	Mixed modes	P-dominated modes	Suppressed modes	Unidentified
Binaries	18	~ 39%	~ 50%	~ 11%	0%
APOKASC	6604	~ 85%	~ 4%	~ 7 %	~ 4%

DOI: 10.1051/epjconf/201716005010

Asteroseismic modelling of the Binary HD 176465

B. Nsamba[1,2,⋆], *M. J. P. F. G. Monteiro*[1,2], *T. L. Campante*[3,5], *D. R. Reese*[4,3], *T. R. White*[5,6,7], *A. García Hernández*[1,2], *C. Jiang*[1,2]

[1]*Instituto de Astrofísica e Ciência do Espaço, Universidade do Porto, Rua das Estrelas, PT4150-762 Porto, Portugal*
[2]*Departamento de Física e Astronomia, Faculdade de Ciências da Universidade do Porto, Portugal*
[3]*School of Physics and Astronomy, University of Birmingham, Birmingham B15 2TT, UK*
[4]*LESIA, Observatoire de Paris, PSL Research University, CNRS, Sorbonne Université, UPMC Univ. Paris 06, Univ. Paris Diderot, Sorbonne Paris Cité, 92195 Meudon, France*
[5]*Stellar Astrophysics Centre, Department of Physics and Astronomy, Aarhus University, Ny Munkegade 120, DK-8000 Aarhus C, Denmark*
[6]*Institut für Astrophysik, Georg-August-Universität Göttingen, Friedrich-Hund-Platz 1, 37077 Göttingen, Germany*
[7]*Max-Planck-Institut für Sonnensystemforschung, Justus-von-Liebig-Weg 3, 37077 Göttingen, Germany*

Abstract. The detection and analysis of oscillations in binary star systems is critical in understanding stellar structure and evolution. This is partly because such systems have the same initial chemical composition and age. Solar-like oscillations have been detected by Kepler in both components of the asteroseismic binary HD 176465. We present an independent modelling of each star in this binary system. Stellar models generated using MESA (Modules for Experiments in Stellar Astrophysics) were fitted to both the observed individual frequencies and complementary spectroscopic parameters. The individual theoretical oscillation frequencies for the corresponding stellar models were obtained using GYRE as the pulsation code. A Bayesian approach was applied to find the probability distribution functions of the stellar parameters using AIMS (Asteroseismic Inference on a Massive Scale) as the optimisation code. The ages of HD 176465 A and HD 176465 B were found to be 2.81 ± 0.48 Gyr and 2.52 ± 0.80 Gyr, respectively. These results are in agreement when compared to previous studies carried out using other asteroseismic modelling techniques and gyrochronology.

Key words. Asteroseismology – Stars: fundamental parameters – Stars: individual: HD 176465

1 Introduction

The accomplishments of the CoRoT and Kepler space telescopes have tremendously transformed asteroseismology of solar-like oscillations into an effective tool for characterising stars [7, 20]. Apart from the observation of solar-like oscillations in several hundreds of solar-type stars, a few binary systems with solar-like oscillations detected separately in both components have also been observed, e.g., 16 Cyg [18], α Cen [3, 15]. Binary star systems play an important role in the understanding of stellar structure and evolution. This is because such systems share the same formation history, i.e., formed from the same molecular cloud therefore having the same initial chemical composition and approximately the same age.

Different tools are currently employed in asteroseismic modelling of stars to derive fundamental stellar parameters [26]. The asteroseismic inference is achieved by matching model parameters to observed individual oscillation frequencies or ratios of characteristic frequency separations, and spectroscopic parameters such as effective temperature and metallicity [19].

The individual frequencies and spectroscopic parameters for the asteroseismic binary HD 176465 obtained with high-quality photometric observations using Kepler are presented in [30]. In this article, we adopt this available data set to estimate a robust set of fundamental properties for both components in this system by using the state-of-the-art AIMS[1] optimisation tool based on a grid of evolutionary models generated using MESA[2].

This paper is organised as follows. Section 2 gives a brief description of our target and defines the observables to be matched with our generated models, while our model grid constituents and optimisation procedures are discussed in Section 3. The results are discussed in Section 4 including a comparison with results from other asteroseismic fitting algorithms. Our conclusions are given in Section 5.

2 Target and data

HD 176465 is a nearby binary system displaying solar-like oscillations in both components. The Kepler space tele-

⋆e-mail: benard.nsamba@astro.up.pt

[1]http://bison.ph.bham.ac.uk/spaceinn/aims/
[2]http://mesa.sourceforge.net/

 DOI: 10.1051/epjconf/201716005010

scope observed the binary system in short-cadence mode (SC; 58.85s sampling), for 30 days during the asteroseismic survey phase (20-07-2009 to 19-08-2009, i.e., Quarter 2.2), and continuously from the end of the survey (37 months, Quarters 5 - 17). Refer to [14] and [10] for details on how the SC time series were prepared from raw observations and corrected to remove outliers and jumps. Both components in the binary system have overlapping frequencies which have been extracted through detailed modelling of the power spectrum and analysis of the échelle diagram (see [30] for details).

Table 1: Spectroscopic parameters of HD 176465 A and HD 176465 B

Parameter	HD 176465 A	HD 176465 B
[Fe/H](dex)	-0.30 ± 0.06	-0.30 ± 0.06
T_{eff} (K)	5830 ± 90	5740 ± 90

In order to increase the precision of asteroseismic inference, we adopted both the spectroscopic parameters (i.e., metallicity (Fe/H) and effective temperature (T_{eff}) in Table 1) and the seismic parameters (i.e., the observed individual frequencies in [30]). A spectrum of HD 176465 was obtained from the ESPaDOnS spectrograph on the 3.6-m Canada - France - Hawaii Telescope (See [5] and [21] for details on how the spectrum was analysed, and the fundamental parameters and element abundances derived).

3 Modelling

MESA version 7624 [22–24] was used to construct a grid of stellar models for each of the individual component in the binary system HD 176465. The stellar models used opacities from OPAL tables [13] for temperatures above 10^4 K. At lower temperatures, tables from [8] were used. Nuclear reaction rates were obtained from tables provided by the NACRE collaboration [1]. Specific rates for $^{14}N(p,\gamma)^{15}O$ were described by [12] and for $^{12}C(\alpha,\gamma)^{16}O$ by [16]. The standard grey Eddington atmosphere was used to integrate the atmospheric structure from the photosphere to an optical depth of $\tau = 10^{-4}$. The mixing length theory [4] was used to describe convection. No convection overshooting or semiconvection was included, however diffusion was included to cater for gravitational settling of heavy elements according to the method of [28].

The stellar grid contained a mass range M ∈ [0.7, 1.15] in steps of 0.05 $M_\odot$, a mixing length parameter range α ∈ [1.5, 2.2] in steps of 0.1 and a metal abundance range Z ∈ [0.005, 0.021] in steps of 0.001. For every combination of M, Z, and α, stellar evolution tracks were evolved by MESA from the pre-main sequence to the end of the main sequence. Over 60 models at different ages along each evolution track were generated. The oscillation frequencies for each model were calculated using GYRE[3] [29]. The surface effects in the model frequencies were corrected using the two-term correction described by [2] and implemented in AIMS.

AIMS uses a Bayesian approach to find the probability distribution functions (PDFs) of the stellar parameters. By interpolating in the grid of models calculated by MESA, AIMS generates a representative sample of models using Monte Carlo Markov Chain (MCMC) approach based on the Python package emcee.py[4] [9]. The results of the derived stellar parameters are presented in Figures 1 and 2, and Tables 2 and 3.

4 Discussion

Our results are compared with findings obtained using other modelling procedures and tools. These include the

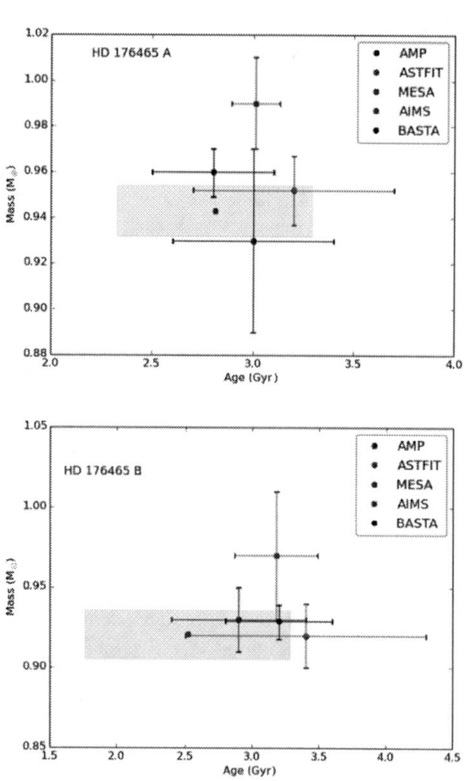

Figure 1: Comparison of our results with those obtained by several pipelines in [30]. The red box corresponds to 1σ-error interval in our results.

Asteroseismic Modelling Portal (AMP), the Aarhus STellar Evolution Code (ASTEC) Fitting method (ASTFIT), the Bayesian Stellar Algorithm (BASTA), and MESA code. Refer to [30] and references therein for details on the different tools. MESA can also be used for optimisation purposes because it contains a module capable of fitting a set of asteroseismic constraints (results in Tables 2 and

[3] https://bitbucket.org/rhdtownsend/gyre/

[4] http://dan.iel.fm/emcee/current/

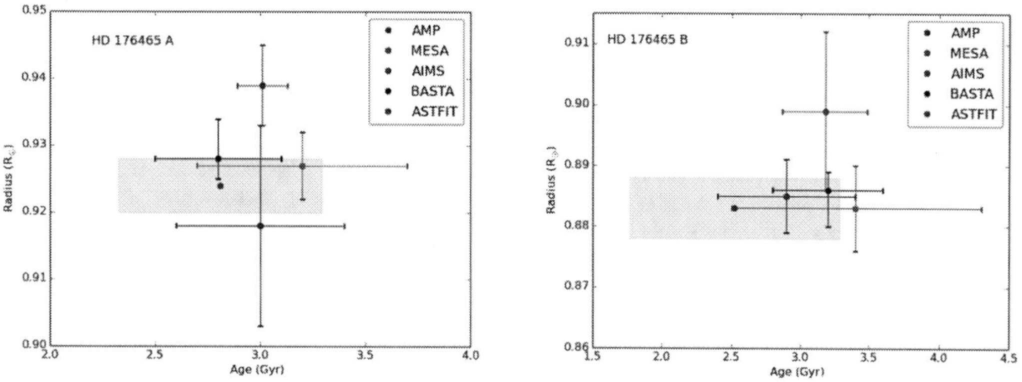

Figure 2: Comparison of our results with those obtained by several pipelines in [30]. The red box corresponds to 1σ-error interval in our results.

Table 2: Comparison of our results (AIMS) with those obtained using other modelling strategies (AMP, ASTFIT, BASTA, and MESA [30]) for HD 176465 A

Parameter	AMP	ASTFIT	BASTA	MESA	AIMS
Mass ($M_\odot$)	0.930± 0.04	0.952 ± 0.015	$0.960^{+0.010}_{-0.011}$	0.99 ± 0.02	0.943 ± 0.011 (± 0.001)
Radius ($R_\odot$)	0.918 ± 0.015	0.927 ± 0.005	$0.928^{+0.006}_{-0.003}$	0.939 ± 0.006	0.924 ± 0.004 (± 0.0003)
Age (Gyr)	3.0 ± 0.4	3.2 ± 0.5	2.8 ± 0.3	3.01 ± 0.12	2.81 ± 0.48 (± 0.6)
Z	0.0085 ± 0.0010	0.0103 ± 0.0010	0.011 ± 0.004	0.0094 ± 0.0009	0.012 ± 0.0010 (± 0.0004)
Y_i	0.258 ± 0.024	0.262 ± 0.003	0.265 ± 0.002	0.23 ± 0.02	——
α	1.90 ± 0.18	1.80	1.791	1.79	1.78 ± 0.11(± 0.01)

Note: —— The surface helium abundance (Y_i) is not determined in AIMS.

Table 3: Comparison of our results (AIMS) with those obtained using other modelling strategies (AMP, ASTFIT, BASTA, and MESA [30]) for HD 176465 B.

Parameter	AMP	ASTFIT	BASTA	MESA	AIMS
Mass ($M_\odot$)	0.930± 0.02	0.92 ± 0.02	$0.929^{+0.010}_{-0.011}$	0.97 ± 0.04	0.921 ± 0.015 (± 0.002)
Radius ($R_\odot$)	0.885 ± 0.006	0.883 ± 0.007	$0.886^{+0.003}_{-0.006}$	0.899 ± 0.013	0.883 ± 0.005 (± 0.001)
Age (Gyr)	2.9 ± 0.5	3.4 ± 0.9	3.2 ± 0.4	3.18 ± 0.31	2.52 ± 0.80 (± 0.1)
Z	0.0085 ± 0.0007	0.0096 ± 0.0011	0.011 ± 0.004	0.0122 ± 0.0011	0.011 ± 0.0012 (± 0.0002)
Y_i	0.246 ± 0.013	0.262 ± 0.003	0.265 ± 0.002	0.24 ± 0.04	——
α	1.94 ± 0.12	1.80	1.791	1.79	1.79 ± 0.11 (± 0.01)

Note: —— The surface helium abundance (Y_i) is not determined in AIMS.

3, under column MESA). In this study, MESA was only used to generate a grid of stellar models as described in Section 3. The input physics used varies across the set of modelling tools used. Furthermore, each modelling technique uses different stellar evolution codes, pulsation codes and/or optimization techniques. The results in Tables 2 and 3 include only statistical uncertainties returned by each pipeline.

The error bars in brackets under column AIMS (Tables 2 and 3) were obtained using an unreduced χ^2 value in the likelihood function. These errors are consistent from a statistical point of view, i.e., they reflect how the observational errors propagate to the derived stellar parameters. The main error bars used were obtained using the overall reduced χ^2 which is inconsistent from a statistical point of view. This option, however, generates error bars comparable to those obtained using other optimization tools.

The different modelling tools converge on models for HD 176465 A with a mean mass of 0.95 ± 0.02 $M_\odot$ and a mean radius of 0.93 ± 0.01 $R_\odot$. For HD 176465 B, they converge with a mean mass of 0.93 ± 0.02 $M_\odot$ and a mean radius of 0.89 ± 0.01 $R_\odot$. Figures 1 and 2 show that results from MESA are slightly higher regarding the stellar masses and radii compared to results from other modelling tools. The initial helium abundance of the best-fitting MESA models are lower than the primordial value from standard Big Bang nucleosynthesis (Y = 0.2482 ± 0.0007)[27]. This may explain why masses and radii of MESA models are larger compared to other modeling tools.

The age of HD 176465 A and HD 176465 B was found to be 2.81 ± 0.48 Gyr and 2.52 ± 0.80 Gyr respectively. This is consistent within 1σ-error interval. All models agree on the ages of both stars within the uncertainties even when no restrictions were made on the ages during the different modelling procedures. This is expected assuming both stars were formed at approximately the same time and from the same molecular cloud. The gyrochronology relations based on [11] calibrations were used in [30] and the age of the binary was found to be 3.2 $^{+1.2}_{-0.8}$ Gyr. This is in good agreement with results obtained using the asteroseismic modelling tools.

The metal abundances for HD 176465 A and HD 176465 B are 0.012 ± 0.0010 and 0.011 ± 0.0012, respectively. This is consistent within 1σ-error interval. This is expected assuming that both stars were formed from the same molecular cloud with approximately the same chemical composition.

5 Conclusions

We derived precise fundamental stellar parameters using state of the art modelling tools for the asteroseismic binary system. The obtained results were compared with those available from other modeling tools [30]. The methods agree on the stellar properties in almost all cases.

Binary star systems provide important tests of stellar structure and evolution. They can be used to provide important constraints in the asteroseismic modelling techniques especially when orbital parameters are known. For HD 176465, position measurements were made in the Washington Double Star Catalog [17]. Unfortunately the fraction of the orbit covered is insufficient to constrain the total mass [30].

Only a handful of main sequence binary systems have been made available from space missions such as CoRoT and Kepler. Future missions such as TESS [6] and PLATO [25] are expected to increase asteroseismic data for similar systems. This will help in setting tight constraints on the physics used in stellar evolution and asteroseismology.

Acknowledgements

We thank the MESA community for the engaging conversations about the MESA project. B. Nsamba acknowledges support from Fundação para a Ciência e a Tecnologia (FCT, Portugal) under Grants PD/BD/113744/2015 (Program PhD::SPACE) and UID/FIS/04434/2013. TLC acknowledges the support of the UK Science and Technology Facilities Council (STFC). Furthermore, A. García Hernández acknowledges support from Fundação para a Ciência e a Tecnologia (FCT, Portugal) through the fellowship SFRH/BPD/80619/2011 and from the EC Project SPACEINN (FP7-SPACE-2012-312844). GYRE has been developed with support from the Astronomy and Astrophysics Grants (AAG) program of the National Science Foundation, through awards AST-0908688 and AST-0904607. The "Asteroseismic Inference on a Massive Scale" (AIMS) project was developed at the University of Birmingham by Daniel R. Reese as one of the deliverables for the SPACEINN network. The SPACEINN network is funded by the European Community Seventh Framework Programme (FP7/2007-2013) under grant agreement no. 312844.

References

[1] Angulo, C., Arnould, M., et al. 1999, Nuclear Physics A, **656**, 3
[2] Ball, W. H & Gizon, L. 2014, A&A, **568**, A123
[3] Bedding, T. R., Kjeldsen, H., et al. 2004, ApJ, **614**, 380
[4] Böhm - Vitense, E. 1958, Zeit. Astrophys., **46**, 108
[5] Bruntt, H., Basu, S., et al. 2012, MNRAS, **423**, 122
[6] Campante, T. L., Schofield, M., et al. 2016, ApJ, **830**, 138
[7] Chaplin, W. J., & Miglio A., 2013, ARA&A , **51**, 353
[8] Ferguson, J. W., Alexander, D. R., et al. 2005, ApJ, **623**, 585
[9] Foreman - Mackey, D., et al. 2013, PASP, **125**, 306 - 312
[10] Garcia, R. A., Hekker, S., et al. 2011, MNRAS, **414**, L6
[11] Garcia, R. A., Ceillier, T., et al. 2014, A&A, **572**, A34
[12] Imbriani, G., Costantini, H., et al. 2005, European Physical Journal A, **25**, 455
[13] Iglesias C. A. & Rogers, F. J. 1996, ApJ, **464**, 943
[14] Jenkins, J. M., Caldwell, D. A., et al. 2010, ApJ, **713**, L87
[15] Kjeldsen, H., Bedding, T. R., et al. 2005, ApJ, **635**, 1281
[16] Kunz, R., Fey, M., et al. 2002, ApJ, **567**, 643
[17] Mason, B. D., Wycoff, G. L., et al. 2001, AJ, **122**, 3466
[18] Metcalfe, T. S., Chaplin, W. J., et al. 2012, ApJ, **748**, L10
[19] Metcalfe, T. S., Creevey, O. L., et al. 2014, ApJS, **214**, 27
[20] Michel, E., Baglin, A., et al. 2008, Science, **322**, 558
[21] Molenda-Zakowicz, J., Sousa, S. G., et al. 2013, MNRAS, **434**, 1422
[22] Paxton, B., Bildsten, L., et al. 2011, ApJS, **192**, 3
[23] Paxton, B., Cantiello, M., et al. 2013, ApJS, **208**, 4
[24] Paxton, B., Marchant, P., et al. 2015, ApJS, **220**, 15
[25] Rauer, H., et al. 2014, Experimental Astronomy, **38**, 249
[26] Silva, V. A., et al. 2015, MNRAS, **452**, 2127-2148
[27] Steigman, G. 2010, J. Cosmology Astropart. Phys., **4**, 029
[28] Thoul, A. A., Bahcall, J. N., & Loeb, A. 1994, ApJ, **421**, 828
[29] Townsend, R. H. D & Teitler, S. A. 2013, MNRAS, **435**, 3406 - 3418
[30] White T. R., et al. 2016, A&A, **601**, A82

EPJ Web of Conferences **160**, 05011 (2017) DOI: 10.1051/epjconf/201716005011
Seismology of the Sun and the Distant Stars II

Stellar magnetic activity and exoplanets

A. A. Vidotto[1,*]

[1]*School of Physics, Trinity College Dublin, the University of Dublin, Dublin-2, Ireland*

Abstract. It has been proposed that magnetic activity could be enhanced due to interactions between close-in massive planets and their host stars. In this article, I present a brief overview of the connection between stellar magnetic activity and exoplanets. Stellar activity can be probed in chromospheric lines, coronal emission, surface spot coverage, etc. Since these are manifestations of stellar magnetism, these measurements are often used as proxies for the magnetic field of stars. Here, instead of focusing on the magnetic proxies, I overview some recent results of magnetic field measurements using spectropolarimetric observations. Firstly, I discuss the general trends found between large-scale magnetism, stellar rotation, and coronal emission and show that magnetism seems to be correlated to the internal structure of the star. Secondly, I overview some works that show evidence that exoplanets could (or not) act as to enhance the activity of their host stars.

1 Introduction

Stellar activity can manifest itself in the form of, e.g., surface spot coverage, emissions from the chromosphere, transition region and corona [e.g., 1, 2]. These manifestations are all indicators of magnetic activity and are, therefore, often used as proxies for stellar magnetism. As stellar magnetism is responsible for driving stellar winds, flares and coronal mass ejections, stellar magnetic fields are the main driver of space weather on (exo)planets. Therefore, understanding the host star magnetism is a key step towards understanding how stellar activity affects exoplanets.

The host star magnetism can have effects on planetary mass-loss [3, 4], habitability [5, 6], and on the magnetospheres of exoplanets [7], to name a few. It has also been suggested that exoplanets, in particular those orbiting close to the star, can also affect stellar activity, through tidal interactions and/or magnetic reconnection events [8–12].

In this article, I present a brief review of stellar surface magnetism (Section 2) and some proposed effects that exoplanets might have on the activity of their host stars (Section 3).

2 Stellar surface magnetism: brief overview of its latest results

Magnetic fields are responsible for driving the winds of cool, main-sequence stars. These outflows carry away angular momentum from the star, slowing down its rotation. Since rotation is intimately related to the generation of stellar magnetic fields, with age, stars spin down and the magnetic fields they generate decrease in magnitude [13].

One way to map surface magnetism is through the tomographic imaging technique Zeeman Doppler imaging (ZDI) [14]. This technique has now successfully mapped the large-scale surface field of hundreds of cool, pre-main-sequence and main-sequence stars [e.g., 15]. Due to the increasing sample of stars with mapped fields, several trends are emerging [13, 15–19]. For example, it has been shown that the topology of surface magnetic fields depends on the internal structure, where fully convective stars present mainly poloidal fields [17, 18] as compared to stars that have a radiative core. This change in topology might be caused by the development of the tachocline [17, 18].

Evolutionary speaking, magnetic fields of main-sequence stars are also observed to decrease with age t. [13] showed that the average magnetic field intensity $\langle|B|\rangle$ is a function of $t^{-0.66\pm0.05}$. This age decay has a power law consistent to the power-law of -0.5 between rotation rate $\Omega_\star$ and age [1]: $\Omega_\star \propto t^{-1/2}$. This similarity could be an indication that the dynamo operating inside the star is of a linear type (i.e., the magnetic field is linearly proportional to stellar rotation rate), as already suggested by [1]. Indeed, ZDI observations have revealed that, for stars with Rossby numbers Ro $\gtrsim 0.1$, the average large-scale magnetic field decays with $\mathrm{Ro}^{-1.38\pm0.14}$ [13] (see bottom panel in Figure 1). Since Rossby number is defined as $\mathrm{Ro} = 2\pi(\Omega_\star\tau_{conv})^{-1}$, where τ_{conv} is the convective turnover time, this relation indicates an approximately linear decay of magnetism with rotation rate.[1]

*e-mail: Aline.Vidotto@tcd.ie

[1]A decay is also observed between magnetism and Rossby number for active evolved (single) G – K giants, albeit with a different slope: [20] showed that the line-of-sight magnetic field strength B_l decays with $\mathrm{Ro}^{-0.68}$. A recent study proposes that planet engulfment could be enhancing the magnetic field of giant stars [21]. I will return to this point in section 3.

DOI: 10.1051/epjconf/201716005011

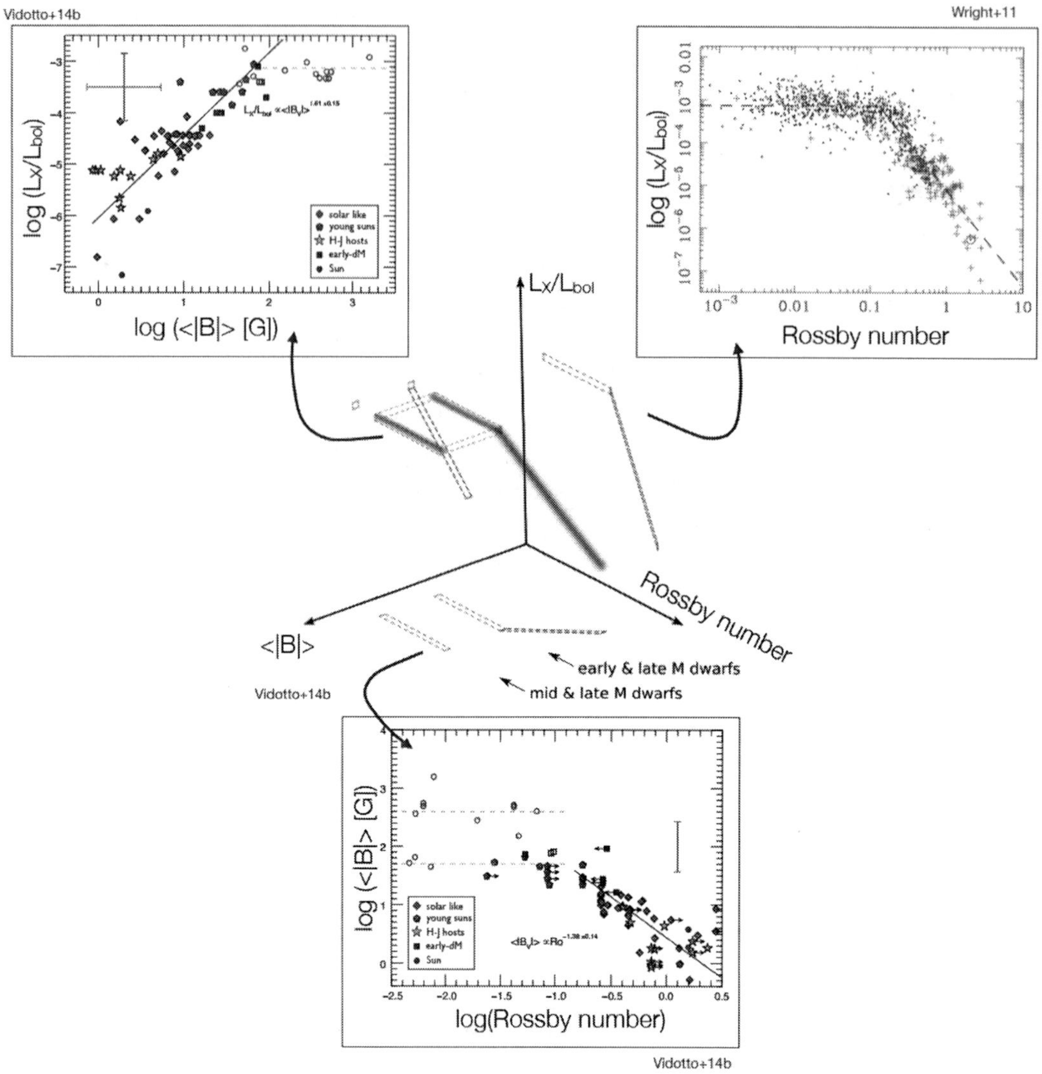

Figure 1: Stellar activity is a complex function of many variables, such as age, mass, rotation, and magnetism. This figure illustrates how magnetic fields are related to X-ray emission (left-most panel) and to Rossby number (bottom panel). It also shows the more traditional activity-rotation relation between X-ray emission and Rossby number (right-most panel). Each of these panels are projections of a 3D distribution of $\{L_x/L_{\rm bol}, {\rm Ro}, \langle|B|\rangle\}$ (middle panel). This figure is based on the plots presented in [13] and [22].

Stars with Ro $\lesssim$ 0.1 are *saturated*, i.e., their activity does not vary with Rossby number. This saturation has been identified, for example, in the relations between $L_x/L_{\rm bol}$ and Ro [22–24], where L_x is the X-ray luminosity and $L_{\rm bol}$ the stellar bolometric luminosity (see right-most panel in Figure 1). When investigating the behaviour of stellar magnetism for a sample of saturated early- and mid-M dwarf stars, [25, 26] noticed that these stars saturated at different magnetic field strengths, whereby early-M dwarfs presented a lower saturation threshold than mid-M dwarfs (see dashed lines in the bottom panel in Figure 1). These authors suggested that this could be caused by different dynamo efficiencies at producing small- and large-scale magnetic fields.

Since rotation is linked to magnetic field strength and X-ray emission is linked to rotation, then the immediate question we asked ourselves is how magnetic field strength can be related to X-ray emission. The left-most panel of Figure 1 shows precisely this investigation, where we note that X-ray emission increases with average magnetic field strength $\langle|B|\rangle^{1.61\pm0.15}$. These three diagrams, illustrated in Figure 1, are actually projections of a 3D distribution of $\{L_x/L_{\rm bol}, {\rm Ro}, \langle|B|\rangle\}$ (sketch shown in the middle panel). In fact, stellar activity is a function of many other variables,

 DOI: 10.1051/epjconf/201716005011

such as age and stellar mass, with the sketch presented in Figure 1 representing a cut of a multi-dimensional function.

A common characteristic of the plots regarding stellar activity are their relatively large spread. Part of this spread is due to intrinsic variability of stellar activity and not only due to observational uncertainties. The Sun, for instance, varies in magnetic field intensity and X-ray emission along its cycle (see left-most and bottom panels of Figure 1, showing two points for the Sun, one during activity minimum and the other during maximum of activity). Other stars present, like our Sun, variations on their surface magnetic fields, some of which are associated to well-defined magnetic cycles [27–30]. This medium term variability (on the order of years) acts as to increase the spread in rotation-magnetic-activity relations. Figure 2 illustrates the case of 61 Cyg A [29]. This star shows polarity reversals on the large-scale field with a period of a full magnetic cycle (i.e., two reversals) of around 7 years. The reversals are in phase with the chromospheric and coronal cycle, similar to what is observed for the Sun.

3 Can planets influence the activity of their host stars?

That stellar activity could affect exoplanets is not hard to imagine. After all, the Earth and the planets in the solar system regularly experience the effects of solar energetic particles interacting with their magnetospheres/atmospheres. Several works have studied the effects of the stars and their winds on their surrounding exoplanets [e.g. 7, 31–38, just to name a few]. But could the opposite also occur? Can planets influence the activity of their host stars?

Motivated by the increase in chromospheric and coronal activity observed in interacting binary stars, [8] proposed that close-in giant planets (or brown dwarfs) could lead to enhancement of dynamo activity. These planet-star interactions were proposed to be of two types: magnetospheric or tidal interactions. In the first case, reconnection events between the magnetic field lines of an exoplanet and of the host star would accelerate particles that would travel along stellar magnetic field lines, interacting with the outer layers of the stellar atmosphere. This interaction would then generate enhanced heat near the magnetic field footpoints. This anomalous activity would be modulated by the orbital period of the planet, as opposed to the 'normal' stellar activity, which is modulated by the stellar rotation period. In the second case, tides raised on the star by the revolving planet would cause repeated expansion and contraction of tidal bulges, generating waves that can dissipate and deposit energy in the upper atmosphere of the star (i.e., transition region, chromosphere). Like in the first scenario, this anomalous activity increase would also be cyclic, but the modulation would take place at half orbital period.

Searches for anomalous activity have since been on going. In a series of works, [9, 39] presented evidence of anomalous activity on a few planet-hosting stars. These anomalies were modulated with the orbital period of their close-in giant planets, possibly indicating a signature of planet-star interaction. Repeated observations (at different epochs), however, failed to detect such modulations [40], leading the authors to suggest that activity enhancement is intermittent and could depend, among others, on the configuration of the star's magnetic field. Searching for chromospheric/transition region signatures of planet-star interaction has developed into an active field of research, yielding positive (or tentatively positive) [41–43] and negative [44–48] detections.

In a recent work, for instance, [43] conducted X-ray observations of HD 17156, the host of an eccentric planet, and reported enhanced X-ray stellar emission a few hours after the periastron passage. This emission, happening near the closest approach of the planet where planet-star interactions are expected to be enhanced, could have been mediated by magnetic reconnection or accretion of planetary material towards the star [e.g., 49]. Following the same idea, i.e., that eccentric planets can trigger stellar activity at orbital phases near the periastron passage, [48] investigated if the activity level of HD 80606 could have been influenced by its eccentric planet. For that, they used optical spectroscopy and photometric monitoring of the star during the periastron passage and close to the apastron passage. They could not find any evidence that the stellar activity level was modified between these two observing epochs. An interpretation (the simplest one) is that this planet does not trigger activity enhancement on the star, or if does, this enhancement would be below the detection sensitivity. Another possibility is that the enhanced activity would appear with a certain phase lag outside the observing window.

If close-in exoplanets can indeed enhance stellar magnetic activity, it might also be possible to verify that in magnetic images of planet-hosting stars. Indeed, among the ZDI sample shown in Figure 1, several of the stars are known to host close-in giant planets (i.e., hot-Jupiters). These stars are shown as yellow star symbols in Figure 1. The close-in, massive planets are expected to cause the largest observational signature of its influence on the large-scale magnetic field of the host star. In terms of magnetic characteristics (topology and intensity), however, there has not been any measurable difference between the hot-Jupiter hosts and the remaining stars of the sample [13, 50, 51]. In other words, the yellow star symbols shown in Figure 1 do not seem to occupy a particular position in these diagrams with respect to other symbols in these plots (i.e., the stars not known to host hot Jupiters).

Another possibility is that the interaction between a close-in planet and the host star might be able to affect stellar *cycles*. The famous planet-host star τ Boo, has been monitored with ZDI for about a decade [27, 28, 30, 50, 52]. This star shows flips in magnetic field polarity every year, making it the star with the shortest full (2 years) magnetic cycle mapped with ZDI. Since τ Boo is a F7V star, its convective envelope is quite thin and it is estimated to have a mass similar to that of the hot Jupiter. The idea is that, through tidal interactions, the planet and the stellar con-

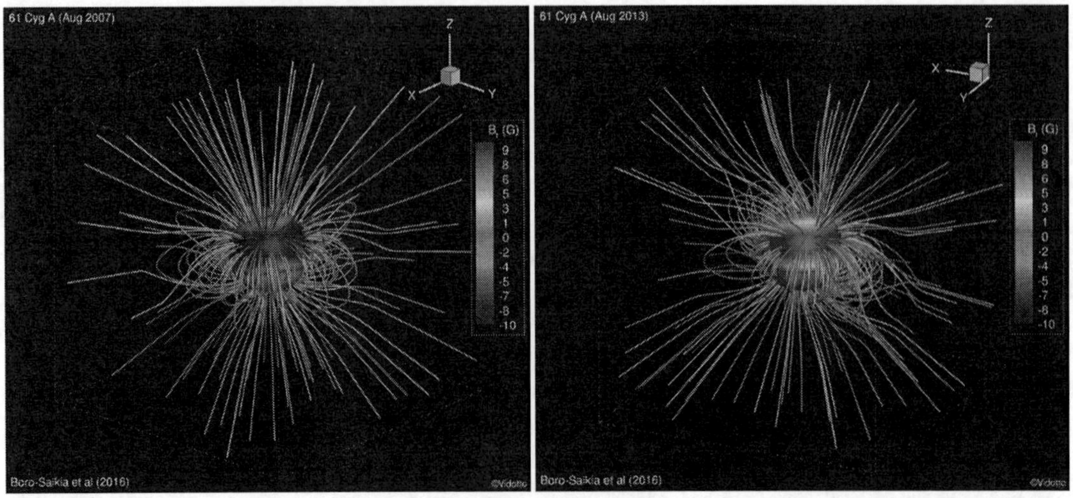

Figure 2: The star 61 Cyg A shows polarity reversals on the large-scale field with a period of a full magnetic cycle (i.e., two reversals) of around 7 years. The reversals are in phase with the chromospheric and coronal cycle. Figure based on the ZDI maps of [29] for Aug 2007 (left) and Aug 2013 (right). The magnetic field lines are extrapolated using a potential field source surface method, which takes into account the radial component of the observed field from [29].

vective envelope would have been synchronised, but the interior radiative core of the star would still be rotating at a different rate. The shear created by different rotation rates between the radiative interior and the convective envelope could be the cause of the short magnetic cycle. According to this interpretation, the planet would have indirectly contributed to altering the generation of magnetic fields in the star. An alternative interpretation involves the high level of surface differential rotation derived in ZDI studies [28].

In a series of works, [12, 21, 53] proposed that planets (initially at large separations) could be engulfed by stars, when their hosts evolve off of the main sequence. With the expansion of the stars at the red giant branch, tidal interactions between a Jupiter-like planet and the star trigger planetary migration, leading the planet to eventually be engulfed by the star. During the migration process, orbital angular momentum are transferred to stellar spin. Therefore, these red giant branch stars would present higher rotation rates, which could not be explained by any reasonable model for single star evolution [12, 53]. From the observational sample of [54], [12, 21, 53] identified some fast rotating red giants that are candidates of having engulfed their planets in a recent past. Since rotation and magnetic fields are closely related [20], [21] then further proposed, that these fast rotating red giants would also show a higher magnetic field. This idea can, of course, be directly tested with spectropolarimetric observations.

4 Summary and Conclusions

This article was based on the talk I presented at the conference "Seismology of the Sun and the Distant Stars 2016" on the topic of "stellar-planet activity". First, I would like to remind the reader that the material I presented during my talk, and that was described in this article, is far from a comprehensive overview of the theme. I hope I could provide some starting references for an interested reader to proceed on their own.

In particular, I had the (hard) task to discuss synergies between "stellar-planet activity" and asteroseismology. The first immediate synergy one can identify is on the relation between the internal structure of stars (probed with asteroseismology) and the global properties of stellar magnetic fields (probed in ZDI studies, Section 2). In the second part of my talk, I focused on how exoplanets might or not enhance stellar magnetic activity. I showed that, although there is some evidence that planets can trigger magnetic activity on their host stars, this evidence seems to be intermittent, making it more difficult to precise the nature of the planet-star interaction. In fact, there is an ongoing debate in the community regarding the detectability of the signature of planet-star interaction. Would there be a way that asteroseismology could contribute to this debate?

Acknowledgements

I would like to warmly thank the organisers of "Seismology of the Sun and the Distant Stars 2016" for the financial support offered for me to attend this conference. Obrigada!

References

[1] A. Skumanich, ApJ, **171**, 565 (1972)
[2] R.W. Noyes, L.W. Hartmann, S.L. Baliunas, D.K. Duncan, A.H. Vaughan, ApJ, **279**, 763 (1984)
[3] H. Lammer, M. Guedel, Y. Kulikov, I. Ribas, T.V. Zaqarashvili, M.L. Khodachenko, K.G. Kislyakova, H. Gröller, P. Odert, M. Leitzinger et al., Earth, Planets, and Space **64**, 179 (2012)

[4] J.E. Owen, F.C. Adams, MNRAS, **444**, 3761 (2014)
[5] A.A. Vidotto, M. Jardine, J. Morin, J.F. Donati, P. Lang, A.J.B. Russell, A&A, **557**, A67 (2013)
[6] I. Ribas, E. Bolmont, F. Selsis, A. Reiners, J. Leconte, S.N. Raymond, S.G. Engle, E.F. Guinan, J. Morin, M. Turbet et al., ArXiv e-prints (2016), 1608.06813
[7] V. See, M. Jardine, A.A. Vidotto, P. Petit, S.C. Marsden, S.V. Jeffers, J.D. do Nascimento, A&A, **570**, A99 (2014), 1409.1237
[8] M. Cuntz, S.H. Saar, Z.E. Musielak, ApJL, **533**, L151 (2000)
[9] E. Shkolnik, G.A.H. Walker, D.A. Bohlender, P.G. Gu, M. Kürster, ApJ, **622**, 1075 (2005), arXiv:astro-ph/0411655
[10] A.F. Lanza, A&A, **505**, 339 (2009), 0906.1738
[11] K. Poppenhaeger, S.J. Wolk, A&A, **565**, L1 (2014), 1404.1073
[12] G. Privitera, G. Meynet, P. Eggenberger, A.A. Vidotto, E. Villaver, M. Bianda, A&A, **593**, A128 (2016), 1606.08027
[13] A.A. Vidotto, S.G. Gregory, M. Jardine, J.F. Donati, P. Petit, J. Morin, C.P. Folsom, J. Bouvier, A.C. Cameron, G. Hussain et al., MNRAS, **441**, 2361 (2014), 1404.2733
[14] J.F. Donati, S.F. Brown, A&A, **326**, 1135 (1997)
[15] J. Donati, J.D. Landstreet, ARA&A, **47**, 333 (2009), 0904.1938
[16] P. Petit, B. Dintrans, S.K. Solanki, J.F. Donati, M. Aurière, F. Lignières, J. Morin, F. Paletou, J. Ramirez Velez, C. Catala et al., MNRAS, **388**, 80 (2008), 0804.1290
[17] J. Morin, J. Donati, P. Petit, X. Delfosse, T. Forveille, M.M. Jardine, MNRAS, **407**, 2269 (2010), 1005.5552
[18] V. See, M. Jardine, A.A. Vidotto, J.F. Donati, C.P. Folsom, S. Boro Saikia, J. Bouvier, R. Fares, S.G. Gregory, G. Hussain et al., MNRAS, **453**, 4301 (2015), 1508.01403
[19] É.M. Hébrard, J.F. Donati, X. Delfosse, J. Morin, C. Moutou, I. Boisse, MNRAS, **461**, 1465 (2016)
[20] M. Aurière, R. Konstantinova-Antova, C. Charbonnel, G.A. Wade, S. Tsvetkova, P. Petit, B. Dintrans, N.A. Drake, T. Decressin, N. Lagarde et al., A&A, **574**, A90 (2015), 1411.6230
[21] G. Privitera, G. Meynet, P. Eggenberger, C. Georgy, S. Ekström, A.A. Vidotto, M. Bianda, E. Villaver, A. ud-Doula, A&A, **593**, L15 (2016), 1608.08893
[22] N.J. Wright, J.J. Drake, E.E. Mamajek, G.W. Henry, ApJ, **743**, 48 (2011), 1109.4634
[23] N. Pizzolato, A. Maggio, G. Micela, S. Sciortino, P. Ventura, A&A, **397**, 147 (2003)
[24] R.D. Jeffries, R.J. Jackson, K.R. Briggs, P.A. Evans, J.P. Pye, MNRAS, **411**, 2099 (2011), 1010.2152
[25] J. Donati, J. Morin, P. Petit, X. Delfosse, T. Forveille, M. Aurière, R. Cabanac, B. Dintrans, R. Fares, T. Gastine et al., MNRAS, **390**, 545 (2008), 0809.0269
[26] J. Morin, J. Donati, P. Petit, X. Delfosse, T. Forveille, L. Albert, M. Aurière, R. Cabanac, B. Dintrans, R. Fares et al., MNRAS, **390**, 567 (2008), 0808.1423
[27] J.F. Donati, C. Moutou, R. Farès, D. Bohlender, C. Catala, M. Deleuil, E. Shkolnik, A. Collier Cameron, M.M. Jardine, G.A.H. Walker, MNRAS, **385**, 1179 (2008), 0802.1584
[28] R. Fares, J. Donati, C. Moutou, D. Bohlender, C. Catala, M. Deleuil, E. Shkolnik, A.C. Cameron, M.M. Jardine, G.A.H. Walker, MNRAS, **398**, 1383 (2009), 0906.4515
[29] S. Boro Saikia, S.V. Jeffers, J. Morin, P. Petit, C.P. Folsom, S.C. Marsden, J.F. Donati, R. Cameron, J.C. Hall, V. Perdelwitz et al., A&A, **594**, A29 (2016), 1606.01032
[30] M.W. Mengel, R. Fares, S.C. Marsden, B.D. Carter, S.V. Jeffers, P. Petit, J.F. Donati, C.P. Folsom, BCool Collaboration, MNRAS, **459**, 4325 (2016)
[31] W.H. Ip, A. Kopp, J.H. Hu, ApJL, **602**, L53 (2004)
[32] M.L. Khodachenko, I. Ribas, H. Lammer, J. Grießmeier, M. Leitner, F. Selsis, C. Eiroa, A. Hanslmeier, H.K. Biernat, C.J. Farrugia et al., Astrobiology **7**, 167 (2007)
[33] M. Jardine, A.C. Cameron, A&A, **490**, 843 (2008), 0808.2889
[34] A. Strugarek, A.S. Brun, S.P. Matt, V. Réville, ApJ, **795**, 86 (2014), 1409.5268
[35] L. Tu, C.P. Johnstone, M. Guedel, H. Lammer, A&A, **577**, L3 (2015), 1504.04546
[36] A.A. Vidotto, R. Fares, M. Jardine, C. Moutou, J.F. Donati, MNRAS, **449**, 4117 (2015), 1503.05711
[37] K.G. Kislyakova, L. Fossati, C.P. Johnstone, M. Holmström, V.V. Zaitsev, H. Lammer, ApJL, **799**, L15 (2015), 1503.06955
[38] B.A. Nicholson, A.A. Vidotto, M. Mengel, L. Brookshaw, B. Carter, P. Petit, S.C. Marsden, S.V. Jeffers, R. Fares, the BCool Collaboration, MNRAS, **459**, 1907 (2016)
[39] E. Shkolnik, G.A.H. Walker, D.A. Bohlender, ApJ, **597**, 1092 (2003)
[40] E. Shkolnik, D.A. Bohlender, G.A.H. Walker, A. Collier Cameron, ApJ, **676**, 628 (2008), 0712.0004
[41] L. Gurdemir, S. Redfield, M. Cuntz, Publications of the Astronomical Society of Australia, **29**, 141 (2012), 1202.3612
[42] K. France, R.O. Parke Loyd, A. Youngblood, A. Brown, P.C. Schneider, S.L. Hawley, C.S. Froning, J.L. Linsky, A. Roberge, A.P. Buccino et al., ApJ, **820**, 89 (2016), 1602.09142
[43] A. Maggio, I. Pillitteri, G. Scandariato, A.F. Lanza, S. Sciortino, F. Borsa, A.S. Bonomo, R. Claudi, E. Covino, S. Desidera et al., ApJL, **811**, L2 (2015), 1509.00662
[44] R. Fares, J. Donati, C. Moutou, M.M. Jardine, J. Grießmeier, P. Zarka, E.L. Shkolnik, D. Bohlender,

C. Catala, A.C. Cameron, MNRAS, **406**, 409 (2010), 1003.6027
[45] B.P. Miller, E. Gallo, J.T. Wright, A.K. Dupree, ApJ, **754**, 137 (2012), 1206.0746
[46] G. Scandariato, A. Maggio, A.F. Lanza, I. Pagano, R. Fares, E.L. Shkolnik, D. Bohlender, A.C. Cameron, S. Dieters, J.F. Donati et al., A&A, **552**, A7 (2013), 1301.7748
[47] B.L. Canto Martins, M.L. Das Chagas, S. Alves, I.C. Leão, L.P. de Souza Neto, J.R. de Medeiros, A&A, **530**, A73 (2011), 1103.5332
[48] P. Figueira, A. Santerne, A. Suárez Mascareño, J. Gomes da Silva, L. Abe, V.Z. Adibekyan, P. Bendjoya, A.C.M. Correia, E. Delgado-Mena, J.P. Faria et al., A&A, **592**, A143 (2016), 1606.05549
[49] I. Pillitteri, A. Maggio, G. Micela, S. Sciortino, S.J. Wolk, T. Matsakos, ApJ, **805**, 52 (2015), 1503.05590
[50] R. Fares, C. Moutou, J.F. Donati, C. Catala, E.L. Shkolnik, M.M. Jardine, A.C. Cameron, M. Deleuil, MNRAS, **435**, 1451 (2013), 1307.6091
[51] M.W. Mengel, S.C. Marsden, B.D. Carter, J. Horner, R. King, R. Fares, S.V. Jeffers, P. Petit, A.A. Vidotto, BCool Collaboration, MNRAS, **submitted** (2016)
[52] C. Catala, J.F. Donati, E. Shkolnik, D. Bohlender, E. Alecian, MNRAS Letters, **374**, L42 (2007), arXiv:astro-ph/0610758
[53] G. Privitera, G. Meynet, P. Eggenberger, A.A. Vidotto, E. Villaver, M. Bianda, A&A, **591**, A45 (2016), 1604.06005
[54] J.K. Carlberg, K. Cunha, V.V. Smith, S.R. Majewski, ApJ, **757**, 109 (2012), 1208.1775

EPJ Web of Conferences **160**, 05012 (2017)
Seismology of the Sun and the Distant Stars II DOI: 10.1051/epjconf/201716005012

Seismology of Giant Planets: General Overview and Results from the *Kepler* K2 Observations of Neptune

Patrick Gaulme[1,2,3,★]

[1]Department of Astronomy, New Mexico State University, P.O. Box 30001, MSC 4500, Las Cruces, NM 88003-8001, USA
[2]Apache Point Observatory, 2001 Apache Point Road, P.O. Box 59, Sunspot, NM 88349, USA
[3]Physics Department, New Mexico Institute of Mining and Technology, 801 Leroy Place, Socorro, NM 87801, USA

Abstract. For this invited contribution, I was asked to give an overview about the application of helio and asteroseismic techniques to study the interior of giant planets, and to specifically present the recent observations of Neptune by *Kepler* K2. Seismology applied to giant planets could drastically change our understanding of their deep interiors, as it has happened with the Earth, the Sun, and many main-sequence and evolved stars. The study of giant planets' composition is important for understanding both the mechanisms enabling their formation and the origins of planetary systems, in particular our own. Unfortunately, its determination is complicated by the fact that their interior is thought not to be homogeneous, so that spectroscopic determinations of atmospheric abundances are probably not representative of the planet as a whole. Instead, the determination of their composition and structure must rely on indirect measurements and interior models. Giant planets are mostly fluid and convective, which makes their seismology much closer to that of solar-like stars than that of terrestrial planets. Hence, helioseismology techniques naturally transfer to giant planets. In addition, two alternative methods can be used: photometry of the solar light reflected by planetary atmospheres, and ring seismology in the specific case of Saturn. The current decade has been promising thanks to the detection of Jupiter's acoustic oscillations with the ground-based imaging-spectrometer SYMPA and indirect detection of Saturn's f-modes in its rings by the NASA Cassini orbiter. This has motivated new projects of ground-based and space-borne instruments that are under development. The K2 observations represented the first opportunity to search for planetary oscillations with visible photometry. Despite the excellent quality of K2 data, the noise level of the power spectrum of the light curve was not low enough to detect Neptune's oscillations. The main results from the K2 observations are the clear detection of the well-known differential rotation of Neptune, measured for the first time through the rotational modulation of its photometry, and the detection of the Sun's oscillations, for the first time in an indirect way in intensity measurements.

1 Interior Structure

The deep internal structures of Jupiter and Saturn, particularly the amount and the radial distribution of heavy elements, are important diagnostics to the formation and evolution of planetary systems in general. Two scenarios are envisioned for the formation of giant planets. The first assumes that cores of rocks and ices are formed by the accumulation of solid planetesimals. When these solid cores grow to about 10 $M_\oplus$ (Earth masses), a phase of rapid H-He gas accretion follows, which leads to the current structure of the planets [1]. In this class of models, the resulting planets are generally enriched in heavy elements. In the second scenario, giant planets form by gravitational instability of the gas in a massive solar nebula [2], resulting in planets with solar chemical compositions and masses probably larger than that of Jupiter. In this scenario, a solid core may form through sedimentation of solid and vaporized material toward the center, or by subsequent capture of solid planetesimals. The gravitational instability scenario requires a relatively massive and cold protoplanetary nebulae to become unstable, while the nucleated instability mechanism can also operate with less massive and hotter disks. Current uncertainties about Jupiter's and Saturn's deep interior structure and composition are large and prevent us from uniquely distinguishing between these two formation mechanisms [3–5].

Jupiter and Saturn are natural reservoirs of liquid metallic hydrogen and helium at pressures of up to tens of Mbar and temperatures of the order of 10^4 K. Such warm, dense states of matter are very challenging to reproduce in laboratory experiments. Megabar pressures can be achieved in shock compression experiments but usually at temperatures well above those found in giant planets [3]. So far, these experiments have not directly probed the conditions of interest for giant planets. On the other hand, significant progress has been achieved recently in computer simulations of the equations of state of H and He. Oddly, this has not led to a better consensus on the internal structure of Jupiter [4, 5].

★e-mail: gaulme@nmsu.edu

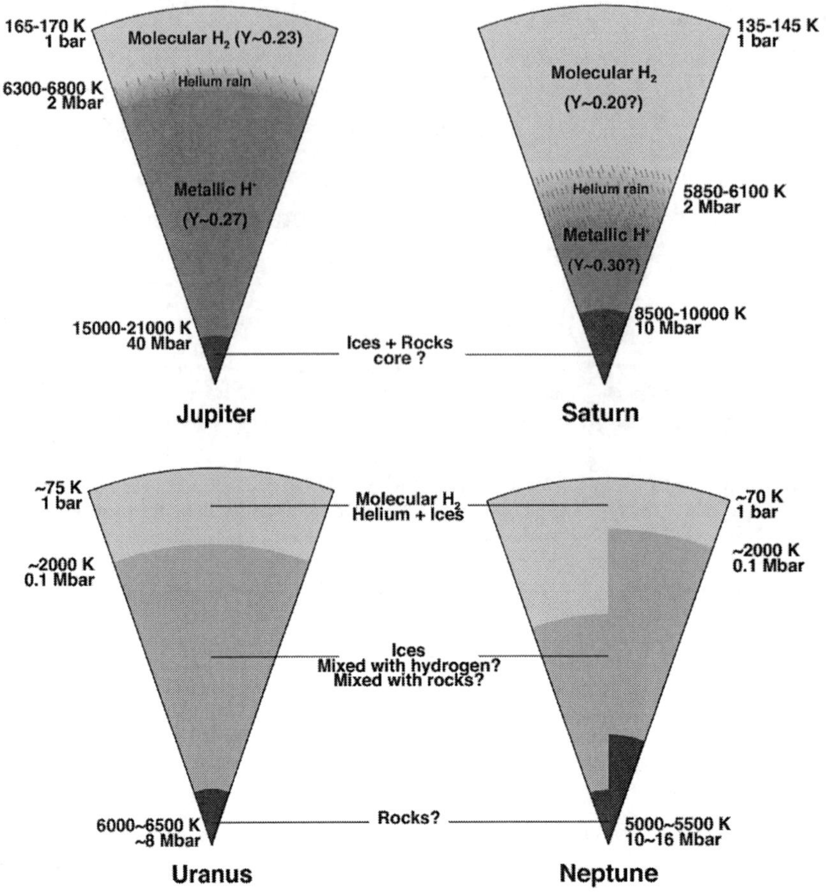

Figure 1: Schematic representation of the interiors of Jupiter, Saturn, Uranus, and Neptune. The range of temperatures for Jupiter and Saturn is for models neglecting the presence of the inhomogeneous region (adapted from [6] and taken from [7]). Helium mass mixing ratios Y are indicated. The size of the central rock and ice cores of Jupiter and Saturn is very uncertain (see text). Similarly, for Uranus and Neptune, considerable uncertainties exist [8]. Recent models of Neptune indicate a wider range of possible solutions of which two representative ones are shown.

Based on our current knowledge of planetary system formation and of equations of state of dense matter, the debate regarding the internal structure and composition of Jupiter and Saturn can be articulated in terms of four main questions [e.g. 9]. i) What is the radial distribution of heavy elements? Are they concentrated in a central core or distributed throughout the H-He envelope? If mixing processes have redistributed the heavy elements after its formation, the thermal evolution of the planet would be profoundly affected [10, 11]. ii) If a dense central core is present, what are its mass and composition? Current estimates of the core mass of Jupiter range from 0 to 15 $M_\oplus$ and the total mass of heavy elements from 10 to 40 $M_\oplus$ [3–5]. iii) Which mechanisms dominate the energy transport from the deep interior to the surface of these planets? iv) Do H-He mixtures separate in giant (exo)planets resulting in a depletion of He in the outer envelope?

2 Seismology and Giant Planets

Starting in 2016, NASA's *Juno* mission will make key contributions to our understanding of Jupiter thanks to precise measurements of its gravity and magnetic fields. Unfortunately, the presence of a denser core of heavy elements only weakly influences even the lowest order (quadrupole) deviation in the gravity field and the core mass will remain essentially model-dependent. On the other hand, seismology, which consists of identifying global acoustic eigenmodes (p-modes), complements *Juno* science by offering a way to directly measure the planet's sound speed profile, and thus its physical properties from the outer envelope to the core.

All of these questions can be addressed with seismology. From an observational point of view, seismology of giant planets is a natural extension of helioseismology. Their common fluid nature is expected to lead to similar oscillations and to the possibility of using similar observa-

 DOI: 10.1051/epjconf/201716005012

Table 1: Comparison of physical and seismic parameters for the four giant planets. Asymptotic $\Delta\nu_{as}$ are obtained as the inverse of the acoustic travel time across the planet, computed from internal structure models ([3, 8, 12] for Jupiter, Saturn, and Neptune respectively, and Nettelmann priv. comm. for Uranus.). $\Delta\nu_{sl}$ and ν_{max} are extrapolated from solar values using asteroseismic scalings with correction of planetary oblateness e. For Jupiter, observed values from [13] are shown for comparison. Table from [7].

Planet	M (M_E)	R_{eq} (km)	e (%)	T_{eff} (K)	$\Delta\nu_{as}$ (μHz)	$\Delta\nu_{sl}$ (μHz)	$\Delta\nu_{obs}$ (μHz)	$\nu_{max,sl}$ (μHz)	$\nu_{max,obs}$ (μHz)	$\delta\nu_{rot}$ (μHz)
Sun	$3\ 10^5$	798 000	0	5777	138				3100	0.4
Jupiter	317.9	71492	6.5	125	152-156	134	155	2030	1250	28
Saturn	95.1	60268	10.2	95	111-115	99		1040		26
Uranus	14.53	25559	2.4	57	~ 172	130		1030		17
Neptune	17.14	24764	2.7	59	198-213	148		1270		19

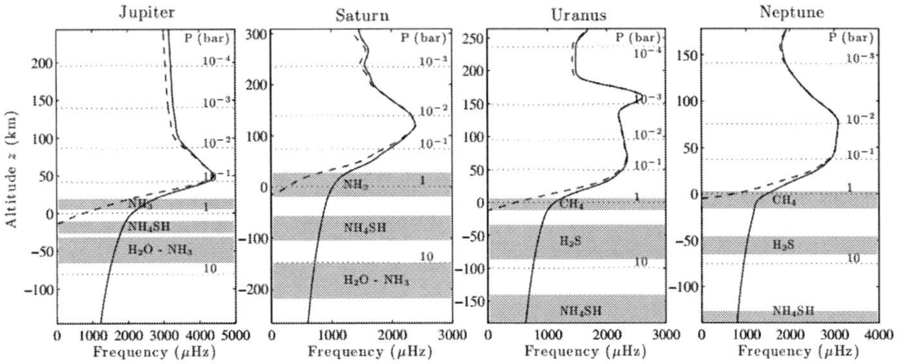

Figure 2: Acoustic cut-off ν_c (plain line) and Brünt-Väisälä $N/2\pi$ (dashed line) frequency profiles in upper atmospheres of the four giant planets. Grey regions indicate cloud locations and their chemical compositions [e.g. 14]. Above the 1-bar level ($z \equiv 0$ km), pressure-temperature profiles come from radio occultation of Voyager spacecrafts [15, and references therein]. Below $z = 0$, pressure-temperature profiles are adiabats, which imposes $N = 0$. From [7].

tional techniques. Theoretical works [16, 17] predict that Jovian global oscillations should have a frequency range of [0.8, 3.5] mHz with 10 to 100 cm s^{-1} amplitude, values that are comparable to those of the Sun. In Table 1, we report the physical and seismic parameters for the four giant planets. Especially, we include the estimates of frequency at maximum amplitude ν_{max} and mean frequency spacing between mode overtones $\Delta\nu$ obtained from the *asteroseismic* scaling relations. Indeed, even though giant planets present significant differences with respect to stars, we still expect ν_{max} to be proportional to the acoustic cut-off frequency (a function of local gravity and pressure scale height), and $\Delta\nu$ to the square root of the mean density of the body. The acoustic cut-off vertical profile of the four giant planets is represented is Fig. 2, compared with the location of cloud and haze layers. For Jupiter and Saturn, waves are similarly trapped close to the ammonia cloud decks, even though Saturn's cut-off is about half Jupiter's. For Uranus and Neptune, waves are trapped at the base of the methane clouds, by keeping in mind that the extension and thickness of cloud coverage suffers from strong uncertainties. We also represent the Brünt-Väisälä frequency profile, which displays values similar to the acoustic cut-off, which means that coupling between acoustic and gravity waves may exist above the tropopause.

Observationally, most of the efforts dedicated to the search of oscillations of giant planets regarded Jupiter, as it is the biggest, closest, and brightest target. There have been several attempts to detect Jovian oscillations using infrared photometry [18], Doppler spectrometry [19–21], and careful searches for excitation of acoustic waves due to the impact of the Shoemaker-Levy 9 comet [22, 23]. In most of these campaigns, the signal-to-noise (SNR) ratio was too low or instrumental artifacts were present that inhibited any positive detection. The fast rotation of Jupiter also limits the precision these instruments were able to obtain.

Jovian seismology had to wait until 2011 to get the first strong evidence of the detection of oscillations using the SYMPA instrument, an imaging spectrometer based on a Mach-Zehnder interferometer at fixed optical path difference [24]. This instrument was designed to overcome some of the earlier limitations by imaging the full planetary disk, similar to solar helioseismic instruments like GONG [25], MDI/SOHO [26], and HMI/SDO [27]. As part of a 10-day observing run in 2005, the SYMPA instrument was able to produce a power spectrum of Jupiter's

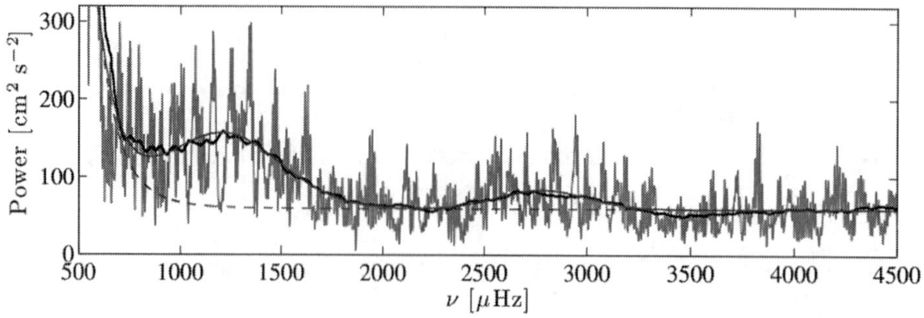

Figure 3: Evidence of the first detection of Jupiter's global modes from the SYMPA instrument, similar to the one to be built in this project. It shows the power spectrum of the mean velocity time series obtained in 2005. Excess oscillation power is detected between 800 and 3400 μHz, as well as a comb-like structure of regularly spaced peaks. The thick lines are smoothed data. From [13].

oscillations shown in Fig. 3 [13, 28]. An excess of acoustic power is observed in the frequency range predicted by theory, as well as the comb-like structure of peaks that is also expected from interior models, thereby confirming Jupiter's global pulsations. Unfortunately, the level of noise in the data is too high to identify individual modes and decisively probe Jupiter's interior.

Regarding Saturn, Marley & Porco (1993) [29] envisionned that "the rings of Saturn may act as a seismograph, recording gravitational perturbations associated with acoustic oscillations modes of the planet". The basic idea is that wave features in Saturn's C rings could be created by resonant interactions with fundamental (f) oscillation modes (i.e., radial order $n = 0$), since these modes perturb the internal density profile and, therefore, the external gravity field. Twenty years later, observations of stellar occultations of stars by the rings made with *Cassini* showed that density-wave structures detected in the C-ring were compatible with resonances due to Saturn f-modes [30–32]. These observations are the *indirect* evidence of these wave forcings [32]. However, observations unexpectedly displayed a splitting of the oscillation frequencies for several azimuthal orders. In a first attempt to explain this, and to retrieve information about the Physics of Saturn's interior, Fuller (2014) [31] proposed that a thermally stable stratified layer might surround the core. This would lead to the existence of a gravity (g) mode cavity, and these g-modes would interfere with Saturn's f-modes and cause the observed split frequencies, somewhat like g-modes do with p-modes in red-giant stars [33].

Seismology of giant planets looks similar to asteroseismology in the 1990s, i.e. with a bright future. The current decade has been promising thanks to the detection of Jupiter's acoustic oscillations with the ground-based imaging-spectrometer SYMPA, and indirect detection of Saturn's f-modes in its rings by the NASA Cassini orbiter. This has motivated new projects of ground-based instruments that are under development, such as the multi-site imaging spectrometer JOVIAL/JIVE (Gonçalves et al., submitted), inherited from SYMPA, which is led by teams from the Observatoire de la Côte d'Azur and New Mexico State University. In the following, we discuss the NASA Kepler "K2" mission, which observed Neptune for about 49 days in early 2015. It also observed Uranus in 2016, but data are not processed yet. Even though it is not sure whether visible photometry is appropriate for giant planet seismology, it was a unique opportunity to test this technique.

3 The Visible Photometry Approach

Several approaches can be used for detecting acoustic modes of giant planets: visible photometry for measuring reflected solar flux changes, infrared photometry for temperature fluctuations, and Doppler spectrometry for radial velocities. The choice of an observation technique is also tightly bounded to the choice of a spectral domain. From the visible up to wavelengths of 5 to 20 μm, the spectra of the four giants is dominated by the solar spectrum that is reflected by clouds and hazes, on top of which are superimposed absorption lines of atmospheric constituents (e.g. H_2, CH_4, NH_3). At larger wavelengths, in the mid infrared, their spectra result from thermal emission. The thermal spectrum dominates the reflected spectrum starting at about 6.5 μm for Jupiter, 10.6 μm for Saturn, 18 μm for Uranus, and 15 μm for Neptune. From Wien's law, their peak emissions are 23 μm for Jupiter, 30 μm for Saturn, and 49 μm for Uranus and Neptune. Obviously, whatever the technique, seismic observations of giant planets follow the same rules as asteroseismic observations: high duty cycle, and long duration (at least 1 week) to reach a frequency resolution sufficient for identifying modes.

A natural approach to perform seismology of the giant planets is to search for variations of the solar light reflected by Jupiter, resulting from distortions of the planetary external radius R by acoustic modes. Mosser (1995) [34] shows that a radial mode $l = 0$ of velocity v leads to a variation of the radius $\delta R = c_s v/g$, where c_s is the sound speed in the upper troposphere and g the gravitation. For Jupiter, $c_s \simeq 800$ m s^{-1}, $g = 24.8$ m s^{-2}, so that a 1-m s^{-1} velocity implies $\delta R = 40$ m, with a corresponding variation of reflected luminosity, $(\delta\Phi/\Phi)_{\rm refl} = 2\delta R/R$ of the order of

1 ppm. We expect about the same level for the four giant planets.

Regarding the photometric approach, Gaulme & Mosser (2005) [35] showed that the photometric fluctuations related to planetary radius variations are likely to be negligible with respect to albedo fluctuations of clouds that are caused by acoustic modes on their thermodynamical equilibrium. They studied Jupiter's cloud response to pressure-temperature perturbations, as acoustic waves, with a one-dimensional model including the thermodynamics and microphysics in a single cloud layer. They demonstrated that acoustic waves drive a shift in the gas/condensate phase equilibrium that defines the cloud, which is damped by the kinetics of condensation. The resulting perturbation of the cloud equilibrium can be converted to albedo variations. For acoustic waves of 1-m s^{-1} amplitudes, photometric fluctuations reach several tens of ppm, which comfortably fits within the photometric performance of current space-based imagers.

The Gaulme & Mosser results are model dependent, and strongly rely on our knowledge of cloud properties of the giant planets. Therefore the photometric approach looks at the same time both promising and uncertain for a space mission. On the one hand, photometric imaging instruments are much simpler and lighter than high-precision spectrometers and their feasibility and accuracy has been demonstrated by the remarkably successful asteroseismic observations of CoRoT and Kepler [36, 37]. Note that such an approach looks more suitable for Jupiter and Saturn rather than for Uranus and Neptune, because a clear cloud structure dominates the reflected solar light of gas giants, while the cloud coverage of ice giants is sparse at levels where sunlight is reflected. On the other hand, the mission K2 represented an exceptional opportunity of testing this technique, because it was mostly free. No much risks and only new information to grab, both in case of detection or non-detection of planetary oscillations. However, K2 could focus only on the ice giants, as Jupiter and Saturn would saturate and move too fast on the chip to allow for precise photometric measurements.

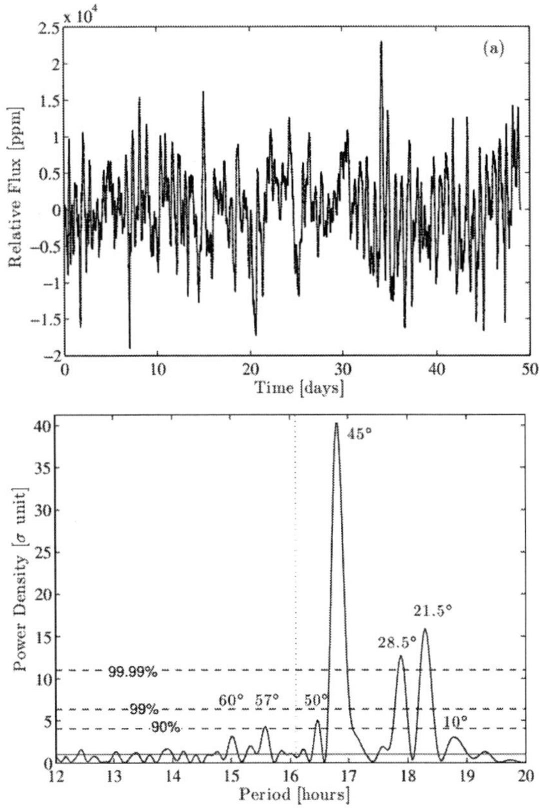

Figure 4: Panel a: K2 Neptune full 49-day light curve, showing normalized brightness variations over time elapsed since December 1, 2014. Panel b: oversampled and whitened power density spectrum as a function of period. The red line corresponds to the noise level of the whitened spectrum and black dashed lines indicate statistical significance levels. Numbers above some peaks indicate the latitudes on Neptune corresponding to that rotation period based on the zonal velocity curve given by Sánchez-Lavega et al. (2015); the features could be in either hemisphere. From [38].

4 The K2 Observations

4.1 Light Curve

Starting on December 1st 2014, *Kepler* K2 observed Neptune continuously for 49 days at a 1-minute cadence. The processing of K2 photometric light-curves is reported by Rowe et al. (submitted). It includes corrections for photometric jumps, intrapixel variations and outliers. The light curve was detrended to remove the observed decrease in flux due to the increasing distance between Neptune and the *Kepler* spacecraft, by subtracting a second-order polynomial (Fig. 4a). Over the 49-day observation window the distance between the *Kepler* spacecraft and Neptune increased by 0.81 AU, which represents a 406-second variation of light travel time. Since we consider physical phenomena on Neptune and on the Sun, we interpolated the data onto a uniform time grid that takes into account the light travel time. The distance variation from the Sun to Neptune is negligible as it corresponds to 0.8 seconds.

The light curve was analyzed in terms of frequencies by computing its power spectral density (PSD) with a Fast Fourier Transform. All short gaps (only a few missing points) were interpolated with a second-order polynomial estimated from the nearby data points. There are no long gaps observed and the overall duty cycle is greater than 98 %. Simon et al. (2016) [38] reported the clear detection of three groups of dominant peaks in between 15 and 17 μHz, 30 and 33 μHz, and 45 and 50 μHz, corresponding to Neptune's rotation and two of its harmonics. Figure 4 shows the periods corresponding to Neptune's rotation (15-17 μHz region). The three major peaks are undoubtedly signal (confidence level larger than 99.99 %), while the neighboring peaks are marginally detected with confidence levels around 90 % (in between 3σ and 6σ). None of the peaks correspond to the periods of Neptune's major moons, nor their harmonics. Horizontal oscillations detected in prior Keck observations [39], potentially linked to tidal forcing by Triton, did not produce a corresponding 7.24 hr signal in our analysis of the photometric light curve. The peaks in the periodogram, if assumed to be created by discrete cloud features, can be used to infer the latitude of those features based on a symmetric zonal wind profile. The most significant peaks roughly correspond to latitudes of 45°, 28.5°, and 21.5° planetographic latitude, respectively. Since the wind profile is symmetric around the equator, these results cannot distinguish between northern or southern features, and we neglect any dispersion in the zonal velocities for the moment. The hemispheric degeneracy can be broken with direct imaging observations of Neptune's cloud locations.

Simultaneous KECK observations allowed to infer that the main feature at 45° is a very bright spot located in the southern hemisphere, which has been observed many times (e.g., [40]). For some smaller features closer to the equator the orientation of the planet make it harder to determine the hemisphere. The dramatic bright spot is long lived and is responsible for the principal component of the variation over a single rotation. However, multiple smaller features both produce irregularities in the light curve and seem to evolve over more rapid timescales, at time as quickly as within a rotation or two, which is compatible with previous observations (e.g., [41]). Without the largest feature, the light curve would be far more irregular, and without the varying smaller spots the rapidity of the evolution would be much less.

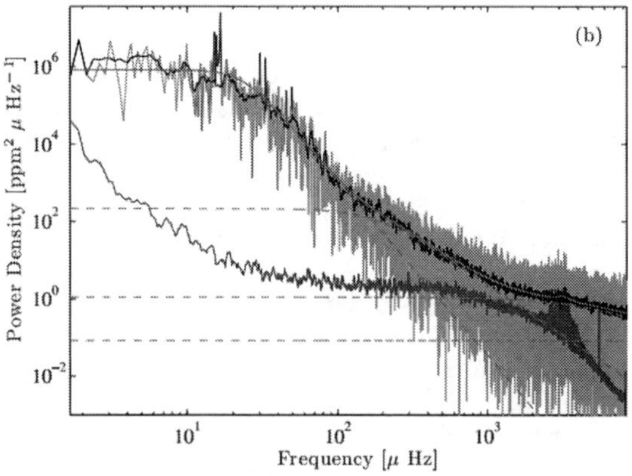

Figure 5: gray line is the power density spectrum of the *Kepler* light curve in the square of parts per million (ppm^2) per μHz, as a function of frequency (μHz). Blue peaks are Neptune's rotation frequencies and harmonics. Black line is the power density smoothed over 100 bins to guide the eye to the mean noise level. The plain red line indicates the noise model plus the mode envelope, which is the sum of three semi-Lorentzians, a Gaussian, and a white noise offset (dashed red lines). The excess power due the solar modes is visible in the bottom right of the plot. The green line is the smoothed (100 bins) power density spectrum of the VIRGO/SPM light curve taken simultaneously with K2 data. From [42].

4.2 Search for Neptune Oscillations

From Fig. 2, Neptune's maximum cut-off frequency for acoustic modes is about 3000 μHz. It means that acoustic modes cannot stand in Neptune at frequencies larger than 3000 μHz. The possible acoustic waves excited at larger frequencies are not trapped by Neptune's troposphere and get dissipated in the stratosphere. More precisely, after increasing monotonically in the upper troposphere, the cut-off frequency reaches a maximum at the tropopause, and decreases to a plateau at about 2000 μHz in the mid-stratosphere. As analyzed by Mosser et al. (1995) [34] for Jupiter, it means that waves with frequencies between 2000 and 3000 μHz can leak into the troposphere by tunneling effect, making their trapping less efficient. Regarding the lower limit, the acoustic cut-off profile shows that waves with frequencies less that 800 μHz are trapped deeper than pressure levels of about 10 bars. Even though radiative transfer in these planets is not fully constrained, it is very likely that optical observations do not probe that deep. Wave at frequencies lower than 800 μHz are thus evanescent at altitudes probed by optical observations. It it therefore reasonable to expect Neptune oscillations' maximum amplitude to be between 800 and 2000 μHz. Note this is perfectly consistent with the $\nu_{\max}$ arising from the asteroseismic relations (Table 1).

Regarding the mean frequency spacing between mode overtones, it is expected to range from 198 to 213 μHz according to internal structure models [7]. This is quite far from the value provided by the asteroseismic scaling relation, likely because of the planet's internal density discontinuities, which affect the distribution of oscillation frequencies. In addition, because of Neptune's fast rotation, acoustic modes of non-radial oscillations with azimuthal order m larger than zero split apart into $+m$ and $-m$ peaks separated by m times the inverse rotation period, i.e. $\approx m \times 17.4$ μHz.

The power spectral density (Fig. 5) does not display any excess power typical of global oscillations in the [800, 2000] μHz range, and nowhere else except for the solar oscillations. We also searched for Neptune's oscillations in the envelope of the autocorrelation (EACF) of the time series, filtered in the expected frequency domain [44, 45]. This approach allows for deriving the mean large separation of a solar-like oscillation spectrum in a blind way without prior information. It has shown to be efficient in cases of low SNR [e.g. 46–48]. The reliability of the result is given by an H_0 test: when the EACF is above a threshold level, the null hypothesis can be rejected, implying that a signal might have been detected.

The envelope of the autocorrelation displays a maximum in between 1.96 and 2.15 h in the frequency range [600, 1400] μHz (Fig. 6). From the H_0 test, the likelihood for this peak to be a signal is about 95 %. This would correspond with a large frequency spacing ranging from 258 to 283 μHz if it is the result of Neptune's oscillations. However, if this excess of power in the autocorrelation di-

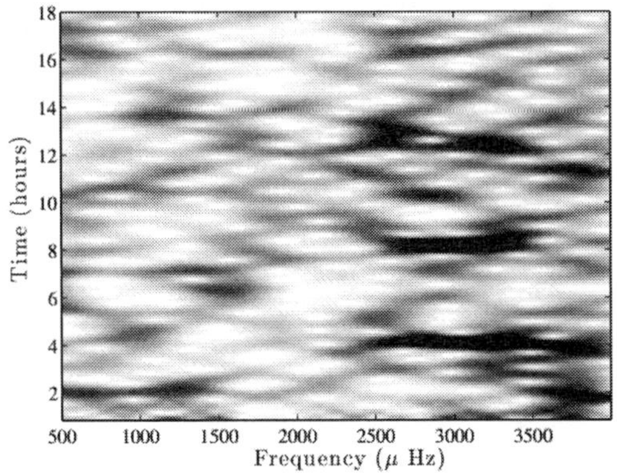

Figure 6: Envelope of the autocorrelation function (EACF) as function of frequency (x-axis) and time (y-axis). The EACF was computed every 20 μHz from 500 to 4000 μHz, by filtering the time series in a frequency bandpass of 500 μHz. The darker a region is, the larger the correlation. Solar modes are clearly visible in between 2500 and 3500 μHz at approximately 4, 8 and 13 hours. From [43].

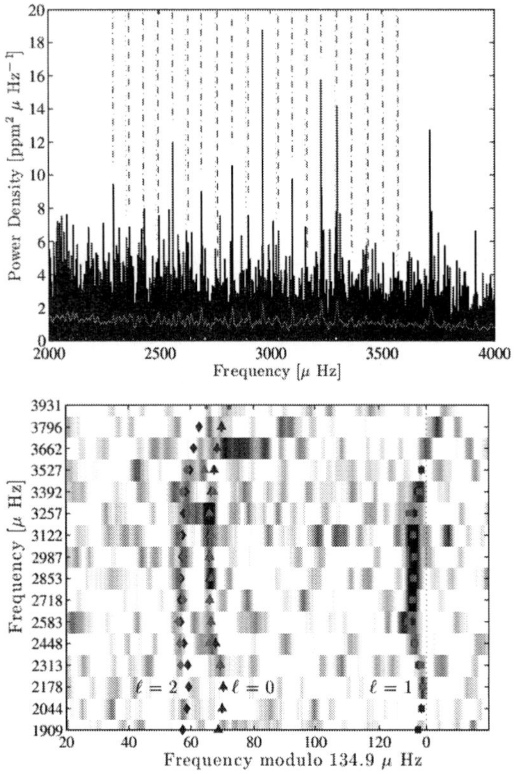

Figure 7: Top panel: K2 Neptune power spectral density on a linear scale (ppm^2 μHz^{-1}) as a function of frequency (μHz). The dot-dashed lines indicate the frequencies Gaulme et al. detected and modeled with the peak-bagging method. Bottom panel: échelle diagram of the power density spectrum. Darker regions correspond to larger peaks in power density. The power density spectrum is smoothed by a weighted moving average over seven bins and folded into 134.9 μHz chunks; each is then stacked on top of its lower-frequency neighbor. Red dots indicate K2 frequencies and the blue dots BiSON's. Error bars are smaller than symbols. From [42].

agram is generated by Neptune's oscillations, we should find peaks corresponding to its rotation period (16-17h) and some harmonics, as modes are split by it. Figure 6 does not exhibit significant maxima at 16 or 8 hours. As a comparison, we displayed the correlation diagram up to frequencies including the solar oscillations. We clearly detect the Sun's large separation, as well as secondary peaks due to the various overtones and the separation between $\ell = 0$ and $\ell = 1$ modes. Finally, we investigated whether a structure could be detected in the échelle diagram corresponding with a $\approx$ 280 μHz and nothing was observed, making it useless to reproduce in the paper. The signal in the [600, 1400] μHz range does not show similar features, and we conclude this 2-hour signal it is likely a spurious signal.

4.3 First Indirect Detection of Solar Oscillations in Intensity Measurements

The 5-minute oscillations in the Sun are routinely used as a reference point for interpreting solar-like oscillations in other stars. This has become increasingly important with the huge amount of high-quality data being returned by NASA's Kepler Mission. Ideally, the measurements of solar oscillations that act as a reference should be observed with the same instrument as the stars.

Observations of Neptune with K2 allowed for a unique opportunity to measure integrated disk seismic properties of the Sun in reflected light and determine fundamental properties (mass, radius) of the Sun as a distant star. Solar oscillations have been measured in radial velocity from the Moon [49], and also from the blue sky in both equivalent width [50] and radial velocity [51]. To our knowledge, Gaulme et al. [42] analysis of K2 photometric observations of reflected solar light from Neptune is the first indirect detection of solar oscillations in intensity.

Another reason for our interest in the K2 observations of Neptune is to calibrate the *amplitudes* of oscillations in the Sun. There has been considerable effort towards understanding how the amplitudes of solar-like oscillations vary from the Sun to other stars, both theoretically [52–60] and observationally [61–68]. A good measurement of the solar amplitude with *Kepler* would serve as an important calibration.

The global asteroseismic approach involves measuring $\Delta\nu$ and $\nu_{\max}$ from the PSD of the light curve. To determine $\nu_{\max}$, one must model background noise in the frequency domain, which is typically dominated by the correlated stellar noise (spots, granulation, meso- and supergranulation). In this case, the background variability arises from sources other than solar spots and granulation. This is obvious when comparing the K2 PSD with simultaneous SOHO/VIRGO (green channel) data (Fig. 4). K2's background overwhelms VIRGO's by up to four orders of magnitude. K2's noise likely results from residuals of effects associated with Neptune's motion across the detector (pixel response non-uniformity, saturation, pointing jitter, etc). It is unlikely that Neptune atmospheric features are responsible for such a noise level given the planet's smooth aspect in the visible, except for isolated cloud structures that appear as outstanding peaks, as we discussed above.

To minimize the bias in estimating the global asteroseismic parameters, Gaulme et al. measured the global helioseismic parameters and individual mode properties (*peak-bagging*) independently with slightly different approaches, by several groups of fitters. The idea was to proceed as one would if this target were one of the many oscillating stars detected by *Kepler*. They obtained a determination of $\Delta\nu$ that is consistent with measurements from SOHO/VIRGO/SPM and BiSON. Differences of about 2 to 3 σ, depending on methods, were observed in the determination of $\nu_{\max}$ relative to the usual solar reference (3100 μHz, e.g. [69]). The application of asteroseismic scaling relations produces a mass and radius of $1.14 \pm 0.06\, M_\odot$ and $1.05 \pm 0.02\, R_\odot$ for the Sun. However, a close look at the simultaneous photometric VIRGO/SPM data indicates that $\nu_{\max}$ was actually larger than the usual solar reference (3160 ± 10μHz), which is not surprising given the stochastic nature of solar oscillations. By taking into account the SNR, the value of $\nu_{\max}$ that is measured from K2 data is consistent with VIRGO, and corresponds to the upper 20 % of the posterior density probability.

The peak-bagging technique leads to mean amplitude and width that match those from VIRGO within error bars. Frequencies are very consistent among fitters and with VIRGO and BiSON. Except for a few peaks with low SNR, all fit within 1 σ. In regards to mode widths, the dispersion is relatively large between fitters, with commonly a factor of two difference, but the error bars are large and mostly overlap. The measured widths from K2 are generally larger than those measured by Gaulme on simultaneous VIRGO data, but agree relatively well with those retrieved from 14 years of VIRGO by Stahn (2010, PhD manuscript) and simultaneous BiSON measurements. As for the amplitudes, mean VIRGO amplitudes (3.6±0.3 ppm) are larger but still compatible with K2 (2.3 ± 0.5 ppm). However, K2's amplitudes are lower than VIRGO's, especially around $\nu_{\max}$, where VIRGO amplitudes are ≈ 4.5 ppm and K2 ≈ 3 ppm, i.e. 1/3 larger. This is presumably due to *Kepler*'s broader and, in particular, redder passband. [70] showed the ratio of the mode amplitudes measured from VIRGO data for the three channels are: blue-to-green ≈ 1.4 and green-to-red ≈ 2, which is consistent with the discrepancies we observe with respect to VIRGO green channel data. Note that BiSON amplitudes are not directly comparable because it is a velocity measurement.

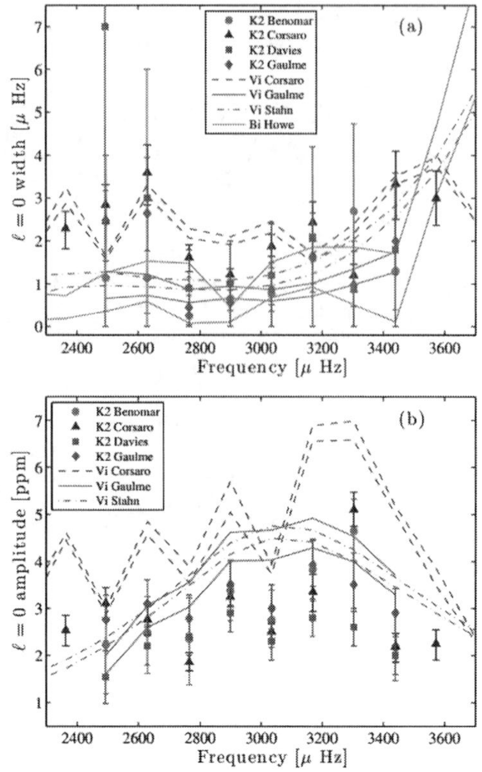

Figure 8: Oscillation width and amplitude as a function of frequency for radial modes. Dashed and dash-dotted lines indicate estimates from VIRGO data (black for simultaneous and magenta for 14 years data). Different colors, symbols and names indicate the results obtained by the different group of fitters from Gaulme et al. (figure extracted from [42]).

5 Conclusions

Seismology applied to giant planets needs a clear recording of the oscillation spectrum of any giant planet, where mode frequencies are unambiguously determined. Due to some peculiarities of giant planets with respect to the Sun or stars (resolved objects, wrapped by clouds, fast rotators), a specific instrumentation is required. Gaulme & Mosser (2005) [35] have explored the possibility of using a space imager as CoRoT for detecting the oscillations of giants planets in the reflected solar flux. They showed that answering to this question depends strongly on the cloud vertical structure and microphysics, which are mostly unknown. The *Kepler* K2 mission represented a unique opportunity of testing this technique on the two ice giants. It observed Neptune continuously for 49 days starting in December 2015 [38], and did the same with Uranus about a year later for 27 days. Neptune's observations have not brought any detection of Neptune's oscillations, but have allowed for detecting the Sun's oscillations reflected on Neptune [42]. K2 Uranus data are currently being processed and it is too early to reject the detection of planetary oscillations, even though they do not seem to be detectable. Nevertheless, solar oscillations are detected at a similar SNR level as for Neptune, and we are keenly interested to see what solar $\nu_{\max}$ the K2 Uranus data will suggest.

Overall, the best instrumental approach to get good SNR oscillations spectra of giant planets are the same as for the Sun, i.e. Doppler imaging. After first the results of the SYMPA instrument, a team led by François-Xavier Schmider (Observatoire de la Côte d'Azur) proposed to place a payload inherited from SYMPA onboard of the JUICE ESA mission to the Jupiter's system. The payload was not selected in the end of a 5-year process, but we were granted enough fundings (CNES, ESA, Thales, NASA) to develop a prototype. The project named JOVIAL (in France) or JIVE (in the US) has now evolved into a three-instrument ground-based network, with sites in France, USA, and Japan, in the same way as helioseismic BiSON, GONG, or asteroseismic SONG. The objective is to observe Jupiter's oscillations and identify its modes up to a degree $\ell = 10$, to definitively answer the question of the existence of a rocky core. A single-site test campaign was led in 2015, and the network will be operational in 2018.

References

[1] G. Magni, A. Coradini, PSS, **52**, 343 (2004)
[2] L. Mayer, T. Quinn, J. Wadsley, J. Stadel, Science **298**, 1756 (2002)
[3] D. Saumon, T. Guillot, ApJ, **609**, 1170 (2004)
[4] B. Militzer, W.B. Hubbard, J. Vorberger, I. Tamblyn, S.A. Bonev, ApJL, **688**, L45 (2008)
[5] N. Nettelmann, A. Becker, B. Holst, R. Redmer, ApJ, **750**, 52 (2012)
[6] T. Guillot, Science **286**, 72 (1999)
[7] P. Gaulme, B. Mosser, F.X. Schmider, T. Guillot, *Seismology of Giant Planets* (2015), pp. 189–202
[8] N. Nettelmann, R. Helled, J.J. Fortney, R. Redmer, PSS, **77**, 143 (2013)
[9] J.J. Fortney, N. Nettelmann, Space Science Reviews, **152**, 423 (2010)
[10] D.J. Stevenson, Icarus, **62**, 4 (1985)
[11] J. Leconte, G. Chabrier, A&A, **540**, A20 (2012)
[12] R. Helled, T. Guillot, ApJ, **767**, 113 (2013)
[13] P. Gaulme, F.X. Schmider, J. Gay, T. Guillot, C. Jacob, A&A, **531**, A104+ (2011)
[14] P.G.J. Irwin, *Giant Planets of Our Solar System* (2009)
[15] G.F. Lindal, AJ, **103**, 967 (1992)
[16] S.V. Vorontsov, V.N. Zharkov, V.M. Lubimov, Icarus **27**, 109 (1976)
[17] D. Bercovici, G. Schubert, Icarus **69**, 557 (1987)
[18] D. Deming, M.J. Mumma, F. Espenak, D.E. Jennings, T. Kostiuk, G. Wiedemann, R. Loewenstein, J. Piscitelli, ApJ, **343**, 456 (1989)
[19] F.X. Schmider, E. Fossat, B. Mosser, A&A, **248**, 281 (1991)
[20] B. Mosser, D. Mekarnia, J.P. Maillard, J. Gay, D. Gautier, P. Delache, A&A, **267**, 604 (1993)
[21] B. Mosser, J.P. Maillard, D. Mékarnia, Icarus **144**, 104 (2000)
[22] C.M. Walter, M.S. Marley, D.M. Hunten, A.L. Sprague, W.K. Wells, A. Dayal, W.F. Hoffmann, M.V. Sykes, L.K. Deutsch, G.G. Fazio et al., Icarus, **121**, 341 (1996)
[23] B. Mosser, P. Galdemard, P. Lagage, E. Pantin, M. Sauvage, P. Lognonne, D. Gautier, F. Billebaud, T. Livengood, H.U. Kaufl, Icarus **121**, 331 (1996)
[24] F.X. Schmider, J. Gay, P. Gaulme, C. Jacob, L. Abe, M. Alvarez, J.A. Belmonte, E. Fossat, B. Gelly, T. Guillot et al., A&A, **474**, 1073 (2007)
[25] J.W. Harvey, F. Hill, R.P. Hubbard, J.R. Kennedy, J.W. Leibacher, J.A. Pintar, P.A. Gilman, R.W. Noyes, A.M. Title, J. Toomre et al., Science **272**, 1284 (1996)
[26] P.H. Scherrer, R.S. Bogart, R.I. Bush, J.T. Hoeksema, A.G. Kosovichev, J. Schou, W. Rosenberg, L. Springer, T.D. Tarbell, A. Title et al., Sol. Phys., **162**, 129 (1995)
[27] P.H. Scherrer, J. Schou, R.I. Bush, A.G. Kosovichev, R.S. Bogart, J.T. Hoeksema, Y. Liu, T.L. Duvall, J. Zhao, A.M. Title et al., Sol. Phys., **275**, 207 (2012)
[28] P. Gaulme, F.X. Schmider, J. Gay, C. Jacob, M. Alvarez, M. Reyes-Ruiz, J.A. Belmonte, E. Fossat, F. Jeanneaux, J.C. Valtier, A&A, **490**, 859 (2008)
[29] M.S. Marley, C.C. Porco, Icarus, **106**, 508 (1993)
[30] M.M. Hedman, P.D. Nicholson, AJ, **146**, 12 (2013)
[31] J. Fuller, Icarus, **242**, 283 (2014)
[32] M.S. Marley, Icarus, **234**, 194 (2014)
[33] P.G. Beck, T.R. Bedding, B. Mosser, D. Stello, R.A. Garcia, T. Kallinger, S. Hekker, Y. Elsworth, S. Frandsen, F. Carrier et al., Science **332**, 205 (2011)
[34] B. Mosser, A&A, **293**, 586 (1995)
[35] P. Gaulme, B. Mosser, Icarus **178**, 84 (2005)
[36] E. Michel, A. Baglin, M. Auvergne, C. Catala, R. Samadi, F. Baudin, T. Appourchaux, C. Barban, W.W. Weiss, G. Berthomieu et al., Science **322**, 558 (2008)
[37] W.J. Borucki, D. Koch, G. Basri, N. Batalha, T. Brown, D. Caldwell, J. Caldwell, J. Christensen-Dalsgaard, W.D. Cochran, E. DeVore et al., Science **327**, 977 (2010)
[38] A.A. Simon, J.F. Rowe, P. Gaulme, H.B. Hammel, S.L. Casewell, J.J. Fortney, J.E. Gizis, J.J. Lissauer, R. Morales-Juberias, G.S. Orton et al., ApJ, **817**, 162 (2016)
[39] S.C. Martin, I. de Pater, P. Marcus, Ap&SS, **337**, 65 (2012)
[40] E. Karkoschka, Icarus, **215**, 439 (2011)
[41] L.A. Sromovsky, Science **254**, 684 (1991)
[42] P. Gaulme, J.F. Rowe, T.R. Bedding, O. Benomar, E. Corsaro, G.R. Davies, S.J. Hale, R. Howe, R.A. Garcia, D. Huber et al., ApJL, **833**, L13 (2016)

[43] J.F. Rowe, P. Gaulme, J.J. Lissauer, M.S. Marley, A.A. Simon, H.B. Hammel, V. Silva Aguirre, T. Barclay, O. Benomar, P. Boumier, Submitted to ApJL (2016)
[44] I.W. Roxburgh, S.V. Vorontsov, MNRAS, **369**, 1491 (2006)
[45] B. Mosser, T. Appourchaux, A&A, **508**, 877 (2009)
[46] B. Mosser, E. Michel, T. Appourchaux, C. Barban, F. Baudin, P. Boumier, H. Bruntt, C. Catala, S. Deheuvels, R.A. García et al., A&A, **506**, 33 (2009)
[47] P. Gaulme, S. Deheuvels, W.W. Weiss, B. Mosser, C. Moutou, H. Bruntt, J. Donati, M. Vannier, T. Guillot, T. Appourchaux et al., A&A, **524**, A47+ (2010)
[48] B. Mosser, K. Belkacem, M.J. Goupil, A. Miglio, T. Morel, C. Barban, F. Baudin, S. Hekker, R. Samadi, J. De Ridder et al., A&A, **517**, A22 (2010)
[49] H. Kjeldsen, T.R. Bedding, R.P. Butler, J. Christensen-Dalsgaard, L.L. Kiss, C. McCarthy, G.W. Marcy, C.G. Tinney, J.T. Wright, ApJ, **635**, 1281 (2005)
[50] H. Kjeldsen, T.R. Bedding, M. Viskum, S. Frandsen, AJ, **109**, 1313 (1995)
[51] H. Kjeldsen, T.R. Bedding, T. Arentoft, R.P. Butler, T.H. Dall, C. Karoff, L.L. Kiss, C.G. Tinney, W.J. Chaplin, ApJ, **682**, 1370 (2008)
[52] J. Christensen-Dalsgaard, S. Frandsen, Sol. Phys., **82**, 469 (1983)
[53] H. Kjeldsen, T.R. Bedding, A&A, **293**, 87 (1995)
[54] H. Kjeldsen, T.R. Bedding, A&A, **529**, L8 (2011)
[55] G. Houdek, N.J. Balmforth, J. Christensen-Dalsgaard, D.O. Gough, A&A, **351**, 582 (1999)
[56] G. Houdek, D.O. Gough, MNRAS, **336**, L65 (2002)
[57] G. Houdek, *Stochastic excitation and damping of solar-like oscillations*, in *Proceedings of SOHO 18/GONG 2006/HELAS I, Beyond the spherical Sun* (2006), Vol. 624 of *ESA Special Publication*, p. 28.1
[58] R. Samadi, D. Georgobiani, R. Trampedach, M.J. Goupil, R.F. Stein, Å. Nordlund, A&A, **463**, 297 (2007)
[59] R. Samadi, K. Belkacem, M.A. Dupret, H.G. Ludwig, F. Baudin, E. Caffau, M.J. Goupil, C. Barban, A&A, **543**, A120 (2012)
[60] K. Belkacem, R. Samadi, M.J. Goupil, Journal of Physics Conference Series **271**, 012047 (2011)
[61] R. Samadi, H.G. Ludwig, K. Belkacem, M.J. Goupil, O. Benomar, B. Mosser, M.A. Dupret, F. Baudin, T. Appourchaux, E. Michel, A&A, **509**, A16 (2010)
[62] D. Huber, T.R. Bedding, D. Stello, S. Hekker, S. Mathur, B. Mosser, G.A. Verner, A. Bonanno, D.L. Buzasi, T.L. Campante et al., ApJ, **743**, 143 (2011)
[63] D. Huber, T.R. Bedding, T. Arentoft, M. Gruberbauer, D.B. Guenther, G. Houdek, T. Kallinger, H. Kjeldsen, J.M. Matthews, D. Stello et al., ApJ, **731**, 94 (2011)
[64] T.L. Campante, R. Handberg, S. Mathur, T. Appourchaux, T.R. Bedding, W.J. Chaplin, R.A. García, B. Mosser, O. Benomar, A. Bonanno et al., A&A, **534**, A6 (2011)
[65] W.J. Chaplin, H. Kjeldsen, T.R. Bedding, J. Christensen-Dalsgaard, R.L. Gilliland, S.D. Kawaler, T. Appourchaux, Y. Elsworth, R.A. García, G. Houdek et al., ApJ, **732**, 54 (2011)
[66] K. Belkacem, M.A. Dupret, F. Baudin, T. Appourchaux, J.P. Marques, R. Samadi, A&A, **540**, L7 (2012)
[67] B. Mosser, Y. Elsworth, S. Hekker, D. Huber, T. Kallinger, S. Mathur, K. Belkacem, M.J. Goupil, R. Samadi, C. Barban et al., A&A, **537**, A30 (2012)
[68] E. Corsaro, H.E. Fröhlich, A. Bonanno, D. Huber, T.R. Bedding, O. Benomar, J. De Ridder, D. Stello, MNRAS, **430**, 2313 (2013)
[69] A.M. Broomhall, W.J. Chaplin, G.R. Davies, Y. Elsworth, S.T. Fletcher, S.J. Hale, B. Miller, R. New, MNRAS, **396**, L100 (2009)
[70] A. Jiménez, T. Roca Cortés, G. Severino, C. Marmolino, ApJ, **525**, 1042 (1999)

AUTHOR INDEX

AUTHOR INDEX